Invertebrates in Freshwater Wetlands

Darold Batzer • Dani Boix
Editors

Invertebrates in Freshwater Wetlands

An International Perspective on their Ecology

Springer

Editors
Darold Batzer
Department of Entomology
University of Georgia
Athens, Georgia, USA

Dani Boix
Institute of Aquatic Ecology
University of Girona
Girona, Spain

ISBN 978-3-319-24976-6 ISBN 978-3-319-24978-0 (eBook)
DOI 10.1007/978-3-319-24978-0

Library of Congress Control Number: 2015959947

Springer Cham Heidelberg New York Dordrecht London

Printed on acid-free paper

Springer International Publishing AG Switzerland is part of Springer Science+Business Media (www.springer.com)

Contents

Contributors

Leandro D. Battirola Universidade Federal de Mato Grosso, Instituto de Ciências Naturais, Humanas e Sociais, Cuiabá, Mato Grosso, Brazil

Darold Batzer Department of Entomology, University of Georgia, Athens, GA, USA

Gustavo Becerra-Jurado School of Biology and Environmental Science, University College Dublin, Dublin, Ireland

Leon Blaustein Community Ecology Lab, Department of Evolutionary and Environmental Ecology and the Institute of Evolution, Faculty of Natural Sciences, University of Haifa, Haifa, Israel

Dani Boix GRECO, Institute of Aquatic Ecology, University of Girona, Girona, Spain

Andrew Boulton Ecosystem Management, School of Environmental and Rural Science, University of New England, Armidale, NSW, Australia

Luc Brendonck Laboratory of Aquatic Ecology, Evolution and Conservation, KU Leuven, Leuven, Belgium

Bryana M. Bush Department of Entomology, University of Georgia, Athens, GA, USA

Steven R. Chipps US Geological Survey, South Dakota Cooperative Fish and Wildlife Research Unit, Department of Natural Resource Management, South Dakota State University, Brookings, SD, USA

Matthew J. Cooper Institute for Great Lakes Research, Central Michigan University, Mt. Pleasant, MI, USA

Jennifer Day Freshwater Research Unit, Department of Zoology, University of Cape Town, Cape Town, South Africa

Claire Duchet Community Ecology Lab, Department of Evolutionary and Environmental Ecology and the Institute of Evolution, Faculty of Natural Sciences, University of Haifa, Haifa, Israel

Sue Eggert USDA Forest Service, Northern Research Station, Grand Rapids, MN, USA

Luis B. Epele CIEMEP (CONICET-UNPSJB), Esquel, Chubut, Argentina

G. Winfield Fairchild Department of Biology, West Chester University, West Chester, PA, USA

Belinda Gallardo Biodiversity Conservation and Ecosystem Restoration Department, Pyrenean, Institute of Ecology (IPE-CSIC), Zaragoza, Spain

Stéphanie Gascón GRECO, Institute of Aquatic Ecology, University of Girona, Girona, Spain

Mark A. Hanson Wildlife Research Unit, Minnesota Department of Natural Resources, Wetland Wildlife Populations & Research Group, Bemidji, MN, USA

Adam K. Janke Department of Natural Resource Management, South Dakota State University, Brookings, SD, USA

Michael J. Jeffries Department of Geography, Northumbria University, Newcastle upon Tyne, UK

Jamie Kneitel Department of Biological Sciences, California State University, Sacramento, CA, USA

Sandro Lanfranco Department of Biology, University of Malta, Msida, MSD, Malta

William F. Loftus Aquatic Research & Communication, LLC, Vero Beach, FL, USA

Mercedes R. Marchese Instituto Nacional de Limnología (CONICET-UNL), Ciudad Universitaria, Santa Fe, Argentina

Marinez I. Marques Universidade Federal de Mato Grosso, Instituto de Ciências Naturais, Humanas e Sociais, Cuiabá, Mato Grosso, Brazil

David Moreno-Mateos Basque Centre for Climate Change (BC3), Bilbao, Spain

IKERBASQUE, Basque Foundation for Science, Bilbao, Spain

Beat Oertli hepia, University of Applied Sciences and Arts Western Switzerland, Geneva, Switzerland

Xavier D. Quintana GRECO, Institute of Aquatic Ecology, University of Girona, Girona, Spain

Julian D. Reynolds Trinity College University of Dublin, Dublin, Ireland

Belinda J. Robson Environmental and Conservation Science, Murdoch University, Murdoch, WA, Australia

Véronique Rosset IRSTEA, UR Maly, Lyon, Villeurbanne, France

Albert Ruhí Julie Ann Wrigley Global Institute of Sustainability, Arizona State University, Tempe, AZ, USA

Jordi Sala GRECO, Institute of Aquatic Ecology, University of Girona, Girona, Spain

Douglas J. Spieles McPhail Center for Environmental Studies, Denison University, Granville, OH, USA

Joshua D. Stafford US Geological Survey, South Dakota Cooperative Fish and Wildlife Research Unit, Department of Natural Resource Management, South Dakota State University, Brookings, SD, USA

Jered M. Studinski Department of Biology, Frostburg State University, Frostburg, MD, USA

Brian Timms Australian Wetlands, Rivers and Landscape Centre, School of Biology, Earth and Environmental Sciences, University of New South Wales, Sydney, NSW, Australia

Joel C. Trexler Department of Biological Science, Florida International University, Miami, FL, USA

Donald G. Uzarski Institute for Great Lakes Research, Central Michigan University, Mt. Pleasant, MI, USA

CMU Biological Station, Central Michigan University, Mt. Pleasant, MI, USA

Csaba F. Vad WasserCluster Lunz, Dr. Carl Kupelwieser Promenade 5, Lunz am See, Austria

Bram Vanschoenwinkel Department of Biology, VUB (Vrije Universiteit Brussel), Brussels, Belgium

Karl M. Wantzen Interdisciplinary Research Center for Cities, Territories, Environment and Society (CNRS UMR CITERES), Université François Rabelais, Tours, France

Elisabeth B. Webb US Geological Survey, Missouri Cooperative Fish and Wildlife Research Unit, Department of Fisheries and Wildlife Sciences, University of Missouri, Columbia, MO, USA

Terry Wheeler Department of Natural Resource Sciences, McGill University, Macdonald Campus, Ste-Anne-de-Bellevue, Quebec, Canada

Matt Whiles Department of Zoology and Center for Ecology, Southern Illinois University, Carbondale, Carbondale, IL, USA

D. Dudley Williams Department of Biological Sciences, University of Toronto Scarborough, Toronto, ON, Canada

Scott A. Wissinger Departments of Biology and Environmental Science, Allegheny College, Meadville, PA, USA

Rocky Mountain Biological Laboratory, Crested Butte, CO, USA

School of Biological Sciences, University of Canterbury, Christchurch, New Zealand

Dale A. Wrubleski Institute for Wetland and Waterfowl Research, Ducks Unlimited Canada, Stonewall, MB, Canada

Haitao Wu Northeast Institute of Geography and Agroecology, Changchun, Jilin, China

Kyle D. Zimmer Department of Biology, University of St. Thomas, St. Paul, MN, USA

Luís Zúñiga Instituto de Biología, Universidad Católica de Valparaíso, Valparaíso, Chile

Chapter 1
An Introduction to Freshwater Wetlands and Their Invertebrates

Darold Batzer and Dani Boix

This text assembles prominent wetland ecologists from across the globe to describe the ecology of the invertebrates residing in the wetlands they each study. Each of their chapters assumes the reader has some basic knowledge about wetland ecology and about invertebrates. Because some may not have this background, we have prepared a brief introductory chapter to familiarize people with some basic aspects of freshwater wetland habitats and provide some foundational information about the invertebrate fauna that exploits freshwater wetlands.

Defining Wetlands

Despite the fact that wetland ecology is now a well-established scientific discipline, what defines a "wetland" habitat remains inconsistent worldwide. Perhaps the most widely used international definition comes from the *Ramsar Convention on Wetlands*, which reads:

> Wetlands are areas of marsh, fen, peatland or water, whether natural or artificial, permanent or temporary, with water that is static or flowing, fresh, brackish or salt, including areas of marine water the depth of which at low tide does not exceed six metres. (www.ramsar.org)

This definition is very broad and includes many habitats, such as shallow lakes and reservoirs, that might not be considered wetlands in many localities. It is primarily meant for a nonscientist audience and lacks the functional mechanistic aspect attractive to ecologists.

In the USA, the history of wetland definition has had a convoluted past (see Sharitz et al. 2014), largely because regulations there confer special protections to and restrictions on wetlands. Thus, what is or is not called a wetland can be controversial. A fairly narrow definition has been adopted, coined by the *US Army Corps of Engineers*, the primary agency charged with regulating US wetlands, which reads:

D. Batzer, D. Boix (eds.), *Invertebrates in Freshwater Wetlands*,
DOI 10.1007/978-3-319-24978-0_1

> The term "wetlands" means those areas that are inundated or saturated by surface or ground water at a frequency and duration sufficient to support, and that under normal circumstances do support, a prevalence of vegetation typically adapted for life in saturated soil conditions. Wetlands generally include swamps, marshes, bogs, and similar areas. (http://el.erdc.usace.army.mil/elpubs/pdf/wlman87.pdf)

This definition has been interpreted to mean that appropriate hydrology, vegetation, and hydric soils *all* need to be present at a site for it to be legally called a wetland and eligible for certain governmental protections. Like the Ramsar definition, it is not an ecological definition, but instead one intended to be used for legal purposes and of course only in the USA. However, it does address how hydrology, vegetation, and soils interact to produce wetland conditions. But for many, the US definition is unnecessarily narrow as it excludes habitats that would be considered wetlands by most ecologists such as non-vegetated mudflats or floodplain areas that routinely flood but still lack hydric soils (e.g., Fig. 1.1).

In Europe, wetland definition is complicated by the diversity of national traditions surrounding the habitats. The most widely accepted definition was developed for the Water Framework Directive (WFD CIS 2003), which reads:

> Wetlands are heterogeneous but distinctive ecosystems which develop naturally or are the product of human activities. Their biogeochemical functions depend notably on a constant or periodic shallow inundation by fresh, brackish or saline water, or saturation at or near the surface of the substrate. They are characterised by standing or slowly moving waters. Common features include hydric soils, micro-organisms, hydrophilous and hygrophilous vegetation and fauna, adapted to chemical and biological processes reflective of periodic or permanent flooding and/or water-logging.

This definition is clearly ecological in nature and expands beyond hydrology, vegetation, and soils to also address microbes and animals. The term *pond* is also widely used in Europe, which is defined as "a waterbody with a maximum depth of no more than 8 m, offering water plants the potential to colonise almost the entire area of the pond" (Oertli et al. 2005). Wetland invertebrate ecologists in Europe have embraced this convention because their organisms of interest are focal components of pond ecosystems. Shallow ponds would be considered wetlands worldwide, but at least in the USA, areas of ponds that are >2 m depth and lack macrophytes would be labeled as deepwater habitats rather than wetlands (Cowardin et al. 1979).

For this book, our goal is to develop an international flavor to the study of wetland invertebrates, and we recognize that what habitats are considered wetlands varies worldwide (even beyond the examples we cite). Thus, we do not impose a specific definition of what constitutes a wetland, nor what constitutes a wetland invertebrate, instead relying on the discretion of the chapter authors. However, most ecologists recognize that wetlands are largely defined by climate, hydrology, and vegetation and that the resident invertebrate faunas are controlled by these factors.

Fig. 1.1 Floodplain of the Oconee River, Georgia. While this habitat floods most years, most of its expanse is not considered *jurisdictional wetland* using criteria of the US Army Corps of Engineers (Environmental Laboratory 1987) because most soils are not hydric and flooding occurs primarily in winter outside the "growing season" for plants. Reprinted with permission from *Ecology of Freshwater and Estuarine Wetlands: Second Edition*, edited by Darold P. Batzer and Rebecca R. Sharitz. © 2014 by the Regents of the University of California. Published by the University of California Press

Defining Wetland Invertebrates

Aquatic invertebrates from a number of animal phyla thrive in wetlands including Turbellaria (flatworms), Rotifera (rotifers), Nematoda (roundworms), Annelida (segmented worms and leeches), Mollusca (snails and clams), and Arthropoda (crustaceans, mites, insects). Ecologists studying aquatic invertebrates in wetlands tend to focus either on microinvertebrates (microturbellarians, rotifers, nematodes, small crustaceans) or macroinvertebrates (large flatworms, annelids, mollusks, large crustaceans, mites, insects). These two size-based groupings have no scientific standing and to some extent simply reflect how the organisms are sampled and habitats they use. By convention a microinvertebrate is <1 mm long, and a macroinvertebrate is >1 mm long, and this metric was largely established by the mesh size of the nets most often used to sample the organisms (coarse mesh for macros and fine mesh (<1 mm) for micros). But obviously, this size-based metric is problematic; for example, an early stage macroinvertebrate (e.g., second instar midge larva) might be <1 mm, but would still be labeled a macro. Possibly more important than actual body size is habitat utilization. Planktonic organisms are mostly microinvertebrates suspended in the water column, where they are sampled using plankton tows or fine-mesh sweep nets. Most macroinvertebrates, in contrast, are associated with various substrates, including the bottom sediments and plant surfaces (benthos) or the water's surface (neuston), where they are usually sampled with corers and coarse-mesh sweep nets. However, numerous exceptions to this dichotomy exist. Microinvertebrates such as nematodes and rotifers are mostly benthic, as are many species of microcrustaceans. On the other hand, several macroinvertebrate species inhabit the water column such as the freshwater jellyfish (*Craspedacusta sowerbii*), fairy shrimps (Anostraca), and phantom midges (Chaoboridae). Considering macroinvertebrates and microinvertebrates separately is purely arbitrary, and efforts addressing both micro- and macroinvertebrates would obviously be optimal. However, such holistic approaches remain rare largely due to more extensive sampling and processing costs (almost double) and also to a lack of taxonomic expertise by most researchers to deal with the full range of organisms.

The Aquatic Macroinvertebrate Fauna

Batzer and Ruhí (2013) recently assembled data on macroinvertebrates from 447 freshwater wetlands from across the globe to assess which taxa tended to dominate these habitats, at least in terms of occurrence. Table 1.1 lists the 40 macroinvertebrate families that occurred in at least 10 % of those 447 wetlands (another 135 less common taxa were also recorded). The list of common taxa includes 25 insects, 5 annelids, 4 crustaceans, 4 molluscans, 1 acarine (water mites), and 1 turbellarian, indicating that insects are by far the most diverse group in wetlands. Among the 25 insect families, 8 families were Diptera (flies), 5 were Hemiptera (water bugs), 4 were Coleoptera (water beetles), 4 were Odonata (damselflies and dragonflies), 2 were Ephemeroptera (mayflies), and 2 were Trichoptera (caddisflies). This is in stark contrast to the aquatic insect faunas in streams, where assemblages are dominated by Ephemeroptera, Plecoptera (stoneflies), and Trichoptera.

Batzer and Ruhí (2013) found that the Chironomidae (midges) and Dytiscidae (predaceous diving beetles) were the only families that were virtually ubiquitous across the 447 wetlands (Table 1.1). Corixidae (water boatmen), Hydrophilidae (water scavenger beetles), and Oligochaeta (aquatic worms) also occurred in most (>50 %) of the wetlands. Remarkably, most macroinvertebrate taxa occurred only in a relatively small subset of available wetlands, although where they occurred, these less-widespread taxa can still be very abundant and ecologically important.

Table 1.1 Forty aquatic macroinvertebrate taxa that had ≥10 % occurrence across a set of 447 wetlands worldwide (from a meta-analysis by Batzer and Ruhí 2013)

Family[a]	Order/class	Percent occurrence	Dry phase strategy	Respiration	Feeding functions: primary/ secondary
Chironomidae	Diptera	97.3	D, M	C	C/P
Dytiscidae	Coleoptera	87.5	D, M	SA	P
Corixidae	Hemiptera	69.1	M	SA	P/C
Hydrophilidae	Coleoptera	67.1	M	SA/C	P
Oligochaeta[a]	Oligochaeta	58.6	D	C	C
Acarina[a]	Acarina	49.2	D, M	C	P/C
Ceratopogonidae	Diptera	46.5	D	C	P/C
Culicidae	Diptera	46.5	D, M	SA	C
Notonectidae	Hemiptera	45.9	M	SA	P
Libellulidae	Odonata	45.2	D, M	G	P
Limnephilidae	Trichoptera	41.6	D	G/C	Sh/P
Haliplidae	Coleoptera	39.6	D, M	SA/C	Sh
Sphaeriidae	Bivalvia	38.9	D	G	P
Physidae	Gastropoda	38.3	D	SA	Sc
Coenagrionidae	Odonata	38.0	M	G/C	P
Planorbidae	Gastropoda	37.6	D	SA	Sc
Baetidae	Ephemeroptera	36.0	M	G	C
Chaoboridae	Diptera	33.8	D, M	C	P
Lestidae	Odonata	29.5	D	G	P
Lymnaeidae	Gastropoda	28.6	D	SA	Sc
Lumbriculidae[b]	Oligochaeta	28.2	D	C	C
Turbellaria[a]	Turbellaria	27.5	D	C	P
Gerridae	Hemiptera	26.8	M	SA	P
Tipulidae/Limoniidae	Diptera	26.8	M	SA	Sh/C
Glossiphoniidae	Hirudinea	22.1	D	C	P
Gyrinidae	Coleoptera	20.4	M	SA/G	P
Aeshnidae	Odonata	19.2	M	G	P
Dixidae	Diptera	18.1	D, M	SA	C
Tubificidae[b]	Oligochaeta	17.9	D	C	C
Asellidae	Malacostraca	17.4	D	C	C/Sh
Tabanidae	Diptera	17.0	D, M	SA	P
Stratiomyidae	Diptera	16.1	D, M	SA	C
Erpobdellidae	Hirudinea	14.8	D	C	P
Dogielinotidae	Malacostraca	13.8	D	G	C/Sh
Caenidae	Ephemeroptera	11.9	M	G	C
Lynceidae	Diplostraca	11.4	D	C	C
Leptoceridae	Trichoptera	10.7	M	C	P/C
Pleidae	Hemiptera	10.5	M	SA	P

(continued)

Table 1.1 (continued)

Family[a]	Order/class	Percent occurrence	Dry phase strategy	Respiration	Feeding functions: primary/ secondary
Crangonyctidae	Malacostraca	10.3	D	G	C/P
Belostomatidae	Hemiptera	10.3	M	SA	P

The last three columns indicate (1) each group's strategy for dealing with drought (*D* desiccation tolerance, *M* migration), (2) each group's primary mode(s) of respiration (*SA* surface air breathers, *C* cutaneous, *G* gills), and (2) each group's primary and secondary feeding functions (*C* collecting, *P* predation, *Sc* scraping, *Sh* shredding)

[a]Oligochaeta, Acarina, and Turbellaria are not families, and these categories include all families for those groups. These higher taxa were used because many authors do not report the families involved

[b]Lumbriculidae and Tubificidae are families in Oligochaeta, so these data also contributed to the table under that classification

The Aquatic Microinvertebrate Fauna

The microinvertebrates of wetlands are studied less than the macroinvertebrates, particularly the meiobenthos. Compositions of microinvertebrates in planktonic and benthic habitats differ markedly. In the water column, planktonic rotifers and microcrustaceans dominate in terms of biomass and species richness. Along wetland substrates, nematodes are the dominant microinvertebrates, although microturbellarian flatworms, Gastrotricha (hairy backs), Tardigrada (water bears), as well as some rotifers and microcrustaceans can be abundant (Rundle et al. 2002) and productive (Anderson et al. 1998). Species richness of nematodes is particularly high, with 605, 327, and 160 species being described in freshwater of Europe, Africa, and North America, respectively (Traunspurger 2002), with many more species as yet undescribed. While often considered planktonic, most rotifers and cladocerans (water fleas) are actually benthic (Margalef 1983; Wetzel 2001).

Ecological roles of planktonic and benthic microinvertebrate communities also differ. Feeding by planktonic microcrustaceans and rotifers can control phytoplankton primary production (Scheffer et al. 1993). In turn, planktivorous and piscivorous fishes can affect zooplankton productivity either directly or indirectly via trophic cascades (sensu Hairston et al. 1960). In this sense, planktonic microinvertebrates become focal to pelagic food webs (Angeler et al. 2003). Feeding by meiobenthos on microbes in biofilms can affect bacterial composition, biomass, or production, again via direct or indirect pathways (Hakenkamp et al. 2002). These impacts on wetland bacteria can affect biochemical processes such as cellulose degradation (Toyohara et al. 2012).

The Terrestrial Invertebrate Fauna

While the aquatic invertebrate fauna has garnered the most research attention, wetland habitats also can support a rich terrestrial invertebrate fauna (myriapods, spiders, mites, beetles). The terrestrial faunas in floodplains (see Chaps. 13 and 14) and peatlands (see Chap. 7) seem especially well developed. The terrestrial fauna is dominated by plant and soil associates. While caterpillars and other herbivorous insects feed on the leaves of wetland macrophytes and trees, the fact that these plants happen to occur in wetlands is not particularly relevant to the ecology of these invertebrates; thus, we do not expand on these invertebrates here. However, the terrestrial invertebrates living in and on the soils of wetlands are subject to periodic flooding and thus must be specifically adapted to tolerate inundation to thrive in wetlands. Just as a habitat-specific aquatic invertebrate fauna exists in wetlands, a habitat-specific terrestrial invertebrate fauna also appears to exist in wetlands (Bright et al. 2010). Many nonaquatic invertebrates in wetlands can survive being underwater for extended periods (Rothenbücher and Schaefer 2006), and some millipedes and spiders have the ability to respire aquatically (Adis 1986; Pedersen and Colmer 2012).

Complete descriptions of terrestrial invertebrate assemblages in wetlands are lacking, and existing research tends to focus on a few groups. Braccia and Batzer (2001) described invertebrate assemblages associated with submersed wood in a floodplain and found terrestrial mites (especially Oribatida), springtails (Collembola), and various wood-associated beetle larvae and adults to be widespread. Mites associated with wood and leaf litter are very widespread in forested floodplains of the Southeastern USA and can readily tolerate flooding. Numerous species of ants (Formicidae) occur in wetlands, and the economically important imported fire ant (*Solenopsis invicta*) of South and now North America is a wetland-associated taxon (Ahrens et al. 2005). In peatlands, ant mounds are focal points for the emissions of greenhouse gases (carbon dioxide and methane) (see Chap. 7; Wu et al. 2013). A diversity of ground beetles (Carabidae) occurs in wetlands, and in Europe carabids are being used as bioindicators of the ecological health of floodplains (Greenwood et al. 1991; Boscaini et al. 2000) and peatlands (Holmes et al. 1993). Spiders are important predators in a host of wetlands (e.g., Jordan et al. 1994; Denno et al. 2002).

Wetland Hydrology and Invertebrates

Because hydrology, controlled by climate, ultimately structures wetland environments, specific wetland types can often be categorized by their water budgets (Jackson et al. 2014). For any particular wetland, knowing which water inputs and water outputs control the hydrology of the habitat can provide useful ecological insights. How water enters and leaves a wetland is summarized in Fig. 1.2 and the following water budget equation:

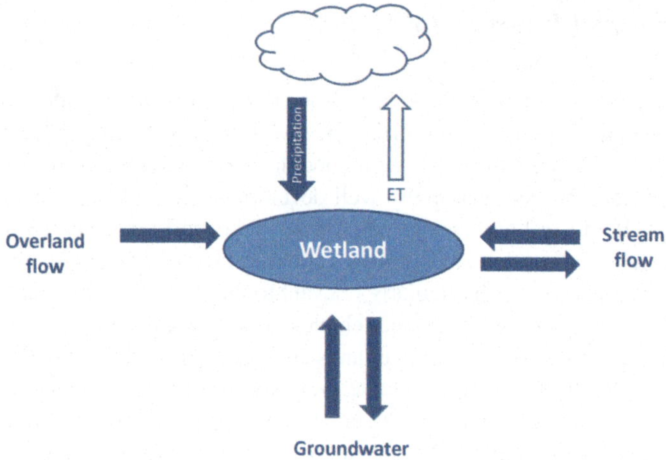

Fig. 1.2 Cartoon depicting the major water inputs and outputs to wetlands

$$P + GW_{in} + \mathrm{OF}_{\mathrm{in}} + \mathrm{SF}_{\mathrm{in}} = \mathrm{ET} + \mathrm{GW}_{\mathrm{out}} + \mathrm{OF}_{\mathrm{out}} + \mathrm{SF}_{\mathrm{out}} + \Delta V$$

where

P = volume of precipitation falling on the wetland
ET = evapotranspiration from the wetland
$\mathrm{GW}_{\mathrm{in}}$ = volume of groundwater flow into the wetland
$\mathrm{GW}_{\mathrm{out}}$ = volume of groundwater flow leaving the wetland
$\mathrm{OF}_{\mathrm{in}}$ = volume of overland flow into the wetland
$\mathrm{SF}_{\mathrm{in}}$ = volume of stream/river flow into the wetland
$\mathrm{SF}_{\mathrm{out}}$ = volume of stream/river flow leaving the wetland
ΔV = change in water volume (or storage) per unit time.

Wetlands in general can be classified into precipitation-, overland flow-, groundwater-, or stream flow-based habitats. Examples of wetlands highlighted in this book that are primarily filled by precipitation include temperate seasonal ponds (Chap. 4), Mediterranean climate ponds (Chap. 5), some peatlands (e.g., bogs, Chap. 7), and rock pools (Chap. 2). Evapotranspiration tends to be the largest avenue of water output from precipitation-based wetlands. Examples of overland flow-based wetlands include alpine wetlands (Chap. 3) and northern seasonal ponds filled by snowmelt. Examples of groundwater-based wetlands include most permanent and semipermanent marshes (Chap. 8), lakeshore marshes (Chap. 9), some peatlands (fens, Chap. 7), turloughs (Chap. 6), and of course groundwater springs and seeps (Chap. 11). Permanently flooded wetlands are typically tied to surficial groundwater aquifers. Examples of wetlands filled by stream or river flow include temperate and tropical floodplains (Chaps. 13 and 14) and beaver wetlands (Chap. 12). Some wetlands defy simple hydrologic categorization such as the Florida Everglades (Chap. 10) where water inputs from direct precipitation, river flow, and groundwater are all important. Managed wetlands such as constructed wetlands (Chap. 15) and managed waterfowl marshes (Chap. 16) often rely on engineered water sources that can include precipitation, groundwater, overland flow, and/or stream/river flow. Even wetland types dominated by one major source of water usually also receive water from secondary sources. Because most aquatic invertebrates congregate in the lowest-lying areas of wetlands where water persists, the longest, secondary sources of water (e.g., groundwater) can often be very important to controlling the ecology of the invertebrate fauna.

Besides water budgets, a wetland's *hydroperiod* (or hydropattern or hydroregime) is an important way to hydrologically categorize habitat (Jackson et al. 2014). Hydroperiod refers to the amount of time surficial, standing water is present in a wetland, regardless of the source. Because invertebrates primarily live in association with the surficial water of wetlands, hydroperiod is an especially important factor controlling them. Wissinger (1999) maintained that five different aspects of hydroperiod combine to control aquatic invertebrate populations and communities, including:

1. Water permanence (permanent vs. semipermanent vs. temporary)
2. Predictability of filling (unpredictably, seasonally, over climatic cycles)
3. Seasonality of filling and drying
4. Duration of wet and dry phases
5. Harshness of wet or dry phases (extremes in temperature and desiccation)

Figure 1.3 summarizes how these various aspects of hydroperiod manifest in different wetland types.

Understanding how each of Wissinger's five aspects of hydroperiod can affect invertebrates yields valuable information about community controls.

1. *Water permanence* is a primary control on invertebrates because if a wetland dries, the permanent water species are eliminated (although they can recolonize) while those with desiccation resistance strategies can persist. Invertebrates in permanent waters can be large, slow-developing taxa, while those in temporary waters must be smaller, fast-developing taxa (Wellborn et al. 1996). Further, water permanence and the presence of invertivorous fish are often correlated.
2. *Predictability* is important to invertebrates because if filling or drying is very unpredictable and brief (e.g., ephemeral wetlands), only highly opportunistic species that can rapidly exploit newly created habitat and develop quickly in brief periods of inundation will occur (e.g., floodwater mosquitoes). But if patterns of drying and filling are very predictable (e.g., in vernal pools), a plethora of invertebrate taxa may adapt their life cycles to match that hydroperiod (e.g., anostracan fairy shrimp, dragonflies and damselflies, limnephilid caddisflies).
3. *Seasonality* is important to invertebrates because as ectotherms they are strongly regulated by temperature. If a wetland fills in summer when temperatures are high, then active flying insects (e.g., odonates, hemipterans) can readily colonize, and all invertebrate types can complete development rapidly. In contrast, if a wetland fills in winter when temperatures are low, few aerial colonists would

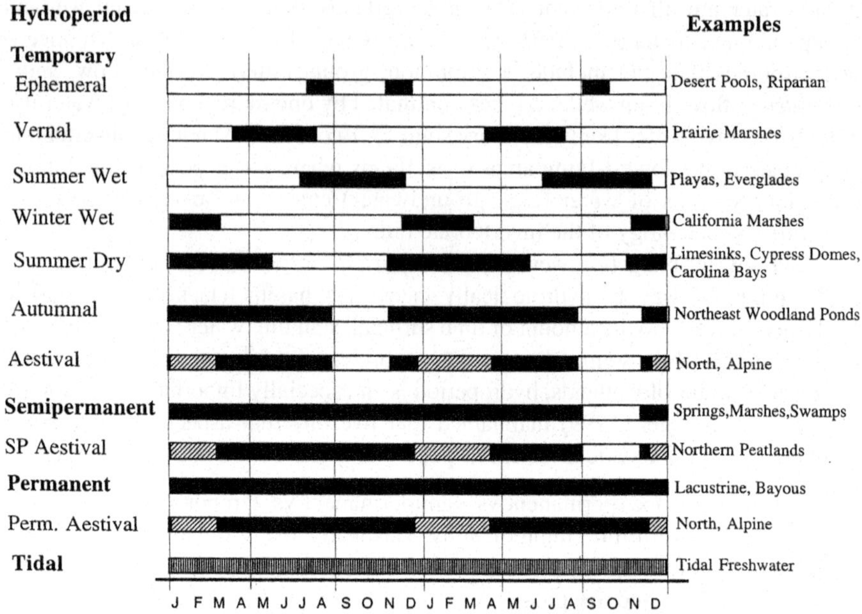

Fig. 1.3 Schematic showing different patterns of flooding and drying in representative wetlands of North America (listed on *right side* of the diagram). Dark bars indicate periods of flooding, and white bars periods of drying over a hypothetical 2 year period. *Hatched bars* indicate when the water column in boreal wetlands might freeze to the bottom in winter. Reprinted with permission from *Ecology of Freshwater and Estuarine Wetlands: Second Edition*, edited by Darold P. Batzer and Rebecca R. Sharitz. © 2014 by the Regents of the University of California. Published by the University of California Press

be active, and development of any invertebrates would be slow, requiring a long duration of flooding for them to succeed.

4. *Duration* of the wet period is important to invertebrates because some species can develop rapidly and exploit even very short duration hydroperiods (e.g., microcrustaceans, fly larvae), while others require many months to develop (e.g., large dragonfly nymphs and beetle larvae) and are only successful in long duration hydroperiods. Similarly, duration of the dry phase is also important because if the dry period is very long (months or years), only a few invertebrate species may be able to persist, but if short (days or weeks), even species poorly adapted to withstand desiccation might still cope.
5. *Harshness* of the wet or dry phase may impose additional constraints on invertebrates. In warm tropical or subtropical wetlands such as the Everglades (see Rader 1999), high water temperatures may stress invertebrates, either directly or via reduced oxygen supplies. In cold climates, invertebrates in seasonally dry wetlands in winter must withstand both desiccation and freezing, and even if a wetland is not dry in winter, the entire water column may still freeze (see Wissinger et al. 1999). In arid or semiarid climates, the substrates of wet-

lands may completely desiccate during dry phases permitting the diapause stages of only a few taxa to persist (Anderson et al. 1999). In contrast, in humid environments, substrates may remain moist even after surface water disappears, and a range of strategies by invertebrates to withstand drying may still be effective (see Ruhí et al. 2013).

The classic paper by Wiggins et al. (1980) categorizes how different aquatic invertebrates exploit annual temporary pools using as criteria the various strategies they employ to deal with wetland drying, including the ways they tolerate or avoid drought, how they disperse (passively vs. active aerially), and the seasonality of dispersal and oviposition. They devised four types:

1. *Overwintering residents*. These organisms can tolerate drought but lack active dispersal and thus occur in the wetlands year-round (in some form). Prominent examples include flightless invertebrates such as mollusks (clams and snails), annelids (worms and leeches), and crustaceans (copepods, cladocerans, ostracods, fairy shrimp, clam shrimp, and tadpole shrimp). As wetlands dry, these organisms can bury into damp substrates to diapause, but more often produce drought-resistant eggs or cysts (that if needed can persist in dry substrates for years).
2. *Overwintering spring recruits*. These organisms can tolerate drought but have adults that emerge from the wetlands and aerially colonize new habitat, laying eggs on the water. Because water in most annual pools is only reliably present in spring, this is the season when oviposition occurs. When wetlands dry, these organisms persist as drought-resistant eggs or nymphs/larvae. Prominent examples include some Dytiscidae beetles and several Chironomidae midges. Some parasitic mites (Hydrachnidia) also fit into this group, and they disperse aerially attached to their insect prey.
3. *Overwintering summer recruits*. These organisms can tolerate drought but have adults that emerge from the wetlands and aerially colonize new habitats, laying eggs on drying wetland substrates. Because drying substrates develop in summer, this is the season when oviposition occurs. (Given the relatively minor differences with Type 2 organisms, some suggest combining Types 2 and 3). Prominent examples include some damselflies and dragonflies (Lestidae, Libellulidae), Limnephilidae caddisflies, *Aedes* and *Ochlerotatus* floodwater mosquitoes, and several Chironomidae.
4. *Non-wintering spring recruits*. These organisms cannot tolerate drought, but instead work to avoid drought. Adults aerially colonize the wetlands after they flood in spring to lay eggs. Immatures then rapidly complete development prior to seasonal drying, emerge, and migrate to other water bodies to spend the winter. Prominent examples include most water bugs (Corixidae, Notonectidae), several beetles (most Dytiscidae and Hydrophilidae beetles), Baetidae mayflies, some Chironomidae, and some dragonflies (Aeshnidae, Libellulidae). These insects mostly spend the winter in nearby aquatic habitats, but some dragonfly species (e.g., *Anax junius*) migrate to warmer areas in winter to produce another generation that then migrate back in spring to oviposit in seasonal ponds as they flood.

While devised for the invertebrate fauna in annual temporary pools of Canada, the Wiggins categories have broader utility to many kinds of nonpermanent wetlands worldwide.

Wissinger (1997) expanded on the Type 4 concept identifying invertebrates he called *cyclic colonizers*, of which some beetles (Dytiscidae) and water bugs (Corixidae) are perhaps the best examples. These organisms cycle predictably between seasonally flooded wetlands (of all kinds) and permanently flooded wetlands (or lakes or rivers). A typical scenario is for reproductive adults to leave permanent water refugia in spring to aerially colonize newly filled seasonal wetlands and lay eggs. In some cases females will then dissolve their flight muscles to provide energy and internal space for additional egg production. After ovipositing, these adults then die. The eggs hatch into a new generation of nymphs or larvae which can exploit the food-rich environments of the seasonal wetlands to develop. In some cases, these immatures develop into one or more generations of flightless short-winged adults that lack flight musculature, diverting that energy into further egg production. As the seasonal wetland begins to dry, a generation of flight-capable adults is produced which leave the site to return to permanent water refugia to spend the dry season. In the following spring, these individuals (or their progeny) then migrate back to the seasonal wetlands to begin a new cycle. Cyclic colonization permits invertebrates to effectively exploit seasonal wetlands despite lacking any ability to tolerate drying. Most cyclic colonizers are predators, and the strategy permits them to access the abundant prey that develop in seasonal wetlands soon after they fill (crustaceans, mosquito and midge larvae). Additionally, because seasonal wetlands are usually fishless, cyclic colonizers can operate there without the threat of fish predation. While migrating to and from seasonal wetlands is likely very risky, the benefits of cyclic colonization clearly outweigh the costs.

Sim et al. (2013) described how climate can influence the relative success of the four Wiggins et al. (1980) strategies. Where temporary wetlands occur in high rain areas, strong dispersers and those that require water for colonization (Types 2 and 4) are favored over weak dispersers and those that lay eggs on dry substrates. In moderately wet climates, most types of taxa can occur, although regular drying facilitates the persistence of drought-adapted taxa. Under low rainfall conditions (arid or semi-arid climates), the Type 1 strategy of desiccation resistance and passive dispersal is favored because colonization from the egg bank may be more efficient than colonization via aerial dispersal. However, under extreme drought conditions, diapausing eggs or cysts of Type 1 organisms may lose viability, reducing their prevalence. Williams (1985) further elaborates on how an arid climate may affect which invertebrates exploit temporary wetlands, maintaining that Type 3 organisms that lay eggs on dry substrates may fair poorly.

Gascón et al. (2008) proposed adding a *Type 5* strategy, consisting of organisms that actively disperse between permanent and temporary water bodies via swimming or crawling, rather than aerially. Lacking desiccation resistance, they (or their progeny) must then migrate back to the source permanent habitats as the seasonally flooded habitat dries. Prominent examples include amphipod crustaceans and snails. In an analogous strategy, leptophlebiid and siphlonurid mayfly nymphs actively swim from river channels into floodplain wetlands during high water events (e.g., Galatowitsch and Batzer 2011). They complete their development on the floodplain, with nymphs often persisting in residual pools of waters long after the hydrologic connection between the river and floodplain is cut. Adult mayflies emerge from the floodplain and fly back to the river channels to lay their eggs, and the cycle repeats.

We categorized the 40 widespread macroinvertebrate taxa in Table 1.1 by whether they were desiccation resistant or instead avoided drying via migration (i.e., were Type 4 taxa sensu Wiggins et al. 1980 or cyclic colonizers sensu Wissinger 1997). Twenty-seven taxa could withstand desiccation and 23 were migratory (with ten families having both desiccation-resistant species and migratory species). While there is a considerable focus on the ability of aquatic macroinvertebrates in wetlands to tolerate drying, it is clear that migratory organisms are also very important constituents (e.g., some Chironomidae and Dytiscidae, most Ephemeroptera, Corixidae, Notonectidae, and Hydrophilidae).

Respiration Strategies of Wetland Invertebrates

The hydrology of wetlands, where shallow standing water occurs in highly organic settings, leads to inherently low levels of dissolved oxygen developing in most wetland waters. Invertebrates from wetlands have developed a particularly wide range of adaptations to acquire oxygen. The unique character of aquatic invertebrate faunas in wetlands vs. streams and rivers is largely dictated by oxygen supplies. Some aquatic insects and mollusks that thrive in well-oxygenated streams, such as Plecoptera stoneflies and non-pulmonate snails, are essentially excluded from many wetlands due to oxygen constraints.

Like invertebrates in other aquatic habitats, some wetland invertebrates still use gills (highly tracheated plates or membranes) to extract oxygen (Fig. 1.4a), including dragonflies and damselflies, mayflies, and some beetles. However, numerous wetland invertebrates rely solely on oxygen exchange across the cuticle and might seem poorly adapted for life in low-oxygen wetland waters. Some of these organisms have long tubular bodies (Fig. 1.4a, b) that yield high surface area to volume ratios to facilitate oxygen transfer (e.g., annelid worms, midge larvae); some beetle larvae (Hydrophilidae, Haliplidae) have lateral extensions of the cuticle to increase surface area. A few taxa have hemoglobin in their hemolymph (e.g., Chironomidae, Tubificidae) that serves a respiratory function (Fig. 1.4b; Resh et al. 2008). Others can switch to anaerobic respiration when oxygen supplies become too low (Mendelssohn et al. 2014).

Many aquatic invertebrates in wetlands do not extract their oxygen needs from the water, but instead directly breathe surface air. Mosquito larvae and most dytiscid beetle larvae have terminal siphons that break the water's surface to access air, often ringed with hydrophobic hairs to prevent flooding. Two genera of mosquitoes

Fig. 1.4 Different adaptations of aquatic insects to extract oxygen for respiration: (**a**) damselfly nymph with three terminal gill plates on its abdomen (photo by M. Galatowitsch); (**b**) tubular chironomid midge larvae, one with reddish color from hemoglobin; (**c**) a *Coquillettidia* mosquito larvae with a siphon inserted into a plant rootlet to access internal oxygen supplies; and (**d**) a dytiscid beetle capturing an air bubble at the water's surface to place under its wings and over its respiratory spiracles. Photos **b–d** reprinted with permission from *Ecology of Freshwater and Estuarine Wetlands: Second Edition*, edited by Darold P. Batzer and Rebecca R. Sharitz. © 2014 by the Regents of the University of California. Published by the University of California Press

(*Coquillettidia*, *Mansonia*) have siphons that are adapted to pierce wetland plant roots or stems to access the air in open aerenchyma space inside the plants (Fig. 1.4c). Adult beetles and water bugs (Corixidae, Notonectidae) capture air bubbles and maintain them under their wings or along their body surfaces (Fig. 1.4d). They directly breathe the oxygen in these bubbles and also use the additional oxygen that tends to diffuse into the bubbles as internal oxygen concentrations are exhausted (this phenomenon is called the physical or compressible gill; Resh et al. 2008). Snails in wetlands are mostly pulmonates (Physidae, Planorbidae, Lymnaeidae, and

Ancylidae), derived from terrestrial ancestors and retaining air-breathing "lungs." Pulmonate snails migrate periodically to the water surface and "gulp" air into lunglike sacs to then use while underwater. Non-pulmonate snails, that dominate non-wetland freshwater habitats, have to extract their oxygen needs from water using internal gills that probably function poorly in many wetlands.

Of the 40 most widespread aquatic macroinvertebrate taxa (Table 1.1), 17 rely of surface air to satisfy their oxygen needs, 17 on respiration across the cuticle, and only 11 on gills (the total is >40 because some groups use multiple strategies). Thus, air breathing is a widely used strategy for invertebrates to exploit oxygen-poor wetland waters. However, the fact that very rudimentary cuticular respiration is also widely used by invertebrates in wetlands is perplexing and as yet unexplained.

Wetland Vegetation and Invertebrates

After hydrology, the dominant vegetation of a wetland is secondarily used for habitat categorization. Wetland floras are comprised of five main categories:

1. Emergent annual macrophytes (e.g., *Bidens*, *Polygonum*)
2. Emergent perennial macrophytes (e.g., *Carex*, *Phragmites*, *Typha*)
3. Submersed macrophytes (e.g., *Myriophyllum*, *Potamogeton*)
4. Woody trees (e.g., *Populus*, *Salix*, *Taxodium*) and shrubs (e.g., *Alnus*)
5. Algae, including cyanobacteria

Each of these plant types is controlled by hydrology. Emergent annual plants thrive in seasonal marshes and wet meadows with short hydroperiods (weeks to months), emergent perennial plants thrive in marshes with intermediate hydroperiods (months to years), and submersed plants thrive in ponds and marshes with long hydroperiods (usually multiple years). Woody vegetation in forested wetlands (often called swamps) tends to occur in short hydroperiod habitat or at least areas that are only flooded in winter and spring when trees and shrubs are largely dormant (e.g., Southeastern US floodplains). Certain wetland trees such as cypress (*Taxodium* spp.) can tolerate long-term flooding (decades), although successful reproduction requires periodic drawdown for seedlings to sprout (Schneider and Sharitz 1988). Because algae establish rapidly and can persist under all hydrologic conditions, algae tend to thrive in most wetland types, except perhaps heavily shaded forested wetlands.

Propagules of most wetland macrophytes persist in substrates as a *seed bank*, which only sprouts after the sediments are exposed during drought events. In the Prairie Pothole Region of North America (van der Valk 1981), a predictable pattern of vegetative succession occurs in response to drought with:

1. Annual plants dominating during dry phases and the early stages of reflooding
2. Perennial emergent hydrophytes dominating in subsequent years, if flooding persists

3. Eventually submersed vegetation and algae dominating with prolonged (multiple year) flooding

Euliss et al. (2004) suggest that invertebrate succession in prairie potholes will track these patterns of hydrologic and vegetative succession.

For wetland invertebrates, plants provide both habitat and food (Batzer and Wissinger 1996). Invertebrates colonize living and dead plant leaves and stems to forage and hide from predators; moist plant litter protects estivating invertebrates from excessive desiccation. Annually senescing emergent macrophytes and trees contribute copious amounts of dead leaves to wetlands, and it has long been assumed that this detritus provides the major trophic base for resident invertebrates. However, only 5 of the 40 widespread macroinvertebrates in wetlands (Table 1.1) are *shredders* (organisms that consume coarse plant matter such as dead leaves and wood). Limnephilidae caddisfly larvae are the only shredders shown to play major ecological roles in wetlands (Díaz-Villanueva and Trochine 2005; Klemmer et al. 2012, Chap. 3).

Instead, *collectors* that consume small particles of organic matter and algae comprise the bulk of the invertebrate primary consumers in wetlands (18 of 40 widespread macroinvertebrates, Table 1.1). Most food web studies in wetlands point toward algae as being the primary food base for resident invertebrates (see Batzer et al. 2014). Besides the macroinvertebrate collectors, snail *scrapers* and most microcrustaceans also feed primarily on algae. A trophic reliance by invertebrates on algae makes ecological sense because algae are an energetically superior food to macrophyte detritus (Fig. 1.5).

Predation and Wetland Invertebrates

Predation can be a pervasive influence on invertebrates in wetlands (Batzer and Wissinger 1996). Wellborn et al. (1996) argue that aquatic animal communities in lentic habitats (lakes and most wetlands) are controlled by two ecological transitions

1. Between temporary and permanent habitats (discussed above)
2. Between fish-bearing and fishless habitats

They maintained that the presence of fish in wetlands would eliminate large, active invertebrates because fish use visual cues to find prey. However, if fish were absent, these large invertebrates (along with amphibians) would become the top predators in the systems.

However, one should not assume that temporary or even ephemeral wetlands are predator-free habitats (Brendonck et al. 2002; Boix et al. 2006). Predators can be particularly important to structuring invertebrate communities of temporary wetlands because many inhabitants are poorly adapted to withstand predation

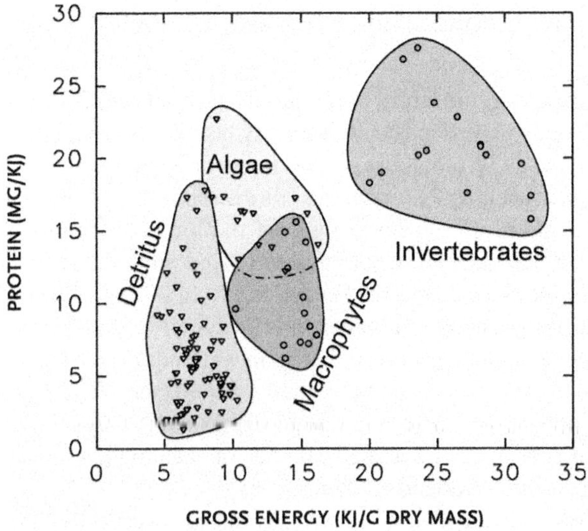

Fig. 1.5 Relative caloric and protein contents of various food resources used by invertebrates in wetlands, showing how detritus would be the lowest-quality foods and other invertebrates the highest-quality foods. Reprinted with permission from *Ecology of Freshwater and Estuarine Wetlands: Second Edition*, edited by Darold P. Batzer and Rebecca R. Sharitz. © 2014 by the Regents of the University of California. Published by the University of California Press

(Wilcox 2001; but see Petrusek et al. 2009). Aerially colonizing insects are common predators in temporary wetlands (Schneider and Frost 1996; Boix et al. 2011), and it has been shown that they can exert significant ecosystem control (Blaustein et al. 1995; Magnusson and Williams 2009).

A case for macroinvertebrate predators being ecologically important in the majority of wetlands is again bolstered by examining which taxa thrive there. In Table 1.1, 23 of 40 widespread macroinvertebrate taxa are either primary or secondary predators. Predation again makes energetic sense because invertebrate prey are high-quality foods (Fig. 1.5) (although often more energetically expensive to acquire than plants). Midge larvae, being the most widespread and typically most abundant invertebrates in wetlands (Table 1.1), are favored foods of virtually every predator that lives in wetlands, whether they be invertebrate (Rasmussen and Downing 1988; Batzer and Resh 1991), amphibian (Bohonak and Whiteman 1999; Wissinger et al. 1999), or fish (Batzer 1998; Batzer et al. 2000). The benefit of predation to invertebrates is evidenced by some limnephilid caddisfly larvae in high alpine wetlands (Chap. 3) that normally consume plant detritus (i.e., serve as shredders), but, as wetlands begin to dry, switch to being predaceous on other insects or even conspecifics in order to use the higher-quality foods to accelerate growth rates.

Importance of Wetland Invertebrates to Society

Wetlands contribute significantly to the biodiversity of the world because so many species occur solely in wetlands or at least rely heavily on wetlands to satisfy important ecological needs. While most people focus on what is often called the *charismatic megafauna*, such as the mammals, birds, reptiles, and plants of wetlands, in terms of sheer numbers of species, most of the biodiversity in many wetlands is comprised of invertebrates. It is not unusual to find more than 50 families of invertebrates in individual wetlands (see Batzer and Ruhí 2013), of which some families might be comprised of numerous genera and species. For the Chironomidae midges, the most widespread family of invertebrates in wetlands (Table 1.1), it is not unusual to find more than 50 species in an individual wetland (see Wrubleski and Rosenberg 1990; Leeper and Taylor 1998; see Chap. 10). In some cases, chironomid species richness might exceed the combined number of mammalian, avian, reptilian, and amphibian species in a wetland habitat.

Additionally, as already discussed, invertebrates play crucial roles in wetland food webs. In many cases, they are the primary trophic link between plants and the charismatic megafauna. Invertebrates feed heavily on living and dead macrophytes and algae and in turn are consumed by wetland fishes, amphibians, and birds (Fig. 1.6). Waterfowl ecologists have come to realize that most ducks consume invertebrates during crucially important periods; nesting hens rely heavily on the protein and lipids in invertebrates for egg production, newly hatched ducklings find invertebrates nutritious and easy to capture prey during their initial weeks of development, molting birds rely on invertebrates for protein for feather growth, and over-wintering migratory ducks focus on invertebrates to fuel flights back to nesting areas (see Chap. 16). Most fishes in wetlands rely on invertebrates as food (see Chaps. 8 and 10), and fishes in lakes and rivers will migrate into wetlands to consume invertebrates (see Chaps. 9, 13 and 14).

Invertebrates have proved to be useful indicators of environmental health in rivers and streams (see Rosenberg et al. 2008), and resident wetland invertebrates may show similar promise (see Chap. 15). In the wetlands of the Great Lakes of North

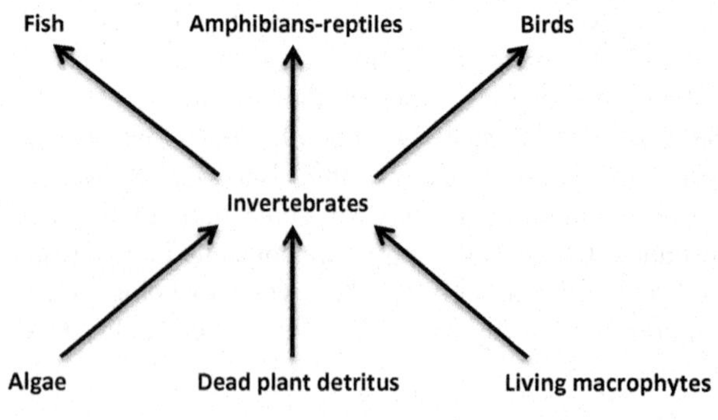

Fig. 1.6 Invertebrates as the primary trophic link between plants and higher animals in wetlands

America (see Chap. 9), invertebrates are widely used as bioindicators in habitat assessment programs. However, the fact that many invertebrates in wetlands are tolerant of harsh environmental conditions (low oxygen levels, high temperatures, fluctuating water levels) might make many taxa of fairly limited use as bioindicators (Batzer 2013). As in streams and rivers (Rosenberg et al. 2008), environmentally sensitive taxa in the insect orders Ephemeroptera, Plecoptera, and Trichoptera could be promising bioindicators in wetlands, although these organisms are not particularly wide spread in many wetlands. A recent meta-analysis suggests that the combined richness of Mollusca, Hemiptera, and Coleoptera might be a useful and easy-to-sample surrogate to predict overall invertebrate taxon richness in wetlands (Ruhí and Batzer 2014). Microcrustaceans have numerous properties that make them useful for water quality assessment (Boix et al. 2005): (a) they are ubiquitous in wetland environments and easily captured; (b) assemblages vary according to

Table 1.2 Significant human diseases associated with invertebrates from wetlands (see Mullen and Durden 2009 for more details)

Human disease	Wetland invertebrate connection
Schistosomiasis	Parasitic fluke (schistosomes) cycle between wetland snails and humans in Africa, Asia, and South America
Malaria	*Anopheles* spp. mosquitoes that breed in marshes worldwide can vector *Plasmodium* protozoan parasites
Filariasis (elephantiasis)	Various *Anopheles*, *Mansonia*, and *Culex* mosquitoes from marshes in Africa and Southeast Asia can vector *Wuchereria* or *Brugia* nematode parasites
West Nile encephalitis	*Culex* spp. mosquitoes that breed in marshes and wet meadows worldwide can vector WNE virus
Japanese encephalitis	*Culex tritaeniorhynchus* mosquitoes that breed in marshes and rice fields of Southeast Asia can vector JE virus
Eastern equine encephalomyelitis	*Culiseta melanura*, *Coquillettidia perturbans*, and *Culex* spp. mosquitoes that breed in forested swamps of the eastern USA can cycle the virus through bird populations (*Culiseta*) or vector EEE virus to humans or horses (other species)
Western equine encephalomyelitis	*Culex tarsalis* mosquitoes that breed in marshes and wet meadows of the western USA and Canada can vector the WEE virus to humans and horses
Murray Valley encephalitis	*Culex* spp. mosquitoes that breed in marshes of Australia can vector MVE virus to humans
St. Louis encephalitis	*Culex* spp. mosquitoes that breed in marshes and wet meadows of the USA can vector SLE virus to humans
Venezuelan equine encephalomyelitis	*Culex* spp. mosquitoes that breed in marshes of northern South America, Central America and Mexico, and south Florida can vector VEE virus to humans and horses
Loiasis (African eyeworm)	Tabanid deer flies that breed in damp soils and wetlands of Africa can vector the parasitic *Loa loa* nematodes to humans
Tularemia (deer fly fever)	Tabanid deer flies that breed in damp soils and wetlands of Utah and Russia can vector the bacteria
Allergies	Bites of wetland breeding mosquitoes, deer and horse flies, and biting midges (no-see-ums) can induce allergic reactions in sensitized people

differences in trophic state; (c) assemblages respond to disturbance gradients; and (d) relationships between microcrustacean assemblages and both phytoplankton and macrophyte communities are well documented. Because resident invertebrates are strongly affected by both temperature and hydroperiods, they may be especially useful bellwethers of the impacts of climate change on wetlands (Ruhí et al. 2013).

Unfortunately, some wetland invertebrates contribute significantly to human suffering. Blood-feeding mosquitoes (Culicidae), biting gnats (Ceratopogonidae), and deer and horse flies (Tabanidae) can plague humans and their livestock. Several important humans diseases, most notably malaria and schistosomiasis, are associated with invertebrates from wetlands (see Table 1.2). Because of their roles as disease vectors, it can be argued that mosquitoes are the most important animals on earth to human well-being. However, it should be noted that not all vector mosquitoes are derived from wetlands (e.g., non-wetland container-breeding species vector yellow fever and dengue viruses) and not all wetland mosquitoes are involved in disease cycles (in fact the vast majority are not).

Conclusion

Invertebrates of wetlands are uniquely adapted to exploit the highly variable and often stressful conditions that develop. Invertebrates comprise much of the biodiversity in wetlands, and invertebrates play focal roles in wetland food webs. A better understanding of the ecology of invertebrate fauna in wetlands will lead to a more complete understanding of overall wetland ecosystem functions. Toward this end, the remainder of the chapters in this book provides detailed and habitat-specific ecological information about invertebrates in wetlands from across the globe.

References

Adis J (1986) An "aquatic" millipede from a Central Amazonian inundation forest. Oecologia 68:247–349

Ahrens ME, Ross KG, Shoemaker DD (2005) Phylogeographic structure of the fire ant *Solenopsis invicta* in its native South American range: roles of natural barriers and habitat connectivity. Evolution 59:1733–1743

Anderson DH, Darring S, Benke AC (1998) Growth of crustacean meiofauna in a forested floodplain swamp: implications for biomass turnover. J N Am Benthol Soc 17:21–36

Anderson CR, Peckarsky BL, Wissinger SA (1999) Tinajas of Southeastern Utah: invertebrate reproductive strategies and the habitat templet. In: Batzer DP, Rader RD, Wissinger SA (eds) Invertebrates in freshwater wetlands of North America: ecology and management. Wiley, New York, pp 791–810

Angeler DG, Chow-Fraser P et al (2003) Biomanipulation: a useful tool for freshwater wetland mitigation? Freshw Biol 48:2203–2221

Batzer DP (1998) Trophic interactions among detritus, benthic midges, and predatory fish in a freshwater marsh. Ecology 79:1688–1698

Batzer DP (2013) The seemingly intractable ecological responses of invertebrates in North American wetlands: a review. Wetlands 33:1–15

Batzer DP, Resh VH (1991) Trophic interactions among a beetle predator, a chironomid grazer, and periphyton in a seasonal wetland. Oikos 60:251–257

Batzer DP, Ruhí A (2013) Is there a core set of organisms that structure macroinvertebrate assemblages in freshwater wetlands? Freshw Biol 58:1647–1659

Batzer DP, Wissinger SA (1996) Ecology of insect communities in nontidal wetlands. Annu Rev Entomol 41:75–100

Batzer DP, Pusateri CR, Vetter R (2000) Impacts of fish predation on marsh invertebrates: direct and indirect effects. Wetlands 20:307–312

Batzer DP, Cooper R, Wissinger SA (2014) Wetland animal ecology. In: Batzer DP, Sharitz RR (eds) Ecology of freshwater and estuarine wetlands, 2nd edn. Univ Calif Press, Berkeley, pp 151–183

Blaustein L, Kotler BP, Ward D (1995) Direct and indirect effects of a predatory backswimmer (*Notonecta maculata*) on community structure of desert temporary pools. Ecol Entomol 20:311–318

Bohonak AJ, Whiteman HH (1999) Dispersal of the fairy shrimp *Branchinecta coloradensis* (Anostraca): effects of hydroperiod and salamanders. Limnol Oceanog 44:487–493

Boix D, Gascón S et al (2005) A new index of water quality assessment in Mediterranean wetlands based on crustacean and insect assemblages: the case of Catalunya (NE Iberian Peninsula). Aquat Conserv Mar Freshw Ecosys 15:635–651

Boix D, Sala J, Gascón S, Brucet S (2006) Predation in a temporary pond with special attention to the trophic role of *Triops cancriformis* (Crustacea: Branchiopoda: Notostraca). Hydrobiologia 571:341–353

Boix D, Magnusson AK et al (2011) Environmental influence on flight activity and arrival patterns of aerial colonizers of temporary ponds. Wetlands 31:1227–1240

Boscaini A, Franceschini A, Maiolini B (2000) River ecotones; carabid beetles as a tool for quality assessment. Hydrobiologia 422(423):173–181

Braccia A, Batzer DP (2001) Invertebrates associated with woody debris in a southeastern U.S. forested floodplain wetland. Wetlands 21:18–31

Brendonck L, Michels E, De Meester L, Riddoch B (2002) Temporary pools are not "enemy-free". Hydrobiologia 486:147–159

Bright EG, Batzer DP, Garnett JA (2010) Variation in invertebrate and fish communities across floodplain ecotones of the Altamaha and Savannah Rivers. Wetlands 30:1117–1128

Cowardin LM, Carter V, Golet FC, LaRoe ET (1979) Classification of wetlands and deepwater habitats of the United States, FWS/OBS-79/31. US Fish Widl Serv, Washington, DC

Denno RF, Gratton C et al (2002) Bottom-up forces mediate natural-enemy impact in a phytophagous insect community. Ecology 83:1443–1458

Díaz-Villanueva V, Trochine C (2005) The role of microorganisms in the diet of Verger cf. limnophilus (Trichoptera: Limnephilidae) larvae in a Patagonian Andean temporary pond. Wetlands 25:473–479

Euliss NH, LaBaugh JW et al (2004) The wetland continuum: a conceptual framework for interpreting biological studies. Wetlands 24:448–458

Galatowitsch ML, Batzer DP (2011) Benefits and costs of *Leptophlebia* (Ephemeroptera) mayfly movements between river channels and floodplain wetlands. Can J Zool 89:714–723

Gascón S, Boix D, Sala S, Quintana XD (2008) Relation between macroinvertebrate life strategies and habitat traits in Mediterranean salt marsh ponds (Emporda wetlands, NE Iberian Peninsula). Hydrobiologia 597:71–83

Greenwood MT, Bickerton MA et al (1991) The use of Coleoptera (Arthropoda: Insecta) for floodplain characterization on the River Trent. Regul Rivers: Res Manage 6:321–332

Hairston NG, Smith FE, Slobodkin LB (1960) Community structure, population control and competition. Am Nat 44:421–425

Hakenkamp CC, Morin A, Strayer DL (2002) The functional importance of freshwater meiofauna. In: Rundle SD, Robertson AL, Schmid-Araya JM (eds) Freshwater meiofauna: biology and ecology. Backhyus, Leiden, pp 321–335

Holmes PR, Boyce DC, Reed DK (1993) The ground beetle (Coleoptera: Carabidae) fauna of Welsh peatland biotopes: factors influencing the distribution of ground beetles and conservation implications. Biol Conserv 63:153–161

Jackson CR, Thompson JA, Kolka RK (2014) Wetland soils, hydrology, and geomorphology. In: Batzer DP, Sharitz RR (eds) Ecology of freshwater and estuarine wetlands, 2nd edn. Univ Calif Press, Berkeley, pp 23–60

Jordan F, Jelks HL, Kitchens WM (1994) Habitat use by the fishing spider *Dolomedes triton* in a northern Everglades wetland. Wetlands 14:239–242

Klemmer AJ, Wissinger SA, Grieg HS, Ostrofsky ML (2012) Nonlinear effects of consumer density on multiple ecosystem processes. J Anim Ecol 81:770–780

Leeper DA, Taylor BE (1998) Insect emergence from a South Carolina (USA) temporary wetlands pond, with emphasis on the Chironomidae (Diptera). J N Am Benthol Soc 17:54–72

Magnusson AK, Williams DD (2009) Top-down control by insect predators in an intermittent pond—a field experiment. Ann Limnol Int J Limnol 45:131–143

Margalef R (1983) Limnología. Omega, Barcelona

Mendelssohn IA, Batzer DP, Holt CR, Graham SA (2014) Abiotic constraints for wetland plants and animals. In: Batzer DP, Sharitz RR (eds) Ecology of freshwater and estuarine wetlands, 2nd edn. Univ Calif Press, Berkeley, pp 61–86

Mullen GR, Durden LA (eds) (2009) Medical and veterinary entomology, 2nd edn. New York, Elsevier

Oertli B, Biggs J et al (2005) Conservation and monitoring of pond biodiversity: introduction. Aquat Conserv Mar Freshw Ecosys 15:535–540

Pedersen O, Colmer TD (2012) Physical gills prevent drowning of many wetland insects, spiders and plants. J Exp Biol 215:705–709

Petrusek A, Tollrian R et al (2009) A "crown of thorns" is an inducible defense that protects *Daphnia* against an ancient predator. Proc Natl Acad Sci U S A 106:2248–2252

Rader RB (1999) The Florida Everglades: natural variability, invertebrate diversity, and foodweb stability. In: Batzer DP, Rader RD, Wissinger SA (eds) Invertebrates in freshwater wetlands of North America: ecology and management. Wiley, New York, pp 25–54

Rasmussen JB, Downing JA (1988) The spatial response of chironomid larvae to the predatory leech *Nephelopsis obscura*. Am Nat 131:14–21

Resh VH, Buchwalter DB, Lamberti GA, Eriksen CH (2008) Aquatic insect respiration. In: Merritt RW, Cummins KW, Berg MB (eds) An introduction to the aquatic insects of North America. Kendall/Hunt, Dubuque IA, pp 39–54

Rosenberg DM, Resh VH, King RS (2008) Use of aquatic insects in biomonitoring. In: Merritt RW, Cummins KW, Berg MB (eds) An introduction to the aquatic insects of North America. Kendall/Hunt, Dubuque IA, pp 123–138

Rothenbücher J, Schaefer M (2006) Submersion tolerance in floodplain arthropod communities. Basic Appl Ecol 7:398–408

Ruhí A, Batzer DP (2014) Assessing congruence and surrogacy among wetland macroinvertebrate taxa towards efficiently measuring biodiversity. Wetlands 34:106–1071

Ruhí A, Boix D et al (2013) Functional and phylogenetic relatedness in temporary wetland invertebrates: current macroecological patterns and implications for future climatic change scenarios. PLoS One 8(11), e81739

Rundle SD, Robertson AL, Schmid-Araya JM (2002) Freshwater meiofauna: biology and ecology. Backhyus, Leiden

Scheffer M, Hosper SH et al (1993) Alternative equilibria in shallow lakes. Trends Ecol Evol 8:275–279

Schneider DW, Frost TM (1996) Habitat duration and community structure in temporary ponds. J N Am Benthol Soc 15:64–86

Schneider RL, Sharitz RR (1988) Hydrochory and regeneration in a bald cypress-water tupelo swamp forest. Ecology 69:1055–1063

Sharitz RR, Batzer DP, Pennings SC (2014) Ecology of freshwater and estuarine wetlands: an introduction. In: Batzer DP, Sharitz RR (eds) Ecology of freshwater and estuarine wetlands, 2nd edn. Univ Calif Press, Berkeley, pp 1–22

Sim LL, Davis JA et al (2013) The influence of changing hydroregime on the invertebrate communities of temporary seasonal wetlands. Freshw Sci 32:327–342

Toyohara H, Park Y, Tsuchiya K, Liu W (2012) Cellulase activity in meiobenthos in wetlands. Fish Sci 78:133–137

Traunspurger W (2002) Nematoda. In: Rundle SD, Robertson AL, Schmid-Araya JM (eds) Freshwater meiofauna: biology and ecology. Backhyus, Leiden, pp 63–104

van der Valk AG (1981) Succession in wetlands: a Gleasonian approach. Ecology 62:688–696

Wellborn GA, Skelly DK, Werner EE (1996) Mechanisms creating community structure across a freshwater habitat gradient. Annu Rev Ecol Syst 27:337–363

Wetzel RG (2001) Limnology: lake and river ecosystems. Academic, San Diego

WFD CIS (2003) The role of wetlands in the Water Framework Directive. Guidance Document, 12. Directorate General Environment of the European Commission, Brussels, ISSN: 1725-1087. ISBN 92-894-6967-6

Wiggins GB, Mackay RJ, Smith IM (1980) Evolutionary and ecological strategies of animals in annual temporary pools. Arch Hydrobiol Suppl 58:97–206

Wilcox C (2001) Habitat size and isolation affect colonization of seasonal wetlands by predatory aquatic insects. Isr J Zool 47:459–475

Williams WD (1985) Biotic adaptations in temporary lentic waters, with special reference to those in semi-arid and arid regions. Hydrobiologia 125:85–110

Wissinger SA (1997) Cyclic colonization and predictable disturbance: a template for biological control in ephemeral crop systems. Biol Contr 10:1–15

Wissinger SA (1999) Ecology of wetland invertebrates: synthesis and applications for conservation and management. In: Batzer DP, Rader RD, Wissinger SA (eds) Invertebrates in freshwater wetlands of North America: ecology and management. Wiley, New York, pp 1043–1086

Wissinger SA, Bohonak AJ, Whiteman HH, Brown WS (1999) Subalpine wetlands in Colorado: habitat permanence, salamander predation, and invertebrate communities. In: Batzer DP, Rader RD, Wissinger SA (eds) Invertebrates in freshwater wetlands of North America: ecology and management. Wiley, New York, pp 757–790

Wrubleski DA, Rosenberg DM (1990) The chironomidae of bone pile pond, Delta Marsh, Manitoba, Canada. Wetlands 10:243–275

Wu HT, Lu XG et al (2013) Ant mounds alter spatial and temporal patterns of CO_2, CH_4 and N_2O emissions from a marsh soil. Soil Biol Biochem 57:884–891

Chapter 2
Invertebrates in Rock Pools

Luc Brendonck, Sandro Lanfranco, Brian Timms, and Bram Vanschoenwinkel

Introduction to Rock Pools: Geology, Hydrology, Environmental Conditions, Flora, and Fauna

Rock Pool Geology and Geography

Rock pools are eroded depressions that occur in a matrix of bedrock (Fig. 2.1). In intertidal zones these habitats contain seawater and house communities of marine organisms; however, further up the rock platform, rainfall input becomes a more important source of water than saltwater spray resulting first in brackish and, further inland, in freshwater supralittoral habitats (Ganning 1971; Egan and Ferrington 2015).

Rock pools are found in different types of bedrock but most often in granite, sandstone, and limestone. Granite is an intrusive igneous rock which forms at high temperatures and under great pressure below the surface. Subsurface weathering can fragment the granite, and only the sturdiest fragments, rich in the more resistant potassium feldspars and quartz, typically remain when they become exposed after erosion of the covering sediments (Campbell 1997). This type of granite outcrop is common in the inland areas of Western Australia. Pools may already be present in these rock slabs (Fig. 2.1a) upon exposure as a result of weathering below the surface (Twidale and Bourne 1975). Once exposed, corrosion by acidic water along near-surface horizontal cracks formed by cycles of hot days and cold nights and rare frosts results in a pool basin, with detritus blown away by wind (Timms and Rankin 2014).

Many sandstone rock pools in Southern Africa (Fig. 2.1c) are formed in a different way. Here, organic concretions that got locked in the sandstone during sedimentation erode faster than the surrounding quartzite. Resulting pool basins in this rock often have near vertical walls, while granite rock pools are most often pan shaped. For the rock pools in the sandstone of the Colorado plateau in

D. Batzer, D. Boix (eds.), *Invertebrates in Freshwater Wetlands*,
DOI 10.1007/978-3-319-24978-0_2

Fig 2.1 Examples of rock pool habitats around the world. (**a**) Inland rock pool on a granite outcrop near Mount Madden, Western Australia; (**b**) 1-m deep pit gnamma in a granite outcrop in Zimbabwe; (**c**) sandstone rock pools on Thaba Phatswa in the Free State, South Africa; (**d**) Rock pool in karstified limestone (Malta); (**e**) man-made rock cut structure at Mosta (Malta) functioning as a rock pool. (Photos a, b, c by B Vanschoenwinkel; d, e by S Lanfranco)

Utah in the USA, it has been suggested that biofilms of cyanobacteria play an important role as they may help dissolve the cement between the sand grains and ensure that pool basins erode faster than the surrounding matrix. Additionally, this film may also help to seal the basins when they hold water (Chan et al. 2005).

Limestone can also hold rock pools (Fig. 2.1d) but since this rock is more susceptible to weathering and erosion, rock pools in limestone presumably have a younger origin and do not persist for millions of years as granite rock pools do. Geological uplifting of ancient limestone reefs, e.g., in coastal areas of the Caribbean, can provide a matrix for rock pools to form (Romanuk and Kolasa 2002).

In the biseasonal Mediterranean climate, the initial stages of pool formation on karstic limestone involve ponding of rainwater and runoff water in small, localized sinks. This promotes solutional erosion of the bedrock gradually forming basin-like depressions. The sides of these basins are generally steep due to undercutting and often display a basal corrosion notch. The gradual enlargement of a solution basin modifies the microtopography in the vicinity and allows the interception of surface water that would otherwise accumulate elsewhere. As such, the formation process tends to develop isolated, relatively large basins surrounded by smaller pits. As the basin grows, adjacent basins may merge giving the lobed margin of several rock pools. Breaching of the walls of these basins arrests their gradual enlargement by forming an overflow channel that drains the pool and provides hydrological connectivity with other basins in the same pool landscape.

Rock pools act as traps for material transported by runoff and wind and therefore accumulate a layer of unconsolidated bottom sediment that mainly comprises clastic mineral particles and organic detritus that, depending of the age of the basin, is mainly from endogenic or exogenic origin. The sediment layer is a fundamental factor in determining the year-to-year persistence of species within these habitats, as it provides a matrix for the accumulation of propagules (seeds and spores) and resting stages (eggs, cysts) of pool species. This creates a "reserve of dormant stages" that could emerge under different conditions (storage effect, cf. Chesson 1985) and may promote long-term species coexistence. The sediment layer also provides a last reserve of moisture for pool biota toward the end of the wet season. Exploiting this reserve of moisture may lengthen the time available for reproductive cycles to be completed and may be the difference between reproductive success and reproductive failure. A study based on 39 pools from six pool landscapes showed that as sediment depth decreased, the number of faunal taxa in the pool decreased too (Briffa et al. 2014).

Freshwater rock pools are known under different names: "oricangas" in South America, "rock pools" in South Africa, "gnammas" in Australia, "lithothelms" in Bulgaria, riverine "potholes" in the USA, "tinajas" in the Western USA and Mexico, and "opferkessel" in Germany (Jocqué et al. 2010a). "Pans or pan gnammas" are terms often used for flat-floored, mostly shallow rock pools and "pit gnammas" are deep rock pools in Australia that may contain water throughout the year, even in very dry regions (White 2009; Timms 2013). Non-coastal, inland rock pool habitats occur worldwide but are most frequently found in arid and semiarid regions where the elements are more likely to expose large rock slabs. On the other end of the climate spectrum, rock pools also exist in arctic and alpine regions where glaciers can expose underlying bedrock (Ghilarov 1967; Meier and Soininen 2014). Rock pools may hold water from several days up to several months depending on pool depth, surface area, and local climatic conditions. In some cases very large or very deep rock pools may keep water year-round. Examples of these are the pit gnammas which can, for instance, be found in Australia (Timms 2013) (Fig. 2.1b) and Zimbabwe (Anusa et al. 2012).

Not all eroded depressions in exposed bedrock evolve into rock pools. Some may fill up with soil and house terrestrial plants. A recent study in South Africa has shown that the most exposed depressions on the edge of a rocky outcrop are more likely to lose any sediment and terrestrial vegetation and become pools, while less exposed basins situated near the center of outcrops can more easily accumulate sediment and evolve into terrestrial "gardens" (Buschke et al. 2012). The balance between sediment capture and sediment loss determines whether the sediment layer is stable or unstable in the long term. If the autochthonous organic component is disregarded, then the rates of sediment capture and loss would mainly depend on wind-borne and waterborne transport of particles. For a given basin, these rates depend on the surface area to morphological volume ratio (*A:V*) of the basin concerned (Lanfranco and Briffa 2012). Pool basins with a large surface area are a larger target and intercept a larger volume of sediment than basins with smaller surface areas. Deeper basins are efficient sediment traps since captured sediment is less likely to be removed. Basins with a low *A:V* may undergo complete infilling and would not function as an aquatic system since the aquatic phase would either be very brief or absent altogether. In these situations, the sediment is colonized by terrestrial macrophytes and further consolidated by the anchorage systems of these plants. Basins with a high surface area and shallow depth would not retain a confluent layer of sediment since they would be inadequately shielded from outward transport by wind and runoff. Such basins would not generally be functional rock pools since the high surface area to volume ratio would also promote high rates of evaporative water loss.

Inland rock pools on granite and sandstone outcrops (inselbergs) typically occur in a particular spatial hierarchical structure (Brendonck et al. 2010). Not only do pools occur in clusters with variable distances between clusters and pools, also the inselbergs are dotted in the landscape and add to the overall spatial structure of "islands (inselbergs) of islands (clusters) of islands (pools)."

A different type of freshwater rock pool can be found in or near rocky riverbeds. Water currents moving rocks can erode cylindrical river potholes within the riverbed (Springer et al. 2006). When the water level of the river drops, these habitats can be isolated from flowing water. Particularly old riverine potholes are often exposed along the edges of rivers and can rely both on riverine and on rainfall input for water. In terms of invertebrate biodiversity, these pools are often species poor, particularly when fish are abundant. Regular flushing by the river can prevent the establishment of egg banks for zooplankton. Additionally, regular colonization by fish may exclude predation-sensitive taxa. In this chapter, we will focus mainly on the endorheic rock pool habitats, i.e., rock pools that rely solely on rainfall for filling. It is these pools that typically contain the most diverse communities including many specialized and endemic taxa.

Finally, a special type of pool that deserves mentioning is man-made rock pools. For instance, a monastery near the town of Montmajour in France has a graveyard of tombs excavated in limestone which now form a cluster of temporary rock pools (Cher 2008). In Malta, in the Mediterranean, Punic tombs excavated in rock also act as temporary pools (Fig. 2.1e).

Rock Pool Hydrology

The hydrology of rock pools is relatively simple. With limestone as a likely exception, most rock pools are not leaky. Rainfall and overflow from neighboring pools are typically the only sources of water input, while it is sensible to assume that evapotranspiration is the most important loss factor. This assumption, however, no longer holds in areas where large mammals or cattle may drink from rock pools. A family of simple models has been developed to reconstruct the long-term inundation regime of rock pools based on long-term climate data and pool morphometry (Hulsmans et al. 2008; Altermatt et al. 2009; Vanschoenwinkel et al. 2009; Lanfranco and Grillas 2010; Tuytens et al. 2014). These models, however, do require some water level time series data to calibrate and validate the models for each pool. Applications of these models include the possibility of accurately quantifying the long-term hydrological selection regime that has helped to shape rock pool communities as well as variation in the evolution of traits in the local populations. For instance, it has been shown that long-term hydrological disturbance regimes can better predict current community structure than the hydroperiod of the pools as observed during sampling (Vanschoenwinkel et al. 2009). New model extensions not only assess water levels and the length of aquatic phases, they also allow estimations of the number of times a pool overflows or receives overflow by neighbors and the volume of water transported. In addition, they generate proxies for the salinity of the water during inundations (Tuytens et al. 2014). The latter is relevant since many rock pool invertebrates use low salinities as a cue for hatching from dormant propagules. Low salinities or conductivities are indicative of heavily diluted water and presumably of a long inundation. It should be emphasized that the total annual rainfall is not necessarily a good predictor of hydroperiod characteristics. A simple hydrological model for rock pools based on 7 years of baseline data (Lanfranco and Grillas 2010) predicted that November to December is the key period for pools in Malta. Rainfall during this period will determine the hydroperiod and, consequently, reproductive success of the pool biota.

Physical and Chemical Environment of Rock Pools

An overview of the chemical and physical conditions in freshwater inland rock pools is summarized in Brendonck et al. (2000) for Southern African rock pools and at a global scale by Jocqué et al. (2010a). In general, basins are filled with rain water, resulting in a highly diluted environment at the start of the inundation with conductivities below 10 μS cm^{-1}, approaching those of distilled water. At later stages of the inundation, conductivity values of up to about 200 μS cm^{-1} were measured. While most western Australian rock pools (gnammas) contain very fresh waters, some are so incised they never overflow and are hyposaline, as in a few on the western Nullarbor Plain (Timms 2012b). If so they support appropriate widespread hyposaline species and not the specialized fauna characteristic of typical gnammas. The generally shallow rock pools also have poor buffering capacity to changes in water quality variables such as pH, temperature, and oxygen. In southern African rock pools, pH values were recorded that varied between 4 and 11, depending on time of the day and phase in the hydrocycle (Brendonck et al. 2000). Pools also closely followed air temperature with values between 10 and 40 °C. Dissolved oxygen also follows a strong dial and seasonal cycle, with concentrations in South African rock pools fluctuating between 5.8 and 7.9 mg L^{-1} (De Vries 1996). Freshwater rock pools in general are rather oligotrophic (Jocqué et al. 2010a). Enrichment happens mainly through bacterial degradation of decaying organisms, fecal input from large (terrestrial) vertebrates, and allochthonous organic material (McLachlan 1981; Osborne and McLachlan 1985). Immediately after filling, dissolved nitrogen and phosphorus concentrations may be quite high, but decline quickly because of nutrient uptake by organisms and a reduced rate of nutrient supply from the sediment (Osborne and McLachlan 1985). Nutrients in the sediment get into the water via bioturbation (e.g., by Notostraca and Spinicaudata) (Osborne and McLachlan 1985). Nutrients are mainly removed from the system by flushing by intense rain, sediment erosion by wind and, to a lesser extent, by terrestrial predators and scavengers removing organisms from the pool basins (Osborne and McLachlan 1985), and possibly also by the emergence of adult insects (Jocqué et al. 2010a). Concentrations of nutrients and dissolved metal ions in freshwater rock pools have only occasionally been measured. Two studies by Baron et al. (1998) and Chan et al. (2005) give measurements of metal concentrations and chemical components in sandstone rock pools in Utah. Most of the values were normal to low for freshwater habitats, but the NO_3-N reached high concentrations (up to 3184 μg L^{-1}) (Chan et al. 2005); this was possibly related to variation in N-fixation by cyanobacterial mats, which lined the potholes in that particular study. Levels of nitrate and phosphate were found to be relatively low in temporary rock pools in Malta, suggesting tight nutrient recycling (Lanfranco 1995).

Flora and Fauna of Rock Pools

Aquatic flora of rock pool systems is often very limited. However, particularly in deeper freshwater rock pools, several groups of submerged, floating, and emergent macrophytes including macroalgae can be found. In many short-lived rock pools in Africa and Australia, mats of the fern *Isoetes* can be found. In Western Australia, the mudwort (*Glossostigma*) is a common rock pool plant. Several rare endemic *Myriophyllum* species are known from the region. Some are known from just one type locality (Geldard 2005). In Botswana rock pools, *Limosella capensis* occurred in more than half of the basins, without clear preference for short- or longer-lived pools (Brendonck 2003, pers. obs.). Mediterranean rock pools are colonized by several species from the genera *Damasonium*, *Ranunculus*, *Elatine*, *Crassula*, and *Callitriche*. Aquatic ferns, such as *Marsilea* and *Pilularia*, are also associated with these habitats. In Australia, Victorian rock pools often support *Callitriche*, *Isoetes*, and *Crassula* (Bayly 2011).

A detailed overview of the fauna found in rock pools worldwide has been assembled by Jocqué et al. (2010a). Although there clearly are biogeographic differences in the species and genera that inhabit rock pools, there are strong similarities in the major phyla, classes, orders, and families of freshwater invertebrates that inhabit rock pools in different parts of the world. The shallowest pools typically lack any aquatic vegetation and house communities of very disturbance-resistant taxa such as oribatid mites. Short-lived pools also often provide good breeding habitats for several groups of dipterans including ceratopogonid and chironomid midges, particularly for those with larvae that can resist desiccation. Slightly deeper pools with hydroperiods typically exceeding 1 week often house specialized temporary pool crustaceans belonging to the class Branchiopoda. Fairy shrimps (Anostraca), clam shrimps (Spinicaudata, Laevicaudata), and water fleas (Anomopoda) form resistant resting eggs and all are relatively common inhabitants of rock pools worldwide. Tadpole shrimps (Notostraca) are also occasionally found in rock pools but less commonly so (Dodson 1987; Vanschoenwinkel et al. 2012). Turbellaria may also hatch shortly after inundation and often become important predators of branchiopod crustaceans (Brendonck et al. 2002). Both branchiopods and flatworms typically hatch early after inundation and quickly reach high population densities. Fairy shrimp and tadpole shrimp tend to hatch first. The more predation-resistant clam shrimp and water fleas presumably can afford to hatch and mature later. Many pools also house diverse assemblages of seed shrimps (Ostracoda). These detritivorous organisms can reach high population densities in the sediment layer of rock pools. Some ostracods have been observed to hatch within a few hours of wetting at the start of the wet season, having survived the summer in a cryptobiotic state (Lanfranco 1995).

Only three species of salamander have been recorded from rock pools: *Ambystoma laterale* and *Ambystoma tigrinum*, both from North America (Smith 1983; Dodson 1987), and *Salamandra salamandra* from Israel (Blaustein et al. 1996, 2004). Fourteen species of toads and frogs belonging to several families have been found in rock pools (Jocqué et al. 2010a).

Overview of Invertebrate Research in Rock Pools

Invertebrate Assemblages of Rock Pools

Several rock pool sites around the world have been the subject of long-term studies. Excellent examples include the coastal rock pools along the Baltic Sea in Europe such as those on the Tvärmine archipelago in Finland (Pajunen and Pajunen 2003). Comparable coastal rock pools can be found on Appledore Island in Maine, USA (Simonis 2013). In Jamaica a large set of coastal rock pools have been studied on a fossil reef in Discovery Bay (Kolasa and Romanuk 2005). Mediterranean rock pools have been studied in Malta (Lanfranco et al. 1991; Lanfranco 1995) and Israël (Blaustein et al. 2004). In arid regions, rock pools have been studied on granite outcrops in Western Australia (Bayly 1982, 1997; Pinder et al. 2000; Weeks et al. 2006; Timms 2012a, b, 2013, 2014a, b), on three granite hills in Botswana (Riddoch et al. 1994), the Korannaberg mountain in central South Africa (Vanschoenwinkel et al. 2007), the rock pools on mountains in Malawi (McLachlan and Ladle 2001), and the rock pools on the Colorado plateau in Utah, USA (Dodson 1987; Graham 1994; Jocqué et al. 2007a, b, c). Studies from tropical and equatorial regions are underrepresented. At least in part this is likely due to the fact that in moister regions, outcropping rock is often covered by vegetation as is the case in the moister SW of Australia. It is also possible that rock pools do not form as readily under tropical conditions. For instance, the rainforests in SE Cameroon house quite a few granite outcrops; however, these do not hold proper rock pools (Vanschoenwinkel pers. obs.). Space restrictions prevent us from providing a complete account of rock pool research. Instead, we highlight different aspects of rock pool ecology, conservation, and evolutionary biology that have been studied in rock pool habitats in different parts of the world.

To fill a gap in integrated knowledge on rock pool ecology, Jocqué et al. (2010a) reviewed more than 150 papers on freshwater rock pools around the world spanning more than 100 years of research. Besides compiling a list of rock pool animal species, in that paper, the authors also present an elaborated and integrated picture of global diversity patterns and the most significant abiotic and biotic processes and also summarize cultural and conservational aspects specifically related to rock pools. As it is not the intention to repeat that rather complete review, we will restrict ourselves here to a summary of the most important patterns in faunal assemblages and bring some updated views on important ecological and evolutionary processes relevant for rock pool populations. About 460 aquatic animal species were listed (Jocqué et al. 2010a), and Appendix provides a truncated list of the invertebrate genera. Approximately 213 of these species were permanent inhabitants without an active migratory stage (i.e., so-called passive dispersers). As the local environmental conditions in rock pools can be highly variable and the flooding regimes unpredictable, this resident fauna is often composed of specialist and endemic species with a high tolerance to stress. A prominent group of rock pool residents are large branchiopods that are especially well represented by anostracans. As these organisms are large and have many pairs of filter-feeding appendages, they are the dominant filter feeders in freshwater rock pools. And although these pools are shallow, clear, and often devoid of vegetation, populations can be surprisingly dense and egg banks vast (Brendonck and Riddoch 2000). Of the six Anostraca genera recorded, *Branchinecta* (USA), *Branchinella* (Australia), *Branchipodopsis* (Africa), and *Branchipus* and *Tanymastix* (Europe) are considered rock pool specialists. *Streptocephalus* and *Linderiella* generally prefer low-transparency mud pools but were occasionally also found in rock pools (De Roeck et al. 2010). *Streptocephalus* species have a slower maturation rate than the typical rock pool species and are therefore less likely to be found in rock pools (Belk 1991; De Roeck et al. 2010). As for the small branchiopods (anomopod cladocerans), six out of the nine families were recorded from rock pools, with Bosminidae, Ophryoxidae, and Acantholeberidae usually absent (Jocqué et al. 2010a). In terms of species richness, Chydoridae, Macrothricidae, and Daphniidae were best represented. Rock pool ostracods are represented by 35 species mostly belonging to the Cyprididae and Limnocytheridae (Martens et al. 2008). The other families are poorly represented, probably due to the lack of drought-resistant resting stages (Martens 1996). Several rock pool taxa are poorly studied and current lists are still an underestimation of the real animal diversity in rock pools. Of the five known Turbellaria species from rock pools, for example, four were described from a single rock pool site in Botswana (Artois et al. 2004).

The remaining taxa listed by Jocqué et al. (2010a), and Appendix, were so-called active dispersers with migration usually restricted to the adult stage. Often these taxa were broadly distributed and also occurred in a wide range of temporary habitats besides rock pools. The Diptera were mainly represented by Chironomidae, Culicidae of the genera *Aedes* and *Culex*, and Ceratopogonidae. Almost all Hemiptera were Corixidae or Notonectidae. Beetles were represented mainly by Dytiscidae and Hydrophilidae.

Jocqué et al. (2010a) compared rock pool species richness among different regions and continents and drew some general patterns. It is no surprise that the highest species richness with a significant endemic component was revealed in southwestern West Australia, thanks to the consistent work on the fauna of freshwater rock pools by Bayly (1997), Pinder et al. (2000), and Timms (2012a,b; 2013, 2014a). At least 230 species have been recorded in the area, with rotifers, branchiopods, ostracods, dytiscids, and chironomids well represented (Pinder et al. 2000). On any rock outcrop, up to about 60–70 species are comprised within one metacommunity, though species lists for individual pools usually number fewer than 30 (Jocqué et al. 2007a, 2010a; Timms 2012a, b). Alpha and gamma diversities decrease northward and northwestward due to shorter and less reliable hydroperiods (Timms 2012a, b; Brendonck et al. 2014); cladocerans and ostracods are particularly affected. Diversity is also lower in the forests of the west coast possibly because of increased acidity due to leaf fall input (Bayly 1982, 1997; Timms 2012b) and the lack of deep pool basins in that area (Vanschoenwinkel pers. obs.). These pools are also more shallow and probably of a younger age. Eastward across the vast waterless Nullarbor Plain, there is a small area of similar inselbergs with gnammas on the upper Eyre Peninsula in South Australia. These are less species rich than the southwestern Australian gnammas, with an average of 18 species present per pool, a diversity per rock outcrop of about 30 species, and an overall fauna of about 60 species (Timms 2014b). The pan gnammas in that region have a limited endemic list mainly of chydorid cladocerans, while pit gnammas have a similar fauna of eurytopic species as in southwestern Australia. The iconic large branchiopods are represented by just two species, *Eulimnadia* n. sp. in the pans and *Lynceus magdaleanae* in the pits. When comparing the shallow Australian rock pools (pan gnammas) with the deeper pit gnammas, it is the shallower gnammas that support the most diverse and characteristic fauna (Pinder et al. 2000). The deeper pit gnammas, while largely seasonal, have longer hydroperiods and a less harsh environment overall and hence tend to have a widespread and ubiquitous fauna (Timms 2014a). Like pan gnammas, they also have iconic branchiopods, in this case two species of the laevicaudatan clam shrimp *Lynceus*.

Metacommunity Dynamics

Several studies showed that species sorting, i.e., the distribution of species based on local environmental conditions, is a dominant structuring process in rock pool communities (Vanschoenwinkel et al. 2007; Ng et al. 2009; Pandit et al. 2009). However, species sorting is imperfect, and within pool clusters there are indications for both mass effects and dispersal limitation. Pajunen (1986) wrote that *Daphnia* populations in certain favorable rock pools were much less prone to go extinct and that these pools are likely to be sources for colonization of less suitable marginal rock pool habitats from which populations regularly go extinct. Spatial community analyses performed in South African rock pools found that nearby and neighboring communities were more similar than would be expected based on similarities in environmental conditions (Vanschoenwinkel et al. 2007). The dispersal that fuels this homogenization over short distances is likely to be mediated by different vectors. First of all, dispersal via temporary overflows or rivulets has been shown to be substantial (Hulsmans et al. 2007; Vanschoenwinkel et al. 2008b; Pellowe-Wagstaff and Simonis 2014), and their homogenizing effect on communities was observed in empirical community patterns (Vanschoenwinkel et al. 2007). On Appledore Island, local differences in the abundance of sea gulls frequenting rock pool clusters were shown to have similar homogenizing effects on communities (Simonis and Ellis 2014). In an inland South African rock pool cluster that is not frequently visited by birds, dispersal fluxes via wind, overflows, and amphibian vectors were compared. Here, wind was shown to be the dominant dispersal agent (Vanschoenwinkel et al. 2008b). Despite high dispersal rates in this system, there was still dispersal limitation. But this was restricted to pools that were both isolated and highly disturbed. In these habitats dispersal can probably not compensate for extinctions mediated by frequent disturbances (Vanschoenwinkel et al. 2013).

Analyses of the genetic structure of fairy shrimp metapopulations in Botswana rock pools revealed that even at the scale of several tens of meters, there was genetic differentiation with an isolation by distance pattern (Brendonck et al. 2000;

Hulsmans et al. 2007). This supports the presence of genetic regionalism even within small pool clusters as also confirmed in other areas (Vanschoenwinkel et al. 2011). Over larger spatial scales—among pool clusters on different mountains—there does not seem to be any gene flow. Instead these fairy shrimp metapopulations seem to be linked by very rare long-distance dispersal events that occur at a timescale of at least tens of thousands of years for mountain populations in South Africa (Vanschoenwinkel et al. 2011) or even up to several millions of years for certain inselberg populations in Western Australia (Zofkova and Timms 2009).

Daphnid Rock Pool Community Ecology

Several researchers have worked on the coexistence of *Daphnia* water flea species and corixid water bugs in Finnish rock pools. Pajunen and Pajunen (2003) assembled data on the occurrence of three *Daphnia* species in 507 rock pools on 16 islands over 17 years. Hanski and Ranta (1983) explained temporal variation in distribution patterns of these species in terms of differences in dispersal and competitive ability. Bengtsson's (1986) experiments showed that competitive exclusion did not occur in the field. He argued that variable responses to biotic and abiotic gradients could be a more likely explanation for variation in species distributions and that, although competition can be intense, closely related *Daphnia* may not be able to outcompete one another in the field. In later experiments he showed that extinctions were more frequent in mesocosms in which all three species were present subscribing the importance of interspecific competition as a driver of extinction (Bengtsson 1989). Building further on the same dataset, Altermatt et al. (2008, 2009) noticed that colonization events occurred more frequently in dry years, when many pools were dry. This claim is supported by earlier observations in African rock pools. Here, it was shown that wind dispersal fluxes were highest immediately after pools dried out (Vanschoenwinkel et al. 2008a, b). Brendonck and Riddoch (2000) also illustrated egg bank erosion in Botswana rock pools that was more significant in shallow pools not protected by vegetation. Altermatt and Ebert (2010) furthermore revealed that small and ephemeral pools contributed to more than 90 % of all dormant stages (ephippia) exposed during desiccation events in the entire metapopulation. They indicated this prominent role of small water bodies as drivers of metapopulation dynamics as an "inverse mainland-island type metapopulation."

Another line of research focused on the microparasites of *Daphnia* species (Bengtsson and Ebert 1998). Among many other findings, it has been shown that gene flow of rock pool *Daphnia* would be beneficial for persistence of the metapopulations because it counteracted the negative effects of inbreeding (Ebert et al. 2002). Besides the long-term data from Finland, another time series dataset exists for the small coastal rock pools (13–200 cm diameter) at Discovery Bay in Jamaica. In this system it was shown that more specious communities exhibited less variation in total abundances over years than less specious ones, although this pattern was initially not observed for individual populations (Romanuk and Kolasa 2002) but could only be confirmed after variation associated with species habitat specialization was taken into account (Kolasa and Li 2003). Additionally, diversity was shown to decrease if ponds dried out more frequently (Therriault and Kolasa 2001). Egan and Ferrington (2015) compared chironomid communities between two zones of freshwater coastal rock pools on Isle Royale (Michigan) in relation to distance from the shore. Pools closer to the shore had significantly higher diversity and housed a totally different community in comparison with more inland pools.

The morphometry of rock pools also has an effect on species diversity. Basin morphometrics favoring several short hydroperiods resulted in more diverse microcrustacean communities (Briffa et al. 2014), while life-form diversity of plants was positively correlated with morphometric heterogeneity and sediment-depth heterogeneity of the basins. While pioneer studies explored the predictive power of using a Levins-type metapopulation perspective with identical patches of rock pools (Hanski and Ranta 1983), more and more evidence has accumulated that even seemingly identical rock pools can be much more heterogeneous than one would expect.

Invertebrate Adaptations for Rock Pool Conditions

A large body of research has focused on life history adaptations of organisms to the temporary rock pool environment. In clusters of rock pools in Botswana, the life history and diapause ecology of the fairy shrimp *Branchipodopsis wolfi* were studied. It was shown that the species hatches at reduced electrical conductivities indicative of fresh rains and well-filled pools (Brendonck et al. 1998). Animals became mature in less than 1 week, depending on temperature and the hydroperiod of the pool (Brendonck et al. 2000). Maturation rate was faster and broods had smaller eggs in the more short-lived pools. A similar pattern was observed for anostracans in South African rock pools by Vanschoenwinkel et al. (2010a). Several parameters were determined that contribute to the egg bank budget in the freshwater rock pool system (Brendonck et al. 1998; Brendonck and Riddoch 2000; Hulsmans et al. 2007; Vanschoenwinkel et al. 2010a, b). It turns out that many populations of this species—with exceptions of those in very ephemeral pools—are remarkably stable and, because of the adaptations to this environment, do not frequently go extinct. Brendonck et al. (1998) tested the capacity of egg banks of *B. wolfi* to buffer against subsequent population crashes due to early drying of the pools. They discovered that the egg bank still served as the source of new cohorts after 16 consecutive wet/dry cycles with premature drying, spread over 2 years.

Although prominent, diapause, of course, is not exclusive to rock pool crustaceans. McLachlan and Ladle (2001) show that diapause also provides important opportunities for certain Diptera to thrive in the most ephemeral rock pools. Many aquatic insects, however, do not diapause and have developed keen senses to actively select suitable rock pools for egg deposition. This has been investigated in detail for mosquitoes in Israeli rock pools (Spencer et al. 2002). Among other observations it was shown that the presence of predatory salamanders affects their choice (Blaustein et al. 1996). In the coastal rock pools in Maine, prey abundance was shown to be a cue for interactive habitat selection in water bugs (Simonis 2013).

As freshwater rock pools usually hold shallow and clear water, it is expected that local populations are exposed to relatively high UV radiation and may require specific adaptations for protection against UV damage (Jocqué et al. 2010a). Adaptations can be behavioral, like the escape behavior observed in *Daphnia* from Scandinavian rock pools that stay close to the sediment during the day but are more evenly distributed during the night (Ranta and Nuutinen 1985). Exposed animals can also show morphological protective adaptations, such as increased pigmentation (Hebert and Emery 1990). Some Copepoda (for instance, *Boeckella opaqua* in Western Australia) have remarkable red pigmentation, while several Cladocera such as *Daphnia jollyi* in Western Australian rock pools, and a chydorid (*Leberis* sp.) in southeastern Botswana, have black pigmentation, not only in the ephippial capsule but also as adults (Jocqué et al. 2010a).

Key Ecological Factors Controlling Invertebrates in Rock Pools

Hydroperiod

Much the same as in other types of temporary aquatic habitats, the *length of the inundation* or *hydroperiod* is an important determinant of the species composition that can be found in a rock pool at a given moment (Fig. 2.2). However, due to strong variation in the lengths of inundations of individual rock pools, the long-term inundation history, including variation in the length of inundations and the onset and the frequency of inundations (often summarized as a multivariate entity known as the hydroregime), may constitute a better predictor. This is particularly so for organisms with dormant propagules that can be considered permanent residents of rock pools, i.e., the Group 1 organisms of Wiggins et al. (1980). For organisms that do not persist in situ during the dry period, past conditions are likely to be much less important than current conditions. For instance, actively dispersing insects obviously use currently available cues to select habitats for oviposition (Spencer et al. 2002), while for organisms with banks of long-lived propagules, current abundances may reflect both successes and demographic catastrophes experienced during an entire sequence of past growing periods. Hydroperiod acts as a life history-based selective filter. For instance, in Malawi, only Diptera with drought-resistant larvae can survive in the most short-lived pools, while those that lack this adaptation require longer hydroperiods (Cantrell and McLachlan 1982).

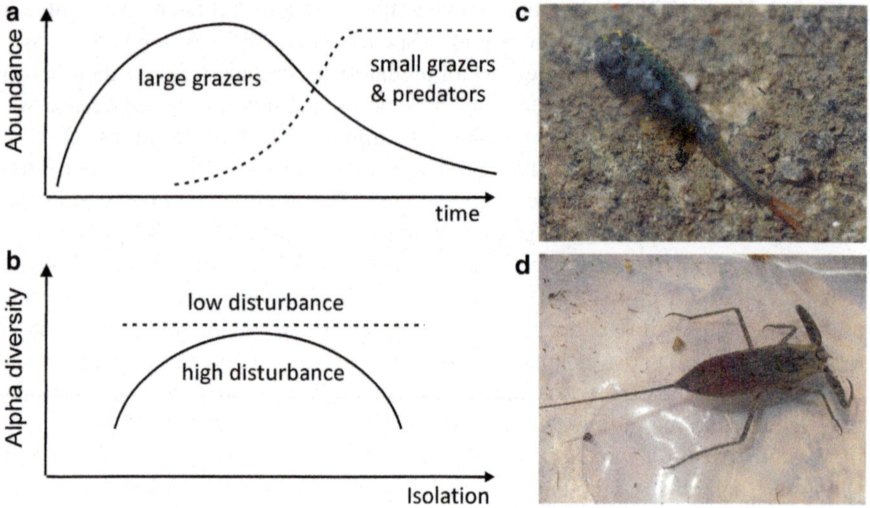

Fig. 2.2 (**a**) Conceptual diagram of succession in temporary rock pools based on observations in Southern Africa (Vanschoenwinkel et al. 2010b). Depending on the length of the hydroperiod, a first group of fast-growing large grazers declines as a result of increasing predation and can be replaced by smaller more predation-resistant grazers. (**b**) A unimodal relationship between isolation and alpha diversity was shown for passive dispersers in frequently disturbed pools in a South African pool cluster. This shows that location of pools within clusters as well as hydrological disturbance regime may determine local diversity patterns. (**c**) *Branchinella longirostris*, a typical early successional large grazing fairy shrimp in Australian rock pools. (**d**) A water scorpion, an occasional late successional actively dispersing inhabitant of rock pools in South Africa. (Photos: B Vanschoenwinkel)

Competition and Predation

There is circumstantial evidence pointing at the importance of a *trade-off between competition and predation tolerance* as an important underlying factor driving succession in these systems (Fig. 2.2). Three studies have documented succession in temporary rock pools in Botswana (30 days) (Jocqué et al. 2007b), South Africa (140 days) (Vanschoenwinkel et al. 2013), and Australia (140 days) (Timms 2012a), respectively. While species abundances do change over time, short-lived pools do not undergo major successional phases over the time period of 1 month (Jocqué et al. 2007b). These communities were dominated by fairy shrimp and clam shrimp as the dominant grazers of algae. In more long-lived pools in South Africa and Australia, a second successional phase could be observed. The arrival and population increase of many flying colonists via active dispersal including many predators (notonectids, beetles, dragonflies) represents the start of a new successional phase (Fig. 2.2b). The increase in predation ensures that populations of the predation-sensitive taxa such as fairy shrimp (Fig. 2.2c) will crash. This may be exacerbated by high densities of predatory flatworms. The large predation-sensitive grazers are replaced by more predation-resistant grazers such as water fleas. Competition trials have shown that these smaller water fleas are competitively inferior to the larger fairy shrimp (Jocqué et al. 2010b), suggesting that a trade-off between competitive ability and predation resistance could facilitate coexistence of these groups. Population dynamics of ostracods and copepods showed variable responses but typically also benefited from longer inundations. While these observations are based on just three field studies, the conclusions might hold for a wider range of rock pools as suggested by observations from other habitats in Australia (Brendonck et al. 2014). Large branchiopods and Turbellaria are typically lacking in the less speciose coastal rock pools where different successional dynamics are expected. Nestedness analyses support that subtropical and semiarid rock pool invertebrate assemblages consist of a core of generalist species that occur in both short-lived pools and in the early successional stages of long-lived pools (Vanschoenwinkel et al. 2013). This group is complemented by a second group of taxa that require longer inundations. The latter include both taxa that are quite predictably present in rock pools but also rarer taxa with longer aquatic life cycles and vagrants that only very rarely disperse into rock pools such as water scorpions (Fig. 2.2d) and Trichoptera.

While temporary rock pools are typically fishless, with exception of pools close to rivers or to the sea (Pajunen and Pajunen 2003), the presence of *keystone predators* such as salamanders impacts habitat selection by mosquitoes. Variable presence of *Notonecta* and *Salamandra* as a top predator had important top-down effects on consumers (Blaustein et al. 1995, 1999) even cascading down to primary producers (Arnér et al. 1998). In a survey of 45 rock pools in southeast Botswana, Brendonck et al. (2002) revealed a negative association between dragonfly nymphs and notonectids on the one hand and the anostracan prey populations on the other hand. In situ predation experiments revealed the predation efficiency of these groups that were indicated as top predators. Turbellarians are also often quite abundant and effective in predating on even much larger prey like fairy shrimp. In addition, they hatch at least as fast from the egg bank as their fairy shrimp prey (Brendonck et al. 2002). De Roeck et al. (2005) showed that besides impact by direct predation, turbellarians also had an indirect impact on the anostracan population by reducing the hatching success of the egg bank. A similar suppression of hatching of zooplankton resting eggs by the presence of predatory salamander larvae was suggested by Blaustein (1997) and confirmed by Spencer and Blaustein (2001).

Spatial Location and Connectivity

Pools near the edge of outcrops are more likely to lose sediment (Buschke et al. 2012). This might increase extinction risk for organisms with dormant egg banks. In rock pools in Malawi, McLachlan (1983) showed that isolation also matters for active dispersers. He found that certain chironomids from isolated pools were larger than their counterparts from pools with lots of proximate neighbors, presumably because of higher dispersal ability. Within a rock pool cluster in South Africa, it was shown that an interaction between pool isolation and long-term disturbance regime determined alpha diversity in these pools, as assessed using four sampling periods (Vanschoenwinkel et al. 2013). A hump-shaped relationship between pool isolation and alpha diversity was only detected in passive dispersers in the most frequently disturbed pools. This suggests that dispersal was not needed to maintain diversity in the least disturbed pools with long hydroperiods (Fig. 2.2b). In isolated disturbed pools it seems dispersal cannot always compensate for local extinctions due to frequent droughts. Lower isolation and increasing connectivity resulted in higher species richness. However richness decreased again in the most connected pools, possibly due to higher abundance of generalist predators in these habitats. Meier and Soininen (2014) report that spatial distances along water connections were more important for explaining similarities in phytoplankton communities than overland distances. Interestingly, spatial patterns in community composition, independent of measured environmental similarities, were even shown for bacteria in coastal pools at small spatial scales, suggesting that history of colonization or localized dispersal dynamics matter even for organisms with such rapid generation times. Using long-term observations of three types of ecosystems, including Jamaican rock pools, Hammond and Kolasa (2014) studied the link between spatial and temporal variation in ecosystem variables. They revealed a strong quantitative link between spatial and regional temporal variation in 136 variables and suggested this as a basis for substituting variables, when long-time series are lacking.

Local Abiotic Conditions

Local abiotic conditions can impose important filters in rock pools that affect community membership. Salinity clearly matters for rock pools in coastal areas (Ganning 1971), but even minor differences in salinity or electrical conductivity can have important consequences for inland temporary pool communities. It has been shown that electrical conductivity was an important hatching cue for temporary pool invertebrates (Brendonck 1996). In a study on the impact of salinity on *Daphnia* communities (three species) in Swedish coastal rock pools, Liao et al. (2015) found that exposure to the local stressor was more important for stress tolerance than genetic diversity or level of isolation in the metacommunity. Little work has been done on the importance of nutrient addition in inland rock pool clusters on rocky outcrops as most rock pools are very oligotrophic. However, as further discussed below, it could be of potential conservation concern. Soininen and Meier (2014) studied patterns of phytoplankton species richness in subarctic Finish rock pools in relation with abiotic conditions and pool size. They found a highly significant positive relation with total P concentrations and a marginally negative relationship with conductivity, while the size of pools was not important. Romanuk and Kolasa (2005) suggested that resource availability (determined by nutrient input and species diversity) may be the principal mechanism determining invasibility at local scales in multi-trophic rock pool communities. Beisner et al. (2006) experimentally tested the invasibility of Jamaican rock pools by a competing ostracod in relation to local productivity and species richness. High resource availability and low species richness increased establishment success of the ostracod species.

Conservation and Management Issues of Importance to Rock Pool Invertebrates

Overall, rock pool habitats are not severely threatened worldwide and are less disturbed by man than many other ecosystems. In part this is likely due to the fact that, unlike many temporary wetlands in lowland areas, the robust nature of the rock slabs makes these habitats difficult to destroy. Additionally, people do not have many incentives to interact with these habitats that are often located away from human settlements. Their occurrence in rocky terrains also makes them largely useless as water sources for agriculture or intensive grazing. Occasionally, local people will build huts on top of the rocky outcrops and use the rock pools to wash their clothes, as is the case in the Gutu region of Zimbabwe. Even those subject to pressure from tourism like Wave Rock in Western Australia have faunas no less diverse than pools on rocks away from the public (Jocqué et al. 2006).

Direct Human Impacts

In the past, native Australian tribes used rock pools as sources of water and marked rock pools, and deeper pit gnammas in particular, on maps they painted under overhanging rocks and in caves (White 2009). For native Australian (aboriginal) people in the arid areas of South Australia, gnammas also facilitated access to a wider area and a larger range of resources. Aboriginal routes in arid areas were largely governed by the occurrence and distribution of rock pools with tracks radiating out from them. Many of the pit gnammas are currently disturbed partly from natural sedimentation due to failure of their former aboriginal custodians to clean them out occasionally and thus maintain their value as a water source. However, in some areas of the country, there is a reconnection of aboriginal communities to their land and appropriate management of their heritage (White 2009). Today, the rainwater that falls on rocky outcrops is still used by farmers in Western Australia. By building walls around an outcrop, they can guide the water, which would otherwise drain into the sediment around the outcrop, to large collector tanks. This practice does not damage rock pools per se since the water in the pools is not harvested. But the walls that are put in place might act as migration barriers for amphibians and reptiles and may prevent some pool habitats that heavily rely on outcrop runoff to fill. Farmers often cover pools to keep wandering sheep from drowning, and damage is done by the decaying infrastructure. In earlier times aboriginals covered some pools to reduce evaporation, and these covers, plus farmer's covers, reduce diversity (Timms 2014a). Quantitative studies of the potential effects of such modifications on biota and pool hydrology are, however, lacking.

Whether rock pools or pool clusters are expendable or not depends on the regional context. In some cases landscapes may be dotted with hundreds of outcrops, each containing clusters of rock pools. However, in other cases small numbers of rocky outcrops with endemic lineages may exist on a single outcrop in a region that lacks similar habitats (Vanschoenwinkel et al. 2011). Community patterns suggest that dispersal among rock pools is often essential to maintain local diversity in rock pool patches (Vanschoenwinkel et al. 2013) and may help to counteract regular extinctions due to disturbance. Therefore it would be recommended to conserve entire clusters rather than single habitats.

Although many rock pool systems are oligotrophic, nutrient enrichment and eutrophication can be problematic in some areas. In the Australian outback, feral goats often gather in substantial numbers around granite outcrops. As a result of wind action, dried goat droppings often accumulate in the rock pools on these outcrops leading to eutrophic conditions or dead pools, without any sign of invertebrate life. During periods of drought in Australia, kangaroos, emus, and assorted reptiles find water in rock pools and can fall in and drown; as a result, their decaying bodies pollute the limited water volume in the pool. Due to the typical oligotrophic state and small volume of inland rock pools, one could argue that even small additions of nutrients could have important effects on biodiversity, but this remains to be studied quantitatively. It has also been hypothesized that nutrient enrichment as a result of mass tourism on the sandstone monolith Uluru could possibly explain the decline of an endemic rock pool crustacean (Timms pers. comm.). Coastal rock pools tend to be more eutrophic. In this case nutrients may be added by visiting sea birds.

In some cases, rock slabs with rock pools are accessible to all-terrain vehicles. Besides damaging the fragile vegetation on the escarpment, vehicles will sometimes drive through rock pools. An experiment by Graham and Wirth (2008) has shown that such physical disturbance of the sediment bank in pools loosens the eggs and seeds contained therein that can hence be blown more easily away by wind. In a South African pool cluster, it was shown that the dried remains of aquatic vegetation that remained during the dry season correlate with the abundance of resting eggs of the fairy shrimp *Branchipodopsis wolfi* that could be maintained in these pools (Vanschoenwinkel et al. 2010a). This was surprising since no link could be found with other factors such as hydroperiod that would directly be linked to reproductive success. Also Brendonck and Riddoch (2000) found egg bank sizes were more stable in pools where eggs were protected by vegetation against wind erosion. These studies indicate that wind erosion during the dry season is likely an important factor affecting long-term persistence of populations.

Climate Change

Given the direct dependence on rainfall and evaporation for filling and on the proportion of inundations that will be suitable for reproduction by different species (Hulsmans et al. 2008; Tuytens et al. 2014), it is no surprise that predicted changes in the earth's climate will ensure that certain rock pool habitats will become less suitable for certain inhabitants. Simulations of realistic changes in rainfall and evaporation for central South Africa revealed that the proportion of inundations suitable for the reproduction of fairy shrimp may decrease by up to 21 %. This is worrying, since among rock pool inhabitants, fairy shrimps are still among the most rapidly reproducing species. Therefore, effects on the reproductive success of most other species are likely to be even more detrimental. Given the strong selection for rapid reproduction in these systems, it is also questionable whether species can adapt to reproduce even faster. It seems more likely that they are already at their physiological limit. In Australia, future climate change may act to reduce hydroperiods, particularly at northern and northwest fringes of the Wheatbelt and Goldfields (Timms 2012a, b). It was sometimes stated that Arctic and subarctic rock pool and pond ecosystems are in general very sensitive to environmental changes because of their small size and high surface area to depth ratios (Smol and Douglas 2007). Tuck and Romanuk (2012), however, conducted an experiment to determine whether thermal variability of +4 °C leads to consistent changes in community structure, temporal dynamics, and ecosystem functioning in laboratory analogues of natural freshwater supralittoral rock pool communities inhabited by meiofauna and zooplankton collected from subarctic, temperate and tropical regions. Subarctic zooplankton communities proved to be more robust to thermal variability than temperate or tropical communities, and this suggests that increasing thermal variability with climate change may have the greatest effects on community structure and function in tropical and temperate regions. Studies on such rock pools give important insights into how future climate changes would affect these sensitive small aquatic ecosystems in polar regions (Soininen and Meier 2014). Altermatt et al. (2008) demonstrated the effect of weather changes on the metacommunity dynamics of three *Daphnia* species in Finnish rock pools. A fourfold increase in colonization rate was observed with warm and dry weather conditions, when egg banks were more exposed to dispersal by animals and winds.

Compared to other temporary aquatic systems, it is likely that given their small water volume, rock pools will be the first to experience the effects of climate warming and changes in the seasonal distribution of water (Hulsmans et al. 2008). As a result they could be considered "sentinel ecosystems." Overall, species with long aquatic life cycles might especially struggle to reproduce in rock pools in the foreseeable future. As an alternative tool for long-term monitoring to track differences in rock pool plant communities, Lanfranco et al. (2015) suggested phylogenetic clustering. According to this method, following the position of pools along a continuum of phylogenetic relatedness may reveal the response of local communities to climate warming.

Although rock pools are usually small and therefore sensitive to climate change, permanent plant and animal inhabitants do show resilience against early drying, mainly through the buffering effect of the egg or seed bank in the sediment. Sediment from Botswana rock pools, for example, still functioned as a source of new hatchlings after 16 subsequent wet/dry cycles with simulated population crashes and no addition of new eggs (Brendonck et al. 1998). Cross et al. (2015) performed a similar experiment with plant seed banks from Australian rock pools and found that a viable seed bank persisted even after ten consecutive wetting events during 3 years.

Conservation Strategies

As freshwater rock pools are rather unique habitats housing a surprisingly high diversity of specialist and endemic species, protection of these habitats is essential. In Mauritania, mountain rock pools are holding a disproportional high richness (32 % of all taxa and 78 % of all Mauritanian endemics) in vertebrates in comparison with the size of the pools (total: 46 ha), while more than 60 % of these pools are currently unprotected (Vale et al. 2015). These pools could function as microrefugia under climate change and expanding dry regions.

Low dispersal among inselbergs, as well as patterns of local endemicity and genetic regionalism, underscore the need to implement conservation strategies at regional rather than local scales (Jocqué` 2006, 2007c, 2010a). Strong links between local hydrological and climatological conditions and diversity patterns also indicate the need to conserve pools with different hydrologies and inselbergs in different climatological regions (Jocqué et al. 2010a). The intensive biological surveys on granite outcrops in Western Australia highlight the need for conservation of these freshwater refuge habitats due to freshwater wetland depletion in the wheat belt from salinization (Jocqué et al. 2007c; White 2009). However, very few gnammas occur in the nation's conservation estate, and if so, it is more by accident than design, i.e., within reserves established for other reasons. Not one gnamma on Eyre Peninsula, South Australia, lies within a National Park, and very few do in Western Australia. Many rock pools, though, lie in reserves set aside for water conservation and so enjoy a measure of protection. Public education is a key process, and already in some local council areas in Western Australia, e.g., at Trayning, tourists are encouraged by roadside signage and brochures to visit extraordinary gnammas. Another interesting rock pool site offering good opportunities to reconcile tourism with conservation of biodiversity is the pilgrimage site of Montmajour near the Camargue in France (Cher 2008). While tourists are in the first place attracted by the Medieval monastery (a world heritage site), they also show interest in the 48 tombs that were cut in the rocks and that became mature rock pools even holding *Triops*. Suggestions were made to the site manager to produce a brochure to inform visitors on the functioning of the rock pool ecosystem.

Appendix

Passive (*left columns*) and active (*right columns*) dispersing invertebrate families and genera recorded from rock pools worldwide (adapted from Jocqué et al. 2010a)

Passive dispersers		Active dispersers	
Higher taxa	Genera	Higher taxa	Genera
Turbellaria		**Insecta**	
Rhabdocoela		**Ephemeroptera**	
Dalyelliidae	*Caliadne*	Baetidae	*Baetis*
	Gieysztoria		*Callibaetis*
Typhloplanidae	*Bothromesostoma*		*Cloeodes*
	Mesostoma		*Cloeon*
	Syrinx	**Odonata**	
Nematoda		Aeshnidae	*Aeshna*
Dorylaimidae	*Dorylaimus*		*Anax*
Mollusca			*Hemianax*
Physidae	*Physella*		*Polycanthagyna*
Planorbidae	*Bulinus*	Coenagrionidae	*Argia*
	Glyptophysa		*Enallagma*
	Isidorella		*Ischnura*
Annelida			*Xanthagrion*
Aeolosomatidae		Corduliidae	*Hemicordulia*
Phreodrilidae	*Antarctodrilus*	Lestidae	*Archilestes*
	Astacopsidrilus		*Austrolestes*
Tubificidae	.	Libellulidae	*Bradinopyga*
Tardigrada			*Libellula*
Macrobiotidae	*Macrobiotus*		*Diplacodes*
Crustacea			*Orthetrum*
Anostraca			*Pantala*
Branchinectidae	*Branchinecta*		*Sympetrum*
Branchipodidae	*Branchipodopsis*		*Trithemis*
	Branchipus	**Hemiptera**	
Chirocephalidae	*Linderiella*	Belostomatidae	*Lethocerus*
Streptocephalidae	*Streptocephalus*	Corixidae	*Agraptocorixa*
Thamnocephalidae	*Branchinella*		*Arctocorisa*
Spinicaudata			*Callicorixa*
Cyzicidae	*Caenestheriella*		*Diaprepocoris*
	Cyzicus		*Graptocorixa*
	Eocyzicus		*Micronecta*
Lepthesteriidae	*Leptestheria*		*Sigara*
Limnadiidae	*Limnadia*	Gerridae	*Aquarius*
	Eulimnadia		*Gerris*
Laevicaudata		Naucoridae	*Ambrysus*

(continued)

(continued)

Passive dispersers		Active dispersers	
Lynceidae	*Lynceus*	Notonectidae	*Anisops*
Notostraca			*Buenoa*
Triopsidae	*Lepidurus*		*Notonecta*
	Triops	Pleidae	*Plea*
Anomopoda		Veliidae	*Microvelia*
Chydoridae	*Allonella*	**Coleoptera**	
	Alona	Dytiscidae	*Agabus*
	Celsinotum		*Allodessus*
	Chydorus		*Deronectes*
	Dunhevedia		*Dytiscus*
	Ephemeroporus		*Eretes*
	Leberis		*Hydroglyphus*
	Monospilus		*Hygrotus*
	Planicirclus		*Hyphydrus*
	Pleuroxus		*Ilybius*
	Plurispina		*Laccophilus*
	Rak		*Lancetes*
Daphniidae	*Ceriodaphnia*		*Liodessus*
	Daphnia		*Megaporus*
Ilyocryptidae	*Ilyocryptus*		*Necterosoma*
Macrothricidae	*Macrothrix*		*Neoclypeodytes*
Moinidae	*Moina*		*Paroster*
Neothricidae	*Neothrix*		*Platynectes*
Simocephalidae	*Simocephalus*		*Potamonectes*
Ostracoda			*Rhantus*
Cyprididae	*Alboa*		*Sternopriscus*
	Amphibolocypris		*Stictotarsus*
	Bennelongia		*Thermonectus*
	Candonocypris		*Uvarus*
	Chlamydotheca	Gyrinidae	*Gyrinus*
	Cypretta	Haliplidae	*Peltodytes*
	Cypricercus	Hydrophilidae	*Berosus*
	Cypriconcha		*Enochrus*
	Cypridopsis		*Hydrochara*
	Cyprinotus		*Hydrophilus*
	Eucyprinotus		*Laccobius*
	Heterocypris		*Limnoxenus*
	Ilyodromus		*Limnoxenus*
	Kapcypridopsis		*Tropisternus*
	Potamocypris	Hydraenidae	*Tympanogaster*
	Strandesia	Limnichidae	

Cypridopsidae	*Sarscypridopsis*	Scirtidae	
Ilyocyprididae	*Ilyocypris*	**Trichoptera**	
Limnocytheridae	*Gomphodella*	Limnephilidae	*Limnephilus*
	Korannacythere	Leptoceridae	*Oecetis*
	Limnocythere		*Triplectides*
Copepoda		**Lepidoptera**	
Cyclopoida		Pyralidae	
Harpacticoida		**Diptera**	
Parastenocarididae		Culicidae	*Aedes*
Canthocamptidae	*Attheyella*		*Anopheles*
	Elaphoidella		*Culex*
	Epactophanes		*Culiciomyia*
	Pindamoraria		*Culiseta*
Centropagidae	*Boeckella*		*Mimomyia*
	Calamoecia		*Ochlerotatus*
Cyclopidae	*Acanthocyclops*		*Opifex*
	Cyclops		*Theobaldia*
	Diacyclops		*Tripteroides*
	Ectocyclops		*Uranotaenia*
	Eucyclops	Ceratopogonidae	*Atrichopogon*
	Macrocyclops		*Bezzia*
	Megacyclops		*Culicoides*
	Metacyclops		*Dasyhelea*
	Microcyclops		*Monohelea*
	Mixocyclops	Chironomidae	*Ablabesmyia*
	Paracyclops		*Allotrissocladius*
	Stolonicyclops		*Alotanypus*
	Tropocyclops		*Apedilum*
Diaptomidae	*Diaptomus*		*Archaeochlus*
	Hesperodiaptomus		*Botryocladius*
	Leptodiaptomus		*Chironomus*
	Lovenula		*Cladopelma*
	Metadiaptomus		*Compterosmittia*
Amphipoda			*Cryptochironomus*
Gammaridae	*Gammarus*		*Dicrotendipes*
Acari			*Gymnometriocnemus*
Eylaidae	*Eylais*		*Kiefferulus*
Oribatidae	*Aquanothrus*		*Microsectra*
	Chudalupia		*Paraborniella*
	Scapheremaeus		*Parakiefferiella*
	Trimalaconothrus		*Paramerina*
Trombidiidae			*Paratendipes*

(continued)

(continued)

Passive dispersers		Active dispersers	
Mesostigmata			*Phaenopsectra*
			Phaenopsectra
			Polypedilum
			Procladius
			Tanytarsus
		Dolichopodidae	
		Ephydridae	
		Muscidae	
		Sciomyzidae	
		Tabanidae	*Tabanus*
		Tipulidae	

References

Altermatt F, Ebert D (2010) Populations in small, ephemeral habitat patches may drive dynamics in a *Daphnia magna* metapopulation. Ecology 91:2975–2982

Altermatt F, Pajunen VI, Ebert D (2008) Climate change affects colonisation dynamics in a metacommunity of three *Daphnia* species. Glob Chg Biol 14:1209–1220

Altermatt F, Pajunen VI, Ebert D (2009) Desiccation of rock pool habitats and its influence on population persistence in a *Daphnia* metacommunity. PLoS One 4

Anusa A, Ndagurwa HGT, Magadza CHD (2012) The influence of pool size on species diversity and water chemistry in temporary rock pools on Domboshawa Mountain, northern Zimbabwe. Afr J Aquat Sci 37:89–99

Arnér M, Koivisto S, Norberg J, Kautsky N (1998) Trophic interactions in rockpool food webs: regulation of zooplankton and phytoplankton by *Notonecta* and *Daphnia*. Freshw Biol 39:79–90

Artois T, Willems W et al (2004) Freshwater Rhabdocoela (Platyhelminthes) from ephemeral rock pools from Botswana, with the description of four new species and one new genus. Zool Sci 21:1063–1072

Baron JS, La Francois T, Kondrateiff BC (1998) Chemical and biological characteristics of desert rock pools in intermittent streams of Capitol Reef National Park, Utah. Great Basin Nat 58:250–264

Bayly IAE (1982) Invertebrate fauna and ecology of temporary pools on granite outcrops in southern Western Australia. Austral J Mar Freshw Res 33:599–606

Bayly IAE (1997) Invertebrates of temporary waters in gnammas on granite outcrops in Western Australia. J Royal Soc W Austral 80:167–172

Bayly IAE (2011) Australia's granite wonderlands. Bas Publishing, Melbourne

Beisner BE, Hovius J et al (2006) Environmental productivity and biodiversity effects on invertebrate community invasibility. Biol Invasions 8:655–664

Belk D (1991) Why only one of 2 common central Texas Anostraca atop Enchanted Rock. Hydrobiologia 212:83–86

Bengtsson J (1986) Life histories and interspecific competition between three *Daphnia* species in rock pools. J Anim Ecol 55:641–655

Bengtsson J (1989) Interspecific competition increases local extinction rate in a metapopulation system. Nature 340:713–715

Bengtsson J, Ebert D (1998) Distributions and impacts of microparasites on *Daphnia* in a rock pool metapopulation. Oecologia 115:213–221

Blaustein L (1997) Non-consumptive effects of larval *Salamandra* on crustacean prey: can eggs detect predators? Oecologia 110:212–217

Blaustein L, Kotler BP, Ward D (1995) Direct and indirect effects of a predatory backswimmer (*Notonecta maculata*) on community structure of desert temporary pools. Ecol Entomol 20:311–318

Blaustein L, Friedman J, Fahima T (1996) Larval *Salamandra* drive temporary pool community dynamics: evidence from an artificial pool experiment. Oikos 76:392–402

Blaustein L, Garb JE, Shebitz D, Nevo E (1999) Microclimate, developmental plasticity and community structure in artificial temporary pools. Hydrobiologia 392:187–196

Blaustein L, Kiflawi M et al (2004) Oviposition habitat selection in response to risk of predation in temporary pools: mode of detection and consistency across experimental venue. Oecologia 138:300–305

Brendonck L (1996) Diapause, quiescence, hatching requirements: what we can learn from large freshwater branchiopods (Crustacea: Branchiopoda: Anostraca, Notostraca, Conchostraca). Hydrobiologia 320:85–97

Brendonck L (2003) Egg banks in freshwater zooplankton: evolutionary and ecological archives in sediment. In: Vandonk E, Boersma M, Spaak P (eds), Proc Vol Recent developments in fundamental and applied plankton research. Hydrobiologia 491:65–84

Brendonck L, Riddoch BJ (2000) Egg bank dynamics in anostracan desert rock pool populations (Crustacea: Branchiopoda). Arch Hydrobiol 148:71–84

Brendonck L, Riddoch BJ, Van de Weghe V, Van Dooren T (1998) The maintenance of egg banks in very short-lived pools—a case study with anostracans (Branchiopoda). In: Brendonck L, De Meester L, Hairston NG (eds), Evolutionary and Ecological Aspects of Crustacean Diapause. Arch Hydrobiol (Spec issue) 52:141–161

Brendonck L, Hamer ML, Riddoch BJ, Seaman MT (2000) *Branchipodopsis* species: specialists of ephemeral rock pools. Afr J Aqua Sci 25:98–104

Brendonck L, Michels E, De Meester L, Riddoch B (2002) Temporary pools are not 'enemy-free'. In Maeda-Martinez AM, Timms BV (eds), Studies on large branchiopod biology 4. Hydrobiologia 486:147–159

Brendonck L, Jocqué M, Hulsmans A, Vanschoenwinkel B (2010) Pools 'on the rocks': freshwater rock pools as model systems in ecological and evolutionary research. Limnetica 29:25–40

Brendonck L, Jocqué M et al (2014) Hydrological stability drives both local and regional diversity patterns in rock pool metacommunities. Oikos. doi:10.1111/oik.01710

Briffa K, Sammut S, Lanfranco S (2014). Morphometric heterogeneity of temporary pools in Malta and its effect on species richness: implications for management and restoration. Proc 6th EPCN Conference, Huesca, Spain, abstract

Buschke FT, Esterhuyse S et al (2012) The dynamics of mountain rock pools—are aquatic and terrestrial habitats alternative stable states? Acta Oecol 47:24–29

Campbell EM (1997) Granite landforms. J Roy Soc W Austral 80:101

Cantrell MA, McLachlan AJ (1982) Habitat duration and dipteran larvae in tropical rain pools. Oikos 38:343–348

Chan M, Moser K et al (2005) Desert potholes: ephemeral aquatic microsystems. Aqua Geochem 11:279–302

Cher O (2008) EPCN Newsl 2:3–4

Chesson PL (1985) Coexistence of competitors in spatially and temporally varying environments: a look at the combined effects of different sorts of variability. Theor Popul Biol 28:263–287

Cross AT, Turner SR et al (2015) Seed dormancy and persistent sediment seed banks of ephemeral freshwater rock pools in the Australian monsoon tropics. Annal Bot 115:847–859

De Roeck ERM, Artois T, Brendonck L (2005) Consumptive and non-consumptive effects of turbellarian (*Mesostoma* sp.) predation on anostracans. Hydrobiologia 542:103–111

De Roeck ER, Waterkeyn A, Brendonck L (2010) Life history traits of *Streptocephalus purcelli* Sars, 1898 (Branchiopoda, Anostraca) from temporary waters with different phenology. Water SA 36:323–328

De Vries CP (1996) Invertebrate community structure and dynamics in Korannaberg rock pools. Univ Orange Free State, Bloemfontein South Africa, MS thesis

Dodson SI (1987) Animal assemblages in temporary desert rock pools: aspects of the ecology of *Dasyhelea sublettei* (Diptera: Ceratopogonidae). J N Am Benthol Soc 6:65–71

Ebert D, Haag C et al (2002) Outbreeding confers a selective advantage to immigrant genes in a metapopulation. Science 295:485–488

Egan AT, Ferrington LC (2015) Zonal stratification and geographic clustering of a species rich chironomid community in freshwater coastal rock pools. Hydrobiologia 751:147–158

Ganning B (1971) Studies on chemical, physical and biological conditions in Swedish rock pool ecosystems. Ophelia 9:51–105

Geldard M (2005) Saving the ridged water-milfoil. Austral Plan Conserv 13:27

Ghilarov AM (1967) Zooplankton of arctic rock pools. Oikos 18:82–95

Graham TB (1994) Predation by dipteran larvae on fairy shrimp (Crustacea: Anostraca) in Utah rock-pools. SW Natural 39:206–207

Graham TB, Wirth D (2008) Dispersal of large branchiopod cysts: potential movement by wind from potholes on the Colorado Plateau. Hydrobiologia 600:17–27

Hammond MP, Kolasa J (2014) Spatial variation as a tool for inferring temporal variation and diagnosing types of mechanisms in ecosystems. PLoS One 9(2), e89245. doi:10.1371/journal.pone.0089245

Hanski I, Ranta E (1983) Coexistence in a patchy environment—three species of *Daphnia* in rock pools. J Anim Ecol 52:263–279

Hebert PDN, Emery CJ (1990) The adaptive significance of cuticular pigmentation in *Daphnia*. Func Ecol 4:703–710

Hulsmans A, Moreau K et al (2007) Direct and indirect measures of dispersal in the fairy shrimp *Branchipodopsis wolfi* indicate a small scale isolation-by-distance pattern. Limnol Oceanogr 52:676–684

Hulsmans A, Vanschoenwinkel B et al (2008) Quantifying the hydroregime of a temporary pool habitat: a modelling approach for ephemeral rock pools in SE Botswana. Ecosystems 11:89–100

Jocqué M, Martens K, Riddoch B, Brendonck L (2006) Faunistics of ephemeral rock pools in southeastern Botswana. Arch Hydrobiol 165:415–431

Jocqué M, Graham T, Brendonck L (2007a) Local structuring factors of invertebrate communities in ephemeral freshwater rock pools and the influence of more permanent water bodies in the region. Hydrobiologia 592:271–280

Jocqué M, Riddoch B, Brendonck L (2007b) Successional phases and species replacements in freshwater rock pools: towards a biological definition of ephemeral systems. Freshw Biol 52:1734–1744

Jocqué M, Timms B, Brendonck L (2007c) A contribution on the biodiversity and conservation of the freshwater fauna of rocky outcrops in the central Wheatbelt of Western Australia. J Royal Soc W Austral 90:137–142

Jocqué M, Vanschoenwinkel B, Brendonck L (2010a) Freshwater rock pools: a review of habitat characteristics, faunal diversity and conservation value. Freshw Biol 55:1587–1602

Jocqué M, Vanschoenwinkel B, Brendonck L (2010b) Anostracan monopolisation of early successional phases in temporary waters? Fund Appl Limnol 176:127–132

Kolasa J, Li BL (2003) Removing the confounding effect of habitat specialization reveals the stabilizing contribution of diversity to species variability. Proc Royal Soc London B Suppl 270:198–201

Kolasa J, Romanuk TN (2005) Assembly of unequals in the unequal world of a rock pool metacommunity. In: Holyoak M, Leibold MA, Holt RD (eds) Metacommunities: spatial dynamics and ecological communities. Univ Chicago Press, Chicago, pp 212–232

Lanfranco S (1995) Ecological succession in Maltese freshwater rock pools. Univ Malta, PhD Dis

Lanfranco S, Briffa K (2012) Temporary freshwater pools of the Maltese Islands. In: Gauci R, Schembri J (eds) Landscapes and landforms of the Maltese Islands, World geomorphological landscapes. Springer, Dordrecht

Lanfranco S, Grillas P (2010) Hydrological characteristics of temporary freshwater pools of the Maltese Islands. 4th Conf European Pond Conserv Network (EPCN) Berlin (Erkner), abstract
Lanfranco S, De Walsche C, Schembri P, Mertens J (1991) Branchiopods (non-cladocerans) of the Maltese Islands (central Mediterranean). Hydrobiologia 212:241–243
Lanfranco S, Grima C et al (2015) Phylogenetic clustering as a tool for long term monitoring of plant communities in rock pools. Mediterranean Temporary Ponds, Paulis Project. Sassari, Italy, Proc Internat Sympo
Liao Y-F, Faulks LK, Östman Ö (2015) Stress tolerance and population stability of rock pool *Daphnia* in relation to local conditions and population isolation. Hydrobiologia 742: 267–278
Martens K (1996) On *Korannacythere* nov. gen. (Crustacea, Ostracoda): a new genus of temporary pool limnocytherids from Southern Africa, with the description of three new species and a generic reassessment of the Limnocytherinae. Bulle Koninklijk Belgisch Instit Natuurwet Biol 66:51–72
Martens K, Schön I, Meisch C, Horne DJ (2008) Global diversity of ostracods (Ostracoda, Crustacea) in freshwater. Hydrobiologia 595:185–193
McLachlan AJ (1981) Food sources and foraging tactics in tropical rain pools. Zool J Linnean Soc 71:265–277
McLachlan A (1983) Habitat distribution and body size in rain-pool dwellers. Zool J Linnean Soc 79:399–407
McLachlan AJ, Ladle R (2001) Life in the puddle: behavioural and life-cycle adaptations in the Diptera of tropical rain pools. Biol Rev 76:377–388
Meier S, Soininen J (2014) Phytoplankton metacommunity structure in subarctic rock pools. Aqua Microbial Ecol 73:81–91
Ng ISY, Carr CM, Cottenie K (2009) Hierarchical zooplankton metacommunities: distinguishing between high and limiting dispersal mechanisms. Hydrobiologia 619:133–143
Osborne PL, McLachlan AJ (1985) The effect of tadpoles on algal growth in temporary, rain-filled rock pools. Freshw Biol 15:77–87
Pajunen VI (1986) Distributional dynamics of *Daphnia* species in a rock-pool environment. Ann Zool Fenn 23:131–140
Pajunen VI, Pajunen I (2003) Long-term dynamics in rock pool *Daphnia* metapopulations. Ecography 26:731–738
Pandit SN, Kolasa J, Cottenie K (2009) Contrasts between habitat generalists and specialists: an empirical extension to the basic metacommunity framework. Ecology 90:2253–2262
Pellowe-Wagstaff KE, Simonis JL (2014) The ecology and mechanisms of overflow-mediated dispersal in a rock-pool metacommunity. Freshw Biol 59:1161–1172
Pinder AM, Halse SA, Shiel RJ, McRae JM (2000) Granite outcrop pools in south Western Australia: foci of diversification and refugia for aquatic invertebrates. J Royal Soc W Austral 83:149–161
Ranta E, Nuutinen V (1985) *Daphnia* exhibit diurnal vertical migration in shallow rock pools. Hydrobiologia 127:253–256
Riddoch BJ, Mpoloka SW, Cantrell M (1994) Genetic variation and localised gene flow in the fairy shrimp, *Branchipodopsis wolfi* in temporary rainwater pools in south-eastern Botswana. In: Beaumont AR (ed) Genetics and evolution of aquatic organisms. Chapman Hall, London, pp 96–102
Romanuk TN, Kolasa J (2002) Environmental variability alters the relationship between richness and variability of community abundances in aquatic rock pool microcosms. Ecoscience 9:55–62
Romanuk TN, Kolasa J (2005) Resource limitation, biodiversity, and competitive effects interact to determine the invisibility of rock pool microcosm. Biol Invasions 7:711–722
Simonis JL (2013) Prey (*Moina macrocopa*) population density drives emigration rate of its predator (*Trichocorixa verticalis*) in a rock-pool metacommunity. Hydrobiologia 715:19–27
Simonis JL, Ellis JC (2014) Bathing birds bias beta-diversity: Frequent dispersal by gulls homogenizes fauna in a rock-pool metacommunity. Ecology 95:1545–1555

Smith DC (1983) Factors controlling tadpole populations of the chorus frog (*Pseudacris triseriata*) on Isle Royale, Michigan. Ecology 64:501–510

Smol J, Douglas MSV (2007) Crossing the final ecological threshold in high Arctic ponds. Proc Natl Acad Sci U S A 104:12395–12397

Soininen J, Meier S (2014) Phytoplankton richness is related to nutrient availability, not to pool size, in a subarctic rock pool system. Hydrobiologia 740:137–145

Spencer M, Blaustein L (2001) Hatching responses of temporary pool invertebrates to signals of environmental quality. Isr J Zool 47:397–418

Spencer M, Blaustein L, Cohen JE (2002) Oviposition habitat selection by mosquitoes (*Culiseta longiareolata*) and consequences for population size. Ecology 83:669–679

Springer GS, Tooth S, Wohl EE (2006) Theoretical modeling of stream potholes based upon empirical observations from the Orange River, Republic of South Africa. Geologija 82:160–176

Therriault TW, Kolasa J (2001) Desiccation frequency reduces species diversity and predictability of community structure in coastal rock pools. Isr J Zool 47:477–489

Timms BV (2012a) Seasonal study of aquatic invertebrates in five sets of latitudinally separated gnanmas in southern western Australian. J Royal Soc W Austral 95:13–28

Timms BV (2012b) Influence of climatic gradients on metacommunities of aquatic invertebrates on granite outcrops in southern western Australia. J Royal Soc W Austral 95:125–135

Timms BV (2013) Geomorphology of pit gnammas in southwestern Australia. J Royal Soc W Austral 96:7–16

Timms BV (2014a) Invertebrates of pit gnammas in Southwest Australia. J Royal Soc W Austral 96:55–67

Timms BV (2014b) Community ecology of aquatic invertebrates in gnammas (rock-holes) of North-Western Eyre Peninsula, South Australia. Trans Royal Soc S Austral 138:147–160

Timms BV, Rankin C (2014) The geomorphology of gnammas (weathering pits) of north-western Eyre Peninsula, South Australia: typology, influence of haloclasty and origins. Trans Royal Soc S Austral

Tuck C, Romanuk TN (2012) Robustness to thermal variability differs along a latitudinal gradient in zooplankton communities. Glob Chg Biol 118:1597–1608

Tuytens K, Vanschoenwinkel B, Waterkeyn A, Brendonck L (2014) Predictions of climate change infer increased environmental harshness and altered connectivity in a cluster of temporary pools. Freshw Biol 59:955–968

Twidale C, Bourne J (1975) Episodic exposure of inselbergs. Geol Soc Am Bull 86:1473–1481

Vale CG, Pimm SL, Brito JC (2015) Overlooked mountain rock pools in deserts are critical local hotspots of biodiversity. PLoS One 10(2), e0118367. doi:10.1371/journal.pone.0118367)

Vanschoenwinkel B, De Vries C, Seaman MT, Brendonck L (2007) The role of metacommunity processes in shaping invertebrate rock pool communities along a dispersal gradient. Oikos 116:1255–1266

Vanschoenwinkel B, Gielen S, Seaman MT, Brendonck L (2008a) Any way the wind blows—frequent wind dispersal drives species sorting in ephemeral aquatic habitats. Oikos 117: 125–134

Vanschoenwinkel B, Gielen S, Vandewaerde H, Brendonck L (2008b) Relative importance of different dispersal vectors for small aquatic invertebrates in a rock pool metacommunity. Ecography 31:567–577

Vanschoenwinkel B, Hulsmans A et al (2009) Community structure in temporary freshwater pools: disentangling the effects of habitat size and hydroregime. Freshw Biol 54:1487–1500

Vanschoenwinkel B, Seaman MT, Brendonck L (2010a) Hatching phenology, life history and egg bank size of a fairy shrimp (Branchiopoda, Crustacea) in relation to the ephemerality of its habitat. Aqua Ecol 44:771–780

Vanschoenwinkel B, Waterkeyn A et al (2010b) Species sorting in space and time-the impact of disturbance regime on community assembly in a temporary pool metacommunity. J N Am Benth Soc 29:1267–1278

Vanschoenwinkel B, Mergeay J et al (2011) Long distance dispersal of zooplankton endemic to isolated mountaintops—an example of an ecological process operating on an evolutionary time scale. PLoS One 6(11), e26730. doi:10.1371/journal.pone.0026730

Vanschoenwinkel B, Pinceel T et al (2012) Toward a global phylogeny of the "Living Fossil" crustacean order of the Notostraca. PLoS One 7(4), e34998

Vanschoenwinkel B, Buschke FT, Brendonck L (2013) Disturbance regime alters the impact of dispersal on alpha and beta diversity in a natural metacommunity. Ecology 94:2547–2557

Weeks SC, Zofkova M, Knott B (2006) Limnadiid clam shrimp distribution in Australia (Crustacea: Branchiopoda: Spinicaudata). J Royal Soc W Austral 89:155–161

White M (2009) Prioritising Rock-holes of Aboriginal and ecological significance in the Gawler Ranges, DWLBC Report 2009/08 Version 1. Gov South Australia, Dept Water Land Biodiver Conserv, Adelaide

Wiggins GB, Mackay RJ, Smith IM (1980) Evolutionary and ecological strategies of animals in annual temporary pools. Arch Hydrobiol Suppl 58:97–206

Zofkova M, Timms BV (2009) A conflict of morphological and genetic patterns in the Australian anostracan *Branchinella longirostris*. Hydrobiologia 635:67–80

Chapter 3
Invertebrate Communities of Alpine Ponds

Scott A. Wissinger, Beat Oertli, and Véronique Rosset

Introduction

Alpine and subalpine ponds and wetlands (hereafter collectively called "alpine ponds") are small standing water bodies situated in high mountain regions at the upper limit of (subalpine) or above (alpine) tree line. Tree line (upper extent of forest) typically marks a transition in the physical environment in mountain ecosystems that is considered a "threshold for high-altitude biota" (Mani 1968). This transition from forest to herbaceous, tundra-like alpine plants is almost always accompanied by a shift in animal communities toward species assemblages adapted to the harsh conditions of alpine environments. The elevation (hereafter called "altitude" to be consistent with the literature on alpine ponds) at which the transition from trees to herbaceous and shrubby, tundra-like vegetation varies considerably around the world depending on latitude, climate, soils, tree taxonomy, and north–south aspect (Körner 2012). In general, tree line occurs at a lower altitude on south-temperate than north-temperate mountains at a given latitude (Körner 1998; Cieraad et al. 2014).

Alpine ponds have been described from nearly every mountainous region in the world (Mani 1968) including the Andes of South America (e.g., Coronel et al. 2004, 2007; Hampel et al. 2010; Maldonado et al. 2011), Europe (e.g., Alps, Pyrenees, Pirin, Tatra; Catalan et al. 2009a, b; Martinez-Sanz et al. 2012; Rosset and Oertli 2011), the Himalayas (Sommaruga 2010; Gardelle et al. 2011), western North America (the Rockies, Sierras: Carlisle and Hawkins 1998; Wissinger et al. 1999a; Knapp et al. 2001), Australasia (Southern Alps in New Zealand, e.g., Wissinger et al. 2009; Chagué-Goff et al. 2010; and Tasmanian, Blue, Snowy in Australia, e.g., Rees and Cwynar 2010; Timms et al. 2013). Although individually small, alpine ponds are often abundant, occur at high densities (e.g., >10 ponds km^{-2}), and like ponds and wetlands at high latitudes, can have a collective surface area that rivals that of deepwater lakes (Downing et al. 2006; Smol and Douglas 2007b; Ewald

D. Batzer, D. Boix (eds.), *Invertebrates in Freshwater Wetlands*,
DOI 10.1007/978-3-319-24978-0_3

et al. 2010). Despite the ubiquity of alpine ponds in mountainous regions around the world, much of the information on benthic invertebrate communities is based on research in Europe and North America with fewer studies from Africa, South America, and Australasia. Research on invertebrate communities in alpine ponds in the Andes is especially focused on zooplanktonic crustaceans (de Los 2005; Declerck et al. 2011; Munoz-Pedreros et al. 2013). There appears to be a dearth of information on benthic invertebrate communities of alpine ponds in the Himalayas, Central Asian plateaus, and Western Asia.

Alpine ponds are typically located in geomorphologic settings that have been influenced by past or current glacial activity. Their hydrology is driven mainly by the accumulation of water from melting snow and ice and from summer precipitation. They can be situated in depressional features in the hummocky terrain of post-glacial or periglacial landscapes (e.g., kettles, swales dammed by moraines), in cirque basins, in riparian wetland complexes, and in alpine glacial valleys as open pools embedded in bog, fen, wet meadow (e.g., Holmquist et al. 2011) and marsh complexes (compare panels in Fig. 3.1). The latter can occur on peatlands formed in glacially scoured bedrock depressions or on unconsolidated glacial deposits (e.g., Chagué-Goff et al. 2010). Supraglacial ponds that form on the ice and rock debris fields of retreating glaciers are among the most ephemeral (in geological time) types of alpine ponds (e.g., Röhl 2008).

In all of these geological settings, there is a size/depth gradient of basins that ranges from large, deep lakes to relatively small, shallow ponds. At the large end of the spectrum, a precise size distinction between “lake” and “pond” is arbitrary with large ponds and shallow lakes sharing many structural and functional characteristics (Søndergaard et al. 2005). The common-language distinction between the English words “pond” vs. “lake” and equivalent words in other languages varies widely across regions of the world, and in some cases a common term is used to describe both types of alpine basins (e.g., “tarn” in Australasia). Hamerlik et al. (2014) find an ecologically relevant threshold related to invertebrate species turnover rates and species–area relationship at about 2 ha in area that has important conservation consequences (see conservation section below). At the shallow, small end of the spectrum, “ponds” are sometimes distinguished from “pools” (with the latter being temporary; De Meester et al. 2005), and in North America, shallow ponds fall within the jurisdictional definition of wetlands (<2 depth). Semantics aside, we focus on small (1 m^2 to a few ha), permanent, or temporary shallow basins with at least seasonal surface water that occur at the upper limits of tree line (subalpine) and above (alpine). Although there has been research on the specialized plant communities of saturated soil wetlands at high altitudes (e.g., Central Asian plateau: Zhang et al. 2010, Northern Andes: Chimner et al. 2010; Cooper et al. 2010), we did not find information on the subsurface, moist-soil invertebrate communities in alpine wetlands that lack surface water. In several instances, we draw from the literature on lakes because (1) in some regions, the only invertebrate data are from small, shallow alpine lakes (e.g., in Australia, Timms et al. 2013); (2) the data from ponds and lakes are combined in the same

Fig. 3.1 Examples of alpine ponds with (**a**) rocky, steep-walled alpine pond in Macun (Swiss National Park), 2660 m.a.s.l. (photo hepia); (**b**) alpine pond with adjacent meadow of cotton sedge beds (Valais, Switzerland), 2500 m.a.s.l. (photo hepia); (**c**) subalpine pond at upper limit of tree line [pond 12 in Fig. 3.2a] in Colorado, 3400 m.a.s.l. (photo by Angus McIntosh); and (**d**) glacial soil and bedrock in tussock meadow in the Southern Alps of New Zealand, 1200 m.a.s.l. (photo by Angus McIntosh)

Fig. 3.1 (continued)

studies, e.g., in Europe, lakes and ponds are grouped together in the data reporting from several extensive surveys including the IBP, international lakes program (Kajak and Hillbricht-Illkowska 1970), and subsequent studies (AL:PE1 and AL:PE2, MOLAR [MOuntain LAke Research], EMERGE [European Mountain lake Ecosystems: Regionalisation, diaGnostic and socio-economic Evaluation];

Marchetto and Rogora 2004); or (3) because invertebrates in the nearshore littoral zones of alpine lakes provide insight into topics that are germane to alpine ponds (e.g., climate change, effects of acidification, fish introductions, permanence gradients, unique alpine taxa).

Physical Characteristics of Alpine Ponds

Substrate conditions in alpine ponds vary depending on their location in the alpine landscape (e.g., Fig. 3.1). Ponds on the highest and steepest slopes of mountains have inclined shorelines with rocky substrates, a narrow littoral zone with minimal benthic habitat zonation, few or no aquatic plants, and low terrestrial plant subsidies (Figs. 3.1a and 3.2b). In contrast alpine pond basins situated in inter-peak meadows, swales, or saddles and on mountainside benches occur in relatively flat settings and have assorted substrates, relatively gradual shoreline profiles that can include zonation in submergent (e.g., *Isoetes* sp.) and emergent wetland plants (e.g., rushes, sedges) and detrital subsidies from plants in adjacent alpine meadows (Fig. 3.1b, c). Temporary or extremely shallow (hence rapid aerobic mineralization) basins typically have a thin organic layer overlying bedrock or glacial soils (Fig. 3.2c), whereas permanently inundated habitats can develop thick accumulations of gyttja and peat (Figs. 3.1c and 3.2d). Clusters of open pond basins that are embedded in alpine fens (e.g., Chagué-Goff et al. 2010) and other peatland habitats can be formed by nivational (under snow) processes associated with the weight and movement of snowpacks (Cooper 1986).

Air temperature is an extremely important parameter that changes with altitude and affects nearly every aspect of the ecology of small ectothermic organisms like aquatic invertebrates. Because of the small overall water volume (hence low heat capacity) and high surface-area-to-volume ratios in alpine ponds, water temperatures respond quickly to radiative and evaporative heat exchange and solar irradiance (Gerten and Adrian 2001). Moreover, because of the short growing season, many aquatic invertebrates have rapid life cycles that are sensitive to small changes in the thermal environment (Adrian et al. 2009). Temperature acts directly on aquatic invertebrates, driving growth and development rates and determining phenology and life-cycle length (Danks 2008).

Because of recent interest in predicting how climate change will affect the ecophysiology and development of species, there has been considerable research on thermal adaptation in aquatic insects (e.g., Nilsson-Örtman et al. 2012; Stoks et al. 2014; Rotvit and Jacobsen 2014). Intuitively, the lower temperatures and shorter growing season at high altitudes should result in slower grow rates and longer development times than at lower altitudes. For example, the development of a mosquito larva (Diptera) requires only 18 days at 20 °C, but 35 days at 16 °C (Bar-Zeev 1958). However, studies across latitudinal gradients have shown that at high latitudes some species exhibit compensatory responses in physiology that match

Fig. 3.2 (**a**) Cluster of 60 alpine (foreground ponds just above tree line at 3450 m.a.s.l.) and subalpine (3400 m.a.s.l.) ponds at the Mexican Cut Nature Reserve in central Colorado; (**b**, upper right) deep organic substrate (gyttja) with sparse submergent vegetation in permanent subalpine pond 5 (photo by Angus McIntosh); (**c**, lower right) rocky substrate in alpine pond 53) (photo by Angus McIntosh); (**d**, lower left) thin organic substrate with sparse submergent vegetation (*Isoetes bolanderi*) in temporary subalpine pond 8 (photo by Angus McIntosh)

their thermal environment, i.e., individuals/ecotypes from high latitudes shift their physiological rates so that they develop faster at lower temperatures (Sniegula et al. 2010; Shama et al. 2011). Although we do not know of comparable studies along altitudinal gradients that compare ecotypes and genotypes, such altitudinal comparisons would control for the photoperiod differences that potentially confound studies on thermal adaptation along latitudinal gradients (Sniegula et al. 2012). Such comparisons could also contribute to our understanding of the degree to which absolute temperature tolerance vs. acclimatory ability limits the distributional range of wetland aquatic invertebrates (Calosi et al. 2010).

The hydrology of alpine ponds varies depending on the dominant sources of inputs (direct precipitation, run off, groundwater, and snow and ice melt) and outputs (surface outflow, groundwater recharge, evaporation). Alpine ponds of all types are typically covered by ice and snow for much of the year, and even during the summer, surface ice can form and melt multiple times on a daily basis.

In winter, shallow alpine ponds often freeze solid, and even relatively deep ponds can form "frazil ice" that creates a hostile environment for invertebrates (Daborn and Clifford 1974; Lee 1989; Oswood et al. 1991). Even permanent alpine ponds that freeze to or into the substrate might be best described as "aestival" in that liquid water is not available for much of the year, and hence overwintering requires some adaptation for aestivation (Daborn 1971; Daborn and Clifford 1974; Danks 2008). Freezing is a major source of mortality for aquatic invertebrates in aestival habitats (Andrews and Rigler 1985; Duffy and Liston 1985), and variation among species in their tolerance to these conditions is likely to act as a filter on community assembly (e.g., Frisbie and Lee 1997; Danks 2008). For example, high-altitude limnephilid caddisflies that live in aestival alpine ponds use an adult diapause to delay oviposition until late autumn, when they lay their eggs under rocks and wood debris at the margins of pond basins. Experimental transfer of eggs to the pond basin results in high mortality compared to those in the adjacent terrestrial environment where they avoid the damaging effects of ice in relatively dry soils insulated by the overlying snowpack (Wissinger et al. 2003). Egg eclosion is triggered by subnivean water during snowmelt, and first instar larvae enter the ponds in the melt water. Species in deep, permanent ponds that oviposit in summer and begin larval development during winter (e.g., the phryganeid, *Agrypnia deflata*) do not survive in adjacent aestival basins (Wissinger et al. 2003). Danks (2008) provides a general summary of the differing conditions and strategies for overwintering in shallow vs. deep ponds at high latitudes and altitudes.

The hydroperiod (schedule of filling and drying) of alpine ponds varies across a gradient from (1) permanent basins that never dry to (2) semipermanent basins that dry late in summer in some but not all years and to (3) temporary basins that dry annually after snowmelt waters have evaporated or drained from locally perched water tables. The physical conditions in temporary alpine ponds can be quite harsh, especially during the latter stages of drying when the high surface-area-to-volume ratio provides little buffering from the dramatic daily temperature swings that occur even in summer at high altitudes. Drying-pond temperatures can fluctuate 30 °C or more daily, so that organisms spend much of the diel cycle at temperatures that are suboptimal (too cold at night and too hot in the day) for growth and development (Jannot et al. 2008; Lund et al. 2016).

The water of alpine ponds is usually clear, and high UV radiation can be harmful for many species, damaging DNA and proteins (Mitchell and Karentz 1993); and, in some instances, causing mortality as observed in the phantom midge *Chaoborus* (Diptera) (Nagiller and Sommaruga 2009). Some species of zooplankton have reduced this problem through UVR avoidance, DNA repair mechanisms, production of mycosporine-like amino acids (MAAs), pigments (carotenoids, melanin), and/or antioxidants (Hairston 1976; Zellmer 1995; Hansson et al. 2007). The relative importance of these traits in alpine zooplankton appears to vary in different mountain regions of the world (Persaud et al. 2007; Sommaruga 2010).

Compared to deep alpine lakes, zooplankton in clear shallow alpine ponds have less opportunity to use diel migration as a strategy for avoiding UVR (Fischer et al. 2015).

Chemical Characteristics of Alpine Ponds

Because most water in alpine wetlands is from snow and ice melt, summer precipitation, and/or short-distance surface runoff, the chemical composition of the water in alpine ponds is primarily determined by the composition and weathering of bedrock, and by soil-forming processes in the context of glacial geomorphology (Marchetto and Rogora 2004). Compared to most lowland ponds and wetlands, total dissolved solids (TDS) are typically low (<100 mg L^{-1}) and vary locally with bedrock composition. For example, in a cluster of alpine ponds in Colorado that occur in glacial depressions in quartzite bedrock, TDS values range from <1 mg L^{-1} during snowmelt to <20 mg L^{-1} in drying pools in late summer. Acid-neutralizing capacity (ANC) ranges from ~10 to 300 milliequivalents L^{-1} (Wissinger and Whiteman 1992), with that 100-fold difference determined by whether ponds do or do not receive inputs from a first-order bedrock stream that crosses a local outcrop of carbonate-rich rock. In general, except in mountainous regions with carbonate-rich sedimentary rocks (e.g., Catalan et al. 2009a, b; Santolaria et al. 2015), alpine ponds should be poorly buffered from acidic inputs (anthropomorphic or natural, e.g., humic or tannic acids) compared to most lowland wetlands (Marchetto and Rogora 2004). Whether that low alkalinity translates into pH that is low enough to affect invertebrate communities will depend on levels of acidic inputs from natural (humic and tannic) and anthropogenic sources (acid deposition—see Human Impacts below).

In contrast to lowland water bodies, alpine lakes and ponds are often oligotrophic and co-limited by both nutrients (N and P) (Elser et al. 2009). Exceptions include those in volcanic mountains with phosphorus-rich rocks and those that receive atmospheric inputs of nutrients, especially nitrogen. In the Canadian Rockies, ponds are more nitrogen limited than lakes, therefore making them more sensitive early warning indicators of the potential effects of anthropogenic nitrogen deposition in remote mountainous regions (Murphy et al. 2010). Oligotrophic alpine ponds typically have low dissolved organic carbon (DOC), especially those that have mineral substrates (bedrock, rocky rubble, glacial till), and lack vegetation. DOC will be relatively high in alpine ponds with extensive littoral or terrestrial subsidies of vascular plant detritus or when associated with peatland soils. Where a particular pond lies on a gradient from low to the high DOC (e.g., brown water alpine ponds embedded in alpine fen) will have cascading effects on water clarity (hence UVR), temperature, primary productivity, and ultimately secondary (invertebrate) productivity. Although many alpine ponds have high oxygen levels associated with cold temperatures and low levels of microbial respiration, boundary layer anoxic conditions should be prevalent in those with organic substrates.

Ecology of Alpine Ponds

Primary Productivity

Compared to low-altitude ponds, area-specific primary productivity in alpine ponds should be, in the absence of human inputs, low and co-limited by the availability of nitrogen and phosphorus (Elser et al. 2009). For example, ponds that occur on quartzite bedrock at 3400 m elev. in Central Colorado are ultra-oligotrophic with extremely low, often undetectable levels of water-column nitrogen and phosphorus (Harte et al. 1985; Wissinger and Whiteman 1992). The overall annual primary productivity of alpine ponds is further constrained by the relatively short growing season, which can range from a few weeks in ponds that only melt for part of the summer to up to 4 months at subalpine altitudes. Aquatic productivity is further reduced in temporary basins where the time between ice melt and drying can be only a few weeks. In arid alpine environments (e.g., Central and Southern Andes), annual primary productivity is more likely to be constrained by drying and hypersaline conditions than by snow cover (e.g., de Los 2005; de Los and Gajardo 2010). Except in "brown water ponds" embedded in peaty wetlands, alpine ponds are often clear with extremely low levels of phytoplankton productivity (e.g., <50 g/cm^3; Wissinger et al. 1999a).

The relative importance of algae vs. detrital sources as primary sources of energy and nutrients should vary widely among alpine ponds depending on the extent to which there are macrophyte communities in the ponds and on the type(s) and productivity of surrounding alpine terrestrial vegetation (compare panels in Figs. 3.1 and 3.2). Although there are few studies to make a formal comparison, one might predict major shifts in the types and diversity of aquatic invertebrates depending on the degree to which phytoplankton, benthic algae, submergent and emergent macrophytes, and/or allochthonous detrital inputs are the dominant types of primary energy being harvested at the base of alpine pond food webs. Open, rocky-bottomed habitats (e.g., Fig. 3.1a) with little or no allochthonous or autochthonous detrital inputs should have very different pathways of energy and nutrient movements than those with detrital inputs (Oertli 1993; Klemmer et al. 2012, Fig. 3.1c). Recent studies in oligotrophic shallow lakes and ponds in the Arctic reveal that benthic algal productivity is nearly tenfold greater than that of phytoplankton (Rautio and Vincent 2006; Cazzanelli et al. 2015). Whether this characterizes patterns of algal productivity in alpine ponds has, to our knowledge, not been studied.

Secondary Productivity

A direct consequence of the generally low primary productivity in alpine ponds is an expected low secondary production of invertebrates (micro- and macroinvertebrates). In general, macroinvertebrate productivity should be low and food-chain length short compared to lowland ponds, both because of low primary productivity and because of metabolic constraints associated with cold temperatures (Plante and Downing 1989; Arim et al. 2007). Because of the extremely low levels of water-column primary productivity, it appears that epibenthic zooplankton rely heavily on benthic resources for growth, development, and survival (Cazzanelli et al. 2012). In general, because of the shallow nature and low water-column productivity in alpine ponds, benthic–pelagic coupling should be especially important for energy flow (Rautio and Vincent 2006). One of the main differences between alpine ponds above tree line and subalpine ponds at the upper edge of tree line is likely to be the additional sources of energy and nutrient subsidies from the surrounding forest, which is a dominant source of energy in ponds embedded in forested landscapes (Oertli 1993).

Invertebrate Community Composition

In large, permanent alpine ponds in Europe (0.2–2 ha), macroinvertebrate communities are often dominated in number and biomass by chironomids (Fureder et al. 2006; Fjellheim et al. 2009). Oligochaetes are also well represented (Laville 1971, 1974), as are mayflies, stoneflies, caddisflies, alderflies, and beetles. Small lakes are dominated by the same taxa (e.g., Oertli et al. 2008), as well as water boatmen, mites, flatworms, hirudineans, bivalves (*Pisidium* sp.), and snails (*Radix* sp.) (see Figs. 3.3, 3.4, 3.5, and 3.6: Appendix). Data surveys in small alpine lakes in the Pyrenees Mountains in France indicate that chironomids are major contributors to biomass and energy flow (Laville 1971, 1974; Le Cren and Lowe-McConnell 1980). Other benthic macroinvertebrates that are major contributors to biomass include caddisflies, bivalves (e.g., *Pisidium* sp.), aquatic beetles, mayflies, megalopterans, stoneflies, oligochaetes, and hirudineans (Le Cren and Lowe-McConnell 1980). The data from small mountain tarns included in lake surveys in New Zealand, Tasmania, and South Temperate mountains of mainland Australia indicate that the dominant contributors to benthic biomass are remarkably similar to each other and to those reported in Europe (Timms 1978; Wissinger et al. 2006b, 2009; Timms et al. 2013). As appears to be the pattern in Europe, deep alpine lake benthic biomass is dominated by chironomids and oligochaetes, whereas that of small alpine lakes and ponds includes a variety of other taxa including caddisflies, bivalves, aquatic beetles, water bugs, and odonates (Wissinger et al. 2009; Timms et al. 2013).

The cold, oxygen-saturated conditions in the highest alpine ponds support a specialized invertebrate fauna that includes Ephemeroptera (mayflies), Plecoptera (stoneflies), and Trichoptera (caddisflies) (EPTs). For example, in Europe, the stonefly *Nemurella picteti* (Plecoptera) and the triclad *Crenobia alpina* (Planaria), which are usually associated with cold well-oxygenated streams, occur in alpine ponds in the Swiss Alps if there are inlet and outlet streams; the predominantly lotic caddisfly *Hesperophylax occidentalis* occurs in flow-through alpine ponds in the Colorado Rockies; and ephemerellid mayflies, chloroperlid stoneflies, and lotic flatworms are more frequently encountered in alpine fens and wet meadows with than without flow in the Sierras (Holmquist et al. 2011) (Fig. 3.3c). Similarly, EPTs

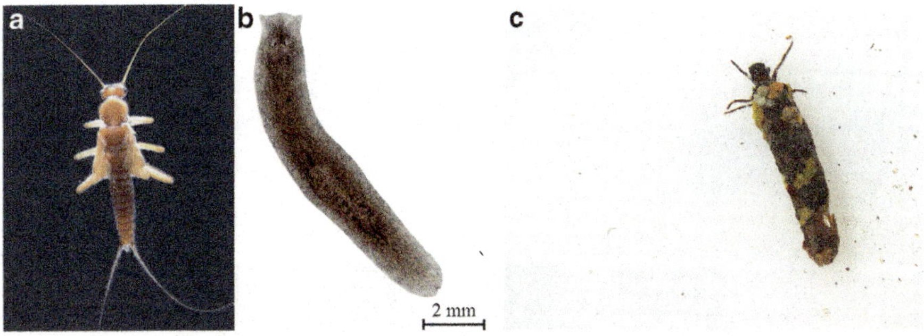

Fig. 3.3 Stenothermal taxa present in well-oxygenated streams but also occurring in alpine ponds: (**a**) Plecoptera: *Nemurella picteti* in the Swiss Alps (photo: J.L. Gatolliat), (**b**) Planaria *Crenobia alpina* in the Swiss Alps (© Biopix), (**c**) Trichoptera *Hesperophylax occidentalis* in Colorado Rocky Mountains (photo by Susan Washko)

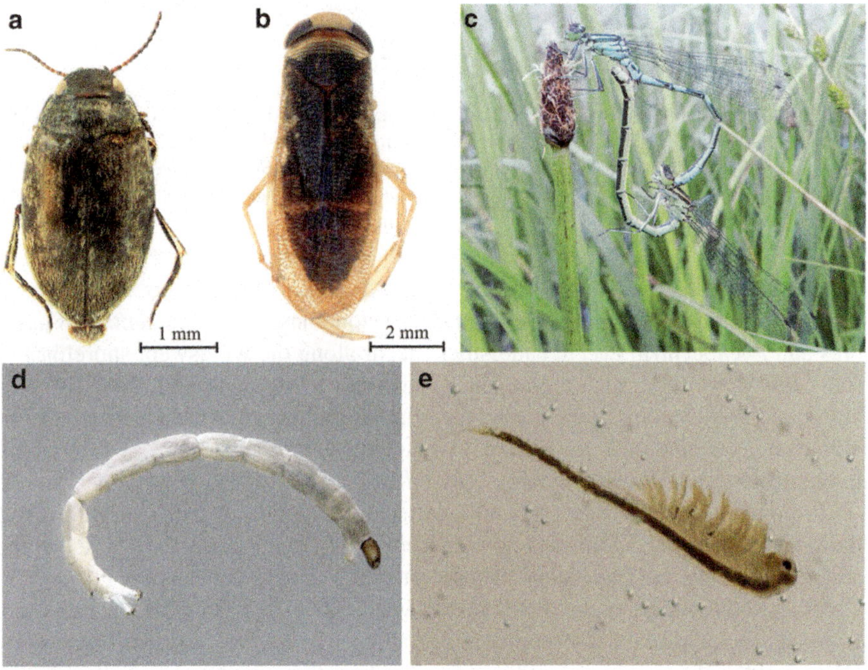

Fig. 3.4 A selection of cold stenotherm species, associated with alpine ponds in Europe: (**a**) the beetle *Hydroporus foveolatus* (Coleoptera) (photo: hepia), (**b**) the water bug *Arctocorisa carinata* (Heteroptera) (photo: hepia), (**c**) the damselfly *Coenagrion hastulatum* (Odonata) (photo: Antonin Jaquet), (**d**) the midge *Pseudodiamesa nivosa* (Chironomidae) (photo: hepia), (**e**) the arctic fairy shrimp *Branchinecta paludosa* (alpine ponds from high Arctic; photo: Markus Lindholm)

Fig. 3.5 Cold stenothermal benthic taxa common in alpine ponds above 3100 m in the Rocky Mountains in Central Colorado: (**a**) larva of the caddisfly *Asynarchus nigriculus* (photo by Scott Wissinger), (**b**) nymph of the mountain emerald dragonfly *Somatochlora semicircularis* (photo by Ellie Irons), (**c**) adult boreal bluet *Enallagma boreale* (photo by Pierre Deviche), (**d**) adult of the water bug *Arctocorisa lawsoni* (photo by Mark Dreiling)

typically found in streams in New Zealand (e.g., *Nesameletus ornatus* and *Deleatidium* mayflies, *Zelandobius furcillatus* stoneflies, and *Pycnocentrodes aureolus* and *Pycnocentria evecta* caddisflies) occur along the windswept shorelines of rocky and gravel-bottomed alpine lakes and ponds (Wissinger et al. 2009). Across altitudinal gradients in alpine zones, the relative abundance of eurythermal taxa in alpine ponds should decrease with cold-adapted stenotherms (Fig. 3.4) dominating in the highest and coldest habitats (Rosset and Oertli 2011).

To compare invertebrate communities in alpine ponds in the Alps (above) to those in the Rocky Mountains, we use data collected from a cluster of >60 subalpine and alpine ponds in The Nature Conservancy's Mexican Cut Nature Preserve in Colorado (3540–3580 m elev.). These data were collected by multiple investigators from the Rocky Mountain Biological Laboratory over the past 40 years (reviews by Dodson 1982; Wissinger et al. 1999a and references therein). As in Europe, chironomids and other small Diptera (e.g., ceratopogonids, culicids) are major contributors to secondary productivity in North American alpine ponds, although low standing stock biomass belies their contribution to productivity except just before pulses of emergence in midsummer. Chironomid diversity (3–15 taxa) increases with pond size and permanence (Wissinger et al. 1999a). Oligochaetes and bivalves (*Pisidium*) are especially abundant in ponds with soft sediments. The mayfly, *Callibaetis ferruginosa*, occurs in permanent subalpine ponds without vertebrate predators (fish, salamanders), but as is often the case in Europe, alpine pond communities do not

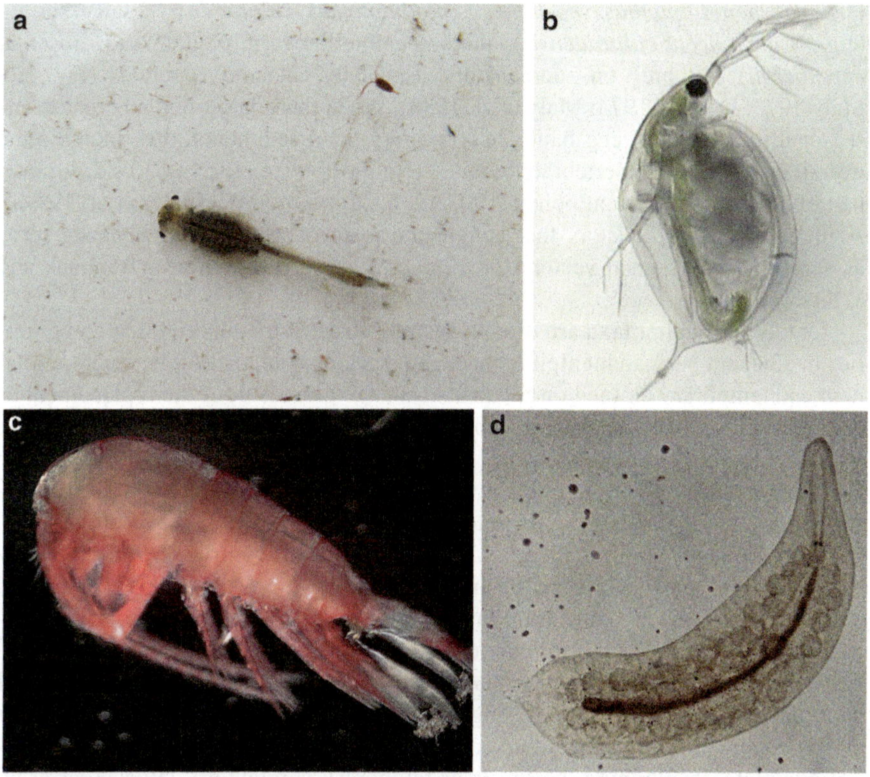

Fig. 3.6 Common water-column invertebrates in alpine ponds in the central Rocky Mountains that lack fish and salamander predators: (**a**) the fairy shrimp, *Branchinecta coloradensis* (photo by Susan Washko), (**b**) large-bodied cladocera, *Daphnia middendorffiana* (photo by James Haney), (**c**) UV protective red-pigmented calanoid copepod, *Hesperodiaptomus shoshone* (photo by Edward Maly), and (**d**) *Mesostoma* flatworm predator on zooplankton in alpine ponds (photo by S.E. Thorpe)

typically have mayflies or stoneflies unless they are flow-through basins with a mixed lotic–lentic fauna. Cased caddisflies (Limnephilidae and Phryganeidae) (Fig. 3.5a) can be extremely abundant and dominate the invertebrate biomass (>50 %) of subalpine ponds at peak abundance during the growing season (Wissinger et al. 2003; Klemmer et al. 2012). Species of odonates, water bugs, and dytiscid beetles with boreal-alpine distributions (Fig. 3.5) (see section "Biogeography of Alpine Pond Invertebrates") are the top predators in fishless alpine ponds.

Alpine ponds typically contain a variety of small-bodied benthic and epibenthic crustaceans including ostracods, bosminids, chydorids, daphniids, and cycloid and calanoid copepods that survive drying and freezing in a resting stage in the substrate (Dodson 1982; Wissinger et al. 1999a). In addition to relatively small species, temporary alpine ponds in the Central Rockies contain a characteristic assemblage of large-bodied epibenthic crustaceans including cladocerans in the *Daphnia pulex* complex (e.g., *D. middendorffiana*), one or more bright-red calanoid copepods (e.g., *Hesperodiaptomus shoshone*, *Leptodiaptomus coloradensis*), fairy shrimp (e.g., *Branchinecta coloradensis*), and epibenthic flatworm predators (*Mesostoma ehrenbergii*) that prey on zooplankters, especially calanoid copepods (Fig. 3.6, Maly 1973; Dodson 1974; Maly et al. 1983). When these large-bodied crustaceans seasonally hatch from egg banks in temporary-pond sediments, they increase the overall area-specific invertebrate biomass to almost twice (1864 ± 630 s.d. mg m^{-2}) that of adjacent permanent ponds (1002 ± 236 s.d. mg m^{-2}; Wissinger et al. 1999a). Adult salamander predators that congregate to feed on this food bonanza serve as across-pond dispersal vectors for these crustacean invertebrates (Bohonak and Whiteman 1999).

Temporary habitat taxa are often a nested subset of the invertebrate communities in adjacent permanant alpine ponds, but species composition between permanent and temporary alpine ponds can also change as a result of the replacement of different, closely related species with alternative adaptive strategies for coping with drying vs. predator avoidance (Table 3.1). For example, species of caddisflies that occur mainly in temporary alpine ponds exhibit a variety of characteristics (high activity and foraging rates, minimal investment in protective cases, intraspecific aggression including cannibalism) that all contribute to their ability to pupate and emerge before basins dry (Wissinger et al. 2006a; Greig and Wissinger 2010). The same traits make those species vulnerable to larval salamander predators, which eliminate them from permanent ponds. In contrast, the behavioral and developmental traits of permanent-pond caddisflies (low overall activity, risk-sensitive foraging, high investment in protective cases) facilitate coexistence with salamanders but extend development to the extent that they cannot complete development before their temporary habitats dry (Wissinger et al. 1999b, 2006a). Similarly, the large-bodied zooplankters described above are pref-

Table 3.1 Distributions of detritivorous caddisflies in Central Colorado

Montane–elevation basins			Subalpine–elevation basins		
Temporary	Semipermanent	Permanent w/ predators	Temporary	Semipermanent	Permanent w/ predators
	***L. externus* (*LE*)**	***L. externus* (*LE*)**		***L. externus* (*LE*)**	***L. externus* (*LE*)**
	**L. sublunatus* (*LS*)	**L. sublunatus* (2000)		*H. occidentalis* (*HO*)	
	***L. picturatus* (*LP*)**	***L. picturatus* (*LP*)**		**L. picturatus* (*1998*)	**L. picturatus* (*1998*)
***L. tarsalis* (*LT*)**	***G. lorretae* (*GL*)**	*G. lorretae* (*GL*)		**G. lorretae* (2009)	**G. lorretae* (2009)
L. secludens	***L. secludens* (*LSC*)**		***A. nigriculus* (*AN*)**	***A. nigriculus* (*AN*)**	

L Limnephilus, A Asynarchus, G Grammotaulius, H Hesperophylax. Ranges based on survey data from 1990 to 2015 (Wissinger et al. 2003; unpublished data). ***Bold*** indicates dominant species. *Asterisk* indicates recent upslope range shift in elevation. Dates show first occurrence at that elevation

erentially preyed upon by fish and salamanders. As a result, these large temporary-pond taxa are replaced by small-bodied daphniid cladoceran and copepods and their chaoborid predators in permanent alpine ponds (Dodson 1970, 1974; Sprules 1972; Maly and Maly 1974). All of these zooplankton species can survive drying; thus, in this case, the replacement pattern across pond types is driven mainly by predators. Such species replacements across "predator-permanence" gradients have been documented for species within a genus, and genera within a family, for nearly every taxonomic group of aquatic invertebrates (Wellborn et al. 1996). However, the details of the proximate mechanisms (different morphologies, physiologies, and behavior) and evolutionary patterns of diversification within taxa are only well understood for odonates (e.g., McPeek and Brown 2000; Stoks and McPeek 2006; McCauley 2008) and caddisflies (see above). For other taxa that exhibit species replacements along "predator-permanence" gradients in alpine ponds (dytiscid beetles, water bugs, chironomids), the degree to which the pattern is driven by predators vs. permanence per se has not been studied experimentally. Across all taxa in a cluster of alpine ponds, it appears that a shift in the top predator along the permanence gradient is more important than hydroperiod (Wissinger et al. 1999a). They used ordination to rank the (dis)similarity in community composition for all taxa among ponds and then structured regression models (path analysis) to tease apart the relative roles of the various factors that might explain patterns of community composition (water chemistry, pond area, depth, permanence, predators). The most important factor that affected pond community composition was the presence or absence of large larval salamander predators, which occur only in relatively deep ponds that remain permanent for multiple years (Whiteman et al. 2012). This is consistent with the prediction that population interactions (competition, predation) should be relatively benign in harsh or frequently disturbed environments like temporary alpine ponds because low prey abundances reduce the likelihood of resource competition and/or because harsh environments ameliorate the efficacy of predation (e.g., Peckarsky 1983). In contrast, Greig et al. (2013) found that top-down predator effects are actually weaker in permanent than temporary high-altitude ponds in New Zealand and attribute this result to nonconsumptive effects of fish that weaken predation by invertebrate predators in the complex food webs of permanent ponds. They suggest that prevalence of generalists (rather than a replacement pattern of specialists) across predator-permanence gradients reflects a balance between strong interactions in fishless permanent habitats and weak predator effects in permanent ponds.

Because alpine ponds often occur in clusters that include multiple basins that differ in harshness and drying regimes, they provide an excellent opportunity for (1) studying interactions between the relative importance of abiotic and biotic filters on community organization (also see Brendonck et al. 2002; Greig and Wissinger 2010), (2) the complicated and unresolved role that disturbance and predators have on local species diversity (see paragraph above and Hughes et al. 2007; Dornelas 2010; Fox 2013; Vanschoenwinkel et al. 2013), and (3) the role of dispersal in metacommunity dynamics at different scales and elevations (e.g., Rundle et al. 2002; Declerck et al. 2011; De Bie et al. 2012; Bertin et al. 2015).

Species Richness in Alpine Ponds

At a given altitude, patterns of species richness at different special scales (alpha, beta, gamma) have been attributed to a variety of non-mutually exclusive habitat characteristics including water chemistry, basin size, vegetation, degree of pond permanence, presence and type of predators, pond clustering, and connectivity (De Meester et al. 2005; reviews by Batzer 2013; Ruhí and Batzer 2014; Batzer et al. 2015). Here we summarize the evidence for the role that these factors play in determining species richness in alpine ponds.

Pond Area

Across all types of taxa, there is evidence that species richness is strongly affected by basin size. For example, in alpine ponds and lakes in the Tatra Mountains in Europe, ponds showed lower local diversity (alpha), higher among site diversity (beta), and similar regional diversity (gamma) than large lakes. These relationships were attributed to a combination of the species–area relationship (SAR) per se, habitat heterogeneity, which often covaries with area, and environmental harshness (Hamerlik et al. 2014). Similarly, in a cluster of ponds (3450–3480 m.a.s.l.) in western North America, within-pond species richness ranges from 5 (0.02 ha) to 75 (4.5 ha) invertebrate taxa (Fig. 3.7). However, these overall trends do not always hold for individual taxa (e.g., Oertli et al. 2002), and patterns observed at low altitude are not necessarily relevant for alpine ponds. For example, in lowland ponds, dragonfly richness is strongly governed by pond area (e.g., Oertli et al. 2002) but that does not appear to be the case in alpine ponds (Hinden et al. 2005; Ilg and Oertli 2014). This difference could be due to the overall decrease in richness with altitude such that it is only at the overall community level that a species–area relationship can be detected in alpine ponds.

Vegetation

Many studies have linked species richness in ponds to the presence and types (submergent vs. emergent; stem architecture) of macrophytes (Hinden et al. 2005; de Mendoza et al. 2012; Ilg and Oertli 2014). Compared to lowland ponds, aquatic macrophyte densities in alpine pines are often relatively sparse, but some basins can have dense stands of both submergent and emergent aquatic plants (Fig. 3.1). There is some evidence that the presence of aquatic vegetation enhances species richness in alpine ponds (Wissinger et al. 2009), but we do not know of any studies that addressed the degree to which that association is related to one of several mechanisms associated with habitat heterogeneity (e.g., niche partitioning, refuge from predators) or some aspect of resource productivity (e.g., plants as a substrate for biofilm or as a source of autochthonous detritus). In alpine fen and wet meadow complexes in the California Sierras, the presence / absence of sheet flow during

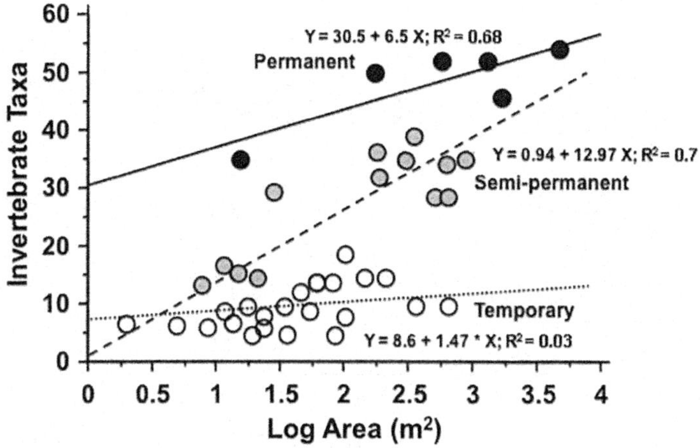

Fig. 3.7 Invertebrate species richness as a function of area in a cluster of alpine ponds in the central Rocky Mountains of Colorado, 3450–3480 m. Permanent ponds (dark circles) do not dry. Semipermanent ponds (gray circles) dry in autumn in some years. Temporary ponds (open circles) dry in early summer in most years (updated from Wissinger et al. 1999a, b to include recent upslope immigrants)

snow melt in spring has a greater effect on aquatic invertebrate communities than did vegetation type (Holmquist et al. 2011).

Permanence

Temporary ponds at low altitudes have lower species richness than permanent ponds (e.g., Schneider and Frost 1996; Tarr et al. 2005; Vanschoenwinkel et al. 2013), and this same pattern is observed in alpine ponds (Fig. 3.7). This result is often confounded by pond area because large ponds are more likely to be permanent than small ponds. However, even when controlling for pond area, species richness is lower in temporary than in permanent basins (Figs. 3.7 and 3.8; Vanschoenwinkel et al. 2009; Wissinger et al. 2009). In alpine ponds, the species that occur in temporary habitats are often a "nested subset" of those in adjacent permanent habitats that have one or more adaptations to cope with drying including (1) the desiccation-tolerant stages of epibenthic crustaceans (cladoceran, copepods, brachiopods) that survive dry basins for frozen basins; (2) specialized oviposition strategies of aquatic insects that facilitate overwintering in dry basins (e.g., endophytic overwintering eggs of lestid damselflies, semiterrestrial eggs of limnephilid caddisflies); (3) rapidly dispersing adults that rapidly colonize newly inundated basins through egg deposition (e.g., chironomid and culicid dipterans); and (4) taxa with aquatic adults that cyclically move back and forth between permanent and nearby temporary habitats (e.g., dytiscid, hydrophilid, and haliplid water beetles and corixid and notonectid aquatic bugs). Species that live in alpine temporary ponds face considerable interannual variation in the time between snow melt and pond drying and, thus, are likely to exhibit

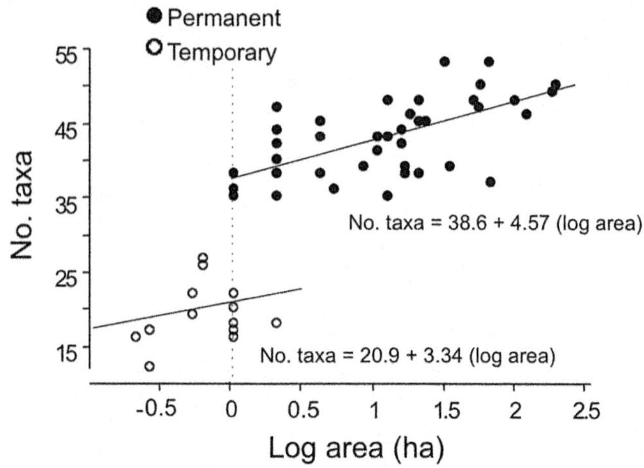

Fig. 3.8 Data from montane and subalpine lakes and ponds on the South Island of New Zealand indicate that independent of area, temporary basins typically have fewer species than permanent ponds (Wissinger et al. 2009)

wide fluctuations in population sizes, with high levels of recruitment in years in which drying occurs relatively late and complete cohort failure when drying occurs relatively early (Greig and Wissinger 2010). Finally, from a landscape diversity perspective, the drying of alpine wetlands can be diversity enhancing because of the seasonal replacement of aquatic by terrestrial invertebrate communities. For example, Holmquist et al. (2011) found that aquatic invertebrate composition and diversity during the wet phase of the hydroperiod differed between fens and wet meadows with and without sheet flow, but that terrestrial invertebrate diversity during the dry phase of the hydroperiod dwarfed that of both types of aquatic communities.

Elevational Gradients in Alpine Ponds

Species Diversity Decreases with Altitude

As with latitude, altitude is a major factor explaining the spatial heterogeneity of the distribution of life. For most biologic groups, species richness decreases with latitude toward the North and South Poles. Similarly, species richness for many taxa decreases with altitude (e.g., Rahbek 1995). The decrease is observed at the regional (regional species pool, gamma richness) and at the local scale (alpha richness). The causes of the decrease in the regional species pool with altitude are related to ecological and evolutionary processes (Graham et al. 2014) that include physical (climate, area, historical perturbations, geology) and/or biological (primary productivity) factors.

Such relations have been described for many ecosystems and taxonomic groups especially for terrestrial biota (e.g., Gaston and Blackburn 2000; Begon et al. 2006) but rarely with pond invertebrates that nevertheless constitute a perfect research model for this aim. Data from alpine ponds from Europe (Switzerland) and North America (Colorado) provide evidence for a lower local species richness above than below tree line for various invertebrate groups (dragonflies, water beetles, caddisflies, and gastropods; Fig. 3.9). Other studies conducted in the North Cascades, Washington, USA (Hoffman et al. 1996), in European alpine ponds and small lakes in the Alps (Fureder et al. 2006; Fjellheim et al. 2009), and in the Pyrenees (de Mendoza and Catalan 2010) all corroborate the sharp decline in taxon richness with increasing altitudes. The latter study cautions that multiple factors underlie these patterns and which factor best explains altitudinal changes in diversity will vary across taxa and locations.

The sharp decline in species richness with altitude is not simply a case of there being a particularly tolerant subset of low-altitude species that also occur in alpine ponds. Although there are some generalists with broad altitudinal ranges that extend into the alpine zone, the change in community composition is often related to species replacements that mirror those observed when moving north or south in latitude. A large proportion of the species from alpine ponds are boreal-alpine species (see section "Biogeography of Alpine Pond Invertebrates"), including many cold stenothermal specialists (Figs. 3.4, 3.5, and 3.10). The proportion of the invertebrate community comprised of stenothermal species increases up to 100 % in the highest alpine ponds. The increasing dominance of boreo-alpine species richness with altitude underlines the uniqueness of invertebrate assemblages and supports the notion that there is indeed a core set of species (sensu Batzer and Ruhí 2013) in alpine pond communities (Appendix).

Life Cycles Shift Along Altitudinal Gradients

Alpine ponds present a unique set of constraints on the ability of organisms to complete their life cycles because of the short growing season and low metabolic rates associated with cold temperatures. Altitudinal shifts in voltinism (generations per year) have been described in several groups of taxa. For example, the water boatmen *Callicorixa audeni* and *Cenocorixa bifida* can complete two or more generations in ponds at montane altitudes but only one per year in subalpine and alpine ponds (Dodson 1975). Similarly, the boreal-alpine dragonfly, *Somatochlora semicircularis*, can complete development in 2 years at montane altitudes (2500–3000 m elev.) in Colorado but takes up to 4 years to complete larval development in alpine (3400–3700 m elev.) ponds (Wiley 1973, 1974; Wissinger et al. 1999a). This change in voltinism can have several effects on population and community ecology. First, it increases the time that multiple year classes of larvae overlap, which is known to affect rates of interference competition, cannibalism, and ultimately cohort survival (Crowley et al. 1987). Secondly, it can result in the redistribution of a species across different types of ponds at a given altitude. For example, at montane altitudes, a species that relies on rapid development to complete larval development in one

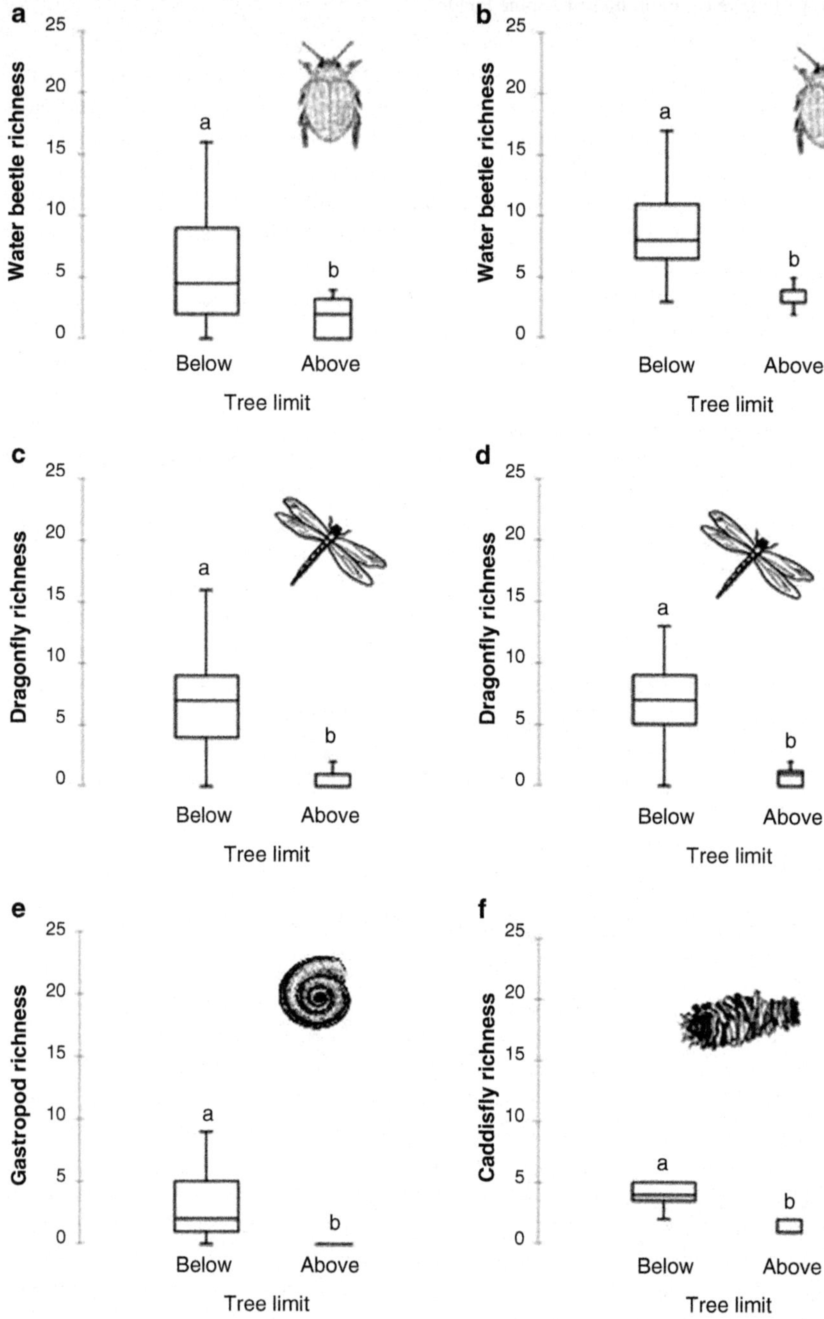

Fig. 3.9 Species richness below and above the tree limit for (**a**) water beetles of Switzerland, (**b**) water beetles of USA, (**c**) adult dragonflies of Switzerland, (**d**) adult dragonflies of USA, (**e**) gastropods of Switzerland, and (**f**) caddisflies of USA. Mann–Whitney test results are shown with different letters above each altitudinal class. Letters denote groupings based on lack of statistical difference among classes. Vertical boxes represent the interquartile range (Q3–Q1), within which the line represents the median. Box size is proportional to the number of sites. Whiskers represent the largest non-outlier values. Drawn after Ilg and Oertli (2014), Oertli et al. (2008), and S. Wissinger (unpublished survey data)

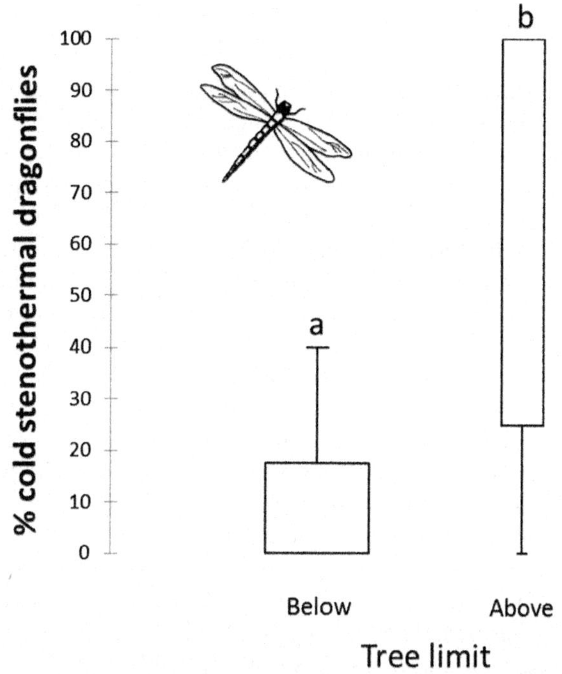

Fig. 3.10 Percentage of cold stenothermal dragonflies present below and above the tree limit in Switzerland. Mann–Whitney test results are shown with different letters above each altitudinal class. Letters denote groupings based on lack of statistical difference among classes. Vertical boxes represent the interquartile range (Q3–Q1), within which the line represents the median. Box size is proportional to the number of sites. Whiskers represent the largest non-outlier values. Drawn after Ilg and Oertli (2014)

summer before ponds dry can exploit both temporary and permanent habitats, whereas at high altitudes, the multiyear development time will restrict the same species to permanent basins. Even in the absence of a change in voltinism, slow growth rates at alpine altitudes can affect whether a species can complete development in habitats that dry in late summer. For example, the cased caddisfly *Limnephilus externus*, which occurs in montane, subalpine, and alpine ponds in the mountains of western North America, can complete development in temporary ponds at montane altitudes but is restricted to semipermanent and permanent ponds at alpine altitudes due to longer development times (Wissinger et al. 2003).

Life-history shifts with altitude are also well described for newts and salamanders that are often the top predators (if there are any) on invertebrates in alpine ponds in Europe and North America (Whiteman et al. 1996; Denoël and Joly 2001). At montane altitudes, these top predators complete larval development in one summer, but in alpine ponds, they require multiple years of growth before they can metamorphose and emerge. At subalpine altitudes, growth rates are so slow that they forgo metamorphosis and remain in ponds as paedomorphic adults that exert strong top-down predator effects on invertebrate communities (Denoël et al. 2007; Whiteman et al. 2012).

Biogeography of Alpine Pond Invertebrates

Arctic/boreal invertebrates that also inhabit alpine ponds at temperate latitudes are of special interest to evolutionary biologists and biogeographers interested in understanding macro- and microevolutionary patterns of persistence and postglacial recolonization (Theissinger et al. 2013). On average, moving upslope 100 m results in a change in climate that is about equivalent to moving 80 km (45 miles or 0.75° of latitude) toward the poles. Another way to view this is to consider that on average, a rise of 122 m in altitude is roughly equal to a 1-°C rise in mean annual air temperature (Danks 1978). Thus, it is not surprising that many of the species that live at low altitudes and high latitudes also inhabit alpine ponds at temperate and even subtropical latitudes. For example, "taiga bluet" (*Coenagrion resolutum*) and "boreal bluet" (*Enallagma boreale*) damselflies (Fig. 3.5c) inhabit lentic habitats across a broad range of altitudes across the entire North American continent above 40°N. However, in the Sierra and Rocky Mountains in western North America, they occur in alpine ponds as far south as northern Mexico, a range extension of nearly 15° in latitude (roughly 1000 km). There appears to be a general biogeographic pattern for alpine pond species in the Northern Hemisphere, especially for taxa with boreal centers of biogeographic origin. Among dragonflies, species of sympetriniid libellulids (e.g., *Leucorrhinia hudsonica, Sympetrum danae, S. costiferum*), corduliids (e.g., *Somatochlora semicircularis*), and aeshnids (e.g., *Aeshna palmata, A. juncea*) are among the top predators in alpine ponds in the Southern Rocky Mountains, far beyond their trans-boreal distributions at lower altitudes (Dubois 2016). Similar patterns are observed for European species in the same genera (e.g., *Somatochlora, Aeshna, Leucorrhinia, Sympetrum*) where postglacial relict and disjunct populations of these boreal-alpine stenotherms occur from the Alps, to the Tatras, and to the Urals (Šácha and Bulánková 2006; Bernard and Daraż 2010; De Knijf et al. 2011). Other arctic–alpine pond invertebrates with broad (often circumboreal) distributions at high latitudes, but restricted alpine distributions further to the south in both Europe and North America, include limnephilid caddisflies (Wissinger et al. 2003; Waringer et al. 2011, 2012) and dytiscid diving beetles (Larson et al. 2000).

Threats and Conservation Issues Associated with Alpine Ponds

Unique Species Assemblages

Ponds in general are considered to have high ecological and conservation value (Boix et al. 2012), and several aspects of alpine pond invertebrate communities present conservation opportunities and challenges. First, although an individual alpine pond is likely to contain fewer species than a large lake, these small basins are often quite abundant and occur across multiple ecological gradients (size, degree of permanence, substrates, productivity) that affect community composition, diversity, and structure (see above). Thus, as with ponds in general (Oertli et al. 2002; De Meester et al. 2005), differences in the fauna among alpine ponds are an important contributor to beta and gamma diversity of aquatic invertebrates in mountainous regions (Oertli et al. 2008; Hamerlik et al. 2014). Secondly, although the overall species richness is often lower than that in comparable (same size, permanence, etc.) low-altitude ponds, alpine pond communities are often dominated by cold stenotherms including endemic or near-endemic species (Oertli et al. 2008). Isolated alpine populations of these species are often distinctive ecotypes (see section "Biogeography of Alpine Pond Invertebrates") and, compared to individuals and populations in the centers of their low-altitude boreal ranges, should be likely to have rare combinations of genes embedded in unique community contexts (e.g., Kaunisto et al. 2015).

Biological conservation of alpine pond invertebrates in Europe has been driven mainly by three flagship groups: aquatic plants, amphibians, and dragonflies. Other invertebrate groups beyond dragonflies are also sometimes considered, especially when they are targeted on local red lists. In Europe, the NGO EPCN (European Pond Conservation Network) identified 30 sites (mostly pond networks) at altitudes >1500 m that are particularly important for pond conservation in the Alpine Arc (Ewald et al. 2010). Their selection was motivated by pond density, services to society, threatened species, and the presence of the three flagship groups described above. Targeted species included five boreo-alpine dragonflies: *Coenagrion hastulatum*, *Leucorrhinia dubia*, *Somatochlora alpestris*, *Aeshna juncea*, and *A. caerulea*.

As exemplified by this EPCN program, the conservation of pond communities has largely focused on networks of different types of ponds and therefore high beta diversity. The high beta and regional diversity associated with pond clusters has been observed in lowland ponds (Oertli et al. 2002), mountain ponds (Martinez-Sanz et al. 2012), and high-altitude bogs (Coronel et al. 2007) and is at least in part enhanced by the stabilizing effects of metapopulation and metacommunity dynamics. Furthermore, high-altitude ponds provide refugia for threatened species because these habitats are often less likely to be directly impacted by human activities than those at lower altitudes (Grant and Samways 2007).

Alpine pond invertebrates, especially boreal-alpine stenotherms, should be especially vulnerable to climate change. Thus, conservation strategies, policies, and priorities will need to account for anticipated shifts in species distributions (Gillson et al. 2013). In the Swiss Alps, Rosset and Oertli (2011) studied a group of alpine pond species potentially endangered by warming and classified them according to their resilience to this perturbation. The dragonfly *Coenagrion hastulatum* (Fig. 3.4c), which is presently widely distributed, has a particularly low resilience index and should be especially vulnerable to future changes in climate. Climate-driven local extinction of species and their geographical retraction have already been observed in other stenothermic taxa, as, for example, with the arctic fairy shrimp *Branchinecta paludosa* (Fig. 3.4e) in alpine ponds from southern Norway (Lindholm et al. 2012).

Sentinels for Regional and Global Atmospheric Human Impacts

Acid Deposition

Alpine ponds, like alpine lakes, are often isolated from local-scale impacts of human activities and therefore often considered to be sentinels of regional and global changes in the environment (e.g., Adrian et al. 2009; Catalan et al. 2009a; Williamson et al. 2009a, b). Early investigations on the effects of humans on alpine ponds and lakes focused on their vulnerability to acidification. Basins situated in unbuffered bedrock settings (igneous, non-carbonate metamorphic, and sedimentary) are vulnerable to even low levels of atmospheric acid deposition (Harte et al. 1985; Camarero et al. 1995). Alpine ponds in mountainous regions that are downwind from major industrial and urban areas have been affected by both sulfuric and nitric acid deposition and, in the latter case, the deposition has affected both pond acidity and nutrient dynamics (Elser et al. 2009; Saros et al. 2011). Similarly, it appears that the melting of glaciers has affected the nutrient status and the planktonic communities of alpine ponds and lakes (Saros et al. 2010), but there appears to be a dearth of information on the impact on benthic invertebrate communities. The pH of relatively unbuffered waters is also vulnerable to increases in temperature (and resulting biological activity), which can enhance rates of acidification (Sommaruga et al. 1997). Acidification reduces DOC concentrations in water bodies allowing increased penetration of solar radiation (Schindler et al. 1996), which has cascading effects on the entire food web. Many invertebrate taxa are sensitive to acidification including microcrustaceans (*Daphnia*, *Hesperodiaptomus*, *Diaptomus*) and aquatic insect larvae (Bradford et al. 1998). Molluscans, malacostracans, caddisflies, and mayflies are typically reduced or absent in ponds where pH is under 5.5 (Friday 1987). In general, species richness is lower in water bodies impacted by acid deposition (Bradford et al. 1998).

Climate Change

Global temperatures are projected to continue to increase during the next century, resulting in a dryer, warmer climate in many parts of the world (IPCC 2014). Climate change impacts on aquatic ecosystems are predicted to be strong at high latitudes and high altitudes where the duration and timing of snow cover and melting can have disproportionately large effects on hydrology and water temperature (Heino et al. 2009; Post et al. 2009; Angeler et al. 2013). The most rapid and dramatic changes have occurred in shallow, temporary basins, because low water volumes and high surface-area-to-depth ratios make them especially vulnerable to changes in snowpack, snowmelt, and evaporation (Barnett et al. 2005; Corcoran et al. 2009). Species that inhabit temporary alpine ponds typically have rapid life cycles that rely on a variety of timing cues such as photoperiod, temperature and heating degree days, water chemistry, and wetting and drying events (Wissinger 1999; Williams 2006). They often exhibit plasticity in these traits in response to seasonal, interannual, and long-term changes in temperature and in the duration of the wet phase of hydroperiods (De Block et al. 2008; Stoks et al. 2014). Thus, it is not clear whether the species that live in temporary alpine ponds will be among the winners or among the losers in the face of a changing climate (Rosset et al. 2009; Rosset and Oertli 2011). For aquatic species that live in temporary habitats that no longer fill annually, the answer is a *fait accompli* (Smol and Douglas 2007a). For species that live in relatively deep alpine ponds that are not likely to dry in near ecological time, climate change impacts on growing season, water temperature, and hence primary and secondary productivity are more difficult to predict. For example, some insect larvae should develop faster and emerge earlier, which in turn should affect voltinism (e.g., semi-voltine species might become univoltine).

Distributional range shifts at the regional scale (latitudinal and altitudinal) are the most frequently observed effect of climate change (Hickling et al. 2006; Chen et al. 2011), and upslope shifts in distribution are expected for species whose ranges are limited by growing season or extreme temperature events. Two separate meta-analyses have estimated average altitudinal boundaries have moved upslope 6.1 m (Parmesan and Yohe 2003) and 11.1 m (Chen et al. 2011) per decade. Whether these range shifts have negative or positive impact on different species of aquatic invertebrates will depend on the outcome of biological interactions in communities with new combinations of species (VanPutten et al. 2010). For example, if competitive outcomes between low-altitude upslope immigrants and boreal-alpine stenotherms are temperature dependent, or if the latter enjoy an altitudinal (temperature) refuge from lower altitude predators or competitors, then those stenotherm taxa could be negatively impacted or even eliminated by such shifts in community composition (Rahel and Olden 2008). Models of the effects of shifting distributions predict a balance of colonization and extinction that will result in a marked increase in local species richness in alpine ponds (Rosset et al. 2010). Research on such topics is still in its infancy and predicting effects on particular species will require understanding what abiotic and biotic constraints limit current distributions along altitudinal gradients, the ability of species to keep pace with a changing climate (Chen et al. 2011), and the degree to which species range expansions are limited by climate sinks (Burrows et al. 2014). For alpine ponds, the availability of yet higher basins becomes a climate-sink barrier for upslope shifts in the distributions of cold stenotherms. It is unclear how upslope shifts in tree line (see review by Parmesan 2006), and alpine plant communities in general, will affect alpine pond invertebrates. Conservation strategies for protecting the unique assemblages of species in alpine ponds will need to be modified to accommodate such shifting distributions (Mawdsley et al. 2009; Gillson et al. 2013). Finally, there is a growing literature on the effects of climate change on the expansion of paraglacial lakes and ponds as a result of the melting and retreat of glaciers (e.g., Gardelle et al. 2011; Komori 2008; Salerno et al. 2014).

Paleolimnological studies from sediment cores provide an important archive of past limnological communities that can be used to reconstruct environmental changes (Smol and Douglas 2007b; Saros 2009; Sayer et al. 2010). Subfossil remains of invertebrates (head shields, postabdomens and claws of cladocerans, ostracod shells, freshwater mollusc shells, bryozoan statoblasts, chironomid head capsules, and preserved chitinous parts of beetles and caddisflies) constitute effective tools for assessing shifts in invertebrate community composition in alpine ponds (Smol et al. 2001; Luoto and Nevalainen 2012).

Predation by Introduced Fish

Fishes, which are often the top predators in lowland permanent ponds and wetlands, were historically absent in many subalpine and alpine ponds because of dispersal barriers. Instead, the aquatic top vertebrate predators were typically limited to adult and larval salamanders [e.g., Pyrenean brook salamander *Calotriton asper* in the European Pyrenees; alpine newts, *Triturus alpestris* (Denoël and Joly 2001) in Europe; and ambystomatid salamander larvae and paedomorphic adults in the mountains of western North America (Whiteman et al. 2012)]. However, over the past two hundred years, nonnative salmonid fishes (trout and salmon) have been introduced into high mountain lakes and permanent ponds for sport fishing throughout the world (MacCrimmon and Marshall 1968; MacCrimmon 1970). The effects of these fishes on alpine aquatic communities have been especially well studied in Europe, western North America, and Australasia (New Zealand and Australia). Impacted invertebrates in alpine water bodies in Europe and North America include large zooplankton species (>1 mm; e.g., *Daphnia* sp.) and large-bodied, conspicuous, and non-burrowing macroinvertebrates (Knapp et al. 2001), e.g., water beetles (Coleoptera, Dytiscidae), caddisflies (Trichoptera), Acari, and water boatmen (Corixidae). Swimming taxa, such as dytiscid beetles, appear particularly sensitive to fish presence (Hinden et al. 2005; de Mendoza et al. 2012). The loss of large-bodied predatory invertebrates appears to benefit the abundance of other taxa including oligochaetes and chironomids.

In Europe and North America, the introduction of nonnative salmonids has also led to the extinction of native amphibians (frogs, toads, newts, salamanders), including those that were the original top predators on invertebrates. The replacement of top native predators by salmonid trout has had cascading effects on ecosystem processes (productivity, energy flow), prey diversity, and relative abundance (Knapp et al. 2001; Tiberti et al. 2014). Not surprisingly, conservation efforts on small lakes and large ponds that support nonnative fishes have been focused on reassessing stocking practices and the removal of fishes (see section on management, below).

Unlike at low altitudes, fish survival in ponds and other shallow lentic habitats at high altitudes will be limited by the degree to which those habitats freeze solid or become anoxic and "winterkill." For large permanent ponds that do not winter-kill, the results from small lakes provide insight into how these introductions have or should affect alpine pond communities. For example, the introduction of trout into subalpine lakes in New Zealand reduces and often eliminates populations of native galaxiid fishes (McDowall 1987) but does not appear to have a major impact on benthic invertebrate communities, perhaps because of the refuge provided by extensive beds of aquatic vegetation (Wissinger et al. 2006a). Thus, relatively shallow permanent alpine ponds and lakes may provide a regional refuge for taxa either because their winterkill or because of the presence of spatial refuges (beds of submergent vegetation and/or emergent vegetation on litoral fringes) that are absent from many alpine lakes.

The effects of fish stocking in alpine ponds and lakes can extend beyond the borders of aquatic basins. For example, in a comparison of fish and fishless lakes in the high Sierra Mountains in California, Empanchin et al. (2010) found that trout-induced reductions in the export of adult aquatic insects (especially mayflies) had a strong negative effect on alpine birds that rely on this food source for feeding nestlings. They recommend that terrestrial recipients of aquatic subsidies be included in conservation and restoration plans for ecosystems that include alpine lakes and ponds.

Domesticated Animal Grazing, Trampling, and Eutrophication

Although many of the types of land use activities that affect lowland ponds and wetlands do not typically threaten alpine ponds, there are several threats that appear to be common across different geographic regions including the development and expansions of ski resorts, pedestrian traffic from hikers and campers, water diversions, and domesticated animal grazing (Brinson and Malvárez 2002). Perhaps the greatest local land use effects of humans on alpine ponds are related to the latter, i.e., cattle, sheep, goats, and yaks, and other domesticated mammals across the world are herded into alpine meadows to take advantage of the flush of terrestrial production during the growing season (Holmquist et al. 2010). This results in the trampling of alpine fen and wet meadow habitats and is of special concern for alpine ponds because many of the core set of species are oligotrophic specialists, and even low levels of grazing can have a major impact on habitat structure (from trampling) and on water quality and water quantity in these small, fragile systems (e.g., Fig. 3.11a). Seasonal grazing in high-altitude meadows should negatively affect the nutrient dynamics of alpine ponds (Enriquez et al. 2014), and trampling by grazing animals can negatively affect riparian plant and animal communities (Wahren et al. 1999). Ponds are often one of the few sources of water for livestock herded into alpine meadows, and herds often frequent or concentrate around these fragile habitats, thus selectively concentrating their nutrient effects in those areas (as in Gusewell et al. 2005).

Cattle grazing and alpine pond conservation can be compatible if the shorelines of existing and newly created alpine ponds (see below) can be protected from trampling by livestock (Fig. 3.11). Although limited trampling can be sustainable (and can even create interesting microhabitats), extensive trampling is destructive, impacting the emergent vegetation and hydrologic stability of the pond (Fig. 3.11a).

Fig. 3.11 Alpine ponds subject of cattle trampling. (**a**) highly eutrophic pond with trampled vegetation (Pyrenees, France) (photo: H.J. Oertli) and (**b**) protected river shore (Alps, Switzerland) (photo: University of Geneva)

The protection of the shore with a simple barrier that encloses the pond (Fig. 3.11b) or a part of it is an inexpensive and efficient management measure that can maintain the ecological integrity of ponds in grazed alpine meadows.

Bioassessment and Monitoring

Sampling Protocols

Bioassessment and monitoring of alpine ponds have to be conducted for various reasons (see section "Management"). Nevertheless tools specific to alpine ponds are missing: Most tools are adopted from lake monitoring programs or have been developed for ponds in lower altitude. For example, sampling of invertebrates in relatively deep ponds is often conducted with grab (e.g., Ekman) or samples of known area collected by snorkeling or SCUBA. For small, shallow ponds, a benthic net appears to be the most common approach, and semiquantitative samples can be obtained by sweeping a known area for a fixed time/number of sweeps. Because of differences in the abundance of large- vs. small-bodied organisms, it is often effective, when collecting quantitative data in shallow ponds (i.e., numbers or biomass per area), to use a relatively large sample device (e.g., hand net across a known area) for large taxa and a small area core for small taxa (microcrustaceans, chironomids, oligochaetes; Wissinger et al. 2009). Complete taxonomic censuses should consider the degree to which samples need to be taken in different microhabitats (e.g., depth, substrate, vegetation cover).

Alpine ponds are often located far from human settlements. Nevertheless, human impacts (acidification, climate change, predator introduction, etc.) necessitate their management and/or monitoring. This management needs to be specific to the alpine environment because the environmental factors driving diversity in alpine ponds differ from that in lowland ponds (Ilg and Oertli 2014). Research on monitoring methods has been conducted on mountain ponds in the Apennines in Italy where Solimini et al. (2008) developed a multimetric biological index based on macroinvertebrates for detecting impairment through eutrophication. Their final index is the summation of seven metrics reflecting different aspects of the structure and function of the macroinvertebrate assemblage: a pollution tolerance metric, three metrics based on taxonomic richness (richness of macroinvertebrate genera, richness of chironomid taxa, and percentage of total richness composed by mayflies, dragonflies, and caddisflies), two metrics based on functional feeding groups (richness of collector gatherer taxa and richness of scraper taxa), and the habit-based metric richness of burrowers. In ponds from Switzerland, a systematic sampling protocol (aquatic macroinvertebrates, adult dragonflies) has been developed and coupled with an assessment through a classification system (Oertli et al. 2005). This approach has been successfully adapted for alpine pond monitoring (Robinson and Oertli 2009).

Monitoring Approaches

The sentinel concept used for alpine lakes and large ponds (Adrian et al. 2009) applies particularly well to ponds (all sizes) because their small water volume responds quickly to perturbation. Although past and present alpine monitoring programs have historically focused on alpine lakes (e.g., Catalan et al. 2009b; Kernan et al. 2009), future program should include ponds and wetlands (0.01–0.1 ha). Because many alpine ponds occur in large clusters, sampling can be done at a relatively low cost and encompass a variety of different habitats (e.g., gradient from temporary to permanent ponds, fish stocked and fishless, with macrophytes and without). Presently there are few monitoring programs targeted per se on alpine ponds. One exception is the long-term monitoring of invertebrates being conducted in the ponds, lakes, and wetlands in the Mexican Cut Nature Reserve in the Colorado Rockies (since 1989) (Wissinger et al. 1999a, b), where monitoring has revealed the arrival of upslope migrants of species previously found only at lower elevations (Wissinger, unpublished data). A second exception is the long-term monitoring in the Swiss National Park that includes a network of 25 alpine ponds that are being surveyed every 2 years for invertebrate diversity (Oertli et al. 2008; Robinson and Oertli 2009). After only 10 years, Oertli and colleagues have observed several significant trends including a colonization of species coming from lower altitude (Ilg et al. 2013) but also a decline in a biotic index based on the water beetle community, which is sensitive to warming (Fig. 3.12, Mavel 2014). Alpine pond monitoring should be coupled with paleolimnological investigations (see details above). The joint information will provide comprehensive patterns of biotic responses to perturbations like climate change.

Restoration and Creation of Alpine Ponds

Removing Nonnative Fish

As described above, the stocking of previously fishless alpine lakes and permanent ponds with salmonids for sport fishing has resulted in changes in the community structure of those habitats. One approach to restoring the native biota in alpine ponds and lakes is to simply remove these fishes. Gill netting has been used to remove fish in high mountain lakes in the Sierra Nevada (Knapp and Matthews 1998), and rotenone is another practice, although this pesticide is also toxic to non-target native species (see also Meronek et al. 1996 for a review on fish-removal methods). Fish removal can be quite successful, with recovery facilitated by the winged adult stages of many benthic macroinvertebrates and by the resting eggs of

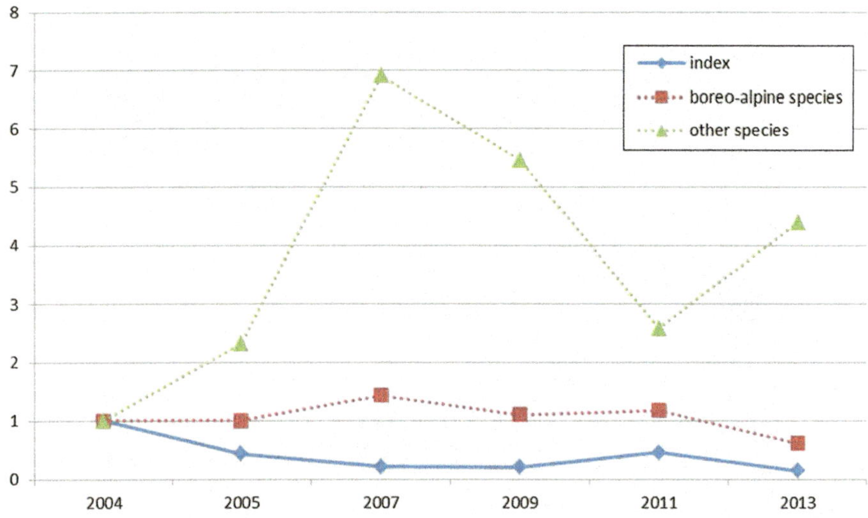

Fig. 3.12 A biotic index (*blue line*) based on the abundance of the water beetles community (Coleoptera, Dytiscidae), separated in boreo-alpine species (*brown line*) and other species (*green line*). The index is the ratio of the two groups and evidences here a declining trend between 2004 and 2013 in the alpine ponds from the Swiss National Park (after Mavel 2014), potentially linked to warming

zooplankton (Knapp et al. 2001). In larger alpine water bodies, fish removal is also coupled with zooplankton reintroduction (McNaught et al. 1999) with the objective of restoring both the planktonic and benthic trophic structure.

Designing and Planning New Alpine Ponds

The motivation for the creation and restoration of alpine ponds is typically focused on the conservation of the unique fauna in these habitats, but there are coupled benefits including water storage (e.g., for cattle or snowmaking at ski resorts) and ecotourism (e.g., as a destination and water source for alpine hiking and camping). The basic principles and multiple benefits for enhancing and creating ponds and wetlands at low altitudes are well established (Fig. 3.13; see constructed and restored wetlands chapter). However these general approaches need to be carefully adapted to alpine ponds because of their unique geomorphological and hydrological setting (e.g., Oertli et al. 2014).

Fig. 3.13 Pond dug in the Swiss Alps (1800 m.a.s.l.). The technical difficulties are often linked to the topography, with machinery having to navigate steep slopes (photos: P.A. Frossard)

Appendix

Representative invertebrate taxa from subalpine and alpine ponds from study sites in the Swiss Alps, Colorado Rockies, and the Southern Alps of New Zealand. Lists are representative and not intended to be comprehensive. Colorado taxa from alpine and subalpine ponds at >3000 m Elk Mountains near Rocky Mountain Biological Laboratory (Dodson 1982; Wissinger et al. 1999a; unpublished data from R. Dubois, W. Brown, S. Wissinger). Swiss Alps data from subalpine and alpine ponds >1500 m elevation (Ilg and Oertli 2014; Oertli et al. 2008; Ilg and Oertli 2014; unpublished data from B. Oertli). New Zealand data from subalpine and alpine ponds >1000 m elevation in central and northern inland ranges of the Southern Alps on the South Island (Wissinger et al. 2006a, b; Wissinger et al. 2009; unpublished data from S. Wissinger, M. Galatowitsch, H. Greig)

	Family	Colorado	Switzerland	New Zealand
Rotifera		*Keratella cochlearis*		*Keratella cochlearis*
		Conchochilus hippocrepis		
		Euchlanis		
Oligochaeta	Enchytraeidae		*Buchholzia*	
			Cernosvitoviella immota	
			Cognettia glandulosa	
			Fridericia	
			Henlea	
	Lumbriculidae	*Lumbriculus variegatus*	*Lumbriculus variegatus*	*Lumbriculus variegatus*
			Stylodrilus heringianus	
	Tubificidae	*Limnodrilus*	*Tubifex tubifex*	*Limnodrilus*
				Aulodrillus
Hirudinea	Erpobdellidae	*Erpobdella obscura*		*Alboglossiphonia multistriata*
	Glossiphoniidae	*Glossiphonia complanata*		
		Helobdella stagnalis		*Placobdelloides maorica*
Platyhelminthes	Planariidae		*Crenobia alpina*	
	Typhlopidae	*Mesostoma ehrenbergii*		*Mesostoma*
Gastropoda	Hydrobiidae			*Potamopyrgus antipodarum*
	Lymnaeidae	*Stagnicola elodes*		*Austropeplea tomentosa*
	Physidae	*Physa (Physella) gyrina*		*Glyptophysa variabilis*
	Planorbidae	*Gyraulus parvus*		*Gyraulus corinna*
Bivalvia	Sphaeriidae	*Musculium*		*Musculium novaezelandiae*
		Pisidium	*Pisidium casertanum*	
		Sphaerium		

(continued)

(continued)

	Family	Colorado	Switzerland	New Zealand
Hydrachnidiae	Arrenuridae	*Arrenurus*		*Arrenurus lacus*
	Eremeidae	*Hydrozetes*		*Hydrozetes*
	Eylaidae	*Eylais*		*Eylais waikawae*
	Hydrachnidiae			*Hydrachna maramauensis*
	Pionidae	*Piona*		*Piona uncata*
Ostracoda		*Cypris palustera*		*Cypretta viridis*
				Cypridopsis vidua
				Cyprinotus incongruens
				Cypricercus sanguineus
Brachiopoda	Brachinectidae	*Branchinecta coloradensis*		
	Bosminidae	*Bosmina longirostris*		*Bosmina meridionalis*
	Chydoridae	*Chydorus sphaericus*		*Chydorus sphaericus*
				Camptocercus australis
	Daphniidae	*Ceriodaphnia quadrangula*		*Ceriodaphnia dubia*
		Daphnia middendorffiana	*Daphnia curvirostris*	
		Daphnia pulex		
		Daphnia rosea		
		Scaphloberis mucronata		
		Simocephalus vetulus		*Simocephalus vetulus*
Copepoda	Calanoida	*Hesperodiaptomus shoshone*		*Boekella*
		Leptodiaptomus coloradensis		
	Cyclopoida	*Acanthocyclops vernalis*		*Acanthocyclops*
Amphipoda	Dogielinotidae	*Hyalella*		
	Gammaridae	*Gammarus*		
Collembola	Poduridae	*Podura aquatica*		

Ephemeroptera	Baetidae	*Callibaetis ferrugineus*	*Cloeon dipterum*	
			Cloeon simile	
	Leptophlebiidae			*Deleatidium*
	Nesameletidae			*Nesameletus*
Odonata	Coenagrionidae	*Coenagrion resolutum*	*Coenagrion puella*	*Xanthocnemis zealandica*
		Enallagma annexum	*Pyrrhosoma nymphula*	
		Enallagma boreale		
		Amphiagrion abbreviatum		
		Ischnura damula		
	Lestidae	*Lestes disjunctus*	*Lestes sponsa*	*Austrolestes colensis*
		Lestes dryas		
	Aeshnidae	*Aeshna interrupta*	*Aeshna caerulea*	
		Aeshna juncea	*Aeshna juncea*	
		Aeshna palmata	*Aeshna cyanea*	
			Anax imperator	
	Corduliidae	*Cordulia shurtleffi*		*Procordulia grayi*
		Somatochlora semicircularis	*Somatochlora alpestris*	
	Libellulidae	*Leucorrhinia hudsonica*		
		Libellula quadrimaculata	*Libellula quadrimaculata*	
		Sympetrum costiferum		
		Sympetrum danae		
Plecoptera	Austroperlidae			*Austroperla cyrene*
	Leuctridae		*Leuctra*	
	Notonemouridae			*Zealandobius furcillatus*
	Nemouridae		*Nemurella picteti*	

(continued)

(continued)

	Family	Colorado	Switzerland	New Zealand
Hemiptera	Corixidae	*Arctocorisa lawsoni*	*Arctocorisa carinata*	*Diaprepocoris nitens*
		Callicorixa audeni		
		Coenocorixa bifida	*Micronecta*	
		Sigara		*Sigara*
	Gerridae	*Aquarius*		
		Gerris gillettei		
	Notonectidae	*Notonecta*		*Anisops*
	Saldidae			*Saldula*
	Veliidae	*Rhagovelia*		*Microvelia*
Trichoptera	Hydroptilidae			*Oxyethira*
				Paroxyethira
	Leptoceridae			*Hudsonema*
				Oecetis
				Triplectides
	Conoesucidae			*Pycnocentria*
				Pycnocentroides
	Limnephilidae	*Asynarchus nigriculus*	*Acrophylax zerberus*	
		Grammotaulius lorretae	*Allogamus uncatus*	
		Hesperophylax occidentalis	*Anabolia nervosa*	
			Consorophylax consors	
		Limnephilus externus	*Limnephilus bipunctatus*	
		Limnephilus picturatus	*Limnephilus coenosus*	
			Pseudopsylopteryx zimmeri	
	Oeconesidae			*Oeconesus*
	Phryganeidae	*Agrypnia deflata*	*Oligotricha striata*	
Lepidoptera				*Hygraula nitens*

Coleoptera	Chrysomelidae	*Plateumaris pusilla*		
	Dytiscidae	*Acilius abbreviatus*		*Antiporus strigosulus*
		Acilius semisulcatus		*Antiporus femoralis*
		Agabus kootenai	*Agabus bipustulatus*	
		Agabus strigulosus	*Agabus lapponicus*	
		Agabus tristis	*Agabus sturmii*	
		Colymbetes sculptilis		*Onychohydrus hookeri*
		Dytiscus dauricus		*Dytiscus semisulcatus*
		Dytiscus alaskensis		
		Hygrotus patruelis		
		Hygrotus suturalis		
		Hydroporus	*Hydroporus sabaudus*	
			Hydroporus discretus	
			Hydroporus foveolatus	
			Hydroporus marginatus	
			Hydroporus memnonius	
			Hydroporus nigrita	
			Hydroporus nivalis	
			Hydroporus palustris	
		Ilybius fraterculus	*Ilybius fuliginosus*	
			Laccophilus	*Lancetes lanceolatus*
		Liodessus obscurellus		*Liodessus plicatus*
		Rhantus binotatus		*Rhantus suturalis*
		Rhantus gutticolis		
		Sanfilippodytes		
		Stictotarsus griseostriatus	*Stictotarsus griseostriatus*	

(continued)

(continued)

	Family	Colorado	Switzerland	New Zealand
	Haliplidae	*Haliplus leechi*	*Haliplus confinis*	
		Haliplus immaculicollis	*Haliplus lineatocollis*	
	Hydrophilidae	*Helophorus paraspelendidus*	*Helophorus glacialis*	*Limnoxenus zealandicus*
		Helophorus eclectus		
		Hydrobius fuscipes		
	Gyrinidae	*Gyrinus affinis*		
	Scirtidae			*Cyphon*
	Staphylinidae	*Stenus*		
Diptera	Ceratopogonidae	*Bezzia*		
		Culicoides		*Paradasyhelea*
		Probezzia		
	Chaoboridae	*Chaoborus americanus*		
		Eucorethra underwoodi		
	Chironomidae	*Chironomus salinarius*		*Chironomus zealandica*
		Chironomus riparius		*Parachironomus cylindricus*
		Cladopelma		*Cladopelma curtivalva*
		Dicrotendipes		*Polypedilum pavidus*
		Endochironomus		*Paucispinigera*
		Microtendipes		
		Pagastiella		
		Pseudodiamesa	*Pseudodiamesa arctica*	
			Pseudodiamesa branickii	
			Pseudodiamesa nivosa	

		Corynoneura	*Corynoneura lobata*	
			Corynoneura scutellata	
		Cricotopus	*Cricotopus sylvestris*	
			Diamesa bertrami	
			Diamesa latitarsis	
			Diamesa steinboecki	
			Diamesa zernyi/cinerella gr.	
		Eukiefferiella	*Eukiefferiella*	
			Pseudokiefferiella parva	
			Pseudosmittia arenaria	
			Heterotrissocladius marcidus	
			Limnophyes asquamatus	
			Metriocnemus eurynotus	
			Metriocnemus ursinus	
			Micropsectra junci	
			Micropsectra radialis	
			Parametriocnemus stylatus	
		Paraphaenocladius	*Paraphaenocladius pseudirritus*	
		Psectrocladius	*Psectrocladius sordidellus*	
		Ablabesmyia		*Ablabesmyia mala*
		Procladius	*Procladius choreus*	
			Orthocladius fuscimanus	
			Protanypus	
		Cladotanytarsus	*Rheocricotopus effusus*	
		Paratanytarsus	*Paratanytarsus austriacus*	
			Parorthocladius nudipennis	

(continued)

(continued)

	Family	Colorado	Switzerland	New Zealand
		Tanytarsus	*Tanytarsus sinuatus*	*Tanytarsus vespertinus*
			Bryophaenocladius	
			Chaetocladius	
			Zavrelimyia melanura	
	Culicidae	*Aedes communis*		
		Aedes pullatus		
		Culex		*Culex pervigilans*
	Dixidae	*Dixella*		
	Stratiomyidae	*Stratiomys*		
	Tipulidae	*Dicranota*		*Limonia*
		Hexatoma		*Zealandotipula*
		Limnophila		

References

Adrian R, O'Reilly CM et al (2009) Lakes as sentinels of climate change. Limnol Oceanogr 54:2283–2297

Andrews D, Rigler FH (1985) The effects of an Arctic winter on benthic invertebrates in the littoral zone of Char Lake, Northwest Territories. Can J Zool 63:2825–2834

Angeler DG, Allen CR, Johnson RK (2013) Measuring the relative resilience of subarctic lakes to global change: redundancies of functions within and across temporal scales. J Appl Ecol 50:572–584

Arim M, Bozinovic F, Marquet PA (2007) On the relationship between trophic position, body mass and temperature: reformulating the energy limitation hypothesis. Oikos 116:1524–1530

Barnett TP, Adam JC, Lettenmaier DP (2005) Potential impacts of a warming climate on water availability in snow-dominated regions. Nature 438:303–309

Bar-Zeev M (1958) The effect of temperature on the growth rate and survival of the immature stages of *Aedes aegypti* (L.). Bull Entomol Res 49:157–163

Batzer DP (2013) The seemingly intractable ecological responses of invertebrates in wetlands: a review. Wetlands 33:1–15

Batzer DP, Ruhí A (2013) Is there a core set of organisms that structure macro-invertebrate assemblages in freshwater wetlands? Freshw Biol 88:1647–1659

Batzer D, Cooper R, Wissinger SA (2015) Ecological roles of animals in wetlands. In: Batzer DP, Sharitz RR (eds) Ecology of freshwater and estuarine wetlands, 2nd edn. Univ Calif Press, Berkley, CA

Begon M, Townsend CR, Harper JL (2006) Ecology: from individuals to ecosystems, 4th edn. Blackwell, Malden

Bernard R, Daraż B (2010) Relict occurrence of East Palaearctic dragonflies in northern European Russia, with first records of *Coenagrion glaciale* in Europe (Odonata: Coenagrionidae). Int J Odonatol 13:39–62

Bertin A, Alvarez E et al (2015) Effects of wind-driven spatial structure and environmental heterogeneity on high-altitude wetland macroinvertebrate assemblages with contrasting dispersal modes. Freshw Biol 60:297–310

Bohonak AJ, Whiteman HH (1999) Dispersal of the fairy shrimp *Branchinecta coloradensis* (Anostraca): effects of hydroperiod and salamanders. Limnol Oceanogr 44:487–493

Boix D, Biggs J et al (2012) Pond research and management in Europe: "Small is Beautiful". Hydrobiologia 689:1–9

Bradford D, Cooper S et al (1998) Influences of natural acidity and introduced fish on faunal assemblages in California alpine lakes. Can J Fish Aquat Sci 55:2478–2491

Brendonck L, Michels E, De Meester L, Riddoch B (2002) Temporary pools are not 'enemy-free'. Hydrobiologia 486:147–159

Brinson MM, Malvárez AI (2002) Temperate freshwater wetlands: types, status, and threats. Environ Conserv 29:115–133

Burrows MT, Schoeman DS et al (2014) Geographical limits to species-range shifts are suggested by climate velocity. Nature 507:492–495

Calosi P, Bilton DT et al (2010) What determines a species' geographical range? Thermal biology and latitudinal range size relationships in European diving beetles (Coleoptera: Dytiscidae). J Anim Ecol 79:194–204

Camarero L, Catalan J et al (1995) Acidification in high Mountain Lakes in Central, Southwest, and Southeast Europe (Alps, Pyrennees, Pirin). Limnologica 25:141–156

Carlisle DJ, Hawkins CP (1998) Relationships between invertebrate assemblage structure, two trout species, and habitat structure in Utah mountain lakes. J N Am Benthol Soc 17:266–300

Catalan J, Barbieri MG et al (2009a) Ecological thresholds in European alpine lakes. Freshw Biol 54:2494–2517

Catalan J, Curtis CJ, Kernan M (2009b) Remote European mountain lake ecosystems: regionalisation and ecological status. Freshw Biol 54:2419–2432

Cazzanelli M, Forsstrom L et al (2012) Benthic resources are the key to *Daphnia middendorffiana* survival in a high arctic pond. Freshw Biol 57:541–551

Chagué-Goff C, Mark AF, Dickinson JJM (2010) Hydrological processes and chemical characteristics of low-alpine patterned wetlands, south-central New Zealand. J Hydrol 385:105–119

Chen I-C, Hill JK et al (2011) Rapid range shifts of species associated with high levels of climate warming. Science 333:1024–1026

Chimner RA, Lemly JM, Cooper DJ (2010) Mountain fen distribution, types and restoration priorities, San Juan Mountains, Colorado, USA. Wetlands 30:763–771

Cieraad E, McGlone MS, Huntley B (2014) Southern Hemisphere temperate tree lines are not climatically depressed. J Biogeogr 41:1441–1630

Cooper DJ (1986) Community structure and classification of Rocky Mountain wetland ecosystems. In: Windell JT (ed) An ecological characterization of Rocky Mountain montane and subalpine wetlands. USDI Fish Wildl Serv Biol Rep, Washington, DC, pp 66–147

Cooper DJ, Wolf E et al (2010) Alpine peatlands of the Andes, Cajamarca, Peru. Arctic Antarctic Alpine Res 42:19–33

Corcoran RM, Lovvorn JR, Heglund PJ (2009) Long-term change in limnology and invertebrates in Alaskan boreal wetlands. Hydrobiologia 620:77–89

Coronel JS, Declerck S et al (2004) Temporary shallow pools in high-Andes 'bofedal' peatlands: a limnological characterization at different spatial scales. Arch Sci 57:85–96

Coronel JS, Declerck S, Brendonck L (2007) High-altitude peatland temporary pools in Bolivia house a high cladoceran diversity. Wetlands 27:1166–1174

Crowley PH, Dillon PM, Johnson DM, Watson CN (1987) Intraspecific interference among larvae in a semivoltine dragonfly. Oecologia 71:447–456

Daborn GR (1971) Survival and mortality of coenagrionid nymphs (Odonata:Zygoptera) from the ice of an aestival pond. Can J Zool 49:569–571

Daborn GR, Clifford HE (1974) Physical and chemical features of an aestival pond in western Canada. Hydrobiologia 44:43–59

Danks HV (1978) Modes of seasonal adaptation in the insects. I. Winter survival. Can Entomol 110:1167–1205

Danks HV (2008) Aquatic insect adaptations to winter cold and ice. In: Lancaster J, Briers RA (eds) Aquatic insects: challenges to populations. CAB Internat, Wallingford, pp 1–19

De Bie T, De Meester L et al (2012) Body size and dispersal mode as key traits determining metacommunity structure of aquatic organisms. Ecol Lett 15:740–747

De Block M, McPeek M, Stoks R (2008) Life history plasticity to combined time and biotic constraints in *Lestes* damselflies from vernal and temporary ponds. Oikos 117:908–916

De Knijf G, Flenker U et al (2011) The status of two boreo-alpine species, *Somatochlora alpestris* and *S. arctica*, in Romania and their vulnerability to the impact of climate change (Odonata: Corduliidae). Int J Odonatol 14:111–126

de Los RP (2005) Richness and distribution of zooplanktonic crustacean species in Chilean Andes Mountains and Southern Patagonia shallow ponds. Polish J Environ Studies 14:817–822

de Los RP, Gajardo G (2010) A null model to explain zooplankton species associations in saline lakes of the South American altiplano. Crustaceana 83:769–777

De Meester L, Declerck S et al (2005) Ponds and pools as model systems in conservation biology, ecology and evolutionary biology. Aquat Conserv Mar Freshw Ecosy 15:715–725

de Mendoza G, Catalan J (2010) Lake macroinvertebrates and the altitudinal environmental gradient in the Pyrenees. Hydrobiologia 648:51–72

de Mendoza G, Rico E, Catalan J (2012) Predation by introduced fish constrains the thermal distribution of aquatic Coleoptera in mountain lakes. Freshw Biol 57:803–814

Declerck S, Coronel JS, Legendre P, Brendonck L (2011) Scale dependency of processes structuring metacommunities of cladocerans in temporary pools of high-Andes wetlands. Ecography 34:296–395

Denoël M, Joly P (2001) Adaptive significance of facultative paedomorphosis in *Triturus alpestris* (Amphibia, Caudata): resource partitioning in an alpine lake. Freshw Biol 46:1387–1396

Denoël M, Whiteman HH, Wissinger SA (2007) Foraging tactics in alternative heterochronic salamander morphs: trophic quality of ponds matters more than water permanency. Freshw Biol 51:1667–1676

Dodson SI (1970) Complementary feeding niches sustained by size-selective predation. Limnol Oceanogr 15:131–137

Dodson SI (1974) Zooplankton competition and predation: an experimental test of the size efficiency hypothesis. Ecology 55:605–613

Dodson SI (1982) Chemical and biological limnology of six west-central Colorado mountain ponds and their susceptibility to acid rain. Am Midl Natur 107:1733–1739

Dodson VE (1975) Distribution and habitat separation of three corixids (Hemiptera: Heteroptera) in western mountains of Colorado. Freshw Biol 5:141–150

Dornelas M (2010) Disturbance and change in biodiversity. Philosophical Trans Royal Soc B Biol Sci 365:3719–3727

Downing JA, Prairie YT et al (2006) The global abundance and size distribution of lakes, ponds, and impoundments. Limnol Oceanogr 5:2388–2397

Dubois RL (2016) Elevational gradients in odonate communities in the central Rocky Mountains

Duffy WG, Liston RC (1985) Survival following exposure to subzero temperatures and respiration in cold acclimatized larvae of *Enallagma boreale* (Odonata: Zygoptera). Freshw Invert Biol 4:1–7

Elser JJ, Kyle M et al (2009) Nutrient availability and phytoplankton nutrient limitation across a gradient of atmospheric nitrogen deposition. Ecology 90:3062–3073

Empanchin PN, Knapp RA, Layler SP (2010) Nonnative trout impact an alpine-nesting bird by altering aquatic insect subsidies. Ecology 91:2406–2415

Enriquez AS, Chimner RA, Cremona MV (2014) Long-term grazing negatively affects nitrogen dynamics in Northern Patagonian wet meadows. J Arid Environ 109:1–5

Ewald N, Nicolet P, et al. (2010) A preliminary assessment of Important Areas for Ponds (IAPs) in the Mediterranean Basin and Alpine Arc. Tech Rep. European Pond Conservation Network (EPCN)

Fischer JM, Olson MH et al (2015) Diel vertical migration of copepods in mountain lakes: the changing role of ultraviolet-radiation across a transparency gradient. Limnol Oceanogr 60:252–262

Fjellheim A, Raddum GG et al (2009) Diversity and distribution patterns of benthic invertebrates along alpine gradients. A study of remote European freshwater lakes. Adv Limnol 62:167–190

Fox JW (2013) The intermediate disturbance hypothesis should be abandoned. Trends Ecol Evol 28:986–992

Friday LE (1987) The diversity of macroinvertebrate and macrophyte communities in ponds. Freshw Biol 18:87–104

Frisbie MP, Lee RE (1997) Inoculative freezing and the problem of winter survival for freshwater macroinvertebrates. J N Am Benthol Soc 16:635–650

Fureder L, Ettinger R et al (2006) Macroinvertebrate diversity in alpine lakes: effects of altitude and catchment properties. Hydrobiologia 562:123–144

Gardelle J, Arnaud Y, Berthier E (2011) Contrasted evolution of glacial lakes along the Hindu Kush Himalaya mountain range between 1990 and 2009. Global Planet Change 75:47–55

Gaston KJ, Blackburn TM (2000) Pattern and process in macroecology, 2nd edn. Blackwell, Oxford, UK

Gerten D, Adrian R (2001) Differences in the persistency of the North Atlantic Oscillation signal among lakes. Limnol Oceanogr 46:448–455

Gillson L, Dawson TP, Jack S, McGeoch MA (2013) Accommodating climate change contingencies in conservation strategy. Trends Ecol Evol 28:135–142

Graham CH, Carnaval AC et al (2014) The origin and maintenance of montane diversity: integrating evolutionary and ecological processes. Ecography 37:711–719

Grant PBC, Samways MJ (2007) Montane refugia for endemic and Red Listed dragonflies in the Cape Floristic Region biodiversity hotspot. Biodiver Conserv 16:787–805

Greig HS, Wissinger SA (2010) Reinforcing abiotic and biotic time constraints facilitate the broad distribution of a generalist with fixed traits. Ecology 91:836–846

Greig HS, Wissinger SA, McIntosh AR (2013) Top-down control of prey increases with drying disturbance in ponds: a consequence of non-consumptive interactions? J Anim Ecol 82:598–607

Gusewell S, Jewell PL, Edwards PJ (2005) Effects of heterogeneous habitat use by cattle on nutrient availability and litter decomposition in soils of an Alpine pasture. Plant and Soil 268:135–149

Hairston NG (1976) Photoprotection by carotenoid pigments in the copepod *Diaptomus nevadensis*. Proc Natl Acad Sci U S A 73:971–974

Hamerlik L, Svitok M et al (2014) Local, among-site, and regional diversity patterns of benthic macroinvertebrates in high altitude waterbodies: do ponds differ from lakes? Hydrobiologia 723:41–52

Hampel H, Cocha J, Vimos D (2010) Incorporation of aquatic ecology to the hydrological investigation of ecosystems in the high Andes. Maskana 1:91–100

Hansson L, Hylander S, Sommaruga R (2007) Escape from UV threats in zooplankton: a cocktail of behavior and protective pigmentation. Ecology 88:1932–1939

Harte J, Lockette GP et al (1985) Acid precipitation and surface water vulnerability on the western slope of the high Colorado Rockies. Water Air Soil Pollut 25:313–320

Heino J, Virkkala R, Toivonen H (2009) Climate change and freshwater biodiversity: detected patterns, future trends and adaptations in northern regions. Biol Rev 89:39–54

Hickling R, Roy DB et al (2006) The distributions of a wide range of taxonomic groups are expanding polewards. Global Change Biol 12:450–455

Hinden H, Oertli B et al (2005) Alpine pond biodiversity: what are the related environmental variables? Aqua Conserv Mar Freshw Ecosys 15:613–624

Hoffman RL, Liss WJ et al (1996) Nearshore distribution of macroinvertebrates in high mountain lakes in the northern Cascades. Arch Hydrobiol 136:363–389

Holmquist JG, Schmidt-Gengenbach J, Haultain SA (2010) Does long-term grazing by pack stock in subalpine meadows result in lasting effects on arthropod assemblages? Wetlands 30:52–362

Holmquist JG, Jones JR, Schmidt-Gengenbach J, Pierotti LF, Love JP (2011) Terrestrial and aquatic macroinvertebrate assemblages as a function of wetland type across a mountain landscape. Arctic Antarct Alpine Res 43: 568–584

Hughes AR, Byrnes JE, Kimbro DL, Stachowicz JJ (2007) Reciprocal relationships and potential feedbacks between diversity and disturbance. Ecol Lett 10:849–864

Ilg C, Oertli B (2014) How can we conserve cold stenotherm communities in warming Alpine ponds? Hydrobiologia 723:53–62

Ilg C, Demierre E, Reymond A-S, Oertli B (2013) Alpen-smaragdlibelle in Weltrekordhöhe. Cratschla 2(13):2–3

IPCC (2014) Climate Change 2014: Synthesis Report. In: Contrib Working Groups I, II and III, Pachauri RK, Meyer LA (eds) 5th Assess Rep Intergovernmental Panel on Climate Change. IPCC, Geneva

Jannot JE, Wissinger SA, Lucas JR (2008) Diet and developmental time constraints alter life history trade-offs in a caddisfly (Trichoptera: Limnephilidae). Biol J Linnean Soc 95:495–504

Kajak Z, Hillbricht-Illkowska A (1970) Productivity problems of freshwaters. Proc IBP-UNESCO symp productivity problems of freshwaters. Kazimierz Dolny Poland, May 6–12, 1970. Varsovie: PWN Polish Sci Publ

Kaunisto KM, Kaunisto P, Vahtera V, Shuhonen J (2015) Populations of the damselfly *Coenagrion hastulatum* at the edge of the species range have fewer gregarine and water mite parasites. Freshw Biol 60:794–801

Kernan M, Ventura M et al (2009) Regionalisation of remote European mountain lake ecosystems according to their biota: environmental versus geographical patterns. Freshw Biol 54:2470–2493

Klemmer AJ, Wissinger SA, Greig HS, Ostrofsky ML (2012) Nonlinear effects of consumer density on multiple ecosystem processes. J Anim Ecol 81:770–780

Knapp RA, Matthews KR (1998) Eradication of nonnative fish by gill netting from a small mountain lake in California. Restor Ecol 6:207–213

Knapp RA, Matthews KR, Sarnelle O (2001) Resistance and resilience of alpine lake fauna to fish introductions. Ecol Monogr 71:401–421

Komori J (2008) Recent expansions of glacial lakes in the Bhutan Himalayas. Quaternary Int 84:177–186

Körner C (1998) Worldwide positions of alpine treelines and their causes. In: Beniston M, Innes JL (eds) The impacts of climate variability on forests. Springer, Heidelberg, pp 221–229

Körner C (2012) Alpine treelines: functional ecology of the global high elevation tree limits. Springer, Basel

Larson DJ, Alarie Y, Roughley RE (2000) Predaceous diving beetles (Coleoptera: Dytiscidae) of the Nearctic Region, with emphasis on the fauna of Canada and Alaska. NRC Research Press, Ottawa

Laville H (1971) Recherches sur les chironomides [Diptera] lacustres du massif de Néouvielle (Hautes-Pyrénées). Int J Limnol 7:335–414

Laville H (1974) Utilisation de substrats artificiels pour l'étude de la faune macrobenthique de la zone littorale rocheuse des lacs de montagne. Int J Limnol 10:163–172

Le Cren ED, Lowe-McConnell RH (1980) The functioning of freshwater ecosystems. Cambridge Univ Press, Cambridge

Lee RE (1989) Insect cold hardiness: to freeze or not to freeze? BioScience 39:308–313

Lindholm M, Stordal F et al (2012) Climate-driven range retraction of an Arctic freshwater crustacean. Freshw Biol 57:2591–2601

Lund JO, Wissinger SA, Peckarsky BL (2016) Caddisfly behavioral responses to drying cues in temporary alpine ponds: Implications for effects of climate change. Freshw Sci. 35:(in press)

Luoto TP, Nevalainen L (2012) Ecological responses of aquatic invertebrates to climate change over the past ~ 400 years in a climatically ultra-sensitive lake in the Niedere Tauern Alps (Austria). Fund Appl Limnol 181:169–181

MacCrimmon HR (1970) World distribution of rainbow trout (*Salmo gairdneri*). J Fish Res Board Can 28:663–674

MacCrimmon HR, Marshall TL (1968) World distribution of brown trout (*Salmo trutta*). J Fish Res Board Can 25:2527–2548

Maldonado M, Maldonado-Campo JA et al (2011) Biodiversity in aquatic systems of the tropical Andes. In: Herzog SK, Martinez R, Jørgensen PM, Tiessen H (eds) Climate change and biodiversity in the tropical Andes. Inter-American Inst Global Change Res, Paris, pp 276–294, http://www.iai.int

Maly EJ (1973) Density, size, and clutch of two high altitude diaptomid copepods. Limnol Oceanogr 18:840–847

Maly EJ, Maly MP (1974) Dietary differences between two co-occurring calanoid alpine copepod species. Oecologia 17:325–333

Maly EJ, Schenholtz S, Arts MT (1983) The influence of flatworm predation on zooplankton inhabiting small alpine ponds. Hydrobiologia 76:233–243

Mani MS (1968) Ecology and biogeography of high altitude insects. Junk, The Hague

Marchetto A, Rogora M (2004) Measured and modelled trends in European mountain lakes: results of fifteen years of cooperative studies. J Limnol 63:55–62

Martinez-Sanz C, Cenzano CSS, Fernandez-Alaez M, Garcia-Criado F (2012) Relative contribution of small mountain ponds to regional richness of littoral macroinvertebrates and the implications for conservation. Aquat Conserv Mar Freshw Ecosys 22:155–164

Mavel N (2014) Détection de l'évolution de la biodiversité aquatique alpine en réponse aux changements climatiques: développement d'un indice biotique intégrant les macroinvertébrés des plans d'eau du Parc National Suisse (Macun, Grisons). HESSO//GE, hepia, Geneva

Mawdsley JR, O'Malley R, Ojima DS (2009) A review of climate-change adaptation strategies for wildlife management and biodiversity conservation. Conserv Biol 22:1080–1089

McCauley SJ (2008) Slow, fast and in between: habitat distribution and behaviour of larvae in nine species of libellulid dragonfly. Freshw Biol 53:253–263

McDowall RM (1987) Impacts of exotic fishes on the native fauna. In: Viner AB (ed) Inland waters of New Zealand. Science Information Publishing Centre, Wellington, NZ, pp 333–347
McNaught AS, Schindler DW et al (1999) Restoration of the food web of an alpine lake following fish stocking. Limnol Oceanogr 44:127–136
McPeek MA, Brown JM (2000) Building a regional species pool: diversification of the *Enallagma* damselflies in eastern North America. Ecology 81:904–920
Meronek TG, Bouchard PM et al (1996) A review of fish control projects. J N Am Fish Manage 16:63–74
Mitchell DL, Karentz D (1993) The induction and repair of DNA photodamage in the environment. In: Young AR, Bjorn LO, Moan J, Nultsch W (eds) Environmental UV photobiology. Plenum Press, New York
Munoz-Pedreros A, De Los RP, Moller P (2013) Zooplankton in Laguna Ljia, a high-altitude Andean shallow lake of the Puna I northern Chile. Crustacean 86:1634–1643
Murphy CA, Thompson PL, Vindebrooke RD (2010) Assessing the sensitivity of alpine lakes and ponds to nitrogen deposition in the Canadian Rocky Mountains. Hydrobiologia 648:83–90
Nagiller K, Sommaruga R (2009) Differential tolerance of UV radiation between *Chaoborus* species and role of photoprotective compounds. J Plankton Res 31:503–513
Nilsson-Örtman V, Stoks R, De Block M, Johansson F (2012) Generalists and specialists along a latitudinal transect: patterns of thermal adaptation in six species of damselflies. Ecology 93:1340–1352
Oertli B (1993) Leaf litter processing and energy flow through macroinvertebrates in a woodland pond (Switzerland). Oecologia 96:466–477
Oertli B, Auderset Joye D et al (2002) Does size matter? The relationship between pond area and biodiversity. Biol Conserv 104:59–70
Oertli B, Auderset Joye D et al (2005) PLOCH: a standardised method for sampling and assessing the biodiversity in ponds. Aquat Conserv Mar Freshw Ecosys 15:665–679
Oertli B, Indermuehle N et al (2008) Macroinvertebrate assemblages in 25 high alpine ponds of the Swiss National Park (Cirque of Macun) and relation to environmental variables. Hydrobiologia 597:29–41
Oertli B, Ilg C et al (2014) Freshwater biodiversity under warming pressure in the Alps: a methodological framework for prioritization of restoration areas for small waterbodies. Eco mont 6:23–34
Oswood MW, Miller LK, Irons JG (1991) Overwintering of freshwater macro- invertebrates. In: Lee RE, Denlinger DL (eds) Insects at low temperature. Chapman Hall, New York, pp 360–375
Parmesan C (2006) Ecological and evolutionary responses to recent climate change. Annu Rev Ecol Evol Syst 37:637–669
Parmesan C, Yohe G (2003) A globally coherent fingerprint of climate change impacts across natural systems. Nature 421:37–42
Peckarsky BL (1983) Biotic interactions or abiotic limitations? A model of lotic community structure. In: Fontaine TD, Bartell SM (eds) Dynamics of lotic ecosystems. Ann Arbor Sci Publ, Ann Arbor, pp 303–323
Persaud AD, Moeller RE, Williamson CE, Burns CW (2007) Photoprotective compounds in weakly and strongly pigmented copepods and co-occurring cladoceran. Freshw Biol 52:2121–2133
Plante C, Downing JA (1989) Production of freshwater invertebrate populations in lakes. Can J Fish Aquat Sci 46:1489–1498
Post E, Forchhammer MC et al (2009) Ecological dynamics across the Arctic associated with recent climate change. Science 325:1355–1358
Rahbek C (1995) The elevational gradient of species richness: a uniform pattern? Ecography 18:200–205
Rahel FJ, Olden JD (2008) Assessing the effects of climate change on aquatic invasive species. Conserv Biol 22:521–533

Rautio M, Vincent WF (2006) Benthic and pelagic food resources for zooplankton in shallow high-latitude lakes and ponds. Freshw Biol 51:1038–1052

Rees ABH, Cwynar LC (2010) Evidence for early postglacial warming in Mount Field National Park, Tasmania. Quartenary Sci Rev 29:443–454

Robinson CT, Oertli B (2009) Long-term biomonitoring of alpine waters in the Swiss National Park. Eco mont 1:23–34

Röhl K (2008) Characteristics and evolution of supraglacial ponds on debris-covered Tasman Glacier, New Zealand. J Glaciol 54:867–880

Rosset V, Oertli B (2011) Freshwater biodiversity under climate warming pressure: identifying the winners and losers in temperate standing waterbodies. Biol Conserv 144:2311–2319

Rosset V, Oertli B, Angélibert S, Indermuehle N (2009) Local diversity of macroinvertebrates in alpine ponds as an indicator of global change: a Gastropod case-study. Verh Int Verein Theor Angewan Limnol 30:482–484

Rosset V, Lehmann A, Oertli B (2010) Warmer and richer? Predicting the impact of climate warming on species richness in small temperate waterbodies. Glob Change Biol 16:2376–2387

Rotvit L, Jacobsen D (2014) Egg development of Plecoptera, Ephemeroptera and Odonata along latitudinal gradients. Ecol Entomol 39:177–185

Ruhí A, Batzer DP (2014) Assessing congruence and surrogacy among wetland macroinvertebrate taxa towards efficiently measuring biodiversity. Wetlands 34:1061–1071

Rundle SD, Foggo A, Choiseul V, Bilton DT (2002) Are distribution patterns linked to dispersal mechanisms? An investigation using pond invertebrate assemblages. Freshw Biol 42:1571–1581

Šácha D, Bulánková E (2006) New records of *Aeshna subarctica elisabethae* (Odonata, Aeshnidae) from the Tatra Mountains (Slovakia). Biologia 18:S221

Salerno F, Gambellis S et al (2014) High alpine ponds shift upwards as average temperatures increase: a case study of the Ortles-Cevedale mountain group (Southern Alps, Italy) over the past 50 years. Global Planet Change 120:81–91

Santolaria Z, Arreubo T et al (2015) Hydrochemistry dynamics in remote mountain lakes and its relation to catchment and atmospheric features: the case study of Sabocos Tarn, Pyrenees. Environ Sci Pollut Res 22:231–247

Saros JE (2009) Integrating neo- and paleolimnological approaches to refine interpretations of environmental change. J Paleolimnol 41:243–252

Saros JE, Rose KC et al (2010) Melting alpine glaciers enrich high-elevation lakes with reactive nitrogen. Environ Sci Technol 44:4891–4896

Saros JE, Clow DW, Blett T, Wolfe AP (2011) Critical nitrogen deposition loads in high-elevation lakes of the western U.S. inferred from paleolimnological records. Water Air Soil Pollut 216:193–202

Sayer CD, Davidson TA, Jones JI, Langdon PG (2010) Combining contemporary ecology and paleolimnology to understand shallow lake ecosystem change. Freshw Biol 55:487–499

Schindler DW, Curtis PJ, Parker BR, Stainton MP (1996) Consequences of climate warming and lake acidification for UV-B penetration in North American boreal lakes. Nature 379:705–708

Schneider DW, Frost TM (1996) Habitat duration and community structure in temporary ponds. J N Am Benthol Soc 15:64–86

Shama LNS, Campero-Paz M et al (2011) Latitudinal and voltinism compensation shape thermal reaction norms for growth rates. Mol Ecol 20:2929–2941

Smol JP, Douglas MSV (2007a) Crossing the final ecological threshold in high Arctic ponds. Proc Natl Acad Sci U S A 104:12395–12397

Smol JP, Douglas MSV (2007b) From controversy to consensus: making the case for recent climate using lake sediments. Front Ecol Environ 5:466–474

Smol J, Birks HJ, Last W (2001) Tracking environmental change using lake sediments. Springer, Dordrecht

Sniegula S, Johansson F (2010) Photoperiod affects compensating developmental rate across latitudes in the damselfly *Lestes sponsa*. Ecol Entomol 35:149–157

Sniegula S, Johansson F, Nilsson-Ortman V (2012) Differentiation in developmental rate across geographic regions: a photoperiod driven latitude compensating mechanism? Oikos 121:1073–1082

Solimini AG, Bazzanti M, Ruggiero A, Carchini G (2008) Developing a multimetric index of ecological integrity based on macroinvertebrates of mountain ponds in central Italy. Hydrobiologia 597:109–123

Sommaruga R (2010) Preferential accumulation of carotenoids rather than of mycosporine-like amino acids in copepods from high altitude Himalayan lakes. Hydrobiologia 648:153–166

Sommaruga R, Wograth S et al (1997) Temperature effects on the acidity of remote alpine lakes. Nature 387:64–67

Søndergaard M, Jeppesen E, Jensen JP (2005) Pond or lake: does it make any difference? Arch Hydrobiol 162:143–165

Sprules WG (1972) Effects of size-selective predation and food competition on high altitude zooplankton communities. Ecology 58:375–386

Stoks R, McPeek MA (2006) A tale of two diversifications: reciprocal habitat shifts to fill ecological space along the pond permanence gradient. Amer Natur 168:S50–S72

Stoks R, Geerts AN, De Meester L (2014) Evolutionary and plastic responses of freshwater invertebrates to climate change: realized patterns and future potential. Evol Appl 7:42–55

Tarr R, Barber MJ, Babbitt KJ (2005) Macroinvertebrate community structure across a wetland hydroperiod gradient in southern New Hampshire, USA. Wetl Ecol Manage 13:321–334

Theissinger K, Bálint M, Feldheim KA (2013) Glacial survival and post-glacial recolonization of an arctic–alpine freshwater insect (*Arcynopteryx dichroa*, Plecoptera, Perlodidae) in Europe. J Biogeogr 40:236–248

Tiberti R, von Hardenberg A, Bogliani G (2014) Ecological impact of introduced fish in high altitude lakes: a case of study from the European Alps. Hydrobiologia 724:1–19

Timms BV (1978) The benthos of seven high elevation lakes in Tasmania. Arch Hydrobiol 81:422–444

Timms BV, Morton J, Green K (2013) Temporal changes in the macroinvertebrate fauna of two glacial lakes, Cootapatamba and Albina, Snowy Mountains, New South Wales. Proc Linnean Soc New South Wales 135:45–54

VanPutten WH, Macel M, Visser M (2010) Predicting species distribution and abundance responses to climate change: why it is essential to include biotic interactions across trophic levels. Philos Trans Royal Soc B 365:2025–2034

Vanschoenwinkel B, Hulsmans A et al (2009) Community structure in temporary freshwater pools: disentangling effects of habitat size and hydroregime. Freshw Biol 54:1487–1500

Vanschoenwinkel B, Buschke F, Brendonck L (2013) Disturbance regime alters the impact of dispersal on alpha and beta diversity in a natural metacommunity. Ecology 94:2547–2557

Wahren CH, Williams RJ, Papst WA (1999) Alpine and subalpine wetland vegetation on the Bogong High Plains, southeastern Australia. Austral J Bot 47:165–188

Waringer J, Graf W, Malicky H (2011) Description of the larvae of *Limnephilus femoratus* (Zetterstedt, 1840) and *Limnephilus subnitidus* McLachlan, 1875, with additional notes on *Limnephilus picturatus* McLachlan, 1875 (Trichoptera: Limnephilidae). Aquat Insects 33:371–380

Waringer J, Graf W, Malicky H (2012) Description of the larvae of *Anabolia lombardi* Ris, 1897 and *Limnephilus sericeus* (Say, 1824), with additional notes on *Limnephilus flavospinosus* 1874 (Trichoptera: Limnephilidae). Aquat Insects 34:93–105

Wellborn GA, Skelly DK, Werner EE (1996) Mechanisms creating community structure across a freshwater habitat gradient. Annu Rev Ecol Syst 27:337–363

Whiteman HH, Wissinger SA, Brown WS (1996) Growth and foraging consequences of facultative paedomorphosis in the tiger salamander, *Ambystoma tigrinum nebulosum*. Evol Ecol 10:433–446

Whiteman HH, Wissinger SA et al (2012) Larval growth in polyphenic salamanders: making the best of a bad lot. Oecologia 168:109–118

Wiley RL (1973) Emergence patterns of the subalpine dragonfly *Somatochlora semicircularis* (Odonata: Corduliidae). Psyche 81:121–133

Wiley RL (1974) Drought resistance in subalpine nymphs of *Somatochlora semicircularis* (Odonata: Corduliidae). Am Midl Natur 87:215–221

Williams DD (2006) The biology of temporary waters, 2nd edn. Oxford Univ Press, New York

Williamson CE, Saros JE, Schindler DW (2009a) Sentinels of change. Science 323:887–888

Williamson CE, Saros JE, Vincent WF, Smol JP (2009b) Lakes as sentinels, integrators, and regulators of climate change. Limnol Oceanogr 54:2273–2282

Wissinger SA (1999) Invertebrate ecology in North American wetlands: a synthesis. In: Batzer DP, Rader RD, Wissinger SA (eds) Invertebrates in freshwater wetlands of North America: ecology and management. Wiley, New York, pp 1043–1075

Wissinger SA, Whiteman HH (1992) Fluctuation in a Rocky Mountain population of salamanders: anthropogenic acidification or natural variation? J Herpetol 26:377–391

Wissinger SA, Bohonak AJ, Whiteman HH, Brown WS (1999a) Subalpine wetlands in central Colorado: roles of habitat permanence and salamander predation in determining invertebrate community composition. In: Batzer DP, Rader RD, Wissinger SA (eds) Invertebrates in freshwater wetlands of north America: ecology and management. Wiley, New York, pp 757–790

Wissinger SA, Whiteman HH et al (1999b) Foraging trade-offs along a predator-permanence gradient in subalpine wetlands. Ecology 80:2102–2116

Wissinger SA, Brown WS, Jannot JE (2003) Caddisfly life histories along permanence gradients in high-altitude wetlands in Colorado (USA). Freshw Biol 48:255–270

Wissinger SA, McIntosh AR, Greig HS (2006a) Impacts of introduced brown and rainbow trout on benthic invertebrate communities in shallow New Zealand lakes. Freshw Biol 51:2009–2028

Wissinger SA, Whissel JC, Eldermire C (2006b) Predator defense along a subalpine permanence gradient: roles of case structure, behavior, and developmental phenology in caddisflies. Oecologia 148:667–678

Wissinger SA, Greig HS, McIntosh AR (2009) Absence of species replacements between permanent and temporary lentic communities in New Zealand. J N Am Benthol Soc 28:12–23

Zellmer ID (1995) UV-B-tolerance of alpine and arctic *Daphnia*. Hydrobiologia 307:153–159

Zhang Y, Wang C et al (2010) Alpine wetlands in the Lhasa River Basin, China. J Geogr Sci 20:375–388

Chapter 4
Invertebrates in Temporary Wetland Ponds of the Temperate Biomes

Michael J. Jeffries, Luis B. Epele, Jered M. Studinski, and Csaba F. Vad

Temperate, Temporary Wetlands: Overlooked but Essential

The world's temperate biomes conjure up iconic and particular visions of landscape, ecology and culture: the pampas grasslands of South America; the prairie plains and verdant broadleaved forests of North America; the fertile lowlands of Western Europe which, eastwards, grade into mighty river basins; and then the steppes of Russia through to central Asia. The very names of these regions suggest horizon wide tracts of grassland or sylvan woodland. Temporary wetlands and ponds do not feature large in any popular images of the temperate biomes, compared, for example, to the flooded forest of the Amazon tropics or bog pools of the Boreal north. This is a shame because temporary wetlands are a consistent and distinct feature of temperate regions. The very nature of temperate biomes with their seasonal rainfall, milder climates and gentler terrain lends itself to the formation of small wetlands. Look closely and each temperate region has characteristic wetlands with evocative names: the mallines of Patagonia; prairie potholes and seasonal woodland ponds of North America; pingos, soda pans and parkland grazing ponds of Europe; and across to the seasonal wetlands of Daurian steppe as Mongolia grades into China.

These wetlands are often biodiversity hotspots, with distinct and rare invertebrates. However, they have been overlooked until the last 20 years or so or, worse, seen as a nuisance. The world's temperate lowlands are under great pressure, often providing the best agricultural land. Our purpose here is to capture the importance of temporary wetlands for invertebrates (with occasional examples from plants and amphibians too where these are revealing) and explore the challenging ecology of these superficially simple systems so that the mallines, potholes, pingos, flashes and slacks can be better appreciated as fundamental to the nature of temperate biomes.

D. Batzer, D. Boix (eds.), *Invertebrates in Freshwater Wetlands*,
DOI 10.1007/978-3-319-24978-0_4

From Diseased Miasmas to Biodiversity Hotspots

Temporary wetlands and their fauna are not a significant feature of the scientific literature before the mid-twentieth century, but the limited studies published were strangely prescient, identifying themes that have come to dominate contemporary research. Much of the early work reflected fears that small wetlands were a source of contagion and disease, from the earliest records in China back to 2700 BC, via the Anglo-Saxon epic poem Beowulf, whose monstrous Grendel lurked in the swamps, to the assault on "marsh fever" and its mosquito vectors in England during the First World War.

Wetlands as the source of disease motivated much of the work, in particular the distribution of mosquitoes. For example, Root (1924) surveyed ponds and wetlands in the state of Georgia. He identified particular types of temporary pond, from limestone sinks, and rain or flood pools, from open fields or wooded habitats. Different species of mosquito showed preferences for different wetlands. The presence or absence of Culicidae in general was attributed to variations in shelter provided by aquatic vegetation or the incidence of *Gambusia* fish. Similar detail informed Macan's (1939) survey of mosquitoes around Cambridge, UK. Small, temporary wetlands dominated the mix of habitats, and Macan identified habitat cues that were likely to affect oviposition, primarily variations in hydrology and the ability of female mosquitoes to gauge these changes. He also identified significant effects of land use, notably the intensive agricultural modification of fenland which had reduced the frequency of some species (e.g. *Aedes cinereus*), had no discernable impact on others and increased populations of some (e.g. *Ochlerotatus rusticus*) due to the creation of ditch habitats.

The search for habitat factors determining mosquito distributions underpins more global reviews too, e.g. wetlands in China and Japan (Lamborn 1922) or Ceylon (Senior-White 1925): although not solely temperate in focus again the role of hydrology, oviposition cues and food resources are made clear, even if the ability to unravel the invertebrate community provokes Senior-White to write "…the problem still remained in the state graphically summed up by Watson (1921) in this preface '…in a single ravine, the various insect inhabitants may come and go in the wondrous fashion of a fairy tale'". Mosquito ecology underpinned later studies too, which introduced new themes. The seasonal sequence of mosquitoes and their beetle predators in vernal ponds in Ontario were recorded by James (1967). Trapping Dytiscidae beetles and Hemiptera bugs revealed a replicable sequence of arrival, with the beetles colonising late in spring when mosquito larvae had already hatched. The sequence was different from that of permanent ponds, the beetles overwintering in permanent sites but then moving into temporary pond to breed. Different species of beetles showed different microhabitat preferences with more taxa amongst the shallow, vegetated edges.

Another disease, liver fluke (*Fasciola hepatica*, a debilitating flatworm primarily infecting sheep), also inspired early studies of temporary wetlands which are the home of its vector, the snail *Galba truncatula*. In the UK, major outbreaks of the fluke were associated with wet years, resulting in *G. truncatula* populations increasing to "catastrophic proportions" (Peters 1938), whilst the snail could go unrecorded in between. Peters' detailed study included permanent ponds down to wet ground pockmarked with hoofprints to identify the snail's ideal wetland. Peters also noted the apparent reluctance of the snail to move far, this limited dispersal begging the general question about the dispersal of temporary wetland invertebrates. The ability of freshwater molluscs to disperse had attracted interest, perhaps because it seems superficially so limited. The ability of snails to hitch a ride on birds and flying invertebrates such as beetles and bugs had attracted particular attention, not least the speed with which they could arrive in newly refilled sites (e.g. Rees 1965).

Many of the key themes that have inspired contemporary research into the invertebrates of temporary wetlands are already to be found in this older literature: habitat preferences, the role of hydrology, predator impacts, dispersal and behavioural responses to microhabitat. The one theme missing was an appreciation of these habitats for their own sake.

Scattered studies, usually of small numbers of ponds and focusing on animals' adaptations, sometimes compared temporary and permanent ponds (Kenk 1949; Barclay 1966). A rare instance of more systematic studies is the work of Mozley who explored the fauna of temporary ponds in several temperate biomes. For example, Mozley (1937) outlined the freshwater habitats of the Kirghiz steppes, part of the vast grasslands of central Asia. In this classically continental landscape, the majority of ponds and wetlands are temporary, filled by snowmelt and dry in summer. He distinguished three types: small depressions within the overall grass plains, noting the characteristic presence of branchiopod shrimps, along with pools on lake beds and intermittent streams, although the latter two are described as having only a meagre fauna. Mozley's studies of temporary wetland habitats in many parts of the world culminated in his eulogy for temporary wetlands and their invertebrates as globally important but overlooked (Mozley 1944). Invertebrates feature in other early studies of temporary ponds, such as Brown's (1951) study of Hemiptera in ponds in southern England and the varying behaviours and wing development of permanent versus temporary pond species.

A sea change in our understanding of temporary ponds and their significance as biodiversity hotspots began in the 1980s. In their review of temporary ponds in Europe, Williams et al. (2001) suggested that our understanding of these habitats "lags at least 50 years behind those of better known water bodies". They provide some telling details. For example, in an otherwise excellent core review of Britain's habitats, produced to underpin conservation policy, Ratcliffe (1977) dismissed temporary ponds as "not an important water body type in Britain". The UK's National Countryside Survey, a periodic nationwide audit of the condition of the landscape and habitats carried out every few years since the 1980s, did not record temporary ponds until 1996, when it was estimated that 40 % of ponds

(82,000) in lowland Britain were of this type. Unfortunately, conservation advice of the time routinely suggested that drying out was a major threat to pond wildlife and that digging out ephemeral wetlands was a good starting point for pond creation.

General surveys of invertebrates in the 1980s and early 1990s focused on permanent ponds (e.g. Barnes 1983; Friday 1987) or did not include temporary sites at all in systematic studies of pond management for conservation (e.g. Gee et al. 1997). Temporary pond invertebrates were featured more in studies of fundamental ecology processes, e.g. seasonal succession and trophic groups in an Ontario vernal pond (Williams 1983) and patterns of community assembly across 30 freshwater pools across Scottish wetlands (Jeffries 1989). These studies made clear the importance of species traits allowing them to exploit the ponds: the invertebrates of William's pond showed variable phenologies and growth rates allowing different species to exploit and survive at different phases of inundation. The Scottish wetland pool fauna was clearly structured by the extent of summer dry phases interacting with distance from potential recolonisation sources.

In the UK the renaissance in our understanding of the importance of ponds, including of temporary ponds, blossomed with the work of Pond Action founded in 1989 and recently renamed the Freshwater Habitats Trust. Pond Action set out to undertake a local pond survey around their Oxford base to develop a methodology that could be used for a national survey. No such inventory had been undertaken before, and, importantly, temporary ponds were included. Collinson et al. (1995) provided the first detailed analyses of the significance of temporary ponds, using Oxfordshire data which included 27 permanent ponds, six that dried occasionally and six that dried every year over the 6 years of the study. Macroinvertebrates were sampled by 3 min of net sweeping, divided proportionately amongst the main habitat types in each pond, e.g. submerged pondweeds and emergent plants, a methodology that underpins all their studies, creating effective comparisons both amongst pond and management types and with other freshwater habitats. The macroinvertebrate data allowed comparisons of total species richness, rarity (in particular, a species rarity index based on national distributions) and heterogeneity in overall communities. In this early study, temporary ponds supported markedly fewer invertebrates, a mean of 17.1 compared to 35–36 in the permanent or occasionally dry sites. However, rarity scores from temporary ponds were as high as those from other ponds; indeed four of the five highest indices were from the small number of temporary sites, including the rarest species found in the survey, the beetle *Haliplus furcatus*, a Red Data Book 1 species, the highest category available. Overall invertebrate communities from temporary versus permanent wetlands were not distinctly different, although there were a few taxa associated only with the temporary sites such as the caddisfly *Limnephilus auricula* and the snail *Anisus leucostoma*. Collinson et al. concluded that we should be careful not to write off temporary ponds. Evidence was accumulating from elsewhere in the UK too. A survey of invertebrates and plants from ponds from Scottish forestry plantations in 1989 also notes that drying out may not be the problem it was so often seen as and that temporary ponds supported specific communities "in their own right" (Jeffries 1992). In the UK, pingo wetlands, essentially shallow, temporary habitats of the Norfolk Brecklands, created by the melting of buried ice lenses left behind as ice age glaciers retreated and perhaps 8000 years old, were identified as internationally important for their rare water beetles (Foster 1993). The invertebrates of "the humble muddy hollow" (Bratton 1990) at last had their defenders, and these changes were made explicit in new advice on pond management, busting the myths that drying out was catastrophic and that changes to water levels should be minimised (Biggs et al. 1994).

In 2001 Pond Action organised a conference specifically to promote the importance of temporary ponds in a European context (Rouen 2001), making plain the significance of these habitats as being common, natural, widespread and persistent but also their vulnerability and the lack of detailed studies. Nicolet (2001) highlighted the growing evidence from Pond Action's national surveys of the importance of temporary ponds for rare species of invertebrate, with rare species found in 82 % of 70 temporary ponds surveyed from seminatural landscapes, although the proportion of rare taxa in permanent and temporary pond faunas was much the same, 16–17 %, but higher than in comparable samples from rivers, 11 %. The same conference emphasised the role of temporary wetlands as home for taxa that tend to be neither wholly terrestrial nor aquatic, such as Dolichopodidae flies which court and hunt on wet mud and Sciomyzidae flies whose larvae predate aquatic snails (Drake 2001) and Carabidae and Staphylinidae beetles (Lott 2001); these taxa often get overlooked by researchers who come from either aquatic or terrestrial disciplines. The beetles provided a particularly striking example of this. From a survey of 30 ponds in Leicestershire, England, the species richness of Carabidae and Staphylinidae was higher than that of the most diverse aquatic taxa such as Dytiscidae and Hemiptera. The amphibious taxa of these transitional liminal habitats remain a poorly understood fauna.

Pond Action developed their work to include a survey specifically of temporary ponds in England and Wales. The ponds were selected from seminatural areas. Invertebrate data from 60 ponds again allowed comparisons of species richness, rarity and overall community patterns; 242 species, ~30 % of the UK freshwater invertebrate list, were found, with beetles dominant, often making up over 50 % of the taxa. Seventy uncommon invertebrates were identified (29 % of the national list), and 82 % of the ponds yielded at least one uncommon species, although many taxa were also known from permanent habitats. The general community types were associated with variations in pH (e.g. acid heaths versus circumneutral grassland) and between bare substrate and vegetated ponds. The days when a seasonally flooded pool would be seen as the ideal target to dig out were gone: "many temporary ponds are best left alone", the authors concluded.

Box 4.1 The Patagonian Mallines of Argentina

The Argentinian Provinces of Neuquén, Río Negro, Chubut, Santa Cruz and Tierra del Fuego make up the Argentinian steppe, primarily grassland and rich in endemic plants and a characteristic invertebrate fauna. The climate is generally dry, cold and windy, and the strong west-to-east rainfall gradient has created two main phytogeographical provinces: the subantarctic forest and Patagonian steppe. On the steppe, precipitation is concentrated in winter and declines from 300 mm in the west to <150 mm in the east, increasing slowly towards the Atlantic coast. This has created abundant small ponds and wetland features called "mallines", an aboriginal Mapuche word, which means swampy area or lowland area where water accumulates. They may be regarded as key components of wider landscapes, providing habitat for terrestrial and aquatic endemic species (Perotti et al. 2005; Kutschker et al. 2014). However, the mallines are also important as a water supply for livestock and can produce 10–20 times more forage than other natural pastures, providing an essential food source for both domestic stock and wildlife (Gaitan et al. 2011). The ecological equilibrium of Patagonia is highly susceptible to human impacts, and the present methods of natural rangeland management are based on extreme overuse in space and time which has induced desertification, one of the main environmental threats to Patagonian biomes. The consequences that these activities are having on Patagonian mallines are not well known.

(continued)

Box 4.1 (continued)

One example of the impact of different levels of livestock pressure on water quality and aquatic invertebrate assemblages comes from a study of 30 mallines from Middle West Patagonia, 15 of them geographically isolated and 15 connected (Pessacq et al. 2011). The analysis of environmental variables showed higher nutrient levels (especially ammonium and soluble reactive phosphorus) and lower aquatic plant coverage at more disturbed mallines, regardless of the mallines' hydrology. Invertebrate assemblages were dominated by Diptera, Coleoptera, Amphipoda and Hemiptera (Appendix), the typical core insect fauna of temperate wetlands around the world. Gamma diversity was 130 taxa. The isolated mallines supported 88 taxa, whilst 89 were recorded from connected sites. Forty-seven taxa were common to both types of malline; however, endemic species were only recorded at isolated wetlands (ten species) including a previously undescribed species of crustacean, *Lynceus mallinensis*. At very disturbed sites, the aquatic invertebrate communities were very species poor. Analysis of the invertebrates from disturbed site invertebrates distinguished assemblages associated with different levels of disturbance (Epele et al. 2013). Oligochaeta (*Lumbriculus variegatus*), Copepoda, Diplostraca, Coleoptera and Diptera (Syrphidae) dominated mallines subjected to high levels of anthropogenic deterioration.

Whilst current management of livestock grazing produces changes in the physicochemical characteristics and aquatic invertebrate assemblages of mallines, the impact appears to be more pronounced at isolated ones. Moreover, given the number of endemic species, these environments would be more vulnerable than those wetlands connected to lotic systems. Some specific actions for conservation of these environments should include controlled grazing areas, alternative sources of water supply and in some cases livestock exclusion.

Temporary Wetlands in the Landscape

The Number of Temporary Wetlands

Auditing the numbers and size of wetlands, even the largest, remains a challenge: the lack of reliable data is itself a threat to the conservation of temporary ponds and stymies opportunities for positive protections. Projects such as the Global Lakes and Wetlands Database and GlobWetlands have capitalised on advances in remote sensing, building on existing databases (see reviews in Lehnner and Döll 2004; Jones et al. 2009). However, the minimum wetland size thresholds in these audits are larger than most temporary wetlands, >1 km^2 (Lehnner and Döll 2004) or 0.5–1 ha (Jones et al. 2009). Estimating the numbers of ponds and wetlands has proven very difficult, even with advanced remote sensing. Small water bodies, permanent or temporary, are routinely missed from audits of global freshwater habitats, which degrades the opportunity for statutory protection and management interventions. The fundamental nature of temporary wetlands, the very characteristics that make them so biodiverse, such as their small size, cryptic and changeable nature, defies easy survey (Fig. 4.1). Recent attempts to estimate numbers have used fine-grained remote sensing and intensive ground survey. Remote sensing uses both visual imagery such as air photographs or spectra in satellite images and geomorphology mapped by airborne radar, e.g. Lidar.

Even the most detailed national or regional databases miss most small, temporary sites. For example, the US Fish and Wildlife Service National Wetlands Inventory (NWI) currently represents one of the finest audits but has a minimum area for sites of between 0.4 and 1.2 ha, varying with pond type and surrounding landscape (Pitt et al. 2012). This results in the omission of at least 50 % of temporary ponds. Pitt et al. tested the difference that local knowledge made to audits of ponds amongst wet woodland in South Carolina, compared to remote sensing. Detailed satellite data provided resolutions down to 1.0 m^2 and identified 4567 wetlands down to 0.0001 ha in area and revealed densities of between 0.33 and 0.53 ponds per km^2. However, when local knowledge was used to search for sites across just 10 % of the study area, an additional 44 wetlands were found, all shallow, all in canopied forest. Their conclusion was that local knowledge was likely to reveal significantly more wetlands than remote sensing alone. Air photos have also been tested by Calhoun et al. (2003), again in wooded habitats, in Maine, in northern USA. They combined air photos of different scales, with NWI data and ground survey. Results varied with the resolution of photos. For their study region, air photos at a 1:4800 scale suggested 516 potential ponds, 93 % of which were correctly identified. However, 1:12,000 photographs suggested only 170 potential ponds, 90 % of which turned out to be correct. Ponds found by ground truthing were significantly shallower than those on the photos. Worse still the density of ponds identified by ground transects was markedly higher, between 23.1 and 49.5 ponds per km^2, varying with precise woodland type, compared to 1.1–13.5 ponds per km^2 from the air photos. The majority ponds were relatively small, two thirds less than 400 m^2, similar to figures reported for other surveys of these American woodland wetlands where the majority of ponds are <1000 m^2.

Results from surveys using diverse technologies all suggest that temporary wetlands are significantly under-recorded. Estimates from air photos of wet woodland in New England suggested that 50 % of the ponds detected in detailed 1:12,000 photos were not in the NWI (Baldwin and de Maynadier 2009). Gala and Melesse (2012) combined Landsat satellite images with Lidar land elevation data to audit wetlands in the Prairie pothole region of central Canada. The combination resulted in detection of substantially more wetlands than recorded on existing maps, the satellite and Lidar combination being more effective than either alone, representing a 200–400 % increase compared to known, mapped numbers. However, there were problems even with the detailed remote sensing, in particular the image pre-processing and de-speckling interfering with detection of the smallest sites.

Fig. 4.1 The impact of local climate variations on wetland inundation. A temporary pond in grazing meadows in northeast England photographed in July 2012 (*above*) and July 2014 (*below*). 2012 was an unusually wet summer, preventing drying out. The wet conditions also resulted in livestock being removed from the field. In 2013 the pond is nearly dry, heavily grazed and trampled

The numbers of temporary ponds and wetlands in temperate biomes remain largely unknown. The limited evidence suggests that existing audits significantly underestimate numbers. More positively, these results show the fundamental nature of these small, cryptic systems within the temperate biomes; woodlands, prairie and steppe and riverine wetlands, each has its own spatial and temporal rhythm of ponds. Identifying what this pattern is for each habitat would be a powerful addition to our understanding of temporary ponds.

Temporary Ponds in Low- and High-Intensity Landscapes

Temporary ponds are a significant feature of anthropogenic landscapes where land use is low intensity, often creating and sustaining a diversity of habitat types. Conversely, many lowland temperate landscapes have long histories of intensive agricultural use and increasing urbanisation, which threaten their characteristic wetlands, e.g. the lowlands of Eastern Europe (Williams et al. 2001).

The high species richness and numbers of rare taxa in temporary ponds in seminatural and natural landscapes were recognised in the first national survey of temporary ponds in the UK, intentionally focusing on minimally impacted ponds (Nicolet et al. 2004). In this survey, mean numbers (and ranges) of plant and macroinvertebrate species were 17 (0–36) and 25 (6–50), respectively, notably higher than results for previous surveys from wider ranges of habitats. The importance of these habitats is made even clearer from detailed surveys of two areas of the UK with abundant, often very small, temporary ponds in amongst low-intensity landscapes: the Lizard Peninsula in Devon, which is predominantly lowland heath and grassland, and the New Forest in Hampshire, a mix of grass, heath and open woodland (Bilton et al. 2009). In both cases, the pools and wetlands are largely anthropogenic, primarily from quarrying activities on the Lizard Peninsula and cattle watering, water supply for agricultural mills, and marl (lime-rich soil) excavation in the New Forest. In both cases, temporary wetlands became characteristic of the landscapes, a form of cultural biodiversity reflecting local economies. On the Lizard Peninsula, conservation management now includes trackway and rut sites, even though they are very small and anthropogenic. A survey of plants and invertebrates identified significant numbers of rare species, 7/119 plants and 25/165 invertebrates. These rare taxa were widespread, 59 % of ponds with at least one uncommon plant and 79 % with an uncommon invertebrate. At the same time, the overall dispersion of species was very patchy with over half the plants and invertebrates found in <10 % of ponds. The overall message is that these ponds are hotspots for overall species richness and rare taxa and that all pond types matter, not just a handful of exceptional sites.

Species with different natural histories may be affected differently by surrounding land use. In a study of invertebrates from playas in the High Plains of Texas, Hall et al. (2004) compared the relative importance of landscape features versus the characteristics within each playa and how these were associated with the presence of (1) drought-resistant residents not relying on dispersal versus (2) transients lacking a drought-resistant stage and capable of active dispersal. Variation in the richness of resident species was significantly related to adjacent land use, highest in natural range land and lowest in agricultural areas, especially early on during the wet phase. However, numbers of transient species did not vary with land use, although they showed positive associations with increased habitat heterogeneity such as emergent plants.

The impacts may depend on the local context. In a detailed analysis of food webs from playas in the Pawnee National Grassland in Colorado, O'Neill and Thorp (2014) expected that sites used by cattle would be impacted due to obvious hoofprints and faeces but could not identify any variations in the structures of food webs (e.g. simplification due to the significant input of organic matter from dung, from grazed versus ungrazed playas). One explanation may be that these wetlands had a history of impacts from large mammals such as bison, and therefore the pond fauna was well adapted to this disturbance and may be even benefitting from microhabitat created by trampling and the transport of propagules from pond to pond via mammal vectors. Arable cropping can produce more subtle effects. Food webs from the playas with corn around in the pond showed more complex food webs, perhaps as the corn provided an additional trophic supplement, but the playas lacking any buffering natural vegetation tended to be more similar to each other than those with some buffer. A surround of natural vegetation resulted in more varied food webs.

Intensively used landscapes can support distinct and valuable temporary wetland biodiversity. But these habitats are often shallow, apparently bare flashes and hollows (Fig. 4.2) and are even more overlooked than temporary wetlands in general,

Fig. 4.2 Shallow flashes and pools in arable fields are often ignored and seen as a problem impacting crop yields. However, these transient habitats support distinct flora and fauna, perhaps most importantly the richness of beetle species when the sites are dry

especially sites that are ploughed and cropped some years. Brose (2003) focused on the diversity of ground beetles (Carabidae) and how this varied with plant communities from shallow field wetlands. Across 30 such wetlands, 138 Carabidae were recorded, ranging between 28 and 58 species per site, along with 92 plant species. The study focused on the relationship between the beetles and the floristic diversity; in particular, did beetle diversity vary with the dominant plant species composition or with the structural complexity of vegetation, regardless of the plant species? The variety of beetles amongst vegetation plots in which the structural complexity was manipulated showed that beetles were more affected by plant architecture than precise plant species. These arable field wetlands represent a disturbed, early successional habitat offering diverse habitat structure as plants responded to variation of inundation and ploughing, promoting local beetle diversity. Although studies from temporary wetlands in arable landscapes are scarce, the results suggest that even the most unprepossessing ponds have their worth. Carabidae are essentially a terrestrial family, a good example of the importance of temporary ponds to nonaquatic taxa.

Evidence of the importance of temporary wetlands for terrestrial beetles was also provided by Lott (2001) who analysed data for Carabidae and the rove beetles (Staphylinidae), another major terrestrial family. In total eight species of Carabidae and 16 Staphylinidae of national conservation status were recorded from 31 temporary ponds out of 112 ponds. The temporary pond beetle fauna was associated more generally with sites with fluctuating water levels in general rather than the flooded ponds themselves; nonetheless, they provide a good example of terrestrial taxa specialising in temporary habitats, often overlooked by freshwater ecologists.

The Community Ecology of Temporary Pond Invertebrates

Where once the invertebrates of temporary ponds were regarded only as the alarming vectors of disease, the last 20 years has seen a revelation in our understanding of the importance of these habitats. The research which identified the significance of temporary ponds for the richness and rarity of their fauna was complemented by insightful studies to explain the complex patterns and processes driving the assembly and dissolution of invertebrate communities. Hydrology dominates these habitats, but there is much more complexity in the detail and the interplay between different species and their varying life histories. In particular, temporary ponds have proved fruitful foci to test different models of animal community assembly. Therefore, we will examine hydrology in detail and then the competing concepts of community assembly which add additional processes into the mix: this split is arbitrary, but we hope it avoids too much muddling of themes.

The Devil in the Hydrological Detail

By the very nature of their habitat, the lives of temporary pond invertebrates are dominated by hydrology: the length of dry and wet phases, seasonal variation, frequency and intensity of the disturbance both from reflooding and drying out. Powerful syntheses of core processes and outcomes are given by Wiggins et al. (1980), Batzer and Wissinger (1996), Wellborn et al. (1996), Schneider and Frost (1996) and Williams (1997). A general paradigm emerges recognising the interaction between the hydrology and various factors, especially predation pressures. The dominant hydrological stress on invertebrates of very short wet phases and annual drying results in life history traits to either resist desiccation or recolonise rapidly. On the other hand, longer wet phases, eventually resulting in habitats that dry out only rarely, impose greater impacts of biotic interactions, especially from predators (fish, amphibians and larger-bodied longer-lived invertebrates).

However, as more studies have accumulated, there has been a growing sense that identifying any coherent, systematic processes that can be reliably applied to most sites has proved difficult: "seemingly intractable" as Batzer (2013) puts it. Outcomes of studies are contradictory, site specific and hard to replicate between places or over time. Even the most obvious likely factor in the lives of wetland invertebrates, hydroperiod length, provides only a coarse guide to species richness: richness increases with hydroperiod length but often a simplistic "very short hydroperiod versus anything longer" step change. Batzer articulated the challenging outcome of this problem. Temporary wetland invertebrates may be largely insensitive to the environmental variation, the inconsistent outcomes of studies therefore being largely trivial, because the animals are highly adapted to thrive in a range of conditions. Alternatively, they may be highly sensitive, responding to their environment in ways we may not be aware of nor can measure or with a hypersensitivity analogous to chaotic systems. This is a dangerous dichotomy if we are to nurture and conserve these vulnerable animals and their habitats. Ultimately, it is the striking variations in species responses and distributions however inconsistent that lead Batzer to conclude that the invertebrates are hypersensitive. "I feel …this review points towards high sensitivity" he writes, which is an unusual choice of words for a scientific review. However, as the many examples cited here show the detail of species responses is compelling and Batzer's "feel" feels justified.

Many recent studies include both field survey and experimental manipulations to untangle the complexity of processes. The timing and frequency of inundation may be as important as the overall length. Kneitel (2014) provides a detailed experimental example using mesocosms to compare the fauna of 16-week-long inundations versus short (8-week) wet phases in either January–February or March–April in vernal pools in California. The richness and density of taxa were markedly higher in the later short wet phase, with the active dispersers (e.g. Diptera) notably increased. Overall length of the inundation was less important. Battle and Golladay (2001) showed similar change over time but with the additional complication of variation amongst habitats. Their study focused on lime-sink wetlands in southwest Georgia, USA, of three types: grass-sedge marsh, cypress savannah and cypress-gum swamp. In the early wet phase, the water chemistry was similar in all three. However, the cypress-gum swamp waters became increasingly nutrient rich, deoxygenated and stained from the rich leaf litter. The grass marsh supports an average of 19.1 taxa in samples at the start compared to 17.0 later, whilst the cypress-gum swamp yielded 14.1, declining to 10.3 later.

Box 4.2 The Pannonian Plains of Hungary

The Pannonian Plain is the lowland heart of the Pannonian Basin, a biogeographically distinctive region of central Europe dominated by the rivers Danube and Tisza and including parts of Hungary, Romania, Croatia and Serbia. The plain is characterised by diverse aquatic habitats. The Kiskunság region is an area particularly rich in temporary waters. Boven et al. (2008a) identified a number of temporary pool types, such as wheel tracks, clay pits, meadow pools, pools in arable land and orchard pools, covering a wide range of hydroperiods. Astatic soda pans are also present in this region. These are shallow saline and alkaline wetlands, hosting unique invertebrate communities, mainly dominated by crustaceans (Horváth et al. 2013a, 2014; Boros et al. 2014; Tóth et al. 2014; Appendix). They also represent important breeding and foraging habitats for water birds (Horváth et al. 2013b). Considering the high diversity of temporary habitats, it is not surprising that the Kiskunság was identified as a "hotspot" for large branchiopods (Boven et al. 2008b).

(continued)

Box 4.2 (continued)

An unusual type of temporary aquatic habitats is bomb crater ponds, often forming very dense clusters; in Northern Kiskunság, a cluster exists that consists of more than 100 bomb craters, between 3 and 12 m in diameter, within an area of approximately 1 km^2 in a sodic meadow. They were created by mistargeted bombing of the nearby airport during the Second World War. The sodic soil results in sodic water. Despite being situated in close proximity, they exhibit a range of salinities from hypo- to mesosaline, along with a wide range of turbidity, hydroperiod, nutrients and vegetation cover.

The communities of invertebrates reflect this environmental heterogeneity, with a high overall regional (gamma) diversity but marked differences between individual wetlands. The most diverse invertebrate group is the coleopterans with close to 60 taxa. Amongst the most widespread species, some typical pond and temporary water species occur such as the Odonata *Lestes barbarus* and *Aeshna affinis*. The ponds are also home to rare species, e.g. amongst fairy shrimps, *Chirocephalus carnuntanus*, a Pannonian endemism, and *Eubranchipus grubii*. Ponds with the highest salinities are less diverse, and some of them have communities typical of soda waters, dominated by the calanoid copepod *Arctodiaptomus spinosus*. In Europe soda waters are rare so this unusual fauna (Appendix) is of high conservation value. Therefore, despite their anthropogenic and ominous origin, these ponds make an important contribution to regional and continental aquatic biodiversity.

Thousands of other bomb crater ponds are present on other parts of the Plain created by military exercises, particularly in the Hortobágy region of the Great Hungarian Plain. Because of their origins, they are often considered wartime scars, and grassland rehabilitation measures have been applied often without a proper evaluation of their contemporary biodiversity. This loss is even more significant because, before river regulation, the major part of these grasslands comprised temporarily flooded areas, creating different water bodies each with their own natural dynamics. Considering the serious losses in such aquatic habitats in central European flood plains, the crater ponds do not deserve their negative reputation which eclipses their value for conservation. They are unusual and distinctive hotspots for temporary water biodiversity, a replacement for the lost wetlands of the old flood plains.

Because the bomb craters are of the same age and in close proximity, any climate and soil-related differences are reduced. Therefore, these areas can be considered as "natural labs" with a sufficient number of replicates exhibiting a wide array of environmental gradients. This makes them excellent model systems for community ecology studies.

Although the loss of water from temporary wetlands was once seen as a problem, unusually long wetting may be as destructive a disturbance to temporary ponds as the loss of water from normally permanent ponds (Jones 2013). Golladay et al. (1997) provide a neat example, again from the lime-sink wetlands of Georgia. These are seasonal wetlands in woodland, usually wet from late spring into early summer. A study of invertebrates from three wetlands in 1994–1995 coincided with unusually heavy rain in the summer of 1994 so that the sites stayed wet. The extended wetting resulted in lower densities of the normally dominant Amphipoda, Isopoda and smaller zooplankton. Conversely, Chironomidae midge larvae and freshwater sponges thrived on the submerged woody debris, perhaps benefitting from a lack of predation. A similar impact occurred in small experimental pools monitored from 1995 to 1999 in Northumberland, UK. In 1995 and 1996, the pools had a single, sustained summer dry phase. In 1997 just as the ponds started to dry very heavy rain, a 1 in 300-year event overfilled them and they stayed wet for 28 months. Some widespread invertebrates were lost (e.g. the caddis *Limnephilus vittatus*), although more permanent pond taxa colonised. The pools also became choked with filamentous algae which may also have reduced the habitat quality. Summer dry phases resumed in 1999, the algal blooms were gone when the pools refilled, and the temporary pool invertebrates reappeared (Jeffries 2005).

These studies exemplify the complexity of the apparently simple wet-phase/dry-phase disturbance. Studies increasingly try to include multiple stressors in the mix to identify their relative importance. Bischof et al. (2013) review some of the uncertainties, for example, the variation between nearby sites, with similar hydrology that may be driven by other environmental influences and the problem of effects only becoming apparent a year or more after the hydrological disturbance. They addressed these challenges by monitoring the invertebrates of 16 seasonal ponds in a forested landscape of Minnesota, sampling every 7 days throughout a 7-week period across 2008 and 2009. In addition to measuring the length of hydroperiod, the precise hydrological types (e.g. recharge, perched) were characterised using wells and other variables such as canopy cover, depth and area included. The fauna was dominated by Diptera such as Culicidae and Chironomidae, along with small Diplostraca and larger shrimps (e.g. *Eubranchipus*). Once again the fine-grained detail matters. There was a distinct seasonal variation of taxa, such as early season detritivore Culicidae. Canopy cover and depth were all significant variables in models to explain the patterns of invertebrate communities. Results varied between years; for example, *Eubranchipus* was positively associated with increased depth in 2008 but the opposite in 2009. Similar outcomes came from a study of snail communities from 43 lakes and ponds in Michigan with a range of hydrologies and varying canopy covers, surface area and the presence of fish (Hoverman et al. 2011). Species richness increased with hydroperiod, the classic general pattern, but also with wetland area and pH, but decreased with increasing canopy cover. A distinctive temporary pond fauna (e.g. *Gyraulus circumstriatus* and *Aplexa elongata*) was identified; these species can aestivate but also are vulnerable to fish predation due to thin shells. This study also focused on wider questions in community ecology, specifically the contest between paradigms that emphasise the role of the niche for sorting out which species live where, versus models of nature where dispersal is more important, and the precise species and their supposed niches unimportant.

These questions lie at the heart of community ecology, and temporary ponds have played a significant role in research about the nature of nature.

Community Ecology and the Contested Explanations of Pattern and Process

The distribution, diversity and abundance of animals in temporary ponds are a specific example of the wider enquiry to identify patterns and process in ecological communities. To reiterate Robert MacArthur's (1972) famous dictum, "to do science is to search for repeated patterns". The trouble is that community ecology, which focuses on describing and explaining the richness and variety of life at individual sites and times, can seem hopelessly lost in outcomes which are contingent on particular local processes and exact scales and "an unsettling morass of theoretical models" (Vellend 2010), making it hard to decipher any systematic rules (Lawton 1999).

In spite of or perhaps because of this, temporary ponds have proven exemplary habitats for community ecology (Wilbur 1997). They are widespread, numerous and often small. They can be recreated by digging out new ponds (e.g. Jeffries 1994, 2011) or using substitutes such as troughs (e.g. Fernando and Galbraith 1973) or pond liner pools (Croel and Kneitel 2011), allowing replication and examination of multiple factors which drive local community assembly in the field. Field conditions can be replicated in finely controlled experiments in laboratory aquaria (e.g. Anholt et al. 2000). Temporary pond faunas are diverse enough to provide a richness of interactions, species' traits are reasonably well documented in many countries, and assemblages can be readily manipulated in experiments. Much of the research focuses on apparently alternative processes driving the development of the community, e.g. deterministic (following recognisable rules) versus stochastic models (dominated by chance) or niche-based (where the traits of individual species are significant) versus neutral models (which treat all species involved as effectively similar in their interaction and fitness). However, there are increasingly effective syntheses of these different versions of how communities work, in particular the metacommunity concept (Leibold et al. 2004) which brings together local effects such as the interactions within a pond and processes that work across the wider landscape such as dispersal.

Leibold et al. (2004) identify four broad metacommunity paradigms that apply to temporary ponds:

1. *Patch dynamics*. This concept assumes that the ponds (i.e. patches) are identical, and, potentially, all the ponds can support the species from the local species pool. The presence or absence of a species is driven largely by local colonisations and extinctions.

2. *Species sorting*. In this case the individual ponds are sufficiently heterogeneous and the species sufficiently sensitive to these differences to cause differences in the presence/absence of taxa. This paradigm emphasises the importance of species' niches, with dispersal playing a role as species move across the landscape tracking gradients and the diversity of pond types.
3. *Mass effects*. The role of immigration and emigration dominates this model. The numbers of individuals dispersing are sufficient to obscure the effects of any local interactions or conditions within a pond. Again, niche differences are important, but the dispersal overrides their outcomes, e.g. immigration topping up populations in poor-quality patches.
4. *Neutral model*. In this case the differences between individual species traits are regarded as insignificant; all species are effectively equivalent in their responses and needs. Their presence/absence is the outcome of random probabilities of emigration, immigration, extinction and speciation.

Temporary ponds are patches in both time and space, requiring their fauna to disperse across the landscape. Many of the individual studies focus on either interactions within a pond or dispersal processes. Keep the metacommunity paradigms in mind to navigate the detail.

Interactions in the Ponds

Within ponds, intra- and interspecific interactions are powerful and complex drivers of species richness. Interactions can be broadly divided between exploitation (e.g. predation, parasitism, herbivory), competition, mutualisms and much less well-understood processes such as ecological engineering and commensalism. Predation and competition have been demonstrated in many studies, amongst many taxa. The small size of ponds and the race to mature before ponds dry out seem to magnify these effects, but they seldom act simply or in isolation.

Predation has proven a particularly powerful interaction in temporary wetlands. Predation has been singled out as the key interspecific interaction, with a gradient of predation pressure created by hydrological variation. Permanent ponds are often home to fish which severely impact invertebrate communities. However, the drying out of temporary wetlands often excludes fish, and large invertebrates such as Odonata and dytiscid larvae become the dominant predators. They can significantly reduce populations of prey, even cause local extinctions within individual wetlands, although across the wider landscape prey survive at sites which the predators have not colonised. Predation by invertebrates can be intense and drive complex responses. For example, Anholt et al. (2000) provide a neat example of how different species of ranid frog tadpoles reduce their time spent moving and movement rates in response to the presence of large dragonfly predators and the amount of food provided. However, the leopard (*Lithobates sphenocephalus*) and wood frog (*L. sylvaticus*) tadpoles, species more characteristic of temporary ponds, responded more to the changes in food resources and showed higher growth even at the lowest food provision. Within a species, these processes can produce intraspecific outcomes, such as smaller tadpoles reducing movement due to predator threat but then losing out to competition from larger tadpoles that remain more mobile (Werner and Anholt 1996). Classic processes of competition and predation create subtle interactions between species. If two species share a predator, then the predator's impact on both may be diluted, so-called apparent mutualism, but if the predator increases its population or preferentially chooses one prey item, then the impacts represent apparent competition, although these effects may be hard to detect (e.g. Stav et al. 2005 experimenting with mosquito larvae and Crustacea from desert pool communities).

Competition within a species can be important too, and the patchy nature of pond habitats may magnify this risk due to inbreeding. The natterjack toad, *Bufo calamita*, is a temporary dune pond amphibian with scattered, often isolated populations in the UK. Rowe and Beebee (2005) showed that the tadpoles of an inbred population grew more slowly than those of an outbred group, an outcome made all the worse by more prolonged exposure to heavy predation from dytiscid water beetles.

The striking examples of significant interactions between pond animals have motivated the search for general patterns in pond food webs, often linked to hydrological variation. Pond permanence may affect the proportion of predatory invertebrates in the overall community. In the study of Bilton et al. (2001), 16 ponds on the Lizard Peninsula in southwest England were sampled for invertebrates and the communities compared to variations in pond area and permanence. Overall species richness increased with permanence and so did the proportion of predatory taxa, whilst pond area showed no relationship with the proportion. In the Lizard ponds, many of the invertebrate predators were larger species of diving beetle and dragonflies, species which are relatively k-selected, i.e. larger, longer-lived and therefore requiring a longer wet phase to develop. Schneider (1997) identified a similar outcome from ponds across a wet-phase gradient, with the numbers of predators increasing with the duration of the inundation as well as the absolute numbers of links and links per taxon in the food webs increasing. Conversely, an unusually severe drought had the opposite effect, reducing links and connections in the web; the effects lingered where large predators such as Notonectidae bugs had been wiped out, allowing very vulnerable prey, e.g. fairy shrimp, to flourish. Again, the detail matters. In a study of playa wetlands in Colorado, O'Neill and Thorp (2014) showed that food webs were more complex in playas dominated by insects, compared to those with large branchiopods such as tadpole shrimps, and that insect diversity increased with playa hydroperiod which was a function of depth. Insects had more diverse traits and were more specialised, whilst branchiopods all did essentially the same thing, filter feeding (i.e. so-called trophic redundancy).

The detail can become overwhelming, but the key message from these examples is the sheer complexity, intensity and speed of interactions which can determine local species richness and population within temporary ponds: hydrological changes may dominate and be conspicuous, but much goes on beneath the surface too.

Box 4.3 Forest Wetlands of the Eastern USA

Temporary wetlands are a common and often overlooked component of forest ecosystems in the deciduous temperate forests of the eastern half of the USA. These wetlands are typically small (<1 ha), shallow (<1.5 m deep), nontidal and non-floodplain systems with considerable canopy coverage and often with live trees within the wetland margins. They have a predictable hydrologic regime, usually drying in summer and re-wetting in either fall, winter or spring. Whilst there may be some primary productivity from algae and submerged macrophytes, these wetlands are usually heterotrophic, and the food web is based on the allochthonous leaf litter which supports a variety of macroinvertebrate grazers, collectors and filterers (Appendix). These in turn support macroinvertebrate predators which occur low in abundance but are usually the most taxonomically diverse functional feeding group. Energy enters these wetlands during fall in the form of leaf litter. This energy is returned to the forest ecosystem months later in the form of invertebrates and amphibians which serve as prey to terrestrial forest organisms, enriching the forest ecosystem as a whole, a good example of reciprocal energy subsidies.

The macroinvertebrate communities of forested temporary wetlands are dominated by insects e.g. Chironomidae, Culicidae, Coleoptera and Odonata, along with Asellidae crustaceans in some regions, although molluscs, microcrustacea and annelids can be locally abundant (Golladay et al. 1997;

(continued)

Box 4.3 (continued)

Schneider 1999; Batzer et al. 2005; Studinski and Grubbs 2007; Appendix). Invertebrate communities of these wetlands vary both temporally and across multiple spatial scales. Within the eastern USA, at the very large regional scale, shifts in invertebrate communities can be observed from north to south, possibly responding to differences in pH (higher in the north) or temperature and precipitation (both higher in the south). These physical gradients likely create fundamental ecological differences, e.g. whilst insects and crustaceans dominate these systems throughout the region, molluscs become more common in the north, and taxa such as Asellidae are very common in southern forested temporary wetlands. At the local scale, invertebrate communities appear to be controlled by wetland size, depth and proximity to other aquatic habitats. The application of biogeography and metapopulation dynamics has been somewhat successful in explaining the differences in macroinvertebrate assemblages amongst these wetlands (March and Bass 1995; Brooks 2000). However, hardy generalist taxa, random colonisation events and yet unknown desiccation resistance strategies complicate efforts to accurately predict macroinvertebrate assemblages, and these forest wetlands are a classic example provoking Batzer's (2013) exploration of the "intractable ecology" of wetland invertebrates.

Forested temporary wetlands increase the complexity of forest ecosystems by providing additional habitat, but they are vulnerable partly because they appear to be common place. These fishless wetlands support a variety of invertebrates including temporary wetland specialists and are also critical habitats for many amphibians. Due primarily to their small size and temporary nature, it is difficult to quantify the recent trends in the loss of these wetlands. It is estimated that forested temporary wetlands account for approximately half of the wetland area remaining in the eastern USA (Tiner 1987). There is little conservation effort directed towards forested temporary wetlands, especially when compared to larger freshwater emergent wetlands or to estuarine marshes. The conversion of forested areas for agricultural, residential or commercial purposes is the most obvious threat, but shifts in hydrologic regimes due to climate change or increases in primary productivity due to timber harvest may also impact these systems. Within the eastern USA, best management practices which regulate timber harvest near these forested wetlands vary widely from state to state (Aust 1994). Forested temporary wetlands are the most common yet some of the least understood wetlands in the eastern USA: these intriguing systems deserve further research.

Dispersal: How, When and Where?

All species disperse but the spatial and temporal scatter of temporary wetlands makes dispersal a dominating process in the ecology of many of their inhabitants. Hydrological change is the obvious driver for dispersal, both to escape from sites that are drying out and recolonise those that have reflooded.

Whilst drying and inundation are obvious cues for animals to disperse, there is evidence of more subtle, and surprising, behaviours. A neat example comes from two species of dytiscid water beetle that inhabit temporary prairie ponds in Alberta: *Graphoderus occidentalis* and *Rhantus sericans*. Yee et al. (2009) set up a field experiment using different densities of adult beetles and aquatic plants within baskets placed in ponds. More beetles left the baskets as overall beetle density increased. The density of plants had less effect although beetles dispersed slightly more at lower stem densities. The possibility that many invertebrates can assess their environment at this level of detail and disperse accordingly suggests that the fine-grained detail of species-specific sensitivities is important before any wider environmental drivers such as land use impacts and stresses. Most studies focus on response to larger scale hydrological and seasonal variation but still pick up the interspecific differences. In a careful mark–recapture study that involved gluing tiny number labels onto the wing cases of beetles without hindering their ability to fly, Davy-Bowker (2002) monitored the dispersal of different species across seven temporary ponds over a three-year period in northwest England. Whilst all species could fly, one, *Colymbetes fuscus*, did not move out even when ponds dried, whilst *Dytiscus marginalis* and *Acilius sulcatus* moved into the temporary sites in spring but retreated to permanent ponds when temporary sites dried out in summer. Reviewing these types of species-specific seasonal variations, Boda and Csabai (2013) produced a general scheme characterising beetles from a Hungarian wetland, highlighting spring, summer and autumn dispersal peaks for different species. Dispersal events appear to be driven more by the seasonal changes to the habitats themselves as they dry or reflood, rather than broader atmospheric conditions such as temperature and wind speed (Boix et al. 2011).

Whether an animal chooses to enter a particular pond, or perhaps colonise by laying eggs, is also subject to surprising responses. Many flying insects such as beetles and bugs notoriously land on shiny surfaces such as vehicles, a mistake exploited by Boda and Csabai (2013), who used plastic sheets to lure in airborne insects for their characterisation of seasonal behaviours. The outcomes, especially for oviposition, are more nuanced with evidence of selection based on habitat colour and resources, e.g. darker versus lighter backgrounds and decomposing leaves versus empty tanks (Williams et al. 2007); avoidance of predators, e.g. mosquitoes avoiding the presence of *Notonecta* predatory bugs (Blaustein et al. 2004); and avoidance of competitors, e.g. *Chironomus* midge avoiding oviposition into aquaria with chemical cues suggesting the presence of large numbers of competing larvae (Stevens et al. 2003). The diversity of individual species' sensitivities to habitat (e.g. presence of rivals), their responses to these conditions (e.g. fly or stay) and the changes to habitats will create complex outcomes, but the underlying mechanisms are universal.

Hydroperiod responses, flight behaviour, oviposition cues and predator–prey interactions: it is the variety and strength of the individual species sensitivities that informed Batzer's (2013) gut feeling that the invertebrates are hypersensitive. The invertebrates of temporary ponds remain a fascinating biota for challenging ecological research tackling fundamental questions about global, regional and local biodiversity. Wetlands are also an arena for an increasing number of studies responding to a new threat, climate change.

Temporary Ponds and Climate Change

Given the predominant role of hydrology in the ecology of temporary ponds, it is no surprise that climate change will have a significant impact on these habitats, often harmful, perhaps at times beneficial. Freshwater habitats in general are subject to multiple stressors (reviewed Stendera et al. 2012), many of them long-standing such as nutrient enrichment and land use change, on top of which climate change only adds to the pressures (Heino et al. 2009). Despite the recognition that small ponds and wetlands will be disproportionately vulnerable (Heino et al. 2009), they, along with ephemeral streams, are often omitted from global and national assessments of climate hazard (Brooks 2009), including some of the rare examples of detailed studies of likely impacts on pond fauna (e.g. Rosset et al. 2010). Impacts will not only be the direct changes to hydrology, such as changes to the length of dry and wet phases, and the seasonal phenology and intensity of precipitation or drought events, but also changes to temperature affecting hatching and development, alterations to trophic inputs to and from the surrounding landscape, the breakdown of site networks impacting metapopulation dynamics and possible change to anthropogenic land use. The existing adaptations of temporary pond biodiversity do seem to confer resistance and resilience to changes, at least to drought (Lake 2011), but the threat arises from changes to the range of extremes, with increased variability and increased maxima to precipitation and rainfall, and extremes that reach novel levels which the fauna has not encountered and against which their existing adaptations are ineffective (Jones 2013).

Demonstrating the effects of climate change is problematic, not just because they are complex but also because they may occur across time scales that require the sorts of long-term data that we did not begin to collect before changes started or the impacts are likely in the future but have not happened yet so we cannot directly identify them. Nonetheless, temporary ponds and their fauna are so tied to the local and regional climate that these habitats allow a range of approaches to investigate climate change risks. At least three approaches, each at different scales, have been developed for ponds. First is the large-scale modelling of the extent of habitat or species distributions and predictions of how these will vary under different climate change scenarios. Second is the field studies that monitor pond hydrology and wildlife during extreme climate events that mimic likely changes. Such studies can be in response to an extreme event but also where local weather extremes impact on sites already being used for long-term monitoring. Third is the use of experimental approaches where local conditions, usually hydrology, are purposefully created or adjusted to mimic likely climate change outcomes. Experiments have the advantage of being much more exactly controlled so that cause and effect can be attributed with greater confidence. Their disadvantage is the usually small spatial scale and artificial set up.

A neat example of the third, experimental, approach comes from Vermont in northeastern USA, classic woodland seasonal pond territory. For this experiment, Hart and Gotelli (2011) used mesocosms, essentially small plastic tanks, laid out in deciduous forest. The tanks were small (0.5 m^2 surface area, 33 cm depth), but the experimental approach allowed precise control of two different aspects of the hydrology: the overall water depth and the variation of the depth within each mesocosm. These two aspects were chosen because of predicted climate change within the region, with increased precipitation in winter and less in summer and greater variation with more intense deluges and longer droughts. The tanks were set up with water and detritus from nearby ponds as well as being open to external aerial colonisation. The populations of juvenile mosquitoes and Chironomidae midges were recorded weekly, these two taxa making up ~90 % of the total abundance of invertebrates in the mesocosms and being core groups in the surrounding ponds. The two taxa responded differently to the hydrological manipulations. Numbers of mosquitoes were higher at low water levels, but the Chironomidae were more abundant at high water levels with little variation. The mosquito growth rates increased as levels declined, whilst for the midges the density-dependent interactions were reduced when water level was higher.

The second approach relies on recording the changes associated with extreme weather events which mimic possible climate change outcomes. This strategy requires long-term site monitoring, usually over at least 5–10 years, ideally longer, which is hard to sustain. Nonetheless, revealing examples exist which show both the predictably intimate relationships between temporary ponds and weather and more subtle variations and complexities. Monitoring wetland invertebrates over long enough periods to include measurable climate variation has also demonstrated significant impacts, both from drought and unusual rainfall events which stop temporary ponds from drying out.

Jeffries (1994) used small replicate 1 m^2 ponds (Fig. 4.3) created by the removal of antitank barriers in a wetland at Aberlady Bay in eastern Scotland, comparing the invertebrates in 1986 and 1987 (Appendix) to the surviving communities in 1992 following an unusual period of drought (Appendix). Ponds that had retained water throughout the study accumulated taxa. However, ponds that had stayed wet before the drought but dried during the low rainfall years showed significant loss of species, typically down from 20–30 to 10–20. Ponds which had previously dried out in the years prior to the drought showed even worse attrition with the already limited fauna reduced from 10 to 12 to just one or two very hardy taxa. This approach was repeated with a set of 30 ponds dug specifically to match the Aberlady ponds, a bit

Fig. 4.3 Small pools used for long-term monitoring of invertebrates. These pools were dug in 1994 at Hauxley Nature Reserve in northeast England. The invertebrates and plants have been monitored since then allowing relatively long-term analysis of changes in response to local climate, notably unusually wet or dry years

further south in Northumberland and monitored for 10 years from 1995 to 2004 (Jeffries 2011). All ponds dried in summer in 1995 and 1996 but unusually heavy, unseasonal rainfall in early summer 1997 kept them filled until summer 1999 when dry phases resumed. In addition, early spring dry phases occurred in 2003 and 2004, followed by a brief late spring refill before the main summer dry phase. The 10-year time period allowed subtleties to show up. A pioneer fauna, typically Ostracoda and Dytiscidae, dominated the ponds in 1995 and 1996. The unusual 1997–1999 wet phase resulted in many permanent pond species colonising such as *Chaoborus* midges and *Cloeon* mayflies. With the resumption of summer dry phases, the diverse fauna was significantly reduced although this took 2–3 years. Ponds hit by the advent of early spring dry phases showed further marked reductions with a remnant fauna of amphibious lumbricid worms and *Helophorus* beetles. These results suggest that increased length and frequency of dry phases will degrade pond fauna, including the characteristic temporary pond species whose resilience is not sufficient when faced with novel extremes and events such as the advent of additional spring dry phases in the Northumberland ponds.

Conversely, extreme inundation can have negative impacts on temporary pond wildlife. The 1 in 300-year rainfall event that prevented the ponds undergoing their normal summer dry phase resulted in filamentous algae smothering the pools, the growth of other plants such as common water buttercup (*Ranunculus aquatilis*) was restricted, and the diversity of invertebrates was reduced as temporary pond specialists were lost (Jeffries 2005, 2008). In a similar natural experiment, Golladay et al. (1997) monitored the invertebrates of lime-sink wetlands in Georgia during which time unusually high rainfall resulted in an extended wet phase. The benthic invertebrate community changed significantly, with decreases in density although a few taxa, notably chironomids, were unusually abundant afterwards. The changes were credited to multiple impacts such as the loss of dry-phase hatching cues and reduced oxygen levels in the deeper water.

The possibility exists that some invertebrates can benefit from weather extremes. In 2012 the UK had an unusually wet summer. The impact of this on the number, area and biodiversity of temporary wetlands was demonstrated on a lowland mixed arable and livestock farm in Northumberland, northeast England, which was used to monitor the number of small wetlands and how this changed with local rainfall. The numbers and area of ponds increased markedly in response to the extreme rainfall, partly because core sites did not dry out but also because new pools developed in hollows and tracks. Vegetation was monitored in 20 ponds, all of which held water throughout the wet summer and were luxuriant, primarily because common emergent species grew more extensively and sites were not ploughed or grazed by livestock: Figure 4.1 shows one of the ponds in 2012 with verdant vegetation compared to the 2013 when grazing resumed. Ponds formed in arable fields with a fauna of ostracods, *Daphnia obtusa*, Chironomidae, Hemiptera and smaller Dytiscidae. The following summers of 2013 and 2014 were warmer and drier than the long-term average, and the majority of the 20 focal ponds dried out. No plant species were lost from across the farm as a whole, but sites with livestock were trampled and grazed, and the vegetation cover became sparser and shorter than in 2012. Four ponds were ploughed and planted with cereal crops with almost total loss of wetland species. The extreme wet summer had benefitted the flora of these shallow sites in this intensively used agricultural land.

Similar impacts have been demonstrated for amphibians and plants using field monitoring or modelling distributions, and again the details are worrying. For example, potential habitat range for the fire-bellied toad, *Bombina bombina*, in Brandenburg, northeast Germany, may increase with a warmer climate, so long as the climate changes do not simultaneously promote land use intensification which would reduce available sites (Dolneger et al. 2014). In woodland pools in northeast USA, the wood frog *Lithobates sylvaticus* can show a significant advance in egg laying date as spring seasons warm earlier, but tadpole metamorphosis only occurs one day earlier, exposing tadpoles to cooler temperatures than they would have encountered if spawning were later (Bernard 2014).

The message from all these approaches is the same. Climate change will affect the numbers and hydrology of temperate, temporary wetlands with potentially serious consequences as novel extremes of drought or inundation stress animals beyond their natural resilience. Less obviously, the effects may include increases in habitat, different responses in different pond types and individual species affected in contradictory ways.

Conclusion

Seasonal ponds, pingos, mallines, soda pans, dune slacks, potholes, lime-sinks and subsidence ponds are fundamental habitats of the temperate biomes, easy to overlook amongst the wider landscape or to dismiss because of their small size and ephemeral nature. Their invertebrates are a disproportionately rich and rare mix, whose natural history has challenged ecologists to research space and time. Our hope is that this chapter does them justice and captures something of the charm of fairy shrimps, the fascination of water beetles and the mysteries of midges. Their ecology, Watson's (1921) "wondrous fashion of a fairy tale", is as fascinating as ever.

Appendix

Invertebrate taxa collected from selected temperate, temporary wetlands from South and North America, central Europe and the UK. (Some taxa not recorded from a site may simply reflect that the organism was not effectively collected by the sampling programme employed; other taxa may be absent for biogeographic reasons).

	Northwest Patagonian wetlands[1]	Central Europe soda pans[2]	Forested depressional wetlands, Minnesota[3]	Experimental pond, northern England, during sustained weather extremes[4]	
				Inundation	Drought
Nematoda		X			
Cnidaria				X	
Turbellaria	X	X		X	
Rotifera		X			
Mollusca					
Gastropoda					
Ancylidae	X				
Chilinidae	X				
Hyriidae	X				
Lymnaeidae	X	X	X	X	
Physidae		X	X		
Planorbidae	X	X	X	X	
Succineidae		X			
Valvatidae		X			
Vertiginidae		X			
Bivalvia					
Sphaeriidae	X		X	X	X
Annelida					
Oligochaeta	X				
Alluroididae	X				
Enchytraeidae	X	X			

(continued)

(continued)

	Northwest Patagonian wetlands[1]	Central Europe soda pans[2]	Forested depressional wetlands, Minnesota[3]	Experimental pond, northern England, during sustained weather extremes[4]	
Lumbricidae			X		X
Lumbriculidae	X		X	X	
Naididae	X	X		X	
Tubificidae		X	X	X	
Hirudinea					
Erpobdellidae		X	X	X	
Glossiphoniidae	X	X	X	X	
Hirudinidae		X			
Crustacea					
Branchiopoda					
Bosminidae		X			
Branchinectidae		X			
Chirocephalidae		X	X		
Chydoridae		X		X	
Cyclestheridae	X				
Daphniidae	X	X	X	X	
Macrothricidae		X			
Notostraca			X		
Sididae		X			
Copepoda					
Calanoida	X	X			
Cyclopoida	X	X		X	X
Harpacticoida	X	X		X	

	Northwest Patagonian wetlands[1]	Central Europe soda pans[2]	Forested depressional wetlands, Minnesota[3]	Experimental pond, northern England, during sustained weather extremes[4]	
Ostracoda	X	X	X	X	X
Amphipoda					
Crangonyctidae				X	
Dogielinotidae	X				
Isopoda					
Asellidae		X		X	
Arachnida					
Hydrachnidia	X	**X**	**X**	X	
Insecta					
Ephemeroptera					
Baetidae	X	X	X	X	
Caenidae	X	X		X	
Leptophlebiidae	X				
Nesameletidae	X				
Oniscigastridae	X				
Odonata					
Aeshnidae	X	X			
Coenagrionidae	X	X	X	X	
Lestidae		X	X	X	
Libellulidae		X	X	X	
Plecoptera					
Gripopterygidae	X				
Hemiptera					
Belostomatidae			X		
Corixidae	X	X	X	X	
Gerridae		X	X	X	
Nepidae		X		X	
Notonectidae	X	X	X	X	
Pleidae		X	X		
Veliidae		X			
Neuroptera					
Sialidae		X			
Coleoptera					
Dytiscidae	X	X	X	X	X
Elmidae	X				
Gyrinidae		X	X		

(continued)

(continued)

	Northwest Patagonian wetlands[1]	Central Europe soda pans[2]	Forested depressional wetlands, Minnesota[3]	Experimental pond, northern England, during sustained weather extremes[4]	
Haliplidae	X	X	X	X	X
Hydraenidae	X	X	X	X	X
Hydrophilidae/ Helophoridae	X	X		X	X
Noteridae		X		X	
Scirtidae	X		X		
Staphylinidae			X		
Trichoptera					
Hydrobiosidae	X				
Hydroptilidae	X		X		
Leptoceridae	X	X	X		
Limnephilidae	X	X	X	X	
Polycentropidae			X		
Diptera					
Ceratopogonidae	X	X	X	X	X
Chaoboridae		X	X	X	
Chironomidae		X	X		
(Chironominae)	X			X	X
(Tanypodinae)	X			X	X
(Orthocladiinae)	X			X	X
(Podonominae)	X				
Culicidae	X	X	X	X	X
Dixidae			X	X	
Dolichopodidae				X	
Empididae	X				
Ephydridae	X	X	X	X	
Muscidae	X				
Psychodidae	X			X	
Simuliidae	X				
Stratiomyidae	X	X	X		
Syrphidae	X		X		
Tabanidae		X	X		
Tipulidae/ Limoniidae	X	X	X	X	X

[1]Epele, unpublished
[2]Wolfram et al. (1999), Boros et al. (2013), Tóth et al. (2014)
[3]Batzer et al. (2004)
[4]Jeffries (2011)

References

Anholt BR, Werner E, Skelly D (2000) Effect of food and predators on the activity of four larval ranid frogs. Ecology 81:3509–3521

Aust WM (1994) Best Management Practices for Forested Wetlands in the Southern Appalachian Region. Water Air Soil Pollut 77:457–468

Baldwin RF, de Maynadier PG (2009) Assessing threats to pool breeding amphibian habitat in an urbanizing landscape. Biol Conserv 142:1628–1638

Barclay MH (1966) An ecological study of a temporary pond near Auckland, New Zealand. Austral J Mar Freshw Resourc 17:239–258

Barnes LE (1983) The colonisation of ball-clay ponds by macroinvertebrates and macrophytes. Freshw Biol 13:561–578

Battle J, Golladay SW (2001) Water quality and macroinvertebrate assemblages in three types of seasonally inundated lime-sink wetlands in southwest Georgia. J Freshw Ecol 16:189–207

Batzer DP (2013) The seemingly intractable ecological responses of invertebrates in North American wetlands: a review. Wetlands 33:1–15

Batzer DP, Wissinger SA (1996) Ecology of insect communities in nontidal wetlands. Annu Rev Entomol 41:75–100

Batzer DP, Palik BJ, Buech R (2004) Relationships between environmental characteristics and macroinvertebrate communities in seasonal woodland ponds of Minnesota. J N Am Benthol Soc 23:50–68

Batzer DP, Dietz-Brantley SE, Taylor BE, DeBiase SE (2005) Evaluating regional differences in macroinvertebrate communities from forested depressional wetlands across eastern and central North America. J N Am Benthol Soc 24:403–414

Bernard MF (2014) Warmer winters reduce frog fecundity and shift breeding phenology, which consequently alters larval development and metamorphic timing. Global Change Biol. doi:10.1111/gcb.12720

Biggs J, Corfield A et al (1994) New approaches to the management of ponds. Brit Wildl 5:273–287

Bilton DT, Foggo A, Rundle SD (2001) Size, permanence and the proportion of predators in ponds. Arch Hydrobiol 151:451–458

Bilton DT, McAbdenbroth LC et al (2009) Ecology and conservation status of temporary and fluctuating ponds in two areas of the southern England. Aquat Conserv Freshw Mar Ecosys 19:134–146

Bischof MM, Hanson MA et al (2013) Invertebrate community patterns in seasonal ponds in Minnesota, USA: response to hydrologic and environmental variability. Wetlands 33: 245–256

Blaustein L, Kiflawi M et al (2004) Oviposition habitat selection in response to risk of predation in temporary pools: mode of detection and consistency across experimental venue. Oecologia 138:300–305

Boda P, Csabai Z (2013) When do beetles and bugs fly? A unified scheme for describing seasonal flight behaviour of highly dispersing primary aquatic insects. Hydrobiologia 703:133–147

Boix D, Magnusson AK et al (2011) Environmental influence on flight activity and arrival patterns of aerial colonizers of temporary ponds. Wetlands 31:1227–1240

Boros E, Horváth Z, Boda P, Móra A (2013). Macroinvertebrates. In Boros E, Ecsedi Z, Oláh J (eds.), Ecology and Management of Soda Pans in the Carpathian Basin. Chapter 5.6. Hortobágy Environmental Association, 109–112

Boros E, Horváth Z, Wolfram G, Voros L (2014) Salinity and ionic composition of the shallow astatic soda pans in the Carpathian Basin. Ann Limnol Int J Limnol 50:59–69

Boven L, Stoks R, Forro L, Brendonck L (2008a) Seasonal dynamics in water quality and vegetation cover in temporary pools with variable hydroperiods in Kiskunság (Hungary). Wetlands 28:401–410

Boven L, Vanschoenwinkel B et al (2008b) Diversity and distribution of large branchiopods in Kiskunság (Hungary) in relation to local habitat and spatial factors: implications for their conservation. Mar Freshw Res 59:940–950

Bratton JH (1990) Seasonal pools. An overlooked invertebrate habitat. Brit Wildl 2:22–31

Brooks RT (2000) Annual and seasonal variation and the effects of hydroperiod on benthic macroinvertebrates of seasonal forest ("vernal") ponds in central Massachusetts, USA. Wetlands 20:707–715

Brooks RT (2009) Potential impacts of global climate change on the hydrology and ecology of ephemeral freshwater systems of the forests of the northeastern United States. Clim Change 95:469–483

Brose U (2003) Bottom-up control of carabid beetle communities in early successional wetlands: mediated by vegetation structure or plant diversity? Oecologica 135:407–413

Brown ES (1951) The relation between migration rate and the type of aquatic habitat in aquatic insects, with special reference to certain types of Corixidae. Proc Zool Soc Lond 121:539–545

Calhoun AJK, Walls TE, Stockwell SS, McCollough M (2003) Evaluating vernal pools as a basis for conservation strategies: a Maine case study. Wetlands 23:70–81

Collinson NH, Biggs J et al (1995) Temporary and permanent ponds: an assessment of the effects of drying out on the conservation value of aquatic macroinvertebrate communities. Biol Conserv 74:125–133

Croel RC, Kneitel JM (2011) Ecosystem-level effects of bioturbation by the tadpole shrimp *Lepidurus packardi* in temporary pond mesocosms. Hydrobiologia 665:169–181

Davy-Bowker J (2002) A mark and recapture study of water beetles (Coleoptera: Dytiscidae) in a group of semi-permanent and temporary ponds. Aquat Ecol 36:435–446

Dolneger N, Freudenberger L et al (2014) Projecting current and potential future distribution of the Fire-bellied toad *Bombina bombina* under climate change in north-eastern Germany. Regional Environ Change 14:1063–1072

Drake M (2001) The importance of temporary waters for Diptera (true-flies). Freshw Forum 17:26–39

Epele LB, Miserendino ML, Kutschker A (2013) Patagonian wetlands subjected to different intensities of cattle grazing pressure. Does it modify aquatic invertebrate and macrophytes community assemblages? Soc Wetl Sci Annu Meeting 2013, Duluth, Minnesota

Fernando CH, Galbraith D (1973) Seasonality and dynamics of aquatic insects colonizing small habitats. Verh Int Verein Limnol 18:1564–1575

Foster G (1993) Pingo fens, water beetles and site evaluation. Antenna 17:184–195

Friday LE (1987) The diversity of macroinvertebrate and macrophyte communities in ponds. Freshw Biol 18:87–104

Gaitan JJ, Lopez CR, Bran D (2011) Vegetation composition and its relationship with the environment in mallines of north Patagonia, Argentina. Wetl Ecol Manage 19:121–130

Gala TS, Melesse AM (2012) Monitoring prairie wet area with an integrated LANDSAT ETM+, RADARSAT-1 SAR and ancillary data from LIDAR. Catena 95:12–23

Gee JHR, Smith BD, Lee KM, Griffiths SW (1997) The ecological basis for freshwater pond management for biodiversity. Aquat Conserv Freshw Mar Ecosys 7:91–104

Golladay SW, Taylor BW, Palik BJ (1997) Invertebrate communities of forested limesink wetlands in southwest Georgia, USA: habitat use and influence of extended inundation. Wetlands 17:383–393

Hall DL, Willig MR et al (2004) Aquatic macroinvertebrate diversity of playa wetlands: the role of landscape and island biogeographic characteristics. Wetlands 24:77–91

Hart EA, Gotelli NJ (2011) The effects of climate change on density-dependent population dynamic of aquatic invertebrates. Oikos 120:1227–1234

Heino J, Virkkala R, Toivonen H (2009) Climate change and freshwater biodiversity: detected patterns, future trends and adaptations in northern regions. Biol Rev 84:39–54

Horváth Z, Vad CF, Vörös L, Boros E (2013a) Distribution and conservation status of fairy shrimps (Crustacea: Anostraca) in the astatic soda pans of the Carpathian basin: the role of local and spatial factors. J Limnol 72:103–116

Horváth Z, Vad CF, Vörös L, Boros E (2013b) The keystone role of anostracans and copepods in European soda pans during the spring migration of waterbirds. Freshw Biol 58:430–440
Horváth Z, Vad CF et al (2014) Opposing patterns of zooplankton diversity and functioning along a natural stress gradient: when the going gets tough, the tough get going. Oikos 123:461–471
Hoverman JT, Davis CJ et al (2011) Environmental gradients and the structure of freshwater snail communities. Ecography 34:1049–1058
James HG (1967) Seasonal activity of mosquito predators in woodland pools in Ontario. Mosq News 27:453–457
Jeffries MJ (1989) Measuring Talling's 'element of chance in pond populations'. Freshw Biol 21:383–393
Jeffries MJ (1992) The ecology and conservation value of forestry ponds in Scotland, United Kingdom. Biol Conserv 58:191–211
Jeffries MJ (1994) Invertebrate communities and turnover in wetland ponds affected by drought. Freshw Biol 32:603–612
Jeffries MJ (2005) Local scale turnover of pond insects: intra-pond habitat quality and inter-pond geometry are both important. Hydrobiologia 543:207–220
Jeffries MJ (2008) The spatial and temporal heterogeneity of macrophyte communities in thirty small, temporary ponds over a period of ten years. Ecography 31:765–775
Jeffries MJ (2011) The temporal dynamics of temporary pond macroinvertebrate communities over a 10-year period. Hydrobiologia 661:391–405
Jones I (2013) The impact of extreme events on freshwater ecosystems. Brit Ecol Soc, London
Jones K, Lanthier Y et al (2009) Monitoring and assessment of wetlands using Earth observation: the GlobWetland project. J Environ Manage 90:2514–2569
Kenk R (1949) The animal life in temporary and permanent ponds in southern Michigan. Museum Zool, Univ Michigan, http://deepblue.lib.umich.edu/bitstream/handle/2027.42/56316/MP071.pdf?sequence=1
Kneitel JM (2014) Inundation timing, more than duration, affects the community structure of California vernal pool mesocosms. Hydrobiologia 732:71–83
Kutschker AM, Epele LB, Miserendio ML (2014) Aquatic plant composition and environmental relationships in grazed Northwest Patagonian wetlands, Argentina. Ecol Eng 64:37–48
Lake PS (2011) Drought and aquatic systems. Effects and Responses, John Wiley, Chichester
Lamborn WA (1922) The mosquitos of some ports of China and Japan. Bull Entomol Res 12:401–409
Lawton JH (1999) Are there general rules in ecology? Oikos 84:177–192
Lehnner B, Döll P (2004) Development and validation of a global data base of lakes, reservoirs and wetlands. J Hydrol 296:1–22
Leibold MA, Holyoak M et al (2004) The metacommunity concept: a framework for multi-scale ecology. Ecol Lett 7:60–613
Lott D (2001) Ground beetles and rove beetles associated with temporary ponds in England. Freshw Forum 17:40–53
Macan TT (1939) The Culicidae of the Cambridge District. Parasitology 31:263–269
MacArthur R (1972) Geographical ecology. Patterns in the Distribution of Species, Princeton Univ Press
March F, Bass D (1995) Application of island biogeography theory to temporary pools. J Freshw Ecol 10:83–85
Mozley A (1937) The ponds, lakes and streams of the Kirghiz Steppe. Scot Geogr Mag 53:1–10
Mozley A (1944) Temporary ponds, a neglected natural resource. Nature 144:490
Nicolet P (2001) Temporary ponds in the UK: a critical biodiversity resource for freshwater plants and animals. Freshw Forum 17:16–25
Nicolet P, Biggs J et al (2004) The wetland plant and macroinvertebrate assemblages of temporary ponds in England and wales. Biol Conserv 120:261–278
O'Neill BJ, Thorp JH (2014) Untangling food-web structure in an ephemeral ecosystem. Freshw Biol 59:1462–1473
Perotti MG, Diéguez MC, Jara F (2005) Estado del conocimiento de humedales del norte patagónico (Argentina): aspectos relevantes e importancia para la conservación de la biodiversidad regional. Revista Chilena Histor Natur 78:723–737

Pessacq P, Epele LB, Rogers C (2011) A new species of *Lynceus* (Crustacea: Branchiopoda:Laevicaudata) from Patagonia, with comments on laevicaudatan systematics. Zootaxa 3043:25–32

Peters BG (1938) Habitats of *Limnaea truncatula* in England and Wales during dry seasons. J Helminthol 16:213–260

Pitt AL, Baldwin RF et al (2012) The missing wetlands: using local ecological knowledge to find cryptic ecosystems. Biodiver Conserv 21:51–63

Ratcliffe D (1977) A nature conservation review. Cambridge Univ Press, Cambridge

Rees WJ (1965) The aerial dispersal of mollusc. Proc Malacol Soc Lond 36:269–282

Root FM (1924) Notes on the mosquitoes of Lee County, Georgia. Am J Hygiene 4:449–455

Rosset V, Lehmann A, Oertli B (2010) Warmer and richer? Predicting the impact of climate warming on species richness in small temperate waterbodies. Global Change Biol 16:2376–2387

Rouen K (2001) European temporary ponds: a threatened habitat. Freshw Biol Assoc, Ambleside

Rowe G, Beebee TJC (2005) Intraspecific competition disadvantages inbred natterjack toad (*Bufo calamita*) genotypes over outbred ones in a shared pond environment. J Anim Ecol 74:71–76

Schneider DW (1997) Predation and food web structure along a habitat duration gradient. Oecologia 110:567–575

Schneider DW (1999) Snowmelt ponds in Wisconsin: influence of hydroperiod on invertebrate community structure. In: Batzer DP, Rader RB, Wissinger SA (eds) Invertebrates in freshwater wetlands of North America: ecology and management. John Wiley and Sons, New York, pp 299–318

Schneider DW, Frost TM (1996) Habitat duration and community structure in temporary ponds. J N Am Benthol Soc 15:64–86

Senior-White R (1925) Physical factors in mosquito activity. Bull Entomol Res 16:187–248

Stav G, Blaustein L, Magalit Y (2005) Individual and interactive effects of a predator and controphic species on mosquito populations. Ecol Appl 15:587–598

Stendera S, Adrian R et al (2012) Drivers and stressors of freshwater biodiversity patterns across different ecosystems and scales: a review. Hydrobiologia 696:1–28

Stevens MM, Warren GN, Braysher BD (2003) Oviposition response of *Chironomus tepperi* to nitrogenous compounds and bioextracts in two-choice laboratory tests. J Chem Ecol 29:911–920

Studinski JM, Grubbs SA (2007) Environmental factors affecting the distribution of aquatic invertebrates in temporary ponds in Mammoth Cave National Park, Kentucky, USA. Hydrobiologia 575:511–520

Tiner RWJ (1987) Mid-Atlantic wetlands, a disappearing natural treasure. US Dept Interior, Fish Wildl Serv, Washington, DC

Tóth A, Horváth Z et al (2014) Zooplankton of the European soda pans: Fauna and conservation of a unique habitat type. Int Rev Hydrobiol 99:255–276

Vellend M (2010) Conceptual synthesis in community ecology. Quarterly Rev Biol 85:183–206

Watson M (1921) The prevention of malaria in the Federated Malay States: a record of twenty years progress. E.P. Dutton and Co, New York

Wellborn GA, Skelly DK, Werner EE (1996) Mechanisms creating community structure across a freshwater habitat gradient. Annu Rev Ecol Syst 27:337–363

Werner EE, Anholt BR (1996) Predator induced behavioural indirect effects: consequences to competitive interactions in anuran larvae. Ecology 77:157–169

Wiggins GB, Mackay RJ, Smith LM (1980) Evolutionary and ecological strategies of animals in annual temporary ponds. Archiv Hydrobiol Suppl 58:97–206

Wilbur HM (1997) Experimental ecology of food webs: complex systems in temporary ponds. Ecology 78:2279–2302

Williams DD (1983) The natural history of a Nearctic temporary pond in Ontario with remarks on continental variation in such habitats. Int Rev Gesamten Hydrobiol Hydrogr 68:239–253

Williams DD (1997) Temporary ponds and their invertebrate communities. Aquat Conserv Mar Freshw Ecosyst 7:105–117

Williams P, Biggs J et al (2001) History origins and importance of temporary ponds. Freshw Forum 17:7–15

Williams DD, Heeg N, Magnusson AK (2007) Habitat background selection by colonising intermittent pond invertebrates. Hydrobiologia 592:487–498

Wolfram G, Donabaum K, Schagerl M, Kowarc VA (1999) The zoobenthic community of shallow salt pans in Austria—preliminary results on phenology and the impact of salinity on benthic invertebrates. Hydrobiologia 408(409):193–202

Yee DA, Taylor S, Vamosi SM (2009) Beetle and plant density as cues initiating dispersal in two species of adult predaceous diving beetle. Oecologia 160:25–36

Chapter 5
Invertebrates of Freshwater Temporary Ponds in Mediterranean Climates

Dani Boix, Jamie Kneitel, Belinda J. Robson, Claire Duchet, Luís Zúñiga, Jennifer Day, Stéphanie Gascón, Jordi Sala, Xavier D. Quintana, and Leon Blaustein

Introduction to Mediterranean Temporary Ponds

Long Known But Neglected Until Recently

Temporary ponds have long attracted the attention of Mediterranean people. In "History of Animals," Aristotle described the seemingly spontaneous generation of life from mud or sand using observations from a temporary pond ("fish" probably refers to the aquatic invertebrates, such as fairy shrimp, that he was observing):

> *The great majority of fish, then, as has been stated, proceed from eggs. However, there are some fish that proceed from mud and sand, even of those kinds that proceed also from pairing and the egg. This occurs in ponds here and there, and especially in a pond in the neighbourhood of Cnidos. This pond, it is said, at one time ran dry about the rising of the Dogstar, and the mud had all dried up; at the first fall of the rains there was a show of water in the pond, and on the first appearance of the water shoals of tiny fish were found in the pond.* [Aristotle (350 BC) Book VI. Chapter 15. In: The History of Animals (translated in 1907 by D'Arcy Wentworth Thompson. London: John Bell)]

The lack of scientific publications on Mediterranean temporary ponds does not reflect their ecological importance (Grillas et al. 2010), and research in these environments has increased significantly over the last two decades (Boix et al. 2012). The absence of comparative studies on aquatic ecosystems in regions with a Mediterranean-type climate has been reported (Gasith and Resh 1999) and partially addressed for Mediterranean-climate streams and rivers (Bonada and Resh 2013). However, studies have emphasized the importance, abundance, functioning, and biodiversity of Mediterranean temporary ponds in the last several decades (e.g., Balla and Davis 1995; Witham 1998; Diget and Rioux 1998; Blaustein and Schwartz 2001), highlighting the need to develop conservation plans for these systems (Zedler 2003; Grillas et al. 2004; Fraga et al. 2010; Sancho and Lacomba 2010).

D. Batzer, D. Boix (eds.), *Invertebrates in Freshwater Wetlands*,
DOI 10.1007/978-3-319-24978-0_5

Mediterranean Biome: The Five Mediterranean Regions

The term "Mediterranean" as a biome is not recognized in the same way as other biomes (e.g., boreal forest, desert, or savannah). Although it is named for a particular geographic region, the term is used to designate regions with a similar climate around the world. Mediterranean climates are typically located in a narrow latitudinal band between 30° and 40°, although in the Mediterranean basin itself the biome expands to 45° (Bolle 2003). This habitat is located on five continents (Eurasia, Africa, South and North America, and Australia), and thus five biogeographic regions (Palearctic, Afrotropic, Neotropic, Nearctic, and Australasian) (Fig. 5.1). The concept of a Mediterranean biome remains controversial (Stamou 1998) because some define the habitat with a definite drought period, while others take into account two features: dry summers and cool winters. However, biogeographic studies have clearly identified Mediterranean forests, woodlands, and scrub, as well as wetlands, since they share similar bioclimatic conditions.

The area covered by the Mediterranean biome is quite small compared to other biomes (Olson et al. 2001). It covers only 5 % of the earth's surface (Fig. 5.1), and the subregions vary tremendously in size (Underwood et al. 2009). South and North America and South Africa combined are small, roughly 100,000 km^2 each (11 % of the Mediterranean area). The south-western Australian Mediterranean-climate region covers more than 750,000 km^2 (25 % of this biome) divided into two separate areas. However, the biome has its maximum extent (64 % of the total area covered by this biome) around the Mediterranean Sea, from which it takes its name. The Circum-Mediterranean region forms an incomplete belt around the Mediterranean Sea (Fig. 5.1), with the maximum width of this belt located on the Iberian Peninsula.

Temporary water bodies predominate in Mediterranean regions because of their typically long, hot, dry summers. Their widespread occurrence and abundance therefore make temporary habitats the characteristic wetlands of this biome. In this chapter, we use "Mediterranean temporary ponds" in a broad sense, including all the temporary freshwater wetlands located in Mediterranean regions, and not restricted by any legal definition. For example, the European Habitats Directive (European Directive 92/43/CEE) considers "Mediterranean temporary ponds" a priority habitat to conserve, but the Directive defines this wetland type strictly by the presence of particular plant species (European Commission 2003; Bagella et al. 2007). Curiously, this protected habitat is identified in the United Kingdom (McAbendroth 2004), outside of the Mediterranean-climate region.

Climate: Hot Dry Summers and Mild Winters

Mediterranean climates are transitional between temperate and tropical climates, hence their prevalence between 30° and 40° of latitude (Aschman 1973; Daget 1984). The climate can be summarized as follows:

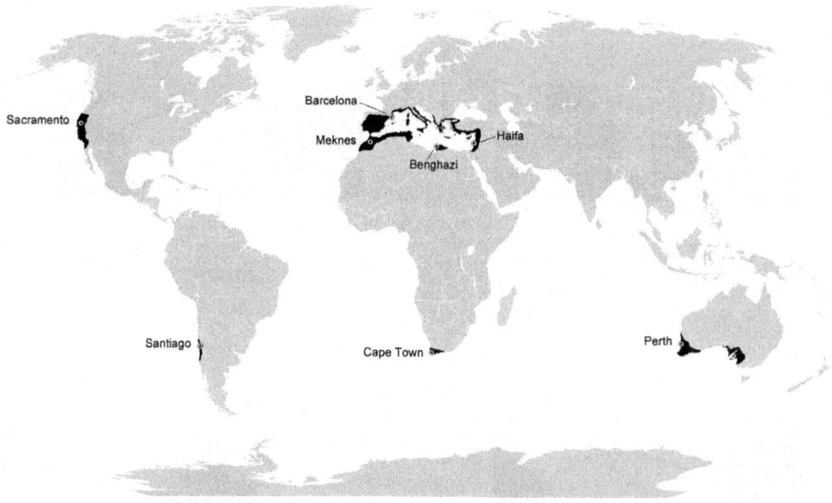

Fig. 5.1 World map showing the location of Mediterranean-climate regions. The sites indicated on the map correspond to the ombrothermic diagrams shown in Fig. 5.2

1. Hot summers which coincide with the dry period
2. Mild, but cool winters
3. High temporal variability (i.e., high intra- and inter-annual differences)

These characteristics determine the hydrology and ecological functioning of the wetlands. The combination of dry weather and high temperatures causes large water-level fluctuations (Álvarez-Cobelas et al. 2005; Beklioglu et al. 2007) and thereby regulates the presence of temporary waters in Mediterranean areas. The mild winters mean that most Mediterranean wetlands remain ice-free (Britton and Crivelli 1993). Freezing of wetlands affects both limnological processes (e.g., thermal or oxygen vertical gradients in the water column) and the availability of habitat and resources for the biota. This particular combination of annual drying and the absence of freezing typify Mediterranean wetlands and distinguish them from wetlands in most other biomes. Moreover, high inter-annual climate variability is also characteristic of Mediterranean regions (Gasith and Resh 1999) (Fig. 5.2).

Despite the narrow latitudinal range and small area of Mediterranean regions, extreme gradients of aridity (i.e., duration of the dry period) and temperature exist (Bolle 2003) (Fig. 5.2). Those gradients have been used to distinguish various climate types and are used to classify Mediterranean climates (Stamou 1998):

1. Duration of the dry period
   (a) Perarid: 11–12 months
   (b) Arid: 9–10 months

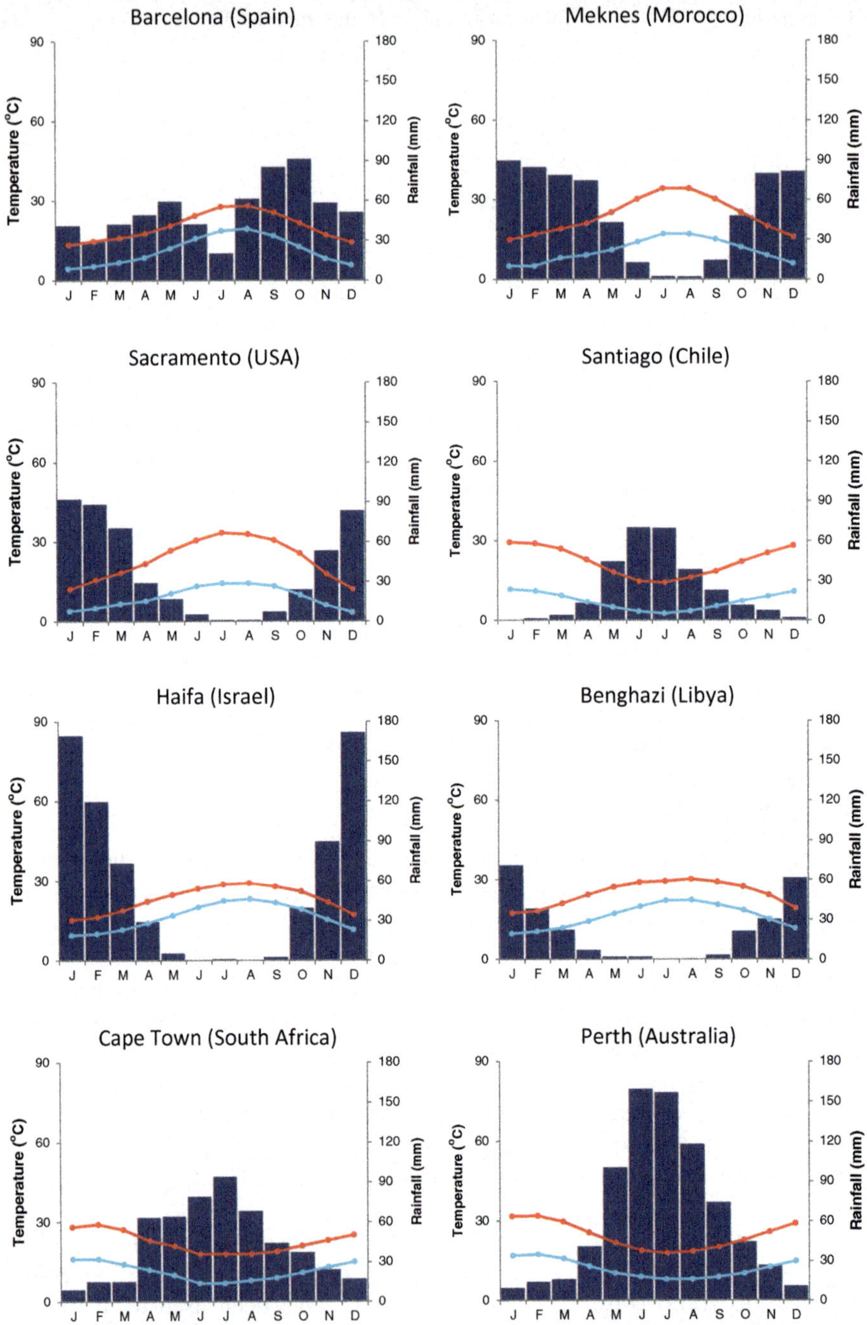

Fig. 5.2 Ombrothermic diagrams for eight locations in Mediterranean-climate regions. *Upper* diagrams correspond to sites in the Mediterranean Basin and *lower* diagrams correspond to sites located in other Mediterranean regions. Monthly mean maximum (*red line*) and minimum (*blue line*) temperature, and monthly mean rainfall (*bars*) were shown. The high intra-annual variability characteristic of mediterranean climates is evident, as well as variability among and within regions (i.e., Mediterranean Basin)

   (c) Semiarid: 7–8 months
   (d) Subhumid: 5–6 months
   (e) Humid: 3–4 months
   (f) Perhumid: 1–2 months

2. Mean temperature of the coldest month

   (a) Warm winter: more than 7 °C (no periods of freezing)
   (b) Mild winter: between 3 and 7 °C (freezing is rare)
   (c) Cold winter: between 0 and 3 °C (freezing is frequent)
   (d) Severe winter: less than 0 °C (freezing may be extensive)

Division of Mediterranean climates into subclasses using a numerical approach has been proposed by Nahal (1981). The use of Ebergers' index of drought (Q) is the most widely accepted approach:

$$Q = \frac{2P}{(M+m)(M-m)}$$

where
P is annual rainfall
M the mean maximum temperature of the hottest month
m the mean minimum temperature of the coldest month
Thus, we can distinguish the following subclasses of Mediterranean climate:

1. Arid: Q values between 20 and 30; annual rainfall between 300 and 500 mm
2. Semiarid: Q values between 30 and 50; annual rainfall between 500 and 700 mm
3. Subhumid: Q values between 50 and 90; annual rainfall between 700 and 1000 mm
4. Humid: Q values more than 90; annual rainfall more than 1000 mm

Mediterranean regions are characterized by an annual precipitation in the range of 275 to 900 mm with over 65% occurring in the 6 colder months (Aschman 1973). However, Mediterranean areas with annual rainfall less than 275 mm do exist (e.g., in Morocco, Spain, and Israel). Two seasonal patterns of precipitation have been described in Mediterranean regions: (1) a single rainfall maximum in winter, and (2) two rainfall maxima, one in spring and the other in autumn. In those Mediterranean regions with strong oceanic influences, winter rainfalls are heavy due to storms moving from high latitudes towards the equator (e.g., southern Australia). In contrast, in areas with less oceanic influence (e.g., coast of the western Mediterranean basin), the storms are weaker, so winter rains are meagre, while spring and autumn rains that are generated by cold air masses are more important to hydrology (Ferrés 1993).

Hydrology

Mediterranean temporary wetlands are usually flooded by rainfall and evaporation is the primary mechanism of drying (Fig. 5.3a). However, there are exceptions. Groundwater-fed temporary wetlands (Fig. 5.3b) are abundant in the southern Australia (Sim et al. 2006a; Horwitz et al. 2009; Boulton et al. 2014) and the African Cape (Mlambo et al. 2011; de Moor and Day 2013) Mediterranean regions, and they are present in some areas of the Mediterranean basin (Rodríguez-Rodríguez 2007). Groundwater-fed temporary wetlands are also found in karstic limestone areas (Boix et al. 2001). Temporary pools or riverine-floodplain wetlands resulting from flash floods (ultimately caused by rains), or by springs that vary in output, resulting in areas that flood (sometimes with fossil water), recede and dry (Issar 1990) (Fig. 5.3c).

Although rain and evaporation are the main inputs and outputs, respectively, freshwater temporary ponds rarely accumulate ions in their basins. Runoff of ions (on the surface or via ground water) must therefore occur from these waterbodies. However, athalassic lakes and ponds (saline water isolated from the sea, sensu Williams 1981) occur in endorheic basins of the more arid locations of Mediterranean climates, such as the Iberian Peninsula, North Africa, California, Australia, and South Africa. In these systems, ions accumulate in the basin because water output occurs only by evaporation. Groundwater-dependent wetlands have water levels that rise and fall seasonally as groundwater tables fluctuate and thus moderate concentrations of ions develop as water levels decline.

Several classifications for temporary ponds have been proposed using hydrological patterns. Duration and predictability of the hydroperiod are frequently used criteria (Comín and Williams 1994; Keeley and Zedler 1998; Boulton et al. 2014; Williams 2006). Our proposal, following Boulton et al. (2014), distinguishes five types of temporary lentic waters:

(a) *Ephemeral*: Filled only after unpredictable rain and by runoff. The flooded area dries out during the days following the flooding and supports low numbers of macroscopic aquatic species.
(b) *Episodic*: Dries in 9 out of 10 years, with rare and irregular flooding (or wet periods) which may last for a few months.
(c) *Intermittent*: Alternating wet and dry periods, but a more irregular frequency of filling than seasonal wetlands. Flooding may persist for months or years.
(d) *Seasonal*: Alternating wet and dry periods annually, in accordance with the season. Usually fill during the wet season of the year, and dry out in a predictable way every year. The flooding lasts for several months, long enough for macroscopic animals and plants to complete their life cycles.
(e) *Near-permanent*: Predictable flooding, though water levels may vary. The annual input of water is greater than the losses (does not dry out) in 9 out of 10 years. The majority of organisms living here cannot tolerate desiccation.

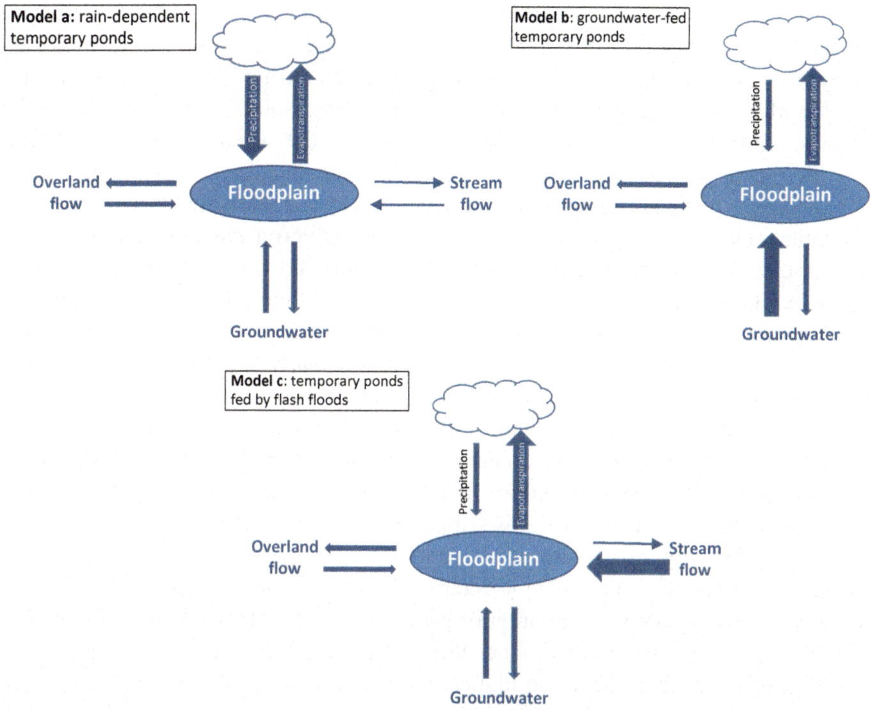

Fig. 5.3 Water budget diagram characteristics of Mediterranean temporary ponds. Model (**a**) represents rain-dependent temporary ponds such as vernal pools; model (**b**) represents groundwater-fed temporary ponds unconnected to rivers commonly found across southern Australia and Cape Mediterranean region, and also present in karst geologies; and model (**c**) shows temporary ponds fed by flash-floods (occurring in some countries of the Mediterranean basin (e.g., Israel))

In this chapter we focus on two of these types, intermittent and seasonal, because these are the types most commonly found in Mediterranean regions. However, these categories represent points along a gradient. For example, some waterbodies dry more often than 1 year in 10 (e.g., 1 year in 3). For these types of ponds, the term "semipermanent" (dry but not every year) is commonly used (e.g., Stewart and Kantru 1972; Collinson et al. 1995; Gascón et al. 2005).

Invertebrate Assemblages of Mediterranean Temporary Ponds

The similar climatic conditions among Mediterranean regions have led to convergent evolution in the flora (Matesanz and Valladares 2014) characterized in part by remarkably high species diversity. However, the regions' isolation and geologic history have resulted in differences in the aquatic fauna (Bonada et al. 2008). For example, fauna groups (e.g., aquatic coleopterans) are characterized by very restricted distributions in the Mediterranean basin region, very likely a consequence of two evolutionary processes: (1) the Mediterranean basin as a refugium during the glacial maxima and (2) the Mediterranean as an area of endemism during isolation in the glacial cycles (Hewitt 2000; Ribera et al. 2003). Similarly, the south-western Australian Mediterranean-climate region has been isolated by surrounding arid zones for millennia, and as a result of this isolation is depauperate in some aquatic fauna, such as stoneflies (Davies and Stewart 2013). Therefore, Mediterranean temporary ponds are characterized by a unique combination of isolation and connectedness at different spatial scales, which can result in the evolution of endemic species (Zedler 2003).

Appendix 1 provides a list of the families of aquatic invertebrates reported from the five Mediterranean regions. The list is geographically and taxonomically biased due to different research efforts among regions and taxonomic groups. The data were obtained in an extensive review of the literature and complemented by unpublished data of the authors (see Appendix 1 for references). Biodiversity (number of genera) in Mediterranean temporary ponds is dominated by arthropods: primarily insects and secondarily crustaceans (Appendix 1). In the case of insects, two families have the highest biodiversity in all Mediterranean regions: Dytiscidae (Coleoptera) and Chironomidae (Diptera). Another six insect families possess at least ten genera (in decreasing order of genera richness): Hydrophilidae, Corixidae, Libellulidae, Ceratopogonidae, Coenagrionidae, and Hydraenidae. Similarly, two families of crustaceans have the highest biodiversity in all Mediterranean regions: Chydoridae (Branchiopoda) and Cyprididae (Ostracoda). These two families are followed by two copepod families: Cyclopidae and Diaptomidae. For non-arthropods, two other families have high genera richness: typhloplanid turbellarians and planorbid gastropods. This pattern of richness among taxonomic groups is in accordance with published comparisons among temporary ponds around the world, regardless of the climate region (e.g., Boix et al. 2001; Boix and Sala 2002; Williams 2006).

Large Branchiopods: Flagship Invertebrate Species of Mediterranean Temporary Ponds

Large branchiopods are a group of crustaceans that almost exclusively inhabit temporary ponds (Hartland-Rowe 1972) and they have a worldwide distribution (Brendonck et al. 2008). In the Mediterranean they are often considered a flagship group of invertebrates for temporary ponds (Belk 1998; Thiéry 2004) and they are promoted to monitoring ecological status of temporary ponds, since they are very sensitive to habitat and landscape degradation (Gascón et al. 2012; van den Broeck et al. 2015a, b). Appendix 2 provides a list (and references) for the large branchiopod species present in Mediterranean ponds to show the high species richness of this group and the high level of endemism. We only included species specifically observed in the temporary ponds located in the Mediterranean biome, below 1500 m.a.s.l., and in fresh water. Species present in the Mediterranean basin but only located at high altitude (e.g., *Chirocephalus algidus*, *C. marchesonii*, *C. ruffoi*, *C. sibyllae* or *C. tauricus*; Cottarelli and Mura 1983; Cottarelli et al. 2010) were not included, nor were saline species (the genus *Parartemia* in Australia has up to 15 species in Australian Mediterranean regions, 11 of them being endemic: Timms 2014 and references therein). Some early-described species are not included in the Appendix because our existing knowledge consists only of very old records, and taxonomic revision is probably needed; this is especially true for the Spinicaudata (Hartland-Rowe 1967; Thiéry 1996) and for some Anostraca of the Mediterranean basin area (such as *Chirocephalus recticornis* and *Chirocephalus festae*; Brauer 1877; Colosi 1922). Only species that appear in the scientific literature during the second half of the twentieth century have therefore been included.

Large branchiopods are known in four of the five Mediterranean regions (no species have been identified from the South American Mediterranean region) and their species richness is high (Appendix 2). The degree of endemism is also quite high, except in South Africa, where only a single endemic species has been recorded. The proportion of endemic species is 55 %, 46 %, and 45 % in the Mediterranean basin, southern Australia, and North America, respectively. At generic level, the anostracan *Linderiella* (California and Mediterranean basin) and the spinicaudatan *Maghrebestheria* (Mediterranean basin) can be considered as endemic to the Mediterranean regions. Some genera have broad distributions across several Mediterranean regions, although endemic species within these genera occur. They represent an example of faunal complexes persisting over millennia with locally adapted endemic species (Keeley and Zedler 1998). This is especially evident in the Notostraca (e.g., *Triops*), Spinicaudata (e.g., *Cyzicus*, *Eulimnadia*), and Laevicaudata (*e.g.,Lynceus*), although it also occurs in some anostracan genera (e.g., *Streptocephalus*, *Branchinecta*). It is interesting to note that some genera are highly speciose in a particular Mediterranean region, such as *Streptocephalus* and *Branchipodopsis* in South Africa, *Branchinecta* in North America, *Branchinella*, *Eulimnadia*, and *Limnadopsis* in southern Australia, and *Chirocephalus*, *Tanymastigites* and *Triops* in the Mediterranean basin.

Dynamics of Invertebrate Assemblages of Mediterranean Temporary Ponds

Key Environmental Factors

Hydroperiod is an important factor determining the faunal composition and structure of Mediterranean temporary ponds (e.g., Boix et al. 2004; Ripley and Simovich 2009; Sim et al. 2013; Kneitel 2014). However, pond size is also considered a determinant factor (e.g., Ebert and Balko 1987; March and Bass 1995; Meintjes 1996; Spencer et al. 1999). Indirect effects of pond size on community structure (i.e., larger ponds have different environmental characteristics than smaller ones) seems to be weak in temporary Mediterranean ponds (Ballón et al. in press). The effects of both hydroregime and habitat size depend on the dispersal modes of the taxa.

Hydroregime has been found to be more important for passive dispersers than for active dispersers (Vanschoenwinkel et al. 2009), for instance, and interactions between inundation length and timing can differentially affect dispersal modes (Kneitel 2014). Further, nested community patterns have been observed in Mediterranean temporary ponds and found to be most associated with environmental variation (hydroperiod and pond size), but the presence of species with poor dispersal abilities can also increase nestedness (Ripley and Simovich 2009; Florencio et al. 2011).

Water quality has also been related to community structure, although it shows an inconsistent pattern over invertebrate successional phases. Water quality fluctuations probably coincide with changes in invertebrate assemblages, rather than causing them (Barclay 1966; Meintjes 1996). The main physical and chemical variables related to species richness and community composition in Mediterranean temporary ponds are salinity (Boix et al. 2008; Waterkeyn et al. 2009; Mlambo et al. 2009), turbidity (Alonso 1998; Mlambo et al. 2009; Ruhí et al. 2014), light (Mokany et al. 2008) and nutrient concentration (Balla and Davis 1995; Mlambo et al. 2009). Although local pond characteristics, and intra- and inter-annual variability are of greater importance than biological factors for shaping the physical and chemical characteristics of temporary ponds, bottom-up and top-down trophic effects are also influential (e.g., Magnusson and Williams 2006), but few studies have specifically addressed this topic in Mediterranean temporary ponds (but see Balla and Davis 1995; Waterkeyn et al. 2013).

Seasonal Succession

Successional Phases

Change in community composition during inundation was the focus of early studies in temperate temporary ponds of the northern hemisphere (e.g., Murray 1911; Mozley 1932; Kenk 1949). These pioneering studies built a general conceptual model, which described the seasonal succession of invertebrate communities. This model has subsequently been improved and/or validated in studies in temperate regions of the southern hemisphere (Barclay 1966; Lake et al. 1989) and in more arid zones with short hydroperiods (e.g., Rzóska 1961; Meintjes 1996; Lahr et al. 1999).

Kenk (1949) identified several different community structures during the hydroperiod of temporary ponds in cool-temperate Michigan (USA) and described them as "stages or phases." Two aquatic phases were described based on water temperature and season: the cold-water and the spring phase. In the southern hemisphere, in New Zealand, Barclay (1966) also observed similar time periods with distinguishable temporary pond communities (autumn-winter stage and spring-early summer stage), but described a third phase during drying (dry stage). Lake et al. (1989) constructed a three-phase conceptual model in which phases were called: "filling," "middle or aquatic" and "drying." The community dynamics of filling and drying phases represent allogenic succession driven by environmental processes. During flooding, resources become abundant, and biotic interactions have limited effect on community composition. During drying, environmental variables change dramatically (e.g., temperature increases, dissolved oxygen fluctuates), and although higher faunal densities intensify biotic interactions, changes in community composition are again mostly driven by environmental conditions. However, the succession observed between these two phases is autogenic, that is, the result of biological processes (i.e., predation and competition). Case studies of faunal community succession in Mediterranean basin ponds are consistent with the three-phase model (Bazzanti et al. 1996; Boix et al. 2004; Florencio et al. 2009; Sahuquillo and Miracle 2010). Additionally, comparisons among the same phases of different hydroperiods suggest that secondary changes in community composition related to season exist in Mediterranean temporary ponds (Yaron 1964; Boix et al. 2004; Culioli et al. 2006). In these studies, invertebrate species compositions for the three successional phases were identified, but species were also identified that only appeared in spring or in autumn-winter hydroperiods. Similarly, in a temporary wetland in the south-western Australian Mediterranean region, Strachan et al. (2014) observed three distinct phases of invertebrate community structure during the process of wetland drying and reflooding, with an almost complete turnover of species between hydrologically defined phases (damp phase, sediment damp but surface water absent; dry phase, groundwater at its lowest level and sediment surface dry; reflooded phase, surface water had returned). These results suggest that groundwater-fed Mediterranean temporary ponds are likely to show very large shifts in community composition over short time periods driven by the marked hydrological changes that occur there.

The study of temporal changes in community structure based on body size is complementary, rather than a replacement for the taxonomic approach (Rodríguez and Magnan 1993). Successional analyses based on taxonomic or functional approaches clearly show temporal changes in the community related to different life-history strategies of the organisms, such as dispersal or resistance to drought (Kenk 1949; Wiggins et al. 1980). In contrast, succession analyses based on body size emphasize the temporal changes in the community related to trophic structure (Quintana et al. 2015). In Mediterranean temporary ponds, differences in biomass-size spectra were observed during succession, while permanent ponds remained more static (Solimini et al. 2005). Size-based approaches have also been used to identify successional phases (Boix et al. 2004).

Mediterranean Versus Cold-Temperate Regions

Although temporary ponds are very important in arid and semiarid areas, such as Mediterranean-climate regions, most ecological knowledge, and the resulting paradigms are biased by studies developed in cold-temperate areas. The general successional models for temporary pond communities may be valid for

Mediterranean ponds, but important distinctions must be recognized. First, in contrast with the mild winters in Mediterranean region, ponds in cold-temperate regions freeze and become snow-covered in winter (e.g., Kenk 1949; Wiggins et al. 1980; Boix et al. 2011) (Fig. 5.4). Consequently, these colder temperate regions may have lower densities of invertebrates and decreases in certain faunal activities, such as aerial dispersal, when compared to Mediterranean temporary ponds (see Ruhí et al. 2012 for an example in created wetlands). Second, the life-history traits and distribution of species differ in cold-temperate and Mediterranean regions (Ruhí et al. 2013a). Species in Mediterranean regions tend to have narrower thermal tolerances and allocate more to reproduction and resistance than species found in colder habitats (Ruhí et al. 2012). Third, precipitation, and hence hydroperiod, is more variable in Mediterranean ponds: annual rainfall vary markedly in some regions across years, and a deviation of 30 % or more from a long-term average is not uncommon (Gasith and Resh 1999; Florencio et al. 2009; Sahuquillo and Miracle 2010; Chester and Robson 2011). Moreover, this inter- and intra-annual variability can result in

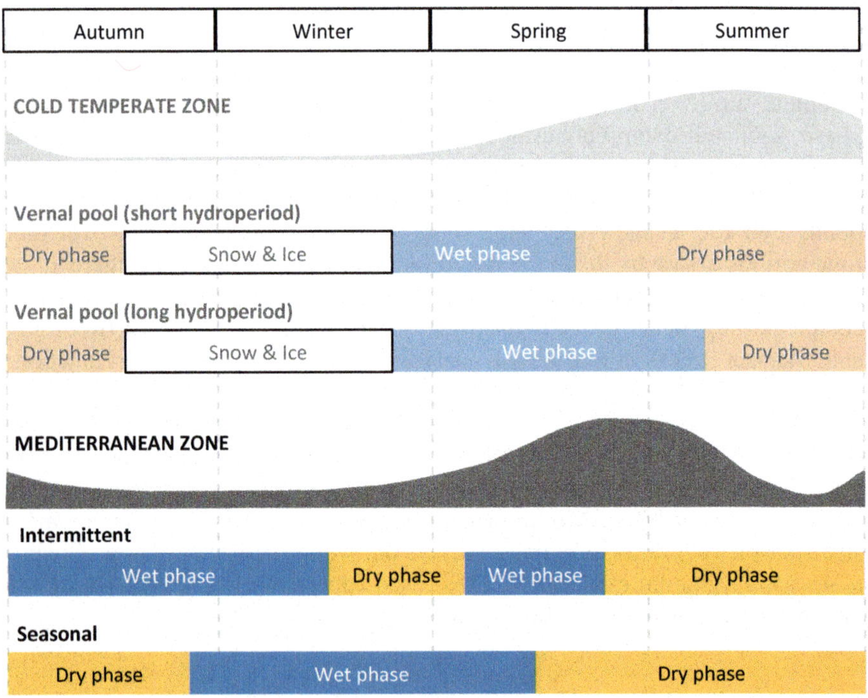

Fig. 5.4 Schematic diagram of temporary pond hydroperiods in cold-temperate and some Mediterranean regions (duration of wet phases can be different depending on the geographical or inter-annual variability of each site). Shaded area indicates favorable environmental conditions for aerial colonizers. In cold-temperate regions, an increase in hydroperiod length implies an increasing number of days with good environmental conditions for aerial dispersers, but this is not the case in all Mediterranean regions

basins flooding in autumn, winter, spring, and occasionally in summer, and ponds can also dry in different seasons, sometimes remaining dry for consecutive years (Fig. 5.4). In contrast, many vernal and autumnal temporary ponds in cold-temperate regions follow a more consistent seasonal pattern, both drying in summer and flooding in spring or autumn, respectively (Wiggins et al. 1980).

Trait-Based Groups During Succession

In cold-temperate temporary ponds, community structure exhibits consistent temporal patterns of functional-feeding groups. Initially, filtering collectors, gathering collectors, and shredders dominate followed by increasing predator densities with increasing hydroperiod (Wiggins et al. 1980; Williams 1983). The majority of predator species cannot persist in the temporary pond during the dry phase. Moreover, predation pressure tends to increase with increasing hydroperiod length (Schneider and Frost 1996). These patterns are commonly associated with changes in resource availability (Wiggins et al. 1980) and the hydrological limitations imposed on certain functional-feeding and life-history groups (Schneider and Frost 1996; Higgins and Merrit 1999). There are similarities between cold-temperate and Mediterranean temporary ponds, but there are many differences.

Temporal patterns of invertebrate composition in temporary Mediterranean ponds can be explained by life-history strategies (Wiggins et al. 1980; see also Chap. 1). Passive dispersers that are desiccation-resistant often dominate at the beginning of the hydroperiod, whereas taxa that are non-desiccation-resistant active dispersers may dominate in both the beginning and in the final stages of the hydroperiod (Culioli et al. 2006; Boix et al. 2009). The dominance of detritivores at the beginning and of predators at the end of the hydroperiod has been observed in many Mediterranean and arid ponds (Lahr et al. 1999; Culioli et al. 2006), although this pattern is not ubiquitous (Bazzanti et al. 1996; Spencer et al. 1999; Boix et al. 2004). Other sequences have been observed, including the dominance of desiccation-resistant taxa at both the beginning (aerially colonizing insects) and in the middle (passive-colonizers) of the hydroperiod (Bazzanti et al. 1996). An absence of a pattern resulting in similar proportions among all life-history groups can also develop (Sim et al. 2013). Predators can be present at the beginning of the hydroperiod after mild winters, or the dominant predators can emerge from pond sediment rather than via flight (i.e., they have drought-resistant stages). For example, when Mediterranean temporary ponds fill in winter, heteropterans (mainly Corixidae) colonize quickly (e.g., Bazzanti et al. 1996; Boix et al. 2001; Florencio et al. 2009). Additionally, it is known communities in which the main predators, such as the notostracan *Triops cancriformis* and the coleopteran *Agabus nebulosus*, are desiccation-resistant and become active during initial flooding (Boix et al. 2006). In these communities the highest predation pressure occurred 20–25 days after flooding with no increase over the rest of the hydroperiod.

High inter-annual variability of hydroperiod length in Mediterranean ponds causes variability in life-history groups (Boix et al. 2009; Sim et al. 2013; Kneitel

2014). Sim et al. (2013) proposed a conceptual model of the relationships between hydroperiod length, proportions of the different life-history groups, and the dominant recolonization strategy (dispersal or egg-bank hatching). The model showed how hydroperiod variability could interact with life-history traits that result in community structural differences (see also Chap. 1). One persistent limitation to progress in understanding these dynamics is the lack of life-history information for many freshwater invertebrates (Robson et al. 2011). Recent studies show that invertebrates might be capable of adapting to variable hydroperiods in ways not previously understood (e.g., Strachan et al. 2015).

Dry Period: Drought Resistance and Dispersal

Drought-tolerance, microrefuges, or high dispersal capacity allow aquatic animals of temporary ponds to survive dry periods. Colonization from other wetlands and persistence in a wetland are not mutually exclusive strategies for some invertebrate taxa (Anderson and Smith 2004). The dry period is considered a constraint for the aquatic fauna, causing lower taxa richness in temporary than in permanent ponds. Studies (e.g., Wiggins et al. 1980; Williams 1996; Boulton et al. 2014; Strachan et al. 2015) have shown, however, that a wide variety of invertebrate groups are adapted to desiccation. For this reason, some authors consider desiccation a mythical constraint rather than a strong ecological filter (Biggs et al. 1994). Although some evidence shows higher species richness in permanent ponds (Della Bella et al. 2005; but see Boix et al. 2008), this may be caused by single sampling events underestimating species richness in temporary wetlands with high temporal turnover in species (Robson and Clay 2005). However, the duration and the predictability of the dry period implies a selection of the fauna, since resistance to desiccation by some groups is related to the duration of the dry period and the existence of suitable refuges (Strachan et al. 2014, 2015). For example, in a temporary pond located in the NE of the Iberian Peninsula, three abundant pioneering macroinvertebrates (*Physa acuta*, *Galba truncatula*, and *Berosus signaticollis*) were always present during previous hydroperiods (dry period length between hydroperiods was less than 6 months), but became absent after the pond was dry for more than 2 years (Boix et al. 2001). Freshwater gastropods can survive short-term exposure to air (Havel et al. 2014), resting in microrefuges such as surface depressions of temporary ponds during the dry period, and some species can also aestivate in the sediment (Strachan et al. 2014). Beetles of the genus *Berosus* rest in the sediment during the metamorphosis from larval instar III to imago while waiting for the pond to reflood (Thiéry 1979; Barbero et al. 1982). Although crustacean eggs may be viable in sediment for long periods (300 years in some copepods; Hairston et al. 1995), surviving numbers decline over time (Jenkins and Boulton 2007). Besides drying stress, aestiviating crustacean eggs can be consumed by predators (Waterkeyn et al. 2011a).

Studies of egg-bank dynamics in Mediterranean temporary ponds are scarce (but see Mura 2004), but the presence of species in ponds has been related to their optimal water temperatures (Nourisson and Aguesse 1961; Waterkeyn et al. 2009). Temperature not only determines the presence of species, but can also determine the presence of different clones, as in the case of the ostracod *Heterocypris incongruens* (Rossi and Menozzi 1990). In laboratory experiments, photoperiod variation can produce clones with different life-history traits (Rossi and Menozzi 1993). In fact, both photoperiod and temperature have been identified as cues for diapause in several species of crustaceans and insects (e.g., Sawchyn and Church 1973; Otero et al. 1998). Temporal hatching patterns, whether bet-hedging or synchronous, can vary for invertebrates. Some crustacean species use bet-hedging in more unstable, temporary habitats, and use synchronous hatching in the more stable places (Simovich and Hathaway 1997; Waterkeyn et al. 2013). Other factors playing a role in the hatching process include salinity (Waterkeyn et al. 2009), light (Pinceel et al. 2013), and predation (Spencer and Blaustein 2001), among others.

Different temporal patterns should be expected among groups that differ in dispersal abilities, such as active vs. passive dispersers, aerial vs. terrestrial dispersers, or large- vs. small-bodied organisms (Bilton et al. 2001; De Bie et al. 2012; Ruhí et al. 2013b). For example, large-bodied species are more dispersal-limited if they are passive dispersers, whereas the opposite is true for active dispersers (De Bie et al. 2012). Small organisms producing resting stages have not been considered dispersal-limited, under the so-called *cosmopolitan paradigm*, but the generality of this paradigm is currently being debated (Incagnone et al. 2015 and references therein). Several vectors for passive dispersers have been described, each acting at different spatial scales: wind (Parekh et al. 2014), insects (Van de Meutter et al. 2008), amphibians (Bohonak and Whiteman 1999), fishes (Beladjal et al. 2007), birds (Frisch et al. 2007), and mammals (Vanschoenwinkel et al. 2008) including human (Valls et al. in press). With animal dispersal, eggs can be transported externally, but there are also cases of dispersal following ingestion and defecation (Bohonak and Whiteman 1999). Different dispersal abilities among active dispersers interact with local and regional factors resulting in different spatiotemporal diversity patterns (Miguel-Chinchilla et al. 2014). Insect flight may be influenced by atmospheric conditions (mainly air temperature, wind speed, air humidity; Boix et al. 2011 and references therein) as well as landscape type, habitat conditions, and biological interactions, such as predation and competition. These factors may also act as cues for the initiation of colonization flights (e.g., Velasco and Millán 1998; Pajunen and Pajunen 2003; Yee et al. 2009), or may be important to insects in their selection of a suitable habitat (Blaustein et al. 2004).

Predation, Competition, and Trophic Webs

Along the hydroperiod gradient, the importance of abiotic and biotic factors both change. In the schematic model proposed by Wellborn et al. (1996), ephemeral habitats were considered as refuges against predation, while increased hydroperiod length led to increased predation pressure. In contrast, permanent wetland invertebrates were viewed as being under higher predation pressure, mainly by fish. This model is widely accepted, but it also generates misunderstandings. First, it is incorrect to assume that predation is absent in temporary ponds, even in those ponds with short hydroperiod length (e.g., Blaustein 1998; Brendonck et al. 2002; Boix et al. 2006; Strachan et al. 2014). Second, in Mediterranean temporary waters, longer hydroperiods are not always associated with the highest predation pressure (Spencer et al. 1999). Wetlands with short spring hydroperiods can show stronger effects of insect predation than wetlands with long autumn-winter hydroperiods (Fig. 5.4). Third, autogenic changes at community level (as described in the Seasonal Succession section) are caused, at least in part, by predation (Higgins and Merrit 1999; Boix et al. 2006). Fourth, indirect effects of predation were also reported in these habitats; for example, bioturbation created by *Triops* negatively affected microcrustaceans by impeding filtering capacities (Waterkeyn et al. 2011a) and altering water physico-chemistry (Croel and Kneitel 2011). Therefore, predation is particularly important in structuring communities in temporary waters because the inhabitants typically lack defences against predation (Wilcox 2001; Petrusek et al. 2009), and the abundance of predaceous insects can be very high in short-duration habitats (Batzer and Wissinger 1996).

From an evolutionary point of view, temporary ponds have been considered faunal refuges from predation (Kerfoot and Lynch 1987). The reduction of the global distribution of branchiopods and the rise of cladocerans (small-sized species) coincided with the increase in fish predation during the Mesozoic. Large branchiopods now almost exclusively inhabit temporary (fish-free) waters (Kerfoot and Lynch 1987). However, some defences against predators, which are exclusive of temporary waters (i.e., *Triops cancriformis*), have been observed in the invertebrate fauna. Morphological changes that increase prey survival (i.e., formation of heart-shaped lobes armed by long spines in the head shield, increased tail spine length, increased body lengths and widths) have been observed in individuals of *Daphnia* spp. when those individuals were incubated with chemical cues released by predatory tadpole shrimp (Petrusek et al. 2009; Rabus et al. 2012).

Other ways of avoiding or reducing the risk of predation also exist. For example, some diel patterns observed in zooplankton species of Mediterranean temporary ponds can be interpreted as adaptations to the effects of predators (Compte et al. in press). In non-Mediterranean fishless ponds, predatory invertebrates that inhabit temporary ponds generate diel responses in zooplankton prey (Neill 1990; Gilbert and Hampton 2001; Trochine et al. 2009). In these studies, the diel pattern varied markedly among zooplankton-prey species, but examples also exist showing that zooplankton diel patterns cannot be explained by predation (Arranz et al. 2015). Other adaptations that reduce exposure to predators also exist in Mediterranean temporary ponds. For example, some culicid (mosquito) females avoid ovipositing in pools that contain predators (Blaustein et al. 2004). Although the cue for oviposition avoidance is generally chemical (Blaustein et al. 2004; Silberbush et al. 2010) mosquitoes appear to use other cues for detecting predators (odonates; Stav et al. 2000), or may not avoid certain kinds of predators (urodeles) when ovipositing (Blaustein et al. 2014).

Predation can also have population-level effects. *Triops*, potentially a keystone predator, may selectively prey upon particular sizes or sexes of prey. Populations of *Daphnia magna* and the mayfly *Callibaetis californicus* exposed to *Triops* predation were size-biased, and characterized by preying on a high proportion of larger individuals (Walton et al. 1991; Rabus et al. 2012). In another study, *Triops* targeted male copepods (*Megacyclops viridis*) (Boix et al. 2006). The role of predation in temporary ponds may therefore be important for population-level, as well as community, dynamics.

Unlike predation, only a few studies have addressed competition in Mediterranean temporary ponds. One interesting feature of these studies is the important effect of predation interacting indirectly with competition through: (1) keystone predation (sensu Paine 1969); and (2) intraguild predation. The most competitively dominant cladoceran genus, *Daphnia*, is also the preferred prey of *Notonecta maculata* (Eitam and Blaustein 2010), so densities of less competitive and smaller cladocerans increase with increasing predator abundance. Consequently species diversity increases, as is typical of keystone predation. Two examples of intraguild predation have been reported: mosquito versus toad competition in a temporary pool in Israel (Blaustein and Margalit 1994), and fairy shrimps versus microcrustaceans in French and Spanish temporary ponds (Sánchez and Angeler 2007; Waterkeyn et al. 2011b). Mosquito larvae (*Culiseta longiareolata*) and toad tadpoles (*Bufotes variabilis*) compete for periphyton food, but late-stage *Culiseta* larvae also prey on *Bufotes* hatchlings. Fairy shrimps compete with and potentially also prey on microcrustaceans.

Analyses of food web structure and top-down and bottom-up dynamics are rare in temporary ponds, and the few existing examples are from studies performed outside the Mediterranean biome (Magnusson and Williams 2009; Schriever and Williams 2013; O'Neill and Thorp 2014). These studies illustrate, however, that (1) food-chains are short (average of 3.3 trophic levels, range of 1.7–4.6) regardless of pond size (Schriever and Williams 2013); (2) food-chain length increases as temporary waterbodies approach the end of the hydroperiod (O'Neill and Thorp 2014); and (3) strong top-down effects generated by insect predators (Odonata and Coleoptera) have been observed, regulating the abundance of dipterans and zooplankton with the effects propagating downwards through the food web to lower trophic levels (i.e., trophic cascades; Magnusson and Williams 2009). Short food-chains and trophic cascades have been also observed in temporary brackish waters in coastal Mediterranean ponds (Compte et al. 2012).

Conservation and Management of Mediterranean Temporary Ponds

From a global perspective, conservation of temporary aquatic environments is precarious, because historically they have been neglected. For example, degradation and disappearance of temporary wetlands and streams progressed continuously during the last century (Holland et al. 1995; Brown 1998), in part due to the negative effects of both intensive (Euliss and Mushet 1999; Barry and Davies 2004; Underwood et al. 2009) and expansive (Robson and Clay 2005; Sim et al. 2006a, b) agricultural practices. Temporary ponds in the Mediterranean basin that were compatible with agricultural activity for thousands of years (Grillas et al. 2004) are now clearly negatively affected by current intensive agriculture (Beja and Alcazar 2003; Parra et al. 2005) as they are in other Mediterranean regions (e.g., southern Australia: Robson and Clay 2005; Sim et al. 2006a, b). Remote sensing and historical comparison studies performed in several Mediterranean regions (e.g., De Roeck et al. 2008; Levin et al. 2009; Gómez-Rodríguez et al. 2010; Rhazi et al. 2012; Tulbure et al. 2014) reveal degradation (i.e., reduced hydroperiod due to human activities) and drastic reduction in the number of temporary ponds. The shallowness and the small size of many Mediterranean temporary ponds have made them very vulnerable to human impacts: they can easily be drained for agriculture, urbanization, tourism, or industrial purposes (Grillas et al. 2004; Zacharias et al. 2007). In other cases, temporary waterbodies have been converted to permanent ones for waste disposal, water storage, or (perceived) aesthetics (e.g., Davis et al. 2001). Thus, Mediterranean temporary ponds are endangered habitats, and consequently the scientific community has emphasized the need to reverse the situation (Giudicelli and Thiéry 1998; Boix et al. 2001; Horwitz et al. 2009; Díaz-Paniagua et al. 2010; Zacharias and Zamparas 2010), so as to restore and preserve these unique and valuable environments.

Negative impacts continue despite the existence of preservation initiatives for Mediterranean temporary ponds and their species (e.g., Europe, European Habitat Directive 92/43/CEE; US Federal Register 2003; but see Zedler 2003). The great value of the flora of Mediterranean temporary aquatic environments has been widely reported (e.g., Holland and Jain 1981; Boutin et al. 1982; Ferchichi-Ben Jamaa et al. 2010; Rhazi et al. 2012), as well as the importance of these aquatic environments for amphibian conservation (e.g., Beja and Alcazar 2003; Gómez-Rodríguez et al. 2009; Ferreira and Beja 2013; Escoriza et al. 2014). Invertebrates have received less attention, with the exception of some crustaceans (e.g., King et al. 1996; Belk 1998; De Roeck et al. 2007). However, the need to protect these environments and even to create new ones for the conservation of endangered invertebrate species has been noted (Baltanás et al. 1992; Valdecasas et al. 1992; Fugate 1998; Chester and Robson 2013). In recent years, public perception, scientific knowledge, and management efforts have improved. For example, the number of scientific symposia and publications for both scientific and general audiences has increased significantly in the last two decades (Witham 1998; Diget and Rioux 1998; Blaustein and Schwartz 2001; Grillas et al. 2004; Fraga 2009). Pioneering examples of vernal pool restoration, mitigation against damage, and conservation activities were developed in California in the 1980s (Black and Zedler 1998; Ferren et al. 1998), and more followed in the other Mediterranean regions. For example, in Europe, local and international projects to establish the value of Mediterranean temporary ponds, and to improve their management, have flourished in the last 15 years (including continental France, Corsica, Minorca, València, southwest of Portugal, Crete and Sardinia; Grillas et al. 2004; Fraga et al. 2010; Sancho and Lacomba 2010). In southern Australia, recent research has focused on understanding the effects of water regime change on temporary wetlands (e.g., Robson and Clay 2005; Chambers et al. 2013; Chester et al. 2013; Sim et al. 2013), the biodiversity and recolonization dynamics of temporary wetland biota (Horwitz et al. 2009; Tuckett et al. 2010; Strachan et al. 2014), and whether artificial wetlands can play a role in conservation as more and more natural wetlands become drier (Chester and Robson 2013; Chester et al. 2013). Globally, several new tools or methods have been proposed for Mediterranean temporary ponds to evaluate the impact of human socioeconomic pressure (Zacharias et al. 2008), estimate their environmental status (Dimitriou et al. 2006), establish their habitat condition or ecological integrity (Sala et al. 2004; Sutula et al. 2006; Chester et al. 2013; van den Broeck et al. 2015a), assess their water quality using invertebrates (Chessman et al. 2002; Boix et al. 2005), and assess the contribution anthropogenic wetlands could make to conservation goals (Chester and Robson 2013; Chester et al. 2013).

Common threats to temporary wetlands are prevalent among Mediterranean regions: habitat loss, hydrological perturbation, disconnection and habitat fragmentation, fire damage, pollution, eutrophication, sedimentation, physical disturbance of the sediment, invasive species, livestock impacts, and climate change (Grillas et al. 2004; Zacharias et al. 2007; Zacharias and Zamparas 2010). However, the relative importance of each differs among countries. For example, increased livestock herd size has caused overgrazing and disturbance of sediments in North Africa (Bouahim et al. 2014, but see Ferchichi-Ben Jamaa et al. 2012). In contrast, the use of livestock has been proposed as a management tool to maintain disturbances that favor rare plant germination and amphibian reproduction in Europe (Grillas et al. 2004), and to reduce invasive plant species and promote natives in California (Marty 2005). Water extraction and diversion can dramatically affect the hydrology of temporary ponds in some areas (Serrano and Serrano 1996; Levin et al. 2009; Sim et al. 2013; Boulton et al. 2014). Habitat loss through drainage appears to be a universal issue across Mediterranean-climate regions (e.g., Hambright and Zohary 1998; Robson and Clay 2005; Horwitz et al. 2009). Moreover, annual rainfall has been declining substantially since 1900 in several Mediterranean regions owing to climate change (IPCC 2007) and already dry periods in rivers and wetlands have been markedly prolonged (Davies 2010; Sim et al. 2013). Many formerly perennial wetlands are now seasonal, and several formerly seasonal wetlands are now rarely inundated. These changes in hydroregime (duration, timing, and frequency of inundation) will imply changes in the populations and metacommunities dynamics, and different patterns between organisms with different dispersion mode or ability are expected (Pyke 2005; Sim et al. 2013; Kneitel 2014; Fig. 5.5).

Although many threats are common among temporary ponds in Mediterranean regions, legal protections and conservation plans differ among locations with very different political and social contexts. However, three fundamental concepts are

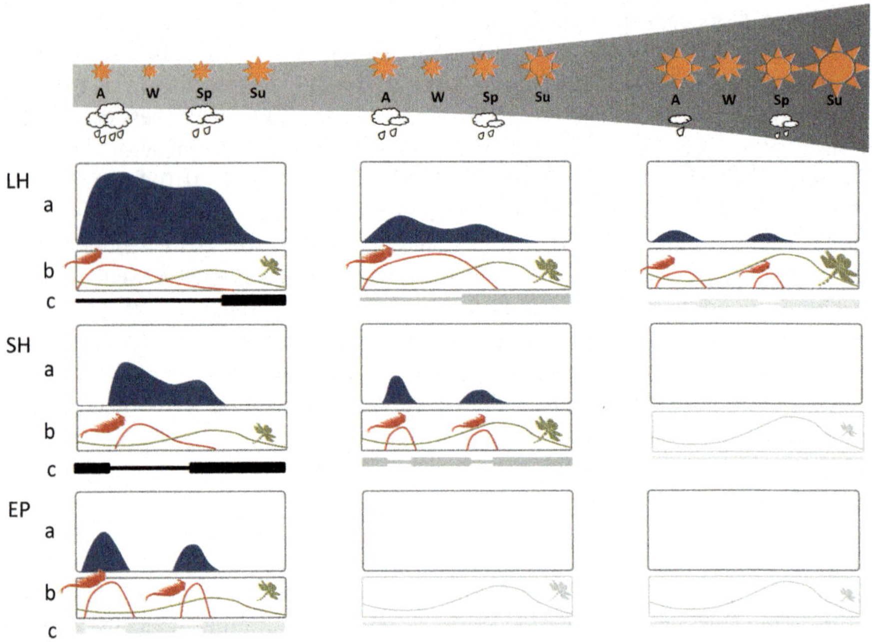

Fig. 5.5 Schematic diagram showing changes in hydroregime and to invertebrates in some future climate change scenarios. Three types of Mediterranean temporary ponds are shown: long hydroperiod (LH); short hydroperiod (SH); and ephemeral (EP). The *left panels* represent the present situation, the *middle panel* shows moderate effects of climate change and the *right-hand panel* shows severe effects of climate change along with the expected changes to hydroperiod length for each pond type. Larger clouds and suns indicate higher rainfall and temperature, respectively. Each panel shows three variables: hydroregime (**a**), organism dynamics (**b**), egg-bank dynamics (**c**). In the hydroregime plots (*a*) the *blue area* shows seasonal water levels (*A* autumn, *W* winter, *Sp* spring, *Su* summer) and hydroperiod length. In the organism dynamic plots (*b*) the bigger the icons (*red* for passive dispersal invertebrates and *green* for the active ones) the higher the population size, and *red* and *green lines* represent the change of population size during the year. Finally, the pattern in (*c*) plots identifies three statuses of the egg-bank: *black*, high density and diversity of propagules; *dark gray*, low density and diversity of propagules; *light gray*, depleted egg-bank. Rainfall reduction and temperature increase cause shorter hydroperiod lengths with decreased life cycle duration and, in turn, a gradual depletion of the egg-bank. For active dispersers, optimal dispersal conditions (i.e., late spring) will be decoupled from hydroperiod since these conditions would occur when ponds would probably be dry. Thus, metapopulation sizes will decrease in time, and taxa presence can be only explained by neighboring ponds (*gray panels*)

shared: (1) habitat loss continues to be the primary challenge to conservation and management; (2) integration of freshwater and terrestrial biodiversity priorities in systematic conservation planning is a major challenge to conservation planners (Amis et al. 2009; Davies and Stewart 2013) and it is especially relevant in areas with a high abundance of temporary ponds (Chester and Robson 2013); and (3) biodiversity protection requires networks of ponds with diverse hydroperiods, where the natural hydrologic regimes are preserved (Beja and Alcazar 2003; Díaz-Paniagua et al. 2010; Chester and Robson 2013; Chester et al. 2013; Florencio et al. 2014). Pond networks in regions under strong human pressure exhibit poorly developed invertebrate metacommunities (Gascón et al. 2012). More knowledge of plants and vertebrates in these environments means that these organisms are sometimes used as surrogates for total biodiversity, but evidence exists that biodiversity patterns and spatiotemporal dynamics of plants, vertebrates, and invertebrates, and even among invertebrate groups, are different (Alexander and Schlising 1998; Gascón et al. 2009; Bagella et al. 2010, 2011; Davies and Stewart 2013; Ruhí et al. 2014; Rouissi et al. 2014). Thus the use of surrogates may be ineffective in Mediterranean-climate regions.

To solve these conservation dilemmas and improve the sustainability of biodiversity and ecosystem function in Mediterranean temporary wetlands, much more research is needed into the dynamics of populations and communities. Comparatively little is known of local and regional patterns of biodiversity, of invertebrate population genetics and dispersal (with the exception of some groups such as large branchiopods; e.g., Aguilar 2011, 2012; Simovich et al. 2013), and of the dynamics of invertebrates in the wide range of habitats and microhabitats present in these wetlands. For example, only recently have invertebrate movements into sediment microrefuges during the drying process been documented in a Mediterranean-climate wetland, showing dynamics not previously observed (Strachan et al. 2014). Finally, progress is being made in elucidating how changes in the hydroregime or in the climate characteristics affect community structure (Ruhí et al. 2014; Kneitel 2014), and further research in this subject should allow improved conservation management of Mediterranean temporary ponds in future scenarios (Fig. 5.5).

Acknowledgement We especially thank Christopher Rogers and Brian Timms for their help with the appendix on large branchiopods, and Valentina Pieri with ostracod genera included in the first appendix. We also thank Laura Serrano for her valuable comments in the preparation of this text.

Appendix 1

Invertebrate taxa observed in Mediterranean temporary ponds of the five world regions (data from published and non-published studies). The figures correspond to the number of identified genera in each family. The symbol + indicates the presence of a certain taxa, but without any genera identified. MED. REG., total number of genera identified in all Mediterranean regions.

		Med. Basin	North America	South America	Southern Australia	South Africa	MED. REG.
NEMATODA							
	Nematoda undet.	+	+	+	+	+	+
Dorylaimida	Dorylaimidae	1	–	–	–	–	1
	Thornenematidae	1	–	–	–	–	1
Enoplida	Tripylidae	1	–	–	–	–	1
Monhysterida	Monhysteridae	1	–	–	–	–	1
Rhabditida	Rhabditida undet.	+	–	–	–	–	+
BRYOZOA							
Plumatellida	Plumatellidae	1	–	1	–	–	1
CNIDARIA							
Hydroida	Hydridae	1	+	1	2	1	2
TURBELLARIA							
	Turbellaria undet.	+	+	+	+	+	+
Catenulida	Stenostomidae	1	–	–	–	–	1
Dalyellioida	Dalyellidae	2	–	–	–	–	2
Rhabdocoela	Polycystidae	1	–	–	–	–	1
	Typhloplanidae	10	–	–	–	1	10
Tricladida	Tricladida undet.	–	–	–	+	–	+
	Dugesiidae	1	–	1	–	–	1
	Planariidae	–	+	–	–	–	+
	Rhynchodemidae	1	–	–	–	–	1
TARDIGRADA							
	Tardigrada undet.	+	–	–	+	+	+
NEMERTEA							
	Nemertea undet.	–	–	–	+	–	+
NEMATOMORPHA							
	Nematomorpha und.	–	–	–	–	+	+

Gordioidea	Gordiidae	1	1	–	–	–	1
ROTIFERA							
	Rotifera undet.	–	–	–	+	+	+
Bdelloidea	Bdelloidea undet.	+	+	+	–	–	+
	Philodinidae	1	–	–	–	–	1
Monogononta	Asplanchnidae	2	–	1	–	–	2
	Atrochidae	1	–	–	–	–	1
	Brachionidae	6	2	3	–	–	6
	Conochilidae	1	–	1	–	–	1
	Euchlanidae	2	–	1	–	–	2
	Filiniidae	1	–	–	–	–	1
	Flosculariidae	1	–	–	–	–	1
	Hexarthridae	1	1	1	–	–	1
	Lecanidae	1	2	2	–	–	2
	Lepadellidae	2	2	2	–	–	3
	Mytilinidae	1	–	–	–	–	1
	Notommatidae	2	–	1	–	–	2
	Proalidae	1	–	–	–	–	1
	Scaridiidae	–	1	1	–	–	1
	Synchaetidae	2	1	1	–	–	2
	Testudinellidae	1	–	–	–	–	1
	Trichocercidae	1	1	1	–	–	1
	Trichotriidae	1	–	1	–	–	1

(continued)

(continued)

		Med. Basin	North America	South America	Southern Australia	South Africa	MED. REG.
MOLLUSCA							
Gastropoda	Ancylidae	2	–	–	1	1	2
	Chilinidae	–	–	1	–	–	1
	Glacidorbidae	–	–	–	1	–	1
	Hydrobiidae	–	–	1	–	–	1
	Lymnaeidae	3	2	–	1	2	5
	Physidae	1	1	1	1	2	2
	Planorbidae	6	–	–	4	2	10
	Pomatiopsidae	–	–	–	1	1	2
	Succineidae	1	–	–	1	–	1
	Tateidae	1	–	–	–	–	1
Bivalvia	Sphaeriidae	2	–	+	2	–	2
ANNELIDA							
Polychaeta	Polychaeta undet.	–	–	+	–	–	+
	Aelosomatidae	1	–	–	1	1	1
Oligochaeta	Oligochaeta undet.	–	+	+	+	–	+
	Enchytraeidae	2	–	–	+	–	2
	Lumbricidae	2	–	–	–	–	2
	Lumbriculidae	+	–	–	–	–	+
	Naididae	3	+	+	3	+	4
	Tubificidae	2	+	+	3	+	5
Hirudinea	Hirudinea undet.	–	–	+	+	–	+
	Erpobdellidae	2	–	–	–	–	2
	Glossiphoniidae	2	–	–	+	–	3
	Hirudinidae	2	–	–	+	–	2
	Richardsonianidae	–	–	–	+	–	+

CRUSTACEA							
Branchiopoda	Bosminidae	1	1	1	2	1	2
	Branchinectidae	1	1	–	–	–	1
	Branchipodidae	1	–	–	1	1	3
	Chirocephalidae	2	2	–	–	–	3
	Chydoridae	16	8	5	16	7	24
	Cyzicidae	2	1	–	1	–	3
	Daphniidae	4	4	3	5	5	6
	Ilyocryptidae	1	–	1	1	–	1
	Leptestheriidae	3	–	–	–	1	3
	Limnadiidae	2	1	–	3	–	5
	Lynceidae	1	1	–	1	–	1
	Macrotrichidae	1	1	–	2	1	2
	Moinidae	1	1	–	1	1	1
	Sididae	2	1	1	2	–	3
	Streptocephalidae	1	1	–	–	1	1
	Tanymastigidae	2	–	–	–	–	2
	Thamnocephalidae	–	–	–	1	–	1
	Triopsidae	2	2	–	2	1	2
Copepoda	Canthocamptidae	4	3	2	1	–	5
	Centropagidae	–	–	–	2	–	2
	Cyclopidae	12	5	5	3	5	13
	Diaptomidae	9	5	–	–	3	15

(continued)

(continued)

		Med. Basin	North America	South America	Southern Australia	South Africa	MED. REG.
Ostracoda	Candonidae	6	2	1	1	2	7
	Cyprididae	14	11	7	15	8	33
	Darwinulidae	1	–	–	1	–	1
	Hemicytheridae	1	–	–	–	–	1
	Ilyocyprididae	1	–	–	1	–	1
	Limnocytheridae	2	2	–	3	1	4
	Notodromadidae	1	–	–	2	–	3
Isopoda	Amphisopidae	–	–	–	1	–	1
	Asellidae	1	–	+	–	–	1
	Scyphacidae	–	–	–	1	–	1
Amphipoda	Ceinidae	–	–	–	1	–	1
	Crangonyctidae	–	1	–	–	–	1
	Hyalellidae	–	–	1	–	–	1
	Perthiidae	–	–	–	1	–	1
	Aeglidae	–	–	1	–	–	1
Decapoda	Cambaridae	1	1	–	–	–	1
	Parastacidae	–	–	1	3	–	4
ARACHNIDA							
Oribatida	Hydrozetidae	1	–	–	–	–	1

Prostigmata	Prostigmata undet.	–	–	+	–	–	+
	Arrenuridae	1	1	–	1	1	1
	Eylaidae	1	1	–	1	1	1
	Hydrachnidae	1	–	–	1	1	1
	Hydrodromidae	1	–	–	1	1	1
	Hydryphantidae	2	–	–	+	3	4
	Hygrobatidae	1	–	–	–	–	1
	Limnesiidae	–	1	–	1	–	1
	Limnocharidae	–	–	–	1	1	1
	Oxidae	–	–	–	1	–	1
	Pionidae	2	–	–	2	1	3
	Teutonidae	1	–	–	–	–	1
	Unionicolidae	–	–	–	2	1	3
Araneae	Pisauridae	–	–	–	+	–	+
INSECTA							
Ephemeroptera	Baetidae	1	1	–	1	1	2
	Caenidae	1	–	1	1	–	2
	Coloburiscidae	–	–	+	–	–	+
	Leptophlebiidae	1	–	1	–	–	2

(continued)

(continued)

		Med. Basin	North America	South America	Southern Australia	South Africa	MED. REG.
Odonata	Aeshnidae	3	2	+	3	1	5
	Austrocorduliidae	–	–	–	1	–	1
	Calopterygidae	1	–	–	–	–	1
	Coenagrionidae	6	3	+	4	3	11
	Corduliidae	1	–	–	2	–	3
	Gomphidae	–	–	–	2	+	2
	Lestidae	3	–	+	1	–	4
	Libellulidae	5	5	+	7	4	13
	Megapodagrionidae	–	–	–	2	–	2
	Petaluridae	–	–	–	1	–	1
	Platycnemididae	1	–	–	–	–	1
	Synthemistidae	–	–	–	2	–	2
	Telephlebiidae	–	–	–	1	–	1
Plecoptera	Capniidae	1	–	–	–	–	1
	Gripopterygidae	–	–	2	–	–	2
	Perlidae	–	–	+	–	–	+
	Notonemouridae	–	–	1	–	–	1

Hemiptera	Belostomatidae	1	1	–	1	1	3
	Corixidae	10	4	+	4	2	13
	Gerridae	2	1	–	–	2	3
	Hebridae	–	–	+	–	–	+
	Hydrometridae	1	–	–	+	–	1
	Mesoveliidae	1	1	+	+	1	1
	Naucoridae	2	–	–	–	–	2
	Nepidae	2	–	–	1	–	2
	Notonectidae	3	2	–	3	2	5
	Pleidae	1	–	–	2	1	2
	Saldidae	1	–	–	–	–	1
	Veliidae	2	–	–	1	–	2
Megaloptera	Sialidae	–	–	1	–	–	1
Neuroptera	Osmylidae	–	–	+	–	–	+
	Sisyridae	–	–	–	1	–	1

(continued)

(continued)

		Med. Basin	North America	South America	Southern Australia	South Africa	MED. REG.
Coleoptera	Alleculidae	+	–	–	–	–	+
	Carabidae	–	–	+	–	–	+
	Chrysomelidae	1	–	–	+	–	1
	Curculionidae	2	+	–	1	–	2
	Dryopidae	1	–	–	–	1	1
	Dytiscidae	27	14	+	23	12	52
	Elmidae	1	–	2	–	–	3
	Erirhinidae	1	–	–	–	–	1
	Georissidae	–	–	–	–	+	+
	Gyrinidae	1	1	–	–	1	2
	Haliplidae	3	3	–	1	1	4
	Helophoridae	1	1	–	–	–	1
	Heteroceridae	–	1	–	–	–	1
	Hydraenidae	3	2	+	2	9	10
	Hydrochidae	1	–	1	1	1	1
	Hydrophilidae	14	6	1	7	11	21
	Hygrobiidae	1	–	–	–	–	1
	Limnichidae	–	–	–	+	–	+
	Noteridae	1	–	–	1	–	2
	Ptiliidae	–	–	–	+	+	+
	Ptilodactylidae	–	–	–	+	–	+
	Scarabaeidae	–	1	–	–	–	1
	Scirtidae	2	–	1	1	+	3
	Spercheidae	–	–	–	–	+	+
	Staphylinidae	–	–	+	–	–	+

Trichoptera	Ecnomidae	–	–	–	+	–	+
	Hydropsychidae	–	–	1	–	–	1
	Hydroptilidae	2	–	1	1	1	3
	Leptoceridae	–	–	2	4	–	5
	Limnephilidae	5	1	–	–	–	5
	Phryganeidae	1	–	–	–	–	1
	Polycentropodidae	1	–	1	–	–	2
	Sericostomatidae	1	–	–	–	–	1
	Stenopsychidae	–	–	+	–	–	+
Lepidoptera	Crambidae	–	–	–	+	–	+
	Pyralidae	1	–	+	+	+	1

(continued)

(continued)

		Med. Basin	North America	South America	Southern Australia	South Africa	MED. REG.
Diptera	Anthomyiidae	–	+	–	+	+	+
	Athericidae	–	–	+	–	–	+
	Ceratopogonidae	6	4	+	8	1	11
	Chaoboridae	1	–	–	1	1	2
	Chironomidae	40	13	+	21	10	49
	Chloropidae	+	–	–	–	–	+
	Culicidae	5	5	+	4	5	7
	Dixidae	2	1	–	–	+	2
	Dolichopodidae	1	+	–	+	–	1
	Empididae	+	–	+	+	–	+
	Ephydridae	4	4	+	+	+	6
	Limoniidae	3	–	+	–	–	3
	Psychodidae	2	1	–	1	1	2
	Rhagionidae	1	–	–	–	–	1
	Scatophagidae	–	1	–	–	–	1
	Sciomyzidae	2	–	–	1	–	2
	Simuliidae	2	+	+	–	–	2
	Stratiomyidae	3	2	–	+	+	4
	Syrphidae	1	2	–	–	–	2
	Tabanidae	1	1	–	+	+	1
	Thaumaleidae	–	–	–	+	–	+
	Tipulidae	3	3	+	+	+	4
MINIMUM NUMBER OF GENERA		394	169	101	253	150	633
MINIMUM NUMBER OF FAMILIES		149	73	77	112	71	204

Main literature sources:

MEDITERRANEAN BASIN—Terzian 1979; Boutin et al. 1982; Metge 1986; Bazzanti et al. 1996; Bazzanti et al. 1997; Chaves 1999; Fahd et al. 2000; Boix et al. 2001, 2005; Eitam et al. 2004; Pieri et al. 2006; Culioli et al. 2006; Carchini et al. 2007; Marrone et al. 2009; Florencio et al. 2009; Sahuquillo and Miracle 2010; Martins et al. 2010; Caramujo and Boavida 2010; Elron and Gafny 2011; Moubayed-Breil et al. 2012; Tornero et al. 2014; Rouissi et al. 2014; Gerecke et al. 2014; Gilbert et al. 2015; Escrivà 2015. NORTH AMERICA—Ebert and Balko 1987; Zedler 1987; King et al. 1996; Rogers 1998, 2014; Belk 1998; Helm 1998; Simovich 1998; de Szalay and Resh 2000; Marchetti et al. 2010. SOUTH AMERICA—Araya and Zúñiga 1985; Villagran-Mella et al. 2006; Figueroa et al. 2009; Correa-Araneda et al. 2014. SOUTHERN AUSTRALIA—Williams 1975; Davis and Christidis 1999; Robson and Clay 2005; Sim et al. 2013; Pinder et al. 2013; Strachan et al. 2014. SOUTH AFRICA –De Roeck et al. 2007; Day et al. 2010; Mlambo et al. 2011.

Appendix 2

Large branchiopods species observed in temporary ponds in four of the five Mediterranean regions (data from published studies; no species have been identified from South American Mediterranean region). **P** means that this species occurs both in and outside the Mediterranean areas, whereas **E** means that it is endemic to one Mediterranean area. For the species inclusion criteria see section "Invertebrate assemblages of Mediterranean temporary ponds?" means that this taxon cannot be included in one of the two previous categories, because taxonomic identity is not at species level.

	Med. Basin	North America	Southern Australia	South Africa
O. NOTOSTRACA				
F. Triopidae				
Lepidurus apus apus	**P**	–	–	–
Lepidurus apus viridis	–	–	**P**	–
Lepidurus couesii	**P**	–	–	–
Lepidurus lubbocki	**P**	–	–	–
Lepidurus packardi	–	**E**	–	–
Triops australiensis	–	–	**P**	–
Triops baeticus	**E**	–	–	–
Triops cancriformis	**P**	–	–	–
Triops emeritensis	**E**	–	–	–
Triops gadensis	**E**	–	–	–
Triops granarius	**P**	–	–	**P**
Triops longicaudatus	–	**P**	–	–
Triops mauritanicus	**E**	–	–	–
Triops simplex	**E**	–	–	–
Triops vicentinus	**E**	–	–	–

(continued)

(continued)

	Med. Basin	North America	Southern Australia	South Africa
O. SPINICAUDATA				
F. Cyzicidae				
Cyzicus bucheti	**E**	–	–	–
Cyzicus californicus	–	**E**	–	–
Cyzicus gihoni	**E**	–	–	–
Cyzicus grubei	**E**	–	–	–
Cyzicus tetracerus	**P**	–	–	–
Eocyzicus saharicus	**P**	–	–	–
Ozestheria mariae	–	–	**E**	–
Ozestheria packardi	–	–	**P**	–
F. Limnadiidae				
Eulimnadia sp.[1]	?	–	–	–
Eulimnadia feriensis	–	–	**E**	–
Eulimnadia datsonae	–	–	**E**	–
Eulimnadia palustera	–	–	**E**	–
Eulimnadia vinculuma	–	–	**E**	–
Eulimnadia texana	–	**P**	–	–
Imnadia yeyetta	**P**	–	–	–
Limnadia lenticularis	**P**	–	–	–
Limnadopsis occidentalis	–	–	**P**	–
Limnadopsis paradoxa	–	–	**E**	–
Limnadopsis tatei	–	–	**P**	–
Paralimnadia badia	–	–	**E**	–
Paralimnadia cygnorum	–	–	**E**	–
Paralimnadia sordida	–	–	**P**	–
F. Leptestheriidae				
Eoleptestheria ticinensis	**P**	–	–	–
Leptestheria dahalacensis	**P**	–	–	–
Leptestheria mayeti	**P**	–	–	–
Leptestheria rubidgei	–	–	–	**P**
Maghrebestheria maroccana	**E**	–	–	–
O. LAEVICAUDATA				
F. Lynceidae				
Lynceus sp.[2]	?	–	–	–
Lynceus baylyi	–	–	**P**	–
Lynceus brachyurus	–	**P**	–	–
Lynceus tatei	–	–	**P**	–
Lynceus macleayanus	–	–	**P**	–
Lynceus magdaleanae	–	–	**P**	–
Lynceus susanneae	–	–	**E**	–

	Med. Basin	North America	Southern Australia	South Africa
O. ANOSTRACA				
F. Streptocephalidae				
Streptocephalus cafer	–	–	–	**P**
Streptocephalus dendyi	–	–	–	**P**
Streptocephalus gracilis	–	–	–	**P**
Streptocephalus ovamboensis	–	–	–	**P**
Streptocephalus papillatus	–	–	–	**P**
Streptocephalus purcelli	–	–	–	**P**
Streptocephalus torvicornis	**P**	–	–	–
Streptocephalus woottoni	–	**E**	–	–
F. Tanymastigidae				
Tanymastix affinis	**E**	–	–	–
Tanymastix stagnalis	**P**	–	–	–
Tanymastix stellae	**E**	–	–	–
Tanymastigites brteki	**E**	–	–	–
Tanymastigites cyrenaica	**P**	–	–	–
Tanymastigites lusitanica	**E**	–	–	–
Tanymastigites perrieri	**P**	–	–	–
F. Branchipodidae				
Australobranchipus parooensis	–	–	**P**	–
Branchipodopsis dayae	–	–	–	**P**
Branchipodopsis hodgsoni	–	–	–	**P**
Branchipodopsis karroensis	–	–	–	**E**
Branchipodopsis wolfi	–	–	–	**P**
Branchipus cortesi	**E**	–	–	–
Branchipus pasai	**E**	–	–	–
Branchipus schaefferi	**P**	–	–	–
F. Thamnocephalidae				
Branchinella affinis	–	–	**P**	–
Branchinella australiensis	–	–	**P**	–
Branchinella basispina	–	–	**E**	–
Branchinella complexidigitata	–	–	**E**	–
Branchinella erosa	–	–	**E**	–
Branchinella kadjikadji	–	–	**E**	–
Branchinella halsei	–	–	**P**	–
Branchinella hattahensis	–	–	**P**	–
Branchinella hearnii	–	–	**E**	–
Branchinella longirostris	–	–	**E**	–
Branchinella lyrifera	–	–	**P**	–
Branchinella nana	–	–	**P**	–
Branchinella occidentalis	–	–	**P**	–
Branchinella papillata	–	–	**P**	–
Branchinella vosperi	–	–	**E**	–

(continued)

(continued)

	Med. Basin	North America	Southern Australia	South Africa
F. Branchinectidae				
Branchinecta campestris	–	**P**	–	–
Branchinecta coloradensis	–	**P**	–	–
Branchinecta conservatio	–	**E**	–	–
Branchinecta dissimilis	–	**P**	–	–
Branchinecta ferox	**P**	–	–	–
Branchinecta gigas	–	**P**	–	–
Branchinecta lindahli	–	**P**	–	–
Branchinecta longiantenna	–	**E**	–	–
Branchinecta lynchi	–	**E**	–	–
Branchinecta mackini	–	**P**	–	–
Branchinecta mesovallensis	–	**E**	–	–
Branchinecta orientalis	**P**	–	–	–
Branchinecta sandiegonensis	–	**E**	–	–
F. Chirocephalidae				
Chirocephalus anatolicus	**E**	–	–	–
Chirocephalus bairdi	**E**	–	–	–
Chirocephalus brteki	**E**	–	–	–
Chirocephalus diaphanus	**P**	–	–	–
Chirocephalus kerkyrensis	**E**	–	–	–
Chirocephalus murae	**E**	–	–	–
Chirocephalus neumanni	**E**	–	–	–
Chirocephalus salinus	**P**	–	–	–
Eubranchipus bundyi	–	**P**	–	–
Eubranchipus oregonus	–	**P**	–	–
Eubranchipus serratus	–	**P**	–	–
Linderiella africana	**E**	–	–	–
Linderiella baetica	**E**	–	–	–
Linderiella massaliensis	**E**	–	–	–
Linderiella occidentalis	–	**E**	–	–
Linderiella santarosae	–	**E**	–	–

[1]The taxonomic identity at species level of this population in Tunisia (Rabet et al. 2015) is not determined

[2]According to Hartland-Rowe (1967) this species is not the ubiquitous *L. brachyurus*

Main literature sources:

MEDITERRANEAN BASIN—Hartland-Rowe 1967; Dimentman 1981; Cottarelli and Mura 1983; Thiéry 1987, 1991, 1996, 2004; Brtek and Thiéry 1995; Alonso 1996; Defaye et al. 1998; Samraoui and Dumont 2002; Brtek and Cottarelli 2006; Cottarelli et al. 2007, 2010; Miracle et al. 2008; Turki and Turki 2010; van den Broeck et al. 2015b; Rabet et al. 2015. NORTH AMERICA - Helm 1998; US Fish and Wildlife Service 2008; Rogers et al. 2010. SOUTHERN AUSTRALIA—Williams 1968; Timms 2002, 2005, 2006, 2008, 2009, 2012, 2013, 2015; Richter and Timms 2005; Timms and Richter 2009; Rogers and Hamer 2012. SOUTH AFRICA—De Roeck et al. 2007.

References

Aguilar A (2011) Weak phylogeographic structure in the endemic western North American fairy shrimp *Branchinecta lynchi* (Eng, Belk and Erickson 1990). Aquat Sci 73:15–20

Aguilar A (2012) Range-wide and local drivers of genetic structure in an endangered California vernal pool endemic crustacean. Conserv Genet 13:1577–1588

Alexander DG, Schlising RA (1998) Patterns in time and space for rare macroinvertebrates and vascular plants in vernal pool ecosystem at the Vina Plains Preserve, and implication for pool landscape management. In: Witham CW (ed) Ecology, conservation, and management of vernal pool ecosystems. California Native Plant Society, Sacramento, pp 161–168

Alonso M (1996) Crustacea, Branchiopoda. In: Ramos MA (ed) Fauna Ibérica, 7. Museo Nacional de Ciencias Naturales, CSIC, Madrid

Alonso M (1998) Las lagunas de la España peninsular. Limnetica 15:1–176

Álvarez-Cobelas M, Rojo C, Angeler DG (2005) Mediterranean limnology: current status, gaps and the future. J Limnol 64:13–29

Amis MA, Rouget M, Lotter M, Day J (2009) Integrating freshwater and terrestrial priorities in conservation planning. Biol Conserv 142:2217–2226

Anderson JT, Smith LM (2004) Persistence and colonization strategies of playa wetland invertebrates. Hydrobiologia 513:77–86

Araya JM, Zúñiga L (1985) Manual taxonómico del zooplancton lacustre de Chile. Boletin Informativo Limnologico. Universidad Austral de Chile 1–110

Arranz I, Sala J et al (2015) Contribution to the knowledge of the distribution of *Chaoborus* species (Diptera: Chaoboridae) in the NE Iberian Peninsula, with notes on the spatial and temporal segregation among them. Limnetica 34:57–68

Aschman H (1973) Distribution and peculiarity in Mediterranean ecosystems. In: Di Castri F, Mooney HA (eds) Mediterranean-type ecosystems: origin and structure. Springer, Berlin, pp 11–19

Bagella S, Caria MC, Farris E, Filigheddu RS (2007) Issues related to the classification of Mediterranean temporary wet habitats according with the European Union Habitats Directive. Fitosociologia 44:245–249

Bagella S, Gascón S et al (2010) Identifying key environmental factors related to plant and crustacean assemblages in Mediterranean temporary ponds. Biodiv Conserv 19:1749–1768

Bagella S, Gascón S et al (2011) Cross-taxon congruence in Mediterranean temporary wetlands: vascular plants, crustaceans, and coleopterans. Comm Ecol 12:40–50

Balla SA, Davis JA (1995) Seasonal variation in the macroinvertebrate fauna of wetlands of differing water regime and nutrient status on the Swan Coastal Plain. Hydrobiologia 299:147–161

Ballón C, Àvila N et al. (in press) Is ecosystem size more important than locality in determining the environmental characteristics of temporary ponds? Limnetica

Baltanás A, Montes C, Marcos AM (1992) Preserving ecological processes, a strategy for biological conservation in fluctuating environments. Conserving and Managing Wetlands for Invertebrates, Council of Europe, Vaduz, In, pp 28–31

Barbero M, Giudicelli J et al (1982) Étude des bicénoses des mares et ruisseaux temporaires a éphémérophytes dominants en région méditerranéenne. Bull d'Écol 13:387–400

Barclay MH (1966) An ecological study of a temporary pond near Auckland, New Zealand. Austral J Mar Freshw Res 17:239–258

Barry MJ, Davies W (2004) Effects of invertebrate predators and a pesticide on temporary pond microcosms used for aquatic toxicity testing. Environ Pollut 13:25–34

Batzer DP, Wissinger SA (1996) Ecology of insect communities in nontidal wetlands. Annu Rev Entomol 41:75–100

Bazzanti M, Baldoni S, Seminara M (1996) Invertebrate macrofauna of a temporary pond in Central Italy: composition, community parameters and temporal succession. Arch Hydrobiol 137:77–94

Bazzanti M, Seminara M, Baldoni S (1997) Chironomids (Diptera: Chironomidae) from three temporary ponds of different wet phase duration in central Italy. J Freshw Ecol 12:89–99

Beja P, Alcazar R (2003) Conservation of Mediterranean temporary ponds under agricultural intensification: an evaluation using amphibians. Biol Conserv 114:317–326

Beklioglu M, Romo S et al (2007) State of the art in the functioning of shallow Mediterranean lakes: workshop conclusions. Hydrobiologia 584:317–326

Beladjal L, Dierckens K, Mertens J (2007) Dispersal of fairy shrimp *Chirocephalus diaphanus* (Branchiopoda: Anostraca) by the trout (*Salmo trutta*). J Crust Biol 27:71–73

Belk D (1998) Global status and trends in ephemeral pool invertebrate conservation: implications for California fairy shrimp. In: Witham CW (ed) Ecology, conservation, and management of vernal pool ecosystems. California Native Plant Society, Sacramento, pp 147–150

Biggs J, Corfield A et al (1994) New approaches to the management of ponds. Brit Wildl 5:273–287

Bilton DT, Freeland JR, Okamura B (2001) Dispersal in freshwater invertebrates. Annu Rev Ecol Syst 32:159–181

Black C, Zedler PH (1998) An overview of 15 years of vernal pool restoration and construction activities in San Diego county, California. In: Witham CW (ed) Ecology, conservation, and management of vernal pool ecosystems. California Native Plant Society, Sacramento, pp 195–205

Blaustein L (1998) Influence of the predatory backswimmer, *Notonecta maculata*, on invertebrate community structure. Ecol Entomol 23:246–252

Blaustein L, Margalit J (1994) Mosquito larvae (*Culiseta longiareolata*) prey upon and compete with toad tadpoles (*Bufo viridis*). J Anim Ecol 63:841–850

Blaustein L, Schwartz SS (eds) (2001) Ecology of temporary pools. Isr J Zool 47:303–528

Blaustein L, Kiflawi M et al (2004) Oviposition habitat selection in response to risk of predation: mode of detection consistency across experimental venue. Oecologia 138:300–305

Blaustein J, Sadeh A, Blaustein L (2014) Influence of fire salamander larvae on among-pool distribution of mosquito egg rafts: oviposition habitat selection or egg raft predation? Hydrobiologia 723:157–165

Bohonak AJ, Whiteman HH (1999) Dispersal of the fairy shrimp *Branchinecta coloradensis* (Anostraca): effects of hydroperiod and salamanders. Limnol Oceanogr 44:487–493

Boix D, Sala J (2002) Riqueza y rareza de los insectos acuáticos de la laguna temporal de Espolla (Pla de l'Estany, Cataluña). Boletín Asoc Española Entomol 26:45–57

Boix D, Sala J, Moreno-Amich R (2001) The faunal composition of Espolla pond (NE Iberian Peninsula): the neglected biodiversity of temporary waters. Wetlands 21:577–592

Boix D, Sala J, Quintana XD, Moreno-Amich R (2004) Succession of the animal community in a Mediterranean temporary pond. J N Am Benthol Soc 23:29–49

Boix D, Gascón S et al (2005) A new index of water quality assessment in Mediterranean wetlands based on crustacean and insect assemblages: the case of Catalunya (NE Iberian peninsula). Aquat Conserv Mar Freshw Ecosyst 15:635–651

Boix D, Sala J, Gascón S, Brucet S (2006) Predation in a temporary pond with special attention to the trophic role of *Triops cancriformis* (Crustacea: Branchiopoda: Notostraca). Hydrobiologia 571:341–353

Boix D, Gascón S et al (2008) Patterns of composition and species richness of crustaceans and aquatic insects along environmental gradients in Mediterranean water bodies. Hydrobiologia 597:53–69

Boix D, Sala J et al (2009) Structure of invertebrate assemblages: contribution to the ecological functioning of the Mediterranean temporary ponds. In: Fraga, P (ed) International Conference on Mediterranean Temporary Ponds: Proceedings & Abstracts, Maó, pp 151–187

Boix D, Magnusson AK et al (2011) Environmental influence on flight activity and arrival patterns of aerial colonizers of temporary ponds. Wetlands 31:1227–1240

Boix D, Biggs J et al (2012) Pond research and management in Europe: "Small is Beautiful". Hydrobiologia 689:1–9

Bolle H-J (2003) Climate, climate variability, and impacts in the Mediterranean area: an overview. In: Bolle H-J (ed) Mediterranean climate. Springer, Berlin, pp 5–86

Bonada N, Resh VH (2013) Mediterranean-climate streams and rivers: geographically separated but ecologically comparable freshwater systems. Hydrobiologia 719:1–29

Bonada N, Rieradevall M et al (2008) Multi-scale assessment of macroinvertebrate richness and composition in Mediterranean-climate rivers. Freshw Biol 53:772–778

Bouahim S, Rhazi L et al (2014) Unravelling the impact of anthropogenic pressure on plant communities in Mediterranean temporary ponds. Mar Freshw Res 65:918–929

Boulton AJ, Brock MA et al (2014) Australian freshwater ecology. Processes and management, 2nd edn. Wiley, Chichester

Boutin C, Lesne L, Thiéry A (1982) Ecologie et typologie de quelques mares temporaires à isoètes d'une région aride du Maroc occidental. Ecol Méditer 8:31–56

Brauer F (1877) Beiträge zur Kenntniss der Phyllopoden. Sitzungsberichte der Kaiserlichen Akademie der Wissenschaften. Mathematisch-Naturwissenschaftliche Classe 75:583–614

Brendonck L, Michels E, De Meester L, Riddoch B (2002) Temporary pools are not enemy-free. Hydrobiologia 486:147–159

Brendonck L, Rogers DC et al (2008) Global diversity of large branchiopods (Crustacea: Branchiopoda) in freshwater. Freshw Anim Div Assess 595:167–176

Britton RH, Crivelli AJ (1993) Wetlands of southern Europe and North Africa: Mediterranean wetlands. In: Whigham DF, Dykyjová D, Hejný S (eds) Wetlands of the world I: inventory, ecology and management. Kluwer Academic, Dordrecht, pp 129–194

Brown KS (1998) Vanishing pools taking species with them. Science 281:626

Brtek J, Cottarelli V (2006) *Chirocephalus murae*, a new species (Branchiopoda, Anostraca) from Turkey. Crustaceana 79:89–98

Brtek J, Thiéry A (1995) The geographic distribution of the European Branchiopods (Anostraca, Notostraca, Spinicaudata, Laevicaudata). Hydrobiologia 298:263–280

Caramujo MJ, Boavida MJ (2010) Biological diversity of copepods and cladocerans in Mediterranean temporary ponds under periods of contrasting rainfall. J Limnol 69:64–75

Carchini G, Della Bella V, Solimini AG, Bazzanti M (2007) Relationships between the presence of odonate species and environmental characteristics in lowland ponds of central Italy. Ann Limnol Int J Limnol 43:81–87

Chambers JM, Nugent G et al (2013) Adapting to climate change: a risk assessment and decision making framework for managing groundwater dependent ecosystems with declining water levels, Development and case studies. National Climate Change Adaptation Research Facility, Gold Coast, 100 p

Chaves ML (1999) Monitorização biológica de lagoas temporárias mediterrânicas—Contribuição para a difinição de estados de referência. Universidade de Lisboa, Estágio de Licenciatura

Chessman BC, Trayler KM, Davis JA (2002) Family and species level biotic indices for macroinvertebrates of wetlands on the Swan Coastal Plain, Western Australia. Mar Freshw Res 53:919–930

Chester ET, Robson BJ (2011) Drought refuges, spatial scale and the recolonization of invertebrates in non-perennial streams. Freshw Biol 56:2094–2104

Chester ET, Robson BJ (2013) Anthropogenic refuges for freshwater biodiversity: their ecological characteristics and management. Biol Conserv 166:64–75

Chester ET, Robson BJ, Chambers JM (2013) Novel methods for managing freshwater refuges against climate change in southern Australia: anthropogenic refuges for freshwater biodiversity. National Climate Change Adaptation Research Facility, Gold Coast, 90 p

Collinson NH, Biggs J et al (1995) Temporary and permanent ponds: an assessment of the effects of drying out on the conservation value of aquatic macroinvertebrate communities. Biol Conserv 74:125–133

Colosi G (1922) Nota sopra alcuni Eufillopodi. Atti della Società Italiana di scienze naturali e del Museo Civico di storia naturale di Milano 61:287–297

Comín FA, Williams WD (1994) Parched continents: Our common future? In: Margalef R (ed) Limnology now. A paradigm of planetary problems. Elsevier, Amsterdam

Compte J, Gascón S, Quintana XD, Boix D (2012) The effects of small fish presence on a species-poor community dominated by omnivores: Example of a size-based trophic cascade. J Exp Mar Biol Ecol 418–419:1–11.

Compte J, Montenegro M, et al (in press) Microhabitat selection and diel patterns of zooplankton 1 in a Mediterranean temporary pond. Hydrobiologia.

Correa-Araneda F, Díaz ME et al (2014) Benthic macroinvertebrate community patterns of Mediterranean forested wetlands and their relation to changes in the hydroperiod. Limnetica 33:361–374

Cottarelli V, Mura G (1983) Anostraci, Notostraci, Concostraci. In: Ruffo S (ed) Guide per il riconoscimento delle specie animali delle acque interne italiane, 18. Consiglio Nazionale delle Richerche

Cottarelli V, Mura G, Sulhi Özkütük R (2007) On two new Anostraca species (Branchiopoda) from Turkey: *Chirocephalus anatolicus* sp. nov. and *Chirocephalus cupreus* sp. nov. Crustaceana 80:455–474

Cottarelli V, Aygen C, Mura G (2010) Fairy shrimps from Asiatic Turkey: Redescription of *Chirocephalus tauricus* Pesta, 1921 and descriptions of *Chirocephalus algidus* sp. nov. and *Chirocephalus brteki* sp. nov. (Crustacea, Branchiopoda, Anostraca). Zootaxa 2528:29–52

Croel R, Kneitel JM (2011) Ecosystem-level effects of bioturbation by the tadpole shrimp *Lepidurus packardi* in temporary pond mesocosms. Hydrobiologia 665:169–181

Culioli JL, Foata J et al (2006) Temporal succession of the macroinvertebrate fauna in a Corsican temporary pond. Vie et Milieu 56:215–221

Daget P (1984) Introduction à une théorie générale de la Méditerranéite. Bull Soc Bot Fr 131:31–36

Davies PM (2010) Climate change implications for river restoration in global biodiversity hotspots. Restor Ecol 18:261–268

Davies PM, Stewart BA (2013) Aquatic biodiversity in the Mediterranean climate rivers of south-western Australia. Hydrobiologia 719:215–236

Davis JA, Christidis F (1999) A guide to wetland invertebrates of southwestern Australia. Western Australian Museum, Perth, WA

Davis JA, Halse SA, Froend RH (2001) Factors influencing biodiversity in coastal plain wetlands of south-western Australia. In: Gopal B, Junk WA, Davis JA (eds) Biodiversity in wetlands: assessment, function and conservation, vol 2. Backhuys Publishers, Leiden, pp 89–100

Day J, Day E, Ross-Gillespie V, Ketley A (2010) The assessment of temporary wetlands during dry conditions. Water Research Commission, Report No. TT 434/09

De Bie T, Meester L et al (2012) Body size and dispersal mode as key traits determining metacommunity structure of aquatic organisms. Ecol Lett 15:740–747

de Moor FC, Day JA (2013) Aquatic biodiversity in the mediterranean region of South Africa. Hydrobiologia 719:237–268

De Roeck E, Vanschoenwinkel B et al (2007) Conservation status of large branchiopods in the Western Cape, South Africa. Wetlands 27:162–173

De Roeck E, Verhoest NE et al (2008) Remote sensing and wetland ecology: a South African case study. Sensors 8:3542–3556

de Szalay FA, Resh VH (2000) Factors influencing macroinvertebrate colonization of seasonal wetlands: responses to emergent plant cover. Freshw Biol 45:295–308

Defaye D, Rabet N, Thiéry A (1998) Atlas et bibliographie des crustacés branchiopodes (Anostraca, Notostraca, Spinicaudata) de France métropolitaine. Patrimoines Naturels, 32. Service du Patrimoine Naturel/IEGB/MNHN, Paris

Della Bella V, Bazzanti M, Chiarotti F (2005) Macroinvertebrate diversity and conservation status of Mediterranean ponds in Italy: water permanence and mesohabitat influence. Aquat Conserva Mar Freshw Ecosyst 15:583–600

Díaz-Paniagua C, Fernández-Zamudio R et al (2010) Temporary ponds from Doñana National Park: a system of natural habitats for the preservation of aquatic flora and fauna. Limnetica 29:41–58

Diget A, Rioux JA (1998) Ecologie et conservation des mares temporaires méditerranéennes: l'exemple des mares de la reserve naturelle de Roque-Haute, Montpellier. Ecol Mediterr 24:105–240

Dimentman C (1981) The rainpool ecosystems of Israel: geographical distribution of freshwater Anostraca (Crustacea). Isr J Zool 30:1–15

Dimitriou E, Karaouzas I, Skoulikidis N, Zacharias I (2006) Assessing the environmental status of Mediterranean temporary ponds in Greece. Ann Limnol 42:33–41

Ebert TA, Balko ML (1987) Temporary pools as islands in space and in time: the biota of vernal pools in San Diego, Southern California, USA. Arch Hydrobiol 110:101–123

Eitam A, Blaustein L (2010) Effects of predator density and duration of predator occupancy on crustacean abundance and diversity in experimental pools. Hydrobiologia 652:269–276

Eitam A, Noreña C, Blaustein L (2004) Microturbellarian species richness and community similarity among temporary pools: relationships with habitat properties. Biodiv Conserv 13:2107–2117

Elron, E, Gafny S (2011) A survey of winter rainpools (ephemeral pools) at the Central Coastal Plain of Israel (March–June, 2010). Report submitted to the Scientific Dept. and the Central District of the Israeli Nature and Parks Authority. 246 pp

Escoriza D, Hassine JB, Boix D (2014) Factors regulating the invasive success of an alien frog: a comparison of the ecology of the native and alien populations. Hydrobiologia 730:127–138

Escrivà A (2015) Ecologia de comnitats d'invertebrats aquatics de la Península Ibèrica, amb especial rellevància als Ostracoda. Ph.D. Thesis, Universitat de València

Euliss NH, Mushet DM (1999) Influence of agriculture on aquatic invertebrate communities of temporary wetlands in the prairie pothole region of North Dakota, USA. Wetlands 19:578–583

European Commission (2003) Interpretation manual of European Union habitats (EUR 25) DG Environment-Nature and Biodiversity, Brussels. http://europa.eu.int/comm/environment/nature/nature_conservation/eu_enlargement/2004/pdf/haitats_im_en.pdf. Accessed 15 Jan 2015

Fahd K, Serrano L, Toja J (2000) Crustacean and rotifer composition of temporary ponds in the Doñana National Park (SW Spain) during floods. Hydrobiologia 436:41–49

Ferchichi-Ben Jamaa H, Muller SD et al (2010) Structures de végétation et conservation des zones humides temporaires méditerranéennes: la région des Mogods (Tunisie septentrionale). C R Biol 333:265–279

Ferchichi-Ben Jamaa H, Muller SD et al (2012) Influence du pâturage sur la structure, la composition et la dynamique de la végétation de mares temporaires méditerranéennes (Tunisie septentrionale). Rev d'Écol 67:3–20

Ferreira M, Beja P (2013) Mediterranean amphibians and the loss of temporary ponds: Are there alternative breeding habitats? Biol Conserv 165:179–186

Ferren WR, Hubbard DM et al (1998) Review of ten years of vernal pool restoration and creation in Santa Barbara, California. In: Witham CW (ed) Ecology, conservation, and management of vernal pool ecosystems. California Native Plant Society, Sacramento, pp 206–216

Ferrés L (1993) Un bioma i un clima a escala humana. In: Folch R (ed) Biosfera, Mediterrànies, vol 5. Enciclopèdia Catalana, Barcelona, pp 17–26

Figueroa R, Suárez ML et al (2009) Caracterización ecológica de humedales de la zona semiárida de Chile Central. Gayana 73:76–94

Florencio M, Serrano L et al (2009) Inter- and intra-annual variations of macroinvertebrate assemblages are related to the hydroperiod in Mediterranean temporary ponds. Hydrobiologia 634:167–183

Florencio M, Díaz-Paniagua C, Serrano L, Bilton DT (2011) Spatio-temporal nested patterns in macroinvertebrate assemblages across a pond network with a wide hydroperiod range. Oecologia 166:469–483

Florencio M, Diaz-Paniagua C, Gomez-Rodriguez C, Serrano L (2014) Biodiversity patterns in a macroinvertebrate community of a temporary pond network. Insect Conserv Div 7:4–21

Fraga P (ed) (2009) International conference on temporary ponds. Proceedings & Abstracts. Consell Insular de Menorca, Maó

Fraga P, Estaún I, Cardona E (eds) (2010) Basses temporals mediterrànies. LIFE BASSES: gestió i conservació a Menorca. Consell Insular de Menorca, Maó

Frisch D, Green AJ, Figuerola J (2007) High dispersal capacity of a broad spectrum of aquatic invertebrates via waterbirds. Aquat Sci 69:568–574

Fugate M (1998) Branchinecta of North America: population structure and its implications for conservation practice. In: Witham CW (ed) Ecology, conservation, and management of vernal pool ecosystems. California Native Plant Society, Sacramento, pp 140–146

Gascón S, Boix D, Sala J, Quintana XD (2005) Variability of benthic assemblages in relation to the hydrological pattern in Mediterranean salt marshes (Empordà wetlands, NE Iberian Peninsula). Arch Hydrobiol 163:163–181

Gascón S, Boix D, Sala J (2009) Are different biodiversity metrics related to the same factors? A case study from Mediterranean wetlands. Biol Conserv 142:2602–2612

Gascón S, Machado M et al (2012) Spatial characteristics and species niche attributes modulate the response by aquatic passive dispersers to habitat degradation. Mar Freshw Res 63:232–245

Gasith A, Resh VH (1999) Streams in Mediterranean climate regions: abiotic influences and biotic responses to predictable seasonal events. Annu Rev Ecol Syst 30:51–81

Gerecke R, Marrone F et al (2014) The water mites (Acari: Hydrachnidia) of the standing waters of Corsica, Sardinia and Sicily: review and new data. Ital J Zool 81:389–408

Gilbert JJ, Hampton SE (2001) Diel vertical migrations of zooplankton in a shallow, fishless pond: a possible avoidance-response cascade induced by notonectids. Freshw Biol 46:611–621

Gilbert JD, De Vicente I et al (2015) A comprehensive evaluation of the crustaceans assemblages in southern Iberian Mediterranean wetlands. J Limnol 74:169–181

Giudicelli J, Thiéry A (1998) La faune des mares temporaires, son originalité et son intérêt pour la biodiversité des eaux continentales méditerranéennes. Ecol Mediterr 24:135–143

Gómez-Rodríguez C, Díaz-Paniagua C et al (2009) Mediterranean temporary ponds as amphibian breeding habitats: the importance of preserving pond networks. Aquat Ecol 43:1179–1191

Gómez-Rodríguez C, Bustamante J, Díaz-Paniagua C (2010) Evidence of hydroperiod shortening in a preserved system of temporary ponds. Remote Sensing 2:1439–1462

Grillas P, Gauthier P, Yavercovski N, Perennou C (2004) Mediterranean temporary pools. Issues relating to conservation, functioning and management, vol 1. Station biologique de la Tour du Valat

Grillas P, Waterkeyn A, Brendonck L, Rhazi L (2010) Basses temporals mediterrànies arreu del món. In: Fraga P, Estaún I, Cardona E (eds) Basses temporals mediterrànies. LIFE BASSES: gestió i conservació a Menorca. Consell Insular de Menorca, Maó, pp 23–40

Hairston NG, Van Brunt RA, Kearns CM, Engstrom DR (1995) Age and survivorship of diapausing eggs in a sediment egg bank. Ecology 76:1706–1711

Hambright K, Zohary T (1998) Lakes Hula and Agmon: destruction and creation of wetland ecosystems in northern Israel. Wetl Ecol Manage 6:83–89

Hartland-Rowe R (1967) An annotated catalogue of phyllopod crustacea recorded from Israel, with a key for their identification. Isr J Zool 16:88–95

Hartland-Rowe R (1972) The limnology of temporary waters and the ecology of Euphyllopoda. In: Clark RB, Wootton RJ (eds) Essays in hydrobiology. University of Exeter, New York, pp 15–31

Havel J, Bruckerhoff LA, Funkhouser MA, Gemberling AR (2014) Resistance to desiccation in aquatic invasive snails and implications for their overland dispersal. Hydrobiologia 741:89–100

Helm BP (1998) Biogeography of eight large branchiopods endemic to California. p. 124–139. In: Witham, CW (ed). Ecology, Conservation, and Management of Vernal Pool Ecosystems—Proceedings from a 1996 Conference. California Native Plant Society, pp 124–139

Hewitt GM (2000) The genetic legacy of the Quaternary ice ages. Nature 405:907–913

Higgins MJ, Merrit RW (1999) Invertebrate seasonal patterns and trophic relationships. In: Batzer D, Rader RB, Wissinger SA (eds) Invertebrates in freshwater. Wetlands of North America. Wiley, New York, pp 279–297

Holland RF, Jain SK (1981) Insular biogeography of vernal pools in the Central Valley of California. Am Nat 117:24–37

Holland CC, Honea J, Gwin SE, Kentula ME (1995) Wetland degradation and loss in the rapidly urbanizing area of Portland, Oregon. Wetlands 15:336–345

Horwitz P, Rogan R et al (2009) Wetland invertebrate richness and endemism on the Swan Coastal Plain, Western Australia. Mar Freshw Res 60:1006–1020

Incagnone G, Marrone F et al (2015) How do freshwater organisms cross the "dry ocean"? A review on passive dispersal and colonization processes with a special focus on temporary ponds. Hydrobiologia 750:103–123

IPCC [Intergovernmental Panel on Climate Change] (2007) Climate Change 2007: The Scientific Basis. Contribution of Working Group I to the Fourth Assessment Report of the Intergovernmental Panel on Climate Change. Cambridge Univ Press, New York

Issar AS (1990) Water shall flow from the rock: hydrogeology and climate in the lands of the Bible. Springer, Berlin

Jenkins KM, Boulton AJ (2007) Detecting impacts and setting restoration targets in arid-zone rivers: aquatic micro-invertebrate responses to reduced floodplain inundation. J Appl Ecol 44:823–832

Keeley JE, Zedler PH (1998) Characterization and global distribution of vernal pools. In: Witham CW (ed) Ecology, conservation, and management of vernal pool ecosystems. California Native Plant Society, Sacramento

Kenk R (1949) The animal life of temporary and permanent ponds in southern Michigan. Miscellaneous Publications Museum of Zoology, University of Michigan 71:1–66

Kerfoot WC, Lynch M (1987) Branchipod communities: associations with planktivorous fish. In: Kerfoot WC, Sih A (eds) Predation: direct and indirect impacts on aquatic communities. University Press of New England, Hanover, pp 367–378

King JL, Simovich MA, Brusca RC (1996) Species richness, endemism and ecology of crustacean assemblages in northern California vernal pools. Hydrobiologia 328:85–116

Kneitel JM (2014) Inundation timing, more than duration, affects the community structure of California vernal pool mesocosms. Hydrobiologia 732:71–83

Lahr J, Diallo AO et al (1999) Phenology of invertebrates living in a sahelian temporary pond. Hydrobiologia 405:189–205

Lake PS, Bayly IAE, Morton DW (1989) The phenology of a temporary pond in western Victoria, Australia, with special reference to invertebrate succession. Arch Hydrobiol 115:171–202

Levin N, Elron E, Gasith A (2009) Decline of wetland ecosystems in the coastal plain of Israel during the 20th century: implications for wetland conservation and management. Landsc Urban Plan 92:220–232

Magnusson AK, Williams DD (2006) The roles of natural temporal and spatial variation versus biotic influences in shaping the physicochemical environment of intermittent ponds: a case study. Arch Hydrobiol 165:537–556

Magnusson AK, Williams DD (2009) Top-down control by insect predators in an intermittent pond–a field experiment. Ann Limnol 45:131–143

March F, Bass D (1995) Application of island biogeography theory to temporary pools. J Freshw Ecol 10:83–85

Marchetti MP, Garr M, Smith ANH (2010) Evaluating wetland restoration success using aquatic macroinvertebrate assemblages in the Sacramento Valley, California. Restor Ecol 18:457–466

Marrone F, Castelli G, Naselli-Flores L (2009) Sicilian temporary ponds: an overview of the composition and affinities of their crustacean biota. In: Fraga P, Estaún I, Cardona E (eds) Basses temporals mediterrànies. LIFE BASSES: gestió i conservació a Menorca. Consell Insular de Menorca, pp 189–202

Martins MJF, Namiotko T et al (2010) Contribution to the knowledge of the freshwater Ostracoda fauna in continental Portugal, with an updated checklist of recent and Quaternary species. J Limnol 69:160–173

Marty JT (2005) Effects of cattle grazing on diversity in ephemeral wetlands. Conserv Biol 19:1626–1632

Matesanz S, Valladares F (2014) Ecological and evolutionary responses of Mediterranean plants to global change. Environ Exp Bot 103:53–67

McAbendroth LC (2004) Ecology and conservation of Mediterranean temporary ponds in the UK. Ph.D. Thesis, University of Plymouth

Meintjes S (1996) Seasonal changes in the invertebrate community of small shallow ephemeral pans at Bain's Vlei, South Africa. Hydrobiologia 317:51–64

Metge G (1986) Etude des écosystèmes hydromorphes (dayas et merjas) de la Méséta Occidentale Marocaine. Ph.D. Thesis, Unive Droit, d'Econ Sci Aix-Marseille

Miguel-Chinchilla L, Boix D, Gascón S, Comín FA (2014) Taxonomic and functional successional patterns in macroinvertebrates related to flying dispersal abilities: a case study from isolated manmade ponds at reclaimed opencast coal mines. Hydrobiologia 732:111–122

Miracle MR, Sahuquillo M, Vicente E (2008) Large branchiopods from freshwater temporary ponds of Eastern Spain. Verhandlungen der Internationale VerTheor Angewan Limnol 30:501–505

Mlambo MC, Reed CC, Day JA (2009) Overview of temporary wetlands' macroinvertebrates assemblages in the Cape Floristic Region, South Africa. In: Fraga, P (ed). International Conference on Mediterranean Temporary Ponds: Proceedings & Abstracts, Maó, pp 203–213

Mlambo MC, Bird MS, Reed CC, Day JA (2011) Diversity patterns of temporary wetland macroinvertebrate assemblages in the south-western Cape, South Africa. Afr J Aquat Sci 36:299–308

Mokany A, Wood JT, Cunningham SA (2008) Effect of shade and shading history on species abundances and ecosystem processes in temporary ponds. Freshw Biol 53:1917–1928

Moubayed-Breil J, Ashe P, Langton PH (2012) *Stygocladius multisetosus* gen. nov., sp. nov., a rheophilic element, inhabiting basaltic and karstic helocrenes in France and Algeria (Diptera: Chironomidae). Fauna Norvegica 31:175–182

Mozley A (1932) A biological study of a temporary pond in Western Canada. Am Nat 66:235–249

Mura G (2004) Structure and functioning of the "egg bank" of a fairy shrimp in a temporary pool: *Chirocephalus ruffoi* from Pollino National Park (Southern Italy) as a case study. Int Rev Hydrobiol 89:35–50

Murray J (1911) The annual history of a periodic pond. Int Rev Gesam Hydrobiol 4:300–310

Nahal I (1981) The Mediterranean climate from a biological view point. In: Di Castri F, Goodal DW, Specht RL (eds) Ecosystems of the world. II. Mediterranean-type shrublands, Elsevier, Amsterdam, pp 63–86

Neill WE (1990) Induced vertical migration in copepods as a defense against invertebrate predation. Nature 345:524–526

Nourisson M, Aguesse P (1961) Cycle annuel des phyllopodes d'une mare temporaire de Camargue. Bull Soc Zool Fr 86:754–762

O'Neill BJ, Thorp JH (2014) Untangling food-web structure in an ephemeral ecosystem. Freshw Biol 59:1462–1473

Olson DM, Dinerstein E et al (2001) Terrestrial ecoregions of the world: a new map of life on earth. BioScience 51:933–938

Otero M, Rossi V, Baltanas A, Menozzi P (1998) Effect of genotype and photoperiod on diapause strategies in *Eucypris virens* (Jurine, 1820)(Crustacea: Ostracoda). Arch Hydrobiol Spec Adv Limnol 52:229–236

Paine RR (1969) The *Pisaster-Tegula* interaction: prey patches, predator food preference and intertidal community structure. Ecology 50:950–961

Pajunen VI, Pajunen I (2003) Habitat selection in rock pool corixids: the effect of local density on dispersal. Hydrobiologia 495:73–78

Parekh PA, Paetkau MJ, Gosselin LA (2014) Historical frequency of wind dispersal events and role of topography in the dispersal of anostracan cysts in a semi-arid environment. Hydrobiologia 740:51–59

Parra G, Jiménez-Melero R, Guerrero F (2005) Agricultural impacts on Mediterranean wetlands: the effect of pesticides on survival and hatching rates in copepods. Ann Limnol Int J Limnol 41:161–167

Petrusek A, Tollrian R et al (2009) A "crown of thorns" is an inducible defense that protects *Daphnia* against an ancient predator. Proc Natl Acad Sci U S A 106:2248–2252

Pieri V, Martens K et al (2006) Distribution of recent ostracods in inland waters of Sicily (Southern Italy). J Limnol 65:1–8

Pinceel T, Vanschoenwinkel B, Uten J, Brendonck L (2013) Mechanistic and evolutionary aspects of light-induced dormancy termination in a temporary pond crustacean. Freshw Sci 32:517–524

Pinder A, Quinlan K, Cale D, Shiel R (2013) Invertebrate communities and hydrological persistence in seasonal claypans of Drummond Nature Reserve, Western Australia. Gov West Austral, Dept Parks Wildl

Pyke CR (2005) Interactions between habitat loss and climate change: implications for fairy shrimp in the Central Valley Ecoregion of California, USA. Clim Change 68:199–218

Quintana XD, Arim M et al (2015) Predation and competition effects on the size diversity of aquatic communities. Aquat Sci 77:45–57

Rabet N, Clarac F et al (2015) Review of the *Eulimnadia* (Branchiopoda: Spinicaudata) from North Africa and adjacent regions, with two new species from Mauritania. Journal of Crustacean Biology 35:461–472

Rabus M, Waterkeyn A et al (2012) Interclonal variation, effectiveness and long-term implications of *Triops*-induced morphological defences in *Daphnia magna* Strauss. J Plank Res 34:152–160

Rhazi L, Grillas P et al (2012) Vegetation of Mediterranean temporary pools: a fading jewel? Hydrobiologia 689:23–36

Ribera I, Foster GN, Vogler AP (2003) Does habitat use explain large scale diversity patterns in European water beetles? Ecography 26:145–152

Richter S, Timms BV (2005) A list of the recent clam shrimps (Crustacea: Laevicaudata, Spinicaudata, Cyclestherida) of Australia, including a description of a new species of *Eocyzicus*. Rec Austral Mus 57:341–354

Ripley BJ, Simovich MA (2009) Species richness on islands in time: variation in ephemeral pond crustacean communities in relation to habitat duration and size. Hydrobiologia 617:181–196

Robson BJ, Clay CJ (2005) Local and regional macroinvertebrate diversity in the wetlands of a cleared agricultural landscape in south-western Victoria, Australia. Aquat Conserv Mar Freshw Ecosys 15:404–414

Robson BJ, Chester ET, Austin CM (2011) Why life history information matters: drought refuges and macroinvertebrate persistence in non-perennial streams subject to a drier climate. Mar Freshw Res 62:801–810

Rodríguez MA, Magnan P (1993) Community structure of lacustrine macrobenthos: do taxon-based and size-based approaches yield similar insights? Can J Fish Aquat Sci 50:800–815

Rodríguez-Rodríguez M (2007) Hydrogeology of ponds, pools, and playa-lakes of southern Spain. Wetlands 27:819–830

Rogers DC (1998) Aquatic macroinvertebrate occurrences and population trends in constructed and natural vernal pools in Folsom, California. In: Witham CW (ed). Ecology, Conservation, and Management of Vernal Pool Ecosystems. California Native Plant Society, pp 224–235

Rogers DC (2014) Anostracan (Crustacea: Branchiopoda) zoogeography I. North American bioregions. Zootaxa 3838:251–275

Rogers DC, Hamer M (2012) Two new species of *Metabranchipus* Masi, 1925 (Anostraca: Branchipodidae). J Crust Biol 32:972–980

Rogers DC, Weeks SC, Hoeh WR (2010) A new species of *Eulimnadia* (Crustacea; Branchiopoda; Diplostraca; Spinicaudata) from North America. Zootaxa 2413:61–68

Rossi V, Menozzi P (1990) The clonal ecology of *Heterocypris incongruens* (Ostracoda). Oikos 57:388–398

Rossi V, Menozzi P (1993) The clonal ecology of *Heterocypris incongruens* (Ostracoda): life-history traits and photoperiod. Funct Ecol 7:177–182

Rouissi M, Boix D et al (2014) Spatio-temporal variability of faunal and floral assemblages in Mediterranean temporary wetlands. Comptes Rendus Biol 337:695–708

Ruhí A, Herrmann J et al (2012) How do early successional patterns in man-made wetlands differ between cold temperate and Mediterranean regions? Limnologica 42:328–339

Ruhí A, Boix D et al (2013a) Functional and phylogenetic relatedness in temporary wetland invertebrates: current macroecological patterns and implications for future climatic change scenarios. PLoS One 8, e81739

Ruhí A, Boix D et al (2013b) Nestedness and successional trajectories of macroinvertebrate assemblages in man-made wetlands. Oecologia 171:545–556

Ruhí A, Chappuis E et al (2014) Environmental filtering determines community patterns in temporary wetlands: a multi-taxon approach. Hydrobiologia 723:25–39

Rzóska J (1961) Observations on tropical rainpools and general remarks on temporary waters. Hydrobiologia 17:265–286

Sahuquillo M, Miracle MR (2010) Crustacean and rotifer seasonality in a Mediterranean temporary pond with high biodiversity (Lavajo de Abajo de Sinarcas, Eastern Spain). Limnetica 29:75–92

Sala J, Gascón S et al (2004) Proposal of a rapid methodology to assess the conservation status of Mediterranean wetlands and its application in Catalunya (NE Iberian Peninsula). Arch Sci 57:141–151

Samraoui B, Dumont HJ (2002) The large branchiopods (Anostraca, Notostraca and Spinicaudata) of Numidia (Algeria). Hydrobiologia 486:119–123

Sánchez B, Angeler DG (2007) Can fairy shrimps (Crustacea: Anostraca) structure zooplankton communities in temporary ponds? Mar Freshw Res 58:827–834

Sancho V, Lacomba JI (eds) (2010) Conservación y restauración de puntos de agua para la biodiversidad. Generalitat Valenciana, Conselleria de Medi Ambient, Aigua, Urbanisme i Habitatge, València

Sawchyn WW, Church NS (1973) The effects of temperature and photoperiod on diapause development in the eggs of four species of *Lestes* (Odonata: Zygoptera). Can J Zool 51:1257–1265

Schneider DW, Frost TM (1996) Habitat duration and community structure in temporary ponds. J N Am Benthol Soc 15:64–86

Schriever TA, Williams DD (2013) Influence of pond hydroperiod, size, and community richness on food-chain length. Freshw Sci 32:964–975

Serrano L, Serrano L (1996) Influence of groundwater exploitation for urban water supply on Temporary Ponds from the Doñana National Park (SW Spain). J Environ Manage 46:229–238

Silberbush A, Markman S et al (2010) Mosquitoes use hydrocarbons to detect larval predators when selecting an oviposition site. Ecol Lett 13:1129–1138

Sim LL, Chambers JM, Davis JA (2006a) Ecological regime shifts in salinised wetland systems. I. Salinity thresholds for the loss of submerged macrophytes. Hydrobiologia 573:89–107

Sim LL, Davis JA, Chambers JM, Strehlow K (2006b) What evidence exists for alternative ecological regimes in salinizing wetlands? Freshw Biol 51:1229–1248

Sim LL, Davis JA et al (2013) The influence of changing hydroregime on the invertebrate communities of temporary seasonal wetlands. Freshw Sci 32:327–342

Simovich MA (1998) Crustacean biodiversity and endemism in California's ephemeral wetlands. In: Witham CW (ed). Ecology, Conservation, and Management of Vernal Pool Ecosystems. California Native Plant Society, pp 107–118

Simovich MA, Hathaway SA (1997) Diversified bet-hedging as a reproductive strategy of some ephemeral pool anostracans (Branchiopoda). J Crust Biol 17:38–44

Simovich MA, Davis KB, Bohonak AJ (2013) Landscape homogenization threatens the genetic integrity of the endangered San Diego fairy shrimp *Branchinecta sandiegonensis* (Branchiopoda: Anostraca). J Crust Biol 33:730–740

Solimini AG, Della Bella V, Bazzanti M (2005) Macroinvertebrate size spectra of Mediterranean ponds with differing hydroperiod length. Aquat Conserv Mar Freshw Ecosyst 15:601–611

Spencer M, Blaustein L (2001) Hatching responses of temporary pool invertebrates to signals of environmental quality. Isr J Zool 47:397–418

Spencer M, Blaustein L, Schwartz S, Cohen J (1999) Species richness and the proportion of predatory animal species in temporary freshwater pools: relationships with habitat size and permanence. Ecol Lett 2:157–166

Stamou GP (1998) Arthropods of Mediterranean-type ecosystems. Springer, Berlin
Stav G, Blaustein L, Margalit Y (2000) Influence of nymphal *Anax imperator* (Odonata: Aeshnidae) on oviposition by the mosquito *Culiseta longiareolata* (Diptera: Culicidae) and community structure in temporary pools. J Vector Ecol 25:190–202
Stewart RE, Kantrud HA (1972) Vegetation of prairie potholes, North Dakota, in relation to quality of water and other environmental factors. US Geol Surv Profess Pap 585-D, Washington
Strachan SR, Chester ET, Robson BJ (2014) Microrefuges from drying for invertebrates in a seasonal wetland. Freshw Biol 59:2528–2538
Strachan SR, Chester ET, Robson BJ (2015) Freshwater invertebrate life history strategies for surviving desiccation. Springer Sci Rev 3:57–75
Sutula MA, Stein ED et al (2006) A practical guide for the development of a wetland assessment method: the California experience. J Am Water Resour Assoc 42:157–175
Terzian E (1979) Ecologie des mares temporaires de l'Isoetion dans la Crau et l'Esterel (France). Univ Droit, Econ Sci Aix Marseille, MS Thesis
Thiéry A (1979) Influence de l'assechement estival sur le peuplement d'insectes aquatiques d'un marais saumatre temporaire en Crau (Bouches-du-Rhone). Ann Limnol 15:181–191
Thiéry A (1987) Les crustacés Branchiopodes Anostraca, Notostraca & Conchostraca des milieu limniques temporaires (Dayas) au Maroc. Taxonomie, biogéographie, ecologie. Ph.D. Thesis, Université Aix-Marseille III
Thiéry A (1991) Multispecies coexistence of branchiopods (Anostraca, Notostraca & Spinicaudata) in temporary ponds of Chaouia plain (western Morocco): sympatry or syntopy between usually allopatric species. Hydrobiologia 212:117–136
Thiéry A (1996) Large branchiopods (Crustacea: Anostraca, Notostraca, Spinicaudata, Laevicaudata) from temporary inland waters of the Arabian Peninsula. Fauna Saudi Arab 15:37–98
Thiéry A (2004) Macrocrustaceans. In: Grillas P, Gauthier P, Yavercovski N, Perennou C (eds) Mediterranean temporary pools. Issues relating to conservation, functioning and management, vol 1. Sta Biologique Tour du Valat, pp 31–33
Timms BV (2002) The fairy shrimp genus *Branchinella* Sayce (Crustacea: Anostraca: Thamnocephalidae) in Western Australia, including a description of four new species. Hydrobiologia 486:71–89
Timms BV (2005) Two new species of *Branchinella* (Anostraca: Thamnocephalidae) and a reappraisal of the *B. nichollsi* group. Mem Queensland Mus 50:441–452
Timms BV (2006) The large branchiopods (Crustacea: Branchiopoda) of gnammas (rock holes) in Australia. J Roy Soc W Austral 89:163–173
Timms BV (2008) Further studies on the fairy shrimp genus *Branchinella* (Crustacea, Anostraca, Thamnocephalidae) in Western Australia, with descriptions of new species. Rec W Austral Mus 24:289–306
Timms BV (2009) A revision of the Australian endemic clam shrimp genus *Limnadopsis* Spencer & Hall (Crustacea: Branchiopoda: Spinicaudata: Limnadiidae). Rec Austral Mus 61:49–72
Timms BV (2012) Further studies on the fairy shrimp genus *Branchinella* (Crustacea: Anostraca: Thamnocephalidae) in Australia, with descriptions of five new species. Zootaxa 3595:35–60
Timms BV (2013) A revision of the Australian species of *Lynceus* Müller, 1776 (Crustacea: Branchiopoda: Laevicaudata, Lynceidae). Zootaxa 3702:501–533
Timms BV (2014) A review of the biology of Australian halophilic anostracans (Branchiopoda: Anostraca). J Biol Res Thessaloniki 21:21
Timms BV (2015) *Eulimnadia* (Branchiopoda: Spinicaudata) in Western Australia: three new species and a description of a rediscovered species. J Crust Biol 35:441–453
Timms BV, Richter S (2009) The clam shrimp *Eocyzicus* (Branchiopoda: Spinicaudata: Cyzicidae) in Australia. J Crust Biol 29:245–253
Tornero I, Sala J et al (2014) Aquatic macrofauna of Vila Nova de Milfontes temporary ponds, with the first record of *Cyphon hilaris* Nyholm, 1944 (Coleoptera: Scirtidae) from Portugal. Boletín Soc Entomol Aragonesa 55:326–330

Trochine C, Modenutti BE, Balseiro EG (2009) Chemical signals and habitat selection by three zooplankters in Andean Patagonian ponds. Freshw Biol 54:480–494

Tuckett RE, Merrit DJ et al (2010) Dormancy, germination and seed bank storage: a study in support of ex situ conservation of macrophytes of southwest Australian temporary pools. Freshw Biol 55:1118–1129

Tulbure MG, Kininmonth S, Broich M (2014) Spatiotemporal dynamics of surface water networks across a global biodiversity hotspot: implications for conservation. Environ Res Lett 9, e114012

Turki S, Turki B (2010) Copepoda and Branchiopoda from Tunisian temporary waters. Int J Biodiver Conserv 2:86–97

Underwood EC, Viers JH et al (2009) Threats and biodiversity in the mediterranean biome. Divers Distrib 15:188–197

US Federal Register (2003) Endangered and threatened wildlife and plants: final designation of critical habitat for four vernal pool crustaceans and eleven vernal pool plants in California and Southern Oregon: final rule. Federal Register 68:46684–46762

US Fish and Wildlife Service (2008) Riverside Fairy Shrimp (*Streptocephalus woottoni*) 5-Year Review: Summary and Evaluation. Carlsbad Office

Valdecasas AG, Camacho AI, Bello E (1992) Small water bodies. A neglected resource for aquatic invertebrate conservation. In: Conserving and Managing Wetlands for Invertebrates, Council Europe, Vaduz, pp 25–27

Valls L, Castillo-Escrivà, A et al. (in press) Human-mediated dispersal of aquatic invertebrates with waterproof footwear. Ambio

Van de Meutter F, Stoks R, De Meester L (2008) Size-selective dispersal of *Daphnia* resting eggs by backswimmers (*Notonecta maculata*). Biol Lett 4:494–496

van den Broeck M, Waterkeyn A et al (2015a) Assessing the ecological integrity of endorheic wetlands, with focus on Mediterranean temporary ponds. Ecol Indic 54:1–11

van den Broeck M, Waterkeyn A, Rhazi L, Brendonck L (2015b) Distribution, coexistence, and decline of Moroccan large branchiopods. J Crust Biol 35:355–365

Vanschoenwinkel B, Waterkeyn A et al (2008) Dispersal of freshwater invertebrates by large terrestrial mammals: a case study with wild boar (*Sus scrofa*) in Mediterranean wetlands. Freshw Biol 53:2264–2273

Vanschoenwinkel B, Hulsmans A et al (2009) Community structure in temporary freshwater pools: disentangling the effects of habitat size and hydroregime. Freshw Biol 54:1487–1500

Velasco J, Millán A (1998) Insect dispersal in a drying desert stream: effects of temperature and water loss. Southw Natur 43:80–87

Villagran-Mella R, Aguayo M, Parra LE, González A (2006) Relación entre características del hábitat y estructura del ensamble de insectos en humedales palustres urbanos del centro-sur de Chile. Revista Chilena de Historia Natural 79:195–211

Walton WE, Tietze NS, Mulla MS (1991) Consequences of tadpole shrimp predation on mayflies in some Californian ponds. Freshw Biol 25:143–154

Waterkeyn A, Grillas P et al (2009) Assemblage structure and dynamics of large branchiopods in Mediterranean temporary wetlands: patterns and processes. Freshw Biol 54:1256–1270

Waterkeyn A, Vanoverbeke J, Van Pottelbergh N, Brendonck L (2011a) While they were sleeping: dormant egg predation by *Triops*. J Plank Res 33:1617–1621

Waterkeyn A, Grillas P et al (2011b) Can large branchiopods shape microcrustacean communities in Mediterranean temporary wetlands? Mar Freshw Res 62:46–53

Waterkeyn A, Van Pottelbergh N et al (2013) Constitutive but no *Triops*-induced differences in bet-hedging strategies for hatching in *Daphnia*. Hydrobiologia 715:29–35

Wellborn GA, Skelly DK, Werner EE (1996) Mechanisms creating community structure across a freshwater habitat gradient. Annu Rev Ecol Syst 27:337–363

Wiggins GB, Mackay RJ, Smith IM (1980) Evolutionary and ecological strategies of animals in annual temporary pools. Arch Hydrobiol Suppl 58:97–206

Wilcox C (2001) Habitat size and isolation affect colonization of seasonal wetlands by predatory aquatic insects. Isr J Zool 47:459–476

Williams DD (1983) The natural history of a Nearctic temporary pond in Ontario with remarks on continental variation in such habitats. Int Rev Gesam Hydrobiol 68:239–253

Williams DD (1996) Environmental constraints in temporary fresh waters and their consequences for the insect fauna. J N Am Benthol Soc 15:634–650

Williams DD (2006) The biology of temporary waters. Oxford Univ Press, Oxford

Williams WD (1968) The distribution of *Triops* and *Lepidurus* (Branchiopoda) in Australia. Crustaceana 14:119–126

Williams WD (1975) A note on the macrofauna of a temporary rainpool in semi-arid Western Australia. Austral J Mar Freshw Res 26:425–429

Williams WD (1981) Inland salt lakes: an introduction. Hydrobiologia 81:1–14

Witham CW (ed) (1998) Ecology, conservation, and management of vernal pool ecosystems. California Native Plant Society, Sacramento

Yaron Z (1964) Notes on the ecology and entomostracan fauna of temporary rainpools of in Israel. Hydrobiologia 24:489–513

Yee DA, Taylor S, Vamosi SM (2009) Beetle and plant density as cues initiating dispersal in two species of adult predaceous diving beetles. Oecologia 160:25–36

Zacharias I, Zamparas M (2010) Mediterranean temporary ponds. A disappearing ecosystem. Biodiv Conserv 19:3827–3834

Zacharias I, Dimitriou E, Dekker A, Dorsman E (2007) Overview of temporary ponds in the Mediterranean region: threats, management and conservation issues. J Environ Biol 28:1–9

Zacharias I, Parasidoy A et al (2008) A "DPSIR" model for mediterranean temporary ponds: European, national and local scale comparisons. Ann Limnol Int J Lim 44(4):253–266

Zedler PH (1987) The ecology of southern California vernal pools: a community profile (No. NP-7900890). San Diego State Univ., CA (USA). Dept. of Biology, Fish and Wildlife Service, Slidell, LA (USA). National Wetlands Research Center

Zedler PH (2003) Vernal pools and the concept of "isolated wetlands". Wetlands 23:597–607

Chapter 6
Invertebrates of Irish Turloughs

Julian D. Reynolds

Introduction to Turloughs

Although turloughs are a recognized European habitat (CEC 2007; NPWS 2013b), almost all identified turloughs occur in Ireland, and the name is an Irish word usually interpreted to mean "dry lake." Published definitions of turloughs differ somewhat, depending on the focus or emphasis of study—geology, hydrology, botany, zoology or even terrestrial agriculture. One such, responding to the EU Water Framework Directive (EC 2000), is "groundwater-dependent terrestrial ecosystems" (Naughton et al. 2012; Kimberley and Coxon 2013).

Limnologically, turloughs are temporary freshwater bodies that form in topographic depressions in karstified limestone, filling from subterranean streams in response to localized rainfall and emptying, partly or completely, to groundwater conduits. The substrate and variable filling pattern of turloughs differentiates them from other forms of temporary waterbody. Figure 6.1 is a schematic diagram of the water budget of turloughs. It indicates that in addition to the main water inputs from groundwater, there may be small additional amounts from precipitation and overland flow. Surface streams may or may not be present. Water is lost by evapotranspiration as well as through sinks into the underlying karst. The extent and duration of flooding depends on seasonality, local weather conditions, and water table. The substrate in a turlough is usually limestone rock or thin soil, sometimes peaty, helping to retain water.

Karstification of limestone was most active in warm or tropical geological periods when sea-level was lower than that at present, and this resulted in development of tunnels and caves within the rock through the action of percolating acidic rainwater, in which streams flow or once flowed. Karstified limestone underlies some 40 % of Ireland. Later erosion and cave collapse has led to their exposure at different levels, streams sometimes meeting the surface as springs, resurgences or sinks. Irish

D. Batzer, D. Boix (eds.), *Invertebrates in Freshwater Wetlands*,
DOI 10.1007/978-3-319-24978-0_6

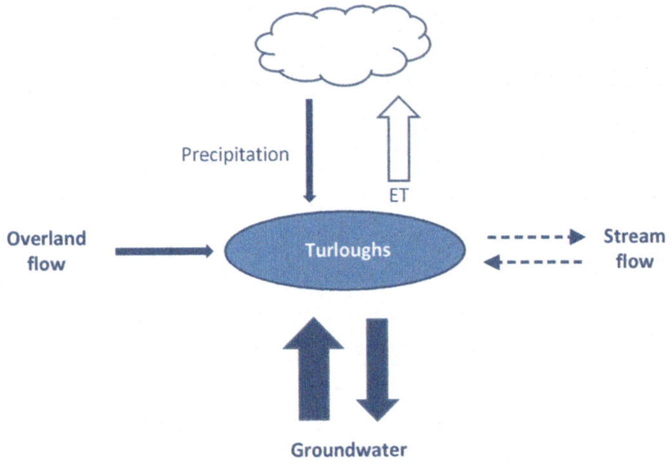

Fig. 6.1 Schematic of turlough water budgets

turloughs apparently all lie within the zone formerly affected by the last glaciation, which stripped off rocks and soil to leave exposed limestone strata.

Even if hydrology and geomorphology are suitable, with a history of glacial erosion, turloughs only develop when climatic conditions permit and rainfall exceeds evapotranspiration for at least part of the year (Fig. 6.1). In Ireland's oceanic climate, turloughs may fill and empty at any time of the year, but aquatic phases are generally found in winter and are therefore cool water, while terrestrial phases tend to occur in the drier, warmer months (Reynolds et al. 2004). Some turloughs at or near sea-level have a lunar tidal pattern influenced by tidal pressure through rock conduits, superimposed on seasonality (Reynolds 2000).

Turloughs are largely confined to the western third of Ireland, from Sligo to Kilkenny and Limerick (Coxon 1987; Goodwillie 1992), but occur most commonly in Mayo, Galway, and Clare, where deep strata of karstified Carboniferous limestone occur in lowlands (NPWS 2013b). However, turlough-like waterbodies have occasionally been described from elsewhere where the environmental conditions are right, with one (Pant-y-llyn) known from Wales (Campbell et al. 1992; Blackstock et al. 1993). Others, such as poljes, large glacial hollows that may or may not flood on an annual basis, and dolines, deep, often funnel-shaped sinkholes, whose extreme manifestations are called "tiankengs," occur in warmer climates (Gunn 2006). Similar karstic wetlands have been described in Slovenia, e.g., the intermittently filling Lake Cerknica (26 km^2) (Kranjc 2006), eastern Spain (Garcia-Gil et al. 1992; Boix et al. 2001), in North America from Mexico to eastern Canada (Coté et al. 1990), and also in China and Australia (Waltham 2006). However, turloughs differ markedly from these other geomorphological features (Coxon 1987; Naughton et al. 2012), and the faunal affinities of Irish turloughs are more northern than Mediterranean or further afield.

A complex waterbody usually referred to as a turlough is Rahasane (220 ha) in County Galway, atypical in that it fills from the adjacent Dunkellin River in flood. The anostracan *Tanymastix stagnalis* has been recorded here (Young 1975, 1976) and in adjacent temporary grassland pools, probably its more typical habitat. The species occurs sporadically in Europe, but is more commonly encountered in Mediterranean regions, and it has not been re-found in recent years in Ireland (Sheehy Skeffington et al. 2006).

In Ireland, temporary "winter-lakes" on flooded grassland show some similarities with turloughs and have been studied with them, principally by Grainger (1966, 1976, 1979), Ali et al. (1987), and Grainger and Holmes (1989). While these Irish habitats differ from turloughs in that they are not groundwater-fed, they have in common a cool-water, winter-filling periodicity and show similarities with Nordic temporary pools on Ordovician limestone Alvar grassland (e.g., Rosén 2006). The first unequivocal Irish record of the calanoid copepod *Diaptomus castor*, previously confused with the widespread planktonic *D. gracilis*, was made by Grainger (1966) in winter-flooded grassland in Mayo. *D. castor* and *D. gracilis* also often occur in turloughs.

Zonation of peripheral turlough vegetation has long been noted (Praeger 1932; Ivimey-Cook and Proctor 1966; Proctor 2010). Most turloughs fill in autumn or early winter. In the aquatic phase, turlough vegetation is limited and mostly annual, but there may also be some perennials in residual water in sinkholes (e.g., *Elodea canadensis*, *Ceratophyllum demersum*), or aquatic mosses resistant to desiccation, such as *Cinclidotus fontinaloides* (Reynolds 2000). Vascular plants of wetlands such as *Persicaria amphibia*, *Mentha aquatica*, and *Potentilla anserina* (e.g., Scannell and Webb 1983; CEC 2007; NPWS 2013a; Reynolds 2014), or the turlough form of *Ranunculus repens*, with finely divided leaves (Lynn and Waldren 2003) are also frequent. Such plants are also characteristic of river floodplains or the drawdown zone of lakes.

As water levels decline, generally in early summer, the substrate becomes exposed and starts to dry. Sparse swards of wetland annual plants are first to appear, although some perennial marsh vegetation such as *Carex nigra* (Williams and Gormally 2009) or reedbeds may persist in parts of the basin where the bottom is sealed by peat or clay (Goodwillie 2003). Depending on time and period of exposure, the terrestrial phase may come to resemble damp limestone grassland, although late-draining turloughs may be dominated by an annual *Chenopodion rubri* community (NPWS 2013b).

Invertebrate Assemblages in Turloughs

Numerous authors have written on turlough invertebrates and their ecology, but as for many other ecosystems and communities in Ireland, research activity on turloughs has been sporadic rather than sustained. Early surveys of Burren vegetation, including turloughs, were carried out by Praeger (1932) and

Ivimey-Cook and Proctor (1966), while Scannell and Webb (1983) and others considered plant distribution and phytosociology, including a seminal report to the NPWS by Goodwillie (1992) identifying and categorizing the larger turloughs in Ireland. Hydrological and geomorphological studies included Coxon (1987) and Drew (1990). These works have been well reviewed, e.g., by Goodwillie (2003).

Turloughs and their invertebrates have been reviewed from different aspects, by Reynolds et al. (1998), Reynolds (2003), Goodwillie and Reynolds (2003), O'Connor et al. (2004), Sheehy Skeffington et al. (2006), Sheehy Skeffington and Gormally (2007), Williams and Gormally (2009), and Porst and Irvine (2009a, b). Some studies reflect an interest in the Carboniferous limestone massif of the Burren and its biodiversity, while others have concentrated on specific taxa or communities. Apart from some pioneering invertebrate studies (Lansbury 1965; Grainger 1976; Donaldson et al. 1979), aquatic and terrestrial invertebrates have been less completely studied than has geomorphology and vegetation.

The best-known of turlough invertebrates are the crustaceans, including over 40 cladocerans and 6 copepods, also gastropod molluscs and certain insect groups including 58 aquatic coleopterans, 25 heteropterans, and 9 odonates (Appendix). Some invertebrate groups found in turloughs have not been investigated, such as the rotifers, nematodes and tardigrades, and the oligochaetes and dipterans of turloughs are, in general, poorly studied. The terrestrial invertebrate fauna has also been neglected. Both aquatic and terrestrial turlough invertebrate assemblages are addressed in this chapter.

Various methods have been used for sampling turlough invertebrates. Sweeping, beating, and pitfall traps are all used for collecting terrestrial phases, while in the aquatic phase dragging a plankton net or sweeping with a benthic net is common (e.g., Porst and Irvine 2009a, b). However a box enclosure method was found to yield more representative samples of the littoral-benthic fauna than use of a sweep net (O'Connor et al. 2004).

Case Studies of Turlough Invertebrates

The following case studies summarize some of the major faunistic findings on invertebrates from turloughs. Appendix lists most taxa recorded from these habitats.

Invertebrates of Burren and Aran Islands Turloughs

Turloughs in the limestone Burren hills of Counties Clare and Galway have received considerable study of their invertebrates. Donaldson et al. (1979) listed the molluscs of three Burren turloughs, and a sustained study of aquatic snails (Byrne 1981), particularly of two *Lymnaea* spp., showed very slow growth in these oligotrophic waters (Byrne et al. 1989). Since then, there appears to be some confusion about the identity of the lymnaeids *L. palustris* and *L. fusca* in turloughs. Tattersfield (1998) summarized work on wetland molluscs from Aran Islands sites, listing 23 wetland species from a total of 34 mollusc species. Through ordination Tattersfield derived a turlough group of molluscs of which *Lymnaea peregra* and *Anisus leucostoma* are most characteristic.

The characteristic microcrustaceans in Burren turloughs have been documented (Reynolds 1982, 1985b) while Duigan (1992) and Duigan and Kovach (1991) published detailed distributional surveys of chydorids in Ireland, including some turloughs. Among typical turlough species is the large chydorid cladoceran *Eurycercus glacialis* of northern and arctic regions, first discovered in Ireland by Duigan (Duigan and Frey 1987a, b) and later in Scotland; its Irish range and ecological requirements were further studied by Reynolds and Marnell (1999).

Byrne (1981) also investigated ephemeropterans and some other insect groups of turloughs, and Lansbury (1965) recorded the heteropterans *Sigara lateralis, S. dorsalis* and *Gerris lacustris* from Burren turlough habitats. In total, Lansbury (1965) and Reynolds (1985a) recorded 29 invertebrate taxa from turloughs and other karstic habitats on the Aran Islands, showing many similarities with the mainland Burren, but with reduced species richness. A maximum of 11 taxa was recorded from turlough sites, compared with four in groundwater-fed wells. *Gammarus duebeni* was found in several habitats, suggesting it may penetrate through fissures in the limestone (Reynolds 1985a).

Gort Turlough Flooding

Following several exceptional and prolonged flooding episodes in the 1990s of a series of turloughs lying in the lowlands north of Gort, County Galway and ranging in size up to 290 ha, a project was initiated to look into causes and remedial action (Southern Water Global 1996, 1997; Tynan et al. 2002). While most attention was paid to hydrology, defining and delimiting the tracks of subterranean waters discharging into Galway Bay, Reynolds (2000) characterized the different turloughs and summarized the characteristic invertebrate fauna of 15 turlough-associated sites, grouped into five districts by their hydrology and apparent drainage relationships. Districts yielded between 36 (eastern sites) and 81 (northern sites) taxa. Species occurring in all five districts included the microcrustaceans *Daphnia pulex, Simocephalus vetulus, Chydorus sphaericus, Polyphemus pediculus, Cyclops agilis* and *Candona candida*, and the macroinvertebrates *Polycelis nigra, Bithynia tentaculata* and *Lymnaea peregra (=Radix balthica)*. The characteristic species *Eurycercus lamellatus, Diaptomus castor* and *Planorbis (Gyraulus) laevis* were not recorded in eastern sites, while *Eurycercus glacialis* was found only in western and southern turloughs. Bond (1997) listed over 240 lepidopterans from the area, relatively few of which had strong turlough links, including *Bactra furfurana, Deltote uncula*, and the scarce *Paraponyx stratiotata*.

Terrestrial Invertebrates of Turloughs

Pioneering work by Speight (1976, 1977) identified carabids and other beetles in the terrestrial phase of turloughs. More recently, Lott and Foster (1990) recorded terrestrial beetles from wetland sites, including turloughs, and insects of the grasslands associated with turloughs have received attention from Good and Butler (2001) and Good (2004). Hydrology and terrestrial phase insect communities have been the focus of attention in a series of studies (Moran et al. 2003, 2008, 2011; Ní Bhriain et al. 2002, 2003; Sheehy Skeffington and Gormally 2007; Williams et al. 2010). Moran et al. (2011) working on the 28 ha Skealogan turlough, designated a Special Area of Protection, studied a *Cirsio-Molinietum* community flooded for approximately three months in the year, and a more aquatic *Ranunculo-Potentillietum anserinae* sward, flooded for about 6 months, and summarized the carabid findings from the two plant communities. *Bembidion aeneum* and *Agonum muelleri* are characteristic carabids of short-duration flooded, grazed swards, with *Nebria brevicollis* and *Chlaenius nigricornis* representative of grazed swards undergoing longer duration flooding. Turlough species survive flooding by their short life cycle, ability to fly, and early breeding followed by hibernation; species of wetter areas tend to be larger.

Williams et al. (2009, 2010) studied the marsh flies (Diptera: Sciomyzidae) of this and other turloughs. They found a fauna of seven species, 91 % dominated by *Ilione albiseta* and other univoltine species, and that microhabitat conditions were important in their survival. Univoltine species with aquatic larvae preyed on pulmonates or bivalves; others survived on damp surfaces, or were fully terrestrial. Most of the group seem to track hydrological regimes and plant communities rather than favoring habitat heterogeneity. In a related study, Williams and Gormally (2009) looked at the role of environmental gradients in turloughs on terrestrial, aquatic, and semiaquatic molluscs.

Ecology and Natural History of Turlough Invertebrates

Aquatic Invertebrates

Aquatic invertebrates of turloughs fall into two groupings; planktonic or semiplanktonic forms (mainly microcrustaceans and dipteran larvae), and littoral-benthic forms, predominantly insects, but also many crustaceans, molluscans, and annelids. In turloughs, development of aquatic invertebrate communities is controlled by a specific range of limnological, chemical, and ecological factors. Whether planktonic or littoral-benthic, aquatic turlough invertebrates experience a suite of often harsh environmental conditions. The schematic in Fig. 6.2 summarizes the influence of some major factors on the aquatic biota of turloughs, with the importance of each factor explained below.

Limnological Controls

Clearly, important limnological factors to turlough invertebrates include hydrography, periodicity of aquatic phase, timing of exposure, and water temperature. Water in turloughs is usually temporary or periodic, most often present in cooler winter months (see Fig. 6.2). As such, the aquatic communities that develop are typically dominated by short-lived taxa, often adapted to cool water such as chydorid and daphniid microcrustaceans and some insect larvae (e.g., *Cloeon dipterum*). The large chydorid microcrustacean *Eurycercus glacialis* (Duigan and Frey 1987a, b), mentioned above, is cold adapted.

Hydroperiod will determine the communities that can develop; short hydroperiods tend to limit longer-lived predators, but too brief flooding may lead to the loss of species before they reach maturity. Turlough aquatic invertebrates must survive periods of desiccation, chiefly in summer, by behavioral or life-history adaptations (e.g., resistant resting stages such as cladoceran ephippia, or terrestrial stages in insects). Drying out in cold seasons can also kill desiccation-resistant resting stages. Because the aquatic phase is unpredictable in its timing and length, the aquatic invertebrate fauna of turloughs is restricted, and may have to recolonize actively or passively (Reynolds et al. 2004; O'Connor, personal communication 2014).

Some longer-lived site-faithful species, such as lymnaeid snails, *Gammarus duebeni* and odonate larvae, may survive dry phases in groundwater or small pools, in mud, or under a felt of drying vegetation. Although most insect colonizers of turloughs are ready flyers, brachypterous or apterous forms, for example among corixids, must be able to survive dry periods in situ (Tobin and McCarthy 2004).

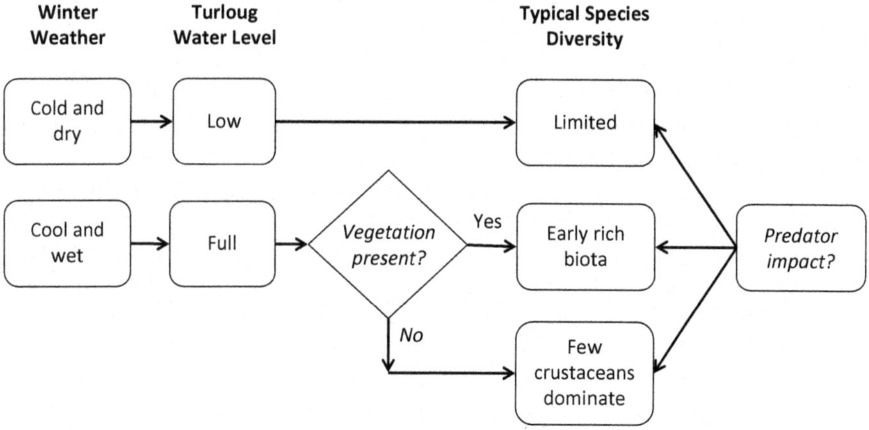

Fig. 6.2 Factors influencing aquatic phase biodiversity

The plankton is chiefly affected by limnological factors such as water movement, depth, area, fetch, and exposure, sometimes causing plankters to drift onshore with risks of stranding. However, the same factors at the correct time can place resting stages such as crustacean ephippia on the shoreline until the next major aquatic period. Active or passive dispersal of temporary or isolated pond macroinvertebrates is of wide interest, with a general review by Bilton et al. (2001a, b) and more specific zooplankton studies by, among others, Louette and De Meester (2005).

Turlough water chemistry is less variable than among many other types of waterbodies, being dominated by calcium and carbonate-based chemistry. However, the considerable variation in nutrient level in turloughs, from ultra-oligotrophic to eutrophic, affects productivity. Dissolved oxygen levels are also important but rarely limiting, except under eutrophic summer-flooding conditions, when algal blooms may decay leading to oxygen deficiency.

Biotic Controls

An important ecological factor for aquatic invertebrates is habitat complexity. Some complexity relates to the presence of irregularities in the turlough basin, where water and aquatic forms may be retained in sinkholes or in marshes. But complexity more often relates to biotic interactions with plants, competitors, and predators.

Microhabitat complexity is primarily related to the presence or absence of vegetation, and its degree or fineness of branching (Eitam et al. 2004). Periphyton grows on plant surfaces, which is food for insect and crustacean grazers, and refuge from predation. Leaves of the turlough form of *Ranunculus repens* and of the ubiquitous *Potentilla anserina* are both finely divided.

Available food webs are initially detritus-based and are quickly exploited by microcrustaceans and insect larvae. Aquatic invertebrate communities are then structured by competition, e.g., between early chydorid colonists for effective dominance (Reynolds et al. 2004), and by predation; both factors result in distinctive faunal species lists and specialist taxa.

Spencer et al. (1999) showed that larger pools tend to have higher proportions of predators, and Bilton et al. (2001a, b) suggested the pattern was due primarily to water permanence. In turloughs invertebrate predators are typically sparse and may be amphibious (such as diving spiders *Argyroneta aquatica*) or able to fly as adults (Coleoptera, Hemiptera, Odonata). Among planktonic predators, *Polyphemus pediculus* is found in many lowland turloughs, where the copepod *Cyclops scutifer* may also be common (Reynolds 2000, 2003). Fish, however, are rarely present in turloughs (Williams et al. 2006), allowing the larger planktonic cladocerans to thrive (Reynolds 1985b; Reynolds and Marnell 1999).

Terrestrial Invertebrates

In their terrestrial phase invertebrate communities of turloughs are structured by physical factors and microclimate, including temperature and desiccation, and by ecological factors (Regan 2005) such as timing of vegetation cover and prevalence of predators. The schematic in Fig. 6.3 indicates how spring weather (wet or dry, early or late), habitat complexity, and presence of predators affects terrestrial phase biodiversity.

As turloughs dry out, insect larvae that survived submergence (e.g., some Coleoptera and Diptera) contribute as adults to a suite of grassland terrestrial detritivores. Chief among terrestrial insect predators are heteropterans and a broad suite of staphylinids (Good 2004), carabids (Moran et al. 2011) and specialized mollusc predators such as sciomyzid fly larvae, which may be terrestrial or aquatic (Ryder et al. 2003; Williams et al. 2009, 2010); others include spiders and mites. These may be controlled by habitat complexity (some turloughs retain patches of standing water or marsh) and by levels of management such as grazing pressure and trampling (Moran et al. 2008, 2011; Ní Bhriain et al. 2002, 2003; Regan 2005; Regan and Moran 2005).

Different plant communities support a characteristic carabid community, e.g., the *Cirsio-Molinietum* community (short duration of flooding), with 9 carabid species in ungrazed habitats, 7 in grazed; and the *Ranunculo-Potentilletum anserinae* community (long duration of flooding), with 10 carabid species in ungrazed and 7 in grazed habitats. Carabids of turloughs are detailed in Moran et al. (2011), arranged by their plant community and indicator importance.

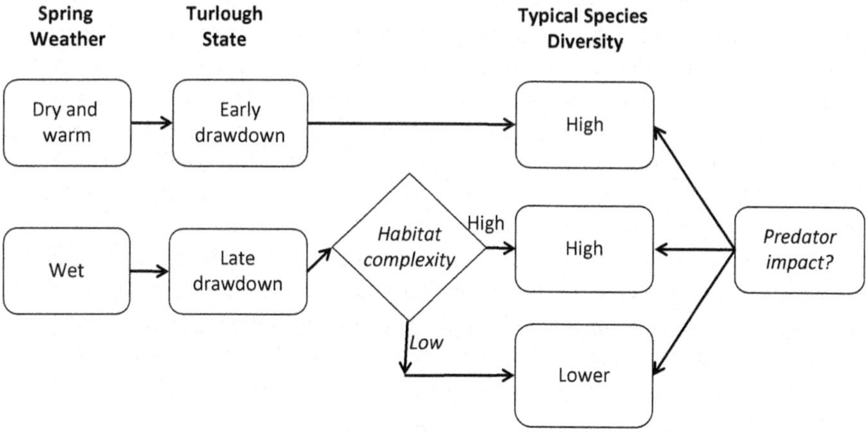

Fig. 6.3 Three main factors influencing terrestrial phase biodiversity

Turlough Invertebrates: Management and Conservation Issues

Turloughs are not uniform, and their management must depend on the primary issues affecting each and their relative importance, having due regard for the livelihood of land owners. Turlough water levels may rise rapidly following local rainfall, e.g., a rise of up to 8 m in 24 h is known in Blackrock Turlough, Co. Galway. Excess water may then damage houses and farm buildings, and inundate roads.

There is now better understanding of the groundwater influences of turloughs and other "groundwater-dependent terrestrial ecosystems" (Tynan et al. 2007; Kimberley and Coxon 2013). County maps of groundwater vulnerability to contaminants from human activities are available (Daly and Warren 1998). Stocking with heavier breeds of cattle may result in trampling of the soil and permanent sealing of the basin, converting turloughs into ponds. Finally, the degree and timing of drying out affects turlough site biodiversity (Collinson et al. 1995; Moran et al. 2008).

Some socioeconomic issues may override conservation priorities; chief among these have been structural damage from flooding (Ní Bhroin 2008) and a growing demand for land for grazing (Ní Bhriain et al. 2002, 2003). Ryder et al. (2005) and Williams et al. (2009) discuss the implications of farmland management and grazing on terrestrial plant and dipteran communities of turloughs. Curtis et al. (2009) stress the importance of avoiding drainage of the basin while managing for good water quality (avoiding ploughing, fertilization or enrichment) and periodicity. Mitigating actions for potentially damaging flood levels, discussed in the Gort Flooding reports (Southern Water Global 1996, 1997) include provision of small moveable dams on feeder streams to hold back or divert excess inflowing water.

Some turlough conservation issues in aquatic and terrestrial phases have been outlined by Reynolds (1996) and NPWS (2013a), and consultancy reports have identified others. Turloughs are a priority habitat in the EU Habitats Directive (EEC 1992) and many turloughs are designated Special Areas of Conservation (SACs) under the Directive—some 70 turlough SACs are listed in Tynan et al. (2007), also giving their trophic sensibilities to nutrient enrichment. Others have less statutory protection, although they may contain protected species. Annex II in the EU Habitats Directive lists priority species, including three species of *Vertigo* snails and the Marsh Fritillary butterfly (*Euphydryas aurinia*); these may occur within turloughs but are not characteristic of them. No characteristic turlough species are protected in this European directive. The only invertebrates apparently limited to turlough habitats in Ireland are the crustacean *Eurycercus glacialis* and the aquatic weevil *Bagous brevis*. Both are considered rare in Ireland and their protection is important.

Biomonitoring, including with invertebrates, is a useful tool to assess turlough ecological health. A large-scale turlough project involving many specialists has been ongoing since 2005, reporting to government via the National Parks and Wildlife Service; some results have been published. Porst and Irvine (2009a, b) listed invertebrates from lowland turloughs and commented on within-turlough habitat variability; this was considered to be less than between turloughs, and Porst and Irvine (2009b) recommended that a single site sample would be an adequate metric for the aquatic community for many purposes. Porst et al. (2012) examined the turlough recolonization process after flooding, highlighting the importance of life-cycle strategies to overcome ecological disturbance. Soils and groundwater pressures were discussed by Kimberley and Coxon (2013) and Kimberley and Waldren (2012).

Water beetles are increasingly seen as useful for characterizing the conservation value of waterbodies, and some detailed surveys of beetles of limestone lakes and turloughs have been carried out. Early surveys by Bilton (1988), Lott and Foster (1990) and Bilton and Lott (1991) identified a suite of beetles associated with the mossy shores of limestone lakes and turloughs. Beetle communities were classified by Foster et al. (1992) and these mossy-edge beetles were considered sensitive to disturbance. A Red List of Irish beetles was produced (Foster et al. 2009), while additions to turlough beetle lists were provided by Bradish et al. (2002), O'Connor et al. (2004), and Reynolds (2014). The beetles *Agabus labiatus, Bagous brevis, B. limosus* and *Berosus stigmaticollis* were found to have particular strongholds in turlough habitats.

Despite an increasing body of knowledge of characteristic turlough species, there is currently no national protection for turlough invertebrates. However, Irish Red Lists, particularly those for water beetles (Foster et al. 2009) and odonates (Nelson et al. 2011) identify species largely restricted to turloughs, and describe their conservation status. It is to be hoped that with such official recognition, some endangered invertebrates of these unique waterbodies will gain appropriate protection at a national level.

Acknowledgments I am very grateful to Áine O'Connor and Brian Nelson of the National Parks and Wildlife Service, Dublin, for much helpful advice and comment. My thanks are also due to many others who have patiently discussed turlough issues with me at various times, in particular Roger Byrne, Catherine Coxon, Catherine Duigan, Garth Foster, Roger Goodwillie, Jack Grainger, Maria Long, Deirdre Lynn, Ferdia Marnell, Mark Murphy, David Nash, Gwendoline Porst, Michael Proctor, Eugenie Regan, Sylvia Reynolds, Cilian Roden, and Micheline Sheehy Skeffington.

Appendix

List of turlough invertebrate taxa. Most invertebrate turlough dwellers are typical of, but not restricted to, turloughs in the wet phase (List A) or in the dry phase (List B). Where species occur in either phase, or where turloughs do not dry out completely, they have been placed in the more characteristic phase habitat. In the list, species in **bold** are typical of turloughs and considered of high conservation value (e.g., Reynolds 2000; NPWS 2007, 2013a; Porst and Irvine 2009a, b). Species in (brackets) have rarely been recorded in turloughs, and records may be erroneous.

Higher classification	Families	Genus-species	References*
A. WET PHASE			
PHYLUM CNIDARIA			
Class Hydrozoa			
Order Anthomedusae	Family Hydridae	*Chlorohydra viridissima*	1, 40
		Hydra sp.	2, 3
PHYLUM PLATY HELMINTHES			
Class Turbellaria			
Order Tricladida	Family Planariidae	*Polycelis nigra/tenuis*	1, 4, 5, 54
		Polycelis nigra	1, 3, 5, 6, 7, 40
		Polycelis sp.	2
PHYLUM ANNELIDA			
Class Oligochaeta			4, 7, 54
Class Hirudinea	Family Glossiphoniidae	*Glossiphonia complanata*	1, 3, 4, 7, 8, 40, 54
		Theromyzon tessulatum	7, 54
		Helobdella stagnalis	40
	Family Haemopidae	*(Haemopis sanguisuga)*	4, 7, 54
PHYLUM MOLLUSCA			
Class Gastropoda			
Order Mesogastropoda	Family Tateidae	*Potamopyrgus antipodarum (jenkinsii)*	1, 3, 4, 7, 53
	Family Lymnaeidae	***Galba truncatula***	1, 4, 7, 9, 52, 54
		(Lymnaea glabra)	4, 54
		Lymnaea (Stagnicola) palustris	3, 7, 8, 11, 12
		Lymnaea peregra (Radix balthica)	1, 3, 4, 7, 8, 9, 11, 40, 53, 54
		Lymnaea palustris	7
		Lymnaea fusca	52, ml
		Lymnaea stagnalis	3, 4, 13, 40
		Myxas glutinosa	4
	Family Physidae	*(Aplexa hypnorum)*	7
		Physa fontinalis	4, 7, 54
	Family Succineidae	*Oxyloma pfeifferi*	9
		Oxyloma elegans	52
		Succinea putris	3, 40, 54
		Succinea sp.	4, 7
	Family Acroloxidae	*(Acroloxus lacustris)*	40

(continued)

Higher classification	Families	Genus-species	References*
	Family Planorbidae	*Anisus leucostoma*	1, 7, 9
		Anisus vortex	40
		Bathyomphalus contortus	3, 4, 40, 54
		Gyraulus crista	4, 8, 52, 54
		Gyraulus albus	7, 40
		Gyraulus laevis	3, 40, 54
		Planorbarius corneus	54
		(Planorbis carinatus)	7
		(Planorbis planorbis)	4, 7, 40, 52
	Family Bithyniidae	***Bithynia tentaculata***	3, 7, 8, 40, 54
	Family Valvatidae	*Valvata cristata*	1, 4, 8, 52
		Valvata piscinalis	8, 40
	Family Ellobiidae	*Carychium minimum*	3
	Family Vertiginidae	*Vertigo antivertigo*	40
	Family Gastrodontidae	*Zonitoides* sp.	4, 7, 54
Class Bivalvia	Family Sphaeriidae		3, 7, 8, 54
		Pisidium obtusale	52, 53
		Pisidium personatum	53
PHYLUM ARTHROPODA			
S/PH CRUSTACEA			
Class Branchiopoda			
Order Anostraca	Family Tanymastigidae	***Tanymastix stagnalis***	14, 15, 16, 17
Order Diplostraca	Family Bosminidae	*Bosmina longirostris*	3, 18, 20, 40
	Family Chydoridae	*Acroperus harpae*	3, 18, 19, 20, 21, 40
		Acroperus elongatus	20
		Alona (Biapertura) affinis	2, 3, 5, 18, 20, 22
		Alona quadrangularis	3, 18, 19, 20, 22, 40
		Alona costata	20, 22
		Alona guttata	19, 20, 22
		Alona rectangula	3, 22, 23, 40
		Alona rustica	5, 22
		Alonella excisa	5, 18, 19, 20, 22
		Alonella exigua	22
		Alonella nana	19, 20, 22
		Alonopsis elongata	5, 22
		Anchistropus emarginatus	22
		Chydorus latus	3
		Chydorus sphaericus	2, 3, 18, 20, 22, 23, 24, 40

(continued)

Higher classification	Families	Genus-species	References*
		Disparalona rostrata	19
		Eurycercus glacialis	2, 3, 5, 18, 22, 25, 26, 40
		Eurycercus lamellatus	2, 3, 18, 20, 22, 27, 40
		Graptoleberis testudinaria	3, 18, 19, 20, 22, 40
		Lathonura rectirostris	22
		Pleuroxus aduncus	22
		Pleuroxus trigonellus	19, 20, 28
		Pleuroxus truncatus	19, 20, 22, 28
		Pleuroxus uncinatus	22
		Picripleuroxus laevis	22, 28
		Pseudochydorus globosus	22
		Rhynchotalona falcata	20
	Family Daphniidae	*Ceriodaphnia dubia*	3, 40
		Ceriodaphnia megops	3, 40
		Ceriodaphnia quadrangula	3, 20, 40
		Daphnia hyalina var lacustris	18
		Daphnia longispina	2, 3, 18, 40
		Daphnia magna	18
		Daphnia obtusa	18
		Daphnia pulex	2, 3, 18, 22, 23, 24, 40
		Daphnia sp.	2, 22
		Scapholeberis mucronata	19
		Simocephalus exospinosus	3
		Simocephalus vetulus	2, 3, 18, 20, 40
		Simocephalus sp.	22
	Family Macrothricidae	*Macrothrix* sp.	22
	Family Ilyocryptidae	*Ilyocryptus sordidus*	20
	Family Polyphemidae	***Polyphemus pediculus***	2, 3, 40
	Family Sididae	*Latona setifera*	20
		Sida crystallina	20, 22
Class Maxillopoda			
S/Cl Copepoda			
Order Harpacticoida	Family Canthocamptidae	*Canthocamptus microstaphylinus*	29
Order Calanoida	Family Diaptomidae	***Diaptomus castor***	2, 3, 5, 30, 31, 40
		Diaptomus cyaneus	16, 30
		Eudiaptomus gracilis	32, 33
		Arctodiaptomus wierzejskii	16, 30

(continued)

Higher classification	Families	Genus-species	References*
Order Cyclopoida	Family Cyclopidae	***Cyclops agilis***	2, 3, 40
		Cyclops scutifer	3, 40
S/Cl Ostracoda		Ostracoda sp.	4, 54
Order Podocopida	Family Candonidae	***Candona candida***	3, 40
	Family Cyprididae	*Cypris puber*	32, 33
Class Malacostraca			
Order Amphipoda	Family Gammaridae	*Gammarus duebeni*	1, 2, 3, 7, 20, 27, 40
		Gammarus lacustris	4, 7, 40, 54
	Family Niphargidae	*Niphargus kochianus irlandicus*	34, 35
Order Isopoda	Family Asellidae	*Asellus aquaticus*	3, 4, 7, 54
		Asellus meridianus	1, 4, 7
		Asellus sp.	3, 40
S/PH CHELICERATA			
Class Arachnida			
S/Cl Acari			4, 7, 54
Order Trombidiformes	Family Hydrodromidae	*Hydrodroma* sp.	3
	Family Pionidae	*Piona conglobata*	3
	Family Unionicolidae	*Unionicola crassipes*	3
Order Araneae	Family Cybaeidae	*Argyroneta aquatica*	2, 4, 54
S/PH HEXAPODA			
Class Insecta			
Order Ephemeroptera	Family Baetidae	***Cloeon dipterum***	3, 4, 7, 36, 40, 54
		Cloeon simile	4, 7, 12, 54
	Family Leptophlebiidae	*Leptophlebia vespertina*	4
	Family Caenidae	*Caenis horaria*	4
	Family Siphlonuridae	*Siphlonurus alternatus (linneanus)*	3, 36, 40
		Siphlonurus armatus	36
Order Odonata	Family Lestidae	***Lestes dryas***	4, 5, 37, 54
		Lestes sponsa	54
		Lestes sp.	4, 54
	Family Coenagrionidae	*Pyrrhosoma nymphula*	13
		Coenagrion pulchellum/ puella	54
		Coenagrion sp.	8, 10, 54
		Ishnura elegans	13, 54
	Family Calopterygidae	*Calopteryx splendens*	13

(continued)

Higher classification	Families	Genus-species	References*
	Family Libellulidae	***Sympetrum sanguineum***	4, 5, 13, 37, 54
		Sympetrum striolatum	13
		Libellula quadrimaculata	1
Order Plecoptera	Family Nemouridae	*Nemoura cinerea*	3, 4, 40
Order Trichoptera	Family Limnephilidae	*Grammotaulius nigropunctatus*	7
		Limnephilus auricula	4, 7, 54
		Limnephilus affinus/incisus	7, 40
		Limnephilus binotatus	7
		Limnephilus centralis	4, 54
		(Limnephilus decipiens)	7, 54
		Limnephilus flavicornis	7, 40
		Limnephilus lunatus	4, 54
		Limnephilus marmoratus	4, 54
		Limnephilus rhombicus	7
		Limnephilus vittatus	7
		Limnephilus sp.	40
	Family Leptoceridae	*Mystacides longicornis*	8, 54
		Phagopteryx brevipennis	4
		Triaenodes bicolor	54
Order Hemiptera	Family Corixidae	*Arctocorisa germari*	7
		Callicorixa praeusta	7, 38, 39, 40, 54
		Corixa affinis	1, 38
		Corixa panzeri	7
		Corixa punctata/iberica	4, 38, 54
		Corixa wollastoni	40
		Cymatia bonsdorffi	39
		Hesperocorixa castanea	38, 39
		Hesperocorixa linnei	39, 54
		Hesperocorixa sahlbergi	39, 54
		Sigara concinna	1, 7, 39, 40
		Sigara dorsalis	1, 7, 39
		Sigara falleni	4, 7, 39, 40
		Sigara lateralis	7, 39, 40
		(Sigara scotti)	7, 40
	Family Gerridae	*Gerris lacustris*	3, 40, 41
		Gerris argentatus	40
		Gerris lateralis	1, 3
		Gerris costai	1
	Family Notonectidae	*Notonecta glauca*	4, 7, 40, 54
	Family Veliidae	*Microvelia reticulata*	7
		Velia caprai	1
	Family Nepidae	*(Nepa cinerea)*	7
	Family Saldidae	*Saldula opacula*	5
		Saldula saltatoria	41

(continued)

Higher classification	Families	Genus-species	References*
Order Coleoptera	Family Chrysomelidae		4, 54
		Donacia clavipes	41
		Donacia sp.	4
		Galerucella lineola	41
		Platyeumaris sericea	41
	Family Curculionidae		4, 54
		Bagous brevis	5, 42
		Bagous limosus	5, 42
	Family Dryopidae	*Dryops luridus*	43
		Dryops similaris	5, 55
		Dryops sp.	4, 7, 54
	Family Dytiscidae	*Agabus bipustulatus*	1, 4, 7
		Agabus labiatus	4, 5, 7, 40, 42, 43, 54
		(Agabus melanocornis)	7
		Agabus nebulosus	4, 5, 7, 40, 43, 54
		Agabus sp.	3, 4, 54
		Colymbetes fuscus	7, 54
		(Dytiscus circumcinctus)	7
		(Dytiscus sulcatus)	7
		Dytiscus sp.	4, 54
		Graptodytes bilineatus	4, 5, 7, 40, 54, 55
		Graptodytes granularis	40
		Hydaticus sp.	4, 54
		Hydroporus erythrocephalus	4, 7, 40, 54
		(Hydroporus memnonius)	7
		Hydroporus palustris	1, 4, 7, 40, 54
		(Hydroporus pubescens)	4, 54
		Hygrotus inaequalis	4, 7, 54
		Hygrotus impressopunctatus	4, 5, 7, 54, 55
		Hygrotus quinquelineatus	4, 7, 54, 55
		Hygrotus sp. larvae	4
		Hyphydrus ovalis	54
		(Ilybius fuliginosus)	7
		Ilybius sp.	4, 54
		Laccophilus minutus	4, 54
		Laccophilus sp.	4
		Porhydrus lineatus	7, 40, 54
		Porhydrus sp.	40
		Rhantus exsoletus	4, 54
		Rhantus frontalis	5, 43, 54
		Rhantus sp.	4, 54
		(Suphrodytes dorsalis)	7

(continued)

Higher classification	Families	Genus-species	References*
	Family Hygrobiidae	*(Hygrobia hermanni)*	7, 54
	Family Elmidae	*Oulimnius sp.*	54
	Family Haliplidae	*Haliplus confinis*	1, 7, 40
		Haliplus fulvus	4, 7
		Haliplus inaequalis	40
		Haliplus lineolatus	40
		Haliplus lineatocollis	7
		Haliplus obliquus	5
		Haliplus ruficollis group	7, 54
		Haliplus 15-lineatus	40
		Haliplus variegatus	5, 7
		Haliplus sp.	4, 40, 54
	Family Helophoridae	*Helophorus aequalis*	43
		Helophorus brevipalpis	1, 4, 7, 43, 54
		Helophorus grandis	7, 43
		Helophorus minutus	5
		Helophorus nanus	5
		Helophorus obscurus	1
		Helophorus sp.	3, 54
	Family Noteridae	*Noterus clavicornis*	54
	Family Hydraenidae	*Ochthebius minimus*	43
		Ochthebius dilatatus	4, 5, 7, 43, 54
		(Ochthebius nilssoni)	44
	Family Hydrophilidae	***Berosus signaticollis***	4, 5, 7, 42, 54, 55
		(Cercyon tristis)	4
		(Hydrobius fuscipes)	1, 4
		Laccobius biguttatus	54
		Laccobius colon	5
		Laccobius minutus	5
	Family Carabidae	*Agonum afrum*	45
		Agonum marginatum	45
		Agonum piceum	5, 45
		Bembidion clarkii agg.	45
		Blethisa multipunctata	5, 45, 46
		Carabus granulatus	45
		Chlaenius nigricornis	5, 45
		Dyschirius globosus	45
		Elaphrus cupreus	45
		Loricera pilicornis	45
		Nebria brevicollis	45
		Pelophila borealis	5, 45, 46
		Pterostichus gracilis	45
		Pterostichus minor	45
	Family Staphylinidae	*Carpelimus impressus*	47
		Philonthus furcifer	5
		Stenus Kiesenwetteri	47

(continued)

Higher classification	Families	Genus-species	References*
Order Diptera	Family Ceratopogonidae		4, 54
	Family Chironomidae		8, 10, 54
		Calopsectra sp.	8
		Cladotanytarsus sp.	8
		Dicrotendipes sp.	8
		Micropsectra sp.	8
		Parachironomus sp.	8
		Paratanytarsus sp.	8
		Phaenopsectra sp.	8
		Polypedilum sp.	8
		Tanypus sp.	8
	Family Culicidae		4, 54
	Family Psychodidae		4, 54
	Family Sciomyzidae	*Colobaea distincta*	5
		Hydromya dorsalis	10
		Ilione albiceta	5, 10, 48, 49
		Ilione lineata	10, 48
		Limnia unguicornis	48
		Pherbellia nana	5, 48
		Pherbina coryleti	5, 10, 48
		Pherbina schoenherri	10
		Sepedon spinipes	10
		Sepedon sphegea	10
		Tetanocera arrogans	10, 48
		Tetanocera hyalipennis	10
		Tetanocera elata	48
		Tetanocera sp.	10
	Family Stratiomyidae		4
	Family Tabanidae		4
	Family Tipulidae		4, 54
Order Lepidoptera	Family Pyralidae	*Acentropus niveus*	8
B: DRY PHASE			
PHYLUM ANNELIDA			
Order Oligochaeta			4
PHYLUM MOLLUSCA			
Class Gastropoda	Family Gastrodontidae	*Zonitoides* sp.	4, 7
	Family Ellobiidae	***Carychium minimum***	3, 52
	Family Valloniidae	*Vallonia pulchella*	52
	Family Agriolimacidae	*Deroceras laeve*	52

(continued)

Higher classification	Families	Genus-species	References*
PHYLUM ARTHROPODA			
Class Insecta			
Order Orthoptera	Family Tetrigidae	*Tetrix subulata*	5, 41
		Tetrix undulata	41
	Family Acrididae	*Chorthippus albomarginatus*	5
Order Hemiptera	Family Saldidae	*Saldula opacula*	5
Order Coleoptera	Family Chrysomelidae		4
		Donacia sp.	4
	Family Curculionidae		4
		Bagous brevis	5, 42
		Bagous limosus	5, 42
	Family Dryopidae	*Dryops* sp. larvae	4
		Dryops similaris	5
	Family Carabidae	*Agonum marginatum*	45
		(Agonum gracile)	45
		Agonum muelleri	5, 45
		Agonum livens	46
		Agonum lugens	5, 46
		Agonum piceum	5, 45
		Agonum viduum	45, 56
		(Agonum thoreyi)	45
		(Acupalpus consputus)	45
		(Amara communis)	45
		(Amara similata)	45
		(Anisodactylus binotatus)	45
		Badister anomalus	47
		Badister meridionalis	5, 46
		Badister peltatus	5, 45
		Bembidion aeneum	5, 45
		Bembidion clarkii	5, 45
		Bembidion doris	45, 56
		Bembidion gutulla	45
		(Bembidion lampros)	45
		(Bembidion mannerheimii)	45
		(Bembidion tetracolum)	45
		Blethisa multipunctata	5, 45, 46, 56
		Carabus granulatus	5, 45
		Chlaenius nigricornis	5, 45
		(Clivina fossor)	45
		Dyschirius globosus	45, 56
		Dyschirius luedersi	45, 56
		Elaphrus cupreus	45, 56

(continued)

Higher classification	Families	Genus-species	References*
		(Harpalus rufipes)	45
		Loricera pilicornis	5, 45, 56
		Pelophila borealis	5, 45, 46
		Platynus livens	5
		(Platynus dorsale)	45
		(Pterostichus anthracinus)	45
		(Pterostichus crenatus)	45
		(Pterostichus versicolor)	45
		Pterostichus diligens	45
		Pterostichus minor	45
		Pterostichus melanarius	45
		Pterostichus niger	45
		Pterostichus nigrita	5, 45
		Pterostichus strenuus	45
		(Stenolophus mixtus)	45
	Family Staphylinidae	*Philonthus furcifer*	5
		Philonthus quisquiliarius	56
		Atheta elongatula	56
		Atheta graminicola	56
		Atheta hygrotopora	56
		Atheta melanocera	56
		Carpelimus rivularis	56
		Gnypeta carbonaria	56
		Stenus binotatus	56
		Stenus boops	56
		Stenus fuscipes	56
		Stenus juno	56
		Stenus umbratilus	56
		Tachyusa atra	56
	Family Silphidae	*Thanatophilus dispar*	5
Order Diptera	Family Sciomyzidae	*Colobaea distincta*	5, 49
		Ilione albiceta	5, 48
		Ilione lineata	48
		Limnia unguicornis	48
		Pherbellia nana	5, 48
		Pherbina coryleti	5, 48
		Tetanocera arrogans	48
		Tetanocera elata	48
	Family Stratiomyidae	*Odontomyia angulata*	50
Order Lepidoptera	Family Glyphipterigidae	*Odontognophos dumentata*	bn
	Family Gelechiidae	*Monochroa lutulentella*	5
	Family Crambidae	*Paraponyx stratiotata*	5, 57

(continued)

Higher classification	Families	Genus-species	References*
	Family Pyralidae	*Acentrotus niveus*	8, 10
		Acentria ephemerella	54
	Family Noctuidae	*Deltote uncula*	5, 57
	Family Pieridae	*Gonepteryx rhamni*	51
	Family Nymphalidae	*Anthocharis cardamines*	51
		Hipparchia semele	51
	Family Tortricidae	*Bactra furfurana*	5, 57

*1. Reynolds (1985a); 2. Reynolds and Marnell (1999); 3. Reynolds (2000); 4. Porst and Irvine (2009a, b); 5. NPWS (2013a); 6. Reynolds (1996); 7. O'Connor et al. (2004); 8. Byrne and Reynolds (1982); 9. Tattersfield (1998); 10. Williams et al. (2010); 11. Byrne et al. (1989); 12. Byrne (1981); 13. Ní Bhroin (2008); 14. Young (1975); 15. Young (1976); 16. Grainger (1979); 17. Grainger (1991); 18. Reynolds et al. (2004); 19. Kane (1903); 20. Reynolds (1985b); 21. Reynolds and Marnell (1999); 22. Duigan (1989); 23. Duigan (1988); 24. Duigan (1987); 25. Duigan and Frey (1987a); 26. Duigan and Frey (1987b); 27. Reynolds (1982); 28. Duigan (1992); 29. O'Connor and Holmes (1990); 30. Ali et al. (1987); 31. Grainger (1966); 32. Grainger in Reynolds (1996); 33. Reynolds et al. (1998); 34. Hazelton (1974); 35. Knight and Penk (2010); 36. Kelly-Quinn and Regan (2012); 37. Nelson et al. (2011); 38. Tully et al. (1991); 39. Tobin and McCarthy (2004); 40. Reynolds (2003); 41. Morris (1966/1967); 42. Foster et al. (2009); 43. Reynolds (2014); 44. O'Callaghan et al. (2009); 45. Moran et al. (2011); 46. CEC (2007); 47. Owen (1994); 48. Williams et al. (2009); 49. Ryder et al. (2003); 50. Gittings (2007); 51. Nash et al. (2012); 52. Williams and Gormally (2009); 53. Tattersfield (1998); 54. Porst and Irvine (2009a, b); 55. Bilton (1988); 56. Lott and Foster (1990); 57. Bond (1997); bn. B. Nelson personal communication (2014); ml. M. Long personal communication (2015).

References

Ali TH, Holmes JMC, Grainger JNR (1987) *Diaptomus cyaneus* Gurney, a freshwater copepod new to Britain and Ireland. Irish Natur J 22:240–241

Bilton DT (1988) A survey of aquatic Coleoptera in central Ireland and in the Burren. Bull Irish Biogeograph Soc 11:77–94

Bilton DT, Lott DA (1991) Further records of aquatic Coleoptera from Ireland. Irish Nat J 23:389–397

Bilton DT, Foggo A, Rundle SD (2001a) Size, permanence and the proportion of predators in ponds. Arch Hydrobiol 151:451–458

Bilton DT, Freeland J, Okamura B (2001b) Dispersal in freshwater invertebrates. Annu Rev Ecol Syst 32:159–181

Blackstock TH, Duigan CA, Stevens DP, Yeo MJM (1993) Vegetation zonation and invertebrate fauna in Pant-y-llyn, an unusual seasonal lake in South Wales, UK. Aqua Conserv Mar Freshw Ecosyst 3:253–268

Boix D, Sala J, Moreno-Amich R (2001) The faunal composition of Espolla pond (NE Iberian Peninsula): the neglected biodiversity of temporary waters. Wetlands 31:577–592
Bond KGM (1997) Insect survey, Lepidoptera. In: Jennings O'D, Southern Water Global (eds) An investigation of the flooding problems in the Gort–Ardrahan area of South Galway. Ecology baseline study. The Office of Public Works, Dublin, pp 1–86
Bradish S, O'Connor Á, Reed T (2002) New records of the water beetles *Berosus signaticollis* (Charpentier) and *Graptodytes bilineatus* (Sturm) in turloughs. Irish Nat J 27:83–84
Byrne RA (1981) Ecological comparisons of three water bodies in the Burren district of County Clare. Dublin, University of Dublin, Department of Zoology M.Sc. Thesis
Byrne RA, Reynolds JD (1982) Preliminary notes on a doline in Lough Gealain, The Burren, Co. Clare. Irish Nat J 20:175–177
Byrne RA, Reynolds JD, McMahon RF (1989) Shell growth, reproduction and life cycles of *Lymnaea peregra* and *L. palustris* (Pulmonata: Basommatophora) in oligotrophic turloughs (temporary lakes) in Ireland. J Zool London 217:321–339
Campbell S, Gunn J, Hardwick P (1992) Pant-y-llyn—the first Welsh turlough? Earth Sci Conserv 31:3–7
CEC (2007) Interpretation manual of European Union habitats. EU27. Euro Commission DG Environ, Brussels
Collinson NH, Biggs J et al (1995) Temporary and permanent ponds: an assessment of the effect of drying out on the conservation value of aquatic macroinvertebrate communities. Biol Conserv 74:125–133
Coté D, Dubois JM, Gwyn H (1990) Mysterious lakes. Geologija 1990(3):20–25
Coxon CE (1987) The spatial distribution of turloughs. Irish Geogr 20:11–23
Curtis T, Donnes G, Ní Chatháin B (2009) The ecological requirements of water-dependent habitats and species designated under the habitats directive. Biol Environ Proc Royal Irish Acad 109B(3):261–319
Daly D, Warren WP (1998) Mapping groundwater vulnerability: the Irish perspective. Geol Soc London Spec Publ 130:179–190
Donaldson E, Donaldson FL, McMillan NF (1979) A note on the Mollusca of three turloughs. Irish Nat J 19:400–401
Drew DP (1990) The hydrology of the Burren, Co. Clare. Irish Geogr 23:69–89
Duigan CA (1987) The Cladocera (Crustacea) of County Offaly, Ireland. Bull Irish Biogeogr Soc 10:81–93
Duigan CA (1988) The Cladocera (Crustacea) of Lough Ree and neighbouring waterbodies in Ireland. Bull Irish Biogeogr Soc 11:100–113
Duigan CA (1989) The taxonomy, ecology and distribution of the littoral freshwater Chydoridae (Crustacea, Cladocera) of Ireland. Dublin: National University of Ireland, Ph.D. Thesis
Duigan CA (1992) The ecology and distribution of the littoral freshwater Chydoridae (Branchiopoda, Anomopoda) of Ireland, with taxonomic comments on some species. Hydrobiologia 241:1–70
Duigan CA, Frey DG (1987a) *Eurycercus glacialis*, a chydorid cladoceran new to Ireland. Irish Nat J 22:180–183
Duigan CA, Frey DG (1987b) *Eurycercus glacialis* in Ireland. Int Rev Gesam Hydrobiol 72:235–249
Duigan CA, Kovach WL (1991) A study of the distribution and ecology of littoral freshwater chydorid (Crustacea, Cladocera) communities in Ireland using multivariate analyses. J Biogeogr 18:267–280
EC (2000) Water Policy Framework Directive 2000/60/EC. Euro Community, Bussels
EEC (1992) Directive on the Conservation of Natural Habitats and of Wild Fauna and Flora, ('Habitats Directive') 92/43/EEC. Euro Econ Community, Brussels
Eitam A, Blaustein L et al (2004) Crustacean species richness in temporary pools: relationships with habitat traits. Hydrobiologia 525:125–130
Foster GF, Nelson BH et al (1992) A classification and evaluation of Irish water-beetle communities. Aquat Conserv Mar Freshw Ecosyst 2:185–208

Foster GN, Nelson BH, O'Connor Á (2009) Ireland Red list No. 1—water beetles. National Parks & Wildlife Service/Department of Environment, Heritage and Local Government, Dublin
Garcia-Gil LJ, Borrego CM et al (1992) Visit to the aquatic ecosystems of Lake Banyoles karstic area. In: Catalan J, Pretus JLI (eds) Mid-Congress Excursions XXV SIL Congress. Univ Barcelona, Barcelona, pp 1–13
Gittings T (2007) *Odontomyia angulata* (Panzer, 1798), a soldierfly (Diptera, Stratiomyidae) new to Ireland. Irish Nat J 28:419
Good J (2004) Calcareous grassland as a habitat for Staphylinidae (Coleoptera) in the Burren National Park, Co. Clare, Ireland. Bull Irish Biogeogr Soc 28:147–162
Good J, Butler FT (2001) Turlough pastures as a habitat for Staphylinidae and Carabidae (Coleoptera) in south-east Galway and north Clare, Ireland. Bull Irish Biogeogr Soc 25:74–94
Goodwillie R (1992) Turloughs over 10 ha: vegetation survey and evaluation. Office of Public Works. Internal Report of National Parks & Wildlife Service, Dublin
Goodwillie R (2003) Vegetation of turloughs. In: Otte ML (ed) Wetlands of Ireland. Distribution, ecology, uses and economic value. Univ College Dublin Press, Dublin, pp 135–144
Goodwillie R, Reynolds JD (2003) Turloughs. In: Otte ML (ed) Wetlands of Ireland. Distribution, ecology, uses and economic value. Univ College Dublin Press, Dublin, pp 130–133
Grainger JNR (1966) *Diaptomus castor* in County Meath. Irish Nat J 15:211
Grainger JNR (1976) Further records for the fairy shrimp *Tanymastix stagnalis* (L.). Irish Nat J 18:326
Grainger JNR (1979) *Diaptomus wierzejskii* in temporary water bodies in Co. Mayo. Irish Nat J 19:372
Grainger JNR (1991) The biology of *Tanymastix stagnalis* (L.) and its survival in small and large temporary water bodies in Ireland. Hydrobiologia 212:77–82
Grainger JNR, Holmes JMC (1989) Further records of *Diaptomus cyaneus* Gurney (Crustacea: Copepoda) from temporary water bodies in Ireland. Irish Nat J 23:160–161
Gunn J (2006) Turloughs and tiankengs—distinctive doline forms. Speleogenesis Evol Karst Aquifers 4(9):4p
Hazelton M (1974) Hypogean fauna recorded from Ireland, 1952–1971. Trans Cave Res Group Great Brit 15:225–252
Ivimey-Cook RB, Proctor MCF (1966) The plant communities of the Burren, Co. Clare. Proc Royal Irish Acad 64B:211–267
Kane WF (1903) A contribution to the knowledge of Irish freshwater Entomostraca—Cladocera. Irish Nat 12:210–218
Kelly-Quinn M, Regan EC (2012) Ireland Red List No 7: Mayflies (Ephemeroptera). National Parks Wildlands Services/Department of Environment, Heritage and Local Government, Dublin
Kimberley S, Coxon C (2013) Evaluating the influence of groundwater pressures on groundwater-dependent wetlands, STRIVE Report Series No 100. Environ Protection Agency, Wexford
Kimberley S, Waldren S (2012) Examinations of turlough soil property spatial variation in a conservation assessment context. Biol Environ Proc Royal Irish Acad 112B(2):193–205
Knight LRFD, Penk MR (2010) Groundwater Crustacea of Ireland: a survey of the stygobytic Malacostraca in caves and springs. Biol Environ Proc Royal Irish Acad 110B(3):211–235
Kranjc A (2006) Seasonal karst lake Lerknica (Slovenia)—2000 years of man versus nature. Helictite 39(2):39–46
Lansbury I (1965) Notes on the Hemiptera, Coleoptera, Diptera and other invertebrates of the Burren, Co. Clare and Inishmore, Aran Islands. Proc Royal Irish Acad 64B:89–115
Lott DA, Foster GN (1990) Records of terrestrial Coleoptera from Wetland sites in 1987, including *Stenus glabellus* Thomson (Staphylinidae) new to the British Isles. Irish Nat J 23(7):280–282
Louette G, De Meester L (2005) High dispersal capacity of cladoceran zooplankton in newly founded communities. Ecology 85:353–359
Lynn D, Waldren S (2003) The turlough form of *Ranunculus repens*. In: Otte ML (ed) Wetlands of Ireland. Univ College Dublin Press, Dublin, pp 157–159

Moran J, Gormally MJ, Sheehy Skeffington M (2003) Records of *Panagaeus crux-major* (L.) and *Perostichus aterrimus* (Herbst) recorded on turloughs, in Cos Clare and Mayo. Irish Nat J 27(8):309–311

Moran J, Sheehy Skeffington M, Gormally M (2008) The influence of hydrological regime and grazing management on the plant communities of a karst wetland (Skealoghan turlough) in Ireland. Appl Veg Sci 11:13–24

Moran J, Gormally M, Sheehy Skeffington M (2011) Turlough ground beetle communities: the influence of hydrology and grazing in a complex ecological matrix. J Insect Conserv 2011:1–19

Morris MG (1966/1967) Weevils (Coleoptera, Curculionoidea) and other insects collected in north-west Clare, with special reference to the Burren Region. Proc Royal Irish Acad 65B:349–71

Nash D, Boyd T, Hardiman D (2012) Ireland's butterflies, a review. Dublin Naturalists' Field Club, Dublin

Naughton O, Johnston PM, Gill LW (2012) Groundwater flooding in Irish karst: The hydrological characterisation of ephemeral lakes (turloughs). J Hydrol 470–471:82–97

Nelson B, Ronayne C, Thompson R (2011) Ireland Red List No 6—Damselflies and Dragonflies (Odonata). National Parks & Wildlife Service/Department of Environment, Heritage and Local Government, Dublin

Ní Bhriain B, Sheehy Skeffington M, Gormally M (2002) Conservation implications of land use practices on the plant and carabid beetle communities of two turloughs in Co. Galway, Ireland. Biol Conserv 105:81–92

Ní Bhriain B, Gormally M, Sheehy Skeffington M (2003) Changes in land-use practices at two turloughs on the east Burren limestones, Co. Galway, with reference to nature conservation. Biol Environ Proc Royal Irish Acad 103B(3):169–176

Ní Bhroin N (2008) Ecological Impact Assessment (EcIA) of the effects of statutory arterial drainage maintenance activities on turloughs (EU Code 3185). Galway: Series of Ecological Assessment of Arterial Drainage Maintenance No. 8. Headford Consulting Office of Public Works

NPWS (2007) Turloughs, Draft Backing Document. In The Status of EU Protected Habitats and Species in Ireland. Backing Documents, Art 17 Forms, Maps. Vol 2, 510–550. Unpubl Rep, National Parks Wildlife Service, Dublin.

NPWS (2013a) Galway Bay Complex SAC (site code 268). Conservation objectives supporting document—Turloughs. National Parks Wildlife Service/Department of Environment, Heritage and Local Government, Dublin

NPWS (2013b) The status of protected habitats and species in Ireland. In: Lynn D (ed) Overview, vol 1. National Parks Wildlife Service/Department of Arts, Heritage & Gaeltacht, Dublin, Unpublished Report

O'Callaghan E, Foster GN, Bilton DT, Reynolds JD (2009) *Ochthebius nilssoni* Herbauer new for Ireland (Hydraenidae, Coleoptera), including a key to Irish *Ochthebius* and *Eniocerus*. Irish Nat J 30:19–23

O'Connor JP, Holmes JMC (1990) A provisional list of the Harpacticoida (Crustacea: Copepoda) of Ireland. Bull Irish Biogeogr Soc 13:44–130

O'Connor Á, Bradish S et al (2004) A comparison of the efficacy of pond net and box sampling methods in turloughs—Irish ephemeral aquatic systems. Hydrobiologia 524:133–144

Owen J (1994) *Badister anomalus* Perris, *Carpelimus impressus* (Lacordaire) and *Stenus Kiesenwetteri* Rosenhauer, Coleoptera new to Ireland. Irish Nat J 24:467–468

Porst G, Irvine K (2009a) Distinctiveness of macroinvertebrate communities in turloughs (temporary ponds) and their response to environmental variables. Aqua Conserv Mar Freshw Ecosyst 19:456–465

Porst G, Irvine K (2009b) Implications of the spatial variability of macroinvertebrate communities for monitoring ephemeral lakes. An example from turloughs. Hydrobiologia 636:421–438

Porst G, Naughton O et al (2012) Adaptation, phenology and disturbance of macroinvertebrates in temporary water bodies. Hydrobiologia 696:47–62

Praeger RL (1932) The flora of the turloughs; a preliminary note. Proc Royal Irish Acad 41B:37–45

Proctor MCP (2010) Environmental and vegetational relationships of lakes, fens and turloughs in the Burren. Biol Environ Proc Royal Irish Acad 110B(1):17–34

Regan EC (2005) Further records of carabid beetles from turloughs. Irish Nat J 28(2):59–61

Regan EC, Moran J (2005) *Thanatophilus dispar* (Herbst) (Silphidae) in turloughs in the west of Ireland. The Coleopterist 14(2):89–91

Reynolds JD (1982) Ecology of turloughs (vanishing lakes) in the Burren, Western Ireland. In: O'Gorman F, Rochford J (eds) Trans International Congress Game Biologists, vol 14. Irish Wildlife Publications, Dublin, pp 183–188

Reynolds JD (1985a) Karstic freshwater habitats of the Aran Islands, Co. Galway, with preliminary notes on their fauna. Irish Nat J 21:430–435

Reynolds JD (1985b) Cladocera from Burren turloughs at Mullagh More, Co. Clare. Bull Irish Biogeogr Soc 9:51–54

Reynolds JD (1996) Turloughs, their significance and possibilities for conservation. In: Reynolds JD (ed) The conservation of aquatic systems. Royal Irish Acad, Dublin, pp 38–46

Reynolds JD (2000) Invertebrate communities of turloughs (temporary lakes) in south-east Galway, Ireland. Verhand Int Ver Limnol 27:1679–1684

Reynolds JD (2003) Fauna of turloughs and other wetlands. In: Otte ML (ed) Wetlands of Ireland. Univ College Dublin Press, Dublin, pp 145–156

Reynolds JD (2014) Preliminary notes on three turloughs (temporary ponds overlying karst) in Co. Limerick, with comments on their vegetation and water beetle fauna. Biol Environ Proc Royal Irish Acad 114B(3):265–270

Reynolds JD, Marnell F (1999) New records of *Eurycercus glacialis* (Cladocera: Chydoridae) in turloughs in south-east Galway. Irish Nat J 26(5/6):177–180

Reynolds JD, Duigan C, Marnell F, O'Connor Á (1998) Extreme and ephemeral water-bodies in Ireland. In: Giller P (ed) Studies in Irish limnology. Marine Institution, Dublin, pp 67–99

Reynolds J, Murphy M, O'Connor Á (2004) Early season cladoceran diversity of Atlantic temporary ponds (turloughs). Arch Sci 57:97–104

Rosén E (2006) Alvar vegetation of Öland—changes, monitoring and restoration. Biol Environ Proc Royal Irish Acad 106B(3):387–399

Ryder CJ, McDonnell RJ, Gormally MJ, Knutson L (2003) *Colobaea distincta* (Meigen): a sciomyzid (Diptera) new to Ireland with a key to the Irish species of the genus. Irish Nat J 27:184–186

Ryder CJ, Moran J, McDonnell RJ, Gormally MJ (2005) Nature conservation implications of grazing practices on the plant and dipteran communities of a turlough in Co. Mayo, Ireland. Biodiv Conserv 14:187–204

Scannell MJP, Webb DA (1983) Flora of Connemara and the Burren. Cambridge Univ Press/Royal Dublin Soc, Cambridge

Sheehy Skeffington M, Gormally M (2007) Turloughs: a mosaic of biodiversity and management systems unique to Ireland. Acta Carsologica 36:217–222

Sheehy Skeffington M, Moran J et al (2006) Turloughs—Ireland's unique wetland habitat. Biol Conserv 133:265–290

Southern Water Global (1996) An investigation of the flooding problems of the Gort-Ardrahan area of south Galway. Interim Factual Rep Office Public Works, Dublin

Southern Water Global (1997) An investigation of the flooding problems of the Gort-Ardrahan area of south Galway. Final Report Office Public Works, Dublin

Speight MCD (1976) *Agonum livens, Asemum striatum* and *Xylota coeruleiviridis:* Insects new to Ireland. Irish Nat J 18(9):274–275

Speight MCD (1977) Notes on three ground beetles (Coleoptera: Carabidae) *Dyschirius luedersi* new to Ireland, *Badister peltatus* and *Chlaenius tristis* reinstated as Irish. Irish Nat J 19(4):116–118

Spencer M, Blaustein L et al (1999) Species richness and the proportion of predatory animal species in temporary freshwater pools: relationships with habitat size and permanence. Ecol Lett 2:157–166

Tattersfield P (1998) Wetland mollusc communities from the Aran Islands. Irish Nat J 26(1/2):8–21

Tobin CM, McCarthy TK (2004) The corixid communities of lowland turloughs in the Galway area and Lough Corrib. Irish Nat J 27(12):450–457

Tully O, McCarthy TK, O'Donnell D (1991) The ecology of the Corixidae (Hemiptera: Heteroptera) in the Corrib catchment, Ireland. Hydrobiologia 210:161–169

Tynan S, Gill M, Johnston P (2002) Water Framework Directive. Development of a methodology for characterisation of a karstic groundwater body with particular emphasis on the linkage with associated ecosystems such as turlough ecosystems. Synthesis Rep Environ Protection Agency, Wexford

Tynan S, Gill M, Johnston P (2007) Water Framework Directive. Development of a methodology for characterisation of a karstic groundwater body with particular emphasis on the linkage with associated ecosystems such as turlough ecosystems. (2002-W-DS-8-M1). Final report of Environment Protection Agency, Wexford

Waltham T (2006) Tiankengs of the world, outside China. Speleogenesis Evol Karst Aquifers 4(1):12

Williams CD, Gormally MJ (2009) Spatio-temporal and environmental gradient effects on mollusc communities in a unique wetland habitat (turloughs). Wetlands 29:854–865

Williams C, Doherty O, Gormally M (2006) *Gasterosteus aculeatus* L. and *Pungitius pungitius* (L.)—two species of stickleback from turloughs in Cos. Mayo and Roscommon. Irish Nat J 28:276–279

Williams CD, Moran J et al (2009) Factors affecting Sciomyzidae (Diptera) across a transect at Skealoghan Turlough (Co. Mayo, Ireland). Aqua Ecol 43:117–133

Williams CD, Gormally MJ, Knutson LV (2010) Very high population estimates and limited movement of snail-killing flies (Diptera: Sciomyzidae) on an Irish turlough (temporary lake). Biol Environ Proc Royal Irish Acad 110B(2):81–94

Young R (1975) *Tanymastix stagnalis* (L.) new to Britain and Ireland. Irish Nat J 18:203–209

Young R (1976) *Tanymastix stagnalis* (L.) in Co. Galway, new to Britain and Ireland. Proc Royal Irish Acad 76B:369–378

Chapter 7
Peatland Invertebrates

Darold Batzer, Haitao Wu, Terry Wheeler, and Sue Eggert

Introduction to Peatlands

Peat can develop in any wetland area where plant production exceeds decomposition. Peatlands are most prevalent in flat landscapes at high latitudes (tundra, boreal zones) where cool temperatures, low evaporation rates, water-logging, and low pH combine to retard plant decomposition (Vitt 1994; Rochefort et al. 2012). Although much less expansive, peatlands can also occur under other climatic conditions provided decomposition is still slow (see the below section on the Okefenokee Swamp). Peatlands are often classified as either bogs or fens, with bogs receiving nutrients almost exclusively from precipitation (i.e., ombrotrophic) and fens also receiving nutrients from surface or subsurface inputs of water (minerotrophic) (Bridgham et al. 1996; Wheeler and Proctor 2000; Rochefort et al. 2012; Fig. 7.1). However, most peatlands are not exclusively ombro- or minerotrophic, and exist along a gradient from bog to fen (e.g., poor fens). Due to the lack of mineral inputs and active acidification by *Sphagnum* mosses, bog-type peatlands tend to be highly acidic (pH $<$ 5). Fens can range from being acidic (pH ~ 5) to circumneutral to basic (pH $>$ 8), depending on hydrology and climate.

The study of invertebrates in peatlands has been unique from other types of wetlands (e.g., other chapters in this book) in that there has been a much stronger emphasis on the terrestrial and semiaquatic fauna (Annelida, Arachnida, Carabidae, brachyceran Diptera, Lepidoptera, Hymenoptera; e.g., Blades and Marshall 1994; Finnamore and Marshall 1994; Marshall et al. 1999; Koponen 2002; Spitzer and Danks 2006) rather than just the aquatic fauna (Odonata, Dytiscidae, nematoceran Diptera, e.g., Rosenberg and Danks 1987). This is probably because dense carpets of vegetation and peat above the waterline provide ample habitat for terrestrial invertebrates (plant and soil dwellers, and their predators), while areas of open, standing water can be limited in many peatlands. The

D. Batzer, D. Boix (eds.), *Invertebrates in Freshwater Wetlands*,
DOI 10.1007/978-3-319-24978-0_7

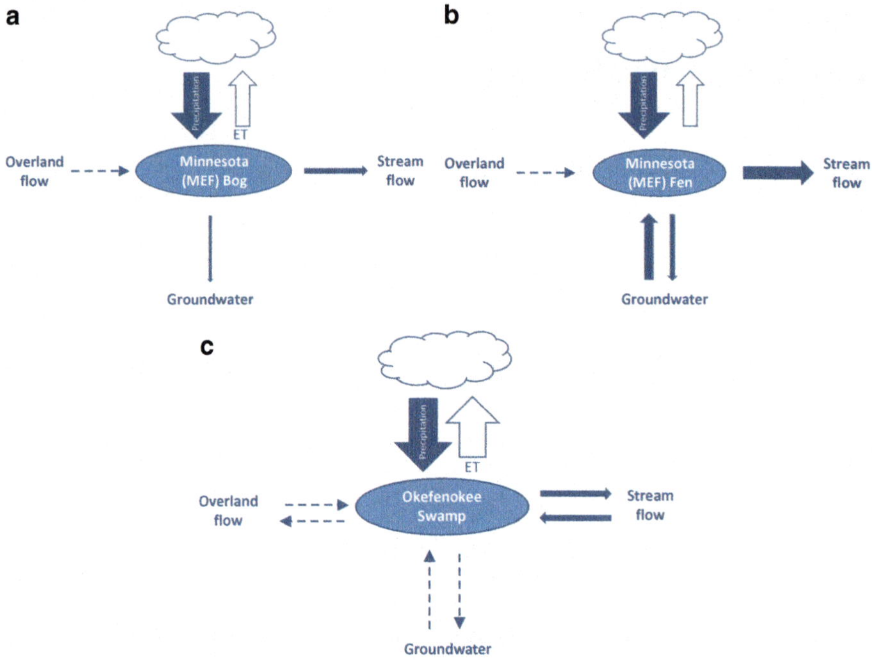

Fig. 7.1 Examples of water budgets quantified from different peatlands: (**a**) Minnesota bog (Bay 1968, 1969; Boelter and Verry 1977); (**b**) Minnesota fen (Verry and Boelter 1975; Boelter and Verry 1977); (**c**) Okefenokee Swamp, Georgia (Rykiel 1984)

terrestrial invertebrate fauna of a peatland and the adjacent upland can overlap to some extent, but many terrestrial taxa are unique to peatland habitats (Spitzer et al. 1999).

For aquatic invertebrates, peatlands can be challenging places to live. On one hand, peatlands rarely dry completely due to close proximity of water tables, whether perched (bogs) or groundwater (fens), or to water retention by peat soils (Rochefort et al. 2012), and thus desiccation is not a strong constraint on peatland invertebrates. On the other, peatland waters tend toward anoxia, and in the case of bogs can be highly acidic. These chemical conditions prevent certain aquatic organisms from becoming well established across peatlands habitats (Mendelssohn et al. 2014). In this chapter, we first review peatland areas across the globe where invertebrate ecology has been a major focus of research including Europe, Canada, the United States, and China. We then look for emerging themes from that body of research to generalize about how invertebrates function across varying peatland habitats.

Focal Areas of Research for Peatland Invertebrates

European Peatlands

Peatland Habitats of Europe

Most European peatlands occur across the northern and western regions (Scandinavia, Great Britain, Ireland, Netherlands) (Verhoeven 2014). As elsewhere, bog peatlands (often called mires in Europe) are dominated by *Sphagnum* mosses and ericaceous shrubs, with the most acidic and least fertile habits being fairly open. In fens, woody trees (*Betula*, *Salix*, *Alnus*) and sedges and grasses becoming more prevalent as pH and nutrient levels rise (Wheeler and Proctor 2000; Hájek et al. 2006).

In terms of invertebrates, the peatland fauna of the United Kingdom and Ireland is perhaps the best known, but excellent recent work has also been conducted in peatlands of Sweden, Finland, Germany, the Netherlands, the Czech Republic, and Spain. Emerging areas of research emphasis include investigating how hydrologic variation affects invertebrate distributions, and especially how peatland invertebrates can be used as bioindicators of human land-use and a changing global climate. Basic ecological work in Europe focuses mostly on the aquatic fauna, while impact assessment work relies more heavily on terrestrial and semiaquatic groups.

Basic Invertebrate Ecology in European Peatlands

Distinct communities of ground-dwelling invertebrates occur across different peatland sites of northern England (e.g., oligotrophic mires, blanket bogs, mixed moor; Coulson and Butterfield 1985). Similarly the compositions of these communities vary greatly seasonally. Thus, both spatial and temporal factors affect invertebrate faunistic diversity of peatlands (Coulson and Butterfield 1985). Invertebrate community composition can also vary across a single peatland, with distinct invertebrate assemblages in bog margins (i.e., laggs) versus bog centers (Bezděk et al. 2006; Mieczan et al. 2014).

As is the case for most kinds of wetland, hydrologic variation is considered a primary control on the aquatic invertebrate fauna in European peatlands. Downie et al. (1998), Standen (1999), and Hannigan and Kelly-Quinn (2012) all compared the aquatic invertebrate faunas in permanent pools and temporary pools/hollows of peatlands (Scotland and Ireland), and found taxonomically richer communities in the permanent water sites. Larger predatory invertebrates (species of Odonata, Hemiptera, Dytiscidae) were restricted to the permanent pools, and Hannigan and Kelly-Quinn (2012) surmised that their presence or absence may serve to structure overall invertebrate communities in peatland pools. Carroll et al. (2011) found that low soil moisture levels, induced by peatland drainage, decreased the abundance of soil-dwelling tipulidae crane fly larvae. Nilsson and Svensson (1995) looked at forested and open (logged) Arctic swamp pools, and found the open pools to be considerably warmer and to support a richer and more abundant dytiscidae and culicid fauna than the forested pools. The study of Hannigan and Kelly-Quinn (2012) addressed mostly acidic bogs, but they also sampled a fen habitat; the community in that fen was dramatically different from the bog-type habitats, being much more taxonomically rich (see taxa lists in the Appendix, and discussion below).

Carrera et al. (2009, 2011) focused on how enchytraeid worms in peatlands were controlled by temperature and moisture levels in the peat, and how in turn the activities of the worms controlled peat breakdown and carbon flux (see also below consideration of climate change). In one study (Carrera et al. 2011), they conducted laboratory incubations of soils from a Spanish peatland under ambient moisture and temperature conditions crossed with elevated temperature or reduced moisture conditions. They further introduced enchytraeid worms to half of the replicates from each treatment. Neither temperature nor moisture level by itself affected CO_2 flux from these soils. However, under moist conditions, higher temperatures induced worm populations to increase, which resulted in greater loss of dissolved organic carbon from the soils, suggesting an important role of enchytraeids in peat decomposition. A companion study focusing solely on temperature (Carrera et al. 2009) also pointed to the importance of the worms to peat breakdown, and suggested that worm activity resulted in a lower release of H+ ions, possibly reducing the effects of acidity in limiting peat decomposition.

Conservation and Invertebrates of European Peatlands

Invertebrates are being increasingly used in Europe to monitor the ecosystem health of peatlands. Groups considered useful as bioindicators include Annelida (Carrera et al. 2009, 2011), Arachnida (Scott et al. 2006; Więcek et al. 2013), Acarina (Więcek et al. 2013; Lehmitz 2014), Collembola (Krab et al. 2013, 2014), Odonata (Drinan et al. 2013), aquatic Hemiptera (Downie et al. 1998; Drinan et al. 2013), Lepidoptera (Spitzer et al. 1999), Tipulidae (Carroll et al. 2011), Formicidae (Vepsalainen et al. 2000), and various Coleoptera (e.g., Carabidae, Dytiscidae: Nilsson and Svensson 1995; Downie et al. 1998; Spitzer et al. 1999; Drinan et al. 2013), as well as whole invertebrate assemblages (Standen 1999; van Duinen et al. 2003; Hannigan and Kelly-Quinn 2012).

Drinan et al. (2013) assessed impacts of forestry practices on macroinvertebrate assemblages in blanket bog lakes. The combination of conifer planting and clear-cut harvesting affected invertebrates most, presumably due to eutrophication. Peatland lakes affected by clear-cuts supported larger dytiscid beetle species and greater corixid water boatmen abundance than the natural, more-oligotrophic lakes. Nilsson and Svensson (1995), working in northern Sweden found a similar Dytiscidae response to tree harvest. Vepsalainen et al. (2000) found that clear-cutting and drainage of Finnish bogs increased overall ant species richness, but these practices apparently eliminated habitat for a few bog-specialist ant species.

Peat extraction and drainage (e.g., Fig. 7.2) have significantly impacted many European bogs, and efforts to restore the habitats are being conducted, using invertebrate response as a metric of success (van Duinen et al. 2003; Hannigan et al.

Fig. 7.2 Ditched peatland at Marcell Experimental Forest, Minnesota. Photo by Sue Eggert, USDA Forest Service, Northern Research Station

2011; Więcek et al. 2013). In some rehabilitated raised bogs in the Netherlands, van Duinen et al. (2003) found limited evidence that the restoration strategies being used were enhancing rare and bog-characteristic invertebrate species. In contrast, Hannigan et al. (2011) found that aquatic invertebrate communities were very similar between pools of a restored bog, where some limited peat extraction and ditching had been conducted, and a largely intact bog, suggesting that here significant progress towards reestablishing natural conditions had been achieved. Więcek et al. (2013) similarly found that water mites, which have very complex life cycles (parasitic larvae, predaceous nymphs and adults), making them especially useful bioindicators, had made significant reestablishment progress in some German peatlands where natural hydrology had been restored after past ditching and peat extraction. Finally, integrating microbial and aquatic invertebrate metrics may provide more insight into environmental responses to anthropogenic impacts than invertebrates alone (Whatley et al. 2014).

It is predicted that climate change may dramatically affect European peatlands through desiccation and warming. Invertebrate responses may be useful in detecting changes, and invertebrate responses may in some ways contribute to any changes in the ecological structure and function of affected peatlands. As mentioned above, Carrera et al. (2009, 2011) found that enchytraeid worms will be harmed by drying, but enhanced by warming conditions, and changes in worm populations may functionally alter carbon cycles in some peatlands. Krab et al. (2013) found that experimental warming of a Swedish subarctic peatland reduced densities of collembolan springtails, another important decomposer, potentially reducing the importance of these invertebrates to peat breakdown. Similarly, drying of a Welsh peatland resulted in declines in tipulidae crane fly larvae populations (Carroll et al. 2011). Invertebrates in peatlands should respond to climate change, although the ecological consequences of invertebrate change will probably be complex.

Canadian Peatlands

Canadian Peatland Habitats

Canada has more peatland habitat than any other country (Vitt 1994). Peatlands occupy 12 % of Canada's landmass and the great majority (97 %) is in the boreal and subarctic regions (Tarnocai 2006), although there are isolated temperate outliers as far south as 42° in Ontario. Most northern peatlands in Canada are fens, usually dominated by brown mosses, *Sphagnum*, and sedges (Vitt 1994). Some arctic wet tundra meadows have many characteristics of fens although they are underlain by permafrost and have shallower accumulations of peat (Vitt 1994). Bogs dominated by *Sphagnum* mosses, ericaceous shrubs, and spruce (*Picea*)/tamarack (*Larix*) forest are widespread in the mid to southern boreal, as well as temperate outliers (Vitt 1994; Rochefort et al. 2012). Because most of Canada was glaciated during the Pleistocene, most extant Canadian peatlands date from the early to mid Holocene, within the last 9000–6000 years (e.g., Vitt 1994; Lavoie et al. 1997).

Diversity and Ecology of Canadian Peatland Invertebrates

Although most Canadian peatlands are boreal or subarctic, much of the research on peatland invertebrates has focused on temperate outliers in southern Canada, especially in Ontario and Quebec. This is likely partly due to accessibility, but also to the perceived higher conservation relevance of these southern sites relative to more extensive and contiguous northern peatlands. Arthropods have received considerably more attention than other invertebrate taxa.

Most research on Canadian peatland arthropods has been species inventories. The most intensive early research was a series of natural history studies of Byron Bog, in London, Ontario by W.W. Judd, beginning in the late 1950s (Judd 1957) and continuing with a long series of papers, many of which focused on insects, over subsequent decades (e.g., Judd 1975). Judd's efforts resulted in probably the most comprehensive inventory of a Canadian peatland at that time.

To address the lack of baseline data, the Biological Survey of Canada launched an initiative in 1981 to document Canadian peatland arthropods. This effort produced two volumes on aquatic (Rosenberg and Danks 1987) and terrestrial arthropods

(Finnamore and Marshall 1994). Many of the chapters were inventories of selected taxa or sites, although there were some focused ecological studies.

Rosenberg and Danks (1987) summarized knowledge of aquatic arthropod taxa in Canadian peatlands and marshes, with species lists and ecological overviews of water mites (Smith 1987), Ephemeroptera, Trichoptera (Flannagan and Macdonald 1987), Odonata (Hilton 1987), Hemiptera (Scudder 1987), Coleoptera (Larson 1987), biting flies (Lewis 1987) and Chironomidae (Diptera) (Wrubleski 1987). Overall, the diversity of aquatic insects in peatlands was considered low, with approximately 11 % of the 4000 aquatic species recorded in Canada found in these habitats, although in many cases it was not possible to distinguish peatland-restricted from peatland-associated or generalist species. Many of the chapters emphasized the need for additional research on geographic distribution and natural history of the taxa.

A subsequent volume on terrestrial arthropods (Finnamore and Marshall 1994) included species-level inventories of selected taxa, often in more geographically limited regions. Aitchison-Benell (1994), Dondale and Redner (1994), and Koponen (1994) documented arachnids, primarily spiders, in Manitoba, Ontario, and Quebec peatlands; Cannings and Cannings (1994) reviewed Odonata of northwestern cordilleran peatlands; Finnamore (1994) provided a checklist and analysis of Hymenoptera in Wagner Fen in central Alberta. Behan-Pelletier and Bissett (1994) and Marshall (1994) provided Canada-wide overviews of peatland oribatid mites and sphaerocerid flies, respectively. Blades and Marshall (1994) summarized results of a broader taxonomic survey from isolated peatlands in southern Ontario.

In contrast to the aquatic fauna, species richness of terrestrial arthropods in peatlands is high. Finnamore (1994) recorded 1410 species of Hymenoptera from Wagner Fen, Alberta and Blades and Marshall (1994) recorded more than 2000 species of arthropods from southern Ontario peatlands. Savage et al. (2011) and Grégoire Taillefer and Wheeler (2012) identified 381 and 699 species of higher Diptera, respectively, from bogs in southern Quebec. Despite the fact that these sites are small habitat fragments, they clearly support high species diversity. One challenge to drawing conclusions about peatland biodiversity from these inventories is that comparable efforts in sampling and identification are often lacking for other habitats in the same regions. Thus it is difficult to determine which species are peatland-specialists or primarily peatland-associated, especially in taxa for which ecological knowledge at the species-level is lacking, such as Diptera (Blades and Marshall 1994; Spitzer and Danks 2006; Savage et al. 2011). Based on available knowledge, the percentage of terrestrial arthropods that are peatland-specialists (10 %) is higher than in the aquatic fauna (1 %) (Marshall and Finnamore 1994).

Several papers, cited previously, in Rosenberg and Danks (1987) and Finnamore and Marshall (1994) discussed ecological aspects of focal taxa in addition to presenting species checklists. This treatment was, in most cases, more developed in aquatic taxa, probably because of a longer history of ecological studies in aquatic entomology, but also because lower diversity, more defined habitats, and higher available taxonomic resolution make community-level analyses more tractable.

Conservation and Invertebrates of Canadian Peatlands

Some recent ecological studies of terrestrial peatland arthropods in Canada have focused on applied questions in conservation and land-use. Peatland conservation efforts, especially in southern Canada where remaining peatlands are small remnants, focus, appropriately, on habitat-level conservation. However, some arthropods restricted to peatlands have been the focus of species-level assessments under federal species-at-risk legislation. Examples include the Bogbean Buckmoth (*Hemileuca* sp., Lepidoptera: Saturniidae) in eastern Ontario, which is listed as Endangered (COSEWIC 2009) and the Georgia Basin Bog Spider (*Gnaphosa snohomish*, Araneae: Gnaphosidae) in southern British Columbia, which is a species of Special Concern (COSEWIC 2012).

There is a need for studies of peatland biodiversity in the context of climate change, especially on boreal and subarctic peatlands, where impacts may be particularly pronounced (Tarnocai 2006). Most subarctic and arctic peatlands overlie permafrost and thawing may have major impacts on hydrology, carbon sequestration and, in turn, biodiversity. Much of the current research on northern peatlands focuses on their roles as carbon sinks and landscape elements, but research on peatland species lags behind. Ongoing research on arthropods in wet tundra (TA Wheeler, unpublished data) shows that taxonomic and ecological diversity of peatland arthropods in the arctic is much higher than documented. Given the suitability of arthropods as bioindicators (McGeoch 1998), further studies of arctic peatland arthropods may provide valuable insights into climate change impacts in the north.

Horticultural peat extraction is one of the major threats to temperate peatlands in eastern Canada. Drainage ditches (e.g., Fig. 7.2) are dug to lower the water table, herbaceous vegetation is removed, and a thin upper layer of dried peat is removed by vacuuming each year (Gorham and Rochefort 2003). Once the usable supply of peat has been exhausted the site is usually abandoned but restoration efforts have been implemented for some sites that are no longer being used for industrial extraction. The restoration process involves restoring the hydrological balance and seeding plant fragments and propagules from nearby undisturbed peatlands, along with mulching and fertilization (Gorham and Rochefort 2003). Restoration success in these sites has primarily been assessed using plants, but some recent studies have also examined the recolonization of insects.

Mazerolle et al. (2006) assessed recolonization of aquatic arthropods in bog pools created as part of restoration efforts and found that species diversity was lower than in natural pools. Assisted restoration of vegetation had a positive effect on arthropod colonization, but aquatic insect diversity was still considerably lower in newly created pools 4 years after restoration. However, most of the species that did colonize successfully were peatland-associated species that are probably adapted to dispersal between isolated bog pools.

Grégoire Taillefer and Wheeler (2012) focused on the response of terrestrial Diptera to restoration of peatland sites in the lower St. Lawrence region of Quebec. They compared higher Diptera diversity in three treatments: natural bogs, abandoned bogs that had been used for peat extraction and left to recover on their own, and bogs that had been restored 7 years earlier. Although overall community structure in restored sites approached that of natural bogs, some functional groups of Diptera (small species, some trophic groups) had not yet recovered. These results suggested that assisted recolonization may be required for small insects in restored sites, much as for plants. Grégoire Taillefer and Wheeler (2013) subsequently found that the usual method for gathering and preparing plant material for restoration (chopping and spreading) did not introduce significant numbers of insects, so that other methods of collecting and introducing arthropods from natural donor sites may be necessary. These studies demonstrate that, despite high species diversity in peatland terrestrial insects, these assemblages may not be resilient to environmental perturbations, and that recolonization to restored and damaged sites may be difficult because of dispersal limitation of the arthropods and fragmented distribution of peatland sites.

Grégoire Taillefer and Wheeler (2010) assessed the role of drainage ditches originally excavated to lower water tables in Johnville Bog in southern Quebec. The presence of ditches (e.g., Fig. 7.2) significantly altered the community structure of terrestrial Diptera at the scale of a few meters from the ditch, suggesting that fine-scale heterogeneity and habitat alteration may have an impact on insect diversity in small peatlands. On a larger scale, Savage et al. (2011) examined the effect of peatland size, vegetation, and surrounding land-use on higher Diptera in six isolated bogs in southern Quebec and northern Vermont. Although peatland size (ranging from 12 to 900 ha) had no measureable impact on community structure, vegetation cover in the sites and surrounding land-use patterns exerted significant influences on the structure of insect assemblages in the peatland. The presence of a forest buffer surrounding these southern sites seemed especially important in maintaining species diversity.

Minnesota Peatlands

Peatland Habitats of Minnesota

Northern peatlands of North America extend into the northern tier of the United States, with Minnesota supporting the greatest area (over 3 million ha; MN DNR 1980). Most are found in the northern half of Minnesota in the lower Glacial Lake Agassiz Region, which extends up into the Great Slave/Great Bear Lake region and the Hudson Bay lowlands of Canada (Glaser 1987; Wright et al. 1992). While the hydrology, chemistry, and vegetation of Minnesota's northern peatlands have been extensively studied (e.g., Heinselman 1970; Boelter and Verry 1977; Glaser et al. 1981; Wheeler et al. 1983), faunal studies are limited to amphibians and reptiles (Karns 1992) and mammals (Berg 1920; Niemi and Hanowski 1992; Nordquist 1992). Little is known about the invertebrate communities inhabiting northern Minnesota's peatlands (Gorham 1990; Wright et al. 1992).

Marcell Experimental Forest (MEF) in north central Minnesota has been a focus of peatland research for decades (e.g., Kolka et al. 2011b). Forested bogs and fens at MEF formed from ice-block depressions that filled with peat ranging in depths of 1–8 m in bogs to 1–6 m in fens (Bay 1967; Verry and Janssens 2011). Sedge and forested peats accumulated in regions where calcium-rich groundwater seeped into depressions forming minerotrophic fens (Boelter and Verry 1977). *Sphagnum* peat accumulated in depressions that were influenced more by low ionic precipitation yielding ombrotrophic bogs.

Bogs at MEF are perched above the aquifer and are primarily precipitation driven (e.g., Bay 1968, 1969; Boelter and Verry 1977) (Fig. 7.1a *water budget of MN bog*). Centers of MEF bogs are raised and surrounded by lagg zones (Sebestyen et al. 2011). Water from the uplands flows into the laggs and drains from the bogs through short outlet streams. At the single fen monitored at MEF, the outlet stream is perennial, a result of continuous groundwater inputs and precipitation (e.g., Verry and Boelter 1975; Boelter and Verry 1977) (Fig. 7.1b *water budget of MN fen*). Bogs at MEF are ion-poor and acidic with pH from 3.7 to 4.9, while groundwater-based fens are ion-rich with water pH ranging from 6 to 7.5 (Boelter and Verry 1977).

Topography, hydrology, and water chemistry influence bog and fen vegetation. Bogs are dominated by black spruce (*Picea mariana*), eastern tamarack (*Larix laricina*), and northern white cedar (*Thuja occidentalis*), *Sphagnum* mosses, and ericaceaous shrubs (Sebestyen et al. 2011). Speckled alder (*Alnus incana*) is common in laggs. Fens contain a higher diversity of understory species including speckled alder, sedges (*Carex* spp.), marsh marigold (*Caltha palustris*), ferns, mosses, and trees (black spruce, balsam fir (*Abies balsamea*), northern white cedar, eastern tamarack, and white birch (*Betula papyrifera*)) (Bay 1967).

Invertebrate Diversity and Ecology in Minnesota Peatlands

Current knowledge of the aquatic invertebrate community composition in northern Minnesota peatlands is limited to an inventory conducted at two bogs and one rich fen at MEF. Twenty-four family or higher-order invertebrate taxa were found in fishless MEF bogs (Appendix). Predators (e.g., Dytiscidae: *Dytiscus* spp., and Cordulidae: *Somatochlora* spp.) and collectors (Culicidae: *Culex*, *Ochlerotatus*, and *Aedes* spp., and Chironomidae: *Chironomus* spp.) were more common than other functional feeding groups, although shredder caddisflies (Limnephilidae: *Limnephilus submonifer*, and *L. indivisus*) were locally common in lagg habitats (Fig. 7.3) where speckled alder trees were abundant. Mitchell et al. (2008) described hotspots of methylmercury production in lagg zones that were related to upland runoff of solutes. Nitrogen-rich leaves from speckled alder trees in laggs may also serve as a high-quality food resource for shredder caddisflies and create hotspots of invertebrate productivity in bog laggs (Fig. 7.3). Multiple individuals of *Philarctus quaeris*, a caddisfly species thought to be extirpated from Minnesota (Houghton 2012) were found in the lagg of a MEF bog during the survey. Despite low pH in the bog water, fingernail clams were present in the lagg habitats and their empty shells are used as case building material by *P. quaeris*. Most bog taxa complete their life cycles in a 2–3 month period since surface water runoff usually ends by mid-June, bog water levels drop, and hollows and lagg habitats become dry.

The fen invertebrate community was somewhat more diverse (33 taxa) than in bogs (24 taxa) (Appendix). Invertebrates in the fen were similar in taxonomic and functional composition to those in bogs, with additional crustaceans (*Hyalella*),

Fig. 7.3 Lagg habitat along upland periphery of a Minnesota peatland (Marcell Experimental Forest). Photo by Sue Eggert, USDA Forest Service, Northern Research Station

snails (*Fossaria*, *Armiger*, and *Gyraulus*), leeches, a mayfly (*Leptophlebia*), and a caddisfly (*Ptilostomis*) found in the permanent outlet stream. *Chilostigma itascae*, an endemic and endangered caddisfly species in Minnesota was not found in peatlands at MEF, although it has been found in similar habitats in the region. Phantom crane fly larvae (*Bittacomorpha*) were common in an iron seep associated with the fen. Several fish species [central mudminnow (*Umbra limi*), brook stickleback (*Culaea inconstans*), and fine-scale dace (*Phoxinus neogaeus*)] were present in the outlet stream which was linked to a downstream lake via a tributary stream. The continuous, nutrient-rich groundwater inputs likely allow for longer invertebrate life cycles, higher animal productivity, and more frequent predator interactions in the fen, although studies of invertebrate life history and trophic interactions in fen and bog food webs at MEF are lacking.

Conservation Biology and Invertebrates of Minnesota Peatlands

For invertebrates, the main thrust of conservation research in Minnesota peatlands has focused on impacts of mercury toxicity. Peatlands are sources of mercury to downstream lakes and rivers via export from outlet streams (Grigal et al. 2000; Kolka et al. 2011a). Methylmercury (MeHg) is produced by microbial processes that respond to the availability of sulfate. To determine the effects of increased atmospheric sulfate deposition on rates of methylation of mercury in MEF peatlands, sulfate was added through a network of PVC pipelines and sprinklers encompassing the downstream half of a bog for 5 years and compared to an upstream control section (Jeremiason et al. 2006). Percent MeHg increased in the treatment section of the bog and in stream water (Jeremiason et al. 2006; Coleman-Wasik et al. 2012). After sulfate addition ended, %MeHg declined in the recovery section relative to the treatment section, but remained higher than the control section. Concentrations of total mercury in mosquito larvae collected in each experimental treatment paralleled MeHg levels in bog water of treatment sections. Study results suggest that reductions in sulfate emissions could result in reductions of MeHg contamination in aquatic food webs in the Upper Midwest United States (Coleman-Wasik et al. 2012).

Northeastern China Peatlands

Peatland Habitats of Northeastern China

Northeastern China (from 38° to 53°N, and 115° to 135°E) is one of the most important areas of peatland wetlands distribution, including Heilongjiang, Jilin and Liaoning Provinces, and the northeast portion of the Inner Mongolia Autonomous Region. This high latitude area, located at the southern margin of the permafrost region of Asia, is conducive to peatland formation due to cold temperatures (mean annual air temperature <1 °C), ample precipitation (400–630 mm, mostly falling from July to September), relatively low evaporation rates, and seasonally frozen soils (Jin et al. 2007). Water and soils in the active permafrost layer (from 45 to 50 cm depths) freeze from October to April.

Peatlands in Northeastern China include both bogs and fens. Bogs are mainly distributed in mountainous areas (e.g., Da Hinggan Mountain with 485 km^2, Xiao Hinggan Mountain with 727 km^2, and Changbai Mountain with 463 km^2 of peatland) (Ma 2013). Fens are most widespread across the Sanjiang Plain, with 350 km^2 of remaining peatland (Ma 2013). Many peatlands of Northeastern China are ombrotrophic, being fed primarily by direct precipitation (Fig. 7.1). Permafrost peatlands typically occur in broad valleys, where frozen soils and flat topographies retain rainwater and surface flow, and prevent water from percolating into the substratum underground (Sun et al. 2011). The peat thickness of both bogs and fens typically ranges from 50 to 60 cm above the permafrost layer (Wang et al. 2010), but can be 1–3 m thick in some places.

Mountain peatlands of northeastern China are vegetatively diverse (700 plant species), with larch (*Larix gmelinii*) being the major tree species (Sun et al. 2011) and the understories being dominated by various shrubs (*Betula fruticosa*, *Chamaedaphne calyculata*, *Ledum palustre*, *Vaccinium* spp.), grasses and sedges (*Calamagrostis angustifolia*, *Carex* spp., *Eriophorum vaginatum*), and mosses (*Sphagnum* spp., *Polytrichum*). Peatland surfaces are a mosaic of microforms, including *Sphagnum* hummocks with woody shrubs (see above), mossy hollows (*Polytrichum juniperinum*), and sedge tussocks (*Eriophorum vaginatum*) (Miao et al. 2012). Marsh develops across some peatlands in Sanjiang Plain, where open water is interspersed with sedges (*Carex schmidtii, C. meyeriana, C. appendiculata, C. lasiocarpa, C. appendiculata*), grasses (*Calamagrostis angustifolia*), and other emergent and submersed plants (*Equisetum heleocharis, Menyanthes trifoliate, Potentilla chinensis, Iris laevigata, Utricularia minor*) (Zhao 1999; Wang et al. 2013). Almost 10 % of the total area of fen peatland on the Sanjiang Plain is classified as marsh (Liu and Ma 2000), and given the presence of ample open water, these marshes are likely important habitats for aquatic invertebrates (Fig. 7.4).

Invertebrate Diversity and Ecology in Northeastern China Peatlands

The vast majority of work on invertebrates in Chinese peatlands has focused on the terrestrial and soil faunas, rather than the aquatic fauna. To some extent, this is logical because bogs and fens of the region lack extensive open water, and the dense vegetative cover and often non-saturated surface soils provide ample habitats for a terrestrial and semiaquatic fauna to develop. In fact, the list of taxa (49 families) in the Appendix of this chapter represents the only known community inventory of aquatic invertebrates from Chinese peatlands (collected from nine fen peatlands in the Sanjiang Plain). Obviously, if basic descriptions of the aquatic fauna are lacking, essentially nothing is known about the ecological dynamics of aquatic invertebrates in Chinese peatlands.

Yin et al. (2003), Zhang et al. (2006, 2008), Wu et al. (2008, 2009), and Bao et al. (2009) each provide descriptions of the terrestrial and semiaquatic invertebrate fau-

Fig. 7.4 Marsh-type habitat in a fen peatland of Northeast China (Honghe National Preserve). Photo by Haitao Wu

nas of Chinese peatlands. The soil invertebrates of peatlands are mainly distributed in the surficial layers (Yin et al. 2003; Zhang et al. 2006; Wu et al. 2008), much as they are in terrestrial forests and grasslands. In many cases, this terrestrial fauna is both taxonomically rich and highly abundant. For example, Zhang et al. (2006) found that soil invertebrates in bogs of Da Hinggan Mountain reached densities of 170,000 individuals/m^2. This fauna was comprised of 4 phyla, 7 classes, 23 orders, and 54 families, with Enchytraeidae worms and Formicidae ants being the numerically dominant families (Huang and Zhang 2008a). From a fen wetland of Sanjiang Plain, Wu et al. (2008) collected 5 phyla, 12 classes, 27 orders and 46 families of soil invertebrates, with mites (Acarina), beetles (Coleoptera), and worms (Enchytraeidae) dominating. In the peatlands of the Xiao Hinggan Mountains, Yin et al. (2003) and Wang et al. (2014) also found that worms (Enchytraeidae) and mites (Acarina), plus springtails (Collembola), were numerically dominant. In terms of biomass, Lumbricidae and Enchytraeidae worms contribute the most (Huang and Zhang 2008b; Zhang et al. 2008). In fens of Sanjiang Plain, Wu et al. (2009) found that soil invertebrate densities peaked in spring. Bao et al. (2009) sampled insects living on fen plants (using sweep nets and yellow-pan traps), and found that Diptera and Hemiptera were numerically dominant, followed by Hymenoptera, Thysanoptera, Collembola, Coleoptera, and Orthoptera.

At local scales, the main environmental factors impacting soil invertebrates in peatlands include temperature, water conditions, soil quality, and vegetation (Zhang et al. 2001). Zhang et al. (2014) found that soil temperature was the main factor affecting the distribution of soil invertebrates across different types of tundra peatlands, and further found a relationship between litter biomass, soil organic matter content, and nutrient content and the distribution of soil macrofauna. In the Changbai Mountains, peatland soil invertebrate densities and community complexities decreased with increasing altitude, mirroring vegetative patterns (Wang et al. 2014). Xin et al. (2009) found that densities of the soil mesofauna were higher in forested bogs than wetland meadows. Zhang and Zhang (2006, 2013) showed that the diversity of the soil macrofauna significantly decreased from continuous to patchy tundra, while the mesofauna and microfauna exhibited the opposite pattern. Water conductivity, pH, soil organic matter content, and water depth all significantly influence water beetle communities (Wei et al. 2002; Dong et al. 2008).

Decomposition is a key process in nutrient recycling and energy flow in peatlands, and studies from Northeastern China on the ecosystem functions of peatland soil faunas have focused on their impacts on leaf litter decomposition. Wu et al. (2009) examined the impacts of soil invertebrates on leaf litter decomposition rates and nutrient fluxes at three successional stages of fen wetland using litterbags with different mesh sizes to include or exclude organisms. Overall, litter breakdown by soil invertebrates was 35.4 % of the total. In coarse mesh bags (4 mm) where most invertebrates had access, litter breakdown was 0.3–4.1 times higher than in fine mesh bags (0.06 mm) where most invertebrates were excluded. Breakdown rates varied among litter from different plants, ranging from 32.9 % for *Carex meyeriana* to 38.2 % for *Calamagrostis angustifolia*; prevalence of these plants in regional peatlands changes with successional status. Litter quality as reflected by carbon (C), nitrogen (N), and phosphorus (P) contents, and C:N and C:P ratios also influence breakdown rates by invertebrates. Variation in invertebrate community compositions and season further influenced litter breakdown.

Ants are very prevalent in fens of the Sanjiang Plain (Wu et al. 2010b, 2013b), and their impacts on soil nutrient pools and cycling are excellent examples of how wetland invertebrates can affect ecosystem function. Wu et al. (2010a, 2013a) found that mounds of *Lasius flavus*, *Lasius niger*, and *Formica candida* ants had greater concentrations of organic C, dissolved organic C, total N, NO_3^-, and NH_4^+ than the surrounding peatland soils. Nutrient pools in ant mounds comprised from 5.3 to 7.6 % of the total in peatland soils overall. Importantly, ant mounds increased the spatial heterogeneity of these nutrient pools.

Conservation and Invertebrates in Northeastern China Peatlands

The major focus of conservation-related research on invertebrates in Chinese peatlands has focused on issues of climate change. Ant mounds also alter the spatial and temporal patterns of gas emissions from peatland soils. Wu et al. (2013b) showed that ant mounds in a Sanjiang Plain fen serve as hot spots for CO_2 emissions, convert soils from being CH_4 sources to CH_4 sinks, and amplify seasonal fluctuations for N_2O emissions. Overall, ant mounds contributed measurable amounts to soil gas emissions from the wetland, averaging 7.0 %, −4.3 %, and 3.4 % of total soil CO_2, CH_4, and N_2O emission, respectively. Laboratory studies suggest that altered gas emissions from ant mounds occur both from changed soil conditions and from ant respiration (Wu et al. 2015). Thus, for a complete understanding of peatland C and N cycles and balances, ant mounds should be considered.

Forest fires happen frequently in the mountains of Northeastern China, and in May 1987 a large fire (known as the Black Dragon Fire) swept across Da Hinggan Mountain forests destroying almost 10,000 km^2 of timber. After the fire, the number of Enchytraeidae in peatlands gradually increased until becoming stable after about five years. The meso- and microfauna was restored after about seven years, while Protura populations did not reappear in burned areas for 16 years (Zhang et al. 2006; Zhang and Zhang 2009).

Subtropical Peatlands of the Southeastern United States

Southeastern US Peatlands

Although most prevalent in cold climates, peatlands also exist in tropical and subtropical climates, despite warm temperatures and high evaporation. In the Southeastern United States, peatlands mostly occur on the Coastal Plain, such as pocosins, a handful of Carolina bays, the Okefenokee Swamp, and the Everglades (see chapters in Batzer and Baldwin 2012). Why these particular wetlands retain peat is not clear, as formative processes, hydrology, pH, and plant communities can all vary widely. Of these Southeastern peatlands, the invertebrates have only been investigated in detail in the Okefenokee Swamp and the Everglades (and the Everglades is the subject of its own chapter in this book, and so not covered here).

The 200,000 ha Okefenokee Swamp in southeastern Georgia and northeastern Florida is among the largest freshwater wetlands in North America (Batzer et al. 2012). The name is aboriginal meaning "land of trembling earth," presumably due to the presence of peat. Deposits of peat up to 4.5 m thick occur across much of the Okefenokee (Cohen et al. 1984a), derived mostly from remnant water lily or cypress debris, and to a lesser extent from *Sphagnum* mosses and sedge.

Hydrology in the Okefenokee is typical for an ombrotrophic peatland (Fig. 7.1c). Water input is dominated by direct rainfall (70–90 %), with some minor inflows from small tributary creeks and off uplands (Rykiel 1984; Brook and Hyatt 1985). Water output is dominated by evapotranspiration (~85 %), with most remaining water losses from stream flow-out of the Suwannee and St. Mary's Rivers to the south. Connection to groundwater, either via discharge or recharge, is considered negligible (<3 %). Early efforts to drain the Okefenokee failed (Izlar 1984), and thus water levels and hydrologic variation remain largely natural. Water pH is acidic (3.5–4.5) (Blood 1980). Mineral concentrations (Ca, Mg, Na, K, Cl) are low, even more so than in many northern bogs (Rykiel 1984). Levels of dissolved carbon are high (46–58 mg C L^{-1}) (Bano et al. 1997), due to organic acids from plant decomposition, which gives the water a characteristic "tea" colored appearance.

Major plant communities of the Okefenokee include forested swamp, scrub-shrub thickets, emergent (grasses, sedges) marsh, and water lily (*Nymphaea*, *Nuphar*) beds (McCaffrey and Hamilton 1984; Fig. 7.5). In peatland forests, bald cypress (*Taxodium distichum*), and pond cypress (*T. ascendens*) are dominant overstory trees (Fig. 7.5). Herbaceous wetland (grass, sedge, and water lily marsh) is locally called "prairie." For more detailed overviews of the Okefenokee see Cohen et al. (1984b) and Batzer et al. (2012).

Invertebrate Community Composition and Ecology in the Okefenokee

Kratzer and Batzer (2007) identified 103 aquatic macroinvertebrate taxa across the Okefenokee (see list of 52 families in the Appendix). Chironomid midge larvae by themselves comprised 66 % of abundance. Mollusks were very rare. Most taxa lack seasonality (Kratzer and Batzer 2007), although microcrustaceans decline in winter from cool temperatures, and in summer from fish predation (Schoenberg 1988). Ecological research on invertebrates of the Okefenokee has focused primarily on natural variation across the mosaic of plant communities and trophic relations.

Kratzer and Batzer (2007) also assessed spatial and temporal variation in macroinvertebrate communities across the Okefenokee Swamp, examining communities in forested, scrub-shrub, prairie, and deepwater (lakes and canals) habitats across different sub-watersheds and in different seasons (29 total locations). They, however, discovered remarkably few invertebrate community patterns across the different plant communities, sub-watersheds, or seasons. A handful of individual taxa exhibited some specialization for particular plant types or seasons, although responses were largely unique for each taxon. Kratzer and Batzer (2007) concluded

Fig. 7.5 Marsh prairie, scrub-shrub thickets, and cypress forest of the Okefenokee Swamp, Georgia. Photo by Mark Galatowitsch, Univ Georgia

that the macroinvertebrate communities in the Okefenokee were dominated by generalist taxa able to exploit the full range of habitat available.

Taylor and Batzer (2010) used stable isotope analyses (C, N) to assess diets of midge larvae in forested and prairie habitat of the Okefenokee. These larvae appeared to be generalist feeders, simply focusing on foods as they were available. Cypress wood was important in forested habitats, and algae and herbaceous plants in marsh prairies; sediment was an important midge food in both habitat types.

Aside from midges, the aquatic invertebrate community of the Okefenokee is dominated by a plethora of large predators (odonates, hemipterans, coleopterans) (Kratzer and Batzer 2007). The presence of so many predatory invertebrates seems at odds with the fact that fish productivity in the Okefenokee Swamp is high (Freeman and Freeman 1985); fish typically exclude predatory invertebrates (Wellborn et al. 1996). However, most fish production is from small-bodied species such as killifish (*Fundulus* spp.), pygmy sunfish (*Elassoma* spp.), and mosquitofish (*Gambusia* spp.). These small fishes feed primarily on microcrustaceans and midge larvae (Freeman and Freeman 1985; Oliver 1991), and probably cannot tackle the larger invertebrate taxa. In much of the Okefenokee (deepwater habitats with large fishes excepted), invertebrates may actually hold a higher position in the food web than fishes. For example, dragonfly nymphs in the Okefenokee frequently have small fish in their guts (B. Freeman, unpublished data).

Fig. 7.6 Wildfire in the Okefenokee Swamp (2007). Photo from US Fish Wildl Serv, Okefenokee National Refuge

Conservation Biology and Okefenokee Invertebrates

Fire, integrated with drought, is considered the most important factor controlling the structure of Okefenokee plant communities (Schlesinger 1978). Fire is believed to maintain open prairie habitat, which otherwise would convert into woody vegetation. Deeper "lakes" scattered across the Okefenokee may have developed where fire burned deep into the peat deposits. Large fires occur every few decades (Yin 1993). A particularly large fire occurred in 2007, and burned >75 % of the Okefenokee (Fig. 7.6). Beganyi and Batzer (2011) assessed invertebrate response to that fire, contrasting burned and non-burned prairie, scrub-shrub, and cypress-forest habitats. Only in cypress forest was significant invertebrate response detected, with populations of leptocerid caddisfly larvae (*Oecetis* sp.) and coenagrionidae damselfly nymphs (*Ishnura* sp.) declining, and populations of corixid water boatmen (*Sigara* sp.) increasing in burned cypress. The rest of the invertebrate community in cypress forest did not appear to numerically respond to fire. In prairie and scrub-shrub thickets, no invertebrate responses to fire were detected. In prairies, fire burned quickly through the habitats; flooded and moist sediments did not burn and herbaceous plants grew back rapidly, likely precluding any invertebrate response. While fire may be a crucial control for plant communities in the Okefenokee, it appears to have a much lesser role in controlling invertebrate community structure (mirroring the lack of plant control on invertebrate communities previously described).

Problems with heavy metals are also a concern in the Okefenokee, and human consumption advisories due to mercury exist for some fish. Anoxic, high temperature conditions in sediments are conducive to mercury methylation, the form of the metal that is most toxic and most likely to bioaccumulate. Mercury levels in Okefenokee macroinvertebrates are unusually high, especially in *Crangonyx* amphipods (George and Batzer 2008; Beganyi and Batzer 2011) (levels in alligators, however, are typical for the Southeast; Jagoe et al. 1998). Levels of mercury in invertebrates are similar across the range of available habitats (plant types, subwatersheds) in the Okefenokee, suggesting that mercury is being introduced via aerial deposition (Jackson et al. 2004; George and Batzer 2008). Beganyi and Batzer (2011) assessed whether the 2007 wildfire magnified mercury levels in macroinvertebrates, but did not find this to be the case.

Synthesis

Perhaps the most telling finding of our review was the dearth of detailed information about the ecology of invertebrates and about invertebrate functional roles in peatlands, despite these habitats being the most expansive and potentially most important wetlands on earth. Many of the published studies simply inventory the fauna or describe basic life histories. However, those studies may suggest why so little is known about the ecology of peatland invertebrates. First, unlike many other wetland types, the terrestrial fauna is especially well developed in peatlands. As noted above, Finnamore (1994) reported 1410 species of the single-order Hymenoptera, in a single Canadian peatland (studies of Hymenoptera are essentially nonexistent in any wetland type, except peatlands). If the goal is to establish the overall importance of invertebrates in peatlands, the terrestrial fauna clearly must be considered, a daunting task. Comparatively, the aquatic invertebrate fauna in peatlands is depauperate, although likely still very important. Despite being easier to work with, studies of aquatic invertebrates in peatlands are few; remarkably this chapter provides the first inventories of aquatic invertebrate from peatlands of Minnesota and China, otherwise fairly well-known habitats.

Although our knowledge remains limited, this review provided some valuable preliminary evidence on the primary ecological controls for peatland invertebrate and the major ecological roles invertebrates play in peatlands, and how invertebrates can be useful in assessing emerging environment threats to peatlands.

Invertebrate Community Ecology in Peatlands

In the Appendix, we report 79 aquatic invertebrate groups from peatland habitats across the globe. These aquatic faunas are moderately diverse compared to other types of wetlands (Batzer and Ruhí 2013), with from 24 to 52 families recorded per location. Some of these variations may simply reflect sampling effort, as the most taxonomically rich peatland habitat, the Okefenokee Swamp, was also the most intensively sampled. Faunas are highly variable from place to place, with only six groups being ubiquitous across all habitats: Branchiopoda (specifically cladocerans) and Copepoda microcrustaceans; Dytiscidae and Hydrophilidae/Helophoridae beetles; and Ceratopogonidae and Chironomidae fly larvae. These same aquatic organisms also tend to dominate other kinds of wetland habitat (Batzer and Ruhí 2013). Overall aquatic invertebrate faunas of peatlands seem to be comprised mostly of generalist organisms, i.e., ubiquitous families.

We saw no obvious gradient in the total taxon richness between bogs (pH < 5) or fens (pH > 6). However, mollusks (Gastropoda, Bivlavia) and leeches (Erpobdellidae, Glossophoniidae) were rarely encountered in low pH habitats (see also Wheeler and Proctor 2000; Hájek et al. 2006), while these groups were widely distributed across fen habitats with more circumneutral pH (Appendix). At least for the mollusks, a lack of calcium carbonate for shell development may exclude many taxa from acidic habitats. However, most aquatic insects and crustaceans appeared to be tolerant of low pH conditions in peatlands, at least as reflected by family-level distributions. Perhaps more sensitivity might become evident in these groups if generic or species-level analyses were possible.

For aquatic invertebrate communities in peatlands, as for most wetlands, water permanence influences compositions, with longer hydroperiods promoting greater taxon richness, especially large predators. Although hydrology likely affects plant community compositions, similar aquatic invertebrate assemblages may occur across a diversity of plant communities, further suggesting that habitat generalists prevail. In northern bogs, however, peripheral lagg habitats appear to be hot spots for aquatic invertebrates, perhaps due to influences of upland water runoff, more open water, or the growth of nitrogen-rich plant foods (e.g., alder leaves). For terrestrial invertebrates, damper soils enhance terrestrial diversity and abundance. Variation of invertebrates related to plant community change is more pronounced for terrestrials, probably because the plants themselves serve as food (herbivores) or habitat.

Fire appears a pervasive influence on peatlands. However, the aquatic invertebrate community was minimally affected by a large wildfire in the Okefenokee Swamp, further evidence of a generalist tendency for that fauna. The terrestrial invertebrate fauna appears more dramatically influenced by fire, likely because the ecological influence of fire is more pronounced above the water line.

Decomposition is likely the most important ecosystem process affected by peatland invertebrates. Invertebrates tend to track the quantity and quality of organic matter in peatlands. In northern peatlands, aquatic limnephilid caddisfly larvae are likely important shredders of organic material (leaves). However, the terrestrial invertebrate fauna probably plays the most important role in decomposition, with ants and annelid worms being key. The fact that these organisms can affect emissions of greenhouse gases from peatlands make them perhaps among the most important wetland invertebrates known.

Predaceous invertebrates, both aquatic and terrestrial, are widespread in peatlands. Most of the 1410 species of Hymenoptera reported by Finnamore were parasitic wasps. The prevalence of predators/parasitoids suggests ecological importance, but we found no studies quantifying their impacts.

Conservation Issues and Peatland Invertebrates

Multiple threatened invertebrate species occur in peatlands, suggesting these habitats are at risk. Researchers frequently mentioned climate change as a major threat, given that northern peatlands primarily exist due to regionally cold temperatures. As mentioned, peatland invertebrates may play key roles in enhancing peat decomposition and gas flux, and as ectotherms their activities are controlled by temperature. This combination could lead to them exacerbating the impacts of climate change.

Drainage, logging, and peat mining are pervasive threats to peatlands. Invertebrates, both aquatic and terrestrial, have proven to be useful indicators of environmental impacts and of the success of restoration efforts, especially by workers in Europe. However, the terrestrial fauna might prove to be superior indicators, given that terrestrials tend to be more specialized than aquatics (e.g., associated with particular plants) and the terrestrial fauna appears to affect crucial ecosystem functions (decomposition, gas flux). Peatland habitats are foci of mercury methylation, and bioaccumulation of mercury in aquatic invertebrates may transfer this heavy metal up food chains.

Conclusion

This review establishes that invertebrates are very diverse and ecologically important components of peatland habitats. However, our review also exposes some real inadequacies in our knowledge. The terrestrial invertebrate fauna has been shown to be especially diverse, and numerous taxa appear to be valuable environmental indicators. However, the ecological and functional roles of terrestrial invertebrates remain poorly known; and this kind of knowledge is sorely needed to understand how these organisms affect peatland ecosystems and why this fauna has tangible value for peatland bioassessment. Even less is known about the aquatic fauna. Basic inventories are lacking, the functional importance of the aquatic fauna to ecosystem processes remains essentially unknown, and their value, if any, to bioassessment is still largely undeveloped. The situation in Europe appears somewhat better than for other parts of the world, and the knowledge about arthropods is somewhat better developed than for the non-arthropod fauna. Because of the extent and importance in climate change scenarios, the virtual dearth of knowledge about invertebrates in Arctic peatlands is of special concern.

Appendix

Aquatic invertebrate taxa recorded from peatlands across the globe, arranged from bogs (lower pH) to fens (higher pH). Dark shading of cells indicates ubiquitous taxa, medium shading indicates sporadic occurrence of taxa, and light shading indicates taxa reported from only a single location. Different sampling techniques and sampling intensities were utilized at each location; hence the absence of a taxon from a particular site may be a sampling artifact. (Genus level classifications are available from the applicable references or from S. Eggert or H. Wu for their previously unpublished data.)

	Okefenokee Swamp (Kratzer and Batzer 2007)	Minnesota bogs (Eggert, unpublished)	Irish bogs (Hannigan and Kelly-Quinn 2012; Hannigan et al. 2011)	Japanese poor-fen (Kato et al. 2009)	Chinese fens (Wu, unpublished)	Minnesota rich fen (Eggert, unpublished)	Irish rich fen (Hannigan and Kelly-Quinn 2012)
pH	3.8–4.7	3.7–4.9	4.3–4.9	4.7–6.3	5.8–6.9	6.5–7.0	6.9–8.3
NEMATODA		■				■	
MOLLUSCA							
Gastropoda							
Ancylidae	■				■		■
Bithyniidae							■
Lymnaeidae				■	■	■	■
Planorbidae		■			■	■	■
Succineidae					■		■
Valvatidae					■	■	■
Bivalvia							
Sphaeriidae		■			■	■	■
ANNELIDA							
Oligochaeta	■		■	■			■
Lumbriculidae		■		■	■	■	
Tubificidae	■	■		■	■	■	
Hirudinea							
Erpobdellidae				■	■	■	■
Glossophoniidae				■	■	■	■
CRUSTACEA							
Branchiopoda (Cladocera)	█	█	█	█	█	█	█
Branchiura							
Arguliidae	■						

Copepoda	■	■	■	■	■	■	■
Ostracoda		●		●		●	
Amphipoda							
Crangonyctidae	●						
Gammaridae	●					●	
Isopoda							
Asellidae	●			●	●		●
Decapoda							
Atydidae				○			
Cambaridae	●			●			
Palaemonidae	○						
ACARINA							
Hydrachnidia	●	●	●		●	●	●
INSECTA							
Collembola							
Entomobryidae	●				●		
Isotomidae		●		●	●	●	
Poduridae	●				●	●	
Sminthuridae	●	●			●		
Ephemeroptera							
Baetidae			●	●			●
Caenidae	●				●		
Leptophlebidae			●			●	●
Odonata							
Coenagrionidae	●		●	●	●		●
Lestidae	●		●	●			
Aeshnidae	●		●		●		●
Corduliidae	●	●			●	●	

(continued)

(continued)

	Okefenokee Swamp (Kratzer and Batzer 2007)	Minnesota bogs (Eggert, unpublished)	Irish bogs (Hannigan and Kelly-Quinn 2012; Hannigan et al. 2011)	Japanese poor-fen (Kato et al. 2009)	Chinese fens (Wu, unpublished)	Minnesota rich fen (Eggert, unpublished)	Irish rich fen (Hannigan and Kelly-Quinn 2012)
Libellulidae							
Hemiptera							
Belostomatidae							
Corixidae							
Gerridae							
Hydrometridae							
Mesoveliidae							
Naucoridae							
Nepidae							
Notonectidae							
Pleidae							
Veliidae							
Coleoptera							
Dytiscidae							
Gyrinidae							
Haliplidae							
Hydraenidae							
Hydrophilidae/ Helophoridae							
Noteridae							
Scirtidae							
Neuroptera							
Corydalidae							

Sialidae							
Sisyridae							
Trichoptera							
Hydroptilidae							
Leptoceridae							
Limnephilidae							
Phryganeidae							
Polycentropodidae							
Lepidoptera							
Pyralidae/Crambidae							
Diptera							
Ceratopogonidae							
Chaoboridae							
Chironomidae							
(Chironominae)							
(Tanypodinae)							
(Orthocladiinae)							
Culicidae							
Dixidae							
Dolichopodidae							
Muscidae							
Psychodidae							
Ptychopteridae							
Sciomyzidae							
Simuliidae							
Stratiomyidae							
Syrphidae							
Tabanidae							
Tipulidae/Limoniidae							
TOTAL TAXA	52	24	34	24	49	33	46

References

Aitchison-Benell CW (1994) Bog arachnids (Araneae, Opiliones) from Manitoba taiga. In: Finnamore AT, Marshall SA (eds) Terrestrial arthropods of peatlands, with particular reference to Canada. Mem Entomol Soc Can 169:21–31

Bano N, Moran MA, Hodson RE (1997) Bacterial utilization of dissolved humic substances from a freshwater swamp. Aquat Microb Ecol 12:233–238

Bao X, Lu XG, Zhang F (2009) Biodiversity and seasonal change of insects in annular wetlands in Sanjiang Plain. J NE Forestry Univ 37:100–104 (In Chinese)

Batzer DP, Baldwin AH (eds) (2012) Wetland habitats of North America: ecology and conservation concerns. Univ Calif Press, Berkeley

Batzer DP, Ruhí A (2013) Is there a core set of organisms that structure macroinvertebrate assemblages in freshwater wetlands? Freshw Biol 58:1647–1659

Batzer DP, Day F, Golladay SW (2012) Southeastern swamp complexes. In: Batzer DP, Baldwin AH (eds) Wetland habitats of North America: ecology and conservation concerns. Univ Calif Press, Berkeley, pp 217–230

Bay RR (1967) Ground water and vegetation in two peat bogs in northern Minnesota. Ecology 48:308–310

Bay RR (1968) The hydrology of several peat deposits in northern Minnesota, USA. In: Proc Third International Peat Congress. National Research Council of Canada, Quebec, Canada, pp 212–218

Bay RR (1969) Runoff from small peatland watersheds. J Hydrol 9:90–102

Beganyi SR, Batzer DP (2011) Wildfire induced changes in aquatic invertebrate communities and mercury bioaccumulation in the Okefenokee Swamp. Hydrobiologia 669:237–247

Behan-Pelletier VM, Bissett B (1994) Oribatida of Canadian peatlands. Mem Entomol Soc Can 169:73–88

Berg WE (1920) Large mammals. In: Wright HE Jr, Coffin BA, Aaseng NE (eds) The patterned peatlands of Minnesota. Univ Minn Press, Minneapolis, pp 73–84

Bezděk A, Jaroŝ J, Spitzer K (2006) Spatial distribution of ground beetles (Coletoptera: Carabidae) and moths (Lepidoptera) in the Mrtvý luh bog, Šumava Mts (Central Europe): a test of habitat island community. Biodiv Conserv 15:395–409

Blades DCA, Marshall SA (1994) Terrestrial arthropods of Canadian peatlands: synopsis of pan trap collections at four southern Ontario peatlands. Mem Entomol Soc Can 169:221–284

Blood ER (1980) Surface water hydrology and biogeochemistry of the Okefenokee Swamp watershed. Ph.D. Dissertation, Univ Georgia, Athens

Boelter DH, Verry ES (1977) Peatland and water in the northern lake States, USDA Forest Serv General Techn Rep NC-31, North Central Forest Experimental Station, St. Paul

Bridgham SD, Pastor J et al (1996) Multiple limiting gradients in peatlands: a call for a new paradigm. Wetlands 16:45–65

Brook GA, Hyatt RA (1985) A hydrological budget for the Okefenokee Swamp watershed, 1981–82. Phys Geogr 6:127–141

Cannings SG, Cannings RA (1994) The Odonata of the northern cordilleran peatlands of North America. Mem Entomol Soc Can 169:89–110

Carrera N, Barreal ME, Gallego PP, Briones MJI (2009) Soil invertebrates control peatland C fluxes in response to warming. Funct Ecol 23:637–648

Carrera N, Barreal ME, Rodeiro J, Briones MJI (2011) Interactive effects of temperature, soil moisture and enchytraeid activities on C losses from a peatland soil. Pedobiologia 54:291–299

Carroll MJ, Dennis P, Pearce-Higgins JW, Thomas CD (2011) Maintaining northern peatland ecosystems in a changing climate: effects of soil moisture, drainage and drain blocking on craneflies. Global Change Biol 17:2991–3001

Cohen AD, Andrejko MJ et al (1984a) Peat deposits of the Okefenokee Swamp. In: Cohen AD, Casagrande DJ et al (eds) The Okefenokee swamp: its natural history, geology, and geochemistry. Wetland Surveys, Los Alamos, pp 493–553

Cohen AD, Casagrande DJ et al (eds) (1984b) The Okefenokee swamp: its natural history, geology, and geochemistry. Wetland Surveys, Los Alamos

Coleman-Wasik JK, Mitchell CPJ et al (2012) Methylmercury declines in a boreal peatland when atmospheric sulfate deposition decreases. Environ Sci Technol 46:6663–6671

COSEWIC (2009) COSEWIC assessment and status report on the Bogbean Buckmoth *Hemileuca* sp. in Canada. Committee on the Status of Endangered Wildlife in Canada, Ottawa, www.sararegistry.gc.ca/status/status_e.cfm

COSEWIC (2012) COSEWIC assessment and status report on the Georgia Basin Bog Spider *Gnaphosa snohomish* in Canada. Committee on the Status of Endangered Wildlife in Canada, Ottawa, www.registrelep-sararegistry.gc.ca/default_e.cfm

Coulson JC, Butterfield JEL (1985) The invertebrate communities of peat and upland grasslands in the north of England and some conservation concerns. Biol Conserv 34:197–225

Dondale CD, Redner JH (1994) Spiders (Araneae) of six small peatlands in southern Ontario or southwestern Quebec. Mem Entomol Soc Can 169:33–40

Dong BL, Wang M, Jiang P (2008) Relations between water beetle diversity and environmental factors in northern slope of Changbai Mountain, northeastern China. J Beijing Forestry Univ 1:74–78 (In Chinese)

Downie IS, Coulson JC, Foster GN, Whitfield DP (1998) Distribution of aquatic macroinvertebrates within peatland pool complexes in the flow country, Scotland. Hydrobiologia 377:95–105

Drinan TJ, Foster GN et al (2013) Macroinvertebrate assemblages of peatland lakes: assessment of conservation value with respect to anthropogenic land-cover change. Biol Conserv 158:175–187

Finnamore AT (1994) Hymenoptera of the Wagner natural area, a boreal spring fen in central Alberta. Mem Entomol Soc Can 169:181–220

Finnamore AT, Marshall SA (1994) Terrestrial arthropods of peatlands, with particular reference to Canada. Mem Entomol Soc Can 169:1–289

Flannagan JF, Macdonald SR (1987) Ephemeroptera and Trichoptera of peatlands and marshes in Canada. Mem Entomol Soc Can 140:47–56

Freeman BJ, Freeman MC (1985) Production of fishes in a subtropical blackwater ecosystem: the Okefenokee Swamp. Limnol Oceanogr 30:686–692

George BM, Batzer DP (2008) Spatial and temporal variations of mercury levels in Okefenokee invertebrates: Southeast Georgia. Environ Pollut 152:484–490

Glaser PH (1987) The ecology of patterned boreal peatlands of northern Minnesota: a community profile. US Fish Wildl Serv Rep 85(7.14)

Glaser PH, Wheeler GA, Gorham E, Wright HE Jr (1981) The patterned mires of the Red Lake Peatland, northern Minnesota: vegetation, water chemistry and landforms. J Ecol 69:575–599

Gorham E (1990) Biotic impoverishment in northern peatlands. In: Woodwell GM (ed) The Earth in transition: patterns and processes of biotic impoverishment. Cambridge Univ Press, New York, pp 65–98

Gorham E, Rochefort L (2003) Peatland restoration: a brief assessment with special reference to *Sphagnum* bogs. Wetl Ecol Manage 11:109–119

Grégoire Taillefer A, Wheeler TA (2010) The effect of drainage ditches on Brachycera (Diptera) diversity in a southern Quebec peatland. Can Entomol 142:160–172

Grégoire Taillefer A, Wheeler TA (2012) Community assembly of Diptera following restoration of mined boreal bogs: taxonomic and functional diversity. J Insect Conserv 16:165–176

Grégoire Taillefer A, Wheeler TA (2013) Animal colonization of restored peatlands: inoculation of plant material as a source of insects. Restor Ecol 21:140–144

Grigal DF, Kolka RK, Fleck JA, Nater EA (2000) Mercury budget of an upland-peatland watershed. Biogeochemistry 50:95–109

Hájek M, Horsák M, Hájková P, Dítě D (2006) Habitat diversity of central European fens in relation to environmental gradients and an effort to standardize fen terminology in ecological studies. Prespect Plant Ecol Evol System 8:97–114

Hannigan E, Kelly-Quinn M (2012) Composition and structure of macroinvertebrate communities in contrasting open-water habitat in Irish peatlands: implications for biodiversity conservation. Hydrobiologia 692:19–28

Hannigan E, Mangan R, Kelly-Quinn M (2011) Evaluation of the success of mountain blanket bog pool restoration in terms of aquatic macroinvertebrates. Biol Environ Proc Royal Irish Acad 111B:95–105. doi:10.3318/BIOE.2011.111.08

Heinselman ML (1970) Landscape evolution, peatland types and the environment in the Agassiz peatlands natural area, Minnesota. Ecol Monogr 40:235–261

Hilton DFJ (1987) Odonata of peatlands and marshes in Canada. Mem Entomol Soc Can 140:57–63

Houghton DC (2012) Biological diversity of the Minnesota caddisflies (Insecta, Trichoptera). Zookeys 189:1–389

Huang LR, Zhang XP (2008a) Characteristics of soil fauna in cold temperate region in Northern Daxing'an Mountains. J NE Forestry Univ 10:22–25 (In Chinese)

Huang LR, Zhang XP (2008b) Community characteristics of mid-micro soil animals in cold-temperature zone of the Daxing'an Mountains, China. China J Appl Environ Biol 3:388–393 (In Chinese)

Izlar RL (1984) A history of Okefenokee logging operations. In: Cohen AD, Casagrande DJ et al (eds) The Okefenokee swamp: its natural history, geology, and geochemistry. Wetland Surveys, Los Alamos, pp 5–17

Jackson BP, Winger PV, Lasier PJ (2004) Atmospheric lead deposition to Okefenokee Swamp, Georgia, USA. Environ Pollut 130:445–451

Jagoe CH, Arnold-Hill B et al (1998) Mercury in alligators (*Alligator mississippiensis*) in the southeastern United States. Sci Total Env 213:255–262

Jeremiason JD, Engstrom DR et al (2006) Sulfate addition increases methylmercury production in an experimental wetland. Environ Sci Technol 40:3800–3806

Jin HJ, Yu QH et al (2007) Degradation of permafrost in the Xing'an Mountains, Northeast China. Permafrost Periglacial Process 18:245–258

Judd WW (1957) Studies of the Byron Bog in southwestern Ontario. I. Description of the bog. Can Entomol 89:235–238

Judd WW (1975) Studies of the Byron Bog in southwestern Ontario. 50. Additional observations on insects in the bog. Entomol News 86:51–56

Karns DR (1992) Amphibians and reptiles. In: Wright HE Jr, Coffin BA, Aaseng NE (eds) The patterned peatlands of Minnesota. Univ Minn Press, Minneapolis, pp 131–150

Kato Y, Takemon Y, Hori M (2009) Invertebrate assemblages in relation to habitat types on a floating mat in Mizorogaike Pond, Kyoto, Japan. Limnology 10:167–176

Kolka RK, Mitchell PJC et al (2011a) Mercury cycling in peatland watersheds. In: Kolka RK, Sebestyen SD, Verry ES, Brooks KN (eds) Peatland biogeochemistry and watershed hydrology at the Marcell Experimental Forest. CRC, Boca Raton, pp 349–370

Kolka RK, Sebestyen SD, Verry ES, Brooks KN (eds) (2011b) Peatland biogeochemistry and watershed hydrology at the Marcell Experimental Forest. CRC, Boca Raton

Koponen S (1994) Ground-living spiders, opilionids, and pseudoscorpions of peatlands in Quebec. Mem Entomol Soc Can 169:41–60

Koponen S (2002) Spider fauna of peat bogs in southwestern Finland. In: Toft S, Scharff N (eds) European arachnology 2000. Aarhus Univ Press, Aarhus, pp 267–271

Krab EJ, van Schrojenstein Lantman IM, Cornelissen JHC, Berg MP (2013) How extreme is an extreme climatic event to a subarctic peatland springtail community? Soil Biol Biochem 59:16–24

Krab EJ, Aerts R, Berg MP, van Hal J, Keuper F (2014) Northern peatland Collembola communities unaffected by three summers of simulated extreme precipitation. Appl Soil Ecol 79:70–76

Kratzer EB, Batzer DP (2007) Spatial and temporal variation in aquatic macroinvertebrates in the Okefenokee Swamp, Georgia, USA. Wetlands 27:127–140

Larson DJ (1987) Aquatic Coleoptera of peatlands and marshes in Canada. Mem Entomol Soc Can 140:99–132

Lavoie C, Elias SA, Filion L (1997) A 7000-year record of insect communities from a peatland environment, southern Québec. Écoscience 4:394–403

Lehmitz R (2014) The oribatid mite community of a German peatland in 1987 and 2012—effects of anthropogenic desiccation and afforestation. Soil Organisms 86:131–145

Lewis DJ (1987) Biting flies (Diptera) of peatlands and marshes in Canada. Mem Entomol Soc Can 140:133–140

Liu XT, Ma XH (2000) Influence of large-scale reclamation on natural environment and regional environmental protection in the Sanjiang Plain. Scienta Geogr Sinica 20:14–19

Ma XH (2013) Carbon reserves and emissions of peatlands in China. Chinese Forestry Press, Beijing (In Chinese)

Marshall SA (1994) Peatland Sphaeroceridae (Diptera) of Canada. Mem Entomol Soc Can 169:173–179

Marshall SA, Finnamore AT (1994) Peatland arthropods introduction. Mem Entomol Soc Can 169:3–5

Marshall SA, Finnamore AT, Blades DCA (1999) Canadian peatlands: diversity and habitat specialization of the arthropod fauna. In: Batzer DP, Rader RB, Wissinger SA (eds) Invertebrates in freshwater wetlands of North America: ecology and management. Wiley, New York, pp 383–400

Mazerolle MJ, Poulin M et al (2006) Animal and vegetation patterns in natural and man-made bog pools: implications for restoration. Freshw Biol 51:333–350

McCaffrey CA, Hamilton DB (1984) Vegetation mapping of the Okefenokee ecosystem. In: Cohen AD, Casagrande DJ et al (eds) The Okefenokee swamp: its natural history, geology, and geochemistry. Wetland Surveys, Los Alamos, pp 201–211

McGeoch MA (1998) The selection, testing and application of terrestrial insects as bioindicators. Biol Rev 73:181–201

Mendelssohn IA, Batzer DP, Holt CR, Graham SA (2014) Abiotic constraints for wetland plants and animals. In: Batzer DP, Sharitz RR (eds) Ecology of freshwater and estuarine wetlands, 2nd edn. Univ Calif Press, Berkeley, pp 61–85

Miao Y, Song C et al (2012) Growing season methane emission from a boreal peatland in the continuous permafrost zone of Northeast China: effects of active layer depth and vegetation. Biogeosciences 9:4455–4464

Mieczan T, Tarkowska-Kukuryk M, Plaska W, Rechulicz J (2014) Abiotic predictors of faunal communities in an ombrotrophic peatland lagg and an open peat bog. Isr J Ecol Evol 60(2-4):62–74

Minnesota Department of Natural Resources, Division of Minerals, Peat Inventory Project (1980) Inventory of peat resources. Koochiching County, Minnesota

Mitchell CPJ, Branfireun BA, Kolka RK (2008) Spatial characteristics of net methylmercury production hot spots in peatlands. Environ Sci Technol 42:1010–1016

Niemi GJ, Hanowski JM (1992) Bird populations. In: Wright HE Jr, Coffin BA, Aaseng NE (eds) The patterned peatlands of Minnesota. Univ Minnesota Press, Minneapolis, pp 111–129

Nilsson AN, Svensson BW (1995) Assemblages of dytiscid predators and culicid prey in relation to environmental-factors in natural and clear-cut boreal swamp forest pools. Hydrobiologia 308:183–196

Nordquist GE (1992) Small mammals. In: Wright HE Jr, Coffin BA, Aaseng NE (eds) The patterned peatlands of Minnesota. Univ Minnesota Press, Minneapolis, pp 85–110

Oliver JD (1991) Consumption rates, evacuation rates, and diets of pygmy killifish, *Leptolucania-ommata*, and mosquitofish, *Gambusia-affinis* (Osteichthyes, Atherinformes) in the Okefenokee Swamp. Brimleyana 17:89–103

Rochefort L, Strack M et al (2012) Northern peatlands. In: Batzer DP, Baldwin AH (eds) Wetland habitats of North America: ecology and conservation concerns. Univ Calif Press, Berkeley, pp 119–134

Rosenberg DM, Danks HV (eds) (1987) Aquatic insects of peatlands and marshes in Canada. Mem Entomol Soc Can 140:1–174

Rykiel EJ Jr (1984) General hydrology and mineral budgets for Okefenokee Swamp. In: Cohen AD, Casagrande DJ et al (eds) The Okefenokee swamp: its natural history, geology, and geochemistry. Wetland Surveys, Los Alamos, pp 212–228

Savage J, Wheeler TA, Moores AMA, Grégoire Taillefer A (2011) Effects of habitat size, vegetation cover and surrounding land use on Diptera diversity in temperate Nearctic bogs. Wetlands 31:125–134

Schlesinger WM (1978) Community structure, dynamics and nutrient cycling in the Okefenokee cypress swamp-forest. Ecol Monogr 48:43–65

Schoenberg SA (1988) Microcrustacean community structure and biomass in marsh and lake habitats of the Okefenokee Swamp: seasonal dynamics and responses to resource manipulations. Holarct Ecol 11:8–18

Scott AG, Oxford GS, Selden PA (2006) Epigeic spiders as ecological indicators of conservation value for peat bogs. Biol Conserv 127:420–428

Scudder GGE (1987) Aquatic and semiaquatic hemiptera of peatlands and marshes in Canada. Mem Entomol Soc Can 140:65–98

Sebestyen SD, Dorrance C et al (2011) Long-term monitoring sites and trends at the Marcell Experimental Forest. In: Kolka RK, Sebestyen SD, Verry ES, Brooks KN (eds) Peatland biogeochemistry and watershed hydrology at the Marcell Experimental Forest. CRC, Boca Raton, pp 15–71

Smith IM (1987) Water mites of peatlands and marshes in Canada. Mem Entomol Soc Can 140:31–46

Spitzer K, Danks HV (2006) Insect biodiversity of boreal peat bogs. Annu Rev Entomol 51:137–161

Spitzer K, Bezděk A, Jaros J (1999) Ecological succession of a relict Central European peat bog and variability of its insect biodiversity. J Insect Conserv 3:97–106

Standen V (1999) Quantifying macroinvertebrate taxon richness and abundance in open and forested pool complexes in the Sutherland Flows. Aquat Conserv Marine Freshw Ecosyst 9:209–217

Sun J, Li XZ et al (2011) Latitudinal pattern in species diversity and its response to global warming in permafrost wetlands in the Great Hing'an Mountains, China. Rus J Ecol 42:123–132

Tarnocai C (2006) The effect of climate change on carbon in Canadian peatlands. Global Planetary Change 53:222–232

Taylor AN, Batzer DP (2010) Spatial and temporal variation in invertebrate consumer diets in forested and herbaceous wetlands. Hydrobiologia 651:145–159

van Duinen GJA, Brock AMT et al (2003) Do restoration measures rehabilitate fauna diversity in raised bogs? A comparative study on aquatic macroinvertebrates. Wetl Ecol Manage 11:447–459

Vepsalainen K, Savolainen R, Tianen J, Vilen J (2000) Successional changes of ant assemblages: from virgin and ditched bogs to forests. Ann Zool Fenn 37:135–149

Verhoeven JTA (2014) Wetlands in Europe: perspective for restoration of a lost paradise. Ecol Eng 66:6–9

Verry ES, Boelter DH (1975) The influence of bogs on the distribution of streamflow from small bog-upland watersheds. In: International symposium on the hydrology of marsh-ridden areas. Proceedings of the Minsk symposium, UNESCO/IAHS. pp 1–11

Verry ES, Janssens J (2011) Geology, vegetation, and hydrology of the S2 bog at the MEF: 12,000 years in northern Minnesota. In: Kolka RK, Sebestyen SD, Verry ES, Brooks KN (eds) Peatland biogeochemistry and watershed hydrology at the Marcell Experimental Forest. CRC, Boca Raton, pp 93–134

Vitt DH (1994) An overview of factors that influence the development of Canadian peatlands. Mem Entomol Soc Can 169:7–20

Wang XW, Li XL et al (2010) Effect of temperature and moisture on soil organic carbon mineralization of predominantly permafrost peatland in the Great Hing'an Mountains, Northeastern China. J Environ Sci 22:1057–1066

Wang XW, Song CC et al (2013) Soil carbon and nitrogen across wetland types in discontinuous permafrost zone of the Xiao Xing'an Mountains, northeastern China. Catena 101:31–37

Wang ZH, Yin XQ, Jiang YF (2014) Structure and diversity of soil fauna communities in the tundra of the Changbai Mountains. Acta Ecol Sinica 3:755–765 (In Chinese)

Wei YL, Ji LZ et al (2002) CCA of water beetles' distribution and environmental factors in lentic samples of North Changbai Mountain. Chin J Appl Ecol 1:91–94 (In Chinese)

Wellborn GA, Skelly DK, Werner EE (1996) Mechanisms creating community structure across a freshwater habitat gradient. Annu Rev Ecol Syst 27:337–364

Whatley MH, van Loon EE et al (2014) Linkages between benthic microbial and freshwater insect communities in degraded peatland ditches. Ecol Indicat 46:415–424

Wheeler BD, Proctor MCF (2000) Ecological gradients, subdivisions and terminology for North-West European mires. Brit Ecol Soc 88:187–203

Wheeler GA, Glaser PH et al (1983) Contributions to the flora of the Red Lake Peatland, northern Minnesota, with special attention to *Carex*. Amer Midl Nat 110:62–96

Więcek M, Martin P, Lipinski A (2013) Water mites as potential long-term bioindicators in formerly drained and rewetted raised bogs. Ecol Indicat 34:332–335

Wright HE Jr, Coffin BA, Aaseng NE (eds) (1992) The patterned peatlands of Minnesota. Univ Minnesota Press, Minneapolis

Wrubleski DA (1987) Chironomidae (Diptera) of peatlands and marshes in Canada. Mem Entomol Soc Can 140:141–161

Wu HT, Lu XG et al (2008) The characteristics of soil fauna community structure of a model wetland in Sanjiang Plain, China. Wetl Sci 6:459–465 (In Chinese)

Wu HT, Lu XG, Jiang M (2009) Impacts of soil fauna on litter decomposition at different succession stages of wetland in Sanjiang Plain, China. Chin Geograph Sci 19(3):258–264

Wu HT, Lu XG et al (2010a) Biogenic structures of two ant species *Formica sanguinea* and *Lasius flavus* altered soil C, N and P distribution in a meadow wetland of the Sanjiang Plain, China. Appl Soil Ecol 46:321–328

Wu HT, Wu DH, Lu XG (2010b) Spatial distribution of ant mounds and effects on soil physical properties in wetlands of the Sanjiang plain, China. Acta Ecol Sinica 30(5):270–275

Wu HT, Batzer DP et al (2013a) Contributions of ant mounds to soil carbon and nitrogen pools in a marsh wetland of Northeastern China. Appl Soil Ecol 70:9–16

Wu HT, Lu XG et al (2013b) Ant mounds alter spatial and temporal patterns of CO_2, CH_4 and N_2O emissions from a marsh soil. Soil Biol Biochem 57:884–891

Wu HT, Lu XG, Tong SZ, Batzer DP (2015) Soil engineering ants increase CO_2 and N_2O emissions by affecting mound soil physicochemical characteristics from a marsh soil: a laboratory study. Appl Soil Ecol 87:19–26

Xin WD, Yin XQ et al (2009) Ecogeographical distribution characteristic of soil mesofauna in forest ecosystem of Xiao Hinggan Mountains. Scientia Geogeraphica Sinica 1:129–133 (In Chinese)

Yin ZY (1993) Fire regime of the Okefenokee Swamp and its relation to hydrological and climatic conditions. Int J Wildl Fire 3:229–240

Yin XQ, Wu DH, Han XM (2003) Diversity of soil animal communities in Xiao Hinggan Mountains. Scientia Geographica Sinica 3:316–322 (In Chinese)

Zhang XP, Zhang W (2006) Geo-ecology of soil fauna in different tundras in Da Xinganling Mountains. Acta Pedol Sinica 6:996–1003 (In Chinese)

Zhang SH, Zhang XQ (2009) Soil animal community diversity in the burned areas of the Great Xing'an Mountains, China. Chin J Appl Environ Biol 5:672–676 (In Chinese)

Zhang W, Zhang XP (2013) Characteristics of soil faunal community structure in mires and wet meadows on different tundras in Da Hinggan Mountains. Wetl Sci 1:145–150 (In Chinese)

Zhang XP, Hou WL, Chen P (2001) Soil animal guilds and their ecological distribution in the Northeast of China. China J Appl Environ Biol 4:370–374 (In Chinese)

Zhang XP, Zhang SH, Li JK (2006) Ecogeographical analysis of soil animals in burned forest areas in Greater Hinggan Mountains. Geograph Res 2:328–337 (In Chinese)

Zhang XP, Huang LR, Jiang XQ (2008) Characteristics of macro soil fauna in forest ecosystem of northern Da Hinggan Mountains. Geograph Res 3:509–518 (In Chinese)

Zhang W, Gu CL et al (2014) Wetland soil faunal community characteristics on different types of tundra environment in Daxing'an Mountain. J NE Forestry Univ 5:101–104 (In Chinese)

Zhao KY (1999) Mires in China. Science Press, Beijing (In Chinese)

Chapter 8
Invertebrates in Permanent Wetlands (Long-Hydroperiod Marshes and Shallow Lakes)

Kyle D. Zimmer, Mark A. Hanson, and Dale A. Wrubleski

Introduction to Permanent Wetlands (Long-Hydroperiod Marshes and Shallow Lakes)

Habitat Attributes

Permanent wetlands are probably best viewed as waters existing along a continuum resulting from gradients of biological, chemical, and physical features and occurring between shallow wetlands that periodically dry and deeper lakes. Rigorous global classification of permanent wetlands has proven difficult for many reasons (Finlayson and Van der Valk 1995). Classification systems allowing for regional or even national inventories are often impossible to apply internationally due to broader patterns of habitat variability and unique nature of regional aquatic systems. Also, shallow freshwaters have highly variable origins, geomorphology, nature and extent of ground-water interactions and flow networks, and often comprise portions of larger more complex lake and wetland habitats. Still, it is worthwhile to apply classification systems to link waters defined here, because classification approaches are widely applied by habitat managers, and because we believe it is important that practitioners understand which aquatic habitats support the invertebrate communities we describe.

For this review, we are narrowing our focus to permanently flooded wetlands, and defining these as lentic freshwaters with relatively shallow maximum depth (usually <4 m, Scheffer 2004), yet presence of standing water during all but extreme droughts (such as that experienced in North America during the 1930s). Lakeshore marshes are covered in another chapter of this book, but we do include some examples of habitats that exchange waters with, and comprise portions of, much larger systems, but are of sufficient size or isolation to develop unique characteristics making them distinct from adjacent lakes.

The classification system described by Cowardin et al. (1979) is especially useful here because it recognizes a continuum of waters and has application to

D. Batzer, D. Boix (eds.), *Invertebrates in Freshwater Wetlands*,
DOI 10.1007/978-3-319-24978-0_8

shallow ecosystems worldwide. Our focus here is on palustrine and lacustrine systems, but we limit these to include only waters defined as littoral (not limnetic) in the Cowardin approach (hereafter permanent wetlands). From an international perspective, these are the habitats classified as lacustrine or shallow water marshes by Warner and Rubec (1997) or as permanent fresh marshes/pools following the international Ramsar Convention (Matthews and Townsend 1993). These permanent wetlands include a wide range of substrates from unconsolidated organic matter to rocky bottoms, emergent and submergent aquatic macrophytes, and a wide range of adjacent upland vegetation from grassland to coniferous forest. Our definition also includes a range of conditions with respect to hydrologic exchange with underlying groundwater, but hydrologic relationships are often poorly known, even for permanent wetlands that have received considerable study.

Our habitat definition is also guided by our conviction that what makes biological communities in these waters unique is the influence of three fundamental properties; together, these attributes establish ecosystem characteristics of permanent wetlands. First, these waters are relatively shallow, in many cases mean depths are <1.5 m. This contributes to the second major property; shallow depth prevents stratification so these waters remain polymictic throughout open-water periods in north-temperate regions, and year-round in mid- and southern latitudes. And finally, sunlight reaches large portions of substrates at levels sufficient to stimulate growth of submergent and emergent vascular plants across the majority of the basin. A grouping of waters based on these features results in a habitat category that spans an extraordinarily wide range of size, geomorphology, and origins, and general examples of these are explained in more detail below. However, we emphasize that understanding factors affecting aquatic invertebrate communities in these ecosystems requires a functional classification that relies heavily on depth, polymixis, and potential for colonization by aquatic macrophytes. In a sense, this is a broad permanent wetland classification with ragged ecological edges, but such an approach is necessary in order to elucidate common factors structuring ecological communities in these waters.

We also recognize that, as Wiggins et al. (1980) and Wellborn et al. (1996) suggested, freshwaters may be ranked along a continuum of habitat permanence and predation potential, and that these two factors work concurrently to establish the importance of other environmental and biological variables structuring resident communities. Following the conceptual framework of Wellborn et al. (1996), we further define permanent wetlands as freshwaters with sufficient depth to sustain permanent aquatic communities, yet encompassing a key ecological transition: these areas may or may not support populations of fish. Here then, permanent wetlands are waters supporting aquatic invertebrate communities that are subject to a variety of predation conditions ranging from only invertebrate predators, to habitats that occasionally include fish, to waters that support permanent fish communities. Throughout north-temperate regions, many of these areas are ice-covered for up to 5–6 months each year. Winter conditions favor prolonged periods of under-ice hypoxia and this sometimes reduces, or even eliminates, fish populations (Peterka

1989). Still, fish predation is among the most important properties structuring aquatic invertebrate communities in permanent wetlands (Zimmer et al. 2000; Hanson et al. 2005).

Geographic Distribution of Permanent Wetlands

Freshwater permanent wetlands are widespread, and the scientific literature indicates that these areas comprise a major global habitat resource for aquatic invertebrates. Permanent wetlands in modern landscapes have been formed by a wide variety of physical processes operating at different temporal and spatial scales. Many permanent wetlands are the direct result of anthropogenic activities; some of the best known of these are wetlands formed by flooding in ancient excavations in the Netherlands, or from water filling ditches and abandoned mining sites in the United Kingdom (Moss et al. 1996; Scheffer 2004). On the other hand, naturally occurring permanent wetlands are probably far more widespread and are a result of a variety of natural processes ranging from precipitation and groundwater filling depressions left behind by retreating glaciers in North America (Kantrud et al. 1989) to extreme flooding events that permanently inundated a vast area previously comprising coastal lowlands in southeastern China (Qin et al. 2007).

As with other freshwater wetlands, characterizing permanent wetland features is difficult because regionally unique combinations of climate, water availability, soils and underlying geological substrates, and other factors contribute to extreme variability in biological, chemical, and physical features of these waters worldwide. Comprehensive descriptions of permanent wetlands from many geographical settings are beyond the scope of this chapter, but examples are useful to illustrate variability and to show how environmental variables may structure aquatic invertebrate communities in these habitats. As a starting point, hydrology is a common structuring factor and a key determinant of wetland characteristics. Hydrology, in turn, interacts with other local, regional, and even continental influences. Resulting among-wetland variability is extraordinary and invertebrate communities may differ sharply in response to water quality and chemistry gradients resulting from different hydrologic relationships (Kantrud et al. 1989) (Fig. 8.1).

For example, throughout the Prairie Pothole Region (PPR) of north-central United States and south-central Canada, thousands of permanent wetlands remain within depressions underlain by soils originating from variable depths of till left behind by retreating glaciers (Kantrud et al. 1989). A negative water balance characterizes most of the region, as annual rates of evapotranspiration usually exceed precipitation. Along with highly variable morainic topography, partially impermeable glacial till (up to 250 m deep in some locations) contributes to complex flow networks and intense interactions between surface and underlying groundwater with combinations of recharge, flow-through, and dis-

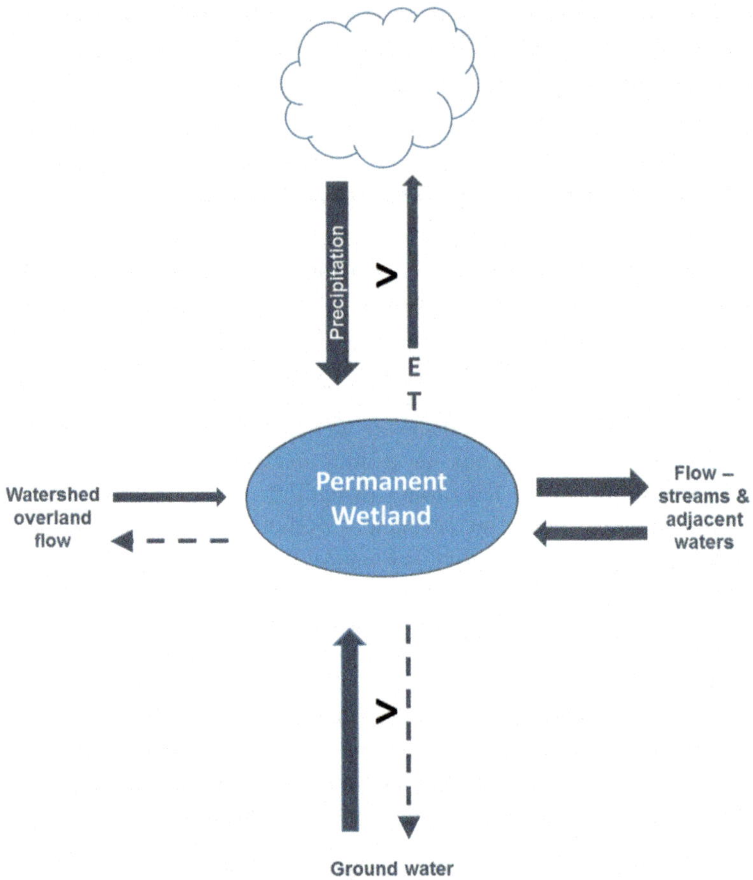

Fig. 8.1 Generalized factors responsible for water movements in permanent wetlands. *Arrow* weights depict relative magnitude of water exchange by individual pathways. *Dashed arrows* indicate relationships of minor importance, but these are often poorly known

charge in shallow waters throughout the region (Winter and Rosenberry 1995). Across the central portion of this area, many small waters were originally isolated and surface connectivity had been a relatively minor factor in water exchange, but this seems to be changing with increasing anthropogenic activity (wetland draining, ditching) and extreme precipitation cycles (Hanson et al. 2005; Herwig et al. 2010).

Vast numbers of permanent wetlands in other regions of the United States and Canada are also of glacial origin. Permanent wetlands are widespread throughout western boreal portions of north-central Alberta, Canada (Bayley et al. 2013). Here, poorly drained outwash plains, moraines, and flat lacustrine plains are prevalent, with wetland water chemistry influenced by relatively thick underlying peat layers in some areas (Bayley et al. 2013). Abundant sediment organic matter and adjacent peatlands contribute to low alkalinity, high levels of total phosphorus (TP) and phytoplankton biomass, and high productivity which often support well-developed communities of plankton and submerged aquatic vegetation (SAV) (Bayley and Prather 2003; Bayley et al. 2013).

Some of the most-studied permanent wetlands (or shallow lakes) in the world are those within lake districts of the northern and western Netherlands and the Norfolk Broads of the United Kingdom. Many of these waters formed following peatland excavations during the seventeenth century (Gulati and Van Donk 2002). Other waters are the result of centuries-old networks of dykes and dams which impounded inland waterways and produced permanent wetlands (or "broads," Moss et al. 1996). Subsequently these wetlands became important habitat for fish production and recreational use, and sometimes supplied water for irrigation of agricultural fields. Prior to the mid-twentieth century, many of these shallow waters exhibited oligotrophic conditions with low nutrient levels, clear water, and lush communities of SAV. With increasing external nutrient loading, many sites transitioned to turbid conditions with poor water quality, further resulting in loss of SAV, declining habitat for piscivorous fish, and diminished recreational value.

Permanent wetlands also occur widely in Mediterranean climates. On the Iberian peninsula in central Spain, permanent wetlands are common. Cladoceran zooplankton communities, especially *Daphnia* spp., have been the focus of studies in these waters due to concerns about lake responses to agricultural chemicals and encroachment, nutrient loading, introduction of non-indigenous fish stocks, and possible effects of climate change (Romo et al. 2004; Fernández-Aláez et al. 2004). Most natural lakes in Greece are shallow waters <1000 ha in size and would be considered permanent freshwater wetlands in our context. These waters have become increasingly valued for recreation and wildlife values, but are threatened by dramatic water-level reductions, eutrophication, and chemical contamination (Coops et al. 2003).

Lake Apopka is a subtropical freshwater wetland in Florida (United States), covering 124 km^2 and averaging 1.7 m depths (Carrick et al. 1993). This wetland overlies clastic, calcareous soils and is characterized by high alkalinity, extreme wind-induced turbidity, and poorly consolidated muck-like sediments. Due to low water clarity, sparse SAV, and declining fish populations, the site has been the subject of one of the most costly rehabilitation efforts in North America (Scheffer 2004). Tropical and subtropical permanent wetlands are well known from South America and Africa. Kosten et al. (2012) surveyed water quality parameters in 83 wetlands along the eastern coast of South America. These waters had a mean depth of 1.9 m, but varied widely in surface area from 0.6 to 27,000 ha. Lake Naivasha near Nairobi, Kenya is a premier destination for bird watching, but its ecological health has declined dramatically due to surrounding urbanization and agriculture, nutrient loading, altered hydrology, and introduction of non-indigenous fish, rusty crayfish (*Procambarus clarkia*), and other invasive species (Harper and Mavuti 2004).

Permanent wetlands are sometimes large and can be prominent landscape features. For example, Lake Arresø is the largest waterbody in Denmark (approximately 40 km^2) and has an average depth of only 3 m. Lake Taihu, the third largest waterbody in China, consists of a vast shallow basin (surface area of 2338 km^2, mean depth = 1.9 m) with extensive submerged, emergent, and floating-leaf vegetation and associated marshlands (Qin et al. 2007). Forty million people live within this lake's watershed (Qin et al. 2010), and the lake provides numerous economic, recreational, and transportation benefits.

Despite extreme variability in formation processes, geomorphology, and hydrology, it is possible to characterize some key features of these habitats, many of which affect aquatic invertebrate communities. These waters are permanently flooded yet shallow enough to permit polymixis throughout open-water periods, and usually allow considerable light penetration to substrates over large areas. Permanent flooding and well-lighted substrates allow for extensive development of emergent and submergent vascular plant communities over a majority of the basin. Thus, permanent wetlands are characterized by a high intensity of physical processes and biological interactions, and biological, chemical, and physical aspects are profoundly affected by complex influences of polymixis, high light availability, and shallow water columns. Organisms (including aquatic invertebrates) often coexist in close proximity with one another and food-web interactions along with water-sediment exchanges are particularly intense (Jeppesen et al. 1997). Physical properties such as light, wind-induced resuspension of sediments, and internal nutrient loading often play larger roles here than in deeper counterparts (Scheffer 2004). As will be elaborated later, permanent wetlands also have a propensity to exhibit alternative stable equilibria (Moss et al. 1996; Romo et al. 2004; Fernández-Aláez et al. 2004; Scheffer 2004; Zimmer et al. 2009). One stable state has low water transparency and sparse SAV (turbid state) while the other has clear water and abundant SAV (clear state). Among freshwaters, permanent wetlands are uniquely affected by this propensity for transitions between clear- and turbid-water states, and aquatic invertebrates both influence, and respond to, these ecological dynamics (Fig. 8.2).

Aquatic Invertebrate Taxa of Permanent Wetlands

The appendix provides a list of the families of aquatic invertebrates reported from a variety of permanent wetland and shallow lake habitats. Permanent wetlands that are covered by other chapters in this book (e.g., lakeshore or riverine marshes, temperate and tropical floodplain wetlands, beaver ponds, Everglades) were not included in this Appendix. Also excluded were taxa from study site descriptions lacking sufficient information to properly define the habitat, or taxa lists that did not distinguish among habitats when several types were sampled. Where more than one habitat type was surveyed (e.g., temporary and permanent wetlands), only

Fig. 8.2 Permanent wetlands exhibit two alternative stable states in many parts of the world. One state is the clear-water condition dominated by submerged aquatic macrophytes with low phytoplankton abundance (background above), and the other state a turbid-water condition dominated by phytoplankton with low abundance of submerged aquatic plants (foreground above). Aquatic invertebrates play a key role in stabilizing both states, and also in inducing shifts from one state to the other. Photo by Brian Herwig (used with permission)

those invertebrates identified as coming from the permanent habitats were included in our list.

A total of 176 families have been reported from permanent wetlands around the world, based on the 27 publications used to assemble our list. Of this number, 75 families were reported from only one continent, with North America and Australia having the largest number of restricted families at 21 each. Only two macroinvertebrate families, Chironomidae and Coenagrionidae, and three microcrustacean families, Cyclopidae, Daphniidae, and Chydoridae, were ubiquitous.

Key Factors Controlling Invertebrates in Permanent Wetlands

Hydrology Controls

In non-permanent wetlands, hydroperiod, along with underlying hydrology, is usually the most fundamental determinant of invertebrate community structure (e.g., how does shortening the inundation period of seasonal wetlands influence invertebrate communities?). In contrast, constant inundation, often with only nominal water-level fluctuations in a single growing season, is typical for many permanent wetlands, where standing water usually persists year-round. In permanent wetlands, research on hydrology and invertebrates has focused on a suite of factors that become more important with extended inundation. Along a hydroperiod continuum, abiotic factors (e.g., desiccation) are most influential in non-permanent wetlands, while biotic interactions (e.g., predation, competition) become increasingly important in permanent wetlands (Wellborn et al. 1996). The roles of predation and other biotic factors have emerged as key proximate drivers, contributing to the high variability often observed among permanent wetlands, even within a given geographical region such as the PPR in North America (Hanson et al. 2005; Anteau et al. 2011). Fish predation is one of the most important determinants of invertebrate community structure in permanent wetlands, and fish presence/absence and community composition are influenced by wetland depth, overland flooding, and connectivity to other waterbodies. Thus, wetland and watershed hydrology have strong indirect influences on invertebrate communities in permanent wetlands via influences on fish distribution and abundance.

At the same time, certain abiotic factors, influenced by hydrology, remain important in permanent wetlands. For example, interactions among underlying soils and geology, ground water, and climate can result in permanent wetlands that range from fresh to saline (Kantrud et al. 1989). Salinity can impact the aquatic invertebrate community indirectly through impacts on aquatic plants (Lacoul and Freedman 2006) or through direct osmoregulatory toxicity (Bayly 1972). Salinity tolerance varies among aquatic invertebrates. Hammer et al. (1990) noted that species richness of macroinvertebrates decreased rapidly at salinities greater than 15 %. Pinder et al. (2004) found salinity to be the primary influence on the distribution of aquatic invertebrates in wetlands of the wheat belt region in Western Australia.

Interactions with Plants and Algae

In wetlands with greater depths and relatively stable water levels, aquatic macrophytes with emergent and submersed growth forms become prominent features and critical habitat for many types of aquatic invertebrates. Aquatic plants increase surface area and habitat complexity, providing additional living space and food within the water column (Cyr and Downing 1988). Both diversity and abundance of aquatic macrophytes increase macroinvertebrate taxon richness and abundance (Zimmer et al. 2000; Hassall et al. 2011), and invertebrate communities often vary among different plant species or communities (Bazzanti et al. 2008). Aquatic plants serve as sites for oviposition, emergence, pupation, attachment, respiration, and as building materials and food (reviewed by Newman 1991). Moreover, presence of aquatic macrophytes contributes to changes in the physical and chemical environment of wetlands, and modifies influences of predator–prey interactions (Carpenter and Lodge 1986).

Given the strong relationship between aquatic plants and invertebrates, grazing of aquatic plants by vertebrate herbivores can indirectly influence invertebrate communities. Muskrats (*Ondatra zibethicus*) are important herbivores in permanent wetlands and shallow lakes (Fritzell 1989). Through their consumption of emergent macrophytes, and harvest of plant material for use in the construction of lodges, muskrats dramatically modify wetland habitats. Such activities have been found to alter abiotic conditions and wetland invertebrate communities (de Szalay and Cassidy 2001). Similar effects have been found for waterfowl that graze submersed macrophytes. Wrubleski (1989) observed contrasting Chironomidae communities in areas where waterfowl had consumed the submersed aquatic macrophytes relative to areas where the plants were protected from grazing.

Research has historically focused on the influence of macrophytes on aquatic invertebrates. However, recent work has shown that invertebrates may conversely also affect submerged macrophytes, indirectly, by controlling abundance of phytoplankton and epiphyton (Scheffer 2004). Epiphyton (and other types of periphyton) and phytoplankton are readily consumed by aquatic invertebrates, and high invertebrate grazing rates can reduce the abundance of both phytoplankton (Hanson and Butler 1994b) and periphyton (Jones and Sayer 2003). Alternatively, low grazing rates of invertebrates can permit periphyton and phytoplankton to accumulate, and resulting light limitation can reduce macrophyte abundance (Sand-Jensen and Borum 1991). Reduced macrophyte abundance, in turn, feeds back to the invertebrate community as described above. Thus, interactions among aquatic invertebrates, epiphyton, phytoplankton, and submersed macrophytes become very complex in permanent wetlands, and play key roles in generating alternative stable states (as elaborated below).

Investigators have also explored the functional importance of invertebrates in processing detritus in wetland food webs, helping to clarify relationships among organic matter, litter decomposition, nutrient cycling, and functional feeding groups of invertebrates in freshwater marshes (Wissinger 1999; Murkin and Ross 2000). Some of this research suggests that macroinvertebrates may play relatively minor roles in nutrient processing (Murkin and Ross 2000), but other studies indicate that zooplankton and macroinvertebrates have potential to translocate sediment-bound nutrients to overlying waters (Fukuhara and Sakamoto 1987), or to excrete nutrients directly to the water column (Vanni 2002).

Interactions with Predators

Fish Predation

Permanent hydroperiods greatly increase the likelihood that wetlands will sustain fish populations, although many remain fishless. The transition from fishless to fish presence represents one of the largest changes in invertebrate community structure in permanent wetlands. Planktivorous and benthivorous fish reduce community-scale biomass, production, and taxon richness of aquatic invertebrates (Batzer 1998; Zimmer et al. 2001b), and also have negative influences on individual taxa via predation or other indirect effects (Zimmer et al. 2000; Hanson et al. 2005). Fish also alter invertebrate behavior, including foraging activity (Johnson 1991), diel migration patterns (Burks et al. 2001), and patterns of oviposition (e.g., females detect chemical cues from fish and avoid laying eggs in basins with fish) (McPeek 1990a; Åbjörnsson et al. 2002). Planktivorous and benthivorous fish also influence invertebrate communities indirectly by favoring shifts to turbid-water states with low abundance of submerged macrophytes and high abundance of phytoplankton.

The importance of fish as a driver of invertebrate community structure in permanent wetlands is supported by both comparative field studies and controlled experiments. For example, Zimmer et al. (2002) compared natural fishless wetlands to wetlands with fathead minnows (*Pimephales promelas*), and found that large-bodied cladocerans and aquatic insects were up to 41-fold and fourfold more abundant in fishless sites, respectively, compared to sites with fish. Hanson and Riggs (1995) also compared invertebrate communities in Minnesota wetlands with and without fathead minnows and reported that invertebrate taxon richness, along with abundance and biomass of aquatic insects and crustaceans, were reduced in the presence of fish. Using experimental mesocosms, Åbjörnsson et al. (2002) reported that presence of fish reduced both biomass and species diversity of aquatic insects, but argued that diversity of large predatory insects may be lower in wetlands with fish because some free-flying insects detect and avoid waters with fish. At the community scale, Zimmer et al. (2000) used a multivariate approach and found that 19 invertebrate taxa (out of 32) occurred more frequently and in higher abundance in fishless wetlands, while only one family (Corixidae) was more abundant in wetlands with fish. Changes in invertebrate behavior between wetlands with and without fish have also been documented in several diverse taxa, with behaviors often focused on minimizing risk of predation by fish. For example, in the absence of fish, damselflies are active, mobile predators, but switch to lie and wait foraging to reduce their predation risk when fish are present (Johnson 1991). *Daphnia* detect chemical cues when fish are present, and undergo diel horizontal migration, moving to macrophyte beds during the day to reduce predation risks, then migrating back to open water at night when the risk of predation is lower (Burks et al. 2001). Several studies have shown that fish effects can exceed influences of abiotic factors (Tangen et al. 2003; Hanson et al. 2012).

Most studies of factors controlling wetland invertebrates have focused on relatively few variables (which is often necessary for large-scale research efforts). However, this limits interpretation and application of results because influences of many factors remain unaccounted for, some of which may be more important than variables included in simple models (Hanson et al. 2012). Relatively little research has assessed concurrent influences of a wide range of factors on wetland invertebrate communities, but available data shed some light on the relative magnitudes of biotic and abiotic influences. For example, Tangen et al. (2003) tested for relationships between land use and invertebrate community structure, but failed to find strong associations. Instead, fish presence/absence showed the strongest relationship to aquatic invertebrate communities. Similarly, Hanson et al. (2012) assessed relationships between invertebrate abundance and several watershed-scale and within-lake variables, and found that characteristics of the fish community were stronger predictors of invertebrate abundance than any measured variable at the watershed scale. Moreover, they also reported that invertebrate abundance was influenced more by simple fish presence/absence than by summed biomass of planktivores and benthivores, suggesting invertebrate communities exhibited a stronger categorical response to fish presence than to density-dependent predation. Further, Hanson et al. (2015) measured community correspondence to relate aquatic invertebrates to a broad suite of environmental variables measured in 104 permanent wetlands within the prairie, parkland, and forested regions of Minnesota. Results indicated that, after controlling for variability attributable to fish abundance, other biotic and abiotic variables were poor predictors of the remaining variation in zooplankton and macroinvertebrates. Overall, available data indicates that permanent wetlands with and without fish are very different ecologically, and that fish presence itself represents a major source of variability in abundance, species composition, and behavior of aquatic invertebrates in permanent wetlands.

It's worth emphasizing that presence of fish in wetlands is naturally variable both temporally and spatially. Many studies emphasize impacts of fish using fish-enclosure experiments in single waterbodies, but this approach may have little relevance to natural fish-bearing systems where fish presence doesn't vary so dramatically. In contrast, studies assessing direct and indirect influences of fish presence–absence on invertebrates across multiple permanent wetlands probably have more ecological relevance because results better mimic natural ecological responses to spatial and temporal variability in communities among these ecosystems (Zimmer et al. 2001a).

We believe there are several major reasons why fish have such strong ecological influences on aquatic invertebrate communities in permanent wetlands. First, in both spatial and temporal contexts, fish presence in permanent wetlands functions as a dynamic categorical variable, with some basins fishless, others with high densities of fish, and still other waterbodies switching from fish to fishless due to colonization or extinction events such as winterkill (Zimmer et al. 2001a; Herwig et al. 2010). Attributes of wetlands and their fish communities may also cause higher fish predation pressure on invertebrates compared to other lentic systems (i.e., deeper lakes). Jeppesen et al. (1997) argued that top-down effects of fish on invertebrates are particularly strong in permanent wetlands because, compared to other systems, (1) fish biomass per unit volume of water is higher, (2) piscivores represent a smaller proportion of the fish community, (3) benthic invertebrate abundance is higher and more accessible to fish, so fish predators rely less on zooplankton prey, and (4) vertical migration of invertebrates to deeper refuge areas is not possible.

Broad diets and high consumption rates also contribute to strong influences of fish on invertebrates in permanent wetlands. Analysis of diets from fish in permanent wetlands confirms that fish consume a wide spectrum of invertebrates, ranging in body size from rotifers to Odonata, and constituting all functional feeding groups (Herwig and Zimmer 2007; Verant et al. 2007). Even fish typically considered piscivorous, such as walleye (*Sander vitreus*), consume numerous invertebrates in permanent wetlands (Ward et al. 2008). The wide spectrum of diet often causes fish predation to have nearly community-wide suppression of invertebrate abundance (Zimmer et al. 2000). Moreover, consumption rates of invertebrates by fish can be very high. Several studies have used bioenergetics modeling to estimate consumption rates of invertebrates by fish, and have found that consumption rates approximate or even exceed production rates of invertebrates during the peak growing season (Duffy 1998; Herwig and Zimmer 2007).

Fish predation on invertebrates is also intensified in permanent wetlands because many invertivorous fish species are not themselves vulnerable to predation by piscivores. In wetlands, benthivorous taxa often outgrow the threat of predation by gape-limited piscivores, and many benthivorous and planktivorous fishes have spines and barbs that defend against predation (e.g., yellow perch *Perca flavescens*). The net result is piscivores are unable to reduce the abundance of invertivores in diverse fish communities, and invertebrate abundance in those wetlands remains low even in presence of piscivores (Friederichs et al. 2011). Finally, fish can influence invertebrate communities indirectly by inducing major changes in abundance of primary producers (Zimmer et al. 2009). High densities of planktivorous and benthivorous fish tend to favor shifts to turbid states with low abundance of submerged macrophytes, which has major impacts on many aquatic invertebrates by changing habitat complexity and the dominance of primary producers (Scheffer 2004).

While many studies from permanent wetlands have reported that fish reduce invertebrate abundance across a variety of taxa, trophic relationships are complex and several studies have reported some apparently contradictory relationships. For example, McParland and Paszkowski (2006) found that gastropods decreased following introduction of fish, while abundance of amphipods and chironomids increased. These authors hypothesized that amphipods and chironomids increased because fish reduced the abundance of their gastropod competitors. Batzer et al. (2000) reported similar results, where fish reduced the abundance of competitors and predators of midge larvae, resulting in a positive overall effect of fish on midge abundance. Moreover, several studies in the PPR of North America have found positive relationships between presence of fish and abundance of Corixidae (Zimmer et al. 2000; Tangen et al. 2003). The mechanisms for this relationship are unknown, but Corixidae may be less sensitive to fish predation, or may benefit from reduced competition when other invertebrates are suppressed by fish. Batzer (1998) suggested that the importance of fish predation on benthic midges varied seasonally, with minimal influences in early summer but more pronounced effects evident during mid-late season. It is perplexing why some studies have found positive effects of fish on select taxa, while others have found consistent negative effects across nearly all taxa. However, variation in fish abundance, the taxonomic composition of fish communities, habitat complexity, and methodological approaches probably all contribute to the range of findings.

Other Predators

Besides fish, other predatory vertebrates may also have important influences on invertebrates, via both direct and indirect effects, especially in fishless habitats. In northern PPR wetlands, gray tiger salamanders (*Ambystoma mavortium diaboli*) can be very abundant, reaching densities of 5000 ha^{-1} (Deutschman and Peterka 1988). These salamanders consume a variety of invertebrates, particularly larger prey such as amphipods and chironomids (Olenick and Gee 1981). Benoy (2008) reported that as tiger salamander abundance increased across 45 PPR wetlands, aquatic insect abundance declined and phytoplankton standing crop increased. These results support the idea that tiger salamanders mimicked the effect of planktivorous fish, inducing cascading effects on the trophic structure of prairie potholes.

In the absence of fish or other vertebrate predators, large-bodied invertebrate taxa such as Odonata, Dytiscidae, and Notonectidae function as top predators (reviewed by Batzer and Wissinger 1996). Many studies have documented the importance of invertebrate predation on invertebrate prey in littoral-type habitats (McPeek 1990b; Åbjörnsson et al. 2002), and invertebrate predators such as *Chaoborus* also have a strong influence on zooplankton communities in fishless water columns (Arnott and Vanni 1993). Similar to fish, invertebrate predators in fishless habitats influence community assemblages (McPeek 1990b), abundances of individual taxa (Åbjörnsson et al. 2002), and the outcomes of competitive interactions (Blois-Heulin et al. 1990).

Invertebrate predators consume not only other invertebrates, but they sometimes alter the presence or behavior of vertebrates. For example, gray treefrog (*Hyla chrysoscelis*) tadpoles reared with predatory dragonfly (*Aeshna umbrosa*) larvae differ in shape and color from tadpoles reared in the absence of dragonflies (McCollum and Leimberger 1997). Smith (1983) reported that in permanent forest pools, dragonfly larvae (*Anax junius*) eliminated chorus frog tadpoles (*Pseudacris triseriata*) when they occurred together in the same pool. While most research on predation in freshwaters has focused on clarifying patterns in response to fish, work to date shows that invertebrate predators likely have important roles structuring invertebrate communities in permanent wetlands, and elucidating those role merits much more study.

A Conceptual Framework for Invertebrate Communities in Permanent Wetlands: Hydrogeomorphology and Alternative Stable States

Permanent wetlands worldwide have been shown to exist in two or more alternative stable states, with the two most common states comprised of a phytoplankton dominated, turbid-water state, versus a submerged macrophyte-dominated, clear-water state (Scheffer 2004) (Fig. 8.2). Both states are relatively stable, although wetlands can shift back and forth between these contrasting conditions (Scheffer et al. 1993). Aquatic invertebrates have been shown to be important for both stabilizing the clear-water state and for inducing shifts between states (Hanson and Butler 1994b; Jones and Sayer 2003). Because of the strong influence of fish on aquatic invertebrates, the importance of invertebrates for stabilizing the clear-water state, and the management emphasis on maintaining clear water (Scheffer et al. 2006), recent research on aquatic invertebrates in permanent wetlands has been focused on understanding the complex relationships among fish, aquatic invertebrates, and alternative stable states at the ecosystem scale.

Aquatic invertebrates stabilize the clear-water state in permanent wetlands two ways. First, zooplankton like *Daphnia* maintain low phytoplankton abundance in spring and early summer while aquatic macrophyte and epiphyton biomass is relatively low (Scheffer 2004). Grazing rates of zooplankton can be very high. For example, Hanson and Butler (1994a) estimated filtration rates of 100–200 % in a large permanent wetland (Lake Christina, MN) following a fish die off. Resulting spring "clear-water phases" may be short-lived, but elevated light levels to wetland sediments are often none-the-less sufficient to trigger growth of submerged macrophytes and epiphyton. This new growth (especially epiphyton) sequesters nutrients from the water column which then helps maintain low phytoplankton abundance for the remainder of the growing season (Scheffer 2004). However, low densities of zooplankton can also induce shifts from clear to turbid states. For example, if densities of planktivorous fish are high during early spring, abundance of zooplankton may be too low to control phytoplankton abundance, leading to high turbidity with low macrophyte abundance, shifting a wetland to the turbid state. Wetlands usually remain in a turbid condition until abundance of planktivorous fish declines to levels low enough to allow zooplankton to again reduce phytoplankton abundance, facilitating a shift back to the clear-water condition (reviewed by Scheffer 2004). As in deeper systems, high densities of planktivorous fish in permanent wetlands induce size-selective predation on zooplankton, resulting in a shift from large-bodied forms like *Daphnia* to small-bodied forms like *Bosmina* (Hanson and Butler 1994a). Although densities of the small-bodied zooplankton can be relatively high, their capacity to control phytoplankton is far lower than large-bodied forms (Lynch and Shapiro 1981) and phytoplankton abundance remains high. Also, if wetlands are sufficiently shallow, submerged macrophytes may persist even in turbid-state conditions because short water columns allow some light penetration to plants growing near the sediment surface.

A second way aquatic invertebrates stabilize clear-water states in wetlands is by exerting sufficient grazing pressure on epiphyton, essentially increasing survival and growth rates of submerged macrophytes by reducing the shading otherwise due to an overgrowth of surface-associated algae. Here the key invertebrates are littoral and benthic forms, especially epiphyton grazers such as gastropods and Ephemeroptera. Jones and Sayer (2003) showed that at high densities invertebrate grazers have potential to reduce epiphyton and prevent shading out of submerged plants. However, sufficiently high densities of insectivorous and molluscivorous fish can reduce epiphyton grazers, allowing epiphyton abundance to increase, favoring large-scale macrophyte declines and triggering transitions to turbid conditions (Jones and Sayer 2003). Thus, the clear-water state in temperate wetlands is stabilized by two different groups of invertebrates, with zooplankton grazing on phytoplankton important in spring and early summer, and littoral-benthic epiphyton grazing becoming more important through the rest of the growing season.

Aquatic invertebrates play a prominent role in inducing state shifts in many permanent wetlands, but other factors can also cause a shift from a clear to turbid state. Nutrient loading can decrease the stability of the clear-water state to the point that a shift occurs to the turbid state, and the turbid becomes the only stable state (Moss et al. 1996). Research has shown that fish are one of the best predictors of shifts to the turbid state, with increasing likelihood of turbid lakes with elevated biomass of planktivorous and benthivorous fish (Zimmer et al. 2009; Nolby et al. 2015). Fish can induce shifts to turbid states via predation effects on invertebrate grazers of phytoplankton and epiphyton, but they can also induce shifts by direct disturbance. Large-bodied benthivores like common carp (*Cyprinus carpio*) uproot submerged macrophytes (Crivelli 1983), and many benthivores may also increase turbidity by disturbing sediments with their feeding activities (Breukelaar et al. 1994), resulting in lower water transparency for submerged plants. Lastly, benthivorous fish may increase internal nutrient loading to the water column by disturbing sediments or by feeding on detritus and translocating nutrients to the water column via excretion (Zimmer et al. 2006). Overall, benthivorous and planktivorous fish favor shifts to the turbid-water state, with subsequent impacts on aquatic invertebrates. Thus, fish impact invertebrate communities both directly via predation and indirectly by favoring the turbid-water state.

Shifts to the turbid-water state influence aquatic invertebrates in multiple ways. First, primary production shifts from submerged macrophytes and epiphyton to phytoplankton and perhaps to sediment-associated algae growing in shallow water with sufficient light (Vadeboncoeur et al. 2003). This shift at the base of the foodweb changes the competitive advantage among invertebrate grazers, shredders, and collectors to species benefitted by high abundance of phytoplankton (Hargeby et al. 1994). Loss of submerged macrophytes also reduces habitat complexity and predation refuge for invertebrates, resulting in increased vulnerability to fish predation (Crowder and Cooper 1982) and altered predator–prey relationships among the invertebrates (Burks et al. 2001).

Fish have strong influences on invertebrates, but not all permanent wetlands have fish, so understanding factors driving fish distributions is necessary for understanding variability in invertebrate communities. A permanent hydroperiod does not ensure a wetland basin will support fish and the status of fish populations in permanent wetlands is quite variable, both within and among wetlands. In the southern PPR of Minnesota, for example, approximately 93 % of permanent wetlands had fish populations (Herwig et al. 2010). On the Canadian side of the PPR, Lawler et al. (1974) found that only 10–20 % of wetlands in southwestern Manitoba supported fish. More recently, Anteau and Afton (2008) reported that 31–45 % of wetlands they sampled in central North Dakota contained fish and that fish populations occurred in 74–84 % of semipermanent and permanent wetlands they sampled in Minnesota and Iowa, along the southeastern margin of the Prairie Pothole Region (PPR) of central North America. These data are consistent with those of Hanson et al. (2005) who suggested that ecological influences of fish in permanent wetlands may increase along a northwest to southeast gradient in the PPR. Even in north-temperate permanent wetlands, fish communities can be surprisingly diverse. Herwig et al. (2010) sampled fish populations in >70 permanent wetlands in Minnesota and reported that 22 fish species occurred in more than one of these sites. They reported that fish species richness averaged 3–4, and that richness was positively correlated with wetland size and watershed area. Large permanent wetlands in North America and elsewhere support diverse, perennial fish populations and several fish feeding guilds (planktivores, benthivores, piscivores) sometimes occur in these systems (Herwig et al. 2010; Friederichs et al. 2011). Although our review indicates that summaries of fish population data from tropical wetlands are limited, fish assemblages in these waters can be much more diverse, especially given effects of repeated introduction of exotic fish species (Jeppesen et al. 2007). While status of fish populations in permanent wetlands is often unknown and probably fluctuates over time, this constitutes a critical ecological threshold that must be considered along with other factors structuring invertebrate communities.

What drives the variability in fish presence among permanent wetlands? At large scales, such as the contrast between the northwestern and southeastern portions of the PPR described above, differences are likely due to variability in climate and wetland morphometry (Hanson et al. 2005). At local scales, classic island biogeography principals appear to be important (Scheffer et al. 2006). Lack of surface-water connectivity and isolation decrease the likelihood that a wetland will be colonized by fish, while reduced wetland depth increases the probability of extinction for existing populations due to winterkill or other factors (Herwig et al. 2010; Nolby et al. 2015). Herwig et al. (2010) studied permanent wetlands along the eastern margin of the PPR in Minnesota and reported that all sites connected to potential sources of fish (e.g., streams and other permanent wetlands) supported fish, as did all unconnected basins with maximum depths greater than 2.15 m. In those wetlands, fish occurred in connected basins because these populations recolonized yearly even if they were eliminated by winterkill. Fish populations also occurred in isolated, but relatively deep basins. This probably reflects the fact that greater water depth reduces frequency (and extent) of winter hypoxia and winterkill, allowing fish populations to persist over long time periods. Fish populations in isolated, deep basins in the PPR may result from rare flooding events allowing periodic colonization in these permanent habitats.

Similar to deeper lakes (Hershey et al. 1999), landscape and wetland basin hydrogeomorphology (wetland depth, connectivity, etc.) appear to have indirect

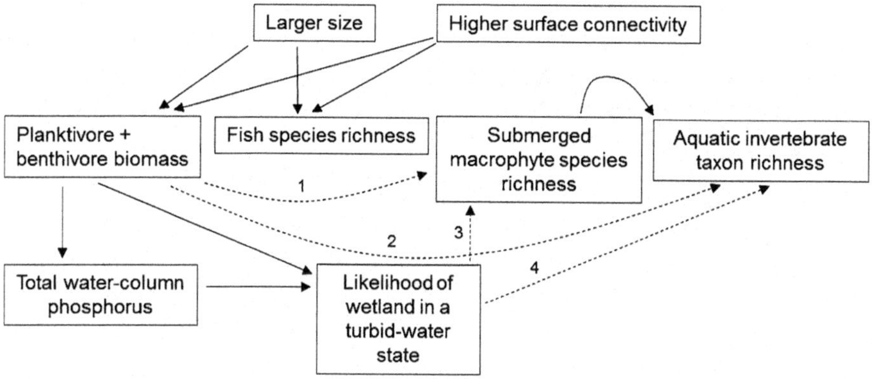

Fig. 8.3 Relationships influencing aquatic invertebrate biodiversity in permanent wetlands (based on Nolby et al. 2015). Solid lines indicate a positive relationship; dashed lines are negative relationships. Lake size and connectivity have a positive influence on planktivore + benthivore biomass. Planktivore + benthivore biomass have negative influence on biodiversity of macrophytes and invertebrates, while total phosphorus and planktivore + benthivore biomass favor turbid states which also has negative impacts on diversity of macrophytes and invertebrates. Although classic island biogeography predicts positive relationships among biodiversity, wetland size, and extent of surface connectivity, influences of planktivorous and benthivorous fish uncouple submerged macrophytes (1) and aquatic invertebrate communities (2) from positive effects of lake size and connectivity. Submerged macrophyte (3) and invertebrate (4) richness are also reduced following transitions to turbid-water states

effects on invertebrate communities in permanent wetlands via influences on fish distribution. Hydrogeomorphology is probably even more influential for invertebrates in wetlands relative to deeper lakes due to intense fish predation and the tendency of planktivorous and benthivorous fish to induce turbid-water states. In complex ways, hydroperiod, maximum depth, and surface connectivity all influence aquatic invertebrate community structure in permanent wetlands through physical processes, but also indirectly by interacting to determine spatial and temporal patterns of fish presence (Fig. 8.3).

Abundance, community composition, and behavior of aquatic invertebrates in permanent wetlands are controlled by a suite of factors that vary in importance from one basin to the next. However, our review suggests that abiotic factors always remain important. Salinity has direct influence on invertebrate communities, but also has indirect influences via limiting distribution of fish and other predators. Wetland depth, wetland connectivity, and nutrient levels are important, but may operate indirectly by influencing the likelihood a wetland will have fish populations, and whether a wetland will be in a turbid or clear-water state. The net result is that invertebrate communities in permanent wetlands are controlled by a series of complex interactions involving both abiotic and biotic variables (Fig. 8.4).

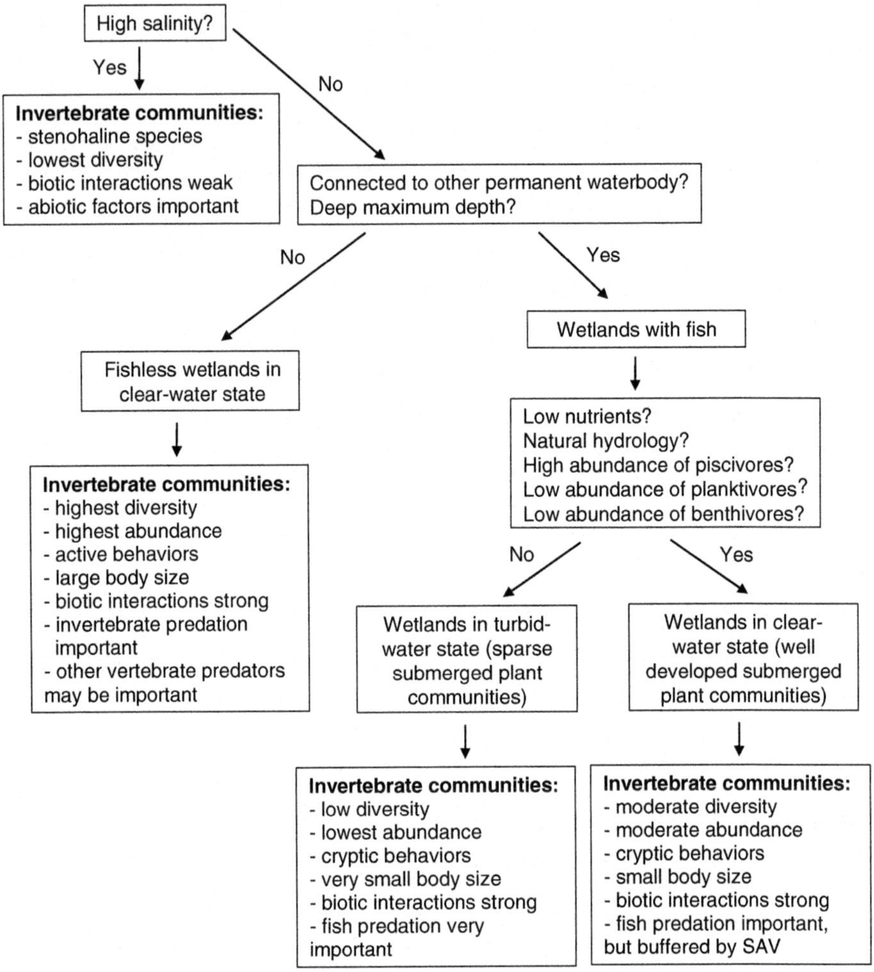

Fig. 8.4 Conceptual model summarizing broad relationships among fish, submerged macrophytes, aquatic invertebrates, and other characteristics of permanent wetlands

Management and Conservation Issues

Current Management and Conservation Issues

Research and management of aquatic invertebrates in permanent wetlands was historically directed toward wetland wildlife, and was mostly focused on understanding invertebrates in the context of their importance as food resources for wetland-dependent vertebrates. For both researchers and managers, we advocate a broader valuation of the roles of aquatic invertebrates in maintaining high water quality and healthy ecological communities in permanent wetlands and shallow lakes.

Perhaps the best conceptual framework illustrating the need for incorporating aquatic invertebrates in wetland management strategies comes from recent work showing the importance of invertebrate communities in the alternative stable state model. As described in detail above, understanding alternative states in permanent wetlands is key to understanding the basic ecology of those ecosystems, and invertebrates play key roles. Many of the conservation issues facing permanent wetlands are important because they influence the stability of alternative stable states and aquatic invertebrates and other wildlife are impacted as wetlands shift states. Management efforts in permanent wetlands often focus on maintaining the clear-water state due to its perceived higher value as habitat for wetland-dependent species (Hanson and Butler 1994a; Scheffer et al. 2006) and greater aesthetic appeal (Moss et al. 1996). Understanding the roles of aquatic invertebrates in controlling abundance of periphyton and phytoplankton, and appreciating the influences of fish on aquatic invertebrates, are key to anticipating whole-ecosystem changes as wetlands transition between clear- and turbid-water states. Studies of invertebrate communities in permanent wetlands identified primary drivers of invertebrate abundance, diversity, and community structure, but also clarified the importance of zooplankton, benthic, and littoral macroinvertebrates in promoting high water clarity and diverse macrophyte communities in the clear-water state. Along with better-known roles in food chain support for waterfowl and other wetland wildlife, this emerging understanding of ecological relationships among water quality, SAV, and alternative stable states points to the need for conservation of aquatic invertebrate communities to favor healthy macrophyte communities and clear-water states in permanent wetlands worldwide (Moss et al. 1996; Scheffer 2004; Zimmer et al. 2009).

Our experience and literature review indicates that several issues are among the most pressing for management of permanent wetlands. Importance of these factors was illustrated by historical research, but subsequent work has clarified the roles of invertebrates in the broader ecological dynamics and alternative ecosystem states of permanent wetlands. Recent efforts also underscore the need for attention to these issues in order to facilitate conservation of aquatic invertebrates and to preserve their functional roles in shallow waters.

First, introduction and proliferation of invasive (albeit sometimes native) fish populations is a critical conservation issue because they often threaten communities of wetland aquatic invertebrates in both north-temperate and tropical regions. The transition from fishless to fish-bearing in permanent wetlands induces a major reorganization of invertebrate behavior and community structure due to direct and indirect effects of predation (Wellborn et al. 1996) and this threshold has been shown to be important in permanent wetlands (Zimmer et al. 2000; Hanson et al. 2012). Both planktivorous and benthivorous fish favor shifts to the turbid state, with subsequent negative effects on most invertebrates due to reduced abundance of SAV (Zimmer et al. 2009; Nolby et al. 2015). Not surprisingly, shallow lake management efforts have traditionally included measures to limit or eliminate dense, undesirable fish populations. Strategies often involve the use of fish toxicants to remove as many fish as possible from deteriorated shallow waters (Hanson and Butler 1994a, b; Zimmer et al. 2001c). In some cases, piscivore stocking has also been used to limit populations of undesirable fish in North America and Europe (Potthoff et al. 2008). These efforts are costly and labor-intensive, but often produce dramatic short-term improvements in water quality and invertebrate communities (Hanson and Butler 1994b; Søndergaard et al. 2007). Unfortunately, target fish populations are rarely eradicated and can recover quickly (Duffy 1998). In our experience, improvements in permanent wetlands following fish removals rarely persist more than 5–10 years. Even when complete fish kills are achieved, migration corridors often allow fish to recolonize permanent wetlands within short periods (Zimmer et al. 2001a).

While management of fish has focused on population control via piscicides or limiting fish access via fish barriers on a basin-by-basin basis, we encourage managers to supplement these tactics with broader approaches where fish dispersion and colonization are addressed at a landscape scale. Installation of culverts and drainage tile, digging of ditches, and integration of wetlands and deeper basins across large spatial scales favors persistent populations of fish in wetlands (Hanson et al. 2005; Herwig et al. 2010). Additionally, increased connectivity also favors dispersal of invasive species among wetland basins, and several studies have documented the impacts of invasive species on invertebrate communities in permanent wetlands (Harper and Mavuti 2004; Rodríguez et al. 2005). Limiting these landscape modifications may help preserve natural variability in community composition among multiple wetland basins, potentially alleviating the need for intensive site-by-site management. Preserving fishless wetlands seems especially important for maintaining aquatic invertebrate diversity at a landscape scale, as many taxa are found more often and in higher abundance in fishless sites relative to those with fish (Wellborn et al. 1996; Nolby et al. 2015). Moreover, fishless wetlands may be uncommon in some parts of the world, and are likely threatened by surface connections to basins with fish. In the PPR of Minnesota (USA), for example, just 7 % of permanent wetlands are fishless, and during a 5-year study of 10 fishless wetlands, two were connected to basins with fish via ditching (Zimmer, personal observation).

The potential for competition between fish and waterfowl for invertebrate prey also highlights the importance of controlling unnatural fish distributions in permanent wetlands. Invertebrates have long been recognized as important prey for many species of waterfowl, especially for females during egg laying (Swanson and Duebbert 1989; Krapu and Reinecke 1992), and interest in waterfowl management stimulated many of the earlier studies on aquatic invertebrates in permanent marshes. Moreover, considerable diet overlaps between fish and many species of waterfowl-spurred-related work to clarify potential for resource competition for invertebrate prey. Early work identified the importance of a wide range of aquatic invertebrates as food for breeding waterfowl and other aquatic birds (Swanson and Duebbert 1989; Krapu and Reinecke 1992). Aquatic insects, crustaceans, snails, and other invertebrates were clearly shown to be required seasonally by breeding waterfowl foraging in temporary or seasonally flooded wetlands. Authors also acknowledged that permanently flooded wetlands were important, especially during drought, when small wetlands are unavailable (Swanson and Duebbert 1989), or as foraging areas for some species like canvasbacks and lesser scaup in North America (Hanson and Butler 1994b). Thus, fish management may be an important component for optimizing waterfowl habitat, especially in areas with limited numbers of temporary and seasonal wetlands.

A second major conservation need is maintenance of natural water-level fluctuations and hydroperiods (duration). Because communities here are comprised of both "wetland" and "lake" species, it might appear that conservation of natural hydroperiods is less critical for these areas; we reject that notion. In North America, altered hydrology usually results in increased maximum depths, stabilized water levels, and in some cases, longer hydroperiods, with permanent flooding of many areas that previously held water only seasonally. Marsh ecologists have shown that vegetation in shallow fresh waters is strongly influenced by flooding depth, frequency, and duration (van der Valk and Davis 1978). Water-level fluctuations, and especially increased water depth, influence virtually all aquatic vascular plants in freshwater wetlands including emergent and submergent forms (van der Valk 2005). Given these relationships, it is not surprising that hydrologic alterations affect vascular plant communities, nutrient cycling, and other properties in permanent wetlands and that natural water-level fluctuations are essential for maintaining wetland processes and biotic communities (Chow-Fraser 2005). Research has shown that permanent wetlands with natural hydroperiods show more diverse plant and animal communities than those with modifications to maintain more stable water levels (van der Valk 2005). Beyond direct implications for aquatic vascular plants, increases in depth and hydroperiod have other interrelated consequences including increased permanence for aquatic organisms (including fish), lower light levels at the sediment surface and ultimately conditions that favor transitions to turbid states (Scheffer 2004). Sustained high-water levels are almost certain to favor loss of SAV and may induce transitions to turbid states, at least in lakes with highly productive populations of benthivorous and/or planktivorous fishes (Coops et al. 2003; Hobbs et al. 2012). Similar to fish distributions, hydroperiod management must be addressed using landscape-level approaches to counter negative consequences of ditching, culvert installation, and consolidation of wetland basins outside an individual wetland's immediate watershed. Despite jurisdictional impediments to wetland drainage and changes to natural water regimes, these practices continue to be widespread throughout North America and elsewhere. Permanent wetlands have been drained less frequently than shallower basins with shorter hydroperiods due to logistical and legal impediments (Oslund et al. 2010), but modifications and even losses still occur in North America and loss rates are much higher in many other parts of the world (Dahl 2014).

Finally, invertebrate communities in many permanent wetlands around the world face severe threats from eutrophication and nutrient loading which affect aquatic invertebrate communities several ways. Higher nutrient levels increase abundance of inedible cyanobacteria (Kosten et al. 2012), favor winter hypoxia in lakes subject to ice cover (Meding and Jackson 2003), and reduce abundance of submerged macrophytes which increases potential for shifts to the turbid state. Nutrients have long been recognized as a primary driver of the stability of alternative states in permanent wetlands, with resilience of turbid states positively related to nutrient levels (Scheffer 2004). Moreover, at higher nutrient levels, greater reductions in fish biomass are needed to induce shifts to clear-water states (Gorman et al. 2014), making management of many turbid waters far more difficult. High external loading of nutrient levels can also have a "legacy" effect on permanent wetlands, where nutrient levels in the water column remain high and the turbid state remains stable long after external loading is reduced due to internal loading of nutrients from wetland sediments (Hobbs et al. 2012). Given the importance of nutrients for inducing transitions to turbid states and the prevalence of internal nutrient loading in permanent wetlands, managers should make control of eutrophication a top priority for conservation of aquatic invertebrates.

Eutrophication seems to be an even greater problem for permanent wetlands at lower latitudes because macrophyte abundance appears to decline at lower levels of nutrients compared to higher latitude wetlands (Kosten et al. 2009). It is important to recognize that trophic relationships among nutrients, water quality, aquatic invertebrates, and macrophytes are more variable and poorly known for tropical wetlands than for north-temperate sites. For example, Bachmann et al. (2002) reported no associations between water column nutrient levels and macrophyte abundance at all but the highest nutrient levels when macrophytes declined and phytoplankton became predominant in permanent wetlands in Florida. Broadly, we think tropical wetland resources need urgent attention from researchers and managers because ecological relationships are poorly known, and because exotic species, eutrophication, demographics, and economics are contributing to extreme pressure on these shallow permanent waters (e.g., Harper and Mavuti 2004). Managing external loading remains paramount yet is difficult due to the ubiquitous use of fertilizers in and around aquatic areas in many regions of the world, and due to the fact that nutrient reduction is a watershed-level issue that often involves multiple stakeholders with conflicting priorities.

Emerging Issues and Information Needs

Several other factors appear to be emerging conservation issues; research to date points to needs for management, yet these issues have received less study than those discussed above. Chemical pollution, defined broadly as pesticides, herbicides, heavy metals, and endocrine disrupting compounds, is an emerging concern for waters at the global scale, yet effects are poorly quantified (Rockström et al. 2009). The same is true for chemical pollution in permanent wetlands, along with impacts on invertebrate communities. Mesocosm and microcosm research have documented the potential toxicity of several commonly used pesticides on aquatic invertebrates found in permanent wetlands (Johnson 1986), and have shown that herbicides may alter relative abundance of primary producers and favor shifts to turbid states in wetlands (Vera et al. 2010). Additionally, researchers have documented pesticide levels in wetlands that exceed thresholds set for maintaining aquatic life by government agencies (Donald et al. 1999). New agricultural chemicals are being developed and those with high efficacy soon are used widely. Pesticides might be especially harmful to wetland invertebrates because applications often occur in close proximity to aquatic habitats, and many target arthropods. Main et al. (2014) recently reported that neonicotinoids are widely used in production of canola, corn, and soybeans across Canada's portion of the PPR. Transport of neonicotinoids into wetlands is likely, where the chemicals may persist for many months. Extensive application of this chemical is relatively new across the North-American prairies, but elsewhere data indicate that toxicity for aquatic invertebrates may be high, especially with prolonged exposure (Main et al. 2014). Additional work on occurrence rates and concentrations of pesticides in wetlands in agricultural areas is clearly needed (Goldsborough and Crumpton 1998).

Research is also needed to clarify the influences of invasive species on wetland invertebrates. Invasive species are known to be a major threat to biodiversity, perhaps second only to habitat destruction (Simberloff et al. 2005), and wetlands are especially prone to colonization by invasive species due to their function as landscape "sinks" (Zedler and Kercher 2004). Information about invasive species effects on aquatic invertebrates in permanent wetlands is relatively scarce, but available evidence suggests the impacts can be large. For example, introduced crayfish in Lake Naivasha, Kenya, reduced macrophyte biomass with impacts cascading to the native invertebrate communities (Harper and Mavuti 2004). More broadly, loss of submerged plant biomass due to invasive species has been shown to induce shifts to turbid states in wetlands, with subsequent impacts on aquatic invertebrates and other organisms (Rodríguez et al. 2005). Threats to conservation of wetland invertebrates from invasive species seem likely to accelerate with an increasingly global society and as urbanization encroaches on remaining permanent wetlands. Urbanization itself also appears to be an emerging threat to conservation of aquatic invertebrates in wetlands. In a study of smaller wetlands spanning a range of wetland types, Holland et al. (1995) found that urbanization and drought eliminated 40 % of wetlands in a rapidly developing area of Portland, USA. Of those remaining, 25 % were severely degraded by human activities. Permanent wetlands may be less vulnerable to draining than basins with shorter hydroperiods, but they are certainly susceptible to degradation from other urban influences.

Lastly, wetlands are especially vulnerable to climate change due to their relatively shallow depths and high evaporation rates (Johnson et al. 2010). Influences of climate change on wetland ecosystems is an active area of research, with studies often using either simulations and model forecasting (Johnson et al. 2010), or comparisons among basins across a latitudinal gradient to mimic changes in climate (Kosten et al. 2012). While these are powerful approaches, we agree with Conly and Van der Kamp (2001) that careful monitoring of individual wetland ecosystems through time is also critically needed in our efforts to assess changes and preserve wetland communities in the face of climate change.

Looking Ahead

Future conservation strategies should include measures to preserve diverse, healthy faunas of aquatic invertebrates in permanent wetlands because this helps to ensure that these areas will continue to provide benefits to wetland wildlife species, but also because invertebrates are necessary in order for permanent wetlands to provide a rich suite of ecosystem services (Hanson et al. 2012). We agree with the broad view of Euliss et al. (2008) who emphasized that future wetland conservation strategies must transition from the traditional focus on wildlife values to comprehensive approaches that strive toward ensuring that permanent wetlands—and invertebrate communities—continue to provide ecosystem services that meet biological, social, political, and even economic needs. We suggest that future management frameworks must retain elements of the traditional emphasis on wildlife, yet be diversified to incorporate the roles of aquatic invertebrate communities in providing ecosystem services not historically associated with invertebrates. This broader paradigm requires a whole-wetland approach where invertebrate communities are better-integrated into studies and management at the scale of the entire wetland. Moreover, several management issues for permanent wetlands, such as increasing distributions of fish and altered hydroperiods, are influenced by factors operating outside wetland watersheds and at landscape scales. Thus, conservation of aquatic invertebrate communities in permanent wetlands will often require that management measures be directed at the wetland-watershed scale.

Our review indicates that increasing demands of agriculture and food production, urbanization, demographics, and economics are contributing to rapid declines in invertebrate communities and other ecological characteristics of permanent wetland habitats. Although permanent wetlands and shallow lakes may be drained and eliminated less frequently than smaller waters, we believe the larger, more permanently flooded wetlands remain vulnerable in temperate and tropical regions worldwide. We challenge wetland scientists and managers to explore new approaches that may help clarify roles of aquatic invertebrate communities in maintaining water quality and ecological integrity of permanent wetlands so these functions will be more widely perceived and valued by an increasingly diverse, urban society.

Appendix

Taxa of Permanent Wetlands

List of aquatic (A) macroinvertebrate and (B) microcrustacean families reported from permanent wetlands. Numbers refer to citations listed in table footnote. Full citations are given in the literature cited. Due to a lack of published information, lists of macroinvertebrates for Africa and microcrustaceans for South America were not compiled.

Taxa	Family	North America	South America	Europe	Asia	Australia
A. Macroinvertebrates						
Mollusca						
Bivalvia	Corbiculidae				22, 23	
	Mycetopodidae		21			
	Pharidae				23	
	Sphaeriidae	1, 2, 4, 8		15	23	26, 27
	Unionidae				23	26
Gastropoda	Acroloxidae			10, 12, 14		
	Ampullariidae		21			
	Ancylidae	8	21	12		25
	Bithyniidae			12, 14, 15	22, 23	
	Cochliopidae			12		
	Ellobiidae			12		
	Hydrobiidae			10, 12, 15		26, 27
	Lymnaeidae	1, 2, 4, 5, 8		10, 14	22, 23	
	Melanopsidae			12		
	Neritidae			12		
	Physidae	1, 2, 4, 5, 8	21	12, 14		26
	Planorbidae	1, 2, 4, 5, 8	21	10, 12, 14, 15	23	25, 26
	Pleuroceridae				23	
	Pomatiopsidae					25, 26, 27
	Stenothyridae				23	
	Valvatidae	2, 4		14, 15		
	Viviparidae			14	22, 23	
Annelida						
Oligochaeta	Aeolosomatidae			12		
	Enchytraeidae			15		25, 27
	Lumbricidae			12		
	Lumbriculidae	4		12		
	Naididae	4		12, 15	23	25, 26
	Phreodrilidae					25
	Tubificidae			12, 15	22, 23	25, 26, 27

(continued)

Taxa	Family	North America	South America	Europe	Asia	Australia
A. Macroinvertebrates						
Hirudinea	Erpobdellidae	2, 4		10, 15		
	Glossiphoniidae	2, 4	21	10, 15	23	25
	Haemopidae	4				
	Piscicolidae			15		
Polychaeta	Nephtyidae				23	
Chelicerata						
Acari	Arrenuridae	3				
	Eylaidae	3				26
	Hydrachnidae	3				
	Hydrodromidae	3				
	Hydryphantidae	3				
	Limnesiidae	3				25
	Limnocharidae	3				25
	Mideopsidae	3				
	Oxidae	3				25
	Pezidae					25
	Pionidae	3				
	Unionicolidae	3				
Crustacea						
Decapoda	Atyidae			12		26
	Hymenosomatidae					26, 27
	Palaemonidae		21	12	23	
	Parastacidae					25
Anaspidacea	Koonungidae					26
Amphipoda	Ceinidae					25, 26, 27
	Corophiidae			12		
	Crangonyctidae			10		
	Dogielinotidae	2, 4	21			
	Gammaridae	2, 4		10, 12, 15	22, 23	25
Isopoda	Anthuridae			12		
	Asellidae	1, 8		10, 15		
	Janiridae					27
	Scyphacidae					27
	Sphaeromatidae			12		
Tanaidacea	Leptocheliidae			12		
Hexapoda						
Collembolla	Entomobryidae	8				
	Isotomidae	1, 8				26
	Sminthuridae	8				
Insecta						

(continued)

Taxa	Family	North America	South America	Europe	Asia	Australia
A. Macroinvertebrates						
Odonata	Aeshnidae	1, 2, 3, 4, 8	19, 21	11		25
	Calopterygidae	1, 8				
	Coenagrionidae	1, 2, 3, 4, 8	19, 20, 21	10, 11, 12, 15	23	25, 26
	Dicteriadidae		21			
	Corduliidae	2, 3, 4		12		25
	Gomphidae	3	19			
	Lestidae	1, 2, 3, 4, 8	19, 21	11, 12		25
	Libellulidae	1, 2, 3, 4, 8	19, 20, 21	11, 12		25
	Perilestidae		21			
	Protoneuridae		21			
Ephemeroptera	Baetidae	1, 2, 3, 4, 8	19, 20, 21	10, 12, 15		25
	Caenidae	1, 2, 3, 4, 8	19, 20, 21	10, 12, 15		27
	Heptageniidae	3				
	Leptohyphidae		21			
	Leptophlebiidae	3	19, 21			
	Polymitarcyidae		19			
	Siphlonuridae	4				
Hemiptera	Belostomatidae	1, 3, 8	18, 20, 21			
	Corixidae	1, 2, 3, 4, 8	18, 20, 21	10, 12, 15		25, 26
	Gerridae	2, 3, 4		10, 12		
	Hebridae	3, 8	18, 21			
	Hydrometridae	3	18			25
	Mesoveliidae	2, 3, 4, 8	21	12		
	Naucoridae	3	21	12		
	Nepidae	3	18, 20			
	Notonectidae	1, 2, 3, 4, 8	18, 20, 21	10, 12		25, 26
	Pleidae	1, 3, 8	18, 20, 21	12		25
	Rhyparochromidae		18			
	Saldidae					25
	Veliidae	1, 3, 4	18	12		25

(continued)

Taxa	Family	North America	South America	Europe	Asia	Australia
A. Macroinvertebrates						
Coleoptera	Chrysomelidae	1, 2, 3, 4, 6, 8	21			
	Curculionidae	2, 3, 4, 6	21			
	Dytiscidae	1, 2, 3, 4, 6, 7, 8	18, 20, 21	10, 12		25, 26, 27
	Dryopidae		20	12		
	Elmidae	2	21			
	Gyrinidae	1, 2, 3, 4, 7, 8		15		
	Haliplidae	1, 2, 3, 4, 6, 7, 8	21	10		25
	Helophoridae	4		10		
	Heteroceridae		18			
	Hydraenidae	2, 3				
	Hydrochidae		18			26
	Hydrophilidae	1, 2, 3, 4, 6, 7, 8	18, 20, 21	10, 12		25, 26, 27
	Hygrobiidae					25
	Lampyridae		21			
	Limnichidae		18			
	Melolonthidae		21			
	Noteridae	3	18, 20, 21			
	Scirtidae	3, 4	18, 21			25
	Staphylinidae		21			26
Neuroptera	Sialidae			10		
Trichoptera	Apataniidae	2				
	Atriplectidae					26
	Brachycentridae	2				
	Calamoceratidae		19			
	Ecnomidae			12		25, 27
	Helicopsychidae					26
	Hydroptilidae	3, 4	19, 21			25
	Leptoceridae	1, 2, 3, 4, 8	19	10, 15		25, 26, 27
	Limnephilidae	2, 3, 4		10, 15		
	Molannidae	3				
	Philorheithridae					26
	Phryganeidae	1, 2, 3, 4, 8				
	Polycentropodidae	3, 4	21	15		

(continued)

Taxa	Family	North America	South America	Europe	Asia	Australia
A. Macroinvertebrates						
Lepidoptera	Crambidae	8				
	Noctuidae		21			
	Pyralidae		21			
Diptera	Ceratopogonidae	1, 2, 3, 4, 8	21	12, 15		25, 26, 27
	Chaoboridae	1, 2, 4, 8	21	15		26
	Chironomidae	1, 2, 3, 4, 8	19, 20, 21	12, 15	22, 23	25, 26, 27
	Culicidae	2, 3, 4	21	15		25
	Dixidae	2, 4	21			
	Dolichopodidae					25
	Empididae	2	21			
	Ephydridae	8	20, 21	12		25, 27
	Muscidae		20, 21			25
	Psychodidae	2, 8				25
	Ptychopteridae			15		
	Scatopsidae					25
	Sciomyzidae	4	21			
	Stratiomyidae	1, 2, 4, 8	21	12		25
	Syrphidae	2	21			26
	Tabanidae	1, 3, 4, 8	21			25
	Tipulidae	2, 4, 8	21			

Taxa	Family	North America	Africa	Europe	Asia	Australia
B. Microcrustaceans						
Diplostraca	Bosminidae	9	16	13	24	
	Chydoridae	9	16, 17	13	24	25
	Daphniidae	9	16, 17	13	24	25
	Ilyocryptidae		17			
	Lynceidae	4				
	Macrothricidae	9	16, 17			25
	Moinidae	9	17		24	
	Polyphemidae	9		13		
	Sididae	9	16, 17	13	24	

(continued)

Taxa	Family	North America	Africa	Europe	Asia	Australia
B. Microcrustaceans						
Copepoda	Ameiridae					25
	Canthocamptidae		16, 17			25
	Canuellidae		16			
	Centropagidae				24	25
	Cletodidae		17			25
	Cyclopidae	9	16, 17	13	24	25
	Diaptomidae	9	16, 17			
	Laophontidae					25
	Oithonidae				24	
	Pseudodiaptomidae				24	
	Sulcanidae					25
Ostracoda	Candonidae		16			25
	Cyprididae		16, 17			25, 26, 27
	Cytherideidae		17			
	Ilyocyprididae		17			25, 26
	Limnocytheridae					25
	Notodromadidae					25

Citations: 1 Hentges and Stewart (2010); 2 Silver et al. (2012); 3 Rosenberg and Danks (1987); 4 Hornung and Foote (2006); 5 Stephen (2006); 6 Lillie (1991); 7 Hanson and Swanson (1989); 8 Maurer (2013); 9 Norlin et al. (2006); 10 Collinson et al. (1995); 11 Carchini et al. (2007); 12 Sahuquillo et al. (2007); 13 Timms and Moss (1984); 14 Brönmark (1985); 15 Mason (1977); 16 Samraoui et al. (1998); 17 Ramdani et al. (2001); 18 Fernández and López Ruf (2006); 19 Maltchik et al. (2012); 20 Fontanarrosa et al. (2013); 21 Krawczyk et al. (2013); 22 Cai et al. (2011); 23 Cai et al. (2012); 24 Guijun et al. (2012); 25 Cale et al. (2004); 26 Khan (2003); 27 Timms (1983).

References

Åbjörnsson K, Brönmark C, Hansson L (2002) The relative importance of lethal and non-lethal effects of fish on insect colonisation of ponds. Freshw Biol 47:1489–1495

Anteau MJ, Afton AD (2008) Amphipod densities and indices of wetland quality across the upper-Midwest, USA. Wetlands 28:184–196

Anteau MJ, Afton AD, Anteau AC, Moser EB (2011) Fish and land use influence *Gammarus lacustris* and *Hyalella azteca* (Amphipoda) densities in large wetlands across the upper Midwest. Hydrobiologia 664:69–80

Arnott SE, Vanni MJ (1993) Zooplankton assemblages in fishless bog lakes—influence of biotic and abiotic factors. Ecology 74:2361–2380

Bachmann RW, Horsburgh CA et al (2002) Relations between trophic state indicators and plant biomass in Florida lakes. Hydrobiologia 470:219–234

Batzer DP (1998) Trophic interactions among detritus, benthic midges, and predatory fish in a freshwater marsh. Ecology 79:1688–1698

Batzer DP, Wissinger SA (1996) Ecology of insect communities in nontidal wetlands. Annu Rev Entomol 41:75–100

Batzer DP, Pusateri CR, Vetter R (2000) Impacts of fish predation on marsh invertebrates: direct and indirect effects. Wetlands 20:307–312

Bayley SE, Prather CM (2003) Do wetland lakes exhibit alternative stable states? Submersed aquatic vegetation and chlorophyll in western boreal shallow lakes. Limnol Oceanogr 48:2335–2345

Bayley S, Wong A, Thompson J (2013) Effects of agricultural encroachment and drought on wetlands and shallow lakes in the Boreal Transition Zone of Canada. Wetlands 33:17–28

Bayly I (1972) Salinity tolerance and osmotic behavior of animals in athalassic saline and marine hypersaline waters. Annu Rev Ecol Syst 3:233–268

Bazzanti M, Grezzi F, Bella VD (2008) Chironomids (Diptera) of temporary and permanent ponds in Central Italy: a neglected invertebrate group in pond ecology and conservation. J Freshw Ecol 23:219–229

Benoy GA (2008) Tiger salamanders in prairie potholes: a “fish in amphibian’s garments?”. Wetlands 28:464–472

Blois-Heulin C, Crowley PH, Arrington M, Johnson DM (1990) Direct and indirect effects of predators on the dominant invertebrates of two freshwater littoral communities. Oecologia 84:295–306

Breukelaar AW, Lammens EHRR, Breteler JGPK, Tátrai I (1994) Effects of benthivorous bream (*Abramis brama*) and carp (*Cyprinus carpio*) on sediment resuspension and concentrations of nutrients and chlorophyll *a*. Freshw Biol 32:113–121

Brönmark C (1985) Freshwater snail diversity: effects of pond area, habitat heterogeneity and isolation. Oecologia 76:127–131

Burks RL, Jeppesen E, Lodge DM (2001) Littoral zone structures as *Daphnia* refugia against fish predators. Limnol Oceanogr 46:230–237

Cai Y, Gong Z, Qin B (2011) Influences of habitat type and environmental variables on benthic macroinvertebrate communities in a large shallow subtropical lake (Lake Taihu, China). Ann Limnol Int J Limnol 47:85–95

Cai Y, Gong Z, Qin B (2012) Benthic macroinvertebrate community structure in Lake Taihu, China: effects of trophic status, wind-induced disturbance and habitat complexity. J Great Lakes Res 38:39–48

Cale DJ, Halse SA, Walker CD (2004) Wetland monitoring in the wheatbelt of south-west Western Australia: site descriptions, waterbird, aquatic invertebrate and groundwater data. Conserv Sci W Austral J 5:20–135

Carchini G, Della Bella V, Solimini AG, Bazzanti M (2007) Relationships between the presence of odonate species and environmental characteristics in lowland ponds of central Italy. Ann Limnol Int J Limnol 43:81–87

Carpenter SR, Lodge DM (1986) Effects of submerged macrophytes on ecosystem processes. Aquat Bot 26:341–370

Carrick HJ, Aldridge FJ, Schelske CL (1993) Wind influences phytoplankton biomass and composition in a shallow, productive lake. Limnol Oceanogr 38:1179–1192

Chow-Fraser P (2005) Ecosystem response to changes in water level of Lake Ontario marshes: lessons from the restoration of Cootes Paradise Marsh. Hydrobiologia 539:189–204

Collinson N, Biggs J et al (1995) Temporary and permanent ponds: an assessment of the effects of drying out on the conservation value of aquatic macroinvertebrate communities. Biol Conserv 74:125–133

Conly FM, Van der Kamp G (2001) Monitoring the hydrology of Canadian prairie wetlands to detect the effects of climate change and land use changes. Environ Monitor Assess 67:195–215

Coops H, Meryem B, Crisman TL (2003) The role of water-level fluctuations in shallow lake ecosystems–workshop conclusions. Hydrobiologia 506:23–27

Cowardin LM, Carter V, Golet FC, LaRoe ET (1979) Classification of wetlands and deepwater habitats of the United States. US Fish Wildlife Service, Washington, DC

Crivelli AJ (1983) The destruction of aquatic vegetation by carp. Hydrobiologia 106:37–41

Crowder LB, Cooper WE (1982) Habitat structural complexity and the interaction between bluegills and their prey. Ecology 63:1802–1813

Cyr H, Downing JA (1988) Empirical relationships of phytomacrofaunal abundance to plant biomass and macrophyte bed characteristics. Can J Fish Aquat Sci 45:976–984

Dahl DE (2014) Status and trends of prairie wetlands in the United States 1997–2009. US Department of International, Fish Wildlife Service, Ecological Service, Washington, DC

de Szalay FA, Cassidy W (2001) Effects of muskrat (*Ondatra zibethicus*) lodge construction on invertebrate communities in a Great Lakes coastal wetland. Am Midl Nat 146:300–310

Deutschman MR, Peterka JJ (1988) Secondary production of tiger salamanders (*Ambystoma tigrinum*) in three North Dakota prairie lakes. Can J Fish Aquat Sci 45:691–697

Donald DB, Syrgiannis J, Hunter F, Weiss G (1999) Agricultural pesticides threaten the ecological integrity of northern prairie wetlands. Sci Total Environ 231:173–181

Duffy WG (1998) Population dynamics, production, and prey consumption of fathead minnows (*Pimephales promelas*) in prairie wetlands: a bioenergetics approach. Can J Fish Aquat Sci 54:15–27

Euliss NH, Smith LM, Wilcox DA, Browne BA (2008) Linking ecosystem processes with wetland management goals: charting a course for a sustainable future. Wetlands 28:553–562

Fernández LA, López Ruf MLL (2006) Aquatic Coleoptera and Heteroptera inhabiting waterbodies from Berisso, Buenos Aires province, Argentina. Rev Biol Trop 54:139–148

Fernández-Aláez M, Fernández-Aláez C et al (2004) A 2-year experimental study on nutrient and predator influences on food web constituents in a shallow lake of north-west Spain. Freshw Biol 49:1574–1592

Finlayson C, Van der Valk A (1995) Wetland classification and inventory: a summary. Vegetatio 118:185–192

Fontanarrosa MS, Collantes MB, Bachmann AO (2013) Aquatic insect assemblages of man-made permanent ponds, Buenos Aires City, Argentina. Neotrop Entomol 42:22–31

Friederichs SJ, Zimmer KD et al (2011) Total phosphorus and piscivore mass as drivers of food web characteristics in shallow lakes. Oikos 120:756–765

Fritzell EK (1989) Mammals in prairie wetlands. In: Van der Valk A (ed) Northern Prairie Wetlands. Iowa State University Press, Ames, pp 302–315

Fukuhara H, Sakamoto M (1987) Enhancement of inorganic nitrogen and phosphate release from lake sediment by tubificid worms and chironomid larvae. Oikos 48:312–320

Goldsborough LG, Crumpton WG (1998) Distribution and environmental fate of pesticides in prairie wetlands. Great Plains Res 8:73–95

Gorman MW, Zimmer KD et al (2014) Relative importance of phosphorus, fish biomass, and watershed land use as drivers of phytoplankton abundance in shallow lakes. Sci Total Environ 466–467:849–855

Guijun Y, Boqiang Q et al (2012) Contrasting zooplankton communities of two bays of the large, shallow, eutrophic Lake Taihu, China: Their relationship to environmental factors. J Great Lakes Res 38:299–308

Gulati RD, Van Donk E (2002) Lakes in the Netherlands, their origin, eutrophication and restoration: state-of-the-art review. Hydrobiologia 478:73–106

Hammer U, Sheard J, Kranabetter J (1990) Distribution and abundance of littoral benthic fauna in Canadian prairie saline lakes. Hydrobiologia 197:173–192

Hanson MA, Butler MG (1994a) Responses to food web manipulation in a shallow waterfowl lake. Hydrobiologia 279(280):457–466

Hanson MA, Butler MG (1994b) Responses of plankton, turbidity, and macrophytes to biomanipulation in a shallow prairie lake. Can J Fish Aquat Sci 51:1180–1188
Hanson MA, Riggs MR (1995) Potential effects of fish predation on wetland invertebrates: a comparison of wetlands with and without fathead minnows. Wetlands 15:167–175
Hanson B, Swanson G (1989) Coleoptera species inhabiting prairie wetlands of the Cottonwood Lake area, Stutsman County, North Dakota. Prairie Nat 21:49–57
Hanson MA, Zimmer KD et al (2005) Biotic interactions as determinants of ecosystem structure in prairie wetlands: an example using fish. Wetlands 25:764–775
Hanson MA, Herwig BR et al (2012) Comparing effects of lake- and watershed-scale influences on communities of aquatic invertebrates in shallow lakes. PLoS One 7:1–9
Hanson, MA, Buelt CB, Zimmer KD, Herwig, BR, Bowe S, Maurer K (2015) Co-correspondence among aquatic invertebrates, fish, and submerged aquatic plants in shallow lakes. Freshwater Science 34:953–964
Hargeby A, Andersson G, Blindow I, Johansson S (1994) Trophic web structure in a shallow eutrophic lake during a dominance shift from phytoplankton to submerged macrophytes. Hydrobiologia 279:83–90
Harper D, Mavuti KM (2004) Lake Naivasha, Kenya: ecohydrology to guide the management of a tropical protected area. Ecohydrol Hydrobiol 4:287–305
Hassall C, Hollinshead J, Hull A (2011) Environmental correlates of plant and invertebrate species richness in ponds. Biodiv Conserv 20:3189–3222
Hentges VA, Stewart TW (2010) Macroinvertebrate assemblages in Iowa prairie pothole wetlands and relation to environmental features. Wetlands 30:501–511
Hershey AE, Gettel GM et al (1999) A geomorphic-trophic model for landscape control of Arctic lake food webs. BioScience 49:887–897
Herwig BR, Zimmer KD (2007) Population ecology and prey consumption by fathead minnows in prairie wetlands: importance of detritus and larval fish. Ecol Freshw Fish 16:282–294
Herwig BR, Zimmer KD et al (2010) Factors influencing fish distributions in shallow lakes in prairie and prairie-parkland regions of Minnesota, USA. Wetlands 30:609–619
Hobbs WO, Hobbs JMR et al (2012) A 200-year perspective on alternative stable state theory and lake management from a biomanipulated shallow lake. Ecol Appl 22:1483–1496
Holland CC, Honea J, Gwin SE, Kentula ME (1995) Wetland degradation and loss in the rapidly urbanizing area of Portland, Oregon. Wetlands 15:336–345
Hornung JP, Foote AL (2006) Aquatic invertebrate responses to fish presence and vegetation complexity in western Boreal wetlands, with implications for waterbird productivity. Wetlands 26:1–12
Jeppesen E, Jensen JP et al (1997) Top-down control in freshwater lakes: the role of nutrient state, submerged macrophytes and water depth. Hydrobiologia 342(343):151–164
Jeppesen E, Meerhoff M et al (2007) Restoration of shallow lakes by nutrient control and biomanipulation—the successful strategy varies with lake size and climate. Hydrobiologia 581:269–285
Johnson BT (1986) Potential impact of selected agricultural chemical contaminants on a northern prairie wetland: a microcosm evaluation. Environ Toxicol Chem 5:473–485
Johnson DM (1991) Behavioral ecology of larval dragonflies and damselflies. Trends Ecol Evol 6:8–13
Johnson WC, Werner B et al (2010) Prairie wetland complexes as landscape functional units in a changing climate. Bioscience 60:128–140
Jones JI, Sayer CD (2003) Does the fish-invertebrate periphyton cascade precipitate plant loss in shallow lakes? Ecology 84:2155–2167
Kantrud HA, Krapu GJ, Swanson GA, Allen JA (1989) Prairie basin wetlands of the Dakotas: a community profile. US Fish Wildlife Service, Jamestown
Khan TA (2003) Limnology of four saline lakes in western Victoria, Australia II biological parameters. Limnologica 33:327–339
Kosten S, Kamarainen A et al (2009) Climate-related differences in the dominance of submerged macrophytes in shallow lakes. Global Change Biol 15:2503–2517

Kosten S, Huszar VL et al (2012) Warmer climates boost cyanobacterial dominance in shallow lakes. Global Change Biol 18:118–126
Krapu GL, Reinecke KJ (1992) Ecology and management of breeding waterfowl. In: Batt BD (ed) Foraging ecology and nutrition. University Minnesota Press, Minneapolis, pp 1–29
Krawczyk ACDB, Baldan LT et al (2013) The invertebrate's community in adjacent Alto Iguacu's anthropic lakes of different environmental factors. Biot Neotrop 13:47–60
Lacoul P, Freedman B (2006) Environmental influences on aquatic plants in freshwater ecosystems. Environ Rev 14:89–136
Lawler G, Sunde L, Whitaker J (1974) Trout production in prairie ponds. J Fish Board Can 31:929–936
Lillie RA (1991) The adult aquatic and semiaquatic Coleoptera of nine northwestern Wisconsin wetlands. Coleopt Bull 45:101–111
Lynch M, Shapiro J (1981) Predation, enrichment, and phytoplankton community structure. Limnol Oceanogr 26:86–102
Main AR, Headley JV et al (2014) Widespread use and frequent detection of neonicotinoid insecticides in wetlands of Canada's Prairie Pothole Region. PLoS One 9, e92821
Maltchik L, Dalzochio MS, Stenert C, Rolon AS (2012) Diversity and distribution of aquatic insects in southern Brazil wetlands: implications for biodiversity conservation in a Neotropical region. Rev Biol Trop 60:273–289
Mason CF (1977) Populations and production of benthic animals in two contrasting shallow lakes in Norfolk. J Anim Ecol 46:147–172
Matthews G, Townsend V (1993) The Ramsar convention on wetlands: its history and development. Ramsar Convention Bureau, Gland
Maurer K (2013) Biophysical interactions in prairie pothole wetlands of Iowa: consequences for macroinvertebrate assemblages and ecosystem condition. MS thesis, Iowa State Univ
McCollum S, Leimberger J (1997) Predator-induced morphological changes in an amphibian: predation by dragonflies affects tadpole shape and color. Oecologia 109:615–621
McParland CE, Paszkowski CA (2006) Effects of small-bodied fish on invertebrate prey and foraging patterns of waterbirds in aspen parkland wetlands. Hydrobiologia 567:43–55
McPeek MA (1990a) Determination of species composition in the *Enallagma* damselfly assemblages of permanent lakes. Ecology 71:83–98
McPeek MA (1990b) Behavioral differences between *Enallagma* species (Odonata) influencing differential vulnerability to predators. Ecology 71:1714–1726
Meding ME, Jackson LJ (2003) Biotic, chemical, and morphometric factors contributing to winter anoxia in prairie lakes. Limnol Oceanogr 48:1633–1642
Moss B, Madgwick J, Phillips G (1996) Guide to the restoration of nutrient-enriched shallow lakes. The Broads Authority, Norwich
Murkin HR, Ross LCM (2000) Invertebrates in prairie wetlands. In: Murkin HR, van der Valk AG, Clark WR (eds) Prairie wetland ecology. Iowa State Univ Press, Ames, pp 201–248
Newman RM (1991) Herbivory and detritivory on freshwater macrophytes by invertebrates: a review. J No Am Benthol Soc 10:89–114
Nolby LE, Zimmer KD, Hanson MA, Herwig BR (2015) Is island biogeography a poor fit to biodiversity patterns in shallow lakes? Freshw Biol 60:870-880
Norlin JI, Bayley SE, Ross LCM (2006) Zooplankton composition and ecology in western boreal shallow-water wetlands. Hydrobiologia 560:197–215
Olenick R, Gee L (1981) Tiger salamanders (*Ambystoma tigrinum*) and stocked rainbow trout (*Salmo gairdneri*): potential competitors for food in Manitoba prairie pothole lakes. Can Field Nat 95:129–132
Oslund FT, Johnson RR, Hertel DR (2010) Assessing wetland changes in the Prairie Pothole Region of Minnesota from 1980 to 2007. J Fish Wildl Manage 1:131–135
Peterka JJ (1989) Fishes in northern prairie wetlands. In: van der Valk A (ed) Northern prairie wetlands. Iowa State University Press, Ames, pp 302–315

Pinder A, Halse McRae SJ, Shiel R (2004) Aquatic invertebrate assemblages of wetlands and rivers in the wheatbelt region of Western Australia. Records W Austral Mus Supple 67:7–37

Potthoff AJ, Herwig BR et al (2008) Cascading food-web effects of piscivore introductions in shallow lakes. J Appl Ecol 45:1170–1179

Qin B, Xu P et al (2007) Environmental issues of lake Taihu, China. Hydrobiologia 581:3–14

Qin B, Zhu G et al (2010) A drinking water crisis in Lake Taihu, China: linkage to climatic variability and lake management. Environ Manage 45:105–112

Ramdani M, Elkhiati N, Flower RJ (2001) Open water zooplankton communities in north African wetland lakes: the CASSARINA project. Aquat Ecol 35:319–333

Rockström J, Steffen W et al (2009) A safe operating space for humanity. Nature 461:472–475

Rodríguez C, Bécares E, Fernández-Aláez M, Fernández-Aláez C (2005) Loss of diversity and degradation of wetlands as a result of introducing exotic crayfish. Biol Invasions 7:75–85

Romo S, Miracle MR et al (2004) Mesocosm experiments on nutrient and fish effects on shallow lake food webs in a Mediterranean climate. Freshw Biol 49:1593–1607

Rosenberg DM, Danks HV (1987) Aquatic insects of peatlands and marshes in Canada. Mem Entom Soc Can 140:1–174

Sahuquillo M, Poquet JM, Rueda J, Miracle MR (2007) Macroinvertebrate communities in sediment and plants in coastal Mediterranean water bodies (Central Iberian Peninsula). Ann Limnol 43:117–130

Samraoui B, Segers H et al (1998) Rotifera, Cladocera, Copepoda, and Ostracoda from coastal wetlands in northeast Algeria. Hydrobiologia 386:183–193

Sand-Jensen K, Borum J (1991) Interactions among phytoplankton, periphyton, and macrophytes in temperate freshwaters and estuaries. Aquat Bot 41:137–175

Scheffer M (2004) Ecology of shallow lakes. Kluwer, Dordrect

Scheffer M, Hosper SH et al (1993) Alternative equilibria in shallow lakes. Trends Ecol Evol 8:275–279

Scheffer M, van Geest GJ et al (2006) Small habitat size and isolation can promote species richness: second-order effects on biodiversity in shallow lakes and ponds. Oikos 112:227–231

Silver CA, Vamosi SM, Bayley SE (2012) Temporary and permanent wetland macroinvertebrate communities: phylogenetic structure through time. Acta Oecol 39:1–10

Simberloff D, Parker IM, Windle PN (2005) Introduced species policy, management, and future research needs. Front Ecol Environ 3:12–20

Smith DC (1983) Factors controlling tadpole populations of the chorus frog (*Pseudacris triseriata*) on Isle Royale, Michigan. Ecology 64:501–510

Søndergaard M, Jeppesen E et al (2007) Lake restoration: successes, failures and long-term effects. J Appl Ecol 44:1095–1105

Stephen, BJ (2006) A survey of the freshwater snails of the prairie pothole region of South Dakota. Report submitted to the South Dakota Department of Game, Fish and Parks

Swanson GA, Duebbert HF (1989) Wetland habitats of waterfowl in the prairie pothole region. In: van der Valk A (ed) Northern prairie wetlands. Iowa State Univ Press, Ames, pp 228–267

Tangen BA, Butler MG, Ell MJ (2003) Weak correspondence between macroinvertebrate assemblages and land use in prairie pothole region wetlands, USA. Wetlands 23:104–115

Timms BV (1983) A study of benthic communities in some shallow saline lakes of western Victoria, Australia. Hydrobiologia 105:165–177

Timms RM, Moss B (1984) Prevention of growth of potentially dense phytoplankton populations by zooplankton grazing, in the presence of zooplanktivorous fish, in a shallow wetland ecosystem. Limnol Oceanogr 29:472–486

Vadeboncoeur Y, Jeppesen E et al (2003) From Greenland to green lakes: cultural eutrophication and the loss of benthic pathways in lakes. Limnol Oceanogr 48:1408–1418

van der Valk AG (2005) Water-level fluctuations in North American prairie wetlands. Hydrobiologia 539:171–188

van der Valk A, Davis C (1978) The role of seed banks in the vegetation dynamics of prairie glacial marshes. Ecology 59:322–335

Vanni MJ (2002) Nutrient cycling by animals in freshwater ecosystems. Annu Rev Ecol Syst 33:341–370

Vera MS, Lagomarsino L et al (2010) New evidences of Roundup® (glyphosate formulation) impact on the periphyton community and the water quality of freshwater ecosystems. Ecotoxicology 19:710–721

Verant ML, Konsti ML, Zimmer KD, Deans CA (2007) Factors influencing nitrogen and phosphorous excretion rates of a fish community in a shallow lake. Freshw Biol 52:1968–1981

Ward MC, Willis DW et al (2008) Consumption estimates of walleye stocked as fry to suppress fathead minnow populations in west-central Minnesota wetlands. Ecol Freshw Fish 17:59–70

Warner BG, Rubec CDA (1997) The Canadian wetland classification system. Wetlands Research Branch, Univ Waterloo

Wellborn GA, Skelly DK, Werner EE (1996) Mechanisms creating community structure across a freshwater habitat gradient. Annu Rev Ecol Syst 27:337–363

Wiggins GB, Mackay RJ, Smith IM (1980) Evolutionary and ecological strategies of animals in annual temporary pools. Arch Hydrobiol Suppl 58:1–206

Winter TC, Rosenberry DO (1995) The interaction of ground water with prairie pothole wetlands in the Cottonwood Lake area, east-central North Dakota, 1979–1990. Wetlands 15:193–211

Wissinger SA (1999) Ecology of wetland invertebrates: synthesis and applications for conservation and management. In: Batzer DP, Rader RB, Wissinger SA (eds) Invertebrates in freshwater wetlands of North America: ecology and management. Wiley, New York, pp 1043–1086

Wrubleski D (1989) The effect of waterfowl feeding on a chironomid (Diptera: Chironomidae) community. Freshwater wetlands and wildlife: perspectives on natural, managed and degraded ecosystems. US Dept Energy Symp 61:691–696

Zedler JB, Kercher S (2004) Causes and consequences of invasive plants in wetlands: opportunities, opportunists, and outcomes. Crit Rev Plant Sci 23:431–452

Zimmer KD, Hanson MA, Butler MG (2000) Factors influencing invertebrate communities in prairie wetlands: a multivariate approach. Can J Fish Aquat Sci 57:76–85

Zimmer KD, Hanson MA, Butler MG (2001a) Effects of fathead minnow colonization and removal on a prairie wetland ecosystem. Ecosystems 4:346–357

Zimmer KD, Hanson MA, Butler MG, Duffy WG (2001b) Size distributions of aquatic invertebrates in two prairie wetlands, with and without fish, with implications for community production. Freshw Biol 46:1373–1386

Zimmer KD, Hanson MA, Butler MG, Duffy WG (2001c) Influences of fathead minnows and aquatic macrophytes on nutrient partitioning and ecosystem structure in two prairie wetlands. Arch Hydrobiol 150:411–433

Zimmer KD, Hanson MA, Butler MG (2002) Effects of fathead minnows and restoration on prairie wetland ecosystems. Freshw Biol 47:2071–2086

Zimmer KD, Herwig BR, Laurich LM (2006) Nutrient excretion by fish in wetland ecosystems and its potential to support algal production. Limnol Oceanogr 51:197–207

Zimmer KD, Hanson MA, Herwig BR, Konsti ML (2009) Thresholds and stability of alternative regimes in shallow prairie-parkland lakes of central North America. Ecosystems 12:843–852

Chapter 9
Invertebrates in Great Lakes Marshes

Matthew J. Cooper and Donald G. Uzarski

Introduction

These habitats form where hydrologic energy sources, such as waves and lake currents, are reduced to the point that macrophytes can establish and persist and where sediment is conducive to macrophyte growth. While marshes form along the shorelines of many large lakes globally, this chapter focuses primarily on marshes of the Laurentian Great Lakes in North America (Fig. 9.1). These are some of the best-studied freshwater coastal wetlands in the world, and much of the research on Great Lakes marsh invertebrate ecology is applicable to other global lakeshore wetlands.

The Laurentian Great Lakes system includes Lakes Superior, Michigan–Huron (hydrologically a single lake), Erie, and Ontario as well as their connecting waterways (Fig. 9.1a). The Great Lakes extend from 41°20′N latitude (the southern shore of Lake Erie) to approximately 49°N latitude (Nipigon Bay on the north shore of Lake Superior), representing approximately 800 km of latitude. The Great Lakes span 1200 km of longitude, from approximately 76°W (eastern Lake Ontario) to 92°W longitude (western Lake Superior). The Great Lakes have over 17,000 km of shoreline, which is greater than the total length of the United States' east and west coasts, combined. This immense freshwater system contains approximately 21 % of the world's surface freshwater supply and 84 % of North America's surface freshwater. Over 2000 coastal wetlands occur along the Great Lakes shoreline (Fig. 9.1b). In this chapter, "lakeshore marsh" refers to a coastal wetland that contains at least some habitat that is dominated by herbaceous vegetation, though these wetlands often contain areas that are dominated by woody vegetation (i.e., swamp) as well.

Lakeshore marshes of the Laurentian Great Lakes are important habitats for fish, amphibians, reptiles, wading birds, and waterfowl (Harris et al. 1983; Jude and Pappas 1992; Prince et al. 1992; Maynard and Wilcox 1997; Weeber and Vallianatos 2000; Uzarski et al. 2005). Invertebrates make up a large component of the diets of these wetland fauna, thus linking algal and detrital energy sources to higher trophic levels. These energy pathways—from primary producers to invertebrate consumers

D. Batzer, D. Boix (eds.), *Invertebrates in Freshwater Wetlands*,
DOI 10.1007/978-3-319-24978-0_9

to fish and other macrofauna—support important functions of coastal wetlands in the broader lake ecosystem (Brazner et al. 2004; Sierszen et al. 2012a). Therefore, because invertebrates represent key trophic linkages in wetland food webs, environmental drivers of invertebrate community structure have ecosystem-level implications.

Climate and Its Influence on Invertebrate Assemblages

In general, the Great Lakes region has a temperate climate with pronounced seasonality. Three primary factors influence the region's climate: air masses that originate in other areas, the continental location of the basin, and the effect of the Great Lakes themselves. In the summer, conditions in the northern portion of the basin are most influenced by cold dry air from the Canadian northwest while the southern portion of the basin receives warm moist air from the Gulf of Mexico. The balance of these different air masses largely dictates local conditions over relatively short timescales. Average July daytime high temperatures are typically around 25 °C in the northern Great Lakes basin and around 30 °C in the southern portion of the basin. In winter, the region is most frequently influenced by Arctic air from the northwest of the continent. Average January nighttime lows are typically around −15 °C in the northern part of the basin and −5 °C in the south. Great Lakes water temperatures continue to drop throughout the winter. Ice frequently covers all of Lake Erie by late winter. The other lakes rarely are fully ice-covered but coastal ice is common. Because coastal wetlands occur in shallow protected bays and inlets, they are usually ice-covered throughout the winter on all of the Great Lakes. Shifting ice along the coast creates an "ice foot" that redistributes sediment and rhizome mats in coastal wetlands. This physical disturbance is an important driver of spatial heterogeneity in lakeshore marsh vegetation communities, which leads to heterogeneity in the resident invertebrate communities (Burton 1985).

Spring and autumn in the Great Lakes region are characterized by highly variable weather. The lakes are slower to warm than the land in the springtime, which tends to keep coastal areas cool well into the spring. In most years, this delays the leafing and blossoming of plants and protects wetland vegetation from late frosts. The lakes are also slow to cool in the autumn, keeping coastal areas warmer than inland regions of the same latitude. These moderating effects of the lakes on coastal climatic conditions also influence wetland invertebrate phenology and the timing of emergence relative to more inland wetlands.

The temperate climate and strong seasonality of the Great Lakes region have significant implications for invertebrate life histories. Most invertebrates cope with the freezing temperatures and ice cover in winter by entering into a dormant stage in the autumn and reemerging the following spring or summer when conditions are once again favorable. Dormancy may involve either quiescence or diapause. Quiescence is the slowed or completely halted development that results as a direct response to the onset of unfavorable conditions, with development resuming when

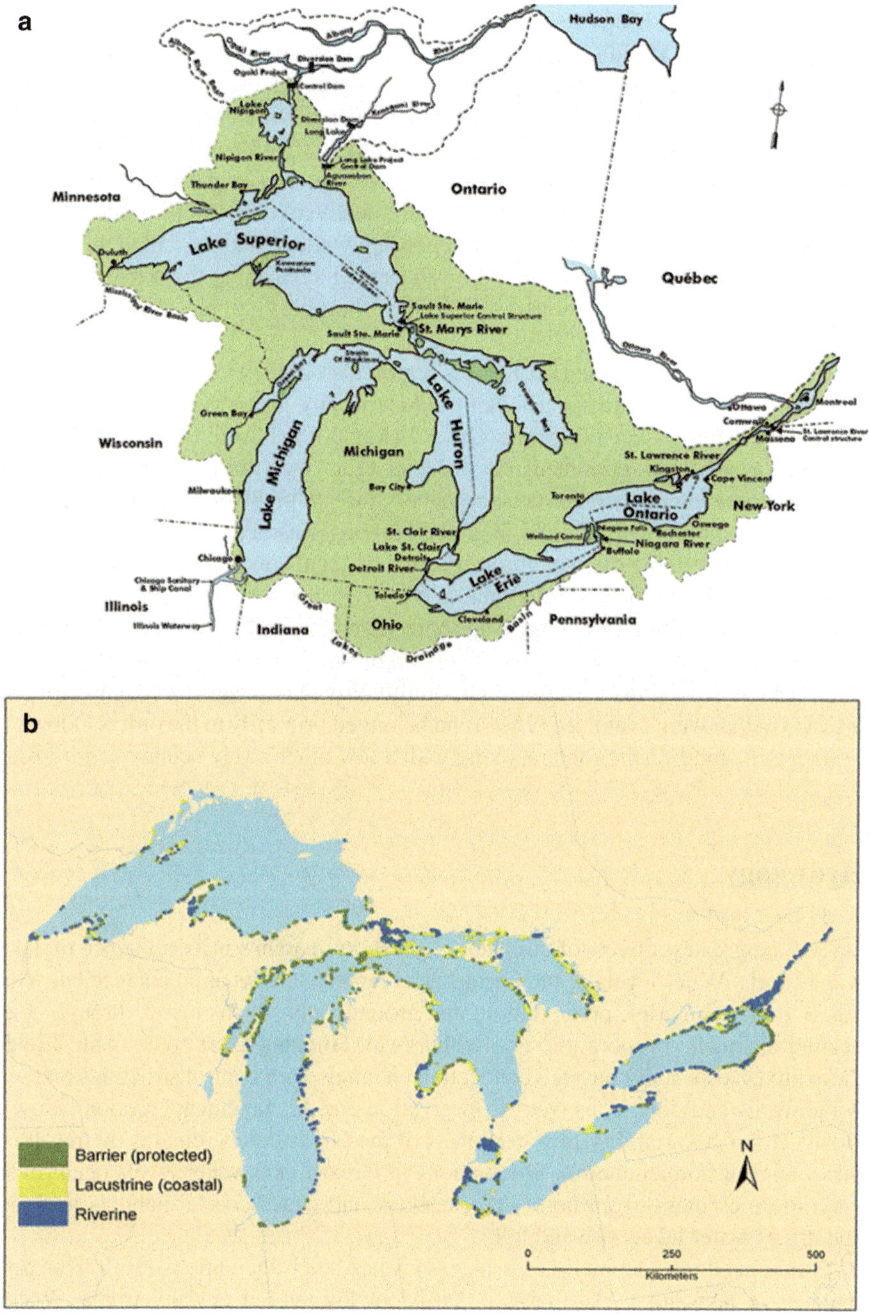

Fig. 9.1 (**a**) The Laurentian Great Lakes and the Great Lakes drainage basin. Select coastal cities, political boundaries, major tributaries, and interconnecting channels are also included (*source*: US Army Corps of Engineers, Detroit District). (**b**) Locations and hydrogeomorphic types of Great Lakes coastal wetlands identified by the Great Lakes Coastal Wetlands Consortium

conditions improve. Diapause is an arrested state of development triggered by specific physiological stimuli. Major environmental cues that induce and/or terminate diapause include temperature, photoperiod, moisture, pH, and changes in dissolved oxygen, among others. The ability to overwinter in a dormant stage and then rapidly recolonize habitats when conditions are favorable is a key adaptation for marsh invertebrates in temperate climates, including those in the Great Lakes. Patrick et al. (2014) noted dramatically increasing invertebrate diversity and density from May to August in a Lake Michigan drowned river mouth wetland and attributed this to taxa coming out of resting stages as conditions became increasingly favorable and to the developing aquatic vegetation communities that provide physical habitat for invertebrates.

Strong seasonality in the Great Lakes region also influences the number of generations that invertebrate species can produce in any given year (i.e., voltinism). Many invertebrate taxa inhabiting Great Lakes coastal wetlands are univoltine, emerging as adults to reproduce only once per year, often during the warm summer months. However, many other taxa, especially insects in the order Diptera, are able to reproduce multiple times throughout the growing season (i.e., multivoltine), with the number and timing of generations dictated by local conditions. Kovalenko et al. (2014) compiled voltinism information for 77 insect taxa collected from Great Lakes littoral habitats (including lakeshore marshes) and found that 48 of these were univoltine, with an additional 10 taxa representing combinations of bi-, uni-, semi-, and merovoltinism. These 58 non-multivoltine taxa represented a majority of insect taxa that were evaluated (75 %) and belonged primarily to the orders Odonata, Trichoptera, and Ephemeroptera, along with a few other rarely occurring groups.

Hydrology

One of the greatest effects of climate on lakeshore marshes is its influence on lake water levels. Water levels of the Great Lakes represent a dynamic balance between inputs from tributaries, precipitation, and groundwater versus losses through connecting channels, evaporation, and withdrawal. Humans exert some control over these fluxes, especially through connecting channels with control structures and via withdrawals and diversions (see "Conservation and Management" section below). However, most variability in water levels of the Great Lakes remains the result of factors out of human control. Fluctuations in Great Lakes water levels occur over varying timescales—from hourly to decadal—and cause coastal habitats to flood and dry as water levels rise and fall.

Pronounced intra-annual (i.e., seasonal) water-level fluctuations result from differences in lake inputs and outputs that occur throughout any given year. Water levels in Lake Michigan–Huron, for example, typically reach an annual maximum in August while Lake Superior reaches its maximum in September, after the previous winter's snowmelt and spring rains have had sufficient time to accumulate.

Annual water-level minima generally occur in late winter when evaporation coincides with reduced tributary inputs. Over broader timescales, variation in basin inputs and outputs from 1 year to the next can cause dramatic interannual water-level fluctuations, often on the order of 1–1.5 m over decadal periods (Fig. 9.2).

Water-level fluctuations are a natural part of the Great Lakes ecosystem. Accordingly, plant and animal species inhabiting lakeshore marshes are uniquely adapted to survive and even flourish in habitats with cyclical wetting and drying (Wilcox 1995; Keough et al. 1999; Mayer et al. 2004; Albert et al. 2005; Uzarski et al. 2009). For example, many wetland plants that cannot establish under permanently flooded conditions are able to germinate in seasonally flooded habitats that maintain a non-flooded, aerobic environment during the spring and early summer. Seedlings of many wetland plants can then survive as water levels rise (Gathman et al. 2005), which results in the productive and diverse vegetation assemblages found in lakeshore marshes. These macrophytes, which often exhibit "zonation" in lakeshore marshes due to physical factors (e.g., depth and wave energy), in turn, form the physical habitat template that invertebrate communities assemble within (Burton et al. 2002; Gathman and Burton 2011).

Superimposed on the seasonal and interannual water-level variation are short-term fluctuations caused by wind-driven or atmospheric pressure-induced seiches. Seiche period depends on basin morphology and wind direction but periods from 2 to 10 h are typical. Seiche amplitudes of 10–20 cm are most common across the Great Lakes (Trebitz 2006), though seiches over 1 m are possible and are generally associated with strong storms. Seiche action causes the shallow marsh habitats at the land–water interface, such as meadow marsh, to cyclically flood and drain several times per day. This is especially pronounced in marshes with gently sloping bathymetry, such as those around Saginaw Bay, Lake Huron. Typical bathymetric slope for Saginaw Bay marshes is approximately 0.25 cm per 1.0 m (M.J. Cooper, unpublished data). Thus, a 15 cm seiche, which is common for Saginaw Bay, will cause the water's edge to move 60 m shoreward and lakeward during a single seiche cycle. The unique inter-seiche meadow marsh habitat is home to invertebrate taxa that are adapted to tolerate such dynamic conditions and exploit the detrital food resources commonly available within these habitats (Burton et al. 2002). For example, collector/gatherer and detritivore crustaceans such as *Gammarus*, *Hyalella*, and *Caecidotea* are often found in high densities within seiche-influenced wet meadow zones (Cardinale et al. 1998; Burton et al. 2002).

The regular water-mixing action induced by seiche activity also helps to distribute nutrients and other dissolved materials within and among lakeshore wetlands (Trebitz 2006). Large storm-driven seiches, particularly when combined with high-energy waves, can serve as a strong physical disturbance in lakeshore marshes, causing sediment redistribution and even destruction of emergent vegetation, with concomitant impacts on resident invertebrate communities. Large-amplitude seiches can also cause vast areas of some marshes to be flooded at highly irregular intervals, especially in marshes with gently sloping bathymetry such as those on Saginaw Bay and western Lake Erie.

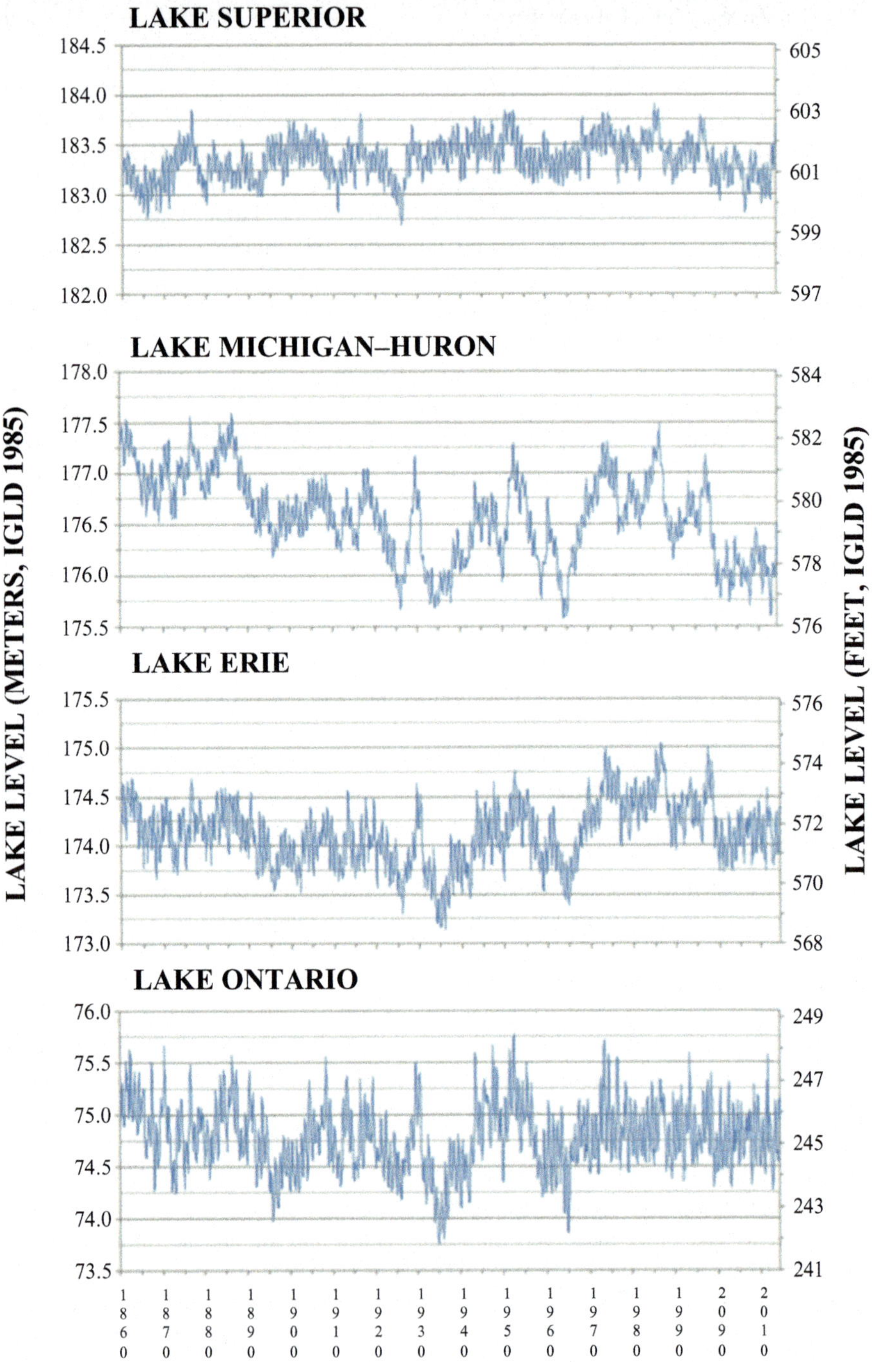

Fig. 9.2 Hydrographs for all five Great Lakes for the period 1860–2013. The figures show the high variability in water levels over the 153 years and the dampening of variability that occurred on Lake Ontario after the Moses–Saunders Power Dam went into operation in 1958. Data were obtained from the National Oceanic and Atmospheric Administration's Center for Operational Oceanographic Products and Services (NOAA CO-OPS; http://www.co-ops.nos.noaa.gov/). Figure courtesy of Douglas A. Wilcox (State University of New York–College at Brockport)

Lakeshore Marsh Hydrogeomorphic Types

In 2002, the Great Lakes Coastal Wetlands Consortium developed a hydrogeomorphic wetland classification system to characterize coastal wetlands of the Laurentian Great Lakes (Albert et al. 2005). The classification system separates wetlands into three broad types—lacustrine, riverine, and barrier-protected—based on geomorphology of the shoreline, primary water source, and hydrologic connectivity to the lake. The scheme includes finer-resolution classification as well and reflects numerous elements of wetland hydrology and geomorphology that collectively influence the structure of floral and faunal communities.

Lacustrine

Lacustrine marshes are adjoined directly to waters of the Great Lakes and are strongly influenced by lake water levels, nearshore currents, and ice scour (Albert et al. 2005). The primary water source for lacustrine marshes is the adjacent lake, though groundwater, tributary streams, and direct precipitation can also contribute to lacustrine marsh hydrology (Fig. 9.3). The main form of water loss is direct outflow to the adjacent lake, though evaporation and evapotranspiration also result in water loss from lacustrine marshes (Fig. 9.3). Geomorphic features along the shoreline such as headlands, embayments, and bathymetry (e.g., sandbars, shallow slope) provide varying degrees of protection from wave energy and coastal currents and allow wetland habitat to develop and persist. Lacustrine marshes can be further subdivided into open and protected embayments, sandspit embayments, and open shoreline wetlands. Invertebrates inhabiting the lakeward margin of lacustrine marshes must be tolerant of wave energy, while those inhabiting wet meadow habitats at the shoreward margin must be tolerant of seiche-induced drying and rewetting cycles.

Riverine

Riverine wetlands occur along the margins of and within tributary streams and rivers and along the margins of large connecting channels between lakes (Albert et al. 2005). Riverine wetlands often occur in deltaic or fluvial habitats at the confluence of rivers and the receiving lake. Water quality, hydraulic processes, and sediment input are controlled in large part by the individual drainages; however, water levels and fluvial processes in these wetlands are directly or indirectly affected by the downstream lake as lake waters flood back into the lower portions of tributary marsh systems. Accordingly, the primary source of water to riverine coastal marshes is direct inputs from tributary streams (Fig. 9.3). The primary

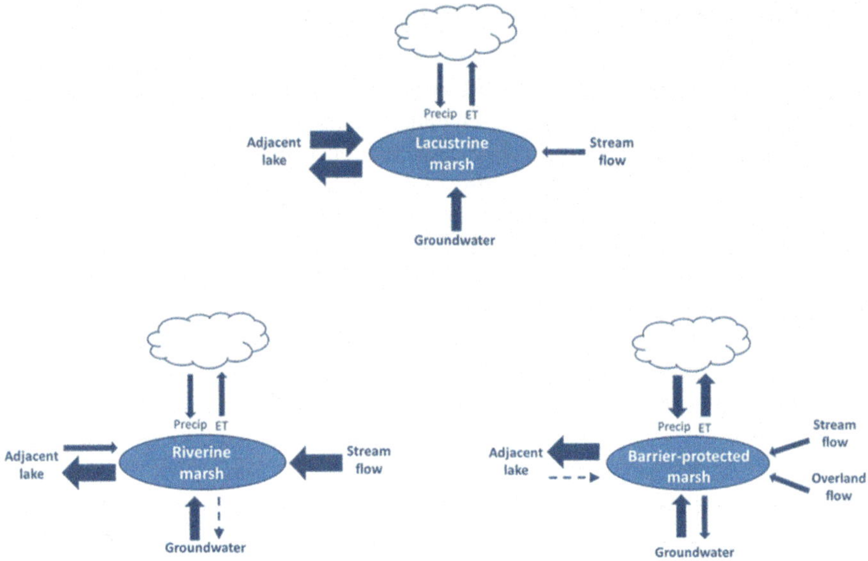

Fig. 9.3 Conceptual models demonstrating the relative magnitudes of water inflows and outflows for the three main hydrogeomorphic types of Great Lakes coastal marshes. *Dashed arrows* represent known but minor flows

outflow is to the receiving lake, though hydrologic inputs and outputs related to atmospheric and groundwater exchange can also occur. Protection from lake waves and coastal currents is provided by sand and gravel bars near river outlets and by channel morphology. Riverine wetlands can be further subdivided into open and barred drowned river mouth wetlands, delta wetlands, and connecting channel wetlands, all of which generally contain herbaceous marsh habitat. Lakeshore riverine wetlands encompass a wide variety of habitat types, from fast-flowing channels to quiet backwater areas with deep organic sediment deposits. Accordingly, a diverse array of invertebrates inhabit these marshes, from rheophilic mayflies in the family Heptageniidae, to sediment-burrowing mayflies in the family Ephemeridae, to grazing snails, shredding and collecting crustaceans (Amphipoda and Asellidae), and surface-dwelling hemipterans in the families Gerridae and Mesoveliidae (Cooper et al. 2007).

Barrier-Protected

Barrier-protected wetlands form as a result of either coastal or fluvial processes that create barriers that separate wetland habitats from the lake (Albert et al. 2005). Barriers may be active or may be the result of some past process that leaves behind the barrier as a relict coastal feature. These wetlands are completely protected from lake waves and currents but may be connected to the lake by one or more channels through the barrier. Water budgets in barrier-protected lakeshore marshes are highly variable and complex, including inputs from precipitation, groundwater, streams, and surface flow and outputs to the atmosphere via evaporation or evapotranspiration as well as outputs to the adjacent lake through temporary channels or shallow subsurface flow (Fig. 9.3). When connected to the lake, water levels reflect those of the adjacent lake because lake water either flows into the marsh or creates sufficient hydraulic head pressure to keep marsh water at the same elevation as the lake. Channels connecting barrier-protected marshes to the lake may be permanent or ephemeral as coastal sediment transport can intermittently close off connecting channels. Invertebrates inhabiting barrier-protected lakeshore marshes must be tolerant of the dramatically fluctuating hydrology that often occurs in these habitats. Strategies to withstand dry periods, such as diapausing eggs or pupae, or the ability to cyclically colonize ephemeral aquatic habitats are common among taxa found in barrier-protected lakeshore marshes (Burton and Uzarski 2009).

Basic Invertebrate Research in Great Lakes Marshes

Our understanding of Great Lakes marsh invertebrate communities has grown considerably in recent years. For example, the structure of these communities has been related to vegetation zonation (Cardinale et al. 1997; Merritt et al. 2002), fetch and wave exposure (Burton et al. 2002, 2004; Cooper et al. 2014), benthic substrate (MacKenzie et al. 2004; Cooper et al. 2007), water-level fluctuation (Gathman and Burton 2011; Cooper et al. 2014), water quality and surrounding land use (King and Brazner 1999; Schneider and Sager 2007; Cooper et al. 2014; Kovalenko et al. 2014; Schock et al. 2014), invasive plants (Kulesza et al. 2008; Holomuzki and Klarer 2010), and habitat fragmentation (Uzarski et al. 2009; Cooper et al. 2012). These drivers are not mutually exclusive of one another, and invertebrate assemblages are often influenced by several of these variables simultaneously.

Indices of Biotic Integrity

In addition to traditional community assembly research, invertebrate-based Indices of Biotic Integrity (IBIs) have been developed and currently are being used throughout the Great Lakes to assess coastal wetland health (Uzarski et al. 2004). The approach leverages the information contained in invertebrate community structure to detect anthropogenic disturbances that may not be discernible with traditional water quality monitoring (Burton et al. 1999; Uzarski et al. 2004). The applicability and performance of IBI-type assessment tools relies on a thorough understanding of invertebrate community responses to both natural and anthropogenic drivers.

An important step in developing wetland IBIs is to partition variability in community structure that is due to natural factors from variability that is due to human disturbance. For Great Lakes coastal wetlands, this has been achieved by developing IBIs for specific wetland types (e.g., lacustrine, riverine, barrier-protected) and vegetation types within wetlands (Burton et al. 1999; Uzarski et al. 2004). Because vegetation structure tends to correlate with hydrology in Great Lakes coastal wetlands (Albert et al. 2005; Gathman et al. 2005), this approach controls for much of the overriding influence of hydrology and macrohabitat structure on IBI metrics. For example, separate sets of IBI metrics have been developed for bulrush-dominated zones and wet meadow zones in Great Lakes lacustrine wetlands (Burton et al. 1999; Uzarski et al. 2004). Stratifying IBIs by vegetation type also allows the protocols to be used at various Great Lakes water levels because vegetation zones move upslope and downslope as water levels fluctuate. Therefore, invertebrate sampling and subsequent IBI metric calculations can "follow" the vegetation zones over time as they move upslope and downslope.

Invertebrate IBI metric identification has been accomplished by comparing community structure in reference wetlands to community structure in impaired wetlands. Attributes of the community that differ between these disturbance categories then have the potential to become IBI metrics. For example, the relative abundance of sphaeriid clams has been shown to decline with increasing human disturbance in Lake Huron lacustrine wetlands, and accordingly, sphaeriid abundance was incorporated into the IBI for these systems (Uzarski et al. 2004). An alternative approach is to quantify anthropogenic disturbance using a multivariate index and then identify invertebrate community metrics that vary predictably along this disturbance gradient. After candidate metrics are identified, metric scoring schemes must be derived to translate metric values into scores for the final IBI determination. Final IBI results are then derived by summing the component metric scores. While the IBI approach is common in lake and stream monitoring and management, it has been used infrequently in wetlands. However, current broadscale monitoring and use of invertebrate-based IBIs in Great Lakes coastal wetlands have become valuable tools for prioritizing wetland restoration projects and tracking restoration outcomes.

Lakeshore Marsh Taxa

Given the immensity of the Great Lakes system, constructing a truly exhaustive list of taxa would be a difficult undertaking. However, several Great Lakes basin-scale invertebrate sampling efforts have occurred or are currently underway in Great Lakes marshes, and these can be used to create a preliminary inventory of taxa. These efforts were conducted in an ecosystem monitoring context, either to develop or test monitoring protocols, or in fully implemented monitoring programs. The Great Lakes Environmental Indicators (GLEI) project (Niemi et al. 2009) sampled invertebrates at 101 coastal wetlands along the US shoreline of the Great Lakes in 2002 and 2003. This program identified 222 invertebrate taxa—most at the genus level (Kovalenko et al. 2014). The Great Lakes Coastal Wetlands Consortium (GLCWC) sampled invertebrates in 67 coastal wetlands in all five Great Lakes in 2002 (Cooper et al. 2014) and identified 215 taxa, mostly at the genus level. In 2011, the GLCWC, along with several researchers from the GLEI group, and others initiated a monitoring program that included sampling invertebrates in lakeshore marshes across the Great Lakes basin. For this effort, over 100 marshes are being sampled each year in the initial 5-year sampling rotation (2011–2015). This monitoring effort is sponsored by the US EPA for the purpose of supporting wetland restoration, protection, and other management activities. More specifically, data and IBI scores are being used to select wetlands for restoration and to track restoration outcomes. While many studies of invertebrate community structure have occurred in Great Lakes coastal marshes in recent decades, the GLCWC effort is the single largest coordinated effort to occur in these systems. In addition to invertebrates, the monitoring program is collecting data on fish, birds, amphibians, vegetation, and water quality in each marsh.

To collect invertebrates, GLCWC researchers use D-frame dip nets to sweep through the water column and vegetation and then "field pick" organisms from the gathered plant matter and detritus. Samples are returned to the laboratory for identification to lowest operational taxonomic unit (usually genus or species) under magnification.

A number of general characteristics of the invertebrate assemblages inhabiting Great Lakes coastal marshes can be gleaned from this large dataset. First, in the initial 3 years of sampling (2011–2013), in which 319 unique marshes were sampled and >270,000 organisms were collected, 331 genera were identified, along with an additional 102 taxa identified at a coarser resolution (Appendix). Second, similar to the finding of Cooper et al. (2014), a small subset of taxa tend to be numerically dominant in the overall assemblage, with the ten most abundant taxa representing 61 % of the organisms collected (Fig. 9.4). Accordingly, the vast majority of taxa could be considered "rare," resulting in a very hollow species-abundance curve (Fig. 9.4). Third, these data reveal that the majority of observed taxa tend to be cosmopolitan, occurring in more than one Great Lake (Appendix).

Environmental Drivers of Invertebrate Communities

Great Lakes Water Levels

The natural fluctuations in water levels of the Great Lakes have important implications for lakeshore marsh invertebrate communities. Perhaps most importantly, intra-annual low-water periods allow macrophyte seeds to germinate and

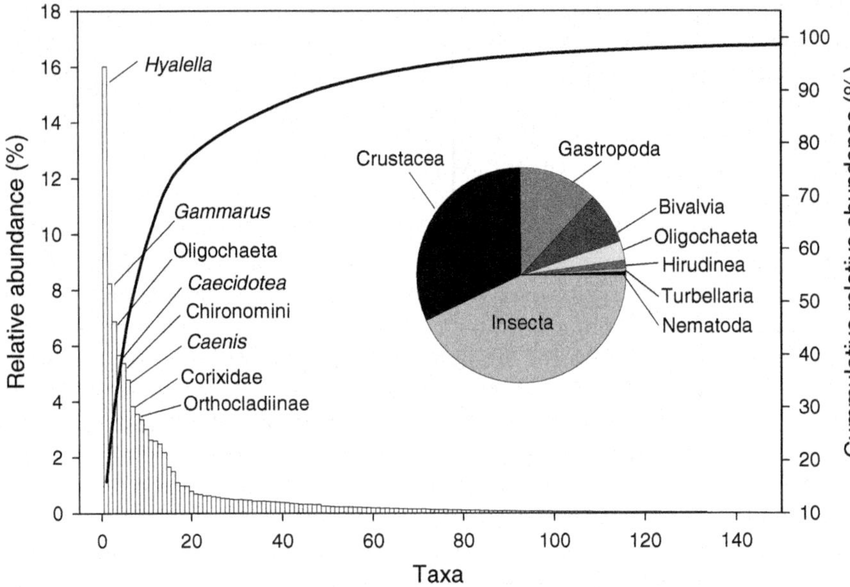

Fig. 9.4 Invertebrate assemblage from 319 Great Lakes marshes sampled by the Great Lakes Coastal Wetlands Consortium from 2011 to 2013. *Vertical bars* represent relative abundances of each taxon from all sites, with the ten most abundant taxa labeled. Only the 135 most abundant taxa (of 433 total taxa collected) are shown on the graph. The *solid line* depicts the asymptotic nature of cumulative relative abundance in the assemblage (right-hand axis). *Pie graph* shows the distribution of invertebrate abundances among major taxonomic groups

seedlings to develop before the wetlands flood again later in the year. Interannual low-water periods (i.e., "low-water years") allow for replenishment of the seedbank. Also during low-water years, the upland portion of meadow marsh is invaded by woody plants, while true marsh (i.e., herbaceous) communities shift lakeward. When water levels rise again, woody species retreat upslope and the emergent marsh and wet meadow communities also move shoreward. Long-term water-level fluctuations, therefore, cause long-term movement and alteration of marsh vegetation zones (Burton 1985; Gathman and Burton 2011). Invertebrate community structure is influenced strongly by structural composition of vegetation (e.g., Voigts 1976; McLaughlin and Harris 1990; Batzer and Resh 1992) as well as sediment characteristics (e.g., Nelson et al. 1990; Cooper et al. 2007). Therefore, maintenance of marsh vegetation structure and sediment characteristics by Great Lakes water levels has a strong influence on invertebrate community structure. Correlations between water levels, marsh vegetation, and invertebrate assemblage structure demonstrate these linkages (Burton et al. 2002; Uzarski et al. 2004; Gathman and Burton 2011).

Lakeshore marsh invertebrate communities appear to also be influenced directly by water levels. Cooper et al. (2014) evaluated a 1997–2012 time series of invertebrate community data from bulrush (*Schoenoplectus* spp.)-dominated habitats at three representative Saginaw Bay wetlands. Their analysis revealed substantial shifts in community structure throughout the period, especially from 2001 through 2004. This period followed a 1 m decline in Lake Huron water levels that occurred between 1997 and 2000. For example, from 2002 to 2004, gastropod relative abundance increased dramatically at all three wetlands, and at one wetland, gastropods increased from just 3 % of the community in 2002 to approximately half of the community in 2004. Over about the same period, insects—especially chironomids—declined substantially at all three wetlands. This decline was particularly evident at one wetland, where chironomids fell from roughly half to just 10 % of the community between 2002 and 2004. Crustaceans declined at all three wetlands beginning in 1999, reaching minima in 2002–2004. Coarse-level community metrics (e.g., % insects, % crustaceans, % gastropods, etc.) correlated with the prior year's water level suggesting a lagged response of communities to the water level decline. Burton et al. (2004) and Uzarski et al. (2004) noted that marsh invertebrate communities in Lake Huron changed surprisingly little during the 1998–2000 water level decline. This observation is consistent with the conclusion of Cooper et al. (2014) that the response to water level was delayed by one to several years following the major decline.

Gathman and Burton (2011) reported changes in invertebrate community structure in a Lake Huron marsh for a 3-year period in which water levels increased approximately 30 cm from year 1 (1996) to year 2 (1997) and then declined again in year 3 (1998). Sampling occurred at fixed stations along transects perpendicular to the shoreline. Multivariate analyses indicated that during the high-water period, assemblages became more homogenized (e.g., wet meadow assemblages resembled emergent marsh assemblages). This was driven by increased dominance by a subset of taxa throughout the marsh, especially *Caecidotea*, Chironomidae, Caenidae, and Amphipoda. Gathman and Burton (2011) also identified four categories of responses to water levels: (1) high-elevation specialists, which were generally restricted to the upper portion of the marsh regardless of the water level; (2) rapid, reversing taxa, which rapidly occupied the wet meadow under high-water conditions, but then retreated back to lower positions as water declined in year 3; (3) time-lagged responders, which expanded upslope as water levels rose, but in a time-lagged nature; and (4) low-elevation specialists, which always remained most common in the emergent marsh, showing little indication of upslope spread with rising water levels. Consistent with Cooper et al. (2014), Gathman and Burton (2011) demonstrate the profound influence of water levels on lakeshore marsh invertebrate communities and provide a framework for evaluating taxonomic responses to interannual water-level fluctuations.

Hydraulic Energy

An important driver of invertebrate community structure in lakeshore marshes is hydraulic energy and its influence on chemical, physical, and biological conditions. Wave energy affects shoreline vegetation by uprooting seedlings, damaging mature plants, and eroding fine sediments around roots and rhizomes (Keddy 1982; Riis and Hawes 2003). Accordingly, plant biomass and wave energy are negatively correlated along most vegetated shorelines, and a threshold exists where rooted vegetation can no longer persist (Keddy 1982; Azza et al. 2007). Effects of wave energy on sediment conditions are complex in lakeshore marshes since the plants themselves attenuate wave energy and affect sedimentation rates (Cooper et al. 2012). In general, however, increased wave energy results in increased particle size and decreased sediment organic content (Keddy 1982; Cooper et al. 2012), which influences basic biogeochemical conditions, including community metabolism (Cooper et al. 2013). Not surprisingly, therefore, exposure to wave and current energy influences faunal community structure in lakeshore marshes, especially for invertebrate communities (Burton et al. 2004; Cooper et al. 2014).

Cooper et al. (2014) analyzed invertebrate data from 67 lakeshore marshes from across the Great Lakes and found that fetch (i.e., potential wave energy) was one of the most important drivers of community structure among 16 candidate variables. The relationship between fetch and community structure was evident in whole-assemblage analyses, though a subset of taxa appeared to drive the observed gradients. For example, Oligochaeta and *Bezzia* were among the dominant taxa in the wave-exposed marshes of Saginaw Bay, while *Gammarus* and *Caecidotea* (both crustaceans) were much less abundant in Saginaw Bay compared to low-fetch marshes such as drowned river mouths of eastern Lake Michigan and protected embayments of northern Lake Huron. These results are largely consistent with Burton et al. (2004) who found that a majority of invertebrate taxa were generalists, occurring in wetlands across varying degrees of exposure, yet subsets of taxa were associated with either low-fetch or high-fetch marshes. Burton et al. (2004) reported higher densities of *Gammarus*, *Crangonyx*, *Caecidotea*, Chironomini, and Tanytarsini in low-fetch wetlands and higher densities of *Sigara*, *Trichocorixa*, Oligochaeta, and *Bezzia* in high-fetch wetlands, which partially overlaps with findings of Cooper et al. (2014).

Specific mechanisms linking wave exposure and invertebrate community structure are unclear; however, a combination of physical disturbance of organisms, the influence of wave energy on sediment organic matter, and the effect of wave-induced turbidity on visual predators are likely all important (Metzler and Sager 1986; Burton et al. 2004; Schneider and Sager 2007). Cooper et al. (2006, 2007) noted that sediment organic content was the best predictor of invertebrate community structure in drowned river mouth wetlands of eastern Lake Michigan. Similarly, MacKenzie et al. (2004) found that in the Peshtigo River wetland, a riverine wetland on the western shore of Lake Michigan, abundances of several invertebrate taxa varied predictably along gradients of sediment organic matter from the river channel into wetland vegetation. Taken collectively, these findings suggest that interactions between wave exposure or other hydraulic forces, sediment organic content, and turbidity are important in structuring invertebrate communities along gradients of hydrologic energy in lakeshore marshes.

Vegetation Zonation

Macrophytes comprise much of the physical habitat that invertebrate communities assemble within. Therefore, differences in vegetation, either different component species or different plant morphotypes, can influence invertebrate community structure. While few studies have investigated the influence of vegetation on invertebrate community structure directly, available evidence suggests that vegetation zonation plays a role in structuring these communities. Burton et al. (1999) suggested that stratifying invertebrate-based indices of biotic integrity by vegetation type would improve the performance of the index. This was later confirmed by Uzarski et al. (2004) who found that stratification by vegetation zone was indeed necessary to account for variation in habitat structure and to allow the index to be used at varying water levels as vegetation zones move upslope and downslope. However, given that vegetation communities are influenced by nearshore hydraulic forces (e.g., wave energy) and Great Lakes water levels, it has been difficult to partition these interacting drivers of invertebrate community structure. For example, Gathman and Burton (2011) found that invertebrate community composition was influenced more by flooding conditions than by vegetation, though vegetation structure is also influenced by flooding regime. Additional experimental research is needed to partition these influences, especially in regard to the effects of nonnative vegetation (e.g., *Phragmites australis* and *Typha* X *glauca*) on invertebrate communities.

Conservation and Management

Lakeshore marshes provide critical habitat for many species of birds, mammals, reptiles, and amphibians (Austen et al. 1994; Hecnar 2004; Hanowski et al. 2007; Wieten et al. 2012). These wetlands also provide essential spawning and nursery areas for many fish species of ecological and economic importance (Chubb and Liston 1986; Klarer and Millie 1992; Uzarski et al. 2005). Additionally, lakeshore marshes trap, process, and remove nutrients from Great Lakes nearshore waters, and their effects on drainage patterns can help recharge groundwater supplies (Burton 1985; Heath 1992). These functions reinforce the notion that conservation and restoration of lakeshore marshes are vital elements of long-term management of the Great Lakes (Sierszen et al. 2012b). Unfortunately, approximately half of the coastal wetland area that was present before European settlement has been converted to other land uses, especially in the lake plains of western Lake Erie and Saginaw Bay where large tracks of wetland were ditched and drained for agriculture and urban development. The majority of remaining wetlands are affected to varying degrees by numerous anthropogenic disturbances.

Water-Level Regulation

Outflow regulation of Lakes Superior and Ontario has altered water-level dynamics within these lakes. On Lake Ontario, sustained deviations from the overall mean water level are noticeably infrequent after the Moses–Saunders Power Dam began its operation in 1958 (Fig. 9.4). The range of fluctuations was approximately 2 m prior to regulation, but this has been reduced to approximately 1 m since regulation began. As a result, cattail (*Typha* spp.) stands spread dramatically in Lake Ontario's marshes, often replacing other more diverse habitat types such as sedge/grass meadow marsh (Wilcox et al. 2008). Invertebrates and other fauna that utilize the dynamic meadow marsh were undoubtedly affected by this change in hydrology and vegetation. Regulation of Lake Superior, which began in the early 1920s, had less of an effect on water levels and Lake Superior reached a near-record high in 1986 and a near-record low in 2007. However, regulation of Lake Superior outflow does dampen seasonal and interannual variability somewhat, with consequences to wetland habitat structure and resident fauna (Ciborowski et al. 2008). Given the importance of natural water-level fluctuations for maintenance of marsh vegetation community structure, it is critical that water-level management policies incorporate natural variation to the greatest extent possible (Ciborowski et al. 2008).

Anthropogenic Nutrient Pollution

Human-derived nitrogen (N) and phosphorus (P) enter aquatic ecosystems from point and nonpoint sources. Runoff from agricultural and urban landscapes is a common source of these nutrients to streams, lakes, and wetlands. Anthropogenic nutrients can impact lakeshore wetlands in dramatic ways, particularly by stimulating excessive primary production (i.e., eutrophication). This production can be in the form of phytoplankton or macrophytes, which can subsequently alter organic matter dynamics as the plants or algae senesce. Organic detritus can be a food resource for invertebrates but excessive organic matter accumulation can cause hypoxic or even anoxic conditions. Thus, nutrient loading has the potential to dramatically alter both the physical habitat and chemical conditions in lakeshore marshes, which can then impact invertebrate communities.

In lakeshore marshes, the response of invertebrates to anthropogenic nutrient loading is perhaps most apparent in relationships between surrounding land use and community structure. Cooper et al. (2014) found that while invertebrate community structure responded most strongly to hydrologic factors (e.g., wave energy, water levels), watershed percent agriculture was also highly correlated with community structure across the Great Lakes basin. These results suggest that at the Great Lakes basin-scale, invertebrate communities respond to the suite of impacts brought about by agricultural runoff, including nutrient loading. Marshes of Saginaw Bay and western Lake Erie, in particular, receive considerable nutrient loads from surrounding agricultural lands (Danz et al. 2007; Dolan and Chapra 2012; He et al. 2013). Others have reported similar relationships between watershed agriculture and coastal wetland invertebrates. For example, Burton et al. (1999) and Uzarski et al. (2004) identified coastal wetland invertebrate community shifts that correlated with surrounding land use in Lake Huron, including Saginaw Bay. Schneider and Sager (2007) reported that agriculturally derived nutrient and sediment loading to Green Bay (Lake Michigan) determined trophic state and light attenuation in Green Bay's coastal wetlands. They further proposed that these variables drove epiphytic invertebrate community structure by influencing food resources. Given the apparent impacts associated with nutrient-laden runoff on lakeshore marsh invertebrates, restoration and protection efforts should identify and ameliorate sources of nutrient pollution when designing projects and programs.

Invasive Species

One of the most serious threats to the biotic integrity of lakeshore marshes is the establishment and spread of nonnative organisms. Pathways of introduction include intentional release, the live bait trade, aquarium trade, ballast water of ships, escape from cultivation, and migration along human corridors such as highways and railroads where natural barriers would have existed otherwise. Because macrophytes form the physical habitat and influence organic matter dynamics in lakeshore marshes, nonnative plant invasions can be particularly detrimental to invertebrates. Examples of invasive macrophytes that dominate in marshes throughout the Great Lakes include submersed aquatic species such as Eurasian water milfoil (*Myriophyllum spicatum*), curly leaf pondweed (*Potamogeton crispus*), and slender naiad (*Najas minor*) and emergent plants such as purple loosestrife (*Lythrum salicaria*), reed canary grass (*Phalaris arundinacea*), and common reed (*Phragmites australis*). It is likely that many invasions around the Great Lakes would not have been successful in healthy marsh ecosystems. However, prior physical habitat disturbances, alteration of natural water-level regimes, and anthropogenic nutrient loading can facilitate the establishment and spread of nonnative plants.

Attempts to control invasive vegetation, especially common reed, often include glyphosate herbicides such as Roundup (Monsanto Corporation) or Glypro (Dow AgroSciences). Effects of these herbicides on non-macrophyte aquatic organisms are not straightforward. Some reports suggest that glyphosate is not harmful to aquatic invertebrates, fish, or algae (USDA 1997; Kulesza et al. 2008) while others show variable toxicity to these taxa (Chen et al. 2004; Relyea 2005). As the use of glyphosate herbicides to treat common reed and other invasive plants continues to increase in lakeshore marshes, additional research on the short- and long-term impacts to nontarget organisms is needed.

Direct impacts of invasive vegetation on invertebrates in lakeshore marshes are equally complex. For example, the extremely high density and biomass of common reed can reduce available nutrients and light and the accumulation of reed detritus can affect system hydrology, can cause sediment anoxia and phytotoxin buildup (e.g., hydrogen sulfide, acetic acid), and can kill the roots of native plants (Armstrong et al. 1996). These effects on native plant communities presumably would impact invertebrate communities as well (Schultz and Dibble 2012). However, Kulesza et al. (2008) and Holomuzki and Klarer (2010) found that *Phragmites* invasion did not adversely affect macroinvertebrate community density and diversity in Lake Erie marshes. Additional factors such as stand age, ambient water quality, and plant community composition prior to invasion likely all influence the degree of impact that nonnative vegetation has on macroinvertebrates.

Nonnative invertebrate species that are now commonly observed in Great Lakes marshes include zebra and quagga mussels (*Dreissena* spp.), rusty crayfish (*Orconectes rusticus*), faucet snails (*Bithynia tentaculata*), Chinese mystery snail (*Cipangopaludina chinensis*), and the amphipod, *Echinogammarus ischnus*. Zebra and quagga mussels are particularly detrimental to native unionids because they colonize unionid shells and outcompete them for food resources (Zanatta et al. 2002, 2015). Additional nonnative species are likely also common in Great Lakes marshes but cryptic identity at the species level impedes detection. Broadscale monitoring and archival of invertebrate collections is an invaluable tool for identifying and tracking range expansions of invertebrate invaders (Peters et al. 2014).

Lakeshore Marsh Restoration

A number of large lakeshore marsh restoration projects have been initiated in the Great Lakes (e.g., Sensiba Wildlife Area in western Green Bay, Cat Island Ecosystem in southern Green Bay, Erie Marsh Preserve in western Lake Erie, Braddock Bay in southern Lake Ontario). Many smaller-scale restoration efforts have also occurred or are planned throughout the basin, often focused on controlling invasive vegetation or reconnecting marsh habitats to the Great Lakes after previous activities such as diking or coastal development has isolated marsh fragments. Significant investment of public and private funds for both large- and small-scale marsh restoration has been made in the region because long-term benefits provided by healthy marshes are believed to outweigh short-term restoration costs.

Effective restoration planning and evaluation require monitoring, and invertebrates can provide important ecological information for this purpose. In Great Lakes marshes, an unprecedented basin-scale ecosystem monitoring program, which includes sampling invertebrates, fish, vegetation, birds, and amphibians, began in 2011 with approximately 1000 marshes scheduled for sampling in the first 5-year rotation. The primary goal of the program is to generate data to prioritize wetland restoration projects and track restoration outcomes. This innovative and strategic approach is being led by the US Environmental Protection Agency with a consortium of wetland researchers who utilize Indices of Biotic Integrity (IBI) and other similar measures to estimate conditions within each wetland. Habitats are targeted for restoration or protection based on the monitoring data. Restoration outcomes are then evaluated over the long term by resampling the restored habitats in subsequent years. A secondary goal of the monitoring program, therefore, is to support adaptive management of restoration techniques as post-restoration monitoring reveals successful and unsuccessful approaches as the biotic communities respond to the restoration activities.

Summary and Conclusions

Lakeshore marshes in the Great Lakes provide habitat for a vast array of invertebrate taxa (well over 400 genera; see "Appendix"). The marshes themselves represent incredible variability in terms of climate, geomorphic types, dominant vegetation, and nutrient conditions. Despite the large number of taxa observed in these habitats, a small subset of taxa tend to be numerically dominant in the overall assemblage, with the ten most abundant taxa representing over 60 % of the organisms collected in basin-scale monitoring programs. Accordingly, most taxa could be considered "rare," resulting in a very hollow species-abundance curve. Invertebrate community structure is driven by a combination of both natural and anthropogenic factors. Important natural drivers include hydrology (e.g., lake water levels), hydraulic forces (e.g., wave energy), and vegetation zonation. Important anthropogenic factors include water quality (e.g., nutrient runoff) and invasive vegetation. Therefore, reducing anthropogenic nutrient loading and controlling the spread of invasive species are important conservation practices. While these marsh-scale impacts warrant attention by managers and policy-makers, the most significant insult to lakeshore marsh invertebrates is the loss of habitat. Approximately half of the original marsh area along the Great Lakes coast has been lost to human development, with even greater losses in some areas. Therefore, restoring severely degraded and previously destroyed marshes and protecting existing intact marshes are key strategies to ensuring the integrity of the Great Lakes coastal ecosystem. Invertebrates inhabiting these critical habitats represent an important nexus as they link primary productivity to higher trophic levels (e.g., fish and waterfowl), facilitate the cycling of wetland nutrients, and provide wetland managers with vital information on ecosystem health.

Appendix

Invertebrate taxa collected as part of the Great Lakes Coastal Wetlands Consortium basin-wide monitoring program (2011–2013). Wetlands were located on Lake Erie (LE), Lake Huron (LH), Lake Michigan (LM), Lake Ontario (LO), and Lake Superior (LS)

Phylum Class **Order** Family	Subfamily, tribe, *genus*, *species*	LE	LH	LM	LO	LS
Cnidaria						
Hydrozoa		X	X	X	X	X
Anthoathecatae		X	X	X	X	X
Hydridae	*Hydra*	X	X	X	X	X
Nematoda		X	X	X	X	X
Nematomorpha		X	X		X	
Platyhelminthes						
Turbellaria		X	X	X	X	X
Annelida						
Clitellata[a]		X	X	X	X	X
Hirudinea		X	X	X	X	X
Euhirudinea[b]		X	X	X	X	X
Arhynchobdellida		X	X	X	X	X
Erpobdellidae		X	X	X	X	X
	Erpobdella	X	X	X	X	
	Mooreobdella	X	X	X	X	X
Rhynchobdellida		X	X	X	X	X
Glossiphoniidae		X	X	X	X	X
	Batracobdella		X	X	X	X
	Desserobdella	X	X		X	X
	Gloiobdella	X	X			
	Glossiphonia	X	X		X	X
	Helobdella	X	X	X	X	X
	Marvinmeyeria		X			
	Placobdella	X	X	X	X	X
	Theromyzon	X	X		X	
Piscicolidae		X	X			X
	Myzobdella					X
Oligochaeta		X	X	X	X	X
Haplotaxida			X	X	X	X
Naididae			X	X	X	X
Tubificidae			X	X		
Lumbriculida					X	
Polychaeta[a]					X	X
Canalipalpata					X	
Sabellidae	*Manayunkia*				X	
Mollusca						
Bivalvia		X	X	X	X	X
Unionoida		X	X			
Veneroida		X	X	X	X	X

Phylum						
Class						
Order						
Family	Subfamily, tribe, *genus*, *species*	LE	LH	LM	LO	LS
Dreissenidae	*Dreissena*	X	X	X	X	X
Sphaeriidae		X	X	X	X	X
	Musculium	X	X	X	X	X
	Pisidium	X	X	X	X	X
	Sphaerium	X	X		X	X
Gastropoda		X	X	X	X	X
Architaenioglossa		X	X	X	X	X
Viviparidae		X	X	X	X	X
	Campeloma		X	X	X	
	Cipangopaludina	X	X	X		X
	Viviparus	X	X	X		X
Basommatophora		X	X	X	X	X
Ancylidae	*Ancylini*	X	X	X	X	X
	Ferrissia		X	X	X	X
	Laevapex	X	X	X	X	X
Lymnaeidae		X	X	X	X	X
	Acella		X		X	
	Bulimnaea		X			X
	Fossaria	X	X	X	X	X
	Lymnaea	X	X	X	X	X
	Lymnaea		X		X	X
	Pseudosuccinea	X	X	X	X	X
	Stagnicola	X	X	X	X	X
Physidae		X	X	X	X	X
	Aplexa		X	X	X	
	Physa	X	X	X	X	X
Planorbidae		X	X	X	X	X
	Armiger		X	X	X	
	Gyraulus	X	X	X	X	X
	Helisoma	X	X	X	X	X
	Menetus		X	X	X	X
	Planorbella	X	X	X	X	X
	Planorbula	X	X	X	X	X
	Promenetus	X	X	X	X	X
Heterostropha		X	X	X	X	X
Valvatidae	*Valvata*	X	X	X	X	X
Mesogastropoda			X	X	X	X
Pomatiopsidae	*Pomatiopsis*		X	X	X	X
Neotaenioglossa						
Bithyniidae		X	X	X	X	X
	Bithynia tentaculata	X	X	X	X	X

(continued)

(continued)

Phylum Class Order Family	Subfamily, tribe, *genus*, *species*	LE	LH	LM	LO	LS
Hydrobiidae		X	X	X	X	X
	Amnicola	X	X	X	X	X
Pleuroceridae		X	X	X	X	X
	Elimia			X	X	
	Goniobasis	X	X		X	
	Pleurocera	X	X	X	X	X
Stylommatophora		X	X	X	X	X
Succineidae	*Succinea*	X	X	X	X	X
Arthropoda						
Arachnida		X	X	X	X	X
Acari[c]		X	X	X	X	X
Malacostraca		X	X	X	X	X
Decapoda		X	X	X	X	X
Cambaridae		X	X	X	X	X
	Cambarus		X	X	X	X
	Orconectes	X	X	X	X	X
Palaemonidae	*Palaemonetes*	X	X	X	X	X
Amphipoda		X	X	X	X	X
Crangonyctidae	*Crangonyx*	X	X	X	X	X
Gammaridae		X	X	X	X	X
	Echinogammarus	X	X	X	X	X
	Gammarus	X	X	X	X	X
Dogielinotidae	*Hyalella azteca*	X	X	X	X	X
Isopoda		X	X	X	X	X
Asellidae		X	X	X	X	X
	Caecidotea	X	X	X	X	X
	Lirceus	X	X	X	X	X
Entognatha		X	X	X	X	X
Collembola		X	X	X	X	X
Isotomidae			X	X	X	
Poduridae	*Podura*		X	X		
Insecta		X	X	X	X	X
Ephemeroptera		X	X	X	X	X
Ameletidae	*Ameletus*		X			
Baetidae		X	X	X	X	X
	Acentrella		X			
	Acerpenna	X				
	Baetis		X	X		X
	Callibaetis	X	X	X	X	X
	Centroptilum	X	X	X	X	X
	Cloeon	X	X	X	X	X
	Procloeon		X	X	X	X
	Pseudocloeon	X		X		

Phylum						
Class						
Order						
Family	Subfamily, tribe, *genus*, *species*	LE	LH	LM	LO	LS
Baetiscidae	*Baetisca*		X		X	X
Caenidae		X	X	X	X	X
	Brachycerus	X	X	X	X	X
	Caenis	X	X	X	X	X
Ephemerellidae		X	X	X	X	X
	Attenella		X			
	Drunella			X		
	Eurylophella		X	X	X	X
	Serratella		X			
	Timpanoga		X			
Ephemeridae		X	X	X	X	X
	Ephemera	X	X	X	X	X
	Hexagenia	X	X	X	X	X
Heptageniidae		X	X	X	X	X
	Macdunnoa		X			
	Stenacron		X			X
	Stenonema		X	X	X	X
Isonychiidae	*Isonychia*		X			
Leptohyphidae	*Tricorythodes*		X		X	X
Leptophlebiidae			X	X		
	Choroterpes			X		
	Leptophlebia		X			
Metretopodidae	*Siphloplecton*					X
Neoephemeridae	*Neoephemera*			X		
Tricorythidae					X	
Odonata		X	X	X	X	X
Anisoptera[d]		X	X	X	X	X
Aeshnidae		X	X	X	X	X
	Aeshna	X	X	X	X	X
	Anax	X	X	X	X	X
	Basiaeschna		X	X		X
	Boyeria		X	X	X	X
Corduliidae		X	X	X	X	X
	Cordulia		X	X	X	X
	Dorocordulia		X	X	X	X
	Epitheca	X	X	X	X	X
	Neurocordulia		X	X		X
	Somatochlora		X	X	X	X

(continued)

(continued)

Phylum Class **Order** Family	Subfamily, tribe, *genus*, *species*	LE	LH	LM	LO	LS
Gomphidae		X	X	X	X	X
	Arigomphus	X			X	
	Dromogomphus			X		
	Gomphus		X	X	X	X
	Hagenius		X			X
	Stylurus					X
Libellulidae		X	X	X	X	X
	Celithemis		X		X	X
	Erythemis	X	X	X	X	X
	Ladona		X			X
	Leucorrhinia	X	X	X	X	X
	Libellula	X	X	X	X	X
	Miathyria		X		X	
	Pantala		X		X	X
	Perithemis	X			X	
	Plathemis		X	X	X	
	Tramea	X	X	X	X	X
Macromiidae			X	X	X	X
	Macromia		X	X		
	Didymops		X		X	X
Zygoptera[d]		X	X	X	X	X
Calopterygidae	*Calopteryx*		X			
Coenagrionidae		X	X	X	X	X
	Amphiagrion		X			
	Argia		X		X	
	Chromagrion		X		X	X
	Coenagrion		X	X	X	
	Enallagma	X	X	X	X	X
	Ischnura	X	X	X	X	X
	Nehalennia	X	X	X		X
Lestidae	*Lestes*	X	X	X	X	X
Plecoptera		X			X	X
Chloroperlidae						X
Perlidae		X			X	
	Neoperla				X	
	Perlesta	X				
Hemiptera		X	X	X	X	X
Belostomatidae		X	X	X	X	X
	Belostoma	X	X	X	X	X
	Lethocerus			X		X

Phylum						
Class						
Order						
Family	Subfamily, tribe, *genus*, *species*	LE	LH	LM	LO	LS
Corixidae		X	X	X	X	X
	Callicorixa		X	X	X	X
	Corisella	X	X	X		X
	Dasycorixa				X	
	Hesperocorixa	X	X	X	X	X
	Neocorixa		X	X	X	
	Palmacorixa	X	X	X	X	X
	Sigara	X	X	X	X	X
	Trichocorixa	X	X	X	X	X
Gerridae		X	X	X	X	X
	Aquarius	X		X		X
	Gerris	X	X	X	X	X
	Limnoporus	X	X	X	X	X
	Metrobates		X		X	X
	Rheumatobates		X	X	X	X
	Trepobates	X	X	X	X	X
Hebridae		X	X	X	X	X
	Hebrus	X	X	X	X	X
	Lipogomphus	X	X	X	X	
	Merragata	X	X	X	X	X
Hydrometridae	*Hydrometra*	X	X	X	X	X
Macroveliidae			X			X
	Macrovelia	X	X			
Mesoveliidae	*Mesovelia*	X	X	X	X	X
Naucoridae	*Pelocoris*	X	X	X	X	
Nepidae	*Ranatra*	X	X	X	X	X
Notonectidae		X	X	X	X	X
	Buenoa	X	X	X	X	X
	Notonecta	X	X	X	X	X
Pleidae	*Neoplea*	X	X	X	X	X
Veliidae		X	X	X	X	X
	Microvelia	X	X	X	X	X
	Steinovelia	X				
Saldidae	*Pentacora*		X			
	Rupisalda		X			
Coleoptera		X	X	X	X	X
Anthicidae		X	X			
Chrysomelidae		X	X	X	X	X
	Donaciinae		X			
Curculionidae		X	X	X	X	X
	Bagous		X		X	
	Lixellus		X			

(continued)

(continued)

Phylum						
Class						
Order						
Family	Subfamily, tribe, *genus*, *species*	LE	LH	LM	LO	LS
Curculionoidea			X			
Dytiscidae		X	X	X	X	X
	Acilius		X	X	X	X
	Agabetes	X	X		X	
	Agabus	X	X	X	X	X
	Celina	X	X	X	X	X
	Colymbetes				X	
	Copelatus	X	X	X	X	
	Coptotomus		X		X	X
	Desmopachria	X	X	X	X	X
	Dytiscus		X	X	X	X
	Graphoderus					X
	Hydaticus	X	X	X	X	X
	Hydroporinae	X	X	X	X	X
	Hydroporus	X	X	X	X	X
	Hydrovatus	X	X	X	X	X
	Hygrotus	X	X	X	X	X
	Ilybius	X	X	X	X	X
	Laccophilus	X	X	X	X	X
	Liodessus	X	X	X	X	X
	Matus	X	X		X	X
	Nebrioporus					X
	Neoporus		X	X	X	X
	Neoscutopterus		X			
	Oreodytes		X		X	
	Rhantus	X				X
	Sanfilippodytes		X			
	Uvarus	X		X		X
Elmidae		X	X	X	X	X
	Dubiraphia	X	X		X	X
	Macronychus				X	X
	Optioservus			X	X	X
	Ordobrevia		X			
	Promoresia				X	
	Stenelmis			X	X	X
Georyssidae	*Georyssus*		X			X
Gyrinidae		X	X	X	X	X
	Dineutus	X	X	X	X	X
	Gyretes			X		X
	Gyrinus	X	X	X	X	X

Phylum						
Class						
Order						
Family	Subfamily, tribe, *genus*, *species*	LE	LH	LM	LO	LS
Haliplidae		X	X	X	X	X
	Haliplus	X	X	X	X	X
	Peltodytes	X	X	X	X	X
Helophoridae	*Helophorus*	X	X	X	X	X
Hydraenidae	*Hydraena*	X	X			X
Hydrochidae	*Hydrochus*	X	X		X	X
Hydrophilidae		X	X	X	X	X
	Anacaena	X	X		X	X
	Berosus	X	X	X	X	X
	Crenitis	X	X	X		X
	Cymbiodyta	X				X
	Enochrus	X	X	X	X	X
	Helochares					X
	Helocombus		X			X
	Hydrobius	X	X	X	X	X
	Hydrochara		X		X	
	Hydrophilus		X	X	X	
	Laccobius	X	X	X		X
	Paracymus	X	X	X	X	X
	Sperchopsis		X			
	Sphaeridiinae	X	X			X
	Tropisternus	X	X	X	X	X
Lampyridae		X	X		X	X
Noteridae	*Hydrocanthus*	X	X	X	X	X
Ptilodactylidae			X	X	X	
	Anchytarsus		X	X		
Scirtidae		X	X	X	X	X
	Cyphon		X	X		
	Elodes					X
	Prionocyphon		X	X		
	Sarabandus		X			
	Scirtes		X			X
Staphylinidae			X	X	X	X
Staphylinoidea[c]			X			X
Neuroptera		X	X	X	X	X
Sisyridae						X
Corydalidae		X	X	X	X	
	Chauliodes	X	X	X	X	
	Nigronia		X			
Sialidae	*Sialis*	X	X	X	X	X

(continued)

(continued)

Phylum Class Order Family	Subfamily, tribe, *genus*, *species*	LE	LH	LM	LO	LS
Trichoptera		X	X	X	X	X
Apataniidae	*Apatania*					X
Brachycentridae	*Brachycentrus*	X	X			
Dipseudopsidae	*Phylocentropus*		X	X		X
Helicopsychidae	*Helicopsyche*	X	X	X	X	X
Hydropsychidae	*Arctopsychinae*			X		
Hydroptilidae		X	X	X	X	X
	Agraylea	X	X	X	X	X
	Hydroptila	X	X	X	X	X
	Neotrichia		X			
	Ochrotrichia	X	X	X	X	
	Orthotrichia	X	X	X	X	X
	Oxyethira	X	X	X	X	X
Lepidostomatidae	*Lepidostoma*		X			X
Leptoceridae		X	X	X	X	X
	Ceraclea	X	X	X	X	X
	Leptocerus	X	X	X	X	X
	Mystacides	X	X	X	X	X
	Nectopsyche	X	X	X	X	X
	Oecetis	X	X	X	X	X
	Setodes	X	X			
	Triaenodes	X	X	X	X	X
	Ylodes		X		X	
Limnephilidae		X	X	X	X	X
	Glyphopsyche		X			X
	Limnephilus		X	X	X	X
	Nemotaulius	X				
	Onocosmoecus				X	
	Psychoglypha		X			
Molannidae			X	X		X
	Molanna		X	X		X
Phryganeidae			X	X	X	X
	Agrypnia					X
	Banksiola		X			X
	Fabria			X		X
	Phryganea		X	X	X	X
Polycentropodidae		X	X	X	X	X
	Cernotina		X	X	X	X
	Neureclipsis	X	X	X		X
	Nyctiophylax					X
	Polycentropus		X	X	X	X

Phylum						
Class						
Order						
Family	Subfamily, tribe, *genus*, *species*	LE	LH	LM	LO	LS
Lepidoptera		X	X	X	X	X
Crambidae		X	X	X	X	X
	Acentria	X	X	X	X	X
	Nymphuliella					X
	Nymphulini				X	
	Petrophila		X			
	Synclita		X	X		X
Noctuidae			X	X		X
	Bellura			X		X
Pyralidae			X	X	X	X
Diptera		X	X	X	X	X
Athericidae	*Atherix*				X	
Ceratopogonidae		X	X	X	X	X
	Leptoconops		X			
	Alluaudomyia		X	X		X
	Atrichopogon		X	X	X	X
	Bezzia	X	X	X	X	X
	Ceratopogon	X	X	X	X	
	Culicoides	X	X	X	X	X
	Dasyhelea			X		X
	Mallochohelea					X
	Palpomyia		X			
	Probezzia	X	X	X	X	X
	Serromyia		X		X	X
	Sphaeromias			X	X	X
	Stilobezzia	X			X	
Chaoboridae		X	X		X	X
	Chaoborus		X			
	Eucorethra		X			
Chironomidae		X	X	X	X	X
	Chironominae	X	X	X	X	X
	Chironomini	X	X	X	X	X
	Tanytarsini	X	X	X	X	X
	Pseudochironomini	X	X	X	X	X
	Orthocladiinae	X	X	X	X	X
	Podonominae	X	X	X	X	X
	Prodiamesinae				X	
	Tanypodinae	X	X	X	X	X
	Coelotanypodini		X			
	Pentaneurini		X	X		
	Tanypodini					X

(continued)

(continued)

Phylum / Class / **Order** / Family	Subfamily, tribe, *genus*, *species*	LE	LH	LM	LO	LS
Culicidae		X	X	X	X	X
	Aedes/Ochlerotatus		X			
	Anopheles	X	X	X	X	X
	Culex	X	X		X	
Dixidae			X			X
	Dixella	X	X	X		X
Dolichopodidae		X	X			X
Empididae		X	X	X	X	X
	Hemerodromia			X	X	X
Ephydridae		X	X	X	X	X
	Ephydra		X			
Phoridae						X
Psychodidae		X	X	X		
	Maruina			X		
	Pericoma		X			
Ptychopteridae			X		X	X
	Bittacomorpha					X
	Ptychoptera		X		X	
Sarcophagidae		X				
Sciomyzidae		X	X	X	X	X
Stratiomyidae		X	X	X	X	X
	Caloparyphus		X			
	Myxosargus	X				
	Odontomyia		X	X	X	
	Hedriodiscus	X	X	X	X	X
	Stratiomys		X		X	
Tabanidae		X	X	X	X	X
	Chrysops	X	X	X	X	X
	Tabanus	X				
Tipulidae		X	X	X	X	X
	Antocha		X			X
	Dicranota			X		
	Erioptera		X			
	Helius	X	X		X	X
	Hexatoma				X	
	Limnophila	X				
	Pilaria		X			
	Prionocera		X			
	Tipula	X	X			X
	Ormosia	X			X	X

[a]Subphylum
[b]Infraclass
[c]Subclass
[d]Suborder

References

Albert DA, Wilcox DA, Ingram JW, Thompson TA (2005) Hydrogeomorphic classification for Great Lakes coastal wetlands. J Gt Lakes Res 31:126–146

Armstrong J, Afreen-Zobayed F, Armstrong W (1996) *Phragmites* die-back: sulphide- and acetic acid-induced bud and root death, lignifications, and blockages with aeration and vascular systems. New Phytol 134:601–614

Austen DJ, Bayley PB, Menzel BW (1994) Importance of the guild concept to fisheries research management. Fisheries 19:12–20

Azza N, van de Koppel J, Denny P, Kansiime F (2007) Shoreline vegetation distribution in relation to wave exposure and bay characteristics in a tropical great lake, Lake Victoria. Trop Ecol 23:353–360

Batzer DP, Resh VH (1992) Macroinvertebrates of a California seasonal wetland and responses to experimental habitat manipulation. Wetlands 12:1–7

Brazner JC, Campana SE, Tanner DK (2004) Habitat fingerprints for Lake Superior coastal wetlands derived from elemental analysis of yellow perch otoliths. Trans Am Fish Soc 133:692–704

Burton TM (1985) The effects of water level fluctuations on Great Lakes coastal marshes. In: Prince HH, D'Itri FM (eds) Coastal wetlands: proceedings of the first great lakes coastal wetlands colloquium. Lewis, Chelsea, pp 3–13

Burton TM, Uzarski DG (2009) Biodiversity in protected coastal wetlands along the west coast of Lake Huron. Aquat Ecosys Health Manage 12:63–76

Burton TM, Uzarski DG et al (1999) Development of a preliminary invertebrate index of biotic integrity for Lake Huron coastal wetlands. Wetlands 19:869–882

Burton TM, Stricker CA, Uzarski DG (2002) Effects of plant community composition and exposure to wave action on invertebrate habitat use of Lake Huron coastal wetlands. Lakes Reserv Res Manage 7:255–269

Burton TM, Uzarski DG, Genet JA (2004) Invertebrate habitat use in relation to fetch and plant zonation in northern Lake Huron coastal wetlands. Aquat Ecosys Health Manage 7:249–267

Cardinale BJ, Burton TM, Brady VJ (1997) The community dynamics of epiphytic midge larvae across the pelagic-littoral interface: do animals respond to changes in the abiotic environment? Can J Fish Aquat Sci 54:2314–2322

Cardinale B, Brady V, Burton T (1998) Changes in the abundance and diversity of coastal wetland fauna from the open water/macrophyte edge towards shore. Wetl Ecol Manag 6:59–68

Chen CY, Hathaway KM, Folt CL (2004) Multiple stress effects of Vision herbicide, pH, and food on zooplankton and larval amphibian species from forest wetlands. Environ Toxicol Chem 23:823–831

Chubb SL, Liston CR (1986) Density and distribution of larval fishes in Pentwater Marsh, a coastal wetland on Lake Michigan. J Gt Lakes Res 12:332–343

Ciborowski JJH, Niemi GJ et al (2008) Ecosystem effects of water level changes in the upper great lakes. International Joint Commission Document

Cooper MJ, Uzarski DG, Burton TM, Rediske RR (2006) Macroinvertebrate community composition relative to chemical/physical variables, land use and cover, and vegetation types within a Lake Michigan drowned river mouth wetland. Aquat Ecosys Health Manage 9:463–479

Cooper MJ, Uzarski DG, Burton TM (2007) Macroinvertebrate community composition in relation to anthropogenic disturbance, vegetation, and organic sediment depth in four Lake Michigan drowned river-mouth wetlands. Wetlands 27:894–903

Cooper MJ, Gyekis KF, Uzarski DG (2012) Edge effects on abiotic conditions, zooplankton, macroinvertebrates, and larval fishes in Great Lakes fringing marshes. J Gt Lakes Res 38:142–151

Cooper MJ, Steinman AD, Uzarski DG (2013) Influence of geomorphic setting on the metabolism of Lake Huron fringing wetlands. Limnol Oceanogr 58:452–464

Cooper MJ, Lamberti GA, Uzarski DG (2014) Spatial and temporal trends in invertebrate communities of Great Lakes coastal wetlands, with emphasis on Saginaw Bay of Lake Huron. J Gt Lakes Res 40(Suppl 1):168–182

Danz NP, Niemi GJ et al (2007) Integrated gradients of anthropogenic stress in the U.S. Great Lakes basin. Environ Manage 39:631–647

Dolan DM, Chapra SC (2012) Great Lakes total phosphorus revisited: 1. Loading analysis and update (1994–2008). J Gt Lakes Res 38:730–740

Gathman JP, Burton TM (2011) A Great Lakes coastal wetland invertebrate community gradient: relative influence of flooding regime and vegetation zonation. Wetlands 31:329–341

Gathman JP, Albert DA, Burton TM (2005) Rapid plant community response to a water level peak in northern Lake Huron coastal wetlands. J Gt Lakes Res 31:160–170

Hanowski J, Danz N et al (2007) Consideration of geography and wetland geomorphic type in the development of Great Lakes coastal wetland bird indicators. Ecohealth 4:194–205

Harris HJ, Milligan MS, Fewless GA (1983) Diversity: quantification and ecological evaluation in freshwater marshes. Bio Conserv 27:99–110

He C, DeMarchi C, Tao W, Johengen TH (2013) Modeling distribution of point and nonpoint sources pollution loadings in the Saginaw Bay watersheds, Michigan. Geospat Tools Urb Water Resour 7:97–113

Heath RT (1992) Nutrient dynamics in Great Lakes coastal wetlands: future directions. J Gt Lakes Res 18(4):590–602

Hecnar SJ (2004) Great Lakes wetlands as amphibian habitats: a review. Aquat Ecosys Health Manage 7:289–303

Holomuzki JR, Klarer DM (2010) Invasive reed effects on benthic community structure in Lake Erie coastal marshes. Wetl Ecol Manag 18:219–231

Jude DJ, Pappas J (1992) Fish utilization of Great Lakes coastal wetlands. J Gt Lakes Res 18:651–672

Keddy PA (1982) Quantifying within-lake gradients of wave energy: interrelationships of wave energy, substrate, particle size, and shoreline plants in Axe Lake, Ontario. Aquat Bot 14:41–58

Keough JR, Thompson TA, Guntenspergen GR, Wilcox DA (1999) Hydrogeomorphic factors and ecosystem responses in coastal wetlands of the Great Lakes. Wetlands 19:821–834

King RS, Brazner JC (1999) Coastal wetland insect communities along a trophic gradient in Green Bay, Lake Michigan. Wetlands 19:426–437

Klarer DM, Millie DF (1992) Aquatic macrophytes and algae at Old Woman Creek estuary and other Great Lakes coastal wetlands. J Gt Lakes Res 18:622–633

Kovalenko KE, Brady JV et al (2014) Congruence of community thresholds in response to anthropogenic stress in Great Lakes coastal wetlands. Freshw Sci 33:958–971

Kulesza AE, Holomuzki JR, Klarer DM (2008) Benthic community structure in stands of *Typha angustifolia* and herbicide-treated and untreated *Phragmites australis*. Wetlands 28:40–56

MacKenzie RA, Kaster JL, Klump JV (2004) The ecological patterns of benthic invertebrates in a Great Lakes coastal wetland. J Gt Lakes Res 30:58–69

Mayer T, Edsall T, Munawar M (2004) Factors affecting the evolution of coastal wetlands of the Laurentian Great Lakes: an overview. Aquat Ecosys Health Manage 7:171–178

Maynard L, Wilcox DA (1997) Coastal wetlands of the Great Lakes. State of the Lakes Ecosystem Conference 1996. Environment Canada and US EPA Report, 905-R-97-015b

McLaughlin DB, Harris HJ (1990) Aquatic insect emergence in two Great Lakes marshes. Wetl Ecol Manag 1:111–121

Merritt RW, Benbow ME, Hudson PL (2002) Wetland macroinvertebrates of Prentiss Bay, Lake Huron, Michigan: diversity and functional group composition. Gt Lakes Entomol 35:149–160

Metzler G, Sager PE (1986) A preliminary study of the macrobenthos of wave-swept and protected sites on the Lake Michigan shoreline at Toft Point Natural Area, Wisconsin. Trans Wis Acad Sci Arts Lett 74:126–132

Nelson JW, Kadlec JA, Murkin HR (1990) Responses by macroinvertebrates to cattail litter quality and timing of litter submergence in a northern prairie marsh. Wetlands 10:47–60

Niemi GJ, Brady VJ et al (2009) Development of ecological indicators for the U.S. Great Lakes coastal region: a summary of applications in Lake Huron. Aquat Ecosys Health Manage 12:77–89

Patrick CJ, Cooper MJ, Uzarski DG (2014) Dispersal mode and ability affect the spatial turnover of a wetland invertebrate metacommunity. Wetlands 34:1133–1143

Peters JA, Cooper MJ et al (2014) Historical changes and current status of crayfish diversity and distribution in the Laurentian Great Lakes. J Gt Lakes Res 40:35–46

Prince HH, Padding PI, Knapton RW (1992) Waterfowl use of the Laurentian Great Lakes. J Gt Lakes Res 18:673–699

Relyea RA (2005) The impact of insecticides and herbicides on the biodiversity and productivity of aquatic communities. Ecol Appl 15:618–627

Riis T, Hawes I (2003) Effect of wave exposure on vegetation abundance, richness and depth distribution of shallow water plants in a New Zealand lake. Freshw Biol 48:75–87

Schneider P, Sager PE (2007) Structure and ordination of epiphytic invertebrate communities of four coastal wetlands in Green Bay, Lake Michigan. J Gt Lakes Res 33:342–357

Schock NT, Murry BA, Uzarski DG (2014) Impacts of agricultural drainage outlets on Great Lakes coastal wetlands. Wetlands 34:297–307

Schultz R, Dibble E (2012) Effects of invasive macrophytes on freshwater fish and macroinvertebrate communities: the role of invasive plant traits. Hydrobiologia 684:1–14

Sierszen ME, Brazner JC et al (2012a) Watershed and lake influences on the energetic base of coastal wetland food webs across the Great Lakes Basin. J Gt Lakes Res 38:418–428

Sierszen ME, Morrice JA, Trebitz AS, Hoffman JC (2012b) A review of selected ecosystem services provided by coastal wetlands of the Laurentian Great Lakes. Aquat Ecosys Health Manage 15:92–106

Trebitz AS (2006) Characterizing seiche and tide-driven daily water level fluctuations affecting coastal ecosystems of the Great Lakes. J Gt Lakes Res 32:102–116

USDA (1997) Glyphosate: herbicide information profile. US Forest Service, Pacific Northwest Region

Uzarski DG, Burton TM, Genet JA (2004) Validation and performance of an invertebrate index of biotic integrity for Lakes Huron and Michigan fringing wetlands during a period of lake level decline. Aquat Ecosys Health Manage 7:269–288

Uzarski DG, Burton TM et al (2005) Fish habitat use within and across wetland classes in coastal wetlands of the five Great Lakes: development of a fish-based index of biotic integrity. J Gt Lakes Res 31:171–187

Uzarski DG, Burton TM, Kolar RE, Cooper MJ (2009) The ecological impacts of fragmentation and vegetation removal in Lake Huron's coastal wetlands. Aqu Ecos Health and Manage 12:45–62

Voigts DK (1976) Aquatic invertebrate abundance in relation to changing marsh vegetation. Am Midl Nat 95:313–322

Weeber RC, Vallianatos M (2000) The Marsh Monitoring Program 1995–1999: monitoring Great Lakes wetlands and their amphibian and bird inhabitants. Bird Studies Canada, Port Rowan, ON

Wieten AC, Cooper MJ, Parker AD, Uzarski DG (2012) Great Lakes coastal wetland habitat use by seven turtle species: influences of wetland type, vegetation, and abiotic conditions. Wetl Ecol Manag 20:47–58

Wilcox DA (1995) Wetland and aquatic macrophytes as indicators of anthropogenic hydrologic disturbance. Nat Areas J 15:240–248

Wilcox DA, Kowalski KP et al (2008) Cattail invasion of sedge/grass meadows and regulation of Lake Ontario water levels: photointerpretation analysis of sixteen wetlands over five decades. J Gt Lakes Res 34:301–323

Zanatta DT, Mackie GL, Metcalfe-Smith JL, Woolnough DA (2002) A refuge for native freshwater mussels (Bivalvia: Unionidae) from impacts of the exotic Zebra Mussel (*Dreissena polymorpha*) in Lake St. Clair. J Gt Lakes Res 28:479–489

Zanatta DT, Bossenbroek JM et al (2015) Distribution of native mussel (Unionidae) assemblages in coastal areas of Lake Erie, Lake St. Clair, and connecting channels, twenty-five years after a dreissenid invasion. NE Natur 22:223–235

Chapter 10
Invertebrates of the Florida Everglades

Joel C. Trexler and William F. Loftus

Introduction

The Everglades is a large karstic wetland located at the southern tip of the Florida, USA, peninsula between the subtropical Western Atlantic and the Gulf of Mexico. Lying within the subtropics, between 25.3°N and 26.7°N, the Everglades experiences mild winters (the average temperature of all months exceeds 17.8 °C) because of the strong influence of the northerly flow of the Florida Current (Gulf Stream) to the east. However, winter temperatures in the interior Everglades are not moderated by the Florida Current and may experience occasional freezes with important effects on the biota, particularly those species derived from the tropics (Duever et al. 1994; Matich and Heithaus 2012; Boucek and Rehage 2014). The ecosystem experiences a seasonal rainfall regime, similar to other tropical/subtropical wetlands that drive annual hydrological cycles, with marked inter-annual variation in the magnitude of marsh drying (Duever et al. 1994). More than 75 % of the annual rainfall of 152 cm is delivered between May and October, the wet season in south Florida, with marked inter-annual variation in the annual deposition resulting from regional climatic drivers (Gaiser et al. 2012).

Peat dating has revealed that the ecosystem is relatively young (approximately 5000 years in its current wetland form). The karstic geology and surficial aquifer yield relatively hard water with high dissolved calcium carbonate (Gleason and Stone 1994), which binds phosphorus and renders the system naturally oligotrophic (McCormick et al. 2002).

The Everglades is recognized widely as a globally iconic ecosystem worthy of conservation, but that has been threatened by drainage, encroaching development, compartmentalization, and nutrient enrichment. It has been designated a World Heritage Site, an International Biosphere Reserve, a Wetland of International Importance, and a specially protected area under the Cartagena Treaty and is the focus of a major state and US Federal restoration effort (NRC 2006). Two observations about the Everglades illustrate the role that aquatic invertebrates play in

D. Batzer, D. Boix (eds.), *Invertebrates in Freshwater Wetlands*,
DOI 10.1007/978-3-319-24978-0_10

its ecology and conservation. The Everglades historically supported large colonies of nesting wading birds (>100,000 pairs/year, among the largest pelicaniform and ciconiiform aggregations recorded) that took advantage of the dry-season concentration of their prey, small fishes and crustaceans, to support their reproductive exuberance (Ogden 1994; Frederick and Ogden 2001; Ogden et al. 2005). How did an oligotrophic ecosystem support seasonally high abundances of apex predators? The Everglades stands out when compared to other aquatic ecosystems in the presence of massive accumulations of primary production in the form of periphyton mats while sustaining low density and biomass of aquatic consumers and, notably, herbivores (Turner et al. 1999; Trexler et al. 2015). Aquatic snails, commonly a key grazer in freshwater ecosystems, are present at extremely low densities and biomass when compared to other aquatic systems worldwide, though the regional species pool is not depauperate (Ruehl and Trexler 2011). Why does so much primary production accumulate in the presence of a robust aquatic-consumer community? In this chapter, we will explore the role of macroinvertebrates in the Everglades food web and use that information to answer these questions. The answers illuminate essential ecological processes that are at risk from water extraction and nutrient enrichment and which are focal goals for a massive restoration program aimed to conserve the unique character of the Everglades (Turner et al. 1999; Gaiser et al. 2012).

The Ecosystem and Its Habitats

Hydrology and Nutrients

The Everglades is the downstream portion of a watershed that extends south of Orlando, Florida to the marine habitats of Florida Bay and the Gulf of Mexico. It has an area of 28,205 km^2 and extends for 449 km north to south and 100 km east to west (Light and Dineen 1994). Prior to the onset of drainage in the late 1800s, the Greater Everglades habitat south of Lake Okeechobee covered an area of 15,000 km^2 (Gaiser et al. 2012), but this has been reduced to the Everglades Agricultural Area (EAA: 3059 km^2), five water conservation areas (WCAs: 3554 km^2), and Everglades National Park (ENP: 4363 km^2) (Light and Dineen 1994; Fig. 10.1). Because the EAA has been drained for agricultural production, only 47 % (7917 km^2) of the historical ecosystem remains as wetlands. Shallow freshwaters in the historical Everglades flowed south from the margins of Lake Okeechobee in a process called sheet flow, with wet-season current speeds believed to have been higher than the contemporary averages of 0.3–1.4 cm s^{-1} (Larsen et al. 2011). Channelization and drainage have greatly diminished the broad, flowing water courses, accompanied by loss of topographic patterning called ridge and slough (Fig. 10.2; McVoy et al. 2011). Sloughs are, on average, 20 cm lower than adjacent sawgrass-dominated (*Cladium jamaicense*) ridges and hold water longer in the dry season. Current velocities of at least 2 cm s^{-1} are typically cited as necessary to suspend and redistribute the flocculent organic layer of detritus that accumulates routinely from bladderwort, periphyton, and vascular plants that are characteristics of Everglades sloughs (Larsen et al. 2011). It has been proposed that water flows of this speed or higher are necessary to maintain the historical organization of the landscape (Larsen et al. 2011; but see Kaplan et al. 2012). The linear (anisotropic) patterning of the landscape parallel to water flow is critical in forming isolated pools of water during dry-season water recession that form early (December) on the system edges and appear progressively later (March–May) toward the center (deeper) areas of the ecosystem. This sequence is critical in providing high-quality foraging patches for wading birds raising chicks between January and May (Gawlik 2002).

Water-depth variation and hydrological patterns are primary drivers of the ecology of the Everglades in general and the lives of aquatic invertebrates in particular. Hydroperiod is defined as the number of days in a year that a site has a water depth exceeding 5 cm, a cutoff selected because of the ubiquitous loose layer of organic flocculent material (floc) that covers the bottom (Trexler et al. 2005). At 5 cm, a floc-filled space remains in the water column, causing fishes and some common macroinvertebrates to expire, presumably from anoxia. Much of the Everglades does not dry annually, but all of the system does dry for short periods of time in relatively dry years. Thus, hydroperiod may be a poor descriptor of hydrological impacts on aquatic ecology. Sites that dry for several days every 2 years (on average) are ecologically quite different from sites that dry for similar lengths of time, but only every 6 or 7 years, though the hydroperiod averaged over multiple years, say 10, would be similar. Everglades sites that dry annually are in a perpetual state of recovery from the event, while sites that dry less frequently may not be, depending on the time required to complete the successional process and the time passed since the most recent drought. Thus, Everglades wetlands are often better characterized by the number of days that have passed since a site last re-flooded (days since dry, DSD). This parameter is correlated with hydroperiod, but also captures time lags inherent in post-drought succession (Trexler et al. 2005). Because seasonal hydrological variation is a key driver of ecology in southern Florida, many workers characterize years by "water year" from the beginning of the wet season (typically considered to be May) to the end of the dry season (typically considered the following April). The transition from wet to dry season typically begins in November, and surface water at the edges of the ecosystem begins to diminish notably by December ("typically" is repeated several times here because marked inter-annual variation is an important element).

Everglades waters are generally described as oligotrophic with low level of total phosphorus (TP), the limiting element for plant growth. Median TP concentration in surface water throughout the Everglades ranges from 4 to 10 ppb, though much higher values routinely occur in some areas (McCormick et al. 2002). Drought conditions can lead to local elevations of about 10 ppb. High values are most common at the inflow points of water coming from the canal system, which receives runoff

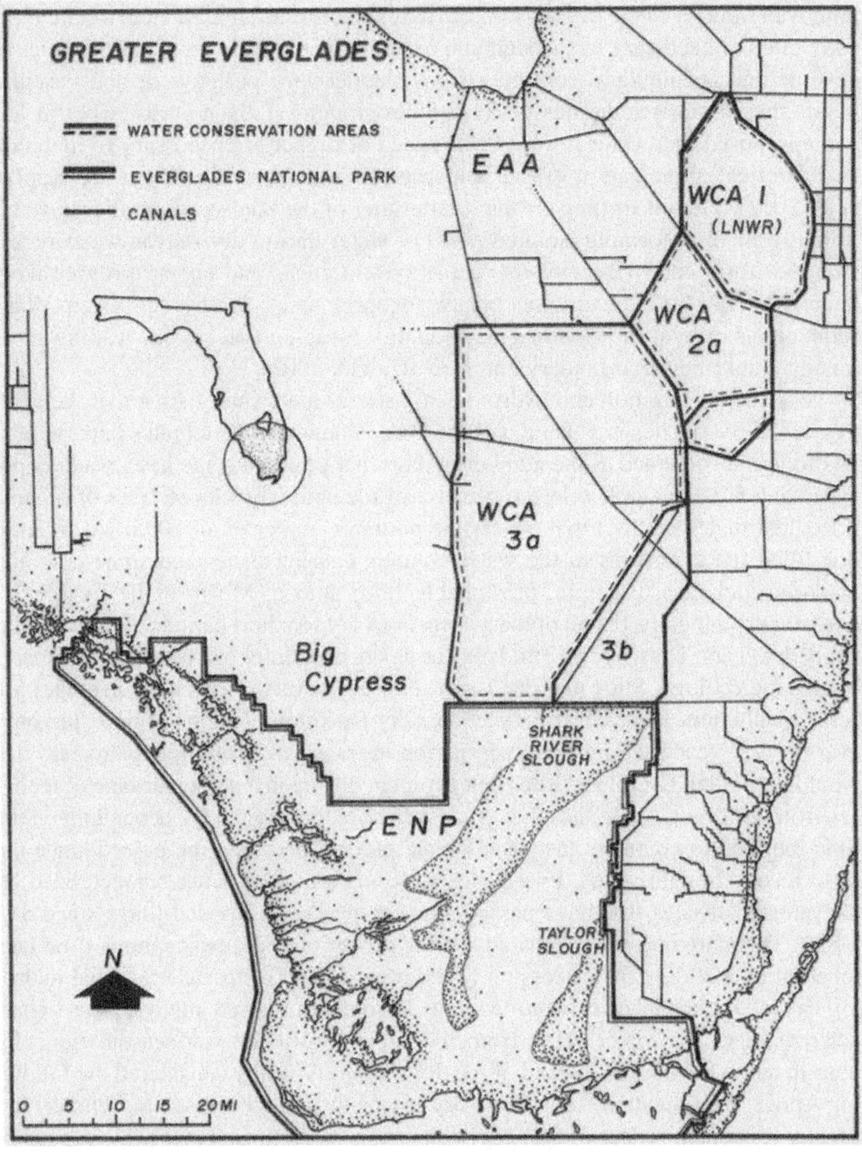

Fig. 10.1 The Greater Everglades ecosystem is located on the southern tip of the Florida peninsula, USA. Shark River Slough and Taylor Slough are bordered by short-hydroperiod marl prairie habitats and delineated on this map for clarity. Unlike other features on this map, they are not encircled by artificial barriers such as canals or levees. Water flows from Lake Okeechobee on the northern extremity of the map (see Florida *inset*) to the south through WCAs and by the canal system to ENP. *EAA* Everglades Agricultural Area, *WCA-1 (LNWR)* Water Conservation Area 1 (Loxahatchee National Wildlife Refuge), *ENP* Everglades National Park

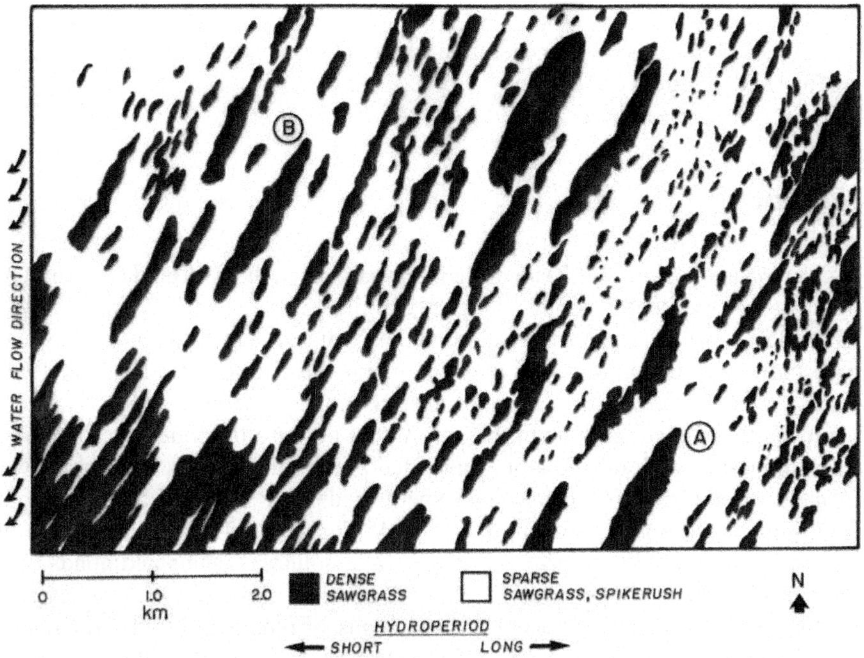

Fig. 10.2 A slice of ridge-and-slough relief from the Shark River Slough, ENP, illustrates landscape characteristic of the Everglades. The long axis of this slice (compass direction N) is approximately 10° to the west or perpendicular to the direction of water flow and captures a short- to long-hydroperiod gradient (west to east). Sawgrass-dominated ridges are modestly higher in elevation than the surrounding sparse sawgrass and spikerush-dominated (*Eleocharis* spp.) sloughs. Point A is located at 25° 38′ 0.79″N, 80° 41′ 40.55W and point B is located at 25° 39′ 30.87″N, 80° 43′ 42.18″W

from agricultural areas. Water-column TP often drops sharply with distance from a canal. Rapid uptake and accumulation of allochthonous TP by bacteria and algae lead to loading that can result in eutrophic conditions (McCormick and O'Dell 1996; Gaiser et al. 2005). The water is also hard, with high levels of calcium carbonate because of the limestone foundation of the Florida peninsula. Phosphorus binds with calcium carbonate, making it biologically unavailable and rendering the ecosystem naturally oligotrophic. The concentration of TP may increase by natural and anthropogenic mechanisms. Natural mechanisms such as accumulation of animal waste beneath wading-bird rookeries or in alligator ponds lead to local patches of enriched conditions, while agricultural runoff has eutrophied large expanses of the ecosystem, notably in northern WCA 2A and much of the perimeter of WCA 1 (LNWR) (Fig. 10.1). Everglades periphyton mats are unusual because in the presence of excess P, the factor limiting their growth at low to moderate concentrations, they disassociate (McCormick and O'Dell 1996; Gaiser et al. 2005). Thus, eutrophied areas of the Everglades generally lack periphyton mats characteristic of areas unaffected by nutrient enrichment.

Habitats

Gunderson and Loftus (1993) identified four habitats in the Everglades ordered by hydroperiod and depth: ponds, sloughs, graminoid wetlands, and forested wetlands (Table 10.1). Canals are an additional habitat added over the past 100 years, with important implications for aquatic animal life (Loftus and Kushlan 1987; Rehage and Trexler 2006; Harvey et al. 2010). Only canals can be considered permanently flooded habitats in this ecosystem because the limestone basement rock limits the depth of alligator-maintained ponds and sloughs to be shallower than the minimum hydrological stage experienced on a decadal scale in all but the most impounded region (southeast WCA 3A; Fig. 10.1). These habitats may be ranked by the diversity of structure in the water column. Canals and ponds typically include areas of open water lacking vegetation; sloughs may have areas of relatively open water, but generally have stems of emergent plants, some submerged aquatic vegetation (SAV), and some periphyton mats; while graminoid wetlands typically have extensive emergent plants, SAV, and luxuriant mats of periphyton.

Most of the ecological studies of Everglades invertebrates have been conducted in slough and wet prairie habitats, with a small amount of work on sawgrass-dominated ridges. There is vertical structure in these habitats that creates a variety of microhabitats for distinct invertebrate communities (Fig. 10.3). Algal production in the Everglades is quite high and may be present as benthic, epipelagic, and epipelic mats covering all surfaces (Gottlieb et al. 2015). Those mats may be quite thick (up to 8 cm) and structured within. The external layers are largely dead algal material, with cyanobacteria dominating the inner living areas. Diatoms, desmids, and green algae, along with heterotrophic bacteria and fungi, make up the interior structure of these mats (Donar et al. 2004). Cyanobacteria that secrete a mucilaginous matrix of extracellular polymeric substances (Thomas et al. 2006; Stewart et al. 2013) are responsible for holding together the mats to form coherent habitats for periphyton infauna. Extracellular polymeric substance is high in protein and polysaccharides (Stewart et al. 2013) and may be a relatively rich food source for heterotrophs. This creates a complex habitat structure, particularly when combined with the diversity of emergent and submerged vascular plants present in much of the ecosystem.

The formation of complex periphyton mats is common in shallow systems where microbes grow attached to substrates because of access to nutrients. It is believed that the more closely related the algal cells are to heterotrophic bacterial ones, the better access they have to enzymes that break organic-bound nutrients,

Table 10.1 Freshwater wetland habitats of the Everglades (modified from Gunderson and Loftus 1993)

A. Canals
B. Ponds—open water lacking substantial submerged aquatic vegetation, e.g., Illinois pondweed (*Potamogeton illinoensis*)
C. Sloughs—deeper area with slow moving water, e.g., white water lily (*Nymphaea*), spatterdock (*Nuphar*)
D. Graminoid wetlands
1. Sawgrass (*Cladium*) marshes; abundant periphyton mats, epiphyton, benthic periphyton
(a) Tall stature
(b) Intermediate stature
2. Wet prairies (peat); abundant periphyton mats, epiphyton, benthic periphyton
(a) *Eleocharis* spp. marshes
(b) *Rhynchospora tracyi* flats
3. Wet prairies (marl); sparse emergent sawgrass, thick benthic periphyton
E. Forested wetlands
1. Bayhead swamp forest (tree island)
2. Pond apple (*Annona*) forests
3. Willow (*Salix*) heads
4. Cypress (*Taxodium*) forests

resulting in increased efficiency of exchange. This requires the microbes to cope with plant allelopathic chemicals and sometimes a reduced light environment (Wetzel 1983). Thus, phytoplankton are relatively uncommon in wetlands unconnected to lakes (Goldsborough and Robinson 1996), so it may not be surprising that waters in the Everglades have high clarity and virtually no phytoplankton. This is partly a result of frequent drying—phytoplanktonic species rarely have adaptations found in benthic species to survive desiccation, so they have trouble establishing in benthic environments (Evelyn Gaiser, personal comm). Most wetland phytoplankton are thought to be benthic species that have become suspended by wind (Goldsborough and Robinson 1996). Periphyton coverage in the Everglades varies seasonally. Floating-mat cover and biomass are lowest in February through April but increase by as much as 30 % and 110 %, respectively, by October (Liston and Trexler 2005). The mats are a critical habitat feature that distinguishes the ecology of the Everglades from other large wetlands in North America (Turner et al. 1999), rendering them more similar to other karstic wetlands throughout the Caribbean (La Hée and Gaiser 2012).

There is a size-structured separation of macroinvertebrate distribution in a typical slough habitat of the Everglades. Smaller invertebrates use the periphyton and benthic floc as a refuge from larger predatory invertebrates and fishes (Dorn et al. 2006; Chick et al. 2008; Trexler et al. 2015), while larger ones graze surfaces or patrol the outside of periphyton capturing prey that venture out. In the laboratory, physical disruption of the mat structure leads to increased consumption of edible

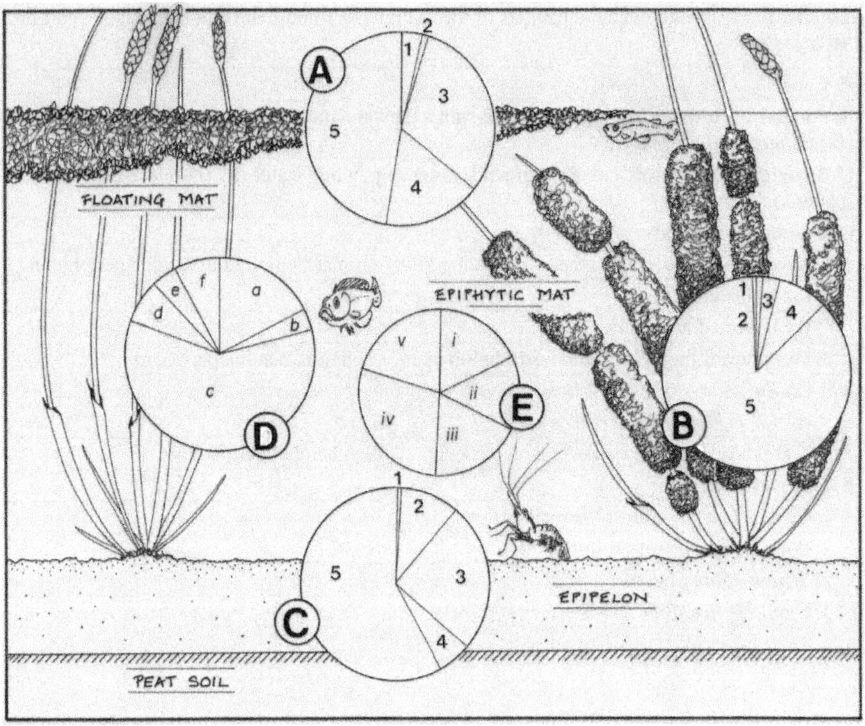

Fig. 10.3 Cross-section of slough habitat illustrating four microhabitats: (**A**) floating mat, (**B**) epiphytic mat encircling emergent plant stem, (**C**) epipelon (benthic organic floc), (**D**) water-column large macroinvertebrates, and (**E**) water-column zooplankton. Pie charts indicate the relative abundance of taxa in each habitat: (*1*) aquatic bladder snails, (*2*) cladocerans + copepods, (*3*) amphipods, (*4*) biting midge (ceratopogonid) larvae, (*5*) midge (chironomid) larvae; (*a*) dragonfly naiads, (*b*) diving beetle larvae, (*c*) riverine grass shrimp, (*d*) creeping water bugs, (*e*) ramshorn snails, (*f*) crayfish, (*i*) flatworms, (*ii*) cladocerans, (*iii*) copepods, (*iv*) ostracods, and (*v*) others

algae and invertebrates that live inside (Geddes and Trexler 2003; Trexler et al. 2015). Floating-mat invertebrates are dominated numerically by bladder snails (physids), amphipods, and biting midges (ceratopogonids) (Fig. 10.3; see Appendix for taxa names). Epiphytic algal mats are dominated numerically by bladder snails, while benthic-floc communities are dominated by midge (chironomid) larvae and amphipods (Fig. 10.3); the density of invertebrates in the mats is substantially higher than in the floc (see next section). The water column is home to larger invertebrates, numerically dominated by riverine grass shrimp (*Palaemonetes paludosus*) and dragonfly naiads (Fig. 10.3); crayfishes (*Procambarus* spp.) are also abundant and dominate invertebrate biomass (not shown) because of the large size of adults. Zooplankton are relatively uncommon in the Everglades water column, particularly during daylight hours (see next section). Zooplankton emerging from the benthos at night are composed equitably of ostracods, copepods, flatworms, and cladocerans (Loftus et al. 1990) (Fig. 10.3).

In the dry season, as the water table drops, water first drains from sawgrass-dominated ridges to the adjacent sloughs. This small topographic heterogeneity has important consequences at the sub-kilometer scale. For example, the recurrent draining leads to a narrow band of nutrient-enriched habitat at the ridge-slough margin that is typically home to elevated abundances of aquatic animals such as riverine grass shrimp. Also, several studies have found that ridges, with dense sawgrass, support elevated densities of crayfish (*P. fallax*) compared to the adjacent sloughs when water level is high enough to flood both habitats (Jelks et al. 1992; Hendrix 2000; Hagerthey et al. 2014). In the dry season, when water drops to expose the surface of ridges, *P. fallax* move to adjacent sloughs where their densities may increase dramatically (Jordan 1996; Cook et al. 2014). In an experimental study, crayfish (*P. fallax*) left ridges as depths dropped below 20 cm, leading to a pulsed increase in their density in recipient sloughs (Cook et al. 2014). The short-hydroperiod dominant species, *P. alleni*, have also been shown to disperse upon reflooding of marshes, but to burrow as water recedes (Acosta and Perry 2001). Habitat shifts and dispersal demonstrated for these crayfish have potential impacts on food availability and habitat use by foraging predators, including wading birds.

Sampling Invertebrates in Everglades Wetlands

Our understanding of aquatic macroinvertebrates is influenced in large part by the methods used to sample them and taxonomic resolution employed to group animals. The standard definition of macroinvertebrates is fauna retained on a 500 μ sieve (Hauer and Resh 2007); those passing through but retained on a 40 μ sieve are meiofauna (Palmer et al. 2007). However, aquatic ecologists commonly term all life stages of aquatic invertebrates with macro-sized adults as macroinvertebrates. In our review, we identified no published Everglades studies that used the term meiofauna.

Most published research on the Everglades invertebrates is either systematic or ecological, conducted on the assemblage of invertebrates obtained by selected sampling devices. The researchers often failed to identify to species those taxa difficult to distinguish. Ecologists often argue that complete taxonomic resolution is unnecessary to document patterns of community structure and invertebrate production relevant to understanding ecological processes. However, King and Richardson (2002, 2008a) provided a compelling case of the benefit of high-resolution taxonomic identification of chironomid larvae for use in biomonitoring in the Everglades. Larvae of the Chironomidae are notoriously difficult to identify to species, requiring mounting of individual specimens and specialized taxonomic knowledge. Because those larvae may be quite abundant, King and Richardson (2002) recommended using a fixed-count method to produce data on relative abundance of species. They demonstrate that some species of midge larvae in the Everglades display habitat specialization that makes them excellent indicators of nutrient enrichment. Jacobsen (2008) developed a key to pupal exuviae of Everglades midges in an effort to simplify taxonomic identification and potentially speed sample processing.

The highly variable and often dense submerged and emergent vegetation of the Everglades creates a challenge for sampling and renders of little use some methods commonly used in other ecosystems. Turner and Trexler (1997) compared eight invertebrate samplers in vegetated habitats by employing them side by side in the Everglades and found that they differed in the number of individuals captured, the number of species captured, and the equitability of species abundances. The methods evaluated were an inverted-funnel trap, a D-frame sweep net, a 1-m^2 throw trap, a stovepipe sampler, a Hester-Dendy artificial substrate, a minnow trap, a benthic corer, and a plankton net. Turner and Trexler (1997) recommended that complementary methods be used to gain a complete representation of the invertebrate assemblage, for example, the funnel trap to capture plankton emerging from the benthos, the D-frame sweep net to capture mat-associated fauna, and the 1-m^2 throw trap to capture larger species such as crayfish and grass shrimp. Though commonly used for bioindicator analysis in other parts of Florida, Turner and Trexler (1997) discouraged the use of Hester-Dendy artificial substrates in the Everglades, because they yielded small collections of animals, they produced a distinctive, non-typical assemblage of animals, and the hard surface had no analogy to indigenous habitats of the region. King and Richardson (2002) recommended the use of a D-frame sweep net deployed in a method similar to Turner and Trexler (1997) and following State of Florida and US Environmental Protection Agency guidelines. The sweeps include brushing across the top of the substrate, followed by sweeping up to the surface with the goal of producing a representative sample of all microhabitats. Liston and Trexler (2005) pointed out that in many areas of the Everglades, this method is problematic because the extensive periphyton mats impede a smooth sweep and may overfill the net bag. The strong integrity of the mats precludes pulling up the net without dragging additional mat from the surrounding area, leading to an overrepresentation of that habitat and possible escape by large mobile macroinvertebrates. Liston and Trexler (2005) recommended sampling periphyton infauna by taking cores (6-cm diameter), which they sorted under magnification to enumerate infauna retained on a 250-mm mesh sieve and with a maximum dimension of 1 mm. This excluded small midge larvae that could be characterized as meiofauna. The authors suggested reporting data as density (number or mass/area sampled) and "crowding" (number or mass/g periphyton). Crowding accounts for the patchy nature of periphyton mats, even when sampled at the small scale of their cores, and represents the encounter rate of animals enumerated within the sample. Later work using this method counted all midges retained on the 205 μm mesh (Sargeant et al. 2011), which increased the number of animals counted, but did not change the spatial or temporal patterns revealed (unpublished data). Dense vegetation typical of areas receiving continuous nutrient enrichment cannot be properly sampled by throw traps because the trap neither settles quickly nor seals effectively on the marsh substrate (Jordan et al. 1997; Turner et al. 1999). Hagerthey et al. (2014) used a portable bottomless lift net to overcome this problem.

There has been limited study of zooplankton in the Everglades. Turner and Trexler (1997) used a conical plankton net to take water-column samples of plankton, but found a number of benthic taxa in the collections. They concluded that the vegetation in the water column made this method ineffective, leading to few animals collected and high inter-sample variation. The Everglades water column is generally very clear, further suggesting that few plankton are present there, at least during the day. Loftus et al. (1990) placed Brakke's (1976) modified Whiteside–Williams (1975) pattern samplers (multiple-funnel trap) on the substrate surface to capture zooplankton as they emerged at night to swim into the water column. To enable more rapid processing of Everglades carbonate-sediment samples, Daraghy et al. (1988) developed a rapid acid-wash method to dissolve carbonate and expose the invertebrates. Bruno et al. (2003) and Bruno and Perry (2004, 2005) studied groundwater copepods with a water suction pump to sample ground water by filtering water drawn from wells reaching into the limestone bedrock with a 40-μm mesh net.

Invertebrate Diversity, Habitat, and Control of Dynamics

Biodiversity

Knowledge of invertebrate communities in the Everglades has greatly increased since it was first reviewed by Rader and Richardson (1992), Gunderson and Loftus (1993), and Rader (1999). Rader (1999) reported that only seven studies were available that described invertebrate communities, and few of those were in the peer-reviewed literature. In a spring 2015 literature search, we found only 20 papers focusing on the ecology of invertebrates in freshwater habitats of the Everglades. Additional papers tangentially provided information on invertebrates as prey of wading birds and snail kites and fish, or as contributing to biogeochemical cycles, particularly of mercury. A small number of book chapters, unpublished theses and dissertations, and technical reports provide yet more information. Little work has focused explicitly on diversity or natural history of aquatic invertebrates living in the Everglades, particularly the diverse aquatic insects. The exception is exemplary work on midge larvae (family Chironomidae) for use as bioindicators (King and Richardson 2002, 2003). The diversity of copepods has also been given careful treatment, described below. Crayfishes and apple snails have received special attention by one or more researchers because of their importance as prey of apex predators (both) and because of the addition of potentially invasive species (snails).

Surveys of Everglades macroinvertebrate biodiversity are limited, but King and Richardson (2002) reported a total of 93 families, 181 genera, and 252 species from their work along a marked gradient of nutrient enrichment in the northern Everglades (WCA 2A). Coleopterans, dipterans, gastropods, odonates, and oligochaetes were the most diverse of the major taxonomic groups. Chironomidae was the most diverse family, represented by 30 and 51 genera and species, respectively. Jacobsen (unpublished technical report) collected approximately 160 species of midges (families Chironomidae and Ceratopogonidae) and 25 species in other dipteran families from ENP. He concluded that his work raised the estimated species richness of the Everglades macroinvertebrate community from 200–250 (Rader 1994, 1999) to over 400 species (summarized in Appendix). Certain groups, particularly the turbellarians and oligochaetes, have been undersampled and understudied and will require a great deal of systematic and ecological work in the future.

Since Rader's (1999) review, Hendrix and Loftus (2000) documented the presence of two species of epigean crayfish in the Everglades and Big Cypress, rather than the one reported in all prior work: Everglades crayfish *P. alleni* (reported by Rader) and slough crayfish *P. fallax*. Hendrix (2003) noted that *P. fallax* ranges throughout Florida, motivating him to reanalyze historical samples from the Shark River Slough. The ecology of these two species is complementary, with *P. fallax* dominating sites that dry infrequently (at least two consecutive years without drying), while *P. alleni* dominate habitats that dry more frequently (Hendrix and Loftus 2000; Dorn and Trexler 2007). Ecological studies carried out without separating the two species tend to find small effects of changing hydrology on crayfish biomass, which overlooks marked species turnover dynamics with possible management implications (Fig. 10.4). VanArman (2011) reviewed the prey and predators of the Everglades crayfishes in the food web.

Bruno and Perry (2004) and Bruno et al. (2001, 2005) surveyed zooplankton, particularly copepods, inhabiting surface and groundwaters of Everglades National Park. From surface waters they recorded a total of 65 taxa of free-living copepods: 9 calanoids, 41 cyclopoids, and 15 harpacticoids (Appendix). Of these, four were newly recorded to the area. They also found 22 species of copepods pumped from groundwater wells, mainly surface-water species.

Cladoceran collections from Everglades National Park were made by Conrow and Loftus during a study described in Loftus et al. (1990), but the data were never published. Identification of specimens was confirmed by David Frey of Indiana University. The samples revealed a diverse fauna of lentic-water Cladocera. A total of 24 genera and at least 42 species were identified in the collections, and chydorids were the most diverse family with 11 genera and 24 species.

The Everglades is home to a diverse anisopteran fauna (Appendix). Urgelles (2010) identified 16 species of dragonfly naiads from Everglades wetlands with hydroperiods of at least 225 days.

In summary, macroinvertebrate biodiversity is not well characterized from the Everglades, particularly in highly diverse groups like the midges, but knowledge has been increasing since the mid-1990s. There is little comparative analy-

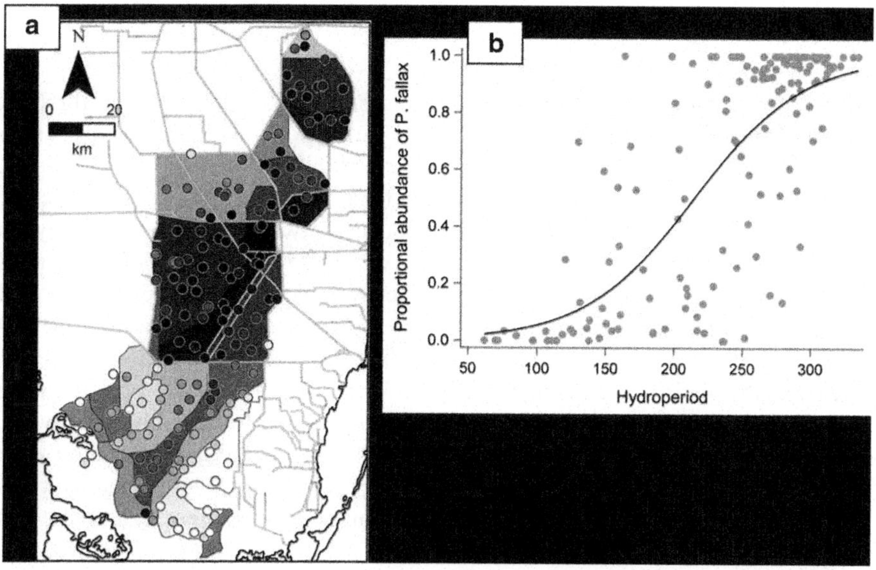

Fig. 10.4 (**a**) Spatial distribution of *Procambarus alleni* and *P. fallax* across the Greater Everglades ecosystem. *Circles* are shaded by the mean proportion of *P. fallax* at a site (i.e., black = 100 % *P. fallax* and 0 % *P. alleni*; white = 0 % *P. fallax* and 100 % *P. alleni*). Landscape sampling units are shaded according to hydroperiod, with longer-hydroperiod sites being *darker* and short-hydroperiod sites being *lighter*. Data collected annually in October through December, 2005–2012 by throw trap with three samples from each of 145 sampling locations. (**b**) Average proportional abundance of *P. fallax* by hydroperiod of sites plotted in (**a**). Response curve was estimated by logistic regression

sis on invertebrate species richness in the Everglades, undoubtedly because identification of specimens in many groups of aquatic invertebrates requires specialized skills that are not widely supported by academic or management organizations. If amphibians and freshwater fish, aquatic taxa better known than invertebrates, are considered, south Florida is species poor (Loftus and Kushlan 1987; Means and Simberloff 1987; Trexler 1995). This has been attributed to a peninsula effect resulting from limited dispersal for obligate freshwater species, the zoogeographical derivation of the aquatic fauna, and the relative youth and lack of diversity of southern Florida aquatic habitats. The karstic wetland yields water chemistry that is hard and alkaline (e.g., the region lacks softwater/blackwater habitats); taxa that require acidic and/or lotic waters may be excluded. Unlike native fishes and amphibians, almost entirely derived from temperate North America, Everglades copepods, cladocerans, and midges have both temperate and tropical affinities. For most invertebrate groups, there has not been a similar state-wide systematic survey of biodiversity to permit a biogeographic analysis. More work is needed!

Periphyton Infauna: Mats, Epiphyton, and Benthos

Floating and epiphytic periphyton mats, and benthic floc, are home to the bulk of small macroinvertebrates, midge larvae, amphipods, and similar-sized taxa that make up a major part of the diets of secondary consumers of the Everglades (Fig. 10.3). Crowding of the most common taxa increased 33–153 % from early to late wet season, and community differences between the two habitat types became more pronounced (Liston and Trexler 2005). Liston (2006) observed that differences in community structure between floating-mat periphyton and epipelon (benthic-floc) microhabitats were greater than any variation attributable to gradients of hydroperiod, P availability, or other spatial factors at the ten sites she sampled from SRS and WCA3A. These studies revealed that floating mats held the highest crowding and density of infauna, followed by epiphyton, with substantially less in epipelon (mat held 6.7 times higher crowding than floc). There were also 1.6 times more taxa per sample in floating-mat samples than in floc, though when the cumulative number of specimens examined was accounted for, the asymptotic species richness was only 17 % higher in mat than floc (35 versus 30).

The drivers of infaunal dynamics appear to differ between floating-mat and epipelon habitats at the same site. Multivariate analyses indicated community structure of epipelon infauna was driven by hydroperiod, although crowding of individual taxa showed no consistent responses to hydroperiod or TP availability (Liston 2006). In contrast, community structure of periphyton mat infauna was driven by the interaction of TP availability and hydroperiod, while densities of mat infauna (no. m^{-2}) were most influenced by hydroperiod (positive correlations). Liston (2006) noted that crowding of mat infauna doubled with P availability in short-hydroperiod marshes, but was constant across the TP gradient in long-hydroperiod marshes. She hypothesized that community structure and density were not different among long-hydroperiod, constantly inundated sites because of the high density of small fish found at these sites. She hypothesized that increased abundance of floating-periphyton mat infauna with P availability at short-hydroperiod sites may result from a release from predation by small fish (Fig. 10.5).

Several studies support the hypothesis that the dynamics of periphyton infauna are closely tied to the fate of the periphyton mat they inhabit. A metacommunity analysis demonstrated that crowding and composition of periphyton infauna are better described as being controlled by "species sorting" than "mass effects," compared to larger invertebrates like crayfish, grass shrimp, dragonfly naiads and creeping water bugs (Naucoridae, *Pelocoris femoratus*), and small fishes (Sokol et al. 2014). The "species sorting" model suggests that dispersal limits local species composition and community-level responses to local environmental conditions, compared to the "mass effects" model that posits species are so mobile that those best matched to the local environment colonize rapidly and saturate the local community (similar to the Baas-Becking hypothesis that

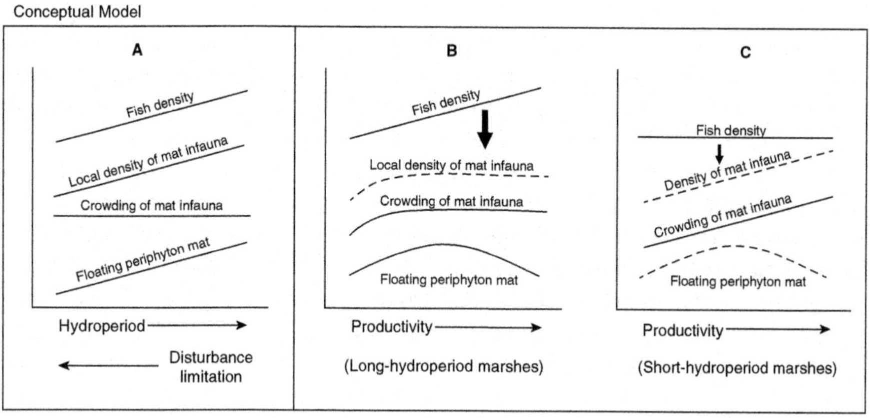

Fig. 10.5 Conceptual model of hypothesized interactive effect of hydroperiod and nutrient addition on the density of periphyton infauna. (**a**) Fish density, periphyton map cover, and infauna density increase with increasing hydroperiod, but infauna crowding is constant because of density-dependent feedback within the mat. (**b**) Fish density increases with increasing primary productivity (TP) at long-hydroperiod sites, but periphyton cover peaks at an intermediate level. Infauna density is flat with decreasing productivity because of the loss of mat cover but stimulation from better quality of food, though at lower abundance; crowding is unaffected because increased food quality is compensated by increased predation. (**c**) Fish density is limited by drying disturbance and does not increase with increasing TP at short-hydroperiod sites. This frees infauna from top-down control to increase crowding in the presence of better quality food, which increases their density in spite of loss of periphyton mat coverage (Redrawn from Liston (2006))

"everything is everywhere, but the environment selects"). Periphyton infaunal communities were modeled best by inclusion of both spatial proximity and local environmental measures, while larger mobile invertebrates were best modeled using just local environmental measures; spatial proximity had the least contribution to explaining landscape patterns of fish community metrics. This result suggests that large invertebrates and fishes are better able to sort themselves in the environment in response to their food availability, physical factors, and predators than are periphyton infauna. Because these large invertebrates and fishes are predators of infaunal invertebrates, the hypothesis of top-down, density-dependent regulation of infaunal invertebrates (Fig. 10.5; Liston 2006) seems reasonable.

Sargeant et al. (2011) used a structural equation modeling (SEM) approach to evaluate competing hypotheses of all bottom-up and mixed bottom-up and top-down control of periphyton mat infauna at 28 sites from Shark River Slough, WCA 3A, WCA 2A, and WCA 1 (LNWR). The best model included both bottom-up and top-down effects among trophic groups and supported top-down control of infauna by omnivores and predators that cascaded to periphyton biomass. The second-best model included bottom-up paths only. Total effects (estimated as the product of all direct and indirect effects) of days-since-dry were negative for all three consumer groups in the study, even when both preferred models suggested positive direct effects for some groups. Total effects of periphyton TP were positive for consumers and generally larger than those of hydrological disturbance and were mediated by paths indicating changes in periphyton community composition and edibility. Finally, Abbey-Lee et al. (2013) used SEM to evaluate models explaining the isotopic niche of Eastern Mosquitofish (*Gambusia holbrooki*), the most abundant intermediate consumer of the Everglades, that has been demonstrated to consume midge larvae, amphipods, cladocerans, and other small macroinvertebrates (Loftus 2000; Taylor et al. 2001; Chick et al. 2008). They reported stable isotope values ($\delta^{15}N$ and $\delta^{13}C$) from samples of at least 17 individual Eastern Mosquitofish collected in the wet season of 2005 from each of the 21 sites located in the Shark River Slough, WCA 3A, WCA 3B, WCA 2A, and WCA 1 (Loxahatchee National Wildlife Refuge). The data indicated that the mosquitofish isotopic niche decreased with increasing density of most types of periphyton infauna (direct effects) and increased as the density of conspecifics and potential competitors increased (indirect effects through periphyton infauna and periphyton edibility). These diet changes are *consistent* with food limitation of Eastern Mosquitofish (the local population became more specialized as food became more available and less specialized as intra- and interspecific competition increased), which could yield a density-dependent, top-down effect on the infauna they consume.

Experimental studies have also indicated the potential of predators to limit the abundance of periphyton infaunal invertebrates. Working in long-hydroperiod marshes of the Shark River Slough, Dorn et al. (2006) compared periphyton, invertebrates, and fish in 1-m^3 cages with 2.54-cm mesh on one side (exclosure cages) to similar cages with a side lacking mesh (control cages) to evaluate the effects of excluding large predators, mostly fishes. After 2 weeks, the exclosure cages held higher densities of intermediate consumers, particularly grass shrimp, than the control cages, suggesting that greater predation risk in the controls caused avoidance. More importantly, small primary consumers (mostly small snails, amphipods, and midges) living in floating periphyton and the flocculent benthos were less abundant in the exclosures, indicative of a trophic cascade. In a similar experiment also conducted in the Shark River Slough, Chick et al. (2008) noted a similar effect of exclosure cages on intermediate consumers, with increased density of dragonflies, crayfish, and grass shrimp inside. Periphyton infauna were not sampled, but inverted-funnel traps were placed in each cage on the last day of the experiment to document zooplankton emergence. Also, similar to the Dorn et al. (2006) study, the treatments had no effect on algal composition or biomass of native periphyton mats placed in the cages, but exclosure cages had less epiphytic algae growing on plastic strips placed in the cages at the start of the study. There were minimal treatment effects on the zooplankton in the presence of elevated density of intermediate consumers, however. Thus, cascading effects on zooplankton and floating-periphyton mats (algal composition and biomass) were small, which Chick et al. (2008) interpreted as support for the hypothesis that periphyton serves as a refuge for infauna. Finally, Liston et al. (2008) observed a dose-dependent increase in periphyton and benthic infauna at low and intermediate TP addition to field mesocosms. Total macroinvertebrate density in periphyton mats increased with intermediate P loads, driven primarily by chironomids and nematodes. However, infaunal crowding in benthic floc decreased with enrichment, driven primarily by loss of chironomids and ceratopogonids (*Dasyhelea*). Thus, macroinvertebrate density increased with TP enrichment until the periphyton mats were lost, after which density decreased markedly. Apparently, mat infauna failed to move into benthic substrates in response to mat loss. These results were noted at nutrient levels too low to yield anoxia and appeared to be linked to the loss of habitat and predation by intermediate consumers present in the mesocosms when the mat-refuge effect was lost.

The recent colonization of the Everglades by a variety of nonnative fishes has been well documented, and their effects have begun to be studied (Kline et al. 2014). Several studies conducted in Everglades field cages and in mesocosms have shown that introduced predatory fishes, particularly the cichlids African Jewelfish (*Hemichromis letourneuxi*) and Mayan Cichlid (*Cichlasoma urophthalmus*), have the potential to affect densities and biomass of aquatic snails and riverine grass shrimp (Porter-Whitaker et al. 2012; Schofield et al. 2013).

Habitat and Size-Structured Predation

Invertebrates in the Everglades live within a dynamic food web affected by the mortality and stress associated with periodic drought. Invertebrates are critical links between primary producers and those apex predators of great ecological, conservation, and economic importance. Furthermore, invertebrates are a key link between the effects of anthropogenic nutrient enrichment and vertebrate species because there are few strictly herbivorous aquatic vertebrates in the Everglades. For example, only two species of the 33 common fish species are primarily herbivorous (Loftus and Kushlan 1987; Loftus 2000), while the vast majority of fishes include invertebrates in their diet. Microbial metabolism of periphyton detritus is an important route for energy flow in this ecosystem (Williams and Trexler 2006; Belicka et al. 2012); mat infaunal consumption of edible algae taxa, extracellular polymeric substances, and heterotrophic bacteria is likely to be a critical link to higher consumers. Many Everglades aquatic invertebrates are predators of other invertebrates (Loftus 2000). Thus, all past reviews of the ecology of invertebrates in the Everglades have speculated about the importance of predation in controlling their dynamics (Rader 1999; King and Richardson 2008b; Trexler et al. 2015). Liston (2006), Sokol et al. (2014), and Trexler et al. (2015) have separated discussion of the controls of invertebrate communities by habitats and size, between (1) taxa that live as periphyton infauna (midge larvae, amphipods, etc.) and may experience a refuge from larger predators and (2) mostly larger invertebrate taxa that cling to the outside of the mats or inhabit the benthos, scavenging for small vulnerable vertebrates and unfortunate infaunal taxa exposed from their refuges.

Dorn and Trexler (2007), Dorn (2008), and Gaiser et al. (2012) pointed out that abundance patterns of some Everglades invertebrates, notably apple snails (Darby et al. 2008), ramshorn snails (Ruehl 2010), and crayfish (Acosta and Perry 2002; Dorn and Trexler 2007), cannot be predicted solely by the time passed since a local site has dried, probably because of their ability to survive mild drying events by burrowing into the sediments and/or rapidly recolonize after droughts. In contrast, riverine grass shrimp, dragonfly naiads, and creeping water bugs are abundant throughout the Everglades, but appear to suffer high mortality during drought events. Using observational data from high-nutrient wetlands, Dorn (2008) suggested that the biomass of crayfish and large predaceous insects could be enhanced indirectly by drying that temporarily eliminated their fish predators (see also Kushlan 1976). Experimental work in low-nutrient constructed wetlands (similar to the Florida Everglades) supported the prediction that crayfish are sensitive to predatory fish (Kellogg and Dorn 2012), but experiments investigating the interactive effects of predators along with other changes associated with drying are lacking. Knorp and Dorn (2014) found that predatory sunfish (Warmouth, *Lepomis gulosus*; Bluespotted Sunfish, *Enneacanthus gloriosus*; and Dollar Sunfish, *Lepomis marginatus*) decreased the density of *P. fallax* but not dragonfly naiads in experimental wetlands. *Procambarus fallax* benefited by simulated marsh drying because it eliminated sunfishes; dragonfly naiad density also decreased. Juveniles of these sunfishes consume macroinvertebrates, particularly odonates, midge larvae, and amphipods. Depending on the season, warmouths display some separation of diet from the other two species, consuming more crayfish and fish as they grow (Loftus 2000; Bransky and Dorn 2013). Gape-size-matched diet overlap of Warmouth with the other two sunfishes was greatest during the wet season, when prey abundance was the greatest. In an 8-year experimental study at the landscape scale (500 km^2 of the Everglades), crayfish densities were positively correlated with the severity of drying (up to 99 days dry) during the preceding dry season (Dorn and Cook 2015). This contrasts with drying effects on small-bodied fishes in the same wetlands, whose densities were seasonally depressed by drying disturbance.

Fire

Fire is an important component of the ecology of the Everglades, both historically and in the present (Ogden et al. 2005). Lightning is responsible for starting fires, primarily in the spring and summer months when water depths may be low or even below ground surface. Many Everglades fires spread in emergent plant stems over shallow standing water, but severe fires on desiccated marshes that consume the peat substrate also occur. Drainage of the Everglades increased the frequency of these peat-consuming fires during the twentieth century, with important effects on the topography, hydropatterns, and biogeochemistry in the aftermath (Gunderson and Snyder 1994; Lockwood et al. 2003; McVoy et al. 2011). There have been few studies of the effects of fire on aquatic invertebrates (or aquatic animals in general) in the Everglades, though relevant impacts have been documented in other ecosystems (e.g., Gresswell 1999; Beganyi and Batzer 2011). A 2015 literature search with key words "Everglades" and "fire" returned 128 papers, but only one of those papers reports studies of the effects of fire on aquatic macroinvertebrates (Venne and Frederick 2013).

The most immediate impacts of fire on aquatic invertebrates are to remove plant cover, increase light penetration into the water column, decrease stem density and habitat structure in the water column, and release nutrients to stimulate primary production (Venne 2012; Venne and Frederick 2013). Some wading birds, notably white ibis (*Eudocimus albus*), are known to include a high frequency of crayfish in their diets and have been observed to forage in and adjacent to wetland areas burning or recently burned (Epanchin et al. 2002). Focusing on wading bird prey, Venne and Frederick (2013) sampled in burned and unburned Everglades sites after controlled burns, but noted few fire-killed prey (minimum water depth in the burned areas was 10 cm). They also found no difference in the density of fishes or macroinvertebrates (grass shrimp, crayfish, belostomatids, dysticids, leeches, odonates, oligochaetes, and creeping water bugs) between burned and adjacent unburned habitats. When they manipulated fire experimentally in plots with shading and vegetation-removal treatments, they also found no treatment effects on macroinvertebrates (Venne 2012). The fire caused a short-lived pulse in nutrients (P) in the water column that was rapidly taken up by bacteria and periphyton in the area, such that it could not be distinguished from unburned control areas within a matter of days.

Conservation and Management Issues for Everglades Invertebrates

Beginning with efforts to regulate hydrology in the Everglades in the twentieth century, phosphorus-laden runoff from agricultural areas has been added to the ecosystem through a network of canals carrying water from Lake Okeechobee and the EAA (Davis 1994; Noe et al. 2001). This has led to eutrophication in several areas (notably northern WCA 2A, southern WCA 1 (LNWR), northern WCA 3A), accompanied by loss of native flora and expansion of cattail (*Typha domingensis*) monocultures (Davis 1994). Cattail invasion is facilitated by drought and fire, which occur more commonly in the modern Everglades that has been reduced in size and suffers from a lowered water table (Newman et al. 1998). Addition of phosphorus leads to the loss of periphyton cover and increase in abundance of nutrient-tolerant macroinvertebrates (Rader and Richardson 1992; King and Richardson 2003, 2008a), changing the ecological character of the ecosystem from one typical of oligotrophic wetlands in the Caribbean basin (Turner et al. 1999). The dense cattails hamper foraging by wading birds (Crozier and Gawlik 2002) and generally diminish the ecological integrity of the affected areas (Sklar et al. 2005). Ecological damage resulting from nutrient enrichment has led to lawsuits and massive investments in public works to remove phosphorus from waters destined for the Everglades. Solving environmental problems resulting from water extraction and nutrient enrichment is the primary goal of a major US Federal and State of Florida partnership to restore the Everglades initiated in 2000 and called the Comprehensive Everglades Restoration Plan (CERP; Sklar et al. 2005; NRC 2006).

In this section, we will focus on four areas of environmental challenges for Everglades restoration and management that are relevant to aquatic invertebrates: phosphorus enrichment, nonnative species invasions, xenobiotics, and trophic dynamics.

Phosphorous Enrichment: The Subsidy-Stress Hypothesis

Anthropogenic nutrient runoff pushes Everglades aquatic communities into a resource state that is enriched beyond what is believed to have been historically present and is absent from areas not receiving enrichment today (Turner et al. 1999). King and Richardson (2007, 2008a) used sweep-net sampling to document a unimodal pattern of invertebrate biomass along a spatial (distance from a canal) and phosphorus (TP in sediment) gradient in an area of the Everglades experiencing extreme P enrichment over a multi-decadal timescale (Fig. 10.1, northern WCA 2A). They interpreted these results in light of the subsidy-stress model first proposed by Odum et al. (1979; Fig. 10.6). In this study, 8 of 12 major taxonomic groups (Amphipoda, Decapoda, Diptera, Ephemeroptera, Gastropoda, Hirudinea, Odonata, Oligochaeta) displayed a unimodal response, three (Coleoptera, Hemiptera, Isopoda) increased monotonically, and one (Trichoptera) decreased monotonically in response to TP. As is typical in the Everglades, periphyton cover was absent at high levels of TP enrichment, and the loss of periphyton is correlated with low invertebrate biomass. The unimodal pattern changed seasonally, however, and was absent early in the wet season following marsh flooding when periphyton is infrequent in all marshes (TP enriched and oligotrophic) because of winter senescence. The authors suggested that nutrient enrichment created an interaction between increased quality and decreased quantity of periphyton, leading to the subsidy-stress patterns observed. An experimental TP-addition study conducted in WCA 2A supported the hypothesis that macroinvertebrates in the Everglades are resource limited (King and Richardson 2008b), supporting the subsidy part of the subsidy-stress hypothesis. McCormick et al. (2004) also observed a decline in abundance and change in species richness to favor nutrient-tolerant taxa at nutrient-enriched sites in WCA 2A. Liston et al. (2008) reported complementary observations in a mesocosm study from the southern Everglades (Taylor Slough); invertebrate density increased with low and intermediate addition of TP, but dropped markedly when experimental nutrient enrichment led to loss of the periphyton mat. It appeared that the loss of periphyton habitat caused the infauna to be vulnerable to predators, leading to their consumption or emigration.

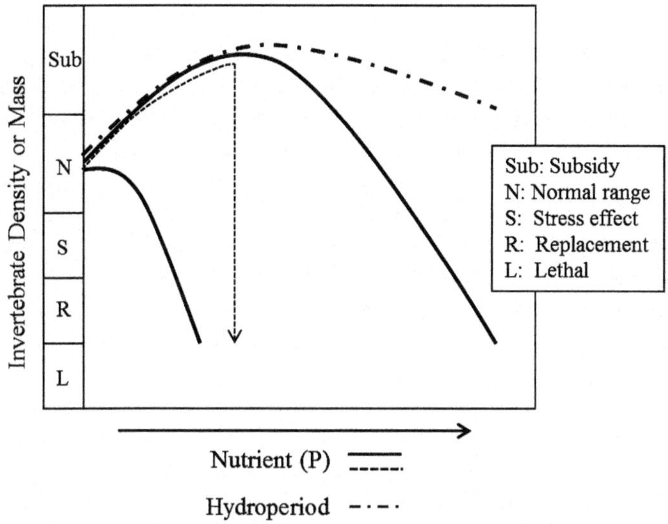

Fig. 10.6 Conceptual diagram of subsidy-stress model redrawn and expanded from King and Richardson (2008a). Two possible patterns are illustrated for TP enrichment, a unimodal subsidy-stress pattern as drawn in King and Richardson (2008a, *solid line*) and an alternative suggested by Liston et al. (2008, *dashed line*). The Liston et al. (2008) relationship illustrates loss of the periphyton mat at a threshold TP concentration with consumption or emigration of all mat inhabitants. A second relationship is illustrated for hydroperiod effects on macroinvertebrates. In this case a subsidy is produced by lengthening hydroperiod until it becomes long enough to permit invasion by carnivorous fishes that crop invertebrate production outside of refuge habitats such as periphyton mats. Note that Everglades marshes not receiving anthropogenic nutrient enrichment seldom pass the inflection point of maximum subsidy because natural processes do not lead to marked elevation in cycling P. Exceptions exist in dry-season alligator ponds and under bird rookeries

Hagerthey et al. (2014) experimentally removed dense cattail and sawgrass from areas experiencing nutrient enrichment to evaluate methods for managing these habitats to regain wading-bird foraging opportunities lost because of the dense habitat cover (Crozier and Gawlik 2002). They found similar species composition in cleared areas and background unenriched sites, but higher density and biomass of macroinvertebrates in the enriched sites, notably crayfish and grass shrimp. Enriched sites with intact vegetation had fewer fishes but more crayfish than cleared areas. Vegetation was so dense in the enriched area that fishes were excluded for lack of space.

The King and Richardson (2007, 2008a) subsidy-stress model for macroinvertebrates is supported for nutrient enrichment as a stressor in the Everglades. The effect of TP enrichment initially subsidizes algal production and changes periphyton species composition to include a higher frequency of edible taxa (green algae and diatoms). This increase in edibility stimulates invertebrate productivity (Sargeant et al. 2011; Trexler et al. 2015). Several studies suggest that this increased productivity is in excess of consumption, and algal mass and macroinvertebrate mass/density all increase with low to intermediate enrichment (King and Richardson 2008b; Liston et al. 2008; Trexler et al. 2015). A few studies provided contrary results, suggesting no response by macroinvertebrates depending on the substrate sampled (Turner et al. 1999; Liston 2006; Liston et al. 2008). Ruehl and Trexler (2015) used a reciprocal transplant of periphyton from enriched and unenriched sites to demonstrate that ramshorn snails (*Planorbella duryi*) do not increase growth rate in areas with high predator density near a canal even when fed nutrient-enriched periphyton. The snails also laid fewer eggs when held in the high-predator area, regardless of the periphyton type provided. In mesocosm experiments, the same snail species increased shell thickness, a predator defense, and decreased growth rate in the presence of crayfish predators feeding on conspecifics (Ruehl and Trexler 2013). Thus, life history trade-offs may slow invertebrate-production responses to nutrient enrichment if predators are also benefited.

We may expect alternative shapes to the subsidy-stress pattern than proposed by King and Richardson (2007, 2008a) depending on the organisms analyzed. Once TP loads became high, periphyton structural coherence was lost and the mats dissociated, exposing the infauna to consumption (Fig. 10.6). This could lead to a shift in production from infaunal taxa to consumers with other predator defenses. Thus, community-wide biomass may not display the threshold decline predicted for mat infauna when the mat disassociates. Hydroperiod and marsh drying is a second important driver in the Everglades. Drying events may lead to high mortality for species lacking adaptations to survive (Gaiser et al. 2012). However, the Everglades is home to several species that can survive moderate-length droughts and may be as abundant or even more abundant following droughts. Dorn and Cook (2015) illustrate that *P. fallax* benefits from drought because of predation release. This suggests a more shallow response to subsidy stress (Fig. 10.6) or a redefinition of stress. However, because drought is very stressful for many taxa, it would be appropriate to be plotted on the *x*-axis of Fig. 10.6 for a community-wide analysis (see Trexler et al. 2005, Fig. 10.6, for an example with fishes). In this case, species turnover and replacement may buffer the community-wide response to the stressor.

Nonnative Species

The Everglades has been invaded by many plant and vertebrate animal species but few aquatic invertebrates. Several nonnative species of mollusks occur in the ecosystem, though only nonnative apple snails have raised conservation concerns. Canals are implicated in playing a role in the survival and dispersal of nonnative invertebrates in the Everglades system (Harvey et al. 2010). Two species of apple snails (Ampullariidae, *Pomacea* spp.) have been introduced to south Florida and now are found in the Everglades or nearby waters. Rawlings et al. (2007) determined that *Pomacea haustrum* were observed in the 1970s in Palm Beach County, Florida but have not spread appreciably in 30 years; however, a disjunct population was identified from the Big Cypress Swamp in the 1990s (Loftus, unpublished data *fide* T. Collins). In contrast, *P. insularum/maculata* was established in

Texas in the late 1980s and spread to Florida by the mid- to late 1990s, where it has continued to expand its range (note that Hayes et al. (2012) recently proposed to group several species of *Pomacea*, including *P. insularum*, under the name *P. maculata*, currently the name that should be used for that introduced species). There is concern about the spread of *P. maculata* into the Everglades and its potential to affect adversely native *P. paludosa* or affect vascular plant density, benthic habitat structure, or water clarity as documented in other systems (Horgan et al. 2014). However, in at least one Florida lake, the highly productive *P. maculata* may provide a valuable food source for the endangered Snail Kite (*Rostrhamus sociabilis*; Cattau et al. 2010). Other nonnative gastropods commonly collected are *Marisa cornuarietis* and *Melanoides tuberculata* (described in Thompson 2004). The bivalve *Corbicula fluminea* is found in canals of south Florida, but not in interior Everglades wetlands.

Mercury, Copper, and Zinc Contamination

There are several xenobiotics that show elevated levels in parts of the Everglades with potential impacts on aquatic invertebrates or on predators that consume them. Copper, in particular, is known to be elevated in three areas that receive runoff from agricultural lands, and herbicide used on those lands is the likely source (Rand and Schuler 2009). Apple snails are sensitive to heavy metals, particularly copper and zinc (Hoang et al. 2009; Hoang and Tong 2015). Though it appears that *P. paludosa* can detoxify stored copper to some extent, it is also clear that exposure increases their mortality rates in experimental settings (Hoang and Rand 2009).

Mercury has been studied extensively in Everglades fishes (Stober et al. 2001), and some data have been gathered for aquatic invertebrates (Cleckner et al. 1998; Loftus 2000). Tissue mercury concentrations in the animals surveyed were well predicted by trophic position estimated by analysis of stable isotopes (Loftus 2000). Thus, all snails tested yielded relatively low tissue concentrations, while fishing spiders, riverine grass shrimp, and some dragonfly naiads had relatively high levels similar to some carnivorous fishes such as *Gambusia holbrooki*.

Trophic Dynamics and Species of Special Concern

The Everglades is home to a number of bird species with protected status that consume macroinvertebrates as their prey. Food limitation appears to be a major driver for the decline of these birds from historical levels; therefore, production of their invertebrate prey is of concern for their management, conservation, and restoration. Snail Kites are specialists on apple snails, and their nest initiation rate is correlated with the availability of these snails as prey (Cattau et al. 2014). Monitoring of Snail

Kite prey and analysis of their habitat requirements are important facets of managing this federally endangered species (Martin et al. 2007). Crayfishes are important diet items of White Ibis and possibly Wood Storks (*Mycteria americana*) (Frederick et al. 2009). Monitoring programs to measure the success of Everglades management and restoration incorporate assessments of crayfish as prey for wading birds (Frederick et al. 2009; Trexler and Goss 2009). Alligators (*Alligator mississippiensis*) are another hallmark charismatic species monitored as an indicator for restoration (Mazzotti et al. 2009). Crayfish, apple snails, and grass shrimp are probably minor components in their diet, but are taken at times. However, aquatic invertebrates sustain the vertebrate prey of alligators and so are also an important part of the food web that supports their health in the environment.

In Everglades management, invertebrates are rarely used as an indicator of general ecosystem health. King and Richardson (2002) discussed the issues required for an invertebrate monitoring program and suggested using midges and other small macroinvertebrates as indicators of nutrient enrichment. However, periphyton was developed to fill this role early in the restoration process, and there has been little interest in funding a second system based on macroinvertebrates.

Conclusions

Aquatic invertebrates are essential elements of the Everglades ecosystem, providing critical links in the food web that sustain animals of great interest to the public. They also contribute to ecosystem function through detrital processing and energy flow. Despite these fundamental roles, their systematics, distribution, and ecology remain poorly understood and documented. More work on biodiversity and ecological processes is needed to understand their roles in the Everglades. Several key taxa are already identified as important contributors to the successful restoration of the Everglades, defined in part as recovering the historical abundance of wading birds and herpetofauna. The glass is certainly half full for aquatic ecology of invertebrates in this internationally important ecosystem. With continued investment in restoration of the Everglades, however, the future for the study of invertebrates there is bright and certain to be rewarding.

Acknowledgments We are grateful for the advice and collaboration of our many colleagues in Everglades research over the years, but especially to Evelyn Gaiser and Tim Collins from Florida International University and Jeff Kline from Everglades National Park for their input and support. We appreciate the help of Melanie Trexler in drawing Figs. 10.1, 10.2, and 10.3. Data presented in this chapter were made possible by funding from Task Agreement CP040130 between the South Florida Water Management District and FIU and Cooperative Agreement W912HZ-11-2-0048 between the US Army Corps of Engineers and FIU. This research was enhanced by collaborations with the Florida Coastal Everglades Long-Term Ecological Research Program (funded by the National Science Foundation, DBI-0620409 and DEB-9910514) and by past funding by the US National Park Service to both authors. The research was conducted in accord with FIU animal care guidelines. This is contribution #740 of the Southeast Environmental Research Center.

Appendix

Aquatic invertebrate taxa recorded from across the Everglades. Data from Conrow and Loftus (unpublished), Rader (1994), Jacobsen (2008), Jacobsen and Perry (2000), Urgelles (2010), Trexler (unpublished), Bruno (unpublished), and Turner and Taylor (1998)

Higher taxa	Family	Genus
Porifera	Spongillidae	*Spongilla*
Cnidaria		
Trachylina	Hydridae	*Hydra*
Rotifera		>50 spp.
Platyhelminthes		*Planaria*
Nemertea		*Prostoma*
Gastrotricha		
Nematoda		
Mollusca		
Gastropoda	Ampullariidae	*Marisa*
		Pomacea
	Hydrobiidae	*Aphaostracon*
		Littoridinops
		Lyogyrus
	Lymnaeidae	*Fossaria*
		Lymnaea
		Pseudosuccinea
	Physidae	*Haitia*
		Physella
	Planorbidae/Ancylidae	*Biomphalaria*
		Drepanotrema
		Ferrissia
		Gyraulus
		Laevapex
		Helisoma
		Micromenetus
		Planorbella
		Planorbula
	Thiaridae	*Melanoides*
Bivalvia	Sphaeriidae	
	Unionidae	*Elliptio*
		Uniomerus
		Villosa
Annelida		
Polychaeta	Nereidae	*Namanereis*

(continued)

(continued)

Higher taxa	Family	Genus
Oligochaeta	Lumbriculidae	*Eclipidrilus*
	Tubificidae	*Allonais*
		Bratislavia
		Dero
		Pristina
		Stylaria
Hirudinea	Erpobdellidae	*Mooreobdella*
	Glossiphoniidae	*Helobdella*
Bryozoa	Plumatellidae	*Pumatella*
Acarina		
Crustacea		
Branchiopoda	Bosminidae	*Bosmina*
		Streblocerus
	Chydoridae	*Alona*
		Alonella
		Camptocercus
		Chydorus
		Dunhevedia
		Ephemeroporus
		Kurzia
		Leydigia
		Pleuroxus
		Pseudochydorus
		Euryalona
	Daphnidae	*Ceriodaphnia*
	Macrothricidae	*Scapholeberis*
		Simocephalus
		Guernella
		Grimaldina
		Ilyocryptus
		Macrothrix
	Moinidae	*Moinodaphnia*
	Sididae	*Diaphanosoma*
		Latanopsis
		Pseudosida
Ostracoda	Cypridae	*Physocypria*
		Scottia

Higher taxa	Family	Genus
Copepoda	Argulidae	*Argulus*
	Centropagidae	*Osphranticum*
	Cletodidae	*Cletocamptus*
	Cyclopidae	*Acanthocyclops*
		Diacyclops
		Ectocyclops
		Eucyclops
		Homocyclops
		Macrocyclops
		Mesocyclops
		Microcyclops
		Paracyclops
		Thermocyclops
		Tropocyclops
	Cletodidaee	*Cletocamptus*
	Lernaeidae	*Lernaea*
	Canthocamptidae	*Canthocamptus*
	Laophontidae	*Onychocamptus*
	Phyllognathopodidae	*Phyllognathopus*
Amphipoda	Crangonyctidae	*Crangonyx*
	Dogielinotidae	*Hyalella*
Decapoda	Cambaridae	*Procambarus*
	Mysidae	*Taphromysis*
	Palaemonidae	*Palaemonetes*
Isopoda	Asellidae	*Caecidotea*
Insecta	Sphaeromatidae	*Sphaeroma*
Collembola	Entomobryidae	*Entomobrya*
	Isotomidae	*Isotomurus*
	Poduridae	
	Sminthuridae	
Ephemeroptera	Baetidae	*Callibaetis*
	Caenidae	*Caenis*
Odonata	Aeshnidae	*Anax*
		Coryphaeschna
		Nasiaecshna
	Coenagrionidae	*Enallagma*
		Ischnura
		Telebasis
	Gomphidae	*Aphylla*
		Arigomphus
	Corduliidae	*Epitheca*
		Epicordulia

(continued)

(continued)

Higher taxa	Family	Genus
	Libellulidae	*Brachymesia*
		Celithemis
		Erythemis
		Erythrodiplax
		Idiataphe
		Libellula
		Macrodiplax
		Pachydiplax
		Pantala
		Perithemis
		Tramea
Hemiptera	Belostomatidae	*Belostoma*
		Lethocerus
	Corixidae	*Palmacorixa*
		Trichocorixa
	Gerridae	*Gerris*
	Hebridae	*Neogerris*
	Hydrometridae	*Merragata*
		Hydrometra
	Macroveliidae	*Oravelia*
	Mesoveliidae	*Mesovelia*
	Naucoridae	*Pelocoris*
	Nepidae	*Ranatra*
	Notonectidae	*Buenoa*
	Veliidae	
Coleoptera	Chrysomelidae	*Donacia*
	Dryopidae	*Pelonomus*
	Dytiscidae	*Agabetus*
		Bidessonotus
		Celina
		Cybister
		Desmopachria
		Hydroporus/ Neoporus
		Hydrovatus
		Ilybius
		Laccophilus
	Gyrinidae	*Gyrinus*
	Haliplidae	*Haliplus*
		Peltodytes

Higher taxa	Family	Genus
	Hydrophilidae	*Berosus*
		Chaetarythria
		Crenitulus
		Derallus
		Enochrus
		Helobata
		Hydrobiomorpha
		Hydrochus
		Paracymus
		Tropisternus
	Helophoridae	*Helophorus*
	Noteridae	*Hydrocanthus*
		Suphis
		Suphisellus
	Psephenidae	
	Scirtidae	*Prionocyphon*
Trichoptera	Hydroptilidae	*Leucotrichia*
		Oxyethira
	Leptoceridae	*Leptocerus*
	Philopotamidae	*Nectopsyche*
		Oecitis
Lepidoptera	Noctuidae	*Simyra*
	Crambidae	*Acentria*
		Paraponyx
Diptera	Ceratopogonidae	*Bezzia/Palpomyia*
		Culicoides
		Dasyhelea
		Forcipomyia
	Chironomidae	*Ablabesmyia*
		Apedilum
		Asheum
		Beardius
		Cantopelopia
		Chironomus
		Cladopelma
		Cladotanytarsus
		Clinotanypus
		Coelotanypus
		Corynoneura
		Cricotopus
		Cryptochironomus
		Cryptotendipes

(continued)

(continued)

Higher taxa	Family	Genus
		Denopelopia
		Dicrotendipes
		Djalmabatista
		Einfeldia
		Endochironomus
		Endotribelos
		Fittkauimyia
		Glyptotendipes
		Goeldichironomus
		Guttipelopia
		Kiefferulus
		Labrundinia
		Larsia
		Limnophyes
		Manoa
		Microchironomus
		Monopelopia
		Nanocladius
		Natarsia
		Nilothauma
		Nimbocera
		Parachironomus
		Parakiefferiella
		Paralauterborniella
		Paramerina
		Paratanytarsus
		Paratendipes
		Phytotelmatocladius
		Polypedilum
		Procladius
		Psectrocladius
		Pseudochironomus
		Pseudosmittia
		Stenochironomus
		Tanypus
		Tanytarsus
		Thienemanniella
		Tribelos
		Xenochironomus
		Zavreliella
	Culicidae	*Aedes*
	Dolichopodidae	
	Ephydridae	*Ephydra*
	Psychodidae	*Pericoma*

Higher taxa	Family	Genus
	Stratiomyidae	*Odontomyia*
	Tabanidae	*Tabanus*
	Tipulidae/Limoniidae	*Elliptera*
		Limonia
		Polymera
		Tipula

References

Abbey-Lee R, Gaiser EE, Trexler JC (2013) Relative role of dispersal dynamics and competition in determining isotopic niche breadth. Freshw Biol 58:780–792

Acosta CA, Perry SA (2001) Impact of hydropattern disturbance on crayfish population dynamics in the seasonal wetlands of Everglades National Park, USA. Aquat Conserv Mar Freshw Ecosys 11:45–57

Acosta CA, Perry SA (2002) Spatio-temporal variation in crayfish production in disturbed marl prairie marshes of the Florida Everglades. J Freshw Ecol 17:641–650

Beganyi SR, Batzer DP (2011) Wildfire induced changes in aquatic invertebrate communities and mercury bioaccumulation in the Okefenokee Swamp. Hydrobiologia 669:237–247

Belicka LL, Sokol ER et al (2012) A molecular and stable isotopic approach to investigate the importance of algal and detrital energy pathways in a freshwater marsh. Wetlands 32:531–542

Boucek RE, Rehage JS (2014) Climate extremes drive changes in functional community structure. Glob Change Biol 20:1821–1831

Brakke DF (1976) Modification of the Whiteside-Williams pattern sampler. J Fish Res Board Can 33:2861–2863

Bransky JW, Dorn NJ (2013) Prey use of wetland benthivorous sunfishes: ontogenetic, interspecific, and seasonal variation. Environ Biol Fish 96:1329–1340

Bruno MC, Perry SA (2004) Exchanges of copepod fauna between surface and ground-water in the Rocky Glades of Everglades National Park (Florida, USA). Arch Hydrobiol 159:489–510

Bruno MC, Perry SA (2005) Temporal and spatial variations in Copepod (Crustacea) communities in groundwater in the Rocky Glades of Everglades National Park (Florida, USA). J Freshw Ecol 20:27–36

Bruno MC, Loftus WF, Perry SA (2001) Preliminary data on microcrustacean communities from ground waters in the southern Everglades. In: Kuniansky EL (ed) USGS Karst interest group proceedings, Water-Resources Investigations Rep 01-401, p 189–197

Bruno MC, Cunningham KJ, Perry SA (2003) Copepod communities from surface and ground waters in the Everglades, South Florida. SE Natur 2:523–546

Bruno MC, Reid JW, Perry SA (2005) A list and identification key for the freshwater, free-living copepods of Florida (USA). J Crustac Biol 25:384–400

Cattau CE, Martin J, Kitchens WM (2010) Effects of an exotic prey species on a native specialist: example of the snail kite. Biol Conserv 143:513–520

Cattau CE, Darby PC, Fletcher RJ, Kitchens WM (2014) Reproductive responses of the endangered snail kite to variations in prey density. J Wildl Manage 78:620–631

Chick JH, Geddes P, Trexler JC (2008) Periphyton mat structure mediates trophic interactions in a subtropical wetland. Wetlands 28:378–389

Cleckner LB, Garrison PJ et al (1998) Trophic transfer of methyl mercury in the northern Florida Everglades. Biogeochemistry 40:347–361

Cook MI, Call EM et al (2014) Seasonal movements of crayfish in a fluctuating wetland: implications for restoring wading bird populations. Freshw Biol 59:1608–1621

Crozier GE, Gawlik DE (2002) Avian response to nutrient enrichment in an oligotrophic wetland, the Florida Everglades. Condor 104:631–642

Daraghy AO, Conrow R, Loftus WF (1988) Acetic acid digestion of high-carbonate substrates: an aid to sorting aquatic invertebrate samples. J Freshw Ecol 4:517–521

Darby PC, Bennetts RE, Percival HF (2008) Dry down impacts on apple snail (*Pomacea paludosa*) demography: implications for wetland water management. Wetlands 28:204–214

Davis SM (1994) Phosphorus inputs and vegetation sensitivity in the Everglades. In: Davis SM, Ogden JC (eds) Everglades: The Ecosystem and Its Restoration. St. Lucie Press, Boca Raton, pp 357–378

Donar CM, Condon KW, Gantar M, Gaiser EE (2004) A new technique for examining the physical structure of Everglades floating periphyton mat. Nova Hedwigia 78:107–119

Dorn NJ (2008) Colonization and reproduction of large macroinvertebrates are enhanced by drought-related fish reductions. Hydrobiologia 605:209–218

Dorn NJ, Cook MI (2015) Hydrological disturbance diminishes predator control in wetlands. Ecology. http://dx.doi.org/10.1890/14-1505.1

Dorn N, Trexler JC (2007) Crayfish assemblage shifts in a large drought-prone wetland: the roles of hydrology and competition. Freshw Biol 52:2399–2411

Dorn NJ, Trexler JC, Gaiser EE (2006) Exploring the role of large predators in marsh food webs: evidence for a behaviorally-mediated trophic cascade. Hydrobiologia 569:375–386

Duever MJ, Meeder JF, Meeder LC, McCollom JM (1994) The climate of South Florida and its role in shaping the Everglades ecosystem. In: Davis SM, Ogden JC (eds) Everglades: The Ecosystem and Its Restoration. St. Lucie Press, Boca Raton, pp 225–248

Epanchin PN, Heath JA, Frederick PC (2002) Effects of fires on foraging and breeding wading birds in the Everglades. Wilson Bull 114:139–141

Frederick PC, Ogden JC (2001) Pulsed breeding of long-legged wading birds and the importance of infrequent severe drought conditions in the Florida Everglades. Wetlands 21:484–491

Frederick P, Gawlik DE et al (2009) The White Ibis and Wood Stork as indicators for restoration of the Everglades ecosystem. Ecol Indicat 9S:S83–S95

Gaiser EE, Trexler JC et al (2005) Exposure to above-ambient phosphorus causes ecosystem state change in the Everglades. J Environ Qual 34:717–723

Gaiser EE, Trexler JC, Wetzel PR (2012) The Florida Everglades. In: Batzer DP, Baldwin AH (eds) Wetland habitats of North America: ecology and conservation concerns. University of California Press, Berkeley, pp 231–252

Gawlik DE (2002) The effects of prey availability on the numerical response of wading birds. Ecol Monogr 72:329–346

Geddes P, Trexler JC (2003) Uncoupling of omnivore-mediated positive and negative effects on periphyton mats. Oecologia 136:585–595

Gleason PJ, Stone P (1994) Age, origin, and landscape evolution of the Everglades peatland. In: Davis SM, Ogden JC (eds) Everglades: The Ecosystem and Its Restoration. St. Lucie Press, Boca Raton, pp 149–197

Goldsborough LG, Robinson GGC (1996) Pattern in wetlands. In: Stevenson RJ, Bothwell ML, Lowe RL (eds) Algal ecology. Freshwater benthic ecosystems. Academic, New York, pp 78–117

Gottlieb AD, Gaiser EE, Lee SS (2015) Changes in hydrology, nutrient loading and conductivity in the Florida Everglades, and concurrent effects on periphyton community structure and function. In: Entry JA, Gottlieb AD, Jayachandrahan K, Ogram A (eds) Microbiology of the Everglades ecosystem. Science Publishers CRC Press, Boca Raton, pp 155–179

Gresswell RE (1999) Fire and aquatic ecosystems in forested biomes of North America. Trans Am Fish Soc 128:193–221

Gunderson LH, Loftus WF (1993) The Everglades. In: Martin WH, Boyce SG, Echternacht AC (eds) Biodiversity of the Southeastern United States. Wiley, New York, pp 199–255

Gunderson LH, Snyder JR (1994) Fire patterns in the Southern Everglades. In: Davis SM, Ogden JC (eds) Everglades: The Ecosystem and Its Restoration. St. Lucie Press, Boca Raton, pp 291–305

Hagerthey SE, Cook MI et al (2014) Aquatic faunal responses to an induced regime shift in the phosphorus-impacted Everglades. Freshw Biol 59:1389–1405

Harvey RG, Loftus WF, Rehage JS, Mazzotti FJ (2010) Effects of canals and levees on Everglades ecosystems. IFAS, University of Florida, Gainesville, Rep WEC204

Hauer FR, Resh VH (2007) Macroinvertebrates. In: Hauer FR, Lamberti GA (eds) Methods in stream ecology, 2nd edn. Academic, New York, pp 435–463

Hayes KA, Cowie RH, Thiengo SC, Strong EE (2012) Comparing apples with apples: clarifying the identities of two highly invasive Neotropical Ampullariidae (Caenogastropoda). Zool J Linnean Soc 166:723–753

Hendrix AN (2000) Population size and life-history parameters of the Everglades crayfish, *Procambarus* spp. M.S. Thesis, University of Washington, Seattle

Hendrix AN (2003) Crayfish (Procambarus spp.) response to hydrologic restoration of the Florida Everglades. Ph.D. Dis, University of Washington, Seattle

Hendrix AN, Loftus WF (2000) Distribution and relative abundance of the crayfishes *Procambarus alleni* (Faxon) and *P. fallax* (Hagen) in southern Florida. Wetlands 20:194–199

Hoang TC, Rand GM (2009) Exposure routes of copper: short term effects on survival, weight, and uptake in Florida apple snails (*Pomacea paludosa*). Chemosphere 76:407–414

Hoang TC, Tong X (2015) Influence of water quality on zinc toxicity to the Florida apple snail (*Pomacea paludosa*) and sensitivity of freshwater snails to zinc. Environ Toxicol Chem 34:545–553

Hoang TC, Schuler LJ et al (2009) Copper release, speciation, and toxicity following multiple floodings of copper enriched agriculture soils: implications in Everglades restoration. Water Air Soil Pollut 199:79–93

Horgan FG, Stuart AM, Kudavidanage EP (2014) Impact of invasive apple snails on the functioning and services of natural and managed wetlands. Acta Oecol 54:90–100

Jacobsen RE (2008) A Key to the Pupal Exuviae of the Midges (Diptera: Chironomidae) of Everglades National Park, Florida. USGS Scientific Investigations Rep 2008-5082

Jacobsen RE, Perry SA (2000) A review of *Beardius* Reiss & Sublette, with description of a new species from Everglades National Park, Florida (Insecta, Diptera, Chironomidae). Spixiana 23:129–144

Jelks HL, Jordan FC, Kitchens WM (1992) Response of wading birds and aquatic macrofauna to hydrological conditions and vegetative structure of Loxahatchee National Wildlife Refuge. Work Order 32, Final Rep. Arthur R. Marshall Loxahatchee National Wildlife Refuge, Rep 1991–4113

Jordan CF Jr (1996) Spatial ecology of decapods and fishes in a northern Everglades wetland mosaic. Ph.D. Dissertation, University of Florida, Gainesville

Jordan CF, Coyne S, Trexler JC (1997) Sampling fishes in heavily vegetated habitats: the effects of habitat structure on sampling characteristics of the 1-m^2 throw trap. Trans Am Fish Soc 126:1012–1020

Kaplan DA, Paudel R, Cohen MJ, Jawitz JW (2012) Orientation matters: patch anisotropy controls discharge. Geophys Res Lett 39, L17401. doi:10.1029/2012GL052754

Kellogg CM, Dorn NJ (2012) Consumptive effects of fish reduce wetland crayfish recruitment and drive species turnover. Oecologia 168:1111–1121

King RS, Richardson CJ (2002) Evaluating subsampling approaches and macro invertebrate taxonomic resolution for wetland bioassessment. J North Am Benthol Soc 21:150–171

King RS, Richardson CJ (2003) Integrating bioassessment and ecological risk assessment: an approach to developing numerical water-quality criteria. Environ Manage 31:795–809

King RS, Richardson CJ (2007) Subsidy-stress response of macroinvertebrate community biomass to a phosphorus gradient in an oligotrophic wetland ecosystem. J North Am Benthol Soc 26:491–508

King RS, Richardson CJ (2008a) Macroinvertebrate responses to a gradient of long-term nutrient additions, altered hydroperiod, and fire. In: Richardson CJ (ed) The Everglades experiments: lessons for ecosystem restoration. Springer, New York, pp 277–319

King RS, Richardson CJ (2008b) Macroinvertebrate and fish responses to experimental P additions in Everglades sloughs. In: Richardson CJ (ed) The Everglades experiments: lessons for ecosystem restoration. Springer, New York, pp 477–502

Kline JL, Loftus WF et al (2014) Recent fish introductions into Everglades National Park: an unforeseen consequence of water-management? Wetlands 34:S175–S187

Knorp NE, Dorn NJ (2014) Dissimilar numerical responses of macroinvertebrates to disturbance from drying and predatory sunfish. Freshw Biol 59:1378–1388

Kushlan JA (1976) Environmental stability and fish community diversity. Ecology 57:821–825

La Hée JM, Gaiser EE (2012) Benthic diatom assemblages as indicators of water quality in the Everglades and three tropical karstic wetlands. Freshw Sci 31:205–221

Larsen L, Aumen N et al (2011) Recent and historic drivers of landscape change in the Everglades ridge, slough, and tree island mosaic. Crit Rev Environ Sci Technol 41(S1):344–381

Light SS, Dineen JW (1994) Water control in the Everglades: a historical perspective. In: Davis M, Ogden JC (eds) Everglades. The ecosystem and its restoration. St. Lucie Press, Boca Raton, pp 47–84

Liston SE (2006) Interactions between nutrient availability and hydroperiod shape macroinvertebrate communities in Florida Everglades marshes. Hydrobiologia 569:343–357

Liston SE, Trexler JC (2005) Spatiotemporal patterns in community structure of macroinvertebrates inhabiting calcareous periphyton mats. J North Am Benthol Soc 24:832–844

Liston SE, Newman S, Trexler JC (2008) Macroinvertebrate community response to eutrophication in an oligotrophic wetland: an in situ mesocosm experiment. Wetlands 28:686–694

Lockwood JL, Ross MS, Sah JP (2003) Smoke on the water: the interplay of fire and water flow on Everglades restoration. Front Ecol Environ 9:462–468

Loftus WF (2000) Accumulation and fate of mercury in an Everglades aquatic food web. Ph.D. Dissertation, Florida International University, Miami, FL

Loftus WF, Kushlan JA (1987) Freshwater fishes of southern Florida. Bull Florida State Mus Biol Sci 31:147–344

Loftus WF, Chapman JD, Conrow R (1990) Hydroperiod effects on Everglades marsh food webs, with relation to marsh restoration efforts. In: Larson G, Soukup M (eds) Fisheries and coastal wetlands research: proceedings of the 1986 conference on science in national parks. US National Park Service and George Wright Society, Fort Collins, pp 1–22

Martin J, Kitchens WM, Hines JE (2007) Importance of well-designed monitoring programs for the conservation of endangered species: case study of the snail kite. Conserv Biol 21:472–481

Matich P, Heithaus MR (2012) Effects of an extreme temperature event on the behavior and age structure of an estuarine top predator, *Carcharhinus leucas*. Mar Ecol Progr Ser 447:165–178

Mazzotti FJ, Best GR et al (2009) Alligators and crocodiles as indicators for restoration of Everglades ecosystems. Ecol Indicat S9:S137–S149

McCormick PV, O'Dell MB (1996) Quantifying periphyton responses to phosphorus in the Florida Everglades: a synoptic-experimental approach. J North Am Benthol Soc 15:450–468

McCormick PV, Newman S et al (2002) Effects of anthropogenic phosphorus inputs on the Everglades. In: Porter JW, Porter KG (eds) The Everglades, Florida Bay, and Coral Reefs of the Florida Keys. An ecosystem sourcebook. CRC, Boca Raton, pp 83–126

McCormick PV, Shuford RBE, Rawlik PS (2004) Changes in macroinvertebrate community structure and function along a phosphorus gradient in the Florida Everglades. Hydrobiologia 529:113–132

McVoy CW, Said WP et al (2011) Landscapes and hydrology of the pre-drainage Everglades. University Press Florida, Gainesville

Means DB, Simberloff D (1987) The peninsula effect: habitat-correlated species decline in Florida's herpetofauna. J Biogeogr 14:551–568

National Research Council (NRC) (2006) Progress toward restoring the Everglades. The first biennial review—2006. National Academies Press, Washington, DC

Newman S, Schuette J et al (1998) Factors influencing cattail abundance in the northern Everglades. Aquat Bot 60:265–280

Noe G, Childers DL, Jones RD (2001) Phosphorus biogeochemistry and the impact of phosphorus enrichment: why is the Everglades so unique? Ecosystems 4:603–624

Odum EP, Finn JT, Franz EH (1979) Perturbation theory and the subsidy-stress gradient. BioScience 29:349–352

Ogden JC (1994) A comparison of wading bird nesting dynamics, 1931–1946 and 1974–1989 as an indication of changes in ecosystem conditions in the southern Everglades. In: Davis S, Ogden JC (eds) Everglades: the ecosystem and its restoration. St. Lucie Press, Boca Raton, pp 533–570

Ogden JC, Davis SM et al (2005) Total system conceptual ecological model. Wetlands 25:955–979

Palmer MA, Strayer DL, Rundle SD (2007) Meiofauna. In: Hauer FR, Lamberti GA (eds) Methods in stream ecology, 2nd edn. Academic, New York, pp 415–433

Porter-Whitaker A, Rehage JS, Loftus WF, Liston SE (2012) Multiple predator effects and native prey responses to two non-native Everglades cichlids. Ecol Freshw Fish 21:375–385

Rader RB (1994) Macroinvertebrates of the northern Everglades: species composition and trophic structure. Florida Sci 57:22–33

Rader RB (1999) The Florida Everglades: natural variability, invertebrate density, and foodweb stability. In: Batzer DP, Rader RB, Wissinger SA (eds) Invertebrates in freshwater wetlands of North America. Ecology and management. Wiley, New York, pp 25–54

Rader RB, Richardson CJ (1992) The effects of nutrient enrichment on algae and macroinvertebrates in the everglades—a review. Wetlands 12:121–135

Rand GM, Schuler LJ (2009) Aquatic risk assessment of metals in sediment from South Florida canals. Soil Sedim Contam 18:155–172

Rawlings TA, Hayes KA, Cowie RH, Collins TM (2007) The identity, distribution, and impacts of non-native apple snails in the continental United States. BMC Evol Biol 7:97

Rehage JS, Trexler JC (2006) Assessing the net effect of anthropogenic disturbance on aquatic communities in wetlands: community structure relative to distance from canals. Hydrobiologia 569:359–373

Ruehl CB (2010) The interactive effects of predators, resources, and disturbance on freshwater snail populations from the Everglades. Ph.D. Dissertation, Florida International University, Miami

Ruehl CB, Trexler JC (2011) Comparisons of snail density, standing stock, and body size among freshwater ecosystems: a review. Hydrobiologia 665:1–13

Ruehl CB, Trexler JC (2013) A suite of prey traits determine predator and nutrient enrichment effects in a tri-trophic food web. Ecosphere 4(6):75

Ruehl CB, Trexler JC (2015) Reciprocal transplant reveals trade-off of resource quality and predation risk in the field. Oecologia 179:117–127

Sargeant BL, Gaiser EE, Trexler JC (2011) Indirect and direct controls of macroinvertebrates and small fish by abiotic factors and trophic interactions in the Florida Everglades. Freshw Biol 56:2334–2346

Schofield PJ, Slone DH, Gregoire DR, Loftus WF (2013) Effects of a non-native cichlid fish (African Jewelfish, *Hemichromis letourneuxi* Sauvage 1880) on a simulated Everglades aquatic community. Hydrobiologia 722:171–182

Sklar FH, Chimney MJ et al (2005) The ecological-societal underpinnings of Everglades restoration. Front Ecol Environ 3:161–169

Sokol ER, Hoch JM, Gaiser EE, Trexler JC (2014) Metacommunity structure along resource and disturbance gradients in Everglades wetlands. Wetlands 34:S135–S146

Stewart TJ, Traber J et al (2013) Characterization of extracellular polymeric substances (EPS) from periphyton using liquid chromatography-organic carbon detection–organic nitrogen detection (LC-OCD-OND). Environ Sci Pollut Res 20:3214–3223

Stober QJ, Thornton K et al (2001) South Florida ecosystem assessment: phase I/II (technical report)—Everglades stressor interactions: hydropatterns, eutrophication, habitat alteration, and mercury contamination. EPA 904-R-01-003. http://www.epa.gov/Region4/sesd/reports/epa904r01003/p2treport.pdf

Taylor RC, Trexler JC, Loftus WF (2001) Separating the effects of intra- and interspecific age-structured interactions in an experimental fish assemblage. Oecologia 127:143–152

Thomas S, Gaiser EE, Gantar M, Scinto LJ (2006) Quantifying the responses of calcareous periphyton crusts to rehydration: a microcosm study (Florida Everglades). Aquat Bot 84:317–323

Thompson FG (2004) An identification manual for the freshwater snails of Florida. Florida Museum of Natural History. https://www.flmnh.ufl.edu/malacology/fl-snail/snails1.htm

Trexler JC (1995) Restoration of the Kissimmee River: a conceptual model of past and present fish communities and its consequences for evaluating restoration success. Restor Ecol 3:195–210

Trexler JC, Goss CW (2009) Aquatic fauna as indicators for Everglades restoration: applying dynamic targets in assessments. Ecol Indicat 9S:S108–S119

Trexler JC, Loftus WF, Perry S (2005) Disturbance frequency and community structure in a twenty-five year intervention study. Oecologia 145:140–152

Trexler JC, Gaiser EE, Kominoski JS, Sanchez J (2015) The role of periphyton mats in consumer community structure and function in calcareous wetlands: lessons from the Everglades. In: Entry JA, Gottlieb AD, Jayachandrahan K, Ogram A (eds) Microbiology of the Everglades Ecosystem. Science Publications, CRC Press, Boca Raton, pp 155–179

Turner PN, Taylor HL (1998) Rotifers new to Florida, USA. Hydrobiologia 387(388):55–62

Turner AM, Trexler JC (1997) Sampling aquatic invertebrates from marshes: evaluating the options. J North Am Benthol Soc 16:694–709

Turner AM, Trexler JC et al (1999) Targeting ecosystem features for conservation: standing crops in the Florida Everglades. Conserv Biol 13:898–911

Urgelles R (2010) Dragonfly assemblage structure in relation to hydroperiod and vegetation gradients in freshwater wetlands of the Florida Everglades. M.S. Thesis, Florida International University, Miami

VanArman PG (2011) Role of native crayfish, *Procambarus alleni* (Faxon) and *Procambarus fallax* (Hagen), in Everglades food webs: a literature review and conceptual model. Florida Sci 74:100–125

Venne LS (2012) Wading bird (Ciconiiformes) response to fire and the effects of fire in the Everglades. Ph.D. Dissertation, University of Florida, Gainesville

Venne LS, Frederick PC (2013) Foraging wading bird (Ciconiiformes) attraction to prescribed burns in an oligotrophic wetland. Fire Ecol 9:78–95

Wetzel RG (1983) Attached algal-substrata interactions: fact or myth, and when and how? In: Wetzel RG (ed) Periphyton of freshwater ecosystems. Develop Hydrobiol 17:207–215

Whiteside MC, Williams JB (1975) A new sampling technique for aquatic ecologists. Verhand Internat Verein Theoreti Angew Limnol 19:1534–1539

Williams AJ, Trexler JC (2006) A preliminary analysis of the correlation of food-web characteristics with hydrology and nutrient gradients in the southern Everglades. Hydrobiologia 569:493–504

Chapter 11
Invertebrates in Groundwater Springs and Seeps

D. Dudley Williams

Introduction to Springs and Seeps

It would be as well to begin with discussion of some definitions of springs and seeps as, today, there is an overabundance of imprecise terms to describe these habitats (e.g. spring source, spring-fed stream, springbrook, headwater spring, spring run, seep, spring/seep). Some definitions are based on the immediate physical and chemical environment, while others focus on the communities of organisms living there. These differences stem, largely, from the disparate backgrounds of the scientists who study them. For example, geologists tend to define thermal springs with respect to temperatures more than 5 °C above the local mean annual air temperature (van Everdingen 1991)—which reflects geothermal warming. Biologists customarily define thermal springs in reference to temperatures above the local annual mean (warm), as well as above the annual mean maximum (relatively hot), and above 40 °C (absolutely hot) (Tuxen 1944). Biologists also sometimes categorize habitat temperatures in a manner that best coincides with the distribution of a given taxon. For example, Smith (1991) defined a "hot" spring as one above 30 °C as none of the Canadian water mites he studied occurred in waters that are hot. However, some insects and other invertebrates can be found in springs exceeding 40 °C, requiring a different cut-off level for "hot" (for further discussion, see Pritchard 1991).

From a simplistic hydrological viewpoint, a spring represents a concentrated point of natural discharge of groundwater that is sufficiently high to maintain flow on the land surface (van Everdingen 1991). The flow of water from a spring often comprises a mixture of water that has infiltrated the subsurface at different times and places. The rate of this recharge varies according to the balance between precipitation, infiltration, run-off and evapotranspiration, and as the contributions from different recharge points change, so too will the output of the spring vary.

D. Batzer, D. Boix (eds.), *Invertebrates in Freshwater Wetlands*,
DOI 10.1007/978-3-319-24978-0_11

Van der Kamp (1995) illustrated such variation in hydrogeology with examples of two spring types: (1) a spring fed by slow seepage through small pore spaces (e.g. sand and rock with fine fractures, typically <0.1 mm wide) with strong filtration which produces stable flow and water devoid of any large particulate matter and (2) a spring fed through cavernous carbonate rock—such "karst" springs typically have strong, but variable, flow and higher particulate loads. In some instances, a spring may be fed by a very shallow or small aquifer resulting in intermittent surface flow. Such contrasting discharge patterns can be predicted to radically affect the nature of the community capable of living in these respective springs. Seeps comprise groundwater that emerges over a more diffuse area, with a rate of flow that is generally insufficient to form an outflow stream—the water typically infiltrating into surrounding soil and/or vegetation, such as mosses. Springs per se have been traditionally subdivided into three broad types: *rheocrenes*, in which the source emerges into a rapidly flowing stream; *limnocrenes*, in which the source first enters a basin with slow flow; and *helocrenes*, in which emerging water percolates into a marshy holding area, frequently rich in mosses (Bornhauser 1913). There are, of course, some springs that do not fit neatly into these definitions, for example, where a limnocrene has sufficient discharge such that the water spills out of the basin forming a fast-flowing stream, which might be termed a "rheolimnocrene".

If we examine a simple water budget for springs and seeps, it is clear that groundwater is their primary source. Precipitation directly on the relatively small surface areas of, for example, most rheocrenes is likely to be low, although that on limnocrenes and larger seeps may be higher. Of the main water losses (outputs), evapotranspiration is likely to be low—but again related to spring/seep surface area, together with the characteristics of local vegetation. Water lost to floodplains may be retained in wetlands or flow down-gradient in springbrook channels. In the case of seeps and helocrenes, water may diffuse away as overland flow into surrounding depressions, soil and vegetation (such as mosses). Depending on local topography and soil infiltration capacity, some spring waters may re-enter shallow downstream aquifers, perhaps to re-emerge at downslope springs or seeps.

Emergence of groundwater, as described above, occurs globally, making springs and seeps commonplace in all but the most arid or cold regions—although even the latter have their own characteristic upwellings (e.g. those found in the Chihuahuan Desert in Texas (Wallace et al. 2005) and on Axel Heiberg Island in the High Arctic (Lay et al. 2013)). Indeed, springs and seeps may exist, or have existed, on other planets in our solar system. For example, there is now evidence of a former groundwater network on Mars from which water may have seeped up to the surface and evaporated (Grossman 2013). Recent, high-resolution imaging has shown dark features extending down steep slopes from bedrock outcrops, often associated with small channels. These features appear to grow incrementally during warm seasons, but fade in cold seasons. Near-surface, liquid brines might explain these observations (McEwen et al. 2011), and it is tempting to make comparisons with the saline springs of the Earth, which support a variety of life.

General Habitat Conditions in Springs

For aquatic invertebrates, it is important to remember that environment and community are intimately linked and that springs and seeps represent a range of habitats from the "classical" highly stable (especially in terms of temperature and discharge) type to those with a more seasonably variable hydroperiod, such as those associated with wetlands (Danks and Rosenberg 1987). Many factors influence habitat conditions in springs and seeps, including those associated with the terrain (e.g. local geology, topography and groundwater reserves) and geographical position (e.g. latitudes and glacial history), alongside smaller-scale influences such as vegetation (submerged, emergent and riparian, especially if it shades the water) (Danks and Williams 1991).

Terrain features dictated by geology and topography determine the supply of water and how it varies, together with the level and variability of its temperature and chemical composition. Thus, large groundwater reservoirs buffer the habitats from erratic drying and other major changes, they tend to enhance the rate of flow but reduce its variation, and they stabilize temperatures. Longer groundwater residence in softer rocks increases mineralization, but depending on the composition of the rocks may either increase or decrease pH (van Everdingen 1991). Permanent springs are fed by groundwater with at least a 1-year residency underground, but the water of many springs is fossil water that emerges from the ground long after it entered, often 10,000 years later (Downing et al. 1977).

Conditions above the ground surface are modified on a broad geographical scale by local climate and other regional elements—such as photoperiod, known to promote the seasonal succession of invertebrate species and algae that occurs in springs despite oftentimes very constant temperature (Teal 1957)—and on a yet smaller scale by the size of the spring, its vegetation and habitat diversity. These influences especially affect the temperature, primary productivity, allochthonous food (dead leaves) supply and other ecological factors. Thus, while a spring may have very uniform abiotic conditions, there may be considerable annual variation in its biotic conditions (Varza and Covich 1995). High mineral content creates mineral springs, including salt springs (Ring 1991); high-temperature groundwater creates thermal springs (Pritchard 1991); and erratic or surficial discharges produce intermittent springs. Discharge has an important effect on spring-bed substrates, which have a major influence on the invertebrate community. For example, constant or intermittent high flow rates remove fine particles thus increasing the mean particle size on the bed and reducing the amount of detrital food materials. In contrast, where the water wells up into a basin, fine particles accumulate. Highly mineralized springs may precipitate flocculent substrates.

In any given spring or seep, all or some of these factors come together to characterize the size, rate and pattern of flow, temperature regime and chemical composition of the habitat. The community of invertebrates that subsequently develops reflects tolerance of and/or adaptation to these factors, together with colonization

Table 11.1 Recommended minimal key descriptors of springs and seeps for biological data

	Status observed		
Key descriptors	Low	Intermediate	High
Nature of source	Helocrene (discharge or seep into marshy/mossy substrate)	Limnocrene (discharge into basin)	Rheocrene (rapid flow into defined channel on discharge)
Discharge at source	Low volume (<0.01 m^3 s^{-1})	Medium volume (0.01–0.5 m^3 s^{-1})	High volume (>0.5 m^3 s^{-1})
Water temperature	Cold (<10°C)	Warm (10–40 °C)	Hot (>40°C)
Chemistry: as total dissolved solids	Freshwater (<1000 ppm)	Mineral (1000–35,000 ppm)	Saline (>35,000 ppm)
Persistence[a] (as approx. interval between major disturbances)	Intermittent (typically annual, but up to 5 years)	Apparently permanent (>5 years, i.e. some signs of inconsistency)	Permanent (>50 years; no disturbance)

Source: Danks and Williams (1991)

[a]Unless historical records are available (e.g. from palaeoecological evidence—see Williams and Williams 1996a, b), interpretation here is likely to be largely subjective, based on observations of physical and biological conditions made at the site

abilities from neighbouring or distant sources. The latter is particularly important in regions where new habitat has been created, such as by shifts in regional geology, affecting the level of the groundwater table, or by the retreat of glaciers during the Pleistocene (Williams and Williams 1999). The role of these factors will be revisited later in the chapter.

As noted, strictly speaking, a spring can be defined only in its proper geological setting, taking into account numerous variables that characterize its nature and potential as a habitat. These variables include some that are easily measured, such as pH, temperature, current speed and dissolved oxygen, and others that are much more difficult to quantify, such as size of underlying aquifer and nature and length of groundwater flow paths (both of which may influence habitat stability). Given limitations in acquiring measurements of some of these variables, and in preference to a rigid classification of springs and seeps, for which there will always be exceptions, Danks and Williams (1991) proposed using a minimum of five key descriptors to characterize each habitat sampled for biological data (Table 11.1). These descriptors are seen as providing sufficient characterization of fundamental elements of springs as habitats and comprise source geometry, rate of water supply, temperature, chemistry and persistence. Thus, for example, the following minimum description provides a useful and comparable characterization: "low volume, warm, freshwater, annual, temporary rheocrene". Where logistically possible, such a basic description should always be supplemented by additional information and measurements.

Spring Invertebrate Faunas

Broad Characteristics of the Fauna

Within the last couple of decades, there have been several attempts to bring together the collective knowledge of spring faunas with the aim of galvanizing further study, synthesis and conservation (Williams and Danks 1991; Ferrington 1995; Hinterlang and Lischewki 1996; Botosaneanu 1998). However, while the spring faunas of some regions (e.g. Europe and North America) and some taxa (e.g. Trichoptera, Diptera and Acari) are becoming better known, the information base for spring communities in most geographic regions of the world is very sparse (Yule 2004a).

In lotic systems, the diversity of invertebrates generally increases downstream. Typically, therefore, the section of channel immediately below the outlet (springbrook) will have a lower diversity than the stream, whereas the spring itself will have the least (Ward and Dufford 1979). However, this is not always the case (Resh 1983), and certain taxa may be particularly species rich at the source. For example, Gathmann and Williams (2006) recorded 38 species of Limoniidae (Diptera) from 30 cold-water springs in Southern Ontario, Canada. Interestingly, of the total of 86 insect species found, 55 % were found at three or fewer sites, and only for very few of these rare species were more than ten individuals m^{-2} collected. Interannual differences were also apparent, with emergence records from a single year, on average, containing around 50 % of the total number of species found in the entire study. Temperature and discharge correlated most strongly with insect community similarity among springs.

Invertebrates collected in or near springs and seeps range from permanent residents to those that are just visiting—to avail themselves of food or use these habitats as "stepping stones" in an attempt to reach other waterbodies. Many of these nonresidents do not breed in springs yet they may dominate the fauna. For example, of the 33 species of aquatic beetle found in a spring in North Dakota, 64 % were nonresident (Roughley and Larson 1991). Resident species are derived from several different sources (Fig. 11.1). At the core are species (termed "crenobionts") that are confined to springs and/or seeps and which exhibit suitable adaptations to cold water and constant abiotic conditions (Danks and Williams 1991). Another important component comprises the phreatic species of turbellarians, micro-crustaceans, mites, plecopterans and other taxa that live in suitable, oxygenated groundwaters and enter the springs that emanate from them. Generalist species from downstream cold-water brooks also enter where there is sufficient flow (rheocrenes). Limnocrenes, helocrenes and seeps may contain some lentic species, although not those that cannot tolerate very cold water (e.g. pond species). Some species are associated with spring margins, especially where there is abundant moss or organic materials, or a distinct splash zone, for example, mites, collembolans and various semiaquatic dipterans. Other species may be derived from the wet soil at a spring's edge (e.g. rotifers, mites, nematodes, snails). Such species living in springs, but not exclusively

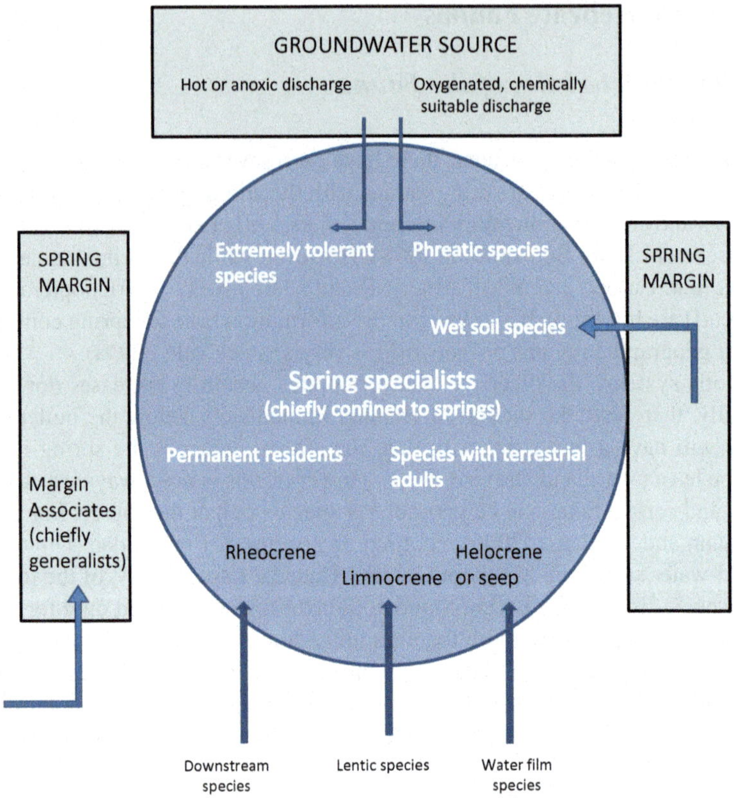

Fig. 11.1 Components of the faunas of springs and seeps (modified after Danks and Williams 1991)

so, are often termed "crenophiles". Hot springs and saline springs are populated by a low diversity of extremely tolerant species, typically in the case of insects, belonging to the Diptera, Coleoptera and Hemiptera (Ring 1991)—additionally, Odonata may occur in hot springs (Pritchard 1991).

Specifics of the Fauna

The Appendix is an attempt to identify, on a global scale, taxa that either are known to occur repeatedly in springs and seeps or that might reasonably be expected to live in them, by virtue of known occurrence in related habitats, such as cold hypogean waters or high-altitude streams. Records are cited mostly at the genus level and include those where a genus is known from one particular geographical region (although not specifically from springs there) but is known from springs in another region—thus, there is the potential for finding it in the former in the future. Where known, habitat details are provided for each genus together with its recorded global distribution and the latter's reference sources. Distributions vary from "widespread" to single zoogeographical regions, such as the "Nearctic", and will be the result of many factors. These include active/passive dispersal and colonization abilities, habitat availability, persistence and connectedness, plate tectonics, the presence/absence/location of barriers and the evolutionary history and chronology of the genus, including opportunities for speciation and endemism. These data are by no means complete, but hopefully will serve as a useful first step towards creating a global framework to which new records of species and their occurrences in springs and seeps can be added.

One of the first observations from the Appendix is that most of the major invertebrate groups found in freshwaters in general are represented in springs and/or seeps. These taxa range from the more sessile forms, such as sponges and hydrozoans, through small motile forms, such as gastrotrichs, rotifers and nematodes, to micro-crustaceans, mites and many insect orders. Groups that have freshwater representatives but appear to be absent from springs and seeps include bryozoans (which tend to be more restricted to warmer lentic and lotic waters; Wood 2010), branchiopods (which are more characteristic of lentic waters that have a drying phase; Williams 2006) and aquatic orthopterans, lepidopterans and hymenopterans. In total, the Appendix lists 379 genera, across all taxa, found in springs per se and 155 genera found in seeps. 100 genera are recorded as occurring in both habitat types, with 55 genera occurring only in seeps. Groups missing, or poorly represented in seeps, include sponges, hydrozoans, gastrotrichs, nematodes, bivalves, tardigrades, cladocerans, isopods and mayflies. Part of these absences could be related to the lower habitat complexity of seeps compared with springs, but also to the lack of study of these taxa in seeps.

Porifera

Freshwater sponges in the genus *Spongilla* are widespread in cold freshwaters, but also occur in thermal springs (Hooper and Van Soest 2002).

Hydrozoa

A few hydra species are known from cold waters in caves in North America, Europe and Australia, and species of the widespread genus *Hydra* have been found in various lentic and lotic freshwaters, including springs. *Velkovrhia enigmatica* (family Bougainvilliidae), found in southwestern Slovenia, is considered to be a freshwater troglobiont (Zagmajster et al. 2011).

Turbellaria

Over 100 species of freshwater planarian are associated with cold waters, with genera such as *Crenobia*, *Polycelis* and *Phagocata* occurring commonly in springs (Reynoldson 1967). The latter, as well as the Nearctic genus *Hymanella*, also have been recorded from seeps (Kenk 1972). Many species are commonly associated with karst habitats (Thorpe and Covich 2010). Habitats of the, largely unidentified, Microturbellaria include springs and underground waters (Kolasa and Tyler 2010).

Gastrotricha

Select freshwater gastrotrichs, primarily from the family Chaetonotidae, occur in interstitial or epibenthic habitats, including springs (Balsamo et al. 2008; Nesteruk 2008).

Rotifera

Rotifer species in the genera *Dipleuchlanis*, *Ptygura* and *Wierzejskiella* have been collected from spring/seep complexes (Segers 2004), and species within the genus *Lindia* are known from saline pools, ditches and hot springs (Wallace et al. 2005).

Nematoda

Species within the orders Monhysterida and Plectida (subclass Chromadoria) are among the most widely reported nematodes from freshwaters (Abebe et al. 2008). The Appendix records indicate that at least eight families of nematode contain species capable of living in hot springs (Poinar 2004, 2010).

Oligochaeta

Around 100 species of aquatic oligochaete are found exclusively in groundwater environments, some of which are deemed important centres of endemism (Martin et al. 2008). A number of tubificid genera (e.g. *Ilyodrilus*, *Limnodrilus* and *Varichaetadrilus*) are associated with karst systems (Webb et al. 1996). Several genera within the subfamily Naidinae have been recorded from springs and/or seeps (e.g. *Allonais*, *Dero*, *Pristina*, *Stylaria*).

Gastropoda

Springs and springbrooks support snail taxa not typically found in larger running waters, although individual sites tend to have low species richness (1–6 species, but at very high densities). Regionally these habitats support highly diverse assemblages of snails—especially of hydrobiids (Strong et al. 2008). Other families known to live in springs include: Melanopsidae, Assimineidae, Moitessieriidae (groundwater systems), Pomatiopsidae (saline springs), Planorbidae and Physidae. The high level of endemism in gastropods occurring in spring and groundwater systems is associated with the extinction of many freshwater snail species as these habitats become degraded or destroyed by human activities, such as water abstraction, trampling of outflows when used for watering livestock and irrigation or mining (Williams et al. 1990; Keleher and Sada 2012).

Bivalvia

While bivalves are important members of the benthic communities of rivers and lakes, their occurrence in springs appears to be much rarer. Indeed, the only records are of the Sphaeriidae genus *Pisidium*, which has a cosmopolitan distribution (Cummings and Graf 2010) and is capable of living in a very wide range of waterbodies.

Tardigrada

Most water bears are "limnoterrestrial", living in the thin films of water on the surfaces of moss, lichens, algae and leaf litter. Some live in wet soils or in aquatic sediments, and others are associated with water margins (Garey et al. 2008). The Appendix lists five, widely distributed genera, from three families, living in springs or at their margins.

Crustacea: Ostracoda

Ostracods are common inhabitants of springs and seeps, where their ecology appears to be a complex product of many environmental variables, including water chemistry, thermal regime and hydrogeology (Forester 1991). Spring-dwelling ostracods are often found in other freshwaters and are well represented by genera belonging to the Cyprididae and Candonidae (see Appendix). A number of genera are known from thermal springs, for example, *Chlamydotheca*, *Cypris*, *Potamocypris* and *Thermopsis*. Other genera contain species adapted to living in the extreme environmental conditions of ephemeral waters, for example, *Cypricercus* (temporary ponds) and *Heterocypris* (ephemeral springs). Indeed, groundwaters and temporary pools, together with Australian salt lakes, are known to be speciation-prone habitats (Martens et al. 2008).

Diplostraca

Cladoceran species richness is greatest in the warm-temperate to subtropical zones of both northern and southern hemispheres, although the Holarctic fauna is both rich and diverse, and Australasia is rich in endemic species (Korovchinsky 2006). Several of the more common genera (*Alona*, *Daphnia*, *Diaphanosoma*) contain species that are associated with hot springs; others live in wet *Sphagnum* moss or temporary ponds. Seven species are permanent inhabitants of subterranean waters (e.g. *Alona phreatica* and *A. smirnovi*; Dumont and Negrea 1996), and some chydorids are semiterrestrial (Forro et al. 2008).

Copepoda

Copepods are important members of most planktonic, benthic and groundwater communities and occur at a wide range of temperatures (glacial meltwater pools to hot springs), habitat sizes (the North American Great Lakes to water-filled leaf axils) and salinities (Boxshall and Defaye 2008). Copepods are most diverse in the Palaearctic, followed by the neotropics, and more than 90 % of species appear endemic to just one zoogeographical region. There are, however, quite a number of highly vagile and cosmopolitan species and genera that live in a range of surface and subterranean waters including springs and seeps (e.g. *Acanthocyclops*, *Diacyclops*, *Eucyclops*, *Attheyella*, *Bryocamptus*—see Appendix). There are also a number of genera that live in groundwater, yet are widely distributed (e.g. *Itocyclops*, *Stygonitocrella*, *Elaphoidella*).

Peracarida: Isopoda

The Asellidae and Stenasellidae contain the greatest number of freshwater species, with the former well represented in epigean lotic and lentic waters and the latter limited to cavernicolous and/or subterranean habitats (see Appendix). Species within the genera *Asellus*, *Caecidotea* and *Lirceus* are also found commonly in cool springs, and *Thermosphaeroma thermophilum* lives in hot springs (Wilson 2008).

Amphipoda

Amphipod diversity is highest in subterranean environments (45 % of species) and especially so in the karst regions of Central and Southern Europe (Niphargidae), North America (Crangonyctidae) and Australia (Paramelitidae) (Vainola et al. 2008). Some genera are widespread, including springs (e.g. *Gammarus*), whereas others live in springs, seeps and a variety of other subterranean waters, such as groundwater and cave pools (e.g. *Stygobromus*) (see Appendix).

Other Peracarida

Among the nine other orders of Peracarida (pouched shrimp), two are associated with subterranean freshwaters. The Spelaeogriphacea comprises four species, one each known from cave waters in Brazil and South Africa and two from aquifers in Australia. The Thermobaenacea consists of 18 species found in cave waters, the interstices of alluvial deposits and in thermo-mineral springs (Jaume 2008).

Decapoda

There appear to be relatively few decapods that inhabit springs per se (e.g. *Palaemonetes*), but there are burrowing crayfishes (e.g. *Fallicambarus*) that inhabit seepage areas where they excavate shafts down to the water table (Williams 2006).

In tropical and subtropical regions, some brachyuran crabs live in a variety of freshwaters, including cave waters (e.g. *Cerberusa*) (Ng 2004; Yeo et al. 2008).

Acari

The Acari comprises two superorders, Parasitiformes and Acariformes, the latter of which contains a wide variety of aquatic mites, those belonging to: the Hydrachnidia (true water mites), the Halacaridae (a primarily marine family within the order Prostigmata) and the Oribatida (a primarily terrestrial order with some aquatic representatives).

Smith (1991) produced a detailed synthesis of the Hydrachnidia from springs in Canada and recorded, at that time, over 115 species in 57 genera and 25 families collected from springs across the country. He divided the water mite fauna into three ecological groups related to the mites' preferences for helocrene, rheocrene or limnocrene habitats. Families with genera very commonly found in springs include the Hydryphantidae, Lebertiidae, Pionidae and Arturidae. The genus *Thermacarus* (Thermacaridae) lives in hot springs.

Of the more than 1000 species of Halacaridae, only around 56 are known to occur in freshwater. These occur in both surface and subterranean waters and have been recorded from springs, wells and the hyporheic zone of rivers. Genera collected from springs in Europe include *Copidognathus* and *Halacarellus* (Benfatti et al. 1989). Halacarids have a very low dispersal ability, yet many genera and even some species are widespread (Bartsch 2008).

Oribatid mites are largely terrestrial, associated with soil, forest litter, mosses and lichens, and only about 90 species, from 10 genera, can be considered truly aquatic. Habitat preferences in many are quite specific and include springs, seeps, temporary and permanent pools, phytotelmata and other water-filled microhabitats (Schatz and Behan-Pelletier 2008). Despite their low diversity, aquatic oribatids can occur at very high densities. Some spring-dwelling species are very ancient, for example, *Mucronothrus nasalis* is thought to predate the breakup of Pangea, about 200 million years ago. The distribution of this species is global, but discontinuous, and seems to be limited by temperature to cold, spring-fed water and cold bogs (Norton et al. 1988).

Collembola

Springtails are particularly abundant in wetlands. In springs and seeps, they are most often encountered on the surfaces of helocrenes and limnocrenes, around the water margins, on saturated soil or on low vegetation (especially mosses) close to the water. Many species live in the water-saturated environments of caves and to considerable depths in soil where they have adopted an interstitial mode of life. Others live under or at the margins of snow fields or glaciers (Deharveng et al. 2008). Cosmopolitan genera commonly found in spring pools include *Isotomurus*, *Podurus* and *Sminthurus* (see Appendix).

Ephemeroptera

Despite being most often found in running waters, mayflies are not abundant or diverse in springs or seeps. Their life cycles are known to be tied closely to changes in water temperature, with small increases, for example, resulting in a different number of nymphal moults and early adult emergence. Perhaps the more constant temperature regimes of many springs lack cues essential to the timing of important mayfly life cycles events.

Odonata

A number of dragonfly and damselfly genera, representing at least ten families, are associated with springs (see Appendix), and, perhaps reflecting their evolution in tropical regions, quite a number of genera support species that live in thermal springs, for example, *Cordulegaster*, *Ophiogomphus*, *Erythemis*, *Libellula*, *Orthemis*, *Pantala*, *Haeterina*, *Amphiagrion*, *Argia*, *Ischnura* and *Lestes*.

Plecoptera

Stonefly nymphs live, typically, in cold, clean running waters but are not particularly common in springs. A number of species (e.g. belonging to the Leuctridae, Nemouridae, Peltoperlidae and Perlodidae) inhabit seeps and splash zones (Stewart and Stark 1993).

Hemiptera

Heteropterans can be found in a range of spring types. For example, corixids, notonectids and belostomatids occur in weedy limnocrenes and helocrenes; ochterids in shallow seeps with exposed rock surfaces; species of *Saldula*, *Micronecta*, *Anisops*, *Ambrysus* and *Limnocoris* in thermal springs; and species of *Diplonychus* in peat mound springs (see Appendix).

Neuroptera

The megalopteran families Corydalidae and Sialidae have fully aquatic larvae that live in a wide variety of lotic and lentic waters (Cover and Resh 2008). These include springs, with species of *Sialis* having been found in slow-flowing limnocrenes in several regions, and the corydalid genus *Neohermes* recorded from seeps in North America (Merritt et al. 2008).

Coleoptera

In Canada, up to 1991, the distribution of the 663 known aquatic beetle species among its major freshwater habitats was lentic 61 %, lotic 23 %, springs 8 %, other <1 % and unknown 8 %. The 63 spring-dwelling species belonged to: Dytiscidae (38 species), Hydrophilidae (9), Hydraenidae (8), Chrysomelidae (subfamily

Donaciinae 6), Haliplidae (1) and Dryopidae (1) (Roughley and Larson 1991). Relatively diverse families deemed absent from the Canadian survey were Gyrinidae, Scirtidae, Curculionidae, Amphizoidae, Elmidae and Psephenidae. The occurrence or absence of these families broadly agrees with the more global findings in the Appendix. Notable exceptions are the occurrence of Hydroscaphidae in seepages and thermal springs, the Scirtidae in seepages adjacent to lotic margins, the elmid genera *Heterelmis* and *Stenelmis* in springs (the latter also in warm springs), *Hydrochus* (Hydrochidae) in mound springs and limnocrenes and *Lutrochus* (Lutrochidae) in mineral springs.

Diptera

Of the 150 families of Diptera, 27 can be considered to be "marginally" to "exclusively" aquatic. The Appendix lists 19 of these as having representatives that inhabit springs and seeps, many of them occurring on a global scale. Among the Nematocera, the following genera contain species that are especially notable for being spring/seep specialists: the tipulids *Brachypremna, Dactylolabis, Pedicia, Thaumastoptera* and *Tipula* and the chironomids Podonominae (*Boreochlus, Paraboreochlus, Parochlus*), Tanypodinae (*Krenopelopia, Macropelopia, Pentaneurella, Zavrelimyia*), Diamesinae (*Diamesa, Pseudokiefferiella, Syndiamesa*), Orthocladiinae (*Antillocladius, Doithrix, Heleniella, Krenosmittia, Lymnophyes, Parachaetocladius, Parametriocnemus, Psilometriocnemus, Stilocladius*) and Chironominae (*Krenopsectra, Neozavrelia, Stempellinella*). Many of the above genera specialize on specific temperature ranges. Alongside these stenotherms are generalist chironomid genera that frequently form part of the spring community: *Ablabesmyia, Procladius, Corynoneura, Cricotopus, Eukiefferiella, Heterotrissocladius, Orthocladius, Psectrocladius, Thienemanniella, Chironomus, Polypedilum, Micropsectra, Paratanytarsus, Stempellina* and *Tanytarsus*. Ferrington (1998) estimated that at least 111 genera of chironomid, comprising at least 185 species, occur in North American springs, representing 19 % of the total described species. For cold springs in Europe, the comparable statistics are 85 genera and over 200 species, representing 20 % of the described fauna (Lindegaard 1995). In springs in the Central High Plains of North America, Blackwood et al. (1995) found that the Orthocladiinae were more prevalent in rheocrenes, while the Chironominae were more abundant in limnocrenes.

Among the Brachycera, the following families contain species associated with springs and/or seeps: Stratiomyidae (*Beris*, spring margins), Empididae (*Hemerodromia, Oreothalia*), Dolichopodidae (raised-peat mound springs), Syrphidae (raised-peat mound springs), Sciomyzidae (snail predators in seeps) and Ephydridae (*Ephydra, Ephydrella, Paracoenia* and *Scatella* in thermal springs).

Trichoptera

Currently, 616 genera and 49 families of caddisfly are known (Trichoptera World Checklist 2015). Of these, at least 21 families, comprising at least 59 genera, have larvae that live in springs and/or seeps, including both cased and net-spinning forms (see

Appendix). Some genera show distinct microhabitat preferences (e.g. *Homoplectra* (rock face springs and seeps in montane areas), *Moselyana* (subalpine forest seeps), *Chilostigma* (seeps in wet meadows), *Pseudostenophylax* (cool spring runs and small intermittent streams)), whereas others encompass a wider range of habitats that include springs (e.g. *Cheumatopsyche, Polycentropus, Lepidostoma, Anabolia, Hesperophylax, Limnephilus, Platycentropus, Helicopsyche* and *Oecetis*). An analysis of the caddisfly larvae living in springs across Canada showed some general trends together with some regional and habitat-related differences (Williams 1991). Notably, the number of species present in the springs increased with habitat diversity, and around 35 % of the species recorded were from the Limnephilidae. Limnocrenes and rheocrenes with a weak current and small substrate particles supported few species, but often large populations. Grazers, shredders and predators were common, but filter-feeders were rare—an exception being the hydropsychid genus *Parapsyche*. Ordination of the data confirmed an east/west difference in caddisfly communities with influential variables including elevation, extent of groundwater source and summer temperature. Factors strongly influencing the composition of the spring communities in both the west and east included riparian vegetation, substrate particle size, current and pH. Springs in which caddisfly larvae were strongly involved in the processing of detritus were dominated by *Homophylax* in the west, but by *Frenesia* and *Lepidostoma* in the east. Predators and scrapers were abundant only in springs having relatively high microhabitat diversity, current speed and pH.

Ecological Controls

Comparisons Among Springs

Most permanent freshwater springs are stable environments, with many of their physical and chemical properties fluctuating less than in streams, rivers and lentic waterbodies. In some springs, however, chemical composition, suspended solids and even discharge and temperature show both seasonal and sudden variation. This is especially evident in seasonal springs discharging from aquifers of limited storage capacity and in temporary springs discharging from shallow systems. However, it also occurs in some perennial springs (Van Everdingen 1991). Springs characterized by such variation have not been studied extensively, but are known to be common in alpine headwaters (Maiolini et al. 2011). Most of the following discussion is aimed at "classical", cold-water springs and seeps, those with low environmental variability.

Although one of the first studies done on springs incorporated standing crop and energy flow models (Silver Springs, Florida; Odum 1957), most of our knowledge on these habitats has come from a taxonomic rather than a quantitative ecological approach. Hence, there persists a weak understanding of the community dynamics of springs, and especially seeps, particularly in terms of how their communities develop, are organized and function. There is also much that is not known of the life histories and physiological properties of spring-dwelling species—adaptations that enable these organisms to live and persist in these habitats.

It is clear that springs support faunas that, collectively, include most of the major taxa of invertebrates capable of living in freshwaters. From sponges, hydras, gastrotrichs, rotifers, nematodes and oligochaetes to molluscs, mites and a wide variety of insects, all have representatives that have adapted to the oftentimes constant physical and chemical environments that springs present. Further, their faunal composition has similarities across much of the world, even to the family and generic level. However, in some regions, global processes have impinged on these patterns. For example, within the last 15,000 years, cold-water springs in some temperate regions have been destroyed by Pleistocene glaciation. Springs reformed since ice retreat often have had insufficient time to repopulate fully such that, in Southern Ontario, Canada, for example, faunas are dominated by the more vagile insects—typically, nemourid stoneflies, chironomids and caddisflies (Williams and Williams 1998). Unlike in non-glaciated parts of North America, molluscs and triclads are not well represented.

At the local regional level, studies show differences in species from spring to spring and also in their abundance and time of adult emergence (e.g. in Plecoptera, Trichoptera and Diptera; Erman 1998; Gathmann and Williams 2006). From a study of the insects living in 75 springs in the south-eastern Alps, Maiolini et al. (2011) concluded that springs have an island-like property, with each spring having its own specific history and abiotic characteristics which select for unique faunal patterns. In a separate study in the same vicinity, Stoch et al. (2011) concluded that, for the meiofauna (oligochaetes, mites and crustaceans), spatial patterns of assemblages at the regional scale were best explained by altitude, water chemistry (related to geology) and water-flow regime—as was earlier found for macro-invertebrate assemblages by Barquin and Death (2006), Ilmonen et al. (2009) and Glazier (1991). To these controlling influences, Myers and Resh (2002) added spring permanence, lack of disturbance and cold-water temperatures as factors responsible for explaining higher species richness of Trichoptera. Ferrington (1998) further added within-spring habitat heterogeneity, operating on a local scale, as a determinant of the taxonomic composition of individual springs. Mattson et al. (1995) provided evidence for spring-bed substrates being among such determinants in that woody substrates in karst-fed springs in Florida were dominated by chironomids, mayflies and caddisflies (with snails and amphipods occasionally abundant), but sandy substrates were populated with other chironomids, snails, bivalves and oligochaetes. In Stoch et al.'s 2011 study, such microhabitat features, together with human disturbance, were less influential, but these authors conceded that where anthropogenic pressure is high (see Sarkka et al. 1997), it can become the foremost influence on spring faunal assemblage structure and distribution. For example, Ferrington (1998) showed that the chironomid genera of two North American springs that had become contaminated by cattle dung came to resemble those found in similarly enriched lower-order streams—the important spring specialists having been lost.

Springs Versus Seeps

An important point to address is the relationship, if any, between the faunas of springs and seeps, particularly whether the latter is a subset of the former. Clearly, while both these habitats are fed by groundwater and may have stable temperature and chemical properties, they can differ markedly in their physical characteristics. For example, springs are likely to have a more defined footprint at their point of issue, deeper water and greater discharge, whereas seeps are shallow (often just a film of water over rock surfaces) and typically discharge more diffusely, often through a carpet of moss. Bearing in mind that seeps have been studied far less than springs, the data from the Appendix show an apparent absence of a few major taxa in the former and, at the genus level, approximately 41 % of the diversity seen in the latter. The two habitat types shared 100 genera, which include many freshwater generalists (e.g. *Ischnura*, *Hydroporus*, *Chironomus* and *Hydroptila*), but there were 55 genera found only in seeps, which include some seep specialists (e.g. *Stolonicyclops*, *Trichothyas*, *Viehoperla*, *Oconoperla*, *Ochterus*, *Antillocladius*, *Oreothalia*, *Chilostigma* and *Clostoeca*). These records suggest that the invertebrate faunas of seeps contain a component that is a subset of the fauna found in springs, but a second component that is comprised of seep specialists.

Specialists Versus Generalists

Another issue to address is what adaptations define spring and seep generalists and specialists. In an analysis of life history traits, Williams (1991) concluded that, in terms of life history theory, the limited information available for spring species showed most support for a deterministic view, based on the predictions of r-, K- and A-selection (Greenslade 1983). Permanent, cold-water springs are habitats characterized by low disturbance and low adversity—assuming that low water temperature is not an adversity for lotic organisms, many of which have their ancestry in cold, headwater streams. According to the predictions, most of the fauna should be K-selected in that: (1) the community should show intermediate to high diversity (see Lindegaard 1995; Ferrington 1998) and consist primarily of sessile grazers/gatherers (see Anderson and Anderson 1995) and filter-feeders with narrow, specialized niches due to heavy competition; (2) productivity and individual growth rate should be high (see Williams and Hogg 1988), although growth may be low in non-cold-adapted species (Iversen 1976); (3) survival rate may be low and length of life, time to maturity and fecundity intermediate (see Tilly 1968); (4) population densities should be near carrying capacity; (5) incidence of dormancy should be low, as should that of parthenogenesis (but see Norton et al. 1988); (6) predators should be rare; and (7) vagility should be low to intermediate.

Community Assembly

Permanent cold-water springs, perhaps more so than seeps, present highly persistent habitats for invertebrates and may contain a disproportionately high number of unusual species for their relatively small size (Erman and Erman 1995). Species presence may be the result of several different entry pathways into the faunal assemblages of individual springs. These pathways result in various species statuses that include (with trichopteran species examples): (1) *habitat specialists* (e.g. *Crunoecia irrorata*, *Anagapetus debilis*) which are largely crenobiontic species arriving from similar habitats that may be adjacent or at some distance, depending on migration abilities; (2) *habitat generalists* (e.g. *Ptilostomis ocellifera*, *Hesperophylax designatus*) including crenophilic species that may arrive similarly, but also by range extension from downstream springbrooks; (3) *relict species* (e.g. *Apatania muliebris*, *Rakiura vernale*) previously abundant species that have undergone range contraction (often through climate change, but increasingly through human activity) and now occur in only one or a few adjacent springs; and (4) *endemic species* (e.g. *Agarodes ziczac*, *Lepidostoma ojanum*) species which originated in a single, stable and persistent spring locality that is ecologically isolated and to which they are now highly adapted. Given these different components to spring faunas, it is important to consider whether spring faunas are stochastic aggregates of habitat (temperature)-limited species or true, coevolved communities. A relevant observation in this regard was made by Myers and Resh (2002), who found that among the caddisfly assemblages found in 28 springs in the Great Basin, USA, although several springs had very similar physicochemical characteristics, none was identical, indeed the assemblage totals ranged from 4 to 18 per spring. Lack of assemblage repeatability suggests a more stochastic colonization process. Further, given the somewhat eclectic nature of the faunal source pathways, it seems unlikely that such a resulting mix of species would be able to function as a true, cohesive and fully interactive community. Clearly, we need many more studies on the development of spring communities in order to resolve these issues. Towards this goal, the main environmental factors thought to shape the composition of these invertebrate assemblages/communities are outlined in Fig. 11.2.

As noted earlier, knowledge of spring invertebrates has arisen from two, largely uncoordinated, approaches: the first based on the taxonomy of particular invertebrate groups whose range of habitats includes springs (e.g. Feldman 1974) and the second comprising a relatively small, unconnected series of population or community studies from single springs (e.g. Teal 1957; Tilly 1968; Winterbourn 1973). Specific, regional surveys of spring faunas have been rare in the past (e.g. Tuxen 1944; Botosaneanu and Negrea 1961), but are now becoming more common and informative. For example, from a study of the mayflies, stoneflies and caddisflies found in springs of the Trentino region of Italy, Maiolini et al. (2011) were able to propose three important functions of

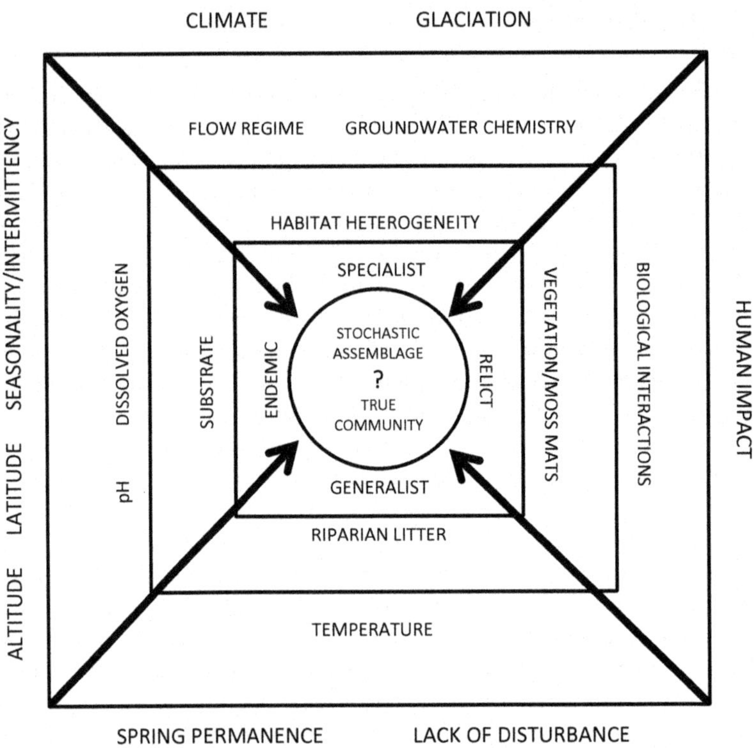

Fig. 11.2 Summary of the main environmental factors thought to shape the composition of invertebrate assemblages/communities in freshwater springs (the arrows indicate the general direction of flow of the factor influences)

cold-water springs in alpine landscapes: (1) in such harsh environments, the mild and stable environmental conditions of springs contribute to maintaining and enhancing the regional biodiversity; (2) springs act as refugia for stream biota, providing more benign conditions during spates and drought—common features in alpine headwaters; and (3) springs provide specialized habitats for strictly crenobiontic species.

Community and Population Ecology

Perhaps the greatest ecological utility that springs and seeps represent is their potential to contribute to our understanding of community ecology and ecosystem function. Indeed, in 1971, Odum proposed that springs are natural, constant-temperature laboratories and pointed out that their study had pioneered development of the trophic dynamics approach to ecosystem studies in general. In 1998, Williams and Williams identified the following features of springs that had the potential to advance community ecology: (1) their communities are naturally less complex than those occurring in most other aquatic habitats, yet, as far as is known, they possess the structural (e.g. a wide range of species representing many major taxa) and functional (e.g. producers, various levels of consumers and decomposers) elements seen in other communities; (2) the natural variation in community structure, and presumably in function, seen among natural springs represents an ideal testing ground for questions of community trophic efficiency, predator–prey dynamics and competition—particularly in the more extreme spring types (thermal and saline) which have very simple community structures (see Collins et al. 1976). A number of population-level research avenues also could be explored using springs, for example, (1) given the high degree of specificity of many spring-dwelling species to this habitat type, alongside the oftentimes isolated nature of individual springs, they could provide ideal habitat island models for the experimental analysis of the metapopulation dynamics theory (Gathmann and Williams 1998); (2) they could be used to compare the population traits of coexisting species that differ widely in their status (e.g. specialists versus generalists), as shown by spring species with very narrow habitat requirements (e.g. the endemic caddisfly *Lepidostoma ermanae*) and the widespread caddisflies *Chyranda centralis*, *Wormaldia occidea* and *Rhyacophila grandis* (Erman and Erman 1995); and (3) using thermal springs to tease apart the influences of important abiotic habitat variables, such as water temperature and chemistry, on populations, effective, natural elimination of thermal differences among adjacent sections of spring or seep complexes (and thus removal of a strong environmental variable) has been proposed as an ideal test condition for the study of the influence of habitat heterogeneity on species richness (Ferrington et al. 1995; see further examples in Williams and Williams 1998).

Conservation Concerns

There are many examples where knowledge of spring faunas has the potential both to address zoogeographical issues and to monitor environmental change. An example of the former is the proposal that for mites, in Canada, springs have played important roles as both refugia and routes for migration. As in Europe, not only have present-day, temperate lowland springs allowed the survival of cold-adapted species after ice retreat, but they also may have allowed the survival of such species in marginal refugia during glacial maxima (through maintenance of interstitial spring habitats in glacial deposits near the southern limits of ice sheets). Refugial survivors then were able to repopulate newly created habitats upon subsequent glacial retreat. It is possible that such "leading-edge" colonization at glacial margins may subsequently have produced a spatial assortment of genomes, perhaps promoting divergence and speciation (Hewitt 1996).

Monitoring environmental change involves the very close relationship, as part of the hydrological cycle, between groundwater and the atmosphere, where changes to the latter (such as a rise in global air temperature) are likely to affect the former. Such a change is likely to have serious consequences for spring discharge, in that permanent springs may become intermittent, and also for the typically stenothermic communities of organisms that live in springs. A study of a spring in which the water temperature was artificially raised by 2–3.5 °C to simulate global warming resulted in a suite of changes to the fauna including: a decrease in total invertebrate numbers, earlier onset of adult insect emergence, increased growth rates and precocial breeding in amphipods, smaller size at maturity in stoneflies and altered sex ratios in caddisflies (Hogg and Williams 1996). Such responses to environmental change make spring faunas very useful in detecting changes in water quality of their source aquifers and hence the potability of these important reservoirs (99.6 % of the world's freshwater resides in this way). Williams et al. (1997) showed that, in a series of springs in Southern Ontario, Canada, there was a strong relationship between the fauna and chloride—a major contaminant of the groundwater in the study area and believed to be derived from the application of road de-icing salt. Several taxa were closely associated with high chloride levels (e.g. the dipteran families Tipulidae and Ceratopogonidae), whereas others occurred only in springs with low chloride (e.g. Turbellaria and *Gammarus pseudolimnaeus*). Not only can spring faunas be used as indicators of contemporary spring health, where sufficient sediment depth occurs coring can yield chitinous remains that may allow reconstruction of the past history of source aquifers and their catchments, especially during the Anthropocene. For example, a core taken from the bed of a Canadian spring showed evidence of changes in land use over a 200-year period. Fossil sclerites of, predominantly, caddisfly and chironomid larvae indicated a shift from pre-European settlement forest, through a land clearing/agricultural phase, to a present-day increase in urban development (Williams and Williams 1996a, b).

Spring faunas provide unique information on endemism and also on post-glacial colonization patterns. Springs are habitats where relict species of these former times have endured, protected from large oscillations in climate. Springs hold a position of importance as study areas that is far out of proportion to their size and number. However, the same cannot be said about their global knowledge base. Some collections of specific taxa from springs exist in the general holdings of national or regional museums, but these collections seldom have sufficient accompanying habitat description or quantification to make them useful except in an introductory capacity. This dearth is occurring at a time when they and their source aquifers have come under extreme pressure from human activities. Springs themselves are being destroyed when, for example, used as stock watering holes, water-bottling sites and spas and as a consequence of logging and transportation routes (Brune 1981)—whereas their water sources are being rendered unfit for use by both invertebrates and humans (Liu and Liptak 1999). Some causes of contamination are localized, while others, such as acid rain, may be widespread. However, contaminants may migrate along subsurface paths, many of which emerge at the surface as springs before flowing into streams, lakes and wetlands. This form of contamination is increasing primarily because the diversity of chemicals used in industry and agriculture is high and many of them persist in the groundwater zone (Bowler 2014). Detection involves costly and repetitive sampling at depth. Biomonitoring of spring faunas has been proposed as a viable alternative, as these organisms are continually subjected to the emerging water and integrate the effects of geology, vegetation and climate over time (as many have at least a 1-year life cycle) (Biological Survey of Canada 1990). Where palaeoecological information is available, it is possible to build up an accurate index of groundwater quality and the history of individual aquifers.

Williams and Danks (1991) drew up a series of recommendations for the protection of springs that comprised: (1) preparing inventories of spring types and their distributions, at both regional and continental levels; (2) surveying their floras and faunas so as to enable the detailed study of representative spring types and regions; (3) making, alongside the biota, detailed descriptions of local geology, hydrology and climate, together with comprehensive analyses of water chemistry; and (4) preserving rare and regionally characteristic spring types and their biotas—the latter to be accomplished through restricting the capping of springs for commercial use; establishing and enforcing protective areas (e.g. woodlands) around springs, including fencing to prevent trampling by livestock; raising public awareness of the importance of springs; and establishing appropriate protective legislation, perhaps most effectively focused on maintaining high-quality groundwater. Groups identified as needing to take responsibility for the implementation of these recommendations include federal and local governments, natural history and other societies, individuals and landowners. Such action is to be strongly encouraged.

Acknowledgments I thank colleagues past and present who have shared my interest in springs and their invertebrate faunas, but most notably the late Dr. Lazare Botosaneanu, Dr. Hugh Danks, Dr. Oliver Gathmann, Dr. Dirk Hinterlang, Dr. Ian Hogg, the late Prof. Noel Hynes, Dr. Nancy Logan, Dr. Ian Smith and Dr. Catherine Yule.

Appendix

Global summary of the genera of invertebrates recorded from, or presumed to live in, springs and seeps

Taxon/family	Genus	Spring	Seep	Genus habitat, where known[a]	Region found[b]	Source[c]
Porifera						
Spongillina	*Spongilla*	√		Occasionally in cold-water and thermal springs	W	1, 2
Hydrozoa						
Hydridae	*Hydra*	√		Various lentic and lotic waters, incl. springs	W	3, 4
Turbellaria						
Microturbellaria	Largely unident.	√		Includes springs and underground waters	W	5
Planariidae				Wide variety of freshwater habitats	W	5, 6
	Crenobia	√		Cool, spring-fed headwater streams	EU	7, 8
	Hymanella	√	√	Includes springs and seeps	NA	9
	Phagocata	√	√	Montane cold streams, springs and seeps	HO, OL	9, 10
	Planaria	√		Lentic and slow lotic waters, incl. springs	HO	8, 9
	Polycelis	√		Cool freshwaters, including springs	HO, OL	9
Kenkiidae	*Sphalloplana*	√		Typically in caves, but also in springs	NA	9, 11
Dendrocoelidae	*Dendrocoelopsis*	√		Eurythermic and stenothermic, incl. springs	HO, AS	9, 12
	Macrocotyla	√		Typically in caves, but also in springs	NA	9, 13
	Procotyla	√		Ponds, lakes, streams and springs	HO	9
Dugesiidae	*Dugesia*	√		Ponds, lakes, streams, springs, cave pools	HO, AT, AU, OL	7
Dimarcusidae	*Mitchellia*	?		Cool cave pools in tropics	OL	14
Gastrotricha				Freshwater and semiterrestrial habitats	W	15
Chaetonotida				Inhabit freshwater and marine environments		
Chaetonotidae	*Chaetonotus*	√		Marine and freshwater, also interstitial forms	W	16–18
	Heterolepidoderma	√		Marine and freshwater, also interstitial forms	W	16–18

Macrodasyida	*Redudasys*	√		Marine, also in deep beach freshwater springs	PA, NT	16, 19
Rotifera				Abundant in interstitial waters, also in bogs	W	20
Proalidae	*Proales*	√		Parasitic in algae (*Vaucheria*)	HO, AT, OL	21, 22
Lindiidae	*Lindia*	√		Variable salinity pools, ditches, hot springs	HO, OL	22–24
Euchlanidae	*Dipleuchlanis*	√	√	Habitats include spring/seep complexes	HO, OL, NT, AT	22, 24
Flosculariidae	*Ptygura*	√	√	Marine and freshwater, spring/seep complexes	HO, OL	22, 24, 25
Dicranophoridae	*Wierzejskiella*	√	√	Marine and freshwater, spring/seep complexes	HO, OL	22, 24, 26
Nematoda				None restricted to cold subterranean waters	W	27
Aphelenchoididae	*Aphelenchoides*	√		Primarily plant pathogens, also in hot springs	W	27, 28
Aphelenchidae	*Aphelenchus*	√		Primarily soil fungus pathogens, hot springs	W	27–30
Dorylaimidae	*Dorylaimus*	√		In soils and freshwater, also in hot springs	W	27, 30
Chromodoridae	*Euchromodora*	√		Many marine species, but also in hot springs	W	27
Monhysteridae	*Monhystera*	√		Range of freshwaters, including hot springs	W	27, 30, 31
Plectidae	*Plectus*	√		Range of freshwaters, including hot springs	W	27, 30, 31
	Tylocephalus	√		Includes thermal soils and hot springs	EU, AU	27, 32
Rhabdolaimidae	*Rhabdolaimus*	√		Includes freshwater mineral and hot springs	W	27, 30, 33
Xyalidae	*Theristus*	√		Marine sediments, also freshwater hot springs	W	27, 34
Oligochaeta				Freshwaters sediments, some in groundwater	W	35
Aeolosomatidae	*Aeolosoma*	√		Includes karst springs	NA, OL	3, 36
Tubificidae	*Ilyodrilus*	√		Lakes and rivers, including karst springs	HO	3
	Limnodrilus	√		Various freshwaters, including karst springs	W	3, 36, 37
	Varichaetadrilus	√		Includes karst springs	HO	3
(Naidinae)	*Allonais*	√		Lentic/lotic waters, also karst springs	W	35, 36
	Dero		√	Lentic and lotic waters, peat mound springs	W	38, 39
	Pristina		√	Lentic and lotic waters, peat mound springs	W	39, 40
	Stylaria	√		Includes karst springs	NA	3
Haplotaxidae	*Haplotaxis*	√	√	Primarily in groundwater	HO, OL	35, 36

(continued)

(continued)

Taxon/family	Genus	Spring	Seep	Genus habitat, where known[a]	Region found[b]	Source[c]
Enchytraeidae	Unident.			Includes karst springs	NA	3
Phreodrilidae	Unident.	√		Includes groundwaters and springs	EQ	36
Lumbriculidae	*Phagodrilus*	√	√	Cool freshwaters, including springs and seeps	NA	35
Gastropoda				Many endemics in springs and groundwater	W	41
Melanopsidae	*Melanopsis*	√		Fresh- and brackish waters, including springs	PA, AT, AU, PAC	41, 42
Assimineidae	Unident.	√		Estuarine and freshwaters, including springs	W	41
Hydrobiidae	*Fontigens*	√		Cool, shallow springs	NA	10
	Various	√	√	Some hydrobiids endemic to springs and seeps	NA	43
Pomatiopsidae	Various	√		Freshwaters, incl. amphibious; saline springs	W	41
Planorbidae	Various	√		Lentic and slow lotic freshwaters, incl. springs	W	41
	Gyraulus		√	Lentic/slow lotic, raised-peat mound springs	HO, AU, OL	39, 41, 44
Physidae	*Physa*	√		Lentic/slow lotic, incl. cold and hot springs	W	3, 41
	Physella	√		Rivers, ponds, lakes, marshes, springs	HO	3, 41
Bivalvia						
Sphaeriidae	*Pisidium*	√		The only sphaeriids in springs	W	10, 45
Tardigrada				Freshwater species not well known	W	46
Halechiniscidae	*Styraconyx hallasi*	√		Marine species, also in a freshwater spring	GR	46
Eohypsibiidae	*Bertolanius*	√		Freshwater sediments, damp bryophytes	W	46
	Eohypsibius	√		Freshwater and terrestrial, including springs	HO, AS, GR	46
	Isohypsibius	√		Soil, leaf litter, moss, also spring margins	HO, AU, PO, OL	46, 47
Hypsibiidae	*Microhypsibius*	√		Freshwater, including springs	HO, GR, AS	46
Murrayidae	*Murrayon*	√		Soil and moss, also in springs	NA, GR, AS, OL	46, 48
Crustacea						
Ostracoda				Significant occurrence in hypogean habitats	W	49, 50

Candonidae	*Candocyprinotus*	√		Common in springs and lacustrine habitats	NA	51
	Candona	√	√	Includes spring/seep habitats	NA, OL	51–53
	Cypria	√		Ponds, lakes, streams and springs	HO, AS, OL	51, 53, 54
	Fabaeformiscandona	√		Lakes, ponds, canals, rice fields, springs	W	51, 55
	Nannocandona	√		Springs supported by deep aquifers	HO	50
	Schellencandona	√		Springs supported by deep aquifers	HO, OL	50, 56
Cyprididae	*Cavernocypris*	√	√	Subterranean waterbodies	HO, AS	52, 57
	Chlamydotheca	√		Warm springs and temporary pools	HO, NT	52, 58
	Cyclocypris	√		Streams, wetlands, springs	HO	50, 51
	Cypricercus	√		Permanent and temporary ponds, springs	HO, NT, OL	50, 51
	Cypridopsis	√		Oligo- to eutrophic waters, springs, wells	W	51, 53, 59
	Cypris	√		Incl. streams, wetlands, hot and cold springs	W	50, 56
	Herpetocypris	√		Stagnant waters, lakes, slow lotic, springs	HO, NT, AT, OL	52, 54
	Heterocypris	√	√	Lakes, temporary ponds, ephemeral springs	HO, OL	52, 54
	Potamocypris	√		Vegetated ponds and pools, also hot springs	HO, AS, AT, OL	50, 54
	Psychrodromus	√		Highest diversity in springs	HO	52, 54
	Scottia	√		Wet leaf litter around spring sources	HO, OL	51, 60
	Thermopsis	√		Hot springs	NA	50
Darwinulidae	*Microdarwinula*	√		Springs, seeps and river gravel; semiterrestrial	HO, AT, OL, NT	51, 61
Ilyocyprididae	*Ilyocypris*	√		Lakes, shallow permanent waters, springs	HO, OL, AU, AT	51, 56
Diplostraca				Lentic and slow lotic waters, also groundwater	W	62
Sididae	*Diaphanosoma*	√		Planktonic in lakes, but also in hot springs	NA, OL, AU, AT	63
	Latonopsis	√		Lakes, ponds, temporary waters, hot springs	HO, OL, AU, AT	63, 64
Chydoridae	*Alona*	√		Lakes, also wet *Sphagnum* mats and hot springs	HO, AS, OL, NT	63
Daphniidae	*Daphnia*	√		Most lentic waters, also hot springs	W	63
Copepoda				Most freshwaters, significant in hypogean	W	49, 65

(continued)

(continued)

Taxon/family	Genus	Spring	Seep	Genus habitat, where known[a]	Region found[b]	Source[c]
Cyclopoida	*Acanthocyclops*	√	√	Ponds, lakes, groundwater and springs/seeps	W	65, 66
	Diacyclops	√	√	Surface and subterranean waters, incl. springs	W	65
	Eucyclops	√		Surface and subterranean waters, incl. springs	W	65, 66
	Itocyclops	√		Stygobionts, also in wells, marshes and springs	HO	65
	Macrocyclops	√		Freshwaters, incl. wetlands and springs	W	65, 66
	Megacyclops		√	Freshwaters, including wells and springs	HO, OL	65, 66
	Mesocyclops	√		Surface and subterranean waters, incl. springs	W	65
	Paracyclops	√		Freshwater, estuaries, bromeliads, springs	W	65, 66
	Stolonicyclops		√	Mountain seeps	NA	65
Harpacticoida				Marine and freshwater, typically benthic		
Phyllognathopodidae	*Phyllognathopus*	√		Eurytopic incl. springs, hyporheic	W	65
Ameiridae	*Stygonitocrella*	√		Freshwater caves, groundwater, wells, springs	HO, NT	65
Canthocamptidae	*Attheyella*	√	√	Surface and groundwaters, incl. hyporheic zone	W	65, 67
	Bryocamptus	√	√	Surface and groundwaters, cave pools, springs	CB, NA	65
	Elaphoidella	√	√	Stygobionts, groundwater and spring-fed pools	W	65
	Maraenobiotus	√		Cold stenothermal, in groundwater and springs	CB, NA	65
	Moraria	√	√	Cave waters, groundwater, springs, wet moss	CB, NA	65
Isopoda: Asellidae	*Asellus*	√		Epigean, hypogean, hyporheic, crenophilic	HO, AS, OL	49, 68
	Caecidotea	√		Epigean, hypogean (groundwater, springs)	NA	3
	Lirceus	√		Lakes, temp. ponds, streams, caves, springs	NA	49
Sphaeromatidae	*Thermosphaeroma*	√		Hot and thermal springs	NA	69
Stenasellidae	*Stenasellus*	?		Subterranean, incl. wells and cave waters	OL	70
Anthuridae	*Stygocyathura*	?		Cave waters	W	70
Amphipoda				Chiefly marine, but ~1900 freshwater species		

Crangonyctidae	*Bactrurus*	√		Springs via predominantly hypogean habitats	NA	71
	Crangonyx	√		Springs via predominantly hypogean habitats	HO	71
	Stygobromus	√	√	Cave pools, groundwater, seeps, springs	HO, EU, AS	71, 72
	Synurella	√		Freshwater subterranean habitats, incl. springs	HO, EU, AS	71, 73
Bogidiellidae	*Indogidiella*	?		Subterranean, including wells and cave waters	OL	74
Gammaridae	*Gammarus*	√		Eurytopic; freshwater and marine, incl. springs	HO	10, 49
Decapoda				Surface waters, subterranean streams and pools	W	75
Palaemonidae	*Palaemonetes*	√		Canals, caves, subterranean waters, springs	HO, AS	75
Cambaridae	*Cambarellus alvarezi*	√		Exterminated from its single spring location	NA	75
	Fallicambarus		√	Semiterrestrial, often within seepage areas	NA	75
Potamidae	*Cerberusa*	?		Cave-dwelling crabs	OL	76
Acari: Hydrachnidia				Commonly found in freshwaters	W	77
Hydrovolziidae	*Hydrovolzia*	√	√	Includes moss around helocrenes	NA, B	77, 78
Piersigiidae	*Piersigia*	√	√	Includes moss around helocrenes	NA, B	77, 78
Apheviderulicidae	*Apheviderulix*	√		Springs	NA	77

(continued)

(continued)

Taxon/family	Genus	Spring	Seep	Genus habitat, where known[a]	Region found[b]	Source[c]
Hydryphantidae	*Almuerzothyas*	√		Springs	NA	77, 78
	Columbiathyas	√		Springs	NA	77, 78
	Cowichania	√		Hyporheic zone	NA	77, 78
	Cyclothyas	√		Interstitial	NA	77, 78
	Euthyas	√	√	Includes vegetation mats around helocrenes	NA	78
	Panisopsis	√	√	Includes vegetation mats around helocrenes	NA, B	77, 78
	Panisus	√	√	Includes vegetation mats around helocrenes	NA, B	77, 78
	Parathyasella	√		Springs and riffles in cold streams	NA, B	77, 78
	Tadjikothyas	√	√	Includes vegetation mats around helocrenes	NA	78
	Tartarothyas	√	√	Includes vegetation mats around helocrenes	NA	77, 78
	Thyas	√	√	Includes vegetation mats around helocrenes	NA	78
	Thyasella	√	√	Includes in helocrene vegetation, interstitial	NA	77, 78
	Thyopsella	√	√	Includes vegetation mats around helocrenes	NA, B	77, 78
	Thyopsis	√	√	Includes vegetation mats around helocrenes	NA	77, 78
	Todothyas	√		Springs and riffles in cold streams	NA, B	77, 78
	Trichothyas		√	Hygropetric and trickle zones	NA	77, 78
	Wandesia	√		Interstitial	NA	77, 78
Thermacaridae	*Thermacarus*	√		Hot springs	NA	77, 78
Sperchontidae	*Sperchon*	√	√	Includes helocrenes	NA	77, 78
Teutoniidae	*Teutonia*	√		Cold stenothermic pools and springs	NA, B	77, 78
Rutripalpidae	*Rutripalpus*	√		Limnocrenes	NA, B	77, 78
Anisitsiellidae	*Bandakia*	√		Includes hyporheic zone	NA, B, OL	77–79
	Bandakiopsis	√	√	Includes vegetation mats around helocrenes	NA	77, 78
	Cookacarus	√		Interstitial	NA	77, 78
	Oregonacarus	√	√	Seepage water habitats	NA	77, 78
	Utaxatax	√	√	Includes interstitial and helocrene habitats	NA	77, 78

Lebertiidae	*Estelloxus*	√		Primarily in cold springs	NA	77
	Lebertia	√		Wide range of freshwater habitats	NA	77, 78
	Scutolebertia	√		Primarily in cold springs	NA	77
Torrenticolidae	*Testudacarus*	√		Includes interstitial	NA, B	78
	Torrenticola	√		Includes interstitial	NA, B, OL	77–79
Limnesiidae	*Tyrrellia*	√	√	Includes helocrenes	NA, B	77, 78
Hygrobatidae	*Atractides*	√		Wide range of freshwater habitats	NA, B	77, 78
	Hygrobetes	√		Wide range of freshwater habitats	NA, B, OL	77–79
Feltriidae	*Feltria*	√	√	Includes helocrenes	NA, B	77, 78
Pionidae	*Forelia*	√		Lentic including limnocrenes	NA, B	78
	Nautarachna	√		Springs and cold streams	NA	77, 78
	Neotiphys	√		Lentic including limnocrenes	NA, B	77, 78
	Piona	√		Lentic including limnocrenes	NA, B	77, 78
	Pionacercus	√		Largely limnocrenes	NA, B	77, 78
	Pionopsis	√		Lentic including limnocrenes	NA, B	77, 78
	Pseudofeltria	√	√	Includes helocrenes	NA	77, 78
	Tiphys	√		Lentic including limnocrenes	NA	77, 78
Arturidae	*Arturus*	√		Surface and hyporheic	NA	77, 78
	Estellacarus	√		Lentic including limnocrenes	NA, B	77, 78
	Ljania	√		Lotic including rheocrenes	NA, B	77, 78
	Neobrachypoda	√		Lentic including limnocrenes	NA, B	78
	Woolastookia	√		Lentic and lotic including springs	NA, B	77, 78
Momoniidae	*Cyclomomonia*	√		Rheocrenes, hyporheic zone	NA	77, 78
	Stygomomonia	√		Rheocrenes, hyporheic zone	NA	77, 78
Nudomideopsidae	*Nudomideopsis*	√	√	Hyporheic, moss mats in helocrenes	NA, B, AU	77, 78
	Paramideopsis	√		Rheocrenes, hyporheic zone	NA, B	77, 78
Mideopsidae	*Mideopsis*	√		Wide range of freshwaters, incl. hyporheic	NA, B	77, 78

(continued)

(continued)

Taxon/family	Genus	Spring	Seep	Genus habitat, where known[a]	Region found[b]	Source[c]
Chappuisididae	*Chappuisides*	√		Hyporheic, rheocrenes	NA	77, 78
Uchidastygacaridae	*Uchidastygacarus*	√		Hyporheic, rheocrenes	NA	77, 78
Neoacaridae	*Neoacarus*	√	√	Helocrenes, rheocrenes, hyporheic	NA, B	77, 78
Acalyptonotidae	*Acalyptonotus*	√	√	Helocrenes, rheocrenes, limnocrenes	NA, B	77, 78
	Paenecalyptonotus	√	√	Helocrenes	NA, B	77, 78
Athienemanniidae	*Chelomideopsis*	√	√	Moss mats in helocrenes, rheocrenes	NA, B	77, 78
	Platyhydracarus	√		Rheocrenes, hyporheic	NA	77, 78
Laversiidae	*Laversia*	√	√	Helocrenes, rheocrenes	NA, B	77, 78
Arrenuridae	*Arrenurus*	√	√	Various freshwater habitats incl. interstitial	NA, AU, OL	39, 77, 78
Halacaridae	*Copidognathus*	√		Cool springs	EU	80
	Halacarellus	√		Cool springs	EU	80
Trhypochthoniidae	*Mucronothrus nasalis*	√		Cold, spring-fed waters and cold bogs	W	81
Collembola				Damp terrestrial habitats and soils	W	82
Isotomidae	*Isotomurus*		√	Still-water surfaces including spring pools	W	82
Poduridae	*Podurus*		√	Still-water surfaces including spring pools	W	82
Sminthuridae	*Sminthurus*		√	Still-water surfaces including spring pools	W	82, 83
Ephemeroptera				Nymphs in virtually all types of freshwater	W	84
Baetidae	*Baetis*	√		Various running waters, including springs	W	85, 86
Ephemerellidae	*Serratella*	√		Swift, rocky streams, also spring headwaters	HO, OL	85, 86
Heptageniidae	*Ecdyonurus*	√		River/stream riffles, also spring headwaters	HO, OL	86, 87
	Epeorus	√		Lotic, occasionally into spring headwaters	HO, OL, NT	86, 87
	Rhithrogena	√		Swift, rocky streams, occasionally in springs	HO, OL	86, 87
	Stenonema	√		Lotic, occasionally into spring headwaters	NA	3
Odonata				Nymphs in most lentic and lotic freshwaters	NT, OL	84
Anisoptera				Approx. 2900 species in various freshwaters		
Cordulegastridae	*Cordulegaster*	√	√	Seeps and springs in marshes, thermal springs	HO	88, 89

Petaluridae	*Tachopteryx*	√	√	Wetland seepage areas, bogs, semiterrestrial	NA	83
Gomphidae	*Lanthus*	√		Clear streams with strong current, springs	NA, OL	83
	Ophiogomphus	√		Rheophilic, also includes thermal springs	HO, AS	88
	Stylogomphus	√		Rocky, spring-fed woodland streams	NA, AS, OL	83
Libellulidae	*Erythemis*	√		Ponds, lakes and marshes; thermal springs	NA, NT	88
	Libellula	√		Ponds, lakes and marshes; thermal springs	HO, AS, OL, NT	88
	Neurothemis		√	Ponds, lakes, streams, peat mound springs	AS, OL, AU	39
	Orthemis	√		Muddy ditches, marshes, thermal springs	NA, NT	88, 90
	Orthetrum		√	Ponds, streams, seepage areas, mound springs	HO, OL, AT, AU	39
	Pantala	√		Ponds, riverine pools, thermal springs	NA, AT, OL, NT	88
	Trapezostigma		√	Raised-peat mound springs	AU	39
Corduliidae	*Somatochlora*	√		Cool waters running through wetlands	HO, OL	83
Hemicorduliidae	*Hemicordulia*		√	Upland lakes and stream pools, mound springs	EU, AU	39
Zygoptera				Approx. 2800 species in various freshwaters		
Calopterygidae	*Hetaerina*	√		Streams and rivers, also thermal springs	HO, NT	88, 91
Coenagrionidae	*Amphiagrion*	√		Lakes, marshes, slow streams, thermal springs	NA	88
	Apanisagrion	√		Slow streams in woodlands	NA, NT	83
	Argia	√		Range of lotic waters, incl. thermal springs	NA, NT	88, 92
	Chromagrion	√		Diverse, often eutrophic, slow-water habitats	NA	83
	Ischnura	√	√	Slow-water habitats, thermal and cold springs	HO, AT, NT, AU	83, 88, 93
Lestidae	*Lestes*	√		Peatland ponds, thermal springs	HO, NT, AT, OL	88, 94
Megapodagrionidae	*Rhinagrion*	√		Includes leaves in trickles from forest springs	OL	95
Plecoptera				Clean, montane or spring-fed lowland streams	W	84, 96
Leuctridae	*Despaxia*	√	√	Spring-fed streams, spring seeps	NA, OL	83
	Megaleuctra	√	√	Spring outflows, seeps	NA, OL	83, 97
Nemouridae	*Amphinemura*	√		Running waters, including spring streams	HO, OL	95, 97

(continued)

(continued)

Taxon/family	Genus	Spring	Seep	Genus habitat, where known[a]	Region found[b]	Source[c]
	Soyedina		√	Spring-fed streams, springs	NA	83
Chloroperlidae	Unident.	√		Rivers and streams, some hyporheic dwellers	HO, OL	10, 98
Peltoperlidae	*Soliperla*	√		Splash zone of small streams and springs	NA	97
	Viehoperla		√	Splash zone of small spring seeps	NA	97
Perlodidae	*Oconoperla*		√	Splash zone of small spring seeps	NA	97
Hemiptera: Saldidae	*Saldula*	√	√	Riparian and littoral, thermal springs, seeps	W	88, 99
Mesoveliidae	*Mesovelia*	√	√	Vegetation in stagnant/slow waters, springs	W	3, 39
Corixidae	*Micronecta*	√	√	Permanent lakes, peat mound and hot springs	PA, OL, AT, AU	39, 100, 101
	Sigara	√		Weedy pools, limnocrenes and helocrenes	W	101, 102
	Trichocorixa	√		Standing/slow waters, karst springs	HO, AT, NT, PAC	3, 103
	Unident.	√		Thermal springs	NA	88
Notonectidae	*Anisops*	√	√	Mound springs, hot springs, vegetated springs	W	39, 100, 104
Naucoridae	*Ambrysus*	√		Clean streams, lake margins, thermal springs	NA, NT	105
	Limnocoris	√		Warm springs, streams and pools in tropics	NA, NT	83, 106
Belostomatidae	*Belostoma*			Weedy ponds, river margins, hot springs	W	107, 105
	Diplonychus		√	Shallow vegetated waters, peat mound springs	AU, AT, OL, AS	39
Ochteridae	*Ochterus*		√	Semiaquatic, seeps on exposed rock surfaces	W	83
Neuroptera						
Corydalidae	*Neohermes*		√	Lotic leaf litter, temporary streams, seeps	NA	83
Sialidae	*Sialis*	√		Slow freshwaters with silty substrates	HO, AS, AU	10, 108
Coleoptera						
Hydroscaphidae	*Hydroscapha*	√	√	Gravel of rivers and seepages, thermal springs	HO, OL, AT NT	83, 109
Haliplidae	*Haliplus*	√		Vegetated ponds, lakes, slow-flowing streams	HO, OL, AU	110

Dytiscidae	*Agabinus* spp.	√		Includes montane streams and coastal springs	NA	110, 111
	Agabus spp.	√		Wetland generalists, springs, small ponds	HO, AT, NT	110, 111
	Allopachria	√		Small forest streams and springs	OL	112,
	Comaldessus	√		Stygobiontic dytiscid found in springs	NA	83, 113
	Eretes		√	Temporary ponds, raised-peat mound springs	W	39
	Hydroglyphus		√	Raised-peat mound springs, pools, wells	AU, AT, OL	39
	Hydroporus spp.	√	√	Generalists, also springs and temporary ponds	HO	110, 111
	Lacconectus	√		Small forested streams and springs	OL	112
	Liodessus	√		Ponds, wetlands, weedy streams, springs	W	110
	Microdytes	√		Small forested streams and springs	OL	112
	Potamonectes spp.	√		Cold waters, high-altitude waters, springs	HO	110, 111
Hydraenidae	*Hydraena* spp.	√		Highly diverse genus, in various freshwaters	HO, OL, AS	112, 113
	Ochthebius spp.	√		Wet mud and pool margins, small waterbodies	W	110
	Rhantus spp.	√		Lentic/lotic waters incl. springs, saline ponds	NA	110
Hydrophilidae	*Berosus* spp.		√	Lentic and slow lotic waters, mound springs	W	39, 114
	Cymbiodyta spp.	√		Shallow waters, stream margins, wet leaf litter	HO	110
	Enochrus	√		Still waters rich in vegetation, small streams	W	110
	Helophorus	√		Vegetated lentic and slow lotic wetland habitats	HO	110
	Paracymus		√	Wetland pools, salt marshes, mound springs	W	39, 115
	Paranacaena		√	Small streams, raised-peat mound springs	AU	39
	Sperchopsis	√		Margins of cold, clear, fast streams	NA	110
	Tropisternus spp.	√		Pioneer species in marshes/flooded grasslands	NA, NT	110
Scirtidae	*Elodes*	√	√	In vegetation alongside rivers and streams	W	83
	Sacodes		√	Various lentic and slow lotic waters	W	83
	Unident.		√	Raised-peat mound springs	AU	39

(continued)

(continued)

Taxon/family	Genus	Spring	Seep	Genus habitat, where known[a]	Region found[b]	Source[c]
Elmidae	*Heterelmis*	√		Riffle areas of shallow streams, springs	NA, NT	116
	Stenelmis	√		Riffle areas of shallow streams, warm springs	HO, OL, AT	112, 117
	Unident.	√		Riffle areas of shallow streams, springs	NA	10
Hydrochidae	*Hydrochus*	√	√	Streams, mound springs, limnocrenes	W	39, 118
Dryopidae	*Helichus*	√		Springs and cold spring-fed streams	NA, NT, OL	110, 119
	Stygoparnus	√		Cave springs	NA	83
Lutrochidae	*Lutrochus*	√		Mineral springs, travertine deposit streams	NA, NT	83, 110
Chrysomelidae	*Donacia*	√		Lakes and wetlands, often with bur-reeds	HO	110
	Plateumaris spp.	√		Wetlands, attached to aquatic plant rhizomes	HO	110
Diptera: Nematocera				Global true flies, larvae are typically aquatic		
Deuterophlebiidae	*Deuterophlebia*	√		Typical of cold, torrential mountain streams	NA, AS	120
Nymphomyiidae	*Nymphomyia*	?		Small, stable temperature (<15 °C) streams	NA, AS	121
Tipulidae	*Brachypremna*	√	√	Soil along stream margins, springs and seeps	NA, NT	83, 122
	Dactylolabis		√	Rocky seeps, splash zones of streams/springs	HO	83, 123
	Leptotarsus	√		Margins and shallow areas of small streams	W	83
	Pedicia	√	√	Stream margins, springs, peatbogs, seepages	HO, OL	83, 124
	Thaumastoptera	√	√	Fine sediments in streams, springs, seeps	W	83, 125
	Tipula		√	Along stream margins and in seepage areas	W	126
Ptychopteridae	Unident.	√	√	Vegetated ponds, slow streams, springs	W, not NT, AU	83, 127
Psychodidae	Unident.		√	Nutrient-rich waters, incl. peat mound springs	W	39, 128
Dixidae	Unident.	√		Lentic waters rich in marginal vegetation	W	10
Chaoboridae	*Eucorethra*	√		Lentic/slow lotic waters, incl. springs and bogs	NA	83, 129
Culicidae	*Anopheles*		√	Habitat incl. groundwater, mound springs	W	39, 130
	Culex	√	√	Groundwater bodies in northern part of range	W	39, 130
Thaumaleidae	Unident.		√	Hygropetric waters (water films on torrents)	W	83, 129

Simuliidae	Unident.	√		Fast waters, incl. cold and thermal springs	W	88
	Simulium urbanum		√	Ditch-like seepages	EU	131
	Simulium costatum	√		Chalk/limestone springs	EU	131
Ceratopogonidae	*Bezzia*	√	√	Shallow, lentic waters; hot and mound springs	W	39, 83
	Culicoides		√	Wetland pool/stream margins, mound springs	W	39, 132
	Dasyhelea		√	Small waterbodies, margins; mound springs	W	39, 133
	Forcipomyia	√		Various standing waters, incl. warm springs	W	83
	Monohelea		√	Wetland pool/stream margins, mound springs	W	39, 133
Chironomidae				>5000 species of chiefly aquatic dipterans		
Podonominae	*Boreochlus*	√	√	Among moss in springs	HO	134, 135
	Paraboreochlus	√		Among moss in springs	HO	134
	Parochlus	√		Includes cool springs	HO, NT, AU	83, 134, 135
	Trichotanypus	√		Among mosses	HO	134, 135
Tanypodinae	*Ablabesmyia*	√		Eurytopic and cosmopolitan, including springs	W	136, 137
	Brundiniella	√		Cool brooks and springs	NA, OL	136, 137
	Conchapelopia	√		Stenothermic-warm eurythermic lotic and lentic	HO, OL, AT, NT	135, 137
	Krenopelopia	√	√	Cold stenothermic, especially springs and seeps	HO, OL	136, 137
	Larsia	√		Cold stenothermic, springs, ditches, littoral	W	136, 137
	Macropelopia	√		Fine sediments in springs, brooks and bogs	HO, AS, NT	136, 137
	Natarsia	√		Cool streams/springs, littoral of cool lakes	HO, OL	136, 137
	Pentaneurella	√	√	Springs and spring-fed streams	EU	136, 137
	Procladius		√	Lentic or slow-flowing waters, mound springs	W	39, 136
	Rheomyia	√		Streams and springs	NA	83, 136
	Trissopelopia	√		Cold stenothermic lotic and lentic waters	HO	136
	Zavrelimyia	√		Cold stenothermic, typically springs	HO, OL, NT	136, 137

(continued)

(continued)

Taxon/family	Genus	Spring	Seep	Genus habitat, where known[a]	Region found[b]	Source[c]
Diamesinae	*Diamesa*	√		Cold adapted, especially springs, hygropetric	HO, OL, AT	137, 138
	Protanypus	√	√	Oligotrophic waters, including springs	HO	135, 138
	Pseudodiamesia	√	√	Various, incl. springs and oligotrophic lakes	HO, OL	135, 137, 138
	Pseudokiefferiella	√		Mountain springs, incl. the hygropetric zone	HO	83, 138
	Sympotthastia	√		Cool flowing water including springs	HO, OL	138
	Syndiamesa	√		Springs and small streams incl. hygropetric	HO, OL	137, 138
Prodiamesinae	*Compteromesa*		√	Small streams and springs	NA, OL	83, 137
	Prodiamesa	√		Shallow freshwaters including springs	HO	135, 137, 139
Orthocladiinae	*Antillocladius*		√	Seepage areas	NA, NT	140
	Brillia	√	√	Cold stenothermic lotic, littoral and hygropetric	HO, NT, OL	135, 140, 141
	Cardiocladius	√		Fast waters, thin water films on waterfalls	W	135, 140, 141
	Chaetocladius	√	√	Frequently semiterrestrial, muddy springs	HO, AT	135, 140, 141
	Compterosmittia		√	Mostly in phytotelmata	HO, AT, OL	83, 137
	Corynoneura	√	√	Eurytopic and cosmopolitan, including springs	W	135, 137, 141
	Cricotopus	√		Wide variety of waterbodies including springs	W	135, 140, 141
	Diplocladius	√	√	Cool waters, including springs	HO	135, 140, 141
	Doithrix	√	√	Margins of streams and seepage areas	HO	83, 135, 137
	Dratnalia	√		Within trichopteran larval/pupal cases, springs	EU	140
	Eukiefferiella	√	√	Eurythermic in flowing waters, also in springs	W	135, 137, 140
	Georthocladius		√	Terrestrial or semiterrestrial, helocrene seeps	HO	137, 140

Orthocladiinae	*Heleniella*	√		Cold stenothermic rheophiles	HO, OL	140, 141
	Heterotrissocladius	√	√	Wide array of waterbodies, including springs	HO, NT	135, 141
	Hydrobaenus	√	√	Various oligotrophic waterbodies, incl. springs	HO	135, 140
	Krenosmittia	√		Springs and hygropetric, also rivers and lakes	HO, AT, OL	135, 137, 140
	Limnophyes	√	√	Semiterrestrial, also springs and seepages	W	135, 137, 140
	Metriocnemus	√	√	Various waters incl. springs and hygropetric	HO, AT, NT	135, 137, 140
	Orthocladius	√	√	Lotic and lentic waters, springs, seepage areas	NH	135, 137, 140
	Parachaetocladius	√		Cold stenothermic, esp. springs and helocrenes	HO	135, 137, 140
	Paracladius	√		Lotic and lentic waters, including springs	HO	135, 137, 140
	Parakiefferiella	√		Lotic and lentic waters, including springs	W	135, 137, 140
	Parametriocnemus	√	√	Cold running waters, especially springs	W	135, 137, 140
	Paraphaenocladius	√	√	Includes damp soil adjacent to springs	HO, AT	135, 137, 140
	Paratrichocladius	√	√	Some are cold stenotherms in running waters	W	135, 140, 141
	Parorthocladius	√		Streams and shallow lentic waters, incl. springs	HO, AT	135, 140, 141
	Platysmittia	√		Running waters	HO	83

(continued)

(continued)

Taxon/family	Genus	Spring	Seep	Genus habitat, where known[a]	Region found[b]	Source[c]
Orthocladiinae	*Plhudsonia*		√	Streams	HO	83
	Psectrocladius	√	√	Lotic and lentic waters, incl. boggy seepages	W	135, 140, 141
	Pseudorthocladius		√	Spring-fed streams often in moss, hygropetric	HO, AT	140
	Pseudosmittia		√	Incl. semiterrestrial species, splash zones	W	141, 140, 135
	Psilometriocnemus	√	√	Damp soil, especially around springs and seeps	NA	140
	Rheocricotopus	√	√	Rheophilic, including springs	W, but not NT	135, 140, 141
	Saetheriella	?		Running waters	HO	83
	Stilocladius	√		Cold running waters, incl. mountain springs	HO, OL	137, 140
	Synorthocladius	√		Various lotic and lentic waters, incl. springs	HO, AT	137, 140
	Thienemannia	√		Often in moss in springs; hygropetric zone	HO	83, 140
	Thienemanniella	√		Most lotic waters, including springs	W	135, 140, 141
	Tvetenia	√		Lotic waters, occasionally in springs	HO, OL, AT	135, 140, 141
	Zalutschia		√	Oligo- and dystrophic lakes, boggy springs	HO	140, 141
Chironominae						
Chironomini	*Chironomus*	√	√	Typical of lentic waters, but incl. springs	W	39, 135, 142
	Einfeldia	√		Primarily dystrophic waters, also in springs	HO, OL	135, 137, 142
	Glyptotendipes	√		Detritus-rich shallow waters, also in springs	HO, OL, AT	135, 142
	Polypedilum spp.	√	√	Most lentic and lotic waterbodies, incl. springs	W	39, 135, 142

Tanytarsini	*Cladotanytarsus*	√		Eurytopic genus, incl. hot springs and brackish	W	142
	Krenopsectra	√		Cold stenothermic in montane regions, springs	HO	142, 143
	Micropsectra	√	√	Wide range; hygropetric and thermal springs	HO	142
	Neostempellina	√		Running waters	HO, AT	83, 137
	Neozavrelia	√		Cold stenothermic, incl. springs and hygropetric	HO, OL, AU	137, 142
	Paratanytarsus	√		Eurytopic genus, occasionally in springs	W	135, 137, 142
	Stempellina	√		Eurytopic genus, incl. cold and thermal springs	W	137, 142
	Stempellinella	√	√	Primarily in springs and small streams	HO, OL, AU	135, 137, 142
	Tanytarsus	√	√	Eurytopic genus, incl. peat mound springs	W	39, 137, 142
	Virgatanytarsus	√		Small rivers and lentic littoral, springs	HO, AT	83, 142
Brachycera				Globally distributed, some have aquatic larvae		
Stratiomyidae	Unident.	√	√	Wetlands, margins, thermal and mound springs	W	39, 88
	Beris	√		In moist moss at spring margins	HO, OL, AS, AU	144
Pelecorhynchidae	Unident.	√		Wetlands, water margins	HO, NT, AU	10, 145
Empididae	*Hemerodromia*		√	Well-oxygenated running waters, incl. springs	W	39, 146
	Oreothalia		√	Water films in seepage areas and on rocks	HO	83
Dolichopodidae	*Liancalus*		√	Algal mats, often on wet rocks at waterfalls	HO	83, 147
	Pelastoneurus		√	Algal mats and mud substrates, various habitats	HO	83, 148
	Unident.		√	Raised-peat mound springs	HO	39
Syrphidae	Unident.		√	Raised-peat mound springs	HO	39
Sciomyzidae	Unident.		√	Snail predators in various waters, incl. seeps	W	10, 149
Ephydridae	*Ephydra*	√		Wetlands, incl. thermal springs and saline pools	HO, OL, AT, NT	83
	Ephydrella	√		Thermal springs	HO, AU	88, 150
	Paracoenia	√		Wetlands, also in alkaline thermal springs	HO	83
	Scatella	√		Thermal springs, incl. hot sulphur springs	HO, OL, AU	83, 151
	Unident.	√		Geysers	HO	152

(continued)

(continued)

Taxon/family	Genus	Spring	Seep	Genus habitat, where known[a]	Region found[b]	Source[c]
Trichoptera						
Rhyacophilidae	*Rhyacophila* spp.	√		Lotic on rocks/cobbles, headwaters and springs	W, not AT, NT	153, 154
Philopotamidae	*Dolophilodes*		√	Fast lotic, also mossy seeps on rock surfaces	NA	83
	Wormaldia	√		Fast-flowing sections of rivers and streams	W	155 ROM[d]
Hydropsychidae	*Arctopsyche*	√		Cold streams, springs esp. boreal/montane	HO, OL	155 ROM
	Cheumatopsyche	√		Rivers and streams, also in karst springs	W	3, 155
	Diplectrona	√		Fast-flowing sections of small, cold streams	W, not AT, NT	153, 156
	Homoplectra	√	√	Rock face springs and seeps, esp. montane	NA	83, 156
	Parapsyche spp.	√		Fast lotic, small, cold streams and springs	NH	153, 155
Polycentropodidae	*Polycentropus*	√		Eurytopic incl. temporary ponds, karst springs	W	3, 155
Xiphocentronidae	*Cnodocentron*	√		Spring-fed ponds	NA	83
Glossosomatidae	*Anagapetus*	√		Cool montane headwaters, alpine springs	NA	83, 156
	Glossosoma spp.	√		On rocks, especially cold rapid streams	HO, OL	153, 155
Hydroptilidae	*Hydroptila*	√	√	Running waters, also lakes	W	83, 155
	Ochrotrichia	√		Running waters, incl. water film on rocks	NA, NT	153, 155
	Palaeagapetus	√	√	Montane, esp. small, cold seepage springs	NA	83, 156
Lepidostomatidae	*Lepidostoma* spp.	√		Cool lotic incl. springs, littoral zone of lakes	NH	83, 155
	Theliopsyche	√		Clean gravel areas of spring runs	NA	156 ROM
Apataniidae	*Allomyia*	√		Cold mountain streams	HO	153, 155
	Apatania	√		Cool lotic, but also lakes at higher altitude	HO, OL	83, 155
	Moselyana	√	√	Subalpine forest seeps	NA	83, 157
Phryganeidae	*Oligostomis*	√		Slow-current areas of cool forest streams	HO	156 ROM
	Ptilostomis	√		Various, incl. temporary pools and cool streams	NA	155 ROM
Brachycentridae	*Amiocentrus*	√		Moderate currents in larger streams	NA	156 ROM
	Eobrachycentrus	√		Moss in very cold spring runs; montane	NA	156 ROM
	Micrasema	√		Small, cold streams, usually in moss on rocks	HO	155 ROM

Limnephilidae	*Anabolia*	√		Slow lotic, marshes, temporary pools	HO	153, 155
	Asynarchus	√		Streams, ponds, temporary pools	HO	155 ROM
	Chilostigma		√	Associated with spring seeps in wet meadows	HO, OL	83, 158
	Chyranda	√		Small cold streams in montane regions	NA	153, 155
	Clostoeca		√	Among leaves in spring seepage areas	NA	156 ROM
	Crenophylax	√		Springbrooks in the White Mountains, Arizona	NA	83
	Cryptochia	√	√	Small, cold spring streams, usually montane	NA	83, 156
	Dicosmoecus	√		Streams of various sizes	NA, AS	155 ROM
	Ecclisocosmoecus	√		Small, cold mountain streams	NA	156 ROM
	Eocosmoecus	√		Small, spring streams, usually montane	NA	83
	Frenesia spp.	√		Cold streams, spring seepage areas	NA	83
	Hesperophylax	√		Range of freshwaters: springs, rivers, lakes	NA	153, 155
	Homophylax	√		Small, cold streams in montane regions	NA	153, 156
	Ironoquia	√		Temporary waters with extremes of temperature	HO	153, 156
	Lenarchus	√		Cool small lakes, marshes, temporary pools	HO	155 ROM
	Limnephilus spp.	√		Range of freshwater habitats, esp. lentic	NH	153, 155
	Neophylax spp.	√		Confined to running waters	HO, OL	153, 156
	Onocosmoecus spp.	√		Cool, slow-flowing streams and lake margins	HO	153, 155
	Platycentropus	√		Eurytopic, from cool streams to warm ponds	NA	153, 156
	Pseudostenophylax	√	√	Cool spring runs, small intermittent streams	OL, HO	83, 156
	Psychoglypha spp.	√		Confined to cool running waters	NA	153, 155
	Pycnopsyche	√		Cool woodland streams and small rivers	NA	153, 156
Goeridae	*Goeracea*		√	On rocks in small, cold spring streams	NA	83, 155
	Lepania	√	√	Cool running waters, incl. seepage springs	NA	83
Rossianidae	*Goereilla*	√		High-gradient headwaters, springs in forests	NA	83
Uenoidae	*Farula*		√	Cold, high-gradient streams, incl. intermittent	NA	83
	Neothremma	√		Rapid, cool mountain streams	NA	153, 155

(continued)

(continued)

Taxon/family	Genus	Spring	Seep	Genus habitat, where known[a]	Region found[b]	Source[c]
Beraeidae	*Beraea*	√	√	In organic matter around spring seepages	HO, AT, AS	83, 156
Sericostomatidae	*Gumaga*	√		Springbrooks and cold springs	NA, OL	83, 156
Helicopsychidae	*Helicopsyche*	√		Various lotic, incl. thermal springs, littoral	W	83, 159
Odontoceridae	*Parthina*	√	√	Small, cold springs	NA	83, 156
Molannidae	*Molanna*	√		Lentic and slowly flowing waters	HO, OL	83, 155
	Molannodes	√	√	Springs and spring seeps	HO, OL	83, 155
Leptoceridae	*Oecetis*	√		Various lotic and lentic, including karst springs	W	3, 155

[a]Habitats have been recorded as including cool/cold-water springs unless otherwise stated

[b]Does not necessarily indicate that the genus has been found specifically in springs or seeps in all of the regions in which it has been recorded. Listings are according to standard biogeographical regions with the addition of geographical locations where authors have given these in preference to the former (e.g. Europe rather than Western Palaearctic and Asia rather than Oriental/southern Palaearctic). These distribution records may well be incomplete for some taxonomic groups, due to insufficient study in certain regions (e.g. Bryozoa and Ephemeroptera in the Oriental). In contrast, other taxonomic groups may appear overly diverse in some regions due to greater intensity of study effort (e.g. Acari and Trichoptera in the Nearctic). Abbreviations are as follows: *ANT* Antarctica, *AS* Asia, *AT* Afrotropical, *AU* Australasia, *B* Boreal, *CB* Circumboreal, *EU* European, *EQ* Equatorial, *GR* Greenland, *HO* Holarctic, *NA* Nearctic, *NH* Northern Hemisphere, *NT* Neotropical, *OL* Oriental, *PA* Palaearctic, *PAC* Pacific Ocean Islands, *PO* polar, *W* widespread

[c]Key to reference sources: (1) Manconi and Pronato (2002); (2) Manconi and Pronato (2008); (3) Webb et al. (1996); (4) Jankowski et al. (2008); (5) Kolasa and Tyler (2010); (6) Schockaert et al. (2008); (7) Roca et al. (1992); (8) Reynoldson (1967); (9) Kenk (1972); (10) Gooch and Glazier (1991); (11) Kenk (1977); (12) Kenk (1973); (13) Kenk (1975); (14) Kawakatsu and Mitchell (2004); (15) Strayer et al. (2010); (16) Kisielwski (1990); (17) Balsamo et al. (2004); (18) Nesteruk (2008); (19) Balsamo et al. (2008); (20) Wallace and Snell (2010); (21) Davis and Gworek (1973); (22) Segers (2004); (23) Tzschachel (1979); (24) Wallace et al. (2005); (25) Edmondson (1940); (26) De Smet (1997); (27) Poinar (2010); (28) Ruess et al. (2000); (29) Giannakis and Sanders (1989); (30) Poinar (2004); (31) Abebe et al. (2008); (32) Yeates (1988); (33) Tahseen et al. (2012); (34) Semprucci (2013); (35) Govedich et al. (2010); (36) Pinder and Ohtaka (2004); (37) Kennedy (1965); (38) Kathman and Brinkhurst (1998); (39) DEC (2009); (40) Rodriquez (2002); (41) Strong et al. (2008); (42) Naser (2006); (43) Brown and Lydeard (2010); (44) Panha and Burch (2004); (45) Cummings and Graf (2010); (46) Nelson et al. (2010); (47) Guidetti and Bertolani (2005); (48) Beasley et al. (2006); (49) Pennak (1978); (50) Smith and Delorme (2010); (51) Taylor (1992); (52) Forester (1991); (53) Victor (2004); (54) Meisch (2000); (55) Escrivá et al. (2012); (56) Karanovic (2012); (57) Marmonier et al. (1989); (58) Diaz and Lorpetto (2011); (59) Matzke-Karasz et al. (2007); (60) Smith et al. (2002); (61) Rossetti and Martens (1998); (62) Dodson et al. (2010); (63) Padhye and Kotov (2010); (64) Silva-Briano and Mirabdullayev (2004); (65) Reid and Williamson (2010); (66) Mirabdullayev (2004); (67) Lopes et al. (2001); (68) Sket (1994); (69) Bowman (1985); (70) Bruce (2004); (71) Covich et al. (2010); (72) Culver et al. (2010); (73) Holsinger (1986); (74) Yule (2004b); (75) Hobbs and Lodge (2010); (76) Ng (2004); (77) Smith et al. (2010);

(78) Smith (1991); (79) Wiles (2004); (80) Benfatti et al. (1989); (81) Norton et al. (1988); (82) Hopkin (1997); (83) Merritt et al. (2008); (84) Williams and Feltmate (1994); (85) Edmunds et al. (1976); (86) Maiolini et al. (2011); (87) Webb and McCafferty (2008); (88) Pritchard (1991); (89) Hager et al. (2012); (90) Garrison (1991); (91) Garrison (1990); (92) Garrison (2004); (93) Tillyard (1905); (94) Corbet and Brooks (2008); (95) Orr et al. (2004); (96) Zwick (1981); (97) Stewart and Stark (1993); (98) Surdick (1985); (99) Polhemus and Polhemus (2006); (100) Hutchinson (1929); (101) Yang et al. (2004); (102) Kurzatkowska and Zawal (2011); (103) Guareschi et al. (2013); (104) Berchi (2011); (105) Harrison (2013); (106) Nieser et al. (2013); (107) Schuh and Slater (1995); (108) Engel (2004); (109) Cai et al. (2012); (110) Roughley and Larson (1991); (111) Yee (2014); (112) Balke et al. (2004); (113) Freitag (2014); (114) Deler-Hernandez (2013); (115) Ruta (2006); (116) Bowles et al. (2003); (117) Schmude (1999); (118) Mart (2009); (119) Rodriguez and Manzo (2012); (120) Courtney (2004a); (121) Courtney (2004b); (122) Brown (2009); (123) Oosterbroek et al. (2007); (124) Petersen (2006); (125) Courtney (2014); (126) De Jong et al. (2008); (127) Ujvarosi et al. (2011); (128) Arnett (2000); (129) Wagner et al. (2008); (130) Cranston et al. (1987); (131) Bass (1998); (132) Mellor et al. (2000); (133) Worth and Grogan (1988); (134) Brundin (1983); (135) Colbo (1991); (136) Fittkau and Roback (1983); (137) Cranston (2010); (138) Oliver (1983); (139) Saether (1983); (140) Cranston et al. (1983); (141) Cranston (1982); (142) Pinder and Reiss (1983); (143) Moubayed and Langdon (1996); (144) Kovac and Rozkosny (2004); (145) Daniels (2014); (146) Plant and Sinclair (2008); (147) Corpus (1986a); (148) Corpus (1986b); (149) Rozkosny (1984); (150) Dumbleton (2008); (151) Mathis and Wirth (1981); (152) Hogue (1993); (153) Williams and Williams (1987); (154) De Moor and Ivanov (2008); (155) Wiggins and Parker (1997); (156) Wiggins (1977); (157) Anderson (1976); (158) Wiggins (1996); (159) Williams et al. (1983)

dBased on information accompanying specimens held at the Royal Ontario Museum, Canada (ROM)

References

Abebe E, Decraemer W, De Ley P (2008) Global diversity of nematodes (Nematoda) in freshwater. Hydrobiologia 595:67–78

Anderson NH (1976) The biology and distribution of the Oregon Trichoptera. Oregon State Univ Agr Exp Sta Tech Bull 134:1–152

Anderson TM, Anderson NH (1995) The insect fauna of spring habitats in semiarid rangelands in Central Oregon. J Kansas Ent Soc 68:65–76

Arnett R (2000) American insects: a handbook of the insects of America north of Mexico. CRC, Boca Raton

Balke M, Jach MA, Hendrich L (2004) Insecta: Coleoptera. In: Yule K, Hoi Sen Y (eds) Freshwater invertebrates of the Malaysian region. Academy of Sciences Malaysia, Kuala Lumpur, pp 555–609

Balsamo M, Todaro MA, Tongiorgi P (2004) Gastrotricha. In: Yule K, Hoi Sen Y (eds) Freshwater invertebrates of the Malaysian region. Academy of Sciences Malaysia, Kuala Lumpur, pp 127–135

Balsamo M, d'Hondt J-L, Kisielewski J, Pierboni L (2008) Global diversity of gastrotrichs (Gastrotricha) in fresh waters. Hydrobiologia 595:85–91

Barquin J, Death RG (2006) Spatial patterns of macroinvertebrate diversity in New Zealand springbrooks and rhithral streams. J North Am Benthol Soc 25:768–786

Bartsch I (2008) Global diversity of halacarid mites (Halacaridae: Acari: Arachnida) in freshwater. Hydrobiologia 595:317–322

Bass J (1998) Last-instar larvae and pupae of the Simuliidae of Britain and Ireland—a key with brief ecological notes. Freshwat Biol Assoc Sci Pub No. 55. Ambleside, Cumbria, UK

Beasley CW, Kaczmarek L, Michalczyk L (2006) New records of tardigrades from China, with zoogeographical remarks. Biol Lett 43:13–20

Benfatti D, Mari M, Morselli I (1989) *Copidognathus dactyloporus*, a new freshwater species (Halacaridae, Acari). Boll Zool 56:99–104

Berchi GM (2011) First record of *Anisops sardeus* (Hemiptera: Heteroptera: Notonectidae) in Romania. Northwest J Zool 7:339–341

Biological Survey of Canada (1990) Freshwater springs: a national heritage. Bull Entomol Soc Can Suppl 22(1):1–9

Blackwood MA, Hall SM, Ferrington LC (1995) Emergence of Chironomidae from springs in the Central High Plains Region of the United States. J Kansas Entomol Soc 68:132–151

Bornhauser K (1913) Die Tierwelt der Quellen in der Umgebung Basels. Int Rev ges Hydrobiol Hydrogr Biol Suppl 5:1–90

Botosaneanu L (ed) (1998) Studies in crenobiology—the biology of springs and springbrooks. Backhuys, Leiden

Botosaneanu L, Negrea S (1961) Une oasis aquatique a faune relique dans la plaine du Danube inferior. Hydrobiologia 18:199–218

Bowler IR (2014) The geography of agriculture in developed market economies. Routledge, Oxford

Bowles DE, Barr CB, Stanford R (2003) Habitat and phenology of the endangered riffle beetle *Heterelmis comalensis* and a coexisting species, *Microcylloepus pusillus*, (Coleoptera: Elmidae) at Comal Springs, Texas, USA. Arch Hydrobiol 156:361

Bowman TE (1985) *Thermosphaeroma cavicaudata* and *Thermosphaeroma macrura*, new Sphaeromatid isopods from Mexican hot springs. Proc Biol Soc Washington 98:1042–1047

Boxshall GA, Defaye D (2008) Global diversity of copepods (Crustacea: Copepoda) in freshwater. Hydrobiologia 595:195–207

Brown BV (2009) Manual of Central American Diptera, vol 1. National Research Council Canada, Ottawa

Brown KM, Lydeard C (2010) Mollusca: Gastropoda. In: Thorpe JH, Covich AP (eds) Ecology and classification of North American freshwater invertebrates. Academic, London, pp 277–307

Bruce NL (2004) Crustacea: Isopoda. In: Yule K, Hoi Sen Y (eds) Freshwater invertebrates of the Malaysian region. Academy of Sciences Malaysia, Kuala Lumpur, pp 298–310

Brundin L (1983) The larvae of Podonominae (Diptera: Chironomidae) of the Holarctic region—keys and diagnoses. Entomol Scand Suppl 19:23–31

Brune G (1981) Springs of Texas, vol 1. Texas A&M University Agriculture Series No. 5

Cai C, Short A, Huang D (2012) The first skiff beetle (Coleoptera: Myxophaga: Hydroscaphidae) from Early Cretaceous Jehol biota. J Paleontol 86:116–119

Colbo MH (1991) A comparison of the spring-inhabiting genera of Chironomidae from the Holarctic with those from natural and man-made springs in Labrador, Canada. Mem Entomol Soc Can 155:169–179

Collins NC, Mitchell R, Wiegert RG (1976) Functional analysis of a thermal spring ecosystem, with an evaluation of the role of consumers. Ecology 57:1221–1232

Corbet PS, Brooks S (2008) Dragonflies. New Naturalist Series, Collins, London

Corpus LD (1986a) Immature stages of *Liancalus similis* (Diptera: Dolichopodidae). J Kansas Entomol Soc 59:635–640

Corpus LD (1986b) Biological notes and descriptions of the immature stages of *Pelastoneurus vagans* Loew (Diptera: Dolichopodidae). Proc Entomol Soc Washington 88:673–679

Courtney GW (2004a) Insecta: Diptera: Deuterophlebiidae. In: Yule K, Hoi Sen Y (eds) Freshwater invertebrates of the Malaysian region. Academy of Sciences Malaysia, Kuala Lumpur, pp 763–768

Courtney GW (2004b) Insecta: Diptera: Nymphomyiidae. In: Yule K, Hoi Sen Y (eds) Freshwater invertebrates of the Malaysian region. Academy of Sciences Malaysia, Kuala Lumpur, pp 769–774

Courtney G (2014) Thaumastoptera. http://www.ent.iastate.edu/dept/research/systematics/thai/tipuloidea/thaumastoptera. Accessed 15 Sept 2014

Cover MR, Resh VH (2008) Global diversity of dobsonflies, fishflies, and alderflies (Megaloptera; Insecta) and spongillaflies, nevrorthids, and osmylids (Neuroptera; Insecta) in freshwater. Hydrobiologia 595:409–417

Covich AP, Thorpe JH, Rogers DC (2010) Introduction to the Subphylum Crustacea. In: Thorpe JH, Covich AP (eds) Ecology and classification of North American freshwater invertebrates. Academic, London, pp 695–723

Cranston PS (1982) A key to the larvae of the British Orthocladiinae (Chironomidae). Freshwat Biol Assoc Sci Pub No. 45. Ambleside, Cumbria, UK

Cranston PS (2010) http://chirokey.skullisland.info/genus/

Cranston PS, Oliver DR, Saether OA (1983) The larvae of Orthocladiinae (Diptera: Chironomidae) of the Holarctic region—keys and diagnoses. Entomol Scand Suppl 19:149–291

Cranston PS, Ramsdale CD, Snow KR, White GB (1987) Adults, larvae and pupae of British mosquitoes (Culicidae). Freshwat Biol Assoc Sci Pub No. 48. Ambleside, Cumbria, UK

Culver DC, Holsinger JR, Christman MC, Pipan T (2010) Morphological differences among eyeless amphipods in the genus *Stygobromus* dwelling in different subterranean habitats. J Crust Biol 30:68–74

Cummings KS, Graf DL (2010) Mollusca: Bivalvia. In: Thorpe JH, Covich AP (eds) Ecology and classification of North American freshwater invertebrates. Academic, London, pp 309–384

Daniels G (2014) Pelecorhynchidae—Australasian/oceanic Diptera catalog—web version. http://hbs.bishopmuseum.org/aocat/peleco.html. Accessed 17 Sept 2014

Danks HV, Rosenberg DM (1987) Aquatic insects of peatlands and marshes in Canada: synthesis of information and identification of needs for research. Mem Entomol Soc Can 140:163–174

Danks HV, Williams DD (1991) Arthropods of springs, with particular reference to Canada: synthesis and needs for research. Mem Entomol Soc Can 155:203–217

Davis SP, Gworek WF (1973) A rotifer parasitizing *Vaucheria* in a Florida spring. Trans Am Microsc Soc 92:135–140

De Jong H, Oosterbroek P et al (2008) Global diversity of craneflies (Insecta, Diptera: Tipulidea or Tipulidae *sensu lato*) in freshwater. Hydrobiologia 595:457–467

De Moor FC, Ivanov VD (2008) Global diversity of caddisflies (Trichoptera: Insecta) in freshwater. Hydrobiologia 595:393–407

De Smet WH (1997) Guides to the identification of the microinvertebrates of the continental waters of the world. 5:1–325

DEC (2009) Resource condition report for a significant Western Australian Wetland: Saunders Spring. Inland aquatic integrity resource condition monitoring (IAI RCM) project. Department of Environment and Conservation, Perth, Australia

Deharveng L, D'Haese CA, Bedos A (2008) Global diversity of springtails (Collembola; Hexapoda) in freshwater. Hydrobiologia 595:329–338

Deler-Hernandez A (2013) A review of the genus *Berosus* Leach of Cuba (Coleoptera, Hydrophilidae). ZooKeys 273:73–106. doi:10.3897/zookeys.273.4591

Diaz AR, Lorpetto EC (2011) The genus *Chlamydotheca* Saussure (Crustacea: Ostracoda) in northeaster Argentina. Nauplius 19:97–107

Dodson SL, Cáceres CE, Rogers DC (2010) Cladocera and other Branchiopoda. In: Thorpe JH, Covich AP (eds) Ecology and classification of North American freshwater invertebrates. Academic, London, pp 773–827

Downing RA, Smith DB et al (1977) The age of groundwater in the Lincolnshire limestone, England and its relevance to the flow mechanism. J Hydrol 33:201–216

Dumbleton LJ (2008) A new species of *Ephydrella* Tonnoir and Malloch (Diptera: Ephydridae) from hot springs and notes on other Diptera from mineralised waters. NZ Entomol 4:38–46

Dumont HJ, Negrea S (1996) A conspectus of the Cladocera of the subterranean waters of the world. Hydrobiologia 325:1–30

Edmondson WT (1940) The sessile Rotatoria of Wisconsin. Trans Am Microsc Soc 59:433–459

Edmunds GF, Jensen SL, Berner L (1976) The Mayflies of North and Central America. University of Minnesota Press, Minneapolis

Engel MS (2004) The alderflies of Kansas (Megaloptera: Sialidae). Trans Kansas Acad Sci 107:119–125

Erman NA (1998) Invertebrate richness and Trichoptera phenology in Sierra Nevada (California, USA) cold springs: sources of variation. In: Botosaneanu L (ed) Studies in crenobiology—the biology of springs and springbrooks. Backhuys, Leiden, pp 95–108

Erman NA, Erman DC (1995) Spring permanence, Trichoptera species richness, and the role of drought. J Kansas Entomol Soc 68:50–64

Escrivá A, Smith RJ et al (2012) Global distribution of *Fabaeformiscandona subacuta*: an exotic invasive Ostracoda on the Iberian Peninsula? J Crustacean Biol 32:949–961

Feldman R (1974) The mollusc fauna of the Bog Springs of the southern Westphalian Uplands. Decheniana 127:135–143

Ferrington LC (ed) (1995) Biodiversity of aquatic insects and other invertebrates in springs. J Kansas Entomol Soc 68:1–165

Ferrington LC (1998) Generic composition of the chironomid fauna in springs of North America. In: Botosaneanu L (ed) Studies in crenobiology—the biology of springs and springbrooks. Backhuys, Leiden, pp 141–155

Ferrington LC, Kavanaugh RG, Schmidt FJ, Kavanaugh JL (1995) Habitat separation among Chironomidae (Diptera) in Big Springs. J Kansas Entomol Soc 68:152–165

Fittkau EJ, Roback SS (1983) The larvae of Tanypodinae (Diptera: Chironomidae) of the Holarctic region—keys and diagnoses. Entomol Scand Suppl 19:33–110

Forester RM (1991) Ostracode assemblages from springs in the western United States: implications for paleohydrology. Mem Entomol Soc Can 155:181–201

Forro L, Korovchinsky NM, Kotov AA, Petrusek A (2008) Global diversity of cladocerans (Cladocera; Crustacea) in freshwater. Hydrobiologia 595:177–184

Freitag H (2014) A mega-diverse water beetle genus (Coleoptera: Hydraenidae: *Hydraena* Kugelann) commonly overlooked in Southeast Asia and its potential use for environmental biomonitoring. Entomol Ornithol Herpetol 3:125. doi:10.4172/2161-0983.1000125

Garey JR, McInnes SJ, Nichols PB (2008) Global diversity of tardigrades (Tardigrada) in freshwater. Hydrobiologia 595:101–106

Garrison RW (1990) A synopsis of the genus *Hetaerina* with descriptions of four new species (Odonata: Calopterygidae). Trans Am Entomol Soc 116:175–259

Garrison RW (1991) A synonymic list of the New World Odonata. Argia 3:1–30

Garrison RW (2004) A synonymic list of the New World Odonata. Website (version Oct. 2004) http://www2.ups.edu/biology/museum/NewWorldOD.html

Gathmann FO, Williams DD (1998) *Inter-site*: a new tool for the simulation of spatially realistic population dynamics. Ecol Model 113:125–139

Gathmann FO, Williams DD (2006) Insect emergence in Canadian coldwater springs: spatial and temporal patterns, and species-environment relationships. Int J Limnol 42:143–156

Giannakis N, Sanders FE (1989) Interactions between mycophagous nematodes, mycorrhizal and other soil fungi. Agric Ecosyst Environ 29:163–167

Glazier DS (1991) The fauna of North American temperate cold springs: patterns and hypotheses. Freshw Biol 26:527–542

Gooch JL, Glazier DS (1991) Temporal and spatial patterns in mid-Appalachian springs. Mem Entomol Soc Can 155:29–49

Govedich FR, Bain BA et al (2010) Annelida (Clitellata): Oligochaeta, Branchiobdellida, Hiurdinida, and Acanthobdellida. In: Thorpe JH, Covich AP (eds) Ecology and classification of North American freshwater invertebrates. Academic, London, pp 385–436

Greenslade PJM (1983) Adversity selection and the habitat template. Am Nat 122:352–365

Grossman L (2013) Mar's youthful skin helps hunt for life. New Scientist 220:10

Guareschi S, Coccia C et al (2013) How far could the alien boatman *Trichocorixa verticalis verticalis* spread? Worldwide estimation of its current and future potential distribution. PLoS One 8(3):e59757. doi:10.1371/journal.pone.0059757

Guidetti R, Bertolani R (2005) Tardigrade taxonomy: an updated check list of the taxa and a list of characters for their identification. Zootaxa 845:1–46

Hager BJ, Kalantari NJ, Scholten VA (2012) The distribution of *Cordulegaster* (Odonata: Cordulegastridae) nymphs in seeps and springs of Nelson Swamp (Madison County, NY). Northeastern Nat 19:67–76

Harrison SP (2013) Plant and animal endemism in California. University of California Press, Berkeley

Hewitt GM (1996) Some genetic consequences of ice ages, and their role in divergence and speciation. Biol J Linnean Soc 58:247–276

Hinterlang D, Lischewki D (eds) (1996) Spring ecology and conservation. Crunoecia 5:1–304

Hobbs HH, Lodge DM (2010) Decapoda. In: Thorpe JH, Covich AP (eds) Ecology and classification of North American freshwater invertebrates. Academic, London, pp 901–967

Hogg ID, Williams DD (1996) Response of stream invertebrates to a global-warming thermal regime: an ecosystem-level manipulation. Ecology 77:395–407

Hogue CL (1993) Latin American insects and entomology. University of California Press, Berkeley

Holsinger JR (1986) Holarctic Crangonyctid amphipods. In: Botosaneanu L (ed) Stygofauna Mundi, a faunistic, distributional, and ecological synthesis for the world fauna inhabiting subterranean waters (including marine interstitial). Brill/Backuys, Leiden

Hooper JNA, Van Soest RWM (2002) Systema Porifera. Guide to the supraspecific classification of sponges and spongiomorphs (Porifera). Plenum, New York

Hopkin SP (1997) Biology of the springtails (Insect: Collembola). Oxford University Press, Oxford

Hutchinson GE (1929) A revision of the Notonectidae and Corixidae of South Africa. Ann S Afr Mus 25:359–474

Ilmonen J, Paasivirta L, Virtanen R, Muotka T (2009) Regional and local drivers of macroinvertebrate assemblages in boreal springs. J Biogeogr 36:822–834

Iversen TM (1976) Life cycle and growth of Trichoptera in a Danish spring. Arch Hydrobiol 78:482–493

Jankowski T, Collins AG, Campbell R (2008) Global diversity of inland water cnidarians. Hydrobiologia 595:35–40

Jaume D (2008) Global diversity of spelaeogriphaceans and thermosbaenaceans (Crustacea; Spelaeogriphacea and Thermosbaenacea) in freshwater. Hydrobiologia 595:219–224

Karanovic I (2012) Recent freshwater ostracods of the world: Crustacea, Ostracoda, Podocopida. Springer, Heidelberg

Kathman RD, Brinkhurst RO (1998) Guide to the freshwater oligochaetes of North America, vol 4. Aquat Resources Center, College Grove, pp 1–264

Kawakatsu M, Mitchell RW (2004) Platyhelminthes: Dugesiidae and Dimarcusidae. In: Yule K, Hoi Sen Y (eds) Freshwater invertebrates of the Malaysian region. Academy of Sciences Malaysia, Kuala Lumpur, pp 55–65

Keleher MJ, Sada D (2012) Desert spring wetlands of the Great Basin. In: Batzer DP, Baldwin AH (eds) Wetland habitats of North America—ecology and conservation concerns. University of California Press, Berkeley, pp 329–341

Kenk R (1972) Freshwater planarians (Turbellaria) of North America. Biota of freshwater ecosystems, identification manual 1, Environ Protection Agency, Washington, p 1–81

Kenk R (1973) Freshwater triclads (Turbellaria) of North America, VI: the genus *Dendrocoelopsis*. Smithsonian Contrib Zool 138:1–16

Kenk R (1975) Freshwater triclads (Turbellaria) of North America, VII: the genus *Macrocotyla*. Trans Am Microsc Soc 94:324–339

Kenk R (1977) Freshwater triclads (Turbellaria) of North America, IX: the genus *Sphalloplana*. Smithsonian Contrib Zool 246:1–38

Kennedy CR (1965) The distribution and habitat of *Limnodrilus* Claparéde (Oligochaeta: Tubificidae). Oikos 16:26–38

Kisielwski J (1990) Origin and phylogenetic significance of freshwater psammic Gastrotricha. Stygologia 5:87–92

Kolasa J, Tyler S (2010) Flatworms: Turbellarians and Nemertea. In: Thorpe JH, Covich AP (eds) Ecology and classification of North American freshwater invertebrates. Academic, London, pp 143–161

Korovchinsky NM (2006) The Cladocera (Crustacea: Branchiopoda) as a relict group. Zool J Linnean Soc 147:109–124

Kovac D, Rozkosny R (2004) Insecta: Diptera, Stratiomyidae. In: Yule CM, Sen YH (eds) Freshwater invertebrates of the Malaysian region. Academy of Sciences Malaysia, Kuala Lumpur, pp 798–804

Kurzatkowska A, Zawal A (2011) *Sigara dorsalis* (Leach, 1817) (Heteroptera: Corixidae)—a new species in Poland and changes in its eastern range extension. Zool Poloniae 56:5–10

Lay C-Y, Mykytczuk NCS, Yergeau E, Lamarche-Gagnon G (2013) Defining the functional potential and active community members of a sediment microbial community in a High-Arctic hypersaline subzero spring. Appl Environ Microbiol 79:3637–3648

Lindegaard C (1995) Chironomidae (Diptera) of European cold springs and factors influencing their distribution. J Kansas Entomol Soc 68:108–131

Liu DHF, Liptak BG (eds) (1999) Groundwater and surface water pollution. CRC, Boca Raton

Lopes RM, Reid JW, Rocha C (2001) Copepoda: developments in ecology, biology and systematics. Hydrobiologia 453(454):1–576

Maiolini B, Carolli M, Silveri L (2011) Ephemeroptera, Plecoptera and Trichoptera in Trentino (south-eastern Alps). J Limnol 70:122–133

Manconi R, Pronato R (2002) Suborder Spongillina subord. Nov.: freshwater sponges. In: Hooper JNA, Van Soest RWM (eds) Systema Porifera: a guide to the classification of sponges. Kluwer, New York, pp 921–1021

Manconi R, Pronato R (2008) Global diversity of sponges (Porifera: Spongillina) in freshwater. Hydrobiologia 595:27–33

Marmonier P, Meisch C, Danielopol DL (1989) A review of the genus *Cavernocypris* Hartmann (Ostracoda, Cypridopsinae): systematics, ecology and biogeography. Bull Soc Naturelle Luxembourg 89:221–278

Mart A (2009) *Hydrochus ibericus* and *Hydrochus nodulifer*, two new records for the Hydrochidae (Coleoptera) fauna of Turkey. Turkish J Zool 33:249–250

Martens K, Schon I, Meisch C, Horne DJ (2008) Global diversity of ostracods (Ostracoda, Crustacea) in freshwater. Hydrobiologia 595:185–193

Martin P, Martinez-Ansemil E et al (2008) Global diversity of oligochaetous clitellates ("Oligochaeta"; Clitellata) in freshwater. Hydrobiologia 595:117–127

Mathis WN, Wirth WW (1981) Studies of Ephydrinae (Diptera: Ephydridae), IV: revision of the Australian species of subgenus *Neoscatella* Malloch. Smithsonian Contrib Zool 325:1–27

Mattson RA, Epler JH, Hein MK (1995) Description of benthic communities in karst, spring-fed streams of north central Florida. J Kansas Entomol Soc 68:18–41

Matzke-Karasz R, Martens K, Schudack M (2007) Ostracodology—linking bio- and geosciences. Hydrobiologia 585:1–278

McEwen AS, Ojha L et al (2011) Seasonal flows on warm Martian slopes. Science 333:740–743

Meisch C (2000) Freshwater Ostracoda of Western and Central Europe. In: Schwoerbel J, Zwick P (eds) Süßwasserfauna von Mitteleuropa. Spektrum Akademischer, Heidelberg, Berlin

Mellor PS, Boorman J, Baylis M (2000) *Culicoides* biting midges: their role as arbovirus vectors. Annu Rev Entomol 45:307–340

Merritt RW, Cummins KW, Berg MB (eds) (2008) An introduction to the aquatic insects of North America. Kendall Hunt, Dubuque

Mirabdullayev IM (2004) Crustacea: Copepoda, Cyclopoidae, Cyclopidae. In: Yule K, Hoi Sen Y (eds) Freshwater invertebrates of the Malaysian region. Academy of Sciences Malaysia, Kuala Lumpur, pp 267–273

Moubayed J, Langdon PH (1996) *Krenopsectra nohedensis* n. sp. and the pupal exuviae of *Micropsectra auvergnensis* Reiss (Diptera: Chironomidae) from the eastern Pyrenees. Br J Ent Nat Hist 9:77–86

Myers MJ, Resh VH (2002) Trichoptera and other macroinvertebrates in springs of the Great Basin: species composition, richness, and distribution. West North Am Nat 62:1–13

Naser MD (2006) Some notes on Melanopsidae (Gastropoda) of Mesopotamia. Iraq Aquat J 2:85–90

Nelson DR, Guidetti R, Rebecchi L (2010) Tardigrada. In: Thorpe JH, Covich AP (eds) Ecology and classification of North American freshwater invertebrates. Academic, London, pp 455–484

Nesteruk T (2008) Assessment of the diversity and density of gastrotrich fauna (Gastrotricha) in bottom sediments of running and standing waters. Teka Komisji Ochrony Kształtowania Środowiska Przyrodniczego—OL. PAN 5:136–143

Ng PKL (2004) Crustacea: Decapoda, Brachyura. In: Yule CM, Sen YH (eds) Freshwater invertebrates of the Malaysian region. Academy of Sciences Malaysia, Kuala Lumpur, pp 311–336

Nieser N, Chen P, Lane de Melo A (2013) A new species and a new synonym in *Limnocoris* (Hemiptera: Heteroptera: Naucoridae) from Brazil. Acta Musei Moraviae, Sci Biol 98:335–346

Norton RA, Williams DD, Hogg ID, Palmer SC (1988) Biology of the oribatid mite *Mucronothrus nasalis* (Acari: Oribatida: Trhypochthoniidae) from a small coldwater springbrook in Eastern Canada. Can J Zool 66:622–629

Odum HT (1957) Trophic structure and productivity of Silver Springs, Florida. Ecol Monogr 27:55–112

Odum EP (1971) Fundamentals of ecology. WB Saunders, Toronto

Oliver DR (1983) The larvae of Diamesinae (Diptera: Chironomidae) of the Holarctic region—keys and diagnoses. Entomol Scand Suppl 19:115–138

Oosterbroek P, Brodo F et al (2007) The Tipulidae and Limoniidae of Greenland (Diptera, Nematocera, Craneflies). Ent Meddr 75:3–33

Orr AG, Butler SG, Hamalainen M, Kemp RG (2004) Insecta: Odonata. In: Yule CM, Sen YH (eds) Freshwater invertebrates of the Malaysian region. Academy of Sciences Malaysia, Kuala Lumpur, pp 409–442

Padhye S, Kotov AA (2010) Cladocera (Crustacea: Branchiopoda) in Indian hot water springs. Invert Zool 7:155–158

Panha S, Burch JB (2004) Mollusca. In: Yule CM, Sen YH (eds) Freshwater Invertebrates of the Malaysian region. Academy of Sciences Malaysia, Kuala Lumpur, pp 207–224

Pennak RW (1978) Freshwater invertebrates of the United States. Wiley, New York

Petersen MJ (2006) A new species of crane fly, *Pedicia* (*Pedicia*) *goldsworthyi* (Diptera: Pediciidae), from eastern North America. Zootaxa 1220:55–61

Pinder AM, Ohtaka A (2004) Annelida: Clitellata, Oligochaeta. In: Yule CM, Sen YH (eds) Freshwater invertebrates of the Malaysian region. Academy of Sciences Malaysia, Kuala Lumpur, pp 162–174

Pinder LCV, Reiss F (1983) The larvae of Chironominae (Diptera: Chironomidae) of the Holarctic region—keys and diagnoses. Entomol Scand Suppl 19:293–435

Plant AR, Sinclair BJ (2008) New species of *Hemerodromia* Meigen from Fiji (Diptera: Empididae). Bishop Mus Occas Pap 98:31–52

Poinar GO (2004) Nematoda. In: Yule CM, Sen YH (eds) Freshwater invertebrates of the Malaysian region. Academy of Sciences Malaysia, Kuala Lumpur, pp 145–156

Poinar GO (2010) Nematoda and Nematomorpha. In: Thorpe JH, Covich AP (eds) Ecology and classification of North American freshwater invertebrates. Academic, London, pp 237–276

Polhemus JT, Polhemus DA (2006) A new Fijian species of *Saldula* Van Duzee (Heteroptera: Saldidae) from the interior of Viti Levu. Bishop Mus Occas Pap 90:43–50

Pritchard G (1991) Insects in thermal springs. Mem Entomol Soc Can 155:89–106

Reid JW, Williamson CE (2010) Copepoda. In: Thorpe JH, Covich AP (eds) Ecology and classification of North American freshwater invertebrates. Academic, London, pp 829–899

Resh VH (1983) Spatial differences in the distribution of benthic macroinvertebrates along a springbrook. Aquat Insects 5:193–200

Reynoldson TB (1967) A key to the British species of freshwater triclads. Freshwat Biol Assoc Sci Pub No. 23. Kendal, Cumbria, UK

Ring RA (1991) The insect fauna and some other characteristics of natural salt springs on Saltspring Island, British Columbia. Mem Entomol Soc Canada 155:51–61

Roca JR, Ribas M, Baguna J (1992) Distribution, ecology, mode of reproduction and karyology of freshwater planarians (Platyhelminthes; Turbellaria; Tricladida) in the springs of the central Pyrenees. Ecography 15:373–384

Rodriguez VA, Manzo V (2012) Redescription and lectotype designation of *Helichus cordubensis* Berg (Coleoptera: Dryopidae). Zootaxa 3168:64–68

Rodriquez P (2002) Benthic and subterranean aquatic oligochaete fauna (Annelida: Oligochaeta) from Coiba Island (Panama) and Cuba. Graellsia 58:3–19

Rossetti G, Martens K (1998) Taxonomic revision of the recent and Holocene representatives of the family Darwinulidae (Crustacea, Ostracoda), with a description of three new genera. Bull l'Institut R Sci Natur Belgique Biol 68:55–110

Roughley RE, Larson DJ (1991) Aquatic Coleoptera of springs in Canada. Mem Entomol Soc Can 155:125–140

Rozkosny R (1984) The Sciomyzidae (Diptera) of Fennoscandia and Denmark. Entomol Scand 14:1–224

Ruess L, Garcia Zapata EJ, Dighton J (2000) Food preferences of a fungal-feeding *Aphelenchoides* species. Nematology 2:223–230

Ruta R (2006) The first record of *Paracymus aeneus* (Germar 1824) (Coleoptera: Hydrophilidae) in Poland with notes on halophilous and halobiotic Hydrophilidae and Hydraenidae in the Polish fauna. Polish J Entomol 75:359–368

Saether OA (1983) The larvae of Prodiamesinae (Diptera: Chironomidae) of the Holarctic region—keys and diagnoses. Entomol Scand Suppl 19:141–147

Sarkka J, Levonen L, Makela J (1997) Meiofauna of springs in Finland in relation to environmental factors. Hydrobiologia 347:139–150

Schatz H, Behan-Pelletier V (2008) Global diversity of oribatids (Oribatida: Acari: Arachnida). Hydrobiologia 595:323–328

Schmude KL (1999) Riffle beetles in the genus *Stenelmis* from warm springs in Southern Nevada: new species, new status, and a key. Entomol News 110:1–12

Schockaert ER, Hooge M et al (2008) Global diversity of free living flatworms (Platyhelminthes, 'Turbellaria') in freshwater. Hydrobiologia 595:41–48

Schuh RT, Slater JA (1995) True bugs of the world (Hemiptera: Heteroptera): classification and natural history. Cornell University Press, Ithaca

Segers H (2004) Rotifera: Monogononta. In: Yule CM, Sen YH (eds) Freshwater invertebrates of the Malaysian region. Academy of Sciences Malaysia, Kuala Lumpur, pp 106–120

Semprucci F (2013) Marine nematodes from the shallow subtidal coast of the Adriatic Sea: species list and distribution. Internat J Biodivers. doi:10.1155/2013/187659

Silva-Briano M, Mirabdullayev IM (2004) Crustacea: Cladocera. In: Yule CM, Sen YH (eds) Freshwater invertebrates of the Malaysian region. Academy of Sciences Malaysia, Kuala Lumpur, pp 274–283

Sket B (1994) Distribution of *Asellus aquaticus* (Crustacea: Isopoda: Asellidae) and its hypogean populations at different geographic scales, with a note on *Proasellus istrianus*. Hydrobiologia 287:39–47

Smith IM (1991) Water mites (Acari: Parasitengona: Hydrachnida) of spring habitats in Canada. Proc Entomol Soc Ontario 118:141–167

Smith AJ, Delorme LD (2010) Ostracoda. In: Thorpe JH, Covich AP (eds) Ecology and classification of North American freshwater invertebrates. Academic, London, pp 725–771

Smith RJ, Matzke-Karasz R, Kamiya T, Ikeda Y (2002) *Scottia birigida* sp. nov. (Cypridoidea: Ostracoda) from western Honshu, Japan and a key to the subfamily Scottiinae Bronstein, 1947. Zootaxa 126:1–20

Smith IM, Cook DR, Smith BP (2010) Water mites (Hydrachnidiae) and other arachnids. In: Thorpe JH, Covich AP (eds) Ecology and classification of North American freshwater invertebrates. Academic, London, pp 485–586

Stewart KW, Stark BP (1993) Nymphs of the North American stonefly genera. University of North Texas Press, Denton

Stoch F, Gerecke R et al (2011) Exploring species distribution of spring meiofauna (Annelida, Acari, Crustacea) in the south-eastern Alps. J Limnol 70:65–76

Strayer DL, Hummon WD, Hochberg R (2010) Gastrotricha. In: Thorpe JH, Covich AP (eds) Ecology and classification of North American freshwater invertebrates. Academic, London, pp 163–172

Strong EE, Gargominy O, Ponder WF, Bouchet P (2008) Global diversity of gastropods (Gastropoda; Mollusca) in freshwater. Hydrobiologia 595:149–166

Surdick RF (1985) Nearctic genera of Chloroperlinae (Plecoptera: Chloroperlidae). Illinois Biol Monogr 54:1–146

Tahseen Q, Sultana R, Khan R, Hussain A (2012) A new genus and species of the family Rhabdolaimidae (Nematoda), with descriptions of two known species and taxonomic discussion. J Nematol 44:302–312

Taylor LC (1992) The response of spring-dwelling ostracodes to intra-regional differences in groundwater chemistry associated with road salting practices in southern Ontario: a test using an urban-rural transect. MS Thesis, University of Toronto

Teal JM (1957) Community metabolism in a temperate cold spring. Ecol Monogr 27:283–302

Thorpe JH, Covich AP (eds) (2010) Ecology and classification of North American freshwater invertebrates. Academic, London

Tilly JM (1968) The structure and dynamics of Cone Spring. Ecol Monogr 27:283–302

Tillyard RJ (1905) On dimorphism in the female of *Ischnura heterosticta*, Burm. Proc Linnean Soc New South Wales 30:302–306

Trichoptera World Checklist (2015) http://entweb.clemson.edu/database/trichop/search.htm

Tuxen SL (1944) The hot springs, their animal communities and their zoogeographical significance. In The zoology of Iceland, vol 1. Einer Munksgard, Copenhagen, p 1–216

Tzschachel G (1979) Marine Rotatoria aus dem interstitial der Nordseeinsel Sylt. Mikrofauna des Meeresbodens 71:1–64

Ujvarosi L, Kolcsar PL, Torok E (2011) An annotated list of the Ptychopteridae (Insecta, Diptera) from Romania, with notes on the individual variability of *Ptychoptera albimana* (Fabricius 1787). Entomol Romanica 16:39–45

Vainola R, Witt JDS et al (2008) Global diversity of amphipods (Amphipoda; Crustacea) in freshwater. Hydrobiologia 595:241–255
Van der Kamp G (1995) The hydrogeology of springs in relation to the biodiversity of spring fauna: a review. J Kansas Entomol Soc 68:4–17
Van Everdingen RO (1991) Physical, chemical, and distributional aspects of Canadian springs. Mem Entomol Soc Canada 155:7–28
Varza D, Covich AP (1995) Population fluctuations within a spring community. J Kansas Entomol Soc 68:42–49
Victor R (2004) Ostracoda. In: Yule CM, Sen YH (eds) Freshwater invertebrates of the Malaysian region. Academy of Sciences Malaysia, Kuala Lumpur, pp 225–253
Wagner R, Bartak M et al (2008) Global diversity of dipteran families (Insecta Diptera) in freshwater (excluding Simuliidae, Culicidae, Chironomidae, Tipulida and Tabanidae). Hydrobiologia 595:489–519
Wallace RL, Snell TW (2010) Rotifera. In: Thorpe JH, Covich AP (eds) Ecology and classification of North American freshwater invertebrates. Academic, London, pp 173–235
Wallace RL, Walsh EJ, Arroyo ML, Starkweather PL (2005) Life on the edge: rotifers from springs and ephemeral waters in the Chihuahuan Desert, Big Bend National Park (Texas, USA). Hydrobiologia 546:147–157
Ward JV, Dufford RG (1979) Longitudinal and seasonal distribution of macroinvertebrates and epilithic algae in a Colorado springbrook-pond system. Arch Hydrobiol 86:284–321
Webb JM, McCafferty WP (2008) Heptageniidae of the world: Part II. Key to the genera. Can J Arthropod Ident. doi:10.3752/cjai.2008.07
Webb DW, Wetzel MJ et al (1996) Karst springs in the sinkhole plain of Illinois: their community diversity and hydrogeology. Technical Report 1996 (11), Illinois Natural History Survey, Centre for Biodiversity, Southern Illinois University, Carbondale
Wiggins GB (1977) Larvae of the North American Caddisfly Genera (Trichoptera). University of Toronto Press, Toronto
Wiggins GB (1996) Larvae of the North American Caddisfly Genera (Trichoptera), 2nd edn. University of Toronto Press, Toronto
Wiggins GB, Parker CR (1997) Caddisflies (Trichoptera) of the Yukon, with analysis of the Beringian and Holarctic species of North America. In: Danks HV, Downes JA (eds) Insects of the Yukon. Biological Survey of Canada, Ottawa, pp 787–866
Wiles PR (2004) Arachnida: Acari, Hydrachnidia. In: Yule CM, Sen YH (eds) Freshwater invertebrates of the Malaysian region. Academy of Sciences Malaysia, Kuala Lumpur, pp 358–368
Williams NE (1991) Geographical and environmental patterns in caddisfly (Trichoptera) assemblages from coldwater springs in Canada. Mem Entomol Soc Can 155:107–124
Williams DD (2006) The biology of temporary waters. Oxford University Press, Oxford
Williams DD, Danks HV (eds) (1991) Arthropods of springs, with particular reference to Canada. Mem Entomol Soc Can 155:1–217
Williams DD, Feltmate BW (1994) Aquatic insects. CAB International, Wallingford
Williams DD, Hogg ID (1988) Ecology and production of invertebrates in a Canadian coldwater spring-springbrook system. Holarct Ecol 11:41–54
Williams DD, Williams NE (1987) Trichoptera from cold freshwater springs in Canada: records and comments. Proc Entomol Soc Ontario 118:13–23
Williams DD, Williams NE (1996a) Paleoecological reconstruction of natural and human influences on groundwater outflows. In: Boon PJ, Howell DL (eds) Freshwater quality: defining the indefinable. HMSO, London, pp 175–183
Williams NE, Williams DD (1996b) Paleoecological reconstruction of natural and human influences on groundwater outflows. In: Boon PJ, Howell DL (eds) Freshwater quality; defining the indefinable? Scottish Natural Heritage/H.M.S.O., London, pp 175–183
Williams DD, Williams NE (1998) Invertebrate communities from freshwater springs: what can they contribute to pure and applied ecology? In: Botosaneanu L (ed) Studies in crenobiology—the biology of springs and springbrooks. Backhuys, Leiden, pp 251–261

Williams DD, Williams NE (1999) Canadian springs: postglacial development of the invertebrate fauna. In: Batzer DP, Rader RB, Wissinger SA (eds) Invertebrates in freshwater wetlands of North America: ecology and management. Wiley, New York, pp 447–467

Williams DD, Read AT, Moore KA (1983) The biology and zoogeography of *Helicopsyche borealis* (Trichoptera: Helicopsychidae): a nearctic representative of a tropical genus. Can J Zool 61:2288–2299

Williams DD, Danks HV et al (1990) Freshwater springs: a national heritage. A brief prepared by the Biological Survey of Canada (Terrestrial Arthropods). Bull Entomol Soc Can 22:1–9

Williams DD, Williams NE, Cao Y (1997) Spatial differences in macroinvertebrate community structure in springs of southeastern Ontario in relation to environmental factors. Can J Zool 75:1404–1414

Wilson GDF (2008) Global diversity of Isopod crustaceans (Crustacea; Isopoda) in freshwater. Hydrobiologia 595:231–240

Winterbourn MJ (1973) Ecology of the Copeland River water springs, South Island, New Zealand. Proc N Z Ecol Soc 20:72–78

Wood TS (2010) Bryozoans. In: Thorpe JH, Covich AP (eds) Ecology and classification of North American freshwater invertebrates. Academic, London, pp 437–454

Worth WW, Grogan WL (1988) The predaceous midges of the world—flora and fauna handbook no. 4. Brill, Leiden, Netherlands

Yang CM, Kovac D, Cheng L (2004) Insecta: Hemiptera, Heteroptera. In: Yule CM, Sen YH (eds) Freshwater invertebrates of the Malaysian region. Academy of Sciences Malaysia, Kuala Lumpur, pp 457–490

Yeates GW (1988) Distribution of Plectidae (Nematoda: Araeolaimida) in pasture soils. N Z J Zool 15:191–202

Yee DA (ed) (2014) Ecology, systematics, and the natural history of predaceous diving beetles (Coleoptera: Dytiscidae). Springer, New York

Yeo DCJ, Ng PKL et al (2008) Global diversity of crabs (Crustacea: Decapoda: Brachyura) in freshwater. Hydrobiologia 595:275–286

Yule CM (2004a) Freshwater environments. In: Yule CM, Sen YH (eds) Freshwater invertebrates of the Malaysian region. Academy of Sciences Malaysia, Kuala Lumpur, pp 1–12

Yule CM (2004b) Crustacea: Pericarida, Amphipoda. In: Yule CM, Sen YH (eds) Freshwater invertebrates of the Malaysian region. Academy of Sciences Malaysia, Kuala Lumpur, pp 307–310

Zagmajster M, Porter M, Fong DW (2011) Freshwater hydrozoans in caves with a report on new records. Speleobiol Notes 3:4–10

Zwick P (1981) Comment. In: Hennig W, Pont C (eds) Insect phylogeny. Wiley, Chichester, p 175

Chapter 12
Invertebrates in Beaver-Created Wetlands and Ponds

Bryana M. Bush and Scott A. Wissinger

Introduction

Eurasian (*Castor fiber*) and North American (*Castor canadensis*) beavers are semi-aquatic mammals that modify the hydrology of streams and other water bodies by constructing dams. The modified aquatic habitats associated with beaver activities were once a ubiquitous feature of the post-Pleistocene landscape throughout the temperate and boreal zones of North America and Europe. By the end of the nineteenth century, trapping and hunting by humans had extirpated beavers across much of their former range (e.g., Johnson and Chance 1974; Danilov et al. 2011). Since then, the recovery and/or reintroduction of populations in North America and Europe (Naiman et al. 1988a, b; Hartman 1994, 1995; Nolet and Rosell 1998; Bluzma 2003; Halley et al. 2012; Law et al. 2014) have led to (1) dramatic changes in the structure and function of headwater and middle orders streams and adjacent riparian zones (Naiman et al. 1988b; Rosell et al. 2005) and (2) the creation and maintenance of wetland habitats within and beyond the boundaries of stream valleys (McCall et al. 1996; Syphard and Garcia 2001).

Research on invertebrate communities in aquatic habitats associated with beaver activities can be divided into general groups of studies: (1) those with a distinctly "running water perspective" that focus on how beaver dams change stream invertebrate communities at multiple scales (reach, stream segment, stream system; Allan 2004) and (2) those with a distinctly wetlands/pond perspective on the plant and animal life that inhabit the many types of shallow lentic habitats outside of stream channels. We first briefly summarize the major themes from the large literature focused on how beavers affect stream invertebrate communities, and then turn to focus on the distinctly lentic invertebrate communities that occur in non-channel wetlands created by beavers.

D. Batzer, D. Boix (eds.), *Invertebrates in Freshwater Wetlands*,
DOI 10.1007/978-3-319-24978-0_12

Stream Ecology Perspective on Beaver Dam Invertebrates

Beaver Reestablishment in Native Range and Changing Streamscapes

Robert Naiman and colleagues established the general paradigm for the effects of beaver activity on stream invertebrates by describing how the presence of beaver dams in stream channels modifies nearly every aspect of the physicochemical (water chemistry, carbon budgets, nutrient spiraling, flow regimes, physical substrates, retention/turnover of organic matter, etc.) and biological (hetero- and auto-trophic microbial assemblages, community metabolism, plants, invertebrates, fish, waterfowl) environment in stream channels and adjacent riparian habitats (Naiman and Melillo 1984; Naiman et al. 1986, 1988a, b; also see Rosell et al. 2005). From the perspective of stream ecologists, hydrologists, and fluvial geomorphologists, the recovery of beaver populations in North America and Eurasia during the past 100 years has prompted a reevaluation of the structure and function of headwater and middle-order streams as compared to when and where beavers had (have) been extirpated (Naiman et al. 1986, 1988b; Cirmo and Driscoll 1993; Devito and Dillon 1993; Hammerson 1994; Pollock et al. 1995; Klotz 1998; Snodgrass and Meffe 1998; Collen and Gibson 2001; Butler and Malanson 2005; Pollock et al. 2007; Burchsted and Daniels 2014; Curran and Cannatelli 2014). The hydrology of beaver ponds in this context is dominated by stream flow inputs and outputs, and the dams can reduce peak channel discharge by temporarily storing water and shunting it to the adjacent riparian zone/floodplain (Fig. 12.1a). This is one of three potential losses of water between channel inflow and outflow in beaver dams. A second is through evapotranspiration because of the increased surface area and residence times, especially in arid environments (Andersen et al. 2011), and a third is through downwelling into the shallow ground water that moves down vallies through unconsolidated sediments. In arid land streams, groundwater recharge from beaver ponds and wetlands can enhance shallow groundwater storage, which later supplements channel flow during low-flow conditions, potentially converting intermittent to perennial streams (Fig. 12.1b, Gibson and Olden 2014).

In steep gradient headwater and middle-order streams, beaver activity in North America and Eurasia leads to the replacement of erosional (riffle) assemblages of invertebrates typical of high-oxygen, turbulent-flow, hard-substrate habitats (e.g., stoneflies, mayflies, riffle beetles, net-spinning caddisflies) with communities that are more typical of depositional environments (pools, runs) that have relatively slow, laminar flow, relatively low oxygen, and a predominance of soft substrates (e.g., chironomids and other dipterans, odonates, dytiscid beetles, hemipterans, annelids, epibenthic crustaceans). At the reach scale, invertebrate biomass is much higher (1.3–11.1 g m^{-2}) in pools behind beaver dams than in adjacent riffles (0.01–0.6 g m^{-2}), but taxonomic diversity between the habitats is similar (McDowell and Naiman 1986; Naiman et al. 1986). However, at the stream-segment or beta

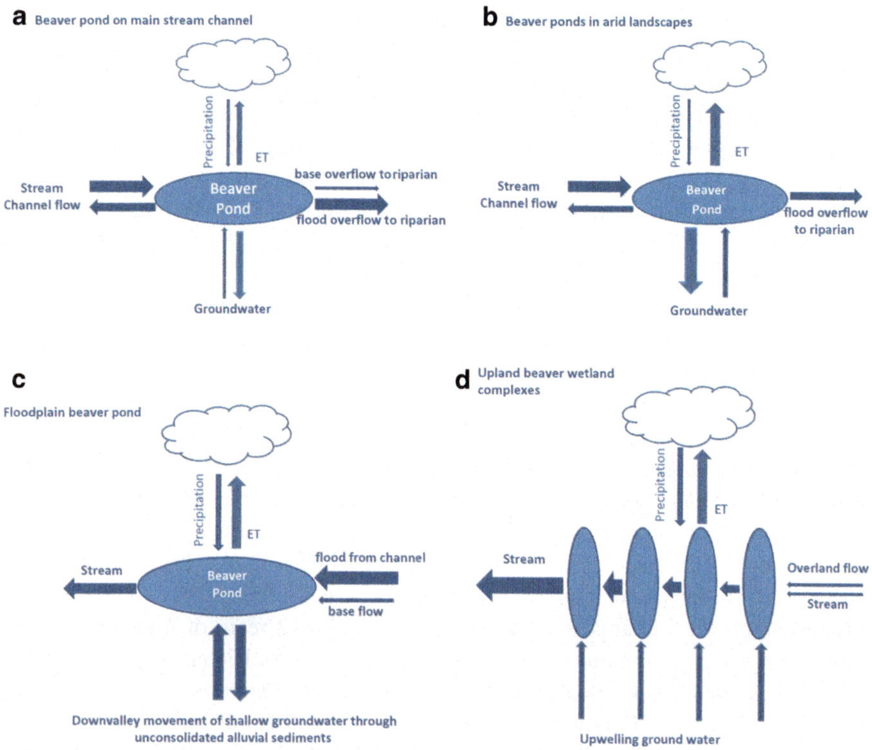

Fig. 12.1 The hydrology of ponds and wetlands associated with beaver activity. (**a**) The hydrology of beaver ponds in main channels will be dominated by stream flow inputs and outputs, and hence be temporally dynamic. Channel flow can be greater than outflow because of (1) bank overflow to the adjacent riparian zone, (2) enhanced evapotranspiration, and/or (3) hyporheic loss with the downstream return flow path dependent on the geomorphologic context (Rosell et al. 2005). (**b**) In arid landscapes, the three losses between stream input and output will be exaggerated because of enhanced evaporation and lateral and vertical losses to shallow groundwater. The temporary storage in that shallow groundwater can be especially important for ameliorating downstream low-flow conditions (Gibson and Olden 2014). (**c**) In beaver ponds and wetlands situated in floodplains away from the main stream channel, the hydrologic budget will be strongly influenced by shallow ground water and lateral hyporheic flow moving down valley through unconsolidated sediments. Proximity to the main channel will determine the frequency and duration of inputs from floods (see Fig. 12.3b). (**d**) Beaver dam wetland complexes in relatively upland landscapes (perched water tables, along valley margins) receive much of their hydrologic input from springs and upwelling groundwater discharge that then leaves the wetland complex as the channel flow of headwater streams. This hydrology model fits beaver meadow complexes in Northeastern North America and those on valley margins in Western North America (see text)

diversity scale (Allan 2004), the longitudinal sequences of these alternating habitat types results in an increase in the overall taxonomic and trophic (functional-feeding-group) diversity as compared to streams lacking beaver dams (e.g., Sprules 1940 [Ontario]; McDowell and Naiman 1986 [Quebec]; Harthun 1999 [Hesse, Germany]; Smith et al. 1991 [New York]; Margolis et al. 2001 [Pennsylvania]; Pliūraitė and Kesminas 2012 [Lithuania]).

The patches of large woody debris associated with dams (Fig. 12.2) and huts can harbor unique assemblages of invertebrate species dominated by grazers and filter feeders (e.g., simuliid larvae; Clifford et al. 1993; Adler and Mason 1997). Rolauffs et al. (2001) found higher invertebrate diversity and higher secondary productivity on coarse woody substrates of dams than in either riffles or the pools created by the dams, perhaps as a result of some combination of the (1) extensive surface area of these complex structures, (2) availability of organic materials (wood substrate with biofilm and flow-through suspended particulates), (3) high organic turnover rate, and (4) aerobic conditions at the water–air interface.

Several studies have compared stream invertebrate communities between comparable habitats above and below beaver dams. In a small, low gradient stream in northeastern North America (New York state), Smith et al. (1991) found that stream invertebrate assemblages below dams are less diverse and have lower densities of Plecoptera, Trichoptera, and filter feeders (taxa not specified). In contrast, Fuller and Peckarsky (2011a) found no systematic differences among functional feeding groups (FFGs) above and below ponds and no differences driven by dam morphology among FFG with the exception of suspension feeders. The abundance of suspension feeders, and especially simuliids, increases below beaver ponds with high hydraulic head dams, which is also typical downstream of man-made reservoirs due to high seston pulses (Mackay and Waters 1986; Richardson and Mackay 1991). However, suspension feeders decrease below ponds with a low hydraulic head dam, and the difference between high- and low-head dams is not driven by algae spillover from dams. Fuller and Peckarsky (2011a) hypothesize that higher abundance of suspension feeders below high head dams could be related to a higher availability of bacterial seston or increased scour downstream of high head dams, but not below low head dams, both of which are favorable for simuliids. Invertebrates were not influenced by any differences in nutrients, algal biomass, and benthic organic matter among stream reaches above and below ponds related to dam morphology.

In a related study, Fuller and Peckarsky (2011b) studied the impact beaver pond morphology had on mayfly life history (Fig. 12.2). They evaluated downstream effects of beaver pond morphology on *Baetis bicaudatus* size and timing of emergence. Reaches downstream of high head, low surface area ponds produced larger females than low head ponds with larger surface area, and females found below the pond were larger than those found above. Male size differences followed similar patterns but were not significantly different. Because large female *B. bicaudatus* are more fecund than small females, Fuller and Peckarsky hypothesize that the next generation could vary in size by +11 to −12 % depending on pond morphology. Larger female size downstream of high head ponds corresponds with colder water temperatures in these areas. Outflow water is colder than pond water, probably as a result of groundwater upwelling below the high hydraulic head dams. Despite temperature differences, pond morphology did not predict timing of emergence of mayflies downstream of dams. In general, where groundwater lost through the hyporheic in beaver ponds resurfaces as channel

Fig. 12.2 Beaver dams on a high gradient stream (West Brush Creek) in the Elk Mountains of Colorado (see Fuller and Peckarsky 2011a, b; photo courtesy of Matt Fuller)

flow (see Fig. 12.1a, b) should have important consequences for how beaver dams affect downstream invertebrate communities.

Impounded reaches of channels in low gradient streams are likely to have an enhanced wetted area of overhanging vegetation and snag habitats along flooded shoreline margins (Johnston and Naiman 1987). The importance of channel-margin overhanging vegetation as substrate for aquatic invertebrates is well described in other stream contexts (e.g., coastal plain rivers—Benke et al. 1985). The secondary production of aquatic invertebrates on these substrates can dwarf that on channel substrates and can be the most important source of production for fisheries in slow-moving, soft-sediment channels (as in Benke et al. 1984). Indeed, literature reviews and meta-analyses of the positive effects of beavers on stream fish cite the high invertebrate productivity in stream habitats associated with beaver activity (pools, wetted margins, dams, huts) as an important positive effect on stream fish abundance, growth, and productivity (reviews by Collen and Gibson 2001; Kemp et al. 2012). Other positive effects of beavers on fish are related to the effects of habitat heterogeneity in the streamscape on overwintering success, juvenile refugia, recruitment, and connectivity between juvenile and adult habitats; whereas barriers to fish movement and increased temperatures (and decreased oxygen) towards upper tolerance thresholds are cited as negative effects in those same reviews. The degree to which beavers have a positive or negative effect on native brook trout, the top predator in high gradient, headwater streams in northeastern North America, appears to vary across locations and geomorphological context (White and Rahel 2008; Niles et al. 2013). From the perspective of our focus here on invertebrates, we did not find any studies that consider how the changes in fish communities associated with beaver activity feed back on invertebrate communities.

There is also a large literature on the positive effects of beaver activity on the growth, survival, and diversity of waterbirds that is attributed to the creation of structurally favorable habitats for breeding and survival (e.g., Brown et al. 1996; McKinstry et al. 2001), and to the high primary and secondary productivity in beaver-created wetlands, including invertebrate production (e.g., Nummi and Hahtola 2008; Nummi and Holopainen 2014). However, it is not clear how increased density, diversity, and production of waterbirds in turn affect beaver pond invertebrate communities.

In relatively flat landscape settings, it appears that the invertebrate communities in the pools that develop upstream of dams are comparable to those typical in standing water habitats. For example, in low gradient streams in Hesse, Germany, the macroinvertebrate communities in beaver ponds are distinctly different from those in unimpounded reaches with high diversity of taxonomic groups (e.g., 11–18 odonates, 11–22 caddisflies including many limnephilids that are typically lentic; Harthun 1999). In contrast, in the Bigoray River in Alberta, Canada, Clifford et al. (1993) found that, although the percent composition varied between habitats, there were seven taxa that were common to both unimpounded and impounded reaches of this slow moving 3rd order stream. For example, Simuliidae represented more than 80 % of the most abundant taxa in the fast water associated with the dams, but less than 3 % in unimpounded sites. Chironomidae made up less than 12 % of the most abundant taxa in dams; however, it comprised more than 48 % in unimpounded reaches of the same streams. In addition, unimpounded sites contained taxa frequently associated with slower reaches (*Pisidium* spp., *Leptophlebia cupida, Caenis* spp.) and both cluster and principal component analysis separated dam sites and stream sites. Thus, it appears that in some hydrologic and geomorphic contexts, beaver dams can be important refuge for lotic taxa in slow moving streams, and in others, are more likely to reduce the available habitat for those taxa. It is possible that in relatively small streams with confined valleys, beaver activities may overwhelm the capacity and competence of low stream discharge to create truly lentic-like habitats, whereas in the context of higher flows and unconfined channels, the redistribution and artificially cascaded nature of channel flows across multiple distributaries may actually enhance the lateral presence of erosional (riffle) and depositional (pool) habitats.

In unconfined geomorphological settings (e.g., broad valleys), the in-channel invertebrate diversity at the stream-segment scale should be complemented by the creation of lateral habitats that support other types of invertebrate assemblages on adjacent shoreline margins (see Johnston and Naiman 1987), and out-of-channel riparian habitats including paleochannels with active and abandoned beaver dams. For example, in relatively flat stream segments in the U-shaped valleys created by mountain glaciers in western North America (Fig. 12.3a), single, meandering channels can be transformed by beavers into valley wide systems of distributaries, each with a complex longitudinal and lateral sequence of habitat types associated with beaver activities including open ponds, systems of channels connecting those ponds, and extensive willow (*Salix* sp.) and sedge (*Carex* sp.) meadow wetland habitats (Fig. 12.3b). The hydrology of floodplain beaver ponds outside of the main channel will vary depending on proximity to the main channel. The hydrology of ponds close to the main channel will be more affected by changes in stream flow conditions than those isolated laterally from the channel. The hydrology of the latter will be dominated by inputs and outputs dominated by the down valley movements of shallow groundwater and lateral hyporheic losses from the main channel (Figs. 12.1c and 12.3b). In this geomorphologic setting, there is likely to

Fig. 12.3 Beaver pond wetlands in the lower East River Valley in the Elk Mountains of central Colorado below the Rocky Mountain Biological Laboratory. (**a**) Overview of meandering river in a glacially widened montane valley in spring (photo by Scott Wissinger). (**b**) *Arrow* indicates location of zoom to floodplain complex of beaver ponds (note beaver hut in pond on *lower left*) and difference in water color between channel (spring runoff) and beaver-created riparian wetlands which include open ponds, channels connecting ponds, and extensive willow-thicket and sedge-meadow wetlands that cover most of the valley bottom (photo by Susan Washko)

be a continuum of invertebrate communities ranging from those dominated by taxa typical of pools in stream channels to those dominated by lentic taxa typical of non-riparian wetlands and ponds (see discussion below of Western Beaver Wetlands; Appendix).

Beaver Impacts Outside of Native Range

Finally, given the transformative effects that beavers can have on nearly every aspect of running water systems, it is not surprising that they are having profound impacts as invasive species on stream ecosystems outside of their native range. In streams of south-temperate South America, Anderson and colleagues studied the impacts of invasive beavers on stream ecosystem structure and function, including the effects on stream invertebrate diversity, community composition, and productivity (Anderson and Rosemond 2007, 2010; Anderson et al. 2009). In a comprehensive review of the impacts of beavers on the physical and biological environments of stream systems in south temperate South America, they concluded that the impacts of beavers as exotic invasive species was of similar magnitude and direction as that observed in studies in the native range of beavers (Anderson et al. 2009). In South America, they compared unimpacted reaches to reaches with beaver ponds to reaches below beaver ponds and found lower taxonomic and FFG diversity in the pools associated with dams than in either upstream or downstream reaches, which did not differ from unimpacted reaches (Anderson and Rosemond 2007). They attributed this difference to the relatively homogenous microhabitat in the soft sediments of the pools, although they did not appear to include other types of habitats (wetted margins, dams, hut). Examining other beaver-associated habitats could be important in obtaining a full picture of invertebrate diversity as these other connected habitats increase habitat heterogeneity, which has increased diversity in beaver-influenced habitats elsewhere. They also found invertebrate abundance, biomass, and secondary production were higher in the pools associated with dams as compared to above or below undammed reaches (Anderson and Rosemond 2007), which was consistent with the literature from North America and Eurasia (Anderson et al. 2009). They tested the hypothesis that this higher productivity was associated with increased production and input of allochthonous detritus using stable isotopes and found a slight increase in reaches with vs. without beaver dams (Anderson and Rosemond 2010). Anderson and colleagues argued that because the in-stream productivity and metabolism in these forested catchments is naturally driven primarily by allochthonous subsidies, beaver impacts are small. They predict that in streams where autochthonous production contributes a larger fraction of the overall energy budget, beavers will have a bigger impact on shifting the metabolism of a stream reach towards autochthonous production (as in Naiman et al. 1986, 1988b).

Lentic Invertebrate Communities in Beaver Wetlands

Beaver-Meadow Wetland Complexes in Northeastern North America

In relatively flat-lying landscapes, beaver activities beyond the main channels of streams can create extensive and persistent wetland complexes that are distinctly lentic in character (Fig. 12.4a). These habitats are variably described as "beaver-pond wetlands," "beaver meadow wetlands," or "valley beaver impoundments" (Burchsted et al. 2010; Polvi and Wohl 2012). As a result of the recolonization of beavers over the past 100 years, these wetland complexes have become a ubiquitous feature of the landscape in northeastern North America (from west to east—Minnesota, Wisconsin, Michigan, Ontario, Ohio, Pennsylvania, New York, New England, Quebec, New Brunswick, and non-urbanized areas of the coastal Atlantic states). Along the northern tier of this region (i.e., southern Canada and border states of the USA), beaver dams that occur beyond the margins of stream courses can transform vast tracts of saturated-soil peatlands into complexes of open ponds, marshes, and shrub swamps that are interconnected by beaver-constructed standing-water canals (Naiman et al. 1986, 1988b; Rebertus 1986; Johnston and Naiman 1990; Woo and Waddington 1990; McCall et al. 1996; Donkor and Fryxell 2000; Ray et al. 2004). Beaver wetland complexes that are not part of peatlands are also common further to the south in glaciated and unglaciated landscapes of Ohio, Pennsylvania, Ontario, and New York, often at the boundary between upland and lowland terrain (Johnston and Naiman 1987; Grover and Baldassarre 1995; Wissinger and Gallagher 1999). The hydrologic budget of beaver wetland complexes is often tied to upwelling areas of groundwater discharge with the outflows below the complexes forming perennial headwater streams (Fig. 12.1d). These wetland complexes are distinctly different from the pools created by beaver dams in large stream channels (Hodkinson 1975a, b; Ray et al. 2001; Burchsted et al. 2010).

Compared to the many detailed studies of how beaver dams change stream invertebrate communities at the streamscape level (see above), there are relatively few studies that describe the invertebrate communities in beaver-meadow wetland complexes. These complexes are ubiquitous in the hummocky glaciated terrain of northwestern Pennsylvania (Fig. 12.4), and the complexes are long-lived, especially where beaver colonies are protected (e.g., PA State Game Lands, Erie National Wildlife Refuge). Wissinger and Gallagher (1999) studied the invertebrate communities in two such complexes (Robinson-South Marsh Complex; and Church-Kiser Marsh Complex in Allegheny College's Environmental Research Reserve). The beaver dams at these sites are located on terraces along the edges of the valley and impound groundwater discharge as it resurfaces at the base of the slopes of uplands. The complex of habitats created in the relatively flat-lying terrain include:

Fig. 12.4 Habitats associated with beaver meadow wetland complex in northwestern Pennsylvania, USA. (**a**) Active beaver pond embedded in a series of ponds and other beaver-affected wetland habitats on Allegheny College's Environmental Research Reserve. Habitats include open water zone in the foreground, lily pad zone, and shrub swamp (plant species given in text) at the forest edge in the background (photo by Ben Plohr). (**b**) Mixed species shrub swamp on the upslope edge of pond, and (**c**) vernal pool in adjacent woodland enhanced by locally raised water table (photo by Scott Wissinger)

1. **Relatively deep (>1 m) ponds** immediately behind the dam structure that have complex plant zonation ranging from an open-water zone of floating pads of spatterdock (*Nuphar variegata*) and submergent vegetation (e.g., *Potamogeton* spp., *Ceratophyllum*) towards shoreline vegetation with deep (e.g., *Typha angustifolia* and *T. latifolia)* and shallow water emergents (e.g., *Scirpus cyperinus* and *Sparganium eurycarpum*) (Fig. 12.4a).
2. **Shallow marshes of emergent vegetation that grade into moist-soil herbaceous communities** that develop on flooded fields adjacent to active dams, and then invade pond basins when dams are abandoned. In addition to the emergent vegetation surrounding the ponds per se (see above), seasonally inundated wet meadow plant assemblages are dominated by rushes and sedges (e.g., *Juncus effusus*, *Carex hystericina*, and *C. lurida*) and herbaceous plants (e.g., *Eupatorium maculatum* and *Verbena hastata*).

3. **Shrub-swamps** along margins of active or abandoned ponds that are dominated by alder (*Alnus rugosa*), wetland dogwoods (*Cornus amomum* and *C. stolonifera*), willows (*Salix* spp.), and buttonbush (*Cephalanthus occidentalis*); and wet meadow shrubby species including meadowsweet (*Spirea alba*) and sweet gale (*Myrica gale*) (Fig. 12.4b).
4. **Back-flooded forests** with dead snags of trees that are intolerant of anaerobic soil conditions soils (e.g., *Prunus serotina*, *Fagus grandifolia*, and *Quercus* spp.).
5. **Living red-maple/hemlock moist-soil swamps with vernal woodland pools** that develop in the depressional micro-topography from the raised water table in back-flooded forests (Figure 15.1 in Wissinger and Gallagher 1999; Fig. 12.4c). Despite the apparent remoteness of these woodland pools to beaver activity, the hydrologic dependence becomes apparent when beaver dams are abandoned, and the adjacent water table falls.

Across all of these subhabitats in these beaver-pond wetland complexes, there is a remarkable diversity of wetland plants (Wissinger et al. 2001) and animals (amphibians, reptiles, fishes, invertebrates) (Wissinger and Gallagher 1999). Wissinger and Gallagher studied the resiliency of the invertebrate communities to short-term drought in the main pond communities by monitoring the multiple pathways by which species recolonized after drought, and experimentally by rehydrating soil cores that were extracted from dried basin sediments. They found that (1) the invertebrate assemblages in semi-permanent basins (abandoned beaver ponds and marginal wetland habitats) were more resilient after drought than those in the permanent basins, (2) the overall rapid recovery of the invertebrate diversity prior to drought (>90 % after 18 months in semi-permanent basins) was attributable to a variety of recolonization modes (e.g., use of micro-refuges in dried basins (see Strachan et al. 2014), desiccation tolerance of eggs, larvae, adults, flexible life history traits, and seasonally timed emergence), and (3) high dispersal rates among habitats with different hydroperiods lead to metapopulation and metacommunity dynamics that stabilize beta diversity across the complex of habitats (Wissinger and Gallagher 1999).

Combining the species identified in the original surveys (see taxonomic list in Wissinger and Gallagher 1999) with subsequent annual surveys (2000–2009; S. Wissinger unpublished data) reveals the presence of >250 invertebrate taxa including 40+ species of odonates, 16 species of caddisflies, 30+ species of beetles, 18 species of water bugs, 50+ dipteran taxa, 11+ molluscs, and 20+ crustaceans (see Appendix for list of families). The taxonomic resolution attainable for odonates and caddisflies (species-level identification of adults and larvae, respectively in the field) provides insight into two levels of habitat heterogeneity that underlie this diversity. First, there is considerable habitat heterogeneity *within* types of subhabitats in the complexes. For example, different species of anisopteran and zygopteran dragonfly larvae are encountered at different depths and in different vegetation zones within main beaver ponds (Table 12.1). This type of spatial niche segregation within ponds (also see Crowley and Johnson 1982; Wissinger 1988; Van de Meutter et al. 2008) is

Table 12.1 Distribution of odonates in different subhabitats within a beaver-meadow wetland complex in northwestern Pennsylvania

Active beaver ponds 1–3 m depth and permanent (pumpkinseed, bluegill, grass pickerel, bass)			Abandoned pond, marsh, shrub swamp <1 m depth, semi-permanent (mudminnow, stickleback, salamander larvae)		Temporary <0.5 m depth (salamanders)
Benthic open basin	Submergent vegetation	Shoreline with emergent vegetation	Marshes—emergent and submergent vegetation	Shrub swamps	Woodland pools/ margins
Libellula lydia	*Libellula luctuosa*	*Libellula luctuosa*	*Libellula pulchella*	*Libellula pulchella*	*Aeshna umbrosa*
Libellula julia	*Libellula incesta*	*Libellula incesta*	*Libellula luctuosa*	*Libellula luctuosa*	
Gomphus exilis	*Erythemis simplicicollis*	*Erythemis simplicicollis*	*Libellula quadrimaculata*	*Aeshna constricta*	
Gomphus spicatus	*Sympetrum vicinum*	*Sympetrum vicinum*	*Anax junius*	*Aeshna canadensis*	
Arigomphus furcifer	*Leucorrhinia intacta*	*Leucorrhinia intacta*	*Sympetrum vicinum*	*Sympetrum ambiguum*	
Perithemis tenera	*Pachydiplax longipennis*	*Pachydiplax longipennis*	*Sympetrum semicinctum*		
Epitheca cynosura	*Celithemis elisa*	*Celithemis elisa*	*Sympetrum obtrusum*		
Epitheca canis	*Celithemis eponina*	*Celithemis eponina*	*Aeshna tuberculifera*		
	Argia fumipennis	*Ischnura verticalis*	*Tramea lacerata*		
	Ischnura verticalis	*Ischnura posita*	*Tramea carolina*		
	Ischnura posita	*Enallagma civile*	*Pantala flavescens*		
	Enallagma signatum	*Enallagma signatum*	*Ischnura verticalis*		
		Enallagma vesperum	*Ischnura posita*		
		Nehalennia irene	*Enallagma civile*		
		Lestes vigilax	*Enallagma boreale*		
			Enallagma aspersum		
			Nehalennia irene		
			Lestes disjunctus		
			Lestes eurinus		
			Lestes congener		

Top vertebrate predators in parentheses (salamanders=*Ambystoma maculatum* and *Notophthalmus viridescens*). Larval occurrences based on Wissinger and Gallagher (1999) and subsequent D-net and aerial surveys taken during May–October between 2000 and 2009. Data are cumulative; i.e., not all species were collected in a given year. Larvae of all species are encountered along shorelines during their emergence periods

not limited to dragonflies—in general, plant zonation is a well-described axis of niche segregation for invertebrates in many types of wetlands (Batzer and Wissinger 1996; Wissinger 1999; De Szalay and Resh 2000; Batzer 2013). Thus, the high diversity of habitat types and distinct plant communities associated with beaver-meadow wetland complexes (Grover and Baldassarre 1995; Wright et al. 2002, 2003) translates into a diverse invertebrate fauna (Hood and Larson 2014). Other microhabitats that create hotspots of diversity within beaver ponds include those associated with the structural complexity of beaver huts and dams (France 1997).

A second scale of heterogeneity in beaver-meadow wetland complexes is related to variation *between* different types of wetland habitats. Although specific patterns are difficult to predict across wetland types (Batzer 2013), for wetlands and ponds of similar size, species richness decreases along a gradient from permanent to temporary habitats, with species in temporary habitats often being a nested subset of those in the permanent habitats. In addition to nestedness patterns in ponds (see review by Batzer and Ruhí 2013), invertebrate community composition can also shift from permanent, relatively deep-water (1–2 m depth) ponds that typically have large-gaped predatory fish that prey on invertebrates (bass, sunfish, pickerel), to semi-permanent marshes and shrub-swamps habitats with small-gaped fish (stickleback, mudminnows) and/or salamander predators, to temporary habitats in which salamander larvae and invertebrates are the top predators (Batzer and Wissinger 1996; Wellborn et al. 1996). In the beaver-wetland complex studied by Wissinger and Gallagher (1999), all of these types of habitats are present and odonates and caddisflies provide evidence for shifts in species composition along predator-permanence gradients. Although there are generalists that occur across habitat types, some species tend to occur mainly at one end (permanent) or the other (temporary) of this gradient (Tables 12.1 and 12.2). Similar differences are observed for beetle assemblages in temporary habitats vs. permanent ponds including beaver ponds beyond stream channels (Fairchild et al. 2000, 2003). Such shifts in species composition are expected for nearly every invertebrate taxon (beetles, bugs, odonates, caddisflies, true flies, crustaceans, molluscs, etc.) associated with wetland habitats; i.e., different combinations of species within genera and different genera within families will be present in different types of basins as a result of differential dispersal and colonization rates combined with the different biological and physico-chemical filters that affect establishment and survival (Batzer and Wissinger 1996; Wellborn et al. 1996; McCauley 2008). The mechanisms that underlie species replacements across permanence gradients (as in Tables 12.1 and 12.2) are well described for odonates and caddisflies, and typically involve tradeoffs between physiological, behavioral, and morphological traits that facilitate coexistence with different types of predators, or tradeoffs between traits that facilitate coexistence with predators and those that expedite the completion of life cycles in temporary habitats (e.g., Stoks and McPeek 2003, 2006; Wissinger et al. 2006; McCauley 2008; McCauley et al. 2010). Patterns of species replacements across habitat types in beaver wetland complexes may be confounded by cycles of dam building and abandonment that lead to legacy effects associated with shifts in permanence and in the presence/absence of large-gaped predatory fish. This temporal variability may be even more likely for beaver ponds and wetlands in the floodplains of major

Table 12.2 Distribution of cased caddisflies (Limnephilidae and Phryganeidae) across subhabitats in two beaver-meadow wetland complexes in northwestern Pennsylvania

Permanent ponds	Semi-perm marshes and shrub swamps	Temporary woodland pools	Seeps and rivulets
Banksiola crotchii	*Nemotaulius hostilis*	*Ptilostomus ocellifera*	*Ironoquia punctatissima*
Platycentropus radiatus	*Limnephilus indivisus*	*Ptilostomus postica*	
Pycnopsyche subfasciata	*Limnephilus submonilifer*	*Ironoquia parvula*	
Agrypnia vestita	*Limnephilus moestus*		
Anabolia consocia	*Banksiola doussaria*		
Fabria inornata	*Bansiola crotchii*		
	Anabolia bimaculata		
	Ptilostomus occellifera		
	Phryganea sp. *(sayi?)*		

Larval occurrences based on Wissinger and Gallagher (1999) and subsequent D-net samples taken during October–November, and April–May from 2000 to 2009. Top predators in (a) permanent ponds (active beaver ponds) are large-gaped fishes (sunfish, bass, grass pickerel); (b) in semi-permanent marshes (herbaceous emergent) and shrubswamps are mudminnows, brook stickleback, and newts; and (c) backflooded temporary habitats and woodland pools are *Ambystoma* salamander predators. Seeps and rivulets are small flowing water habitats between various standing water habitats in the beaver meadow complexes (see Fig. 15.1 Wissinger and Gallagher 1999)

streams because of the stochastic arrival of fishes and unpredictable filling and drying events associated with floods (e.g., Kohler et al. 1999).

In summary, the habitat heterogeneity observed within and between the different types of basins in beaver-meadow wetland complexes combined with the potential for diversity-enhancing metacommunity dynamics associated with dispersal and high connectivity among basins should lead to an overall higher diversity of plants and animals as compared to in structurally simple and isolated wetland basins (Wissinger and Gallagher 1999; Wright et al. 2002, 2003; Caudill 2005; McCauley et al. 2010). Moreover, because the combinations of habitats in beaver-meadow complexes that are part of the cyclic and multi-successional pathways associated with beaver activity (see Naiman et al. 1988b; McMaster and McMaster 2001) are constantly changing, understanding the degree to which assemblage structure and composition at a given point in time at a given location is a result of extant vs. legacy conditions will require long-term and wetlandscape-level study.

Beaver Dam Wetlands in Western North America

Wetland habitats associated with beaver activity in western North America occur in a variety of geomorphological contexts including (1) northern peatlands (e.g., Hood and Bayley 2008a, b, 2009); (2) on rivers of the "High Plains" to the east of the

Rocky Mountains and in arid intermountain basins among the major North American Cordilleran ranges where created ponds and wetlands expand the riparian ecotone of stream systems and create perennial wetland habitats in arid landscapes (e.g., Andersen and Shafroth 2010; Gibson and Olden 2014); (3) on relatively small tributaries in the foothills of mountain ranges (e.g., Hodkinson 1975a, b; Clifford et al. 1993; Morrison et al. 2015); (4) in the riparian zone of montane and subalpine rivers that flow through the U-shaped valleys carved by mountain glaciers in the Rocky and Sierra mountain ranges (Malanson and Butler 1990; Butler and Malanson 1995; Fuller and Peckarsky 2011a, b; Polvi and Wohl 2012; Levine and Meyer 2014) (Figs. 12.3 and 12.5); and (5) on streams flowing on terraces and other valley-side and headwater habitats in montane and alpine valleys (Caudill 2002; Fig. 12.5b).

Beaver Activity Enhances Habitat Heterogeneity in Northern Peatlands

As described for northeastern North America above, beaver activity in the peatlands of western Canada and Alaska enhances existing habitat heterogeneity in these wetland landscapes. In Miquelon Lake Provincial Park, Alberta, Canada, beavers alter existing shallow isolated wetlands via channel digging (Hood and Larson 2014,

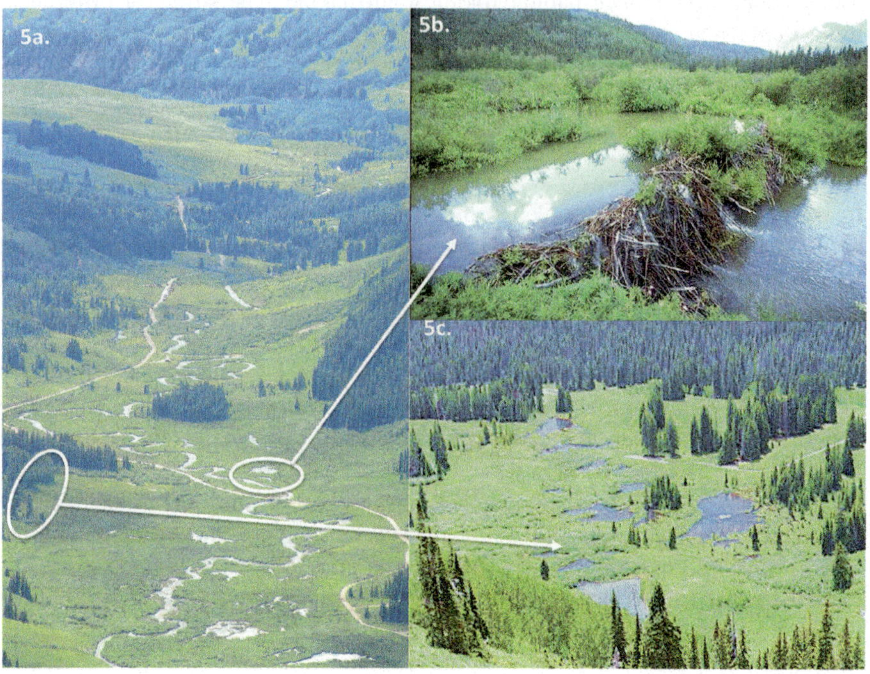

Fig. 12.5 (**a**) Overview of geomorphological settings of beaver dam wetlands in the upper East River Valley in the Elk Mountains of central Colorado (photo by Scott Wissinger), (**b**) valley bottom riparian beaver pond (photo by Chris Caudill), and (**c**) upland beaver wetland complex on valley margin terrace (photo by Susan Washko)

2015). Beavers dig long, deep channels perpendicularly from the wetland edge outward that connect to other wetlands or upland areas, thereby increasing habitat heterogeneity (Hood and Larson 2014) and wetland connectivity (Hood and Larson 2015). Hood and Larson (2014) found that beaver activity increased the amount of vegetated-edge habitat, which had higher species richness, diversity, and evenness than open water and beaver channels. Invertebrate richness, diversity, abundance, and density varied by year and yearly differences were driven by precipitation. Drought resulted in higher densities while higher water levels resulted in more diversity. *Daphnia* spp. were the most abundant taxon regardless of hydrologic conditions. Invertebrates were compared between active and inactive beaver wetlands and between different types of habitats (open water, beaver channels, and vegetated edges) within each category (active, inactive). Predators were the most species-rich group in both active and inactive wetlands, and Chaoboridae larvae were numerically dominant, especially in active beaver channels. Gerridae and Gyrinidae were unique to active channels despite low numbers of individuals. Tabanidae were unique to inactive channels and Culicidae were associated with all three subhabitats in inactive wetlands. Amphipoda were associated with active vegetative edges and Hood and Lawson posit that amphipods are influenced by an increase of organic material brought in by beaver and that beaver maintain deeper water, which may reduce habitat for mosquito larvae. The strongest differences were seen at the within-wetland level with beaver channels and vegetated edges having more functional feeding groups than open water.

Abandoned Beaver Pond Invertebrate Communities

Beaver ponds are notorious sinks for mineral and organic sediments and patterns of accumulation of these various types of sediments should have multiple consequences for the development of benthic invertebrate communities. In general, the sediments at old dam sites contain higher amounts of organic material than those at relatively young sites (Butler and Malanson 1995). Hodkinson (1975a, b) studied the aquatic invertebrates in abandoned beaver ponds in forested landscapes in the foothills of the Rocky Mountains in Alberta, Canada, with a particular focus on understanding patterns of distribution and abundance of dipteran larvae, which were the dominant taxa in the organic-rich sediments in these habitats. Although the invertebrate communities in these abandoned beaver ponds included surface-dwelling Ephemeroptera, Plecoptera, Megaloptera, and Trichoptera, the great majority of taxa listed (67/83 species) were dipterans living in the soft organic-laden sediments in these basins. Tipulid larvae were particularly diverse (26 species), and Hodkinson determined that their distribution and abundance varied among substrate types. Coarse-grained, lotic-like gravel substrates in stream courses were dominated by non-tipulid lotic taxa. The abundance and species composition of the dipteran assemblages in the organic-laden, soft-sediments of the abandoned ponds varied along a gradient that varied in (1) particle size; (2) degree of compaction (flocculent to firm); and (3) amount (mostly organic to mostly mineral) and type (herbaceous, woody deciduous leaves, conifer needles) of detrital plant material. Invertebrate biomass was higher in loose, flocculent, detrital substrates than in relatively compacted, mainly mineral substrates. Dietary analyses by Hodkinson combined with those in previous studies by Pritchard and Hall (1971) and Pritchard and Leischner (1973) revealed that (1) allochthonous vascular plant detritus dominated the diets of most species in these wetlands and (2) habitat partitioning led to dietary partitioning in terms of the type of vascular plant detritus ingested.

Succession in Riverine Floodplains

Beaver dams are frequently breached, rebuilt, relocated, or abandoned in relatively large streamscapes. Malison et al. (2014) studied invertebrates in different successional stages on a large river floodplain in Alaska, USA. Invertebrate communities in beaver ponds differed from flood-channel spring brooks but were similar among early-, mid-, and late-successional ponds despite the fact that early-successional ponds had a greater degree of connectivity to the main channel.

The return of beavers to large river systems in arid landscapes is an area of intensive study because of the potential effects on conservation efforts—both on the positive side of the reestablishment of natural flora and fauna, and on the negative side as an interactor with invasive species (Gibson and Olden 2014). While there is evidence for how beaver reestablishment and subsequent cyclical changes associated with damming and abandonment can influence successional changes in vegetation, there are few data on changes to invertebrate communities (Gibson and Olden 2014).

Metapopulation Dynamics in Montane Beaver Wetland Complexes

Beaver activity on small streams in montane settings often creates step-like complexes of multiple ponds and wetlands that cascade along the relatively flat terrain of mountainside terraces and along the sides of montane valleys (Fig. 12.5b). The lateral development of these complexes creates multi-basin clusters of active and abandoned ponds with hydrologic budgets akin to those described above for the wetland complexes in northeastern North America; i.e., inputs are often dominated by first order streams, springs, and/or groundwater upwellings at valley margins (Fig. 12.1d). The proximity of multiple habitats makes it likely that aquatic insects with even moderate dispersal abilities can move between ponds. In a series of related studies Caudill (2003a, b, 2005) evaluated the dynamics of a mayfly (*Callibaetis ferrugineus hageni*) metapopulation in beaver ponds of the upper East River Valley of Colorado, USA, with and without trout (Fig. 12.5b, c). Caudill found that late instar larval mayflies densities are significantly higher and adult emergence nearly an order of magnitude greater in troutless ponds than in those with trout. Surprisingly, trout ponds with few or no emerging adults subsequently have similar larval recruitment to ponds with high emergence rates (Caudill 2003a). Isotope labeled adult females move between trout and troutless ponds and there is no relationship between oviposition and trout (Caudill 2003b). A comparison of adult emergence compared to larval recruitment rates point to a source–sink population dynamic among beaver ponds, and models based on these empirical data predict that this mayfly cannot persist in ponds with trout in the absence of adult dispersal from neighboring troutless habitats (Caudill 2005).

Beaver-Pond vs. Non-beaver Pond Montane Invertebrate Assemblages

Caudill's beaver-pond study sites are located in the Elk Mountains of Colorado, USA, where Wissinger and colleagues have surveyed the invertebrate community composition of both beaver and non-beaver ponds for the past 25 years (Wissinger et al. 2003; Wissinger, unpublished data). The spatial configurations of the different types of montane wetland and pond habitats in the Elk Mountains are characteristic of many glaciated mountain valleys throughout the central Rocky Mountains—i.e., beavers dam the main stem of the East River as it meanders through a U-shape glacial valley creating wetland complexes with some ponds highly connected to main channel flow (as in Fuller and Peckarsky 2011a, b; Malison et al. 2014), and others that are less directly connected (see Figs. 12.3 and 12.5). Beavers also dam headwater side tributaries where they traverse glacier-formed terraces on valley walls (as in Caudill; Fig. 12.5c). Kettle ponds and other non-beaver dam wetland habitats associated with the glacial landscape also occur in these valleys. A comparison of the invertebrate communities in these various wetland habitats within the same valley reveals several patterns (Table 12.3). First, assemblages in main-stem and valley-floor complexes have a higher number of running-water invertebrates including stream-dwelling mayflies, stoneflies, and caddisflies than those on valley terraces. Inlet and outlet areas of ponds at the upper and lower extent of beaver-pond complexes include some of these stream-dwelling EPTs that, in addition to a distinctly lentic group of organisms (see taxa lists in Appendix; Table 12.3), create a much higher total diversity than in communities in valley-floor complexes. There is considerable overlap in the dominant taxa in upland beaver pond complexes and those in non-beaver kettle ponds, with the former often as a nested subset of the species of the caddisflies, odonates, water bugs, and beetles that dominate (in terms of biomass) the large-bodied invertebrate fauna in non-beaver wetlands (Table 12.3; Fig. 12.6). There are several large-bodied taxa that characteristically dominate the biomass in upland beaver ponds but are rare or absent in non-beaver ponds including (1) *Callibaetis* mayflies (Fig. 12.6b); (2) tipulid flies (6–8 species (Fig. 12.6e)); (3) dixid flies; and (4) amphipod crustaceans. There are also subtle, species-level differences that are consistently observed between beaver- and non-beaver upland wetlands. For example, the water boatman *Callicorixa audeni* and *Cenocorixa bifida* are common in montane kettle ponds, whereas several species of *Hesperocorixa* (a relatively lotic genus) dominate in nearby beaver wetlands (Caudill 2002). Many

Table 12.3 Number of species in relatively lentic vs. lotic aquatic insect groups in ponds and wetlands in the upper East River Valley in the Elk Mountains of Colorado near the Rocky Mountain Biological Laboratory

Order	Family	Valley beaver	Upland beaver	Non-beaver
Ephemeroptera	Total taxa	7	3	2
	Baetidae	1	2	1
	Caenidae		1	1
	Ephemerellidae	1		
	Heptageniidae	3		
	Leptophlebiidae	1		
	Siphlonuridae	1		
Plecoptera	Total taxa	7	2	
	Chloroperlidae	2	1	
	Nemouridae	2	1	
	Perlidae	2		
Odonata	Total taxa	1	11	18
	Aeshnidae	1	3	3
	Coenagrionidae		3	4
	Corduliidae		1	2
	Lestidae		2	3
	Libellulidae		2	6
Coleoptera	Total taxa	7	16	22
	Chrysomelidae		1	1
	Dytiscidae	4	10	15
	Gyrinidae	1	1	1
	Haliplidae	1	1	2
	Helophoridae	1	1	1
	Hydrophilidae		2	2
Hemiptera	Total taxa	3	8	11
	Corixidae	1	2	3
	Gerridae	1	2	3
	Mesoveliidae		1	1
	Notonectidae		1	1
	Saldidae	1	1	2
	Veliidae		1	1
Trichoptera	Total taxa	8	8	10
	Hydropsychidae	1		
	Leptoceridae		1	1
	Limnephilidae	5	6	8
	Phryganeidae		1	1
	Polycentropodidae	2		
	Rhyacophilidae	1		

(1) Valley beaver = ponds with hydrologic connections to the East River (Fig. 12.5b); (2) upland beaver = beaver ponds on small tributaries along the valley sides (Fig. 12.5c); and (3) non-beaver = kettle ponds in mid-valley moraines. Data combined from Caudill (2002), B. Peckarsky (unpub. data), and S. Wissinger (unpub. data)

Fig. 12.6 Frequently encountered abundant invertebrates in upland beaver wetlands in the East River Valley in the Elk Mountains of central Colorado studied by Caudill (2002) and Wissinger (unpublished data): (**a**) larva of the caddisfly, *Limnephilus externus* (photo by Nixie Boddy); (**b**) larva of the mayfly, *Callibaetis ferrugineus hageni* (photo by Chris Caudill); (**c**) adult water boatman, *Hesperocorixa* (photo by Timothy Loh); (**d**) larva of the dragonfly, *Aeshna palmata* (photo by Jim Johnson); (**e**) tipulid fly larva (photo by John Meyer); (**f**) larva of the damselfly, *Coenagrion resolutum* (photo by Susan Washko); (**g**) larva of the meniscus midge *Dixella* (photo by Stephen Luk); and (**h**) adult dytiscid diving beetle, *Agabus tristus* (photo by Susan Washko)

of the dominant taxa in temporary non-beaver ponds are rare or absent in beaver ponds (Wissinger unpublished data). Whether the presence of trout in valley bottom beaver pond complexes explains the rarity of many of the larger bodied lentic taxa (odonates, beetles, water bugs, cased caddisflies; Table 12.3) found in upland habitats (beaver and non-beaver) bears further study.

Finally, beaver dam wetlands have the potential to play a stabilizing role in maintaining beta and regional diversity in wetland habitats in the face of climate change. Wetlands, ponds, and other shallow, temporary basins are considered to be the most vulnerable aquatic habitats to changes in temperature and precipitation regimes, especially at relatively high latitudes and elevations (Barnett et al. 2005; Corcoran et al. 2009; Tuytens et al. 2014). Ponds and other wetland habitats associated with beaver activity are typically permanent because of their hydrological connection to stream courses or their proximity to points of groundwater discharge (Fig. 12.1d). This permanence has the potential for creating refuges for species in habitats that are becoming increasingly temporary (see Smol and Douglass 2007). For example, in the East River Valley, the cased caddisfly, *Limnephilus externus* (Fig. 12.6a), is ubiquitous in beaver and non-beaver ponds and wetlands (also see Hodkinson 1975a, b). The local habitat range of this species is limited by pond drying because of the extended time spent in the final instar during late summer (Wissinger et al. 2003). Censuses of the presence and abundance of this species throughout the valley for 25 years reveal that during the past decade (2005–2015), early pond drying has resulted in complete cohort failures not observed in the previous 15 years (Wissinger, unpublished data). For example, in 2009 and 2012, a combination of a light snow pack, early snow melt, and a dry early summer, led to the disappearance of over 30 populations of this species in temporary wetland habitats in the East River Valley. The only populations of this species that survived to pupate and emerge in the valley in both years were associated with beaver dam wetlands (both main-valley and terrace complexes; see Figs. 12.3 and 12.5), which remained permanent as a result of their landscape position in stream courses. Spatial patterns of recolonization in non-beaver pond basins after these drought events suggest that beaver-pond populations provide a regional haven for this species in drought years and source of colonists for the reestablishment of populations in non-beaver habitats (Wissinger unpublished data).

Beaver Wetlands of the Southeastern USA

Southeastern USA beaver wetlands are typically unstable transitory systems due to regional weather and a history of extreme sedimentation. The Southeastern USA receives more annual precipitation (1300+ mm per year) than most other areas with beaver-created wetlands and is subject to intense tropical and winter storms that create large stream pulses. These pulses can breach many beaver dams, and at least temporarily drain beaver wetlands. In addition, river and stream beds in the

Southeast, particularly the Piedmont region, are unstable (Mukundan et al. 2011) due to poor cotton-era farming practices in the late 1800s and early 1900s that eroded 10–30 cm of topsoil into streams and floodplains (Trimble 1974). At current export rates, Jackson et al. (2005) estimate that 6–10 millennia will be required to export sediment mobilized during the cotton-era from a Georgia Piedmont watershed. The precipitation patterns in the Southeast combined with unstable, sand and silt substrates limit vegetation growth in and around beaver wetlands to relatively simple communities (e.g., *Panicum* grasses) adapted to shifting hydrologic conditions. Exceptions include ponds built to incorporate old roadbeds or those isolated from main channel flows. Overall, however, most beaver wetlands in the Southeast are small, unstable habitats.

Succession in Invertebrate Communities in Southeastern Beaver Wetlands

Beaver wetlands in the Southeastern USA are typically formed from damming small streams that then flood adjacent riparian forest (Fig. 12.1a). In early-stage beaver wetlands, many upland trees persist. The wetlands overall are rather shallow other than the area immediately adjacent to the dam, and in the original stream channel. As beaver wetlands persist, terrestrial vegetation dies under stress from flooding, creating open pond-like wetlands with emergent and submergent vegetation. However, because dams often breach due to frequent and intense storms, these wetlands are frequently abandoned by beaver, and subsequently drain. Abandoned ponds can develop complex braided drainage networks, as the original channel becomes sediment filled and numerous secondary channels develop. Abandoned beaver wetlands usually fill with some seasonal standing water, and upland tree species are slow to reinvade. Typically, abandoned ponds are large open meadow-like wetlands with aquatic, semiaquatic, and terrestrial subhabitats. A few studies have examined invertebrate communities in abandoned wetlands in other regions (Hodkinson 1975a, b; Wissinger and Gallagher 1999; Hood and Larson 2014), but not in southeastern North America.

To compare invertebrate communities of beaver wetlands among three basic stages of habitat succession, invertebrates were sampled in newly formed (created within 2 years; $n=4$), mature (established for >15 years; $n=4$), and abandoned wetlands (breached dams; $n=3$) in October 2013 and May 2014 in Oconee National Forest in Georgia, USA (Bush and Batzer, unpublished data). There were a relatively high number of taxa (>60 families; Appendix) in each wetland type, with strong seasonal variation in invertebrate communities (Fig. 12.7). In October, invertebrate communities differed among all successional stages, while in May only the mature beaver wetland communities differed from newly formed or abandoned ponds (Fig. 12.7). Ostracoda, Copepoda, Branchiopoda (mainly daphniids), Chironomidae, and Ceratopogonidae collectively accounted for 89–95 % of total invertebrates, regardless of condition, with ostracods alone accounting for 49–76 % of all individuals (Fig. 12.8). Ostracods can be particularly abundant in systems with

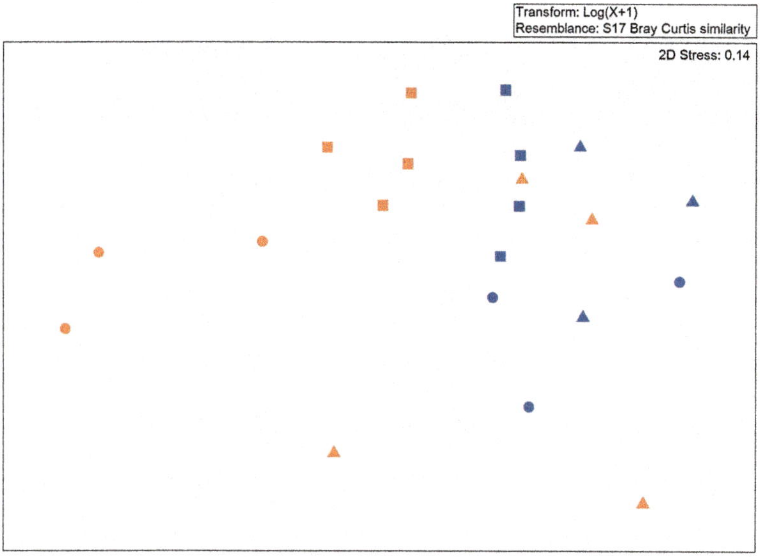

Fig. 12.7 Nonmetric multidimensional scaling plot showing patterns among newly created (*filled triangle*), mature (*filled square*), and abandoned (*filled circle*) beaver wetland invertebrate communities (Bray-Curtis similarity, Kruskal fit scheme 1,25 restarts) in October 2013 (*orange fill*; new vs. mature $R=0.344$, $P=0.029$; new vs. abandoned $R=0.704$, $P=0.029$; mature vs. abandoned $R=0.741$, $P=0.029$) and May 2014 (*blue fill*; new vs. mature $R=0.685$, $P=0.029$; new vs. abandoned $R=0.630$, $P=0.1$; mature vs. abandoned $R=0.556$, $P=0.029$)

copious benthic organic detritus combined with relatively shallow, warm water, which is typical in Southeastern beaver wetlands (Smith and Delorme 2010). This preponderance of small, benthic taxa suggests that fish predation is important in these wetlands (Wellborn et al. 1996), and most beaver wetlands of the Southeastern USA support large populations of *Gambusia* mosquitofish (Poecilidae). Given that all three successional types were dominated by the same five small-bodied taxa, the differences among successional states (Fig. 12.7) are likely driven by rarer, larger-bodied invertebrate taxa.

The greatest differences among successional states were observed in October when terrestrial and semi-aquatic taxa such as springtails (Entomobryidae), scale bugs (Coccoidea), and spiders (Araneae) were among the most abundant macroinvertebrates in abandoned wetlands (Table 12.4). Macroinvertebrates in newly formed and mature beaver wetlands were dominated by common lentic taxa (e.g., Coenagrionidae, Libellulidae, Baetidae, Caenidae), and differences between communities were more likely driven by variation in relative abundance than community composition, per se.

In May, invertebrate communities were similar among successional states, with only the stable mature wetlands exhibiting a unique community structure (Fig. 12.7). This seasonal difference suggests that both seasonal change and

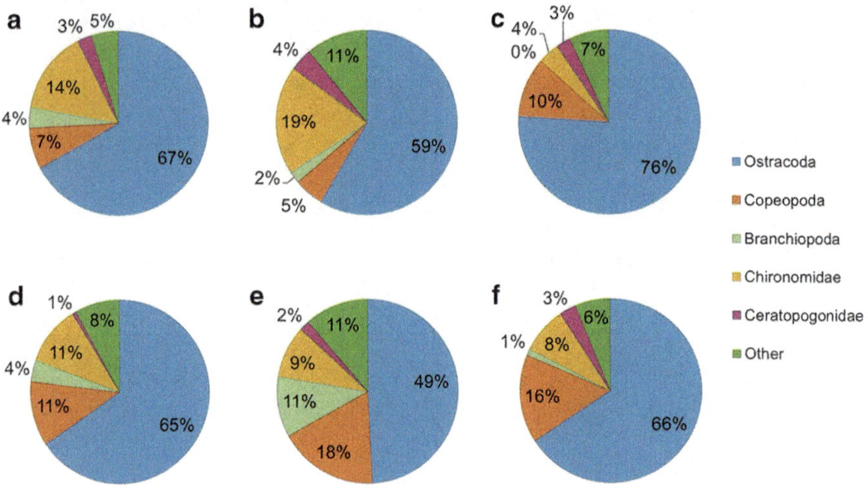

Fig. 12.8 Relative abundance of most dominant taxonomic groups in (**a**) October newly created, (**b**) October mature, (**c**) October abandoned, (**d**) May newly created, (**e**) May mature, and (**f**) May abandoned beaver wetlands of Georgia

Table 12.4 Ten most abundant large-bodied taxa (excluding Ostracoda, Branchiopoda, Copepoda, Chironomidae, and Ceratopogonidae) in new, mature, and abandoned beaver wetlands of Georgia, in October 2013

Newly created	Mature	Abandoned
Oligochaeta	Caenidae	*Entomobryidae*
Dogielinotidae	Dogielinotidae	*Coccoidea*
Coenagrionidae	Coenagrionidae	Oligochaeta
Non-oribatid Acarina	Non-oribatid Acarina	Non-oribatid Acarina
Libellulidae	Libellulidae	*Araneae*
Sphaeriidae	Sphaeriidae	Coenagrionidae
Caenidae	Baetidae	Sphaeriidae
Baetidae	Veliidae	Oribatidae
Dytiscidae	Oribatidae	Corethrellidae
Araneae	Scirtidae	*Delphacidae*

Italicized taxa represent exclusively terrestrial taxa

longer-term succession strongly control invertebrate community structures in these beaver wetlands.

While one might expect a linear successional pattern as a stream changes into a pond and then into a wet meadow (see Naiman et al. 1988b), the succession we observed appears more stochastic. In the Southeastern USA, beaver wetlands are frequently changing from one stage to another, and back again. Dams in new beaver wetlands are frequently breached and abandoned before the wetland ever becomes mature. In our study ponds, two of the newly formed wetlands were abandoned due to dam breaches soon after we sampled, and two newly formed wetlands had been recreated from formerly abandoned sites. Thus, invertebrates in these wetlands have to be able to adapt to constantly changing conditions or be highly mobile colonizers. Where mature beaver wetlands persist, pond-like communities of lentic invertebrates develop that are able to take advantage of both permanent water and high habitat heterogeneity, and can tolerate high fish predation rates (e.g., Benke et al. 1999). Invertebrate communities in abandoned ponds may be (1) former residents of mature wetlands that are able to take advantage of residual channels and seasonal filling, (2) migrants from nearby newly created or mature wetlands in the complex (Hodkinson 1975a; Wissinger and Gallagher 1999), or (3) semi-aquatic or terrestrial residents taking advantage of damp soil conditions or lush vegetation (as in abandoned pond in Hodkinson 1975a, tussock zone of Benke et al. 1999, marsh habitat in Wissinger and Gallagher 1999, and vegetative edges in Hood and Larson 2014).

Zonation and Habitat Heterogeneity Within Beaver Wetlands

As in Northeastern North America (see above), mature beaver wetlands in southeastern North America can be complex heterogeneous habitats with a variety of semi-aquatic, emergent, and submergent vegetation, as well an abundance of woody debris (Benke et al. 1999). Benke and colleagues examined the distribution of invertebrate communities in different habitat zones in a mature beaver wetland in Talladega National Forest, Alabama, USA. This wetland is the largest in a series of beaver-created wetlands on a low gradient small stream in the coastal plain of Alabama, and is sub-divided into three distinct vegetative zones each containing several subhabitats. These zones consist of (1) a small, deep unvegetated area of open water adjacent to the beaver dam, with a thin benthic layer; (2) a moderately shallow area in the middle of the pond dominated by floating white water lily (*Nymphaea odorata*), with a thicker benthic layer and extensive woody debris; and (3) a shallow semi-aquatic region at the edge of the pond dominated by emergent rush (*Juncus effuses*), which had two distinct subhabitats: rivulets and *Juncus* tussocks.

The taxon richness of the invertebrate community in the Talladega beaver pond is tightly coupled with increasing habitat heterogeneity from the open water zone to the *Nymphaea* zone to the *Juncus* zone. The open water has the simplest invertebrate community (Hood and Larson 2014). Copepods are common to both open water and the benthic substrate, while cladocerans dominate the open water. The benthic layer here is dominated by Chironomidae larvae (as was the case for the *Nymphaea* and *Juncus* zones; and like ponds in other regions - e.g. McDowell and Naiman 1986; Clifford et al. 1993; Margolis et al. 2001; Hood and Larson 2014), oligochaetes, and microcrustaceans. Ceratopogonidae larvae are also common (similar to Georgia wetlands above) in all three zones, and are the most important predator by relative abundance in the benthos of the open water zone.

Taxon richness is highest in the structurally complex *Nymphaea* habitat. There is a higher species richness of microcrustaceans in the *Nymphaea* zone compared to open water. Chironomids are the dominant insect in all three subhabitats (vegetation, woody debris, and benthos) of this zone. *Hyalella azteca* is the most common non-insect invertebrate. Overall community structure is similar between the benthos and woody debris, and consisted of many typical lentic taxa. Caenids and baetids were the most abundant mayfly taxa, and dytiscid beetles were the dominant coleopterans. Hydroptilidae and Phryganeidae caddisfly larvae are observed, but in low numbers. The most important predators are Odonata larvae (Coenagrionidae and Libellulidae). The invertebrates on vegetation of the *Nymphaea* zone are similar to woody debris and benthos (although less abundant) with two exceptions: Chrysomelidae beetle larvae (*Donacia* spp.) and Pyralidae moth larvae which are both specifically associated with *Nymphaea* leaves.

The *Juncus* zone has the highest overall taxon richness of all three zones (>100 taxa), which reflects the presence of both aquatic and semi-aquatic subhabitats. The second most abundant taxa (after Chironomidae) are semiaquatic/terrestrial collembolans in both tussocks and rivulets (similar to the abandoned Georgia wetlands). Semiaquatic/terrestrial Carabidae and Staphylinidae beetles, and Lycosidae spiders, along with aquatic Dytiscidae beetles, are the most common predators. Mites are another common predator in the *Juncus* zone, especially in the tussocks. Sciaridae larvae are only found in the *Juncus* zone and are only abundant in the tussocks. While insects are more taxonomically diverse in the *Juncus* zone than in the other zones, there are fewer microcrustacea taxa than in the *Nymphaea* zone, and those present are dominated by copepods and ostracods rather than copepods and cladocerans (as in the *Nymphaea* and open water zones). Curiously, the preponderance of Ostracoda seen in the Georgia beaver wetlands described above (Bush and Batzer, unpublished data) does not develop in the Talladega beaver pond studied by Benke et al. (1999).

Benke et al. (1999) and Stagliano et al. (1998) also studied insect emergence in all three zones of the Talladega beaver pond. Insects emerge in every month of the year, and chironomids are the most frequently collected insects in emergence traps, and the only group collected in the open water zone. Chironomid emergence is highest in the *Nymphaea* zone, which was several times higher than the open water zone even at its lowest point and is annually twice that of the *Juncus* zone. Insect emergence year-round coupled with continuously high chironomid larval abundance likely means that growth continues year round in this warm water wetland. While chironomid emergence is lower in the *Juncus* zone than the *Nymphaea* zone, ceratopogonid emergence is highest in the *Juncus* zone. Sciaridae, Cecidomyiidae, and Lepidoptera also emerge in high numbers from the *Juncus* zone. The majority of emerging insects from the *Juncus* zone are semi-aquatic or terrestrial.

Conclusions

Beaver wetlands have high invertebrate taxon richness. A recent analysis of macroinvertebrate family richness and composition from 447 individual wetlands by Batzer and Ruhí (2013) included five beaver wetlands (four from Wissinger and Gallagher 1999 and one from Benke et al. 1999). These beaver wetlands ranked first, third, fourth, sixth, and ninth overall as supporting the most families out of the 447 sites, not including the semi-terrestrial invertebrate families that are discussed above as being prevalent in abandoned beaver ponds (e.g., beaver complexes in the Southeastern USA; see Table 12.4 and Appendix). Our review suggests that beaver wetlands support high taxon richness primarily due to high habitat heterogeneity (e.g., Benke et al. 1999; Wissinger and Gallagher 1999; Hood and Larson 2014), which has several components:

- Beaver wetlands have a variety of subhabitats including open water, emergent and submergent vegetation, varying water depths, wetted semi-aquatic edges, mud substrates, highly organic benthic layers of varying complexity, and woody debris, which can each support unique organisms (e.g., Benke et al. 1999; Wissinger and Gallagher 1999).
- Beavers increase habitat heterogeneity of existing wetlands by digging long, deep channels that increase connectivity to other wetlands (e.g., Hood and Larson 2014).
- Beavers create complex woody debris structures including lodges (France 1997) and dams the support unique invertebrate assemblages (e.g., Clifford et al. 1993; Rolauffs et al. 2001).
- Beaver activities (damming of streams, building of channels, etc.) create a mosaic of lentic and lotic hydrology that provides habitat for both stream and pond invertebrates (e.g., Table 12.3; Appendix).
- Beaver wetland environments are constantly changing through time, which creates dynamic, frequently nonlinear, multidimensional succession in habitat conditions and invertebrate community structure (Naiman et al. 1988b; Fig. 12.7).
- Beaver dam complexes often include multiple basins that are hydrologically connected and within dispersal distances that foster metapopulation dynamics that enhance alpha and beta diversity (e.g., Caudill 2005).
- Wetted edges, shallow regions, and abandoned beaver wetlands create refuge for many terrestrial and semi-terrestrial taxa (see discussions in Western and Southeastern North America vignettes).
- A variety of predators of invertebrates (other invertebrates, amphibians, fishes, and birds) exploit beaver wetlands, and because predation pressure changes spatially across and temporally within beaver wetlands, heterogeneity should be enhanced. Few studies have considered these potential feedbacks on invertebrate communities in beaver wetlands.
- At the continental scale, conservation efforts to recover and reintroduce beaver populations in their native range (North America and Europe) have been

successful and beaver now have healthy populations that create wetlands across a wide variety of geomorphological, hydrological, and climactic conditions (e.g., Gibson and Olden 2014).

Our review further identifies important ecological values and services to our society that emanate from beaver activities and associated invertebrate communities including:

- Creation and maintenance of new wetlands (e.g., Fustec et al. 2001; Syphard and Garcia 2001; Cunningham et al. 2006; Nummi and Holopainen 2014; Morrison et al. 2015), in the face of ongoing wetland loss in Europe and North America (Zedler and Kercher 2005; Dahl 2011).
- Enhancement of existing wetlands, which aids in increasing wetland density important to conserving wetland dependent organisms as human populations increase (Gibbs 2000).
- Restoration of water quality and quantity in arid lands that has important consequences for regional water management issues and for conservation of plants and animals in those regions (Maret et al. 1987; Gibson and Olden 2014).
- Maintenance of natural flows (Wild 2011) during drought and flood buffering against extreme precipitation events, both of which are likely to become more frequent and severe in the face of climate change (IPCC 2014).
- Creation of refuges for invertebrates during drought, which are then able to recolonize other wetlands post-drought, stabilizing regional diversity (e.g., Wissinger and Gallagher 1999).
- Maintenance of abundant and rich aquatic invertebrate communities that provide important ecosystem services such as the processing of organic matter (e.g., Klemmer et al. 2012; Prather et al. 2013) and linking primary energy sources to wetland fish and waterfowl (Kemp et al. 2012; Nummi and Holopainen 2014).

Beavers and beaver wetlands will likely become especially crucial management partners and resources as climate change and population growth continue to threaten wetlands overall (Wild 2011).

Appendix

Invertebrates recorded (*) in beaver-associated ponds and wetlands in Georgia (Bush and Batzer unpublished), Pennsylvania (Wissinger and Gallagher 1999; Wissinger unpublished), and Colorado, USA (Caudill 2002; B. Peckarsky, unpublished data; S. Wissinger unpublished data).

		SE NA beaver dam wetlands			NE NA beaver pond complexes	Central Colorado Montane beaver dam ponds and wetlands	
	Family	New	Mature	Abandoned	All basins	Valley terrace pond complexes	Stream channel/ riparian distributary
Turbellaria			*		*	*	*
Gastropoda	Lymnaeidae	*	*	*	*	*	*
	Physidae	*	*	*	*	*	*
	Planorbidae	*	*	*	*	*	*
Bivalvia	Sphaeriidae	*	*	*	*	*	*
Oligochaeta		*	*	*	*	*	*
Hirudinea		*	*	*	*	*	*
Araneae		*	*	*	*	*	*
Acari		*	*	*	*	*	*
Pseudoscorpiones		*					
Ostracoda		*	*	*	*	*	
Cladocera	Bosminidae	*	*	*	*	*	*
	Chydoridae				*	*	
	Daphniidae				*	*	
Copepoda	Calanoida	*	*	*	*	*	*
	Cyclopoida					*	
Decapoda	Cambaridae	*	*		*		
Isopoda	Asellidae	*	*		*	*	
Amphipoda	Crangonyctidae	*	*	*	*	*	*
	Dogielinotidae	*	*	*	*	*	*

(continued)

(continued)

		SE NA beaver dam wetlands			NE NA beaver pond complexes	Central Colorado Montane beaver dam ponds and wetlands	
	Family	New	Mature	Abandoned	All basins	Valley terrace pond complexes	Stream channel/ riparian distributary
Collembola	Entomobryidae	*	*	*	*		
	Hypogastruridae	*		*	*		
	Isotomidae	*	*	*			
	Poduridae			*	*	*	*
	Sminthuridae		*	*			
Ephemeroptera	Baetidae	*	*	*	*	*	
	Caenidae	*	*	*	*	*	*
	Ephemeridae		*				
	Ephemerellidae						*
	Heptageniidae						*
	Leptophlebiidae						*
	Siphlonuridae						*
Odonata	Aeshnidae	*		*	*	*	*
	Coenagrionidae	*	*	*	*	*	*
	Gomphidae	*			*	*	*
	Lestidae		*		*	*	*
	Libellulidae	*	*	*	*	*	*
Plecoptera	Chloroperlidae					*	*
	Nemouridae					*	*
	Perlidae						*

Orthoptera	Gryllidae			*			
	Tettigoniidae			*			
Psocoptera				*			
Thysanoptera	Terebrantia			*			
Hemiptera	Aphidae	*	*	*			
	Belostomatidae	*	*	*	*		
	Cercopidae			*			
	Cicadellidae	*	*	*			
	Coccoidea	*	*	*			
	Corixidae	*	*	*	*	*	*
	Delphacidae	*	*	*			
	Gelastocoridae				*		
	Gerridae	*	*		*	*	*
	Hebridae		*	*	*	*	
	Hydrometridae			*	*		
	Mesoveliidae	*	*	*	*	*	
	Miridae	*		*			
	Nepidae	*	*		*		
	Reduviidae		*	*			
	Veliidae	*	*	*	*	*	
Neuroptera	Corydalidae	*	*	*	*		
	Sialidae	*			*		

(continued)

(continued)

		SE NA beaver dam wetlands			NE NA beaver pond complexes	Central Colorado Montane beaver dam ponds and wetlands	
	Family	New	Mature	Abandoned	All basins	Valley terrace pond complexes	Stream channel/ riparian distributary
Coleoptera	Anthicidae		*				
	Carabidae		*	*	*	*	
	Chrysomelidae			*	*	*	
	Coccinellidae	*		*			
	Curculionidae		*	*	*		
	Dytiscidae	*	*	*	*	*	*
	Elateridae			*			
	Elmidae	*		*			*
	Gyrinidae			*	*	*	
	Haliplidae		*	*	*	*	
	Hydraenidae			*	*		
	Hydrophilidae	*	*	*	*	*	
	Lampyridae		*	*			
	Latridiidae			*			
	Noteridae		*	*	*		
	Phalacridae			*			
	Ptilodactylidae		*	*			
	Scirtidae		*	*	*		
	Silvanidae			*			
	Staphylinidae			*	*	*	

Trichoptera	Hydroptilidae	*	*	*			
	Hydropsychidae			*			
	Leptoceridae	*			*	*	
	Limnephilidae				*	*	*
	Phryganeidae				*	*	
	Polycentropodidae	*					*
	Rhyacophilidae	*					*
Lepidoptera	Crambidae	*					
	Noctuidae	*	*	*			
	Tineidae		*				
Diptera	Ceratopogonidae	*	*	*	*	*	*
	Chaoboridae				*	*	
	Chironomidae	*	*	*	*	*	*
	Corethrellidae	*	*	*			
	Culicidae	*	*	*	*	*	*
	Dixidae				*	*	*
	Dolichopodidae				*		
	Empididae		*		*		
	Ephydridae	*	*		*		
	Psychodidae			*	*		
	Ptychopteridae	*	*	*	*		
	Simuliidae			*			*
	Stratiomyidae	*	*	*	*	*	
	Tabanidae	*	*	*	*	*	
	Tipulidae	*	*	*	*	*	*
Hymenoptera	Formicidae		*	*			
	Chalcidoidea			*			
	Mymaridae	*					
	Platygastroidea	*	*	*			

References

Adler PH, Mason PG (1997) Black flies (Diptera: Simuliidae) of East-Central Saskatchewan, with description of a new species and implications for pest management. Can Entomol 129:81–91

Allan JD (2004) Landscapes and riverscapes: the influence of land use on stream ecosystems. Annu Rev Ecol Evol System 35:257–284

Andersen DC, Shafroth PB (2010) Beaver dams, hydrological thresholds, and controlled floods as a management tool in a desert riverine ecosystem, Bill Williams River, Arizona. Ecohydrology 3:325–338

Andersen DC, Shafroth PB, Pritekel CM, O'Neill MW (2011) Managed flood effects on beaver pond habitat in a desert riverine ecosystem, Bill Williams River, Arizona USA. Wetlands 31:195–206

Anderson CB, Rosemond AD (2007) Ecosystem engineering by invasive exotic beavers reduces in-stream diversity and enhances ecosystem function in Cape Horn, Chile. Oecologia 154:141–153

Anderson CB, Rosemond AD (2010) Beaver invasion alters terrestrial subsidies to subantarctic stream food webs. Hydrobiologia 652:349–361

Anderson CB, Pastur GM et al (2009) Do introduced North American beavers *Castor canadensis* engineer differently in southern South America? An overview with implications for restoration. Mammal Rev 39:33–52

Barnett TP, Adam JC, Lettenmaier DP (2005) Potential impacts of a warming climate on water availability in snow-dominated regions. Nature 438:303–309

Batzer DP (2013) The seemingly intractable ecological responses of invertebrates in North American wetlands: a review. Wetlands 33:1–15

Batzer DP, Ruhí A (2013) Is there a core set of organisms that structure macroinvertebrate assemblages in freshwater wetlands? Freshw Biol 58:1647–1659

Batzer DP, Wissinger SA (1996) Ecology of insect communities in nontidal wetlands. Annu Rev Entomol 41:75–100

Benke AC, Arsdall TCV, Gillespie DM, Parrish FK (1984) Invertebrate productivity in a subtropical blackwater river: the importance of habitat and life history. Ecol Monogr 54:25–63

Benke A, Henry R, Gillespie D, Hunter R (1985) Importance of snag habitat for animal production in southeastern streams. Fisheries 10:8–13

Benke A, Ward G, Richardson T (1999) Beaver impounded wetlands of the southeastern coastal plain. In: Batzer D, Rader R, Wissinger S (eds) Invertebrates in freshwater wetlands of North America: ecology and management. Wiley, New York, pp 217–245

Bluzma P (2003) Beaver abundance and beaver site use in a hilly landscape (Eastern Lithuania). Acta Zool Lituanica 13:8–14

Brown DJ, Hubert WA, Anderson SH (1996) Beaver ponds create wetland habitat for birds in mountains of southeastern Wyoming. Wetlands 16:127–133

Burchsted D, Daniels MD (2014) Classification of the alterations of beaver dams to headwater streams in northeastern Connecticut, U.S.A. Geomorphology 205:36–50

Burchsted D, Daniels M, Thorson R, Vokoun J (2010) The river discontinuum: applying beaver modifications to baseline conditions for restoration of forested headwaters. Bioscience 60:908–922

Butler DR, Malanson GP (1995) Sedimentation rates and patterns in beaver ponds in a mountain environment. Geomorphology 13:255–269

Butler DR, Malanson GP (2005) The geomorphic influences of beaver dams and failures of beaver dams. Geomorphology 71:48–60

Caudill CC (2002) Metapopulation biology of the mayfly *Callibaetis ferrugineus hageni* in high-elevation beaver ponds. Ph.D. Dissertation, Cornell University, Ithaca, NY

Caudill CC (2003a) Empirical evidence for nonselective recruitment and a source-sink dynamic in a mayfly metapopulation. Ecology 84:2119–2132

Caudill CC (2003b) Measuring dispersal in a metapopulation using stable isotope enrichment: high rates of sex-biased dispersal between patches in a mayfly metapopulation. Oikos 101:624–630

Caudill CC (2005) Trout predators and demographic sources and sinks in a mayfly metapopulation. Ecology 86:935–946

Cirmo CP, Driscoll CT (1993) Beaver pond biogeochemistry: acid neutralizing capacity generation in a headwater wetland. Wetlands 13:277–292

Clifford HF, Wiley GM, Casey RJ (1993) Macroinvertebrates of a beaver altered boreal stream of Alberta, Canada, with special reference to the fauna on the dams. Can J Zool 71:1439–1447

Collen P, Gibson RJ (2001) The general ecology of beavers (Castor spp.), as related to their influence on stream ecosystems and riparian habitats, and the subsequent effects on fish—a review. Rev Fish Biol Fish 10:439–461

Corcoran RM, Lovvorn JR, Heglund PJ (2009) Long-term change in limnology and invertebrates in Alaskan boreal wetlands. Hydrobiologia 620:77–89

Crowley PH, Johnson DM (1982) Habitat and seasonality as niche axes in an odonate community. Ecology 63:1064–1077

Cunningham JM, Calhoun AJK, Glanz WE (2006) Patterns of beaver colonization and wetland change in Acadia National Park. NE Natural 13:583–596

Curran JC, Cannatelli KM (2014) The impact of beaver dams on the morphology of a river in the eastern United States with implications for river restoration. Earth Surf Process Landf 39:1236–1244

Dahl TE (2011) Status and trends of wetlands in the conterminous United States 2004 to 2009. US Department of the Interior, Fish and Wildlife Service, Washington, DC

Danilov P, Kanshiev V, Fyodorov F (2011) Characteristics of North American and Eurasian beaver ecology in Karelia. In: Sjöberg G, Ball JP (eds) Restoring the European beaver: 50 years of experience. Pensoft, Sofia, pp 55–72

De Szalay FA, Resh VH (2000) Factors influencing macroinvertebrate colonization of seasonal wetlands: responses to emergent plant cover. Freshw Biol 45:295–308

Devito KJ, Dillon PJ (1993) Importance of runoff and winter anoxia to the P and N dynamics of a beaver pond. Can J Fish Aquat Sci 50:2222–2234

Donkor NT, Fryxell JM (2000) Lowland boreal forests characterization in Algonquin Provincial Park relative to beaver (*Castor canadensis*) foraging and edaphic factors. Plant Ecol 148:1–12

Fairchild GW, Faulds AM, Matta JF (2000) Beetle assemblages in ponds: effects of habitat and site age. Freshw Biol 44:523–534

Fairchild GW, Jason C et al (2003) Microhabitat and landscape influences on aquatic beetle assemblages in a cluster of temporary and permanent ponds. J North Am Benthol Soc 22:224–240

France RL (1997) The importance of beaver lodges in structuring littoral communities in boreal headwater lakes. Can J Zool 75:1009–1013

Fuller MR, Peckarsky BL (2011a) Does the morphology of beaver ponds alter downstream ecosystems? Hydrobiologia 668:35–48

Fuller MR, Peckarsky BL (2011b) Ecosystem engineering by beavers affects mayfly life histories. Freshw Biol 56:969–979

Fustec J, Lode T, Le Jacques D, Cormier JP (2001) Colonization, riparian habitat selection and home range size in a reintroduced population of European beavers in the Loire. Freshw Biol 46:1361–1371

Gibbs JP (2000) Wetland loss and biodiversity conservation. Conserv Biol 14:314–317

Gibson PP, Olden JD (2014) Ecology, management, and conservation implications of North American beaver (*Castor canadensis*) in dryland streams. Aquat Conserv Mar Freshw Ecosys 24:391–409

Grover AM, Baldassarre GA (1995) Bird species richness within beaver ponds in south-central New York. Wetlands 15:108–118

Halley D, Rosell F, Saveljev A (2012) Population and distribution of Eurasian beaver (*Castor fiber*). Baltic For 18:168–175

Hammerson GA (1994) Beaver (*Castor canadensis*) ecosystem alterations, management, and monitoring. Nat Areas J 14:44–57
Harthun M (1999) The influence of the European beaver (*Castor fiber albicus*) on the biodiversity (Odonata, Mollusca, Trichoptera, Ephemeroptera, Diptera) of brooks in Hesse (Germany). Limnologica 29:449–464
Hartman G (1994) Long-term population development of a reintroduced beaver (*Castor fiber*) population in Sweden. Conserv Biol 8:713–717
Hartman G (1995) Patterns of spread of a reintroduced beaver *Castor fiber* population in Sweden. Wildl Biol 1:97–103
Hodkinson ID (1975a) A community analysis of the benthic insect fauna of an abandoned beaver pond. J Anim Ecol 44:533–551
Hodkinson ID (1975b) Energy flow and organic matter decomposition in an abandoned beaver pond ecosystem. Oecologia 21:131–139
Hood GA, Bayley SE (2008a) Beaver (*Castor canadensis*) mitigate the effects of climate on the area of open water in boreal wetlands in western Canada. Biol Conserv 141:556–567
Hood GA, Bayley SE (2008b) The effects of high ungulate densities on foraging choices by beaver (*Castor canadensis*) in the mixed-wood boreal forest. Can J Zool 86:484–496
Hood GA, Bayley SE (2009) A comparison of riparian plant community response to herbivory by beavers (*Castor canadensis*) and ungulates in Canada's boreal mixed-wood forest. For Ecol Manage 258:1979–1989
Hood GA, Larson DG (2014) Beaver-created habitat heterogeneity influences aquatic invertebrate assemblages in boreal Canada. Wetlands 34:19–29
Hood GA, Larson DG (2015) Ecological engineering and aquatic connectivity: a new perspective from beaver-modified wetlands. Freshw Biol 60:198–208
IPCC (2014) Climate change 2014: synthesis report. In: Pachauri RK, Meyer LA (eds) Contribution of working groups I, II and III to the fifth assessment report of the intergovernmental panel on climate change. IPCC, Geneva, Switzerland
Jackson CR, Martin JK, Leigh DS, West LT (2005) A southeastern Piedmont watershed sediment budget: evidence for a multi-millennial agricultural legacy. J Soil Water Conserv 60:298–310
Johnson DR, Chance DH (1974) Presettlement overharvest of Upper Columbia River beaver populations. Can J Zool 52:1519–1521
Johnston CA, Naiman RJ (1987) Boundary dynamics at the aquatic-terrestrial interface: the influence of beaver and geomorphology. Landscape Ecol 1:47–57
Johnston CA, Naiman RJ (1990) Aquatic patch creation in relation to beaver population trends. Ecology 71:1617–1621
Kemp PS, Worthington TA et al (2012) Qualitative and quantitative effects of reintroduced beavers on stream fish. Fish Fisher 13:158–181
Klemmer AJ, Wissinger SA, Greig HS, Ostrofsky ML (2012) Nonlinear effects of consumer density on multiple ecosystem processes. J Anim Ecol 81:770–780
Klotz RL (1998) Influence of beaver ponds on the phosphorus concentration of stream water. Can J Fish Aquat Sci 55:1228–1235
Kohler SL, Corti D, Slamecka MC, Schneider DW (1999) Prairie floodplain ponds—mechanisms affecting invertebrate community structure. In: Batzer DP, Rader RD, Wissinger SA (eds) Invertebrates in freshwater wetlands of North America. Wiley, New York, pp 711–730
Law A, Jones KC, Willby NJ (2014) Medium vs. short-term effects of herbivory by Eurasian beaver on aquatic vegetation. Aquat Bot 116:27–34
Levine R, Meyer GA (2014) Beaver dams and channel sediment dynamics on Odell Creek, Centennial Valley, Montana, USA. Geomorphology 205:51–64
Mackay RJ, Waters TF (1986) Effects of small impoundments on hydropsychid caddisfly production in Valley Creek, Minnesota. Ecology 67:1680–1686
Malanson GP, Butler DR (1990) Woody debris, sediment, and riparian vegetation of a subalpine river, Montana, U.S.A. Arctic Alpine Res 22:183–194
Malison RL, Lorang MS, Whited DC, Stanford JA (2014) Beavers (*Castor canadensis*) influence habitat for juvenile salmon in a large Alaskan river floodplain. Freshw Biol 59:1229–1246

Maret TJ, Parker M, Fannin TE (1987) The effect of beaver ponds on the nonpoint source water quality of a stream in Southwestern Wyoming. Water Res 21:263–268

Margolis BE, Raesly RL, Shumway DL (2001) The effects of beaver-created wetlands on the benthic macroinvertebrate assemblages of two Appalachian streams. Wetlands 21:554–563

McCall TC, Hodgman TP, Diefenbach DR, Owen RB (1996) Beaver populations and their relation to wetland habitat and breeding waterfowl in Maine. Wetlands 16:163–172

McCauley SJ (2008) Slow, fast and in between: habitat distribution and behaviour of larvae in nine species of libellulid dragonfly. Freshw Biol 53:253–263

McCauley SJ, Brodin T, Hammond J (2010) Foraging rates of larval dragonfly colonists are positively related to habitat isolation: results from a landscape-level experiment. Am Nat 175(3):E66–E73

McDowell DM, Naiman RJ (1986) Structure and function of a benthic invertebrate stream community as influenced by beaver (Castor canadensis). Oecologia 68:481–489

McKinstry MC, Caffrey P, Anderson SH (2001) The importance of beaver to wetland habitats and waterfowl in Wyoming. J Am Water Resour Assoc 37:1571–1577

McMaster RT, McMaster ND (2001) Composition, structure, and dynamics of vegetation in fifteen beaver-impacted wetlands in western Massachusetts. Rhodora 103(915):293–320

Morrison A, Westbrook CJ, Bedard-Haughn A (2015) Distribution of Canadian Rocky Mountain wetlands impacted by beaver. Wetlands 35:95–104

Mukundan R, Radcliffe DE, Ritchie JC (2011) Channel stability and sediment source assessment in streams draining a Piedmont watershed in Georgia, USA. Hydrol Process 25:1243–1253

Naiman RJ, Melillo JM (1984) Nitrogen budget of a subarctic stream altered by beaver (*Castor canadensis*). Oecologia 62:150–155

Naiman RJ, Melillo JM, Hobbie JE (1986) Ecosystem alteration of boreal forest streams by beaver (*Castor canadensis*). Ecology 67:1254–1269

Naiman RJ, Décamps H, Pastor J, Johnston CA (1988a) The potential importance of boundaries of fluvial ecosystems. J North Am Benthol Soc 7:289–306

Naiman RJ, Johnston CA, Kelley JC (1988b) Alteration of North American streams by beaver. Bioscience 38:753–762

Niles JM, Hartman KJ, Keyser P (2013) Short-term effects of beaver dam removal on brook trout in an Appalachian headwater stream. NE Nat 20:540–551

Nolet BA, Rosell F (1998) Comeback of the beaver *Castor fiber*: an overview of old and new conservation problems. Biol Conserv 83:165–173

Nummi P, Hahtola A (2008) The beaver as an ecosystem engineer facilitates teal breeding. Ecography 31:519–524

Nummi P, Holopainen S (2014) Whole-community facilitation by beaver: ecosystem engineer increases waterbird diversity. Aquat Conserv Mar Freshw Ecosys 24:623–633

Pliūraitė V, Kesminas V (2012) Ecological impact of Eurasian beaver (*Castor fiber*) activity on macroinvertebrate communities in Lithuanian trout streams. Cent Eur J Biol 7:101–114

Pollock MM, Naiman RJ et al (1995) Beaver as engineers: influences on biotic and abiotic characteristics of drainage basins. In: Jones CG, Lawton JH (eds) Linking species & ecosystems. Springer, New York, pp 117–126

Pollock MM, Beechie TJ, Jordan CE (2007) Geomorphic changes upstream of beaver dams in Bridge Creek, an incised stream channel in the interior Columbia River basin, eastern Oregon. Earth Surf Process Landf 32:1174–1185

Polvi LE, Wohl E (2012) The beaver meadow complex revisited—the role of beavers in post-glacial floodplain development. Earth Surface Process Landform 37:332–346

Prather CM, Pelini SL et al (2013) Invertebrates, ecosystem services and climate change. Biol Rev 88:327–348

Pritchard G, Hall HA (1971) An introduction to the biology of craneflies in a series of abandoned beaver ponds, with an account of the life cycle of *Tipula sacra* Alexander (Diptera; Tipulidae). Can J Zool 49:467–482

Pritchard G, Leischner TG (1973) The life history and feeding habits of *Sialis cornuta* Ross in a series of abandoned beaver ponds (Insecta; Megaloptera). Can J Zool 51:121–131

Ray AM, Rebertus AJ, Ray HL (2001) Macrophyte succession in Minnesota beaver ponds. Can J Bot 79:487–499

Ray HL, Ray AM, Rebertus AJ (2004) Rapid establishment of fish in isolated peatland beaver ponds. Wetlands 24:399–405

Rebertus AJ (1986) Bogs as beaver habitat in north-central Minnesota. Am Midl Nat 116: 240–245

Richardson JS, Mackay RJ (1991) Lake outlets and the distribution of filter feeders: an assessment of hypotheses. Oikos 62:370–380

Rolauffs P, Hering D, Lohse S (2001) Composition, invertebrate community and productivity of a beaver dam in comparison to other stream habitat types. Hydrobiologia 459:201–212

Rosell F, Bozser O, Collen P, Parker H (2005) Ecological impact of beavers *Castor fiber* and *Castor canadensis* and their ability to modify ecosystems. Mamm Rev 35:248–276

Smith AJ, Delorme LD (2010) Ostracoda. In: Covich JH, Thorp AP (eds) Ecology and classification of North American freshwater invertebrates, 3rd edn. Academic, San Diego, CA, pp 725–771

Smith ME, Driscoll CT et al (1991) Modification of stream ecosystem structure and function by beaver (*Castor canadensis*) in the Adirondack mountains, New York. Can J Zool 69:55–61

Smol JP, Douglas MSV (2007) Crossing the final ecological threshold in high Arctic ponds. Proc Natl Acad Sci U S A 104:12395–12397

Snodgrass JW, Meffe GK (1998) Influence of beavers on stream fish assemblages: effects of pond age and watershed position. Ecology 79:928–942

Sprules WM (1940) The effect of a beaver dam on the insect fauna of a trout stream. Trans Am Fish Soc 70:236–248

Stagliano DM, Benke AC, Anderson DH (1998) Emergence of aquatic insects from two habitats in a small wetland of the southeastern USA: temporal patterns of numbers and biomass. J North Am Benthol Soc 17:37–53

Stoks R, McPeek MA (2003) Predators and life histories shape *Lestes* damselfly assemblages along a freshwater habitat gradient. Ecology 84:1576–1587

Stoks R, McPeek MA (2006) A tale of two diversifications: reciprocal habitat shifts to fill ecological space along the pond permanence gradient. Am Nat 168(S6):S50–S72

Strachan SR, Chester ET, Robson BJ (2014) Microrefuges from drying for invertebrates in a seasonal wetland. Freshw Biol 59:2528–2538

Syphard AD, Garcia MW (2001) Human- and beaver-induced wetland changes in the Chickahominy River watershed from 1953 to 1994. Wetlands 21:342–353

Trimble SW (1974) Man-induced soil erosion on the southern piedmont. Soil Conservation Society of America, Ankeny, pp 1700–1970

Tuytens K, Vanschoenwinkel B, Waterkeyn A, Brendonck L (2014) Predictions of climate change infer increased environmental harshness and altered connectivity in a cluster of temporary pools. Freshw Biol 59:955–968

Van de Meutter F, Cottenie K, De Meester L (2008) Exploring differences in macroinvertebrate communities from emergent, floating-leaved and submersed vegetation in shallow ponds. Fund Appl Limnol 173:47–57

Wellborn GA, Skelly DK, Werner EE (1996) Mechanisms creating community structure across a freshwater habitat gradient. Annu Rev Ecol System 27:337–363

White SM, Rahel FJ (2008) Complementation of habitats for Bonneville cutthroat trout in watersheds influenced by beavers, livestock, and drought. Trans Am Fish Soc 137:881–894

Wild C (2011) Beaver as a climate change adaptation tool: concepts and priority sites in New Mexico. Seventh Generation Institute, Santa Fe

Wissinger SA (1988) Life history and size structure of larval dragonfly populations. J North Am Benthol Soc 7:13–28

Wissinger SA (1999) Ecology of wetland invertebrates: synthesis and applications for conservation and management. In: Batzer DP, Rader RD, Wissinger SA (eds) Invertebrates in freshwater wetlands of North America: ecology and management. Wiley, New York, pp 1043–1086

Wissinger SA, Gallagher LJ (1999) Beaver pond wetlands in northwestern Pennsylvania. Modes of colonization and succession after drought. In: Batzer D, Rader R, Wissinger S (eds)

Invertebrates in freshwater wetlands of North America: ecology and management. Wiley, New York, pp 333–362

Wissinger SA, Ingmire S, Bogo J (2001) Invertebrate and plant communities in restored wetlands in Pennsylvania. IN: R.R. Rader, D.P. Batzer, and S.A. Wissinger (eds.) Biomonitoring and Management of North American Wetlands. John Wiley & Sons, NY

Wissinger SA, Brown WS, Jannot JE (2003) Caddisfly life histories along permanence gradients in high-altitude wetlands in Colorado (U.S.A.). Freshw Biol 48:255–270

Wissinger SA, Whissel JC, Eldermire C, Brown WS (2006) Predator defense along a permanence gradient: roles of case structure, behavior, and developmental phenology in caddisflies. Oecologia 147:667–678

Woo M-K, Waddington JM (1990) Effects of beaver dams on subarctic wetland hydrology. Arctic 43:223–230

Wright JP, Jones CG, Flecker AS (2002) An ecosystem engineer, the beaver, increases species richness at the landscape scale. Oecologia 132:96–101

Wright JP, Flecker AS, Jones CG (2003) Local vs landscape controls on plant species richness in beaver meadows. Ecology 84:3162–3173

Zedler JB, Kercher S (2005) Wetland resources: status, trends, ecosystem services, and restorability. Annu Rev Environ Resour 30:39–74

Chapter 13
Invertebrates of Temperate-Zone River Floodplains

Darold Batzer, Belinda Gallardo, Andrew Boulton, and Matt Whiles

Introduction

River floodplains are the most widely distributed wetland habitats worldwide, occurring from tropical to polar regions, and from deserts to rainforests (e.g., Junk and Wantzen 2006; Batzer and Baldwin 2012). Given their ubiquity, this book devotes two chapters to floodplains; this one on invertebrates in temperate-zone habitats and the next on invertebrates in tropical floodplains. In simple terms, floodplains are flat areas next to streams or rivers that flood periodically. They range from narrow corridors of deposited sediments through to vast plains that extend many kilometers out from the channel (Junk and Wantzen 2006).

Depending on their specific area of interest, various scientists view floodplains quite differently. Many ecologists perceive floodplains primarily as aquatic ecotonal extensions of river channels, and work on fish and water quality is prominent; most people who work on floodplain invertebrates also come from this tradition. Alternatively, ecologists who work on plants and soils (the criteria typically used to define wetlands, Cole and Somerville 2014) view floodplains as distinct habitats rather than simply ecotones between rivers and uplands; ecologists who study invertebrates are just beginning to apply this perspective (e.g., Bright et al. 2010). In fact, floodplains are a mosaic of sub-habitats—some aquatic, some terrestrial, some wetland—and all perspectives should be incorporated for a holistic examination of their ecology (Ward et al. 2002).

Although floodplains indeed are areas of low topography, they are not "flat," and even modest changes in topography can have important ecological consequences (e.g., creating multiple microhabitats for coexisting plants with different water requirements). Diverse sub-habitats occur across most floodplains, including levees, islands, anabranches or side channels (sloughs), oxbow lakes or billabongs, pools, backswamps, flats, bottomland forests, seasonal and permanent marshes, isolated depressions, and deltas (Fig. 13.1; Ward et al. 2002; King et al. 2012), each potentially

D. Batzer, D. Boix (eds.), *Invertebrates in Freshwater Wetlands*,
DOI 10.1007/978-3-319-24978-0_13

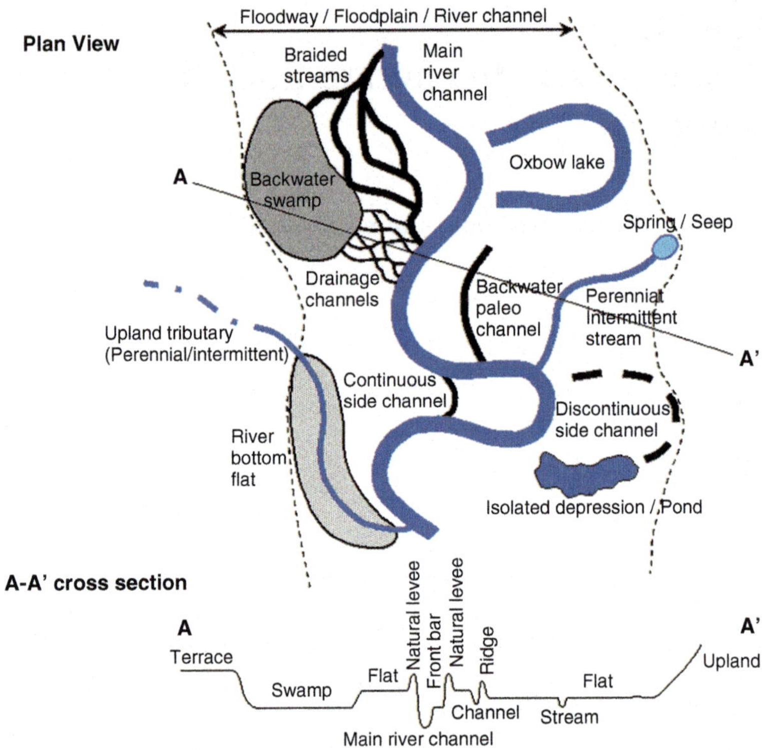

Fig. 13.1 Habitat heterogeneity that can occur across floodplains. Reprinted with permission from *Ecology of Freshwater and Estuarine Wetlands: Second Edition*, edited by Darold P. Batzer and Rebecca R. Sharitz. (c) 2014 by the Regents of the University of California. Published by the University of California Press

supporting a unique biota. Sub-habitat monikers are often regional, and similar features may have different names across the globe.

How floodplains flood is complex (King et al. 2012). While over-bank flooding may contribute the greatest volume of water to floodplain water budgets, groundwater discharge from adjacent uplands or deeper aquifers and direct precipitation onto the floodplain itself are also important contributors (Fig. 13.2). Those unfamiliar with floodplains may assume that these habitats flood only after river-flows over-bank, but in fact we have frequently observed floodplains becoming partly inundated long before the traditional "flood stage" of rivers. Most floodplains inundate seasonally. With the onset of wetter seasons, or seasons with lower evaporation and transpiration, floodplains typically first begin to partially flood from direct precipitation, increased seepage from rising groundwater, the lateral seepage of river water through levees from rising channels, or some combination of all three (Benke et al. 2000; Junk and Wantzen 2006; King et al. 2012). In the driest years, river levels may never rise enough for a direct influx of river water, yet

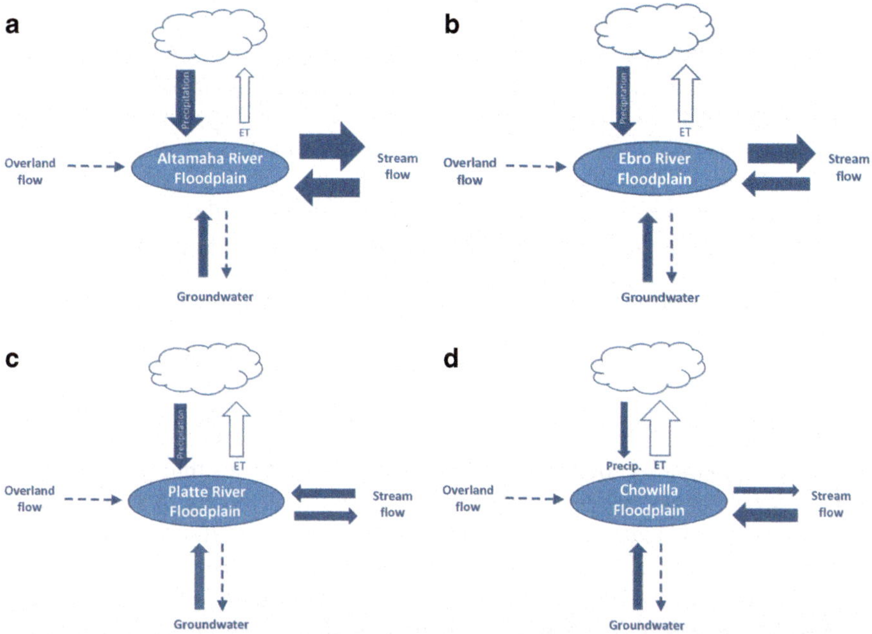

Fig. 13.2 Water budgets for typical temperate-zone floodplains in (**a**) Southeastern USA (Altamaha River, Georgia), (**b**) Southwestern Europe (Ebro River, Spain), (**c**) Central Prairie of the USA (Platte River, Nebraska), and (**d**) Southern Australia (Murray River). This order reflects progressive climatic change with the Southeastern USA being the wettest climate and Southern Australia the driest climate

substantial areas of the floodplain may still become inundated. In moderately dry years, river water may begin to enter the floodplain through natural breaks in the levees, and much of the floodplain may become inundated even if "flood stage" is never reached. In these years, floodplains may function much like tidal wetlands, with water flowing through levee breaks and associated sub-channels onto the floodplain as river levels rise, and then ebbing back out the same pathways as river waters recede (albeit this occurs over days or weeks). In wet years, when river levels top the levees, the entire floodplain may become inundated so that it becomes an extension of the river, with unidirectional downstream flow developing until river levels again drop. Even if the actual influx of water onto a floodplain is brief, the lowest-lying portions of the floodplain may remain inundated for many more weeks, months, or even years, providing important habitat for aquatic organisms.

Floodplains are considered pulsing systems, with the Flood Pulse Concept (FPC) of Junk et al. (1989) being the major paradigm developed to explain their ecological controls. First developed for large tropical rivers, we direct readers to the detailed description of the FPC found in the next chapter. However, the concept also has utility for temperate-zone floodplains (Tockner et al. 2000; Malard et al. 2006; Paillex et al. 2007; Reese and Batzer 2007; Gallardo et al. 2008, 2009a). The FPC maintains that water pulses ecotonally connect the channel and floodplain, with water-borne materials and aquatic organisms moving between systems. Additionally, flood pulses are important controls on the plants and soils that give floodplains their wetland character (González et al. 2010).

Where do invertebrates fit into this emerging body of floodplain research? For the aquatic invertebrate fauna in temperate-zone floodplains, a major thrust of research addresses how connectivity of floodplains with river channels affects invertebrates and habitat quality (e.g., Sheldon et al. 2002; Jenkins and Boulton 2003; Arscott et al. 2005; Gallardo et al. 2008, 2009b, 2014; Starr et al. 2014), and describes the movements of invertebrates between rivers and their floodplain (e.g., Söderström and Nillson 1987; Smock 1994; Galatowitsch and Batzer 2011). However, it has also been established that much of the invertebrate community on floodplains comprises obligate wetland organisms that are not derived from river channels, but instead spend dry seasons aestivating in moist soils or residing in permanent lentic water bodies that persist on the floodplain (e.g., Arscott et al. 2005; Tronstad et al. 2005a; Bright et al. 2010; Starr et al. 2014). While most work focuses on the aquatic invertebrate fauna of floodplains, the terrestrial component is being increasingly recognized (e.g., Braccia and Batzer 2008; Paetzold et al. 2008; Persson Vinnersten et al. 2014). Different floodplains flood, pulse, and dry in different ways, depending on climate, weather, and topography. This variation in flood seasonality, extent, intensity, and duration undoubtedly affects the ecology of the resident invertebrates (Batzer and Wissinger 1996; Ward et al. 1999b; Wissinger 1999; Paetzold et al. 2008; Persson Vinnersten et al. 2014), implying that invertebrates might be useful bioindicators of anthropogenically induced alterations of floodplain hydrology (Paetzold et al. 2008).

Humans pose many threats to temperate floodplain habitats (Brinson and Malvárez 2002). Perhaps the greatest impact on floodplain biota is from flow regulation via dam, weir, or levee construction (Poff et al. 1997; Kingsford 2000; Bunn and Arthington 2002). These structures limit the extent of floodplain inundation, alter the length and amplitude of flood pulses, and sever connections between the river channel and its floodplain (Ward et al. 1999b), which all typically impact floodplain biota.

Key Locations for Floodplain Invertebrate Research

Globally, the invertebrate fauna has been extensively researched only at a few temperate-zone floodplain ecosystems, mainly in the USA, Europe, and Australia. In this section we review examples from each of these areas as model systems (again, readers interested in tropical systems are referred to the next chapter). While acknowledging that useful information has been collected from other floodplain-river complexes besides the ones highlighted in this chapter, we feel the comprehensive nature of the work in each of our model systems permits integrating multiple ecological perspectives within similar settings. We conclude this chapter by using these detailed region-specific case studies to develop a more overarching synthesis of the ecology of invertebrates across temperate-zone floodplains, including implications for their successful management, that hopefully will be broadly useful.

Southeastern US Floodplains (Altamaha, Ogeechee, Satilla, Savannah, Sipsey Rivers)

In terms of hydrology, geomorphology, and plant ecology, floodplains of the Southeastern USA might be the world's most intensively studied (e.g., King et al. 2012), although information on the region's floodplain invertebrates is limited. While the lower Mississippi River is the region's most prominent system, other major rivers include the Apalachicola-Chattahoochee-Flint and the Mobile-Alabama River systems that flow into the Gulf of Mexico and the Altamaha, Savannah, Santee, Pee Dee, Cape Fear, and Roanoke River systems that flow into the Atlantic Ocean (Benke and Cushing 2005). Much of the work on floodplain invertebrates, however, has focused on lesser known rivers of the region (e.g., Coosawhatchie, Ogeechee, Satilla, Sipsey), especially ones whose hydrology is minimally regulated.

River floodplains of the Southeast USA typically flood seasonally in winter and spring (Benke et al. 2000; King et al. 2012). However, apart from the Mississippi, this pattern is not related to seasonal changes in precipitation or snow melt. Instead, rainfall in this region is consistent year-round (~100 mm/month). Floodplain inundation occurs when evapotranspiration rates decline in winter and early-spring, due to low temperatures and slow plant growth, causing channels to fill and spill onto floodplains (Benke et al. 2000); direct precipitation and groundwater discharge onto floodplains also persists longer in winter and early-spring (Fig. 13.2a). Into summer, higher temperatures and active tree growth spur evapotranspiration, and river levels fall, residual water on floodplains evaporates, transpires, and drains, and seasonal floodplain wetlands dry. As Southeastern US floodplains are densely forested, transpiration is a particularly important control on hydrology. Tropical storms can create secondary overbank floods in late summer or early fall, although these floods tend to be brief.

While the majority of the water entering floodplains is derived from large overbanking events, and this source dominates overall water budgets (Fig. 13.2a), the composition of invertebrate communities on Southeastern US floodplain wetlands may be controlled more by water from other sources. Even in drought years when over-bank floods do not occur, productive invertebrate communities still develop on Southeastern floodplains (Reese and Batzer 2007). Long before major floods occur, low-lying areas on floodplains begin to fill from direct rainfall, groundwater discharge off uplands, and lateral seepage from rising river channels. It is in these low-lying areas, whether permanently or seasonally flooded, where most aquatic invertebrates on floodplains reside, feed, and reproduce. In this region, the major effect of large floods on aquatic invertebrates is probably to permit broader access across seasonally flooded habitat, and to provide an influx of nutrients, sediments, and (in some cases) organisms from the river channels.

Floodplain Invertebrate Community Variation and Assembly

Invertebrates that successfully exploit Southeastern US floodplains (Appendix) are derived from four sources: (1) desiccation-resistant aquatic organisms; (2) riverine aquatic organisms; (3) lentic water body aquatic organisms; and (4) flood-tolerant terrestrial organisms (Table 13.1). Some taxa such as certain crustaceans (Table 13.1) exploit multiple strategies. As most research attention has focused on the aquatic fauna (Groups 1–3) and on flooded locations, terrestrial Group 4 invertebrates are likely often overlooked. However, one study (Braccia and Batzer 2001), targeting invertebrates associated with woody debris sampled in both wet and dry periods, found that 43 of 67 families associated with wood were Group 4 organisms. Thirty of these 43 "terrestrial" families were collected living in or on wood that had been completely submersed for extended periods (weeks or months), suggesting that they can function under water.

Table 13.1 Characteristics of invertebrates able to exploit floodplain habitats of the Southeastern USA

Organism character	Example taxa	Relevant references
Group 1: Desiccation-resistant aquatic organisms that spend dry periods in the soil/litter layer	Annelida, Asellidae, Culicidae, Chironomidae	Tronstad et al. (2005a, b), Reese and Batzer (2007)
Group 2: Lotic aquatic organisms that enter from the adjacent river channels	Leptophlebiidae, Siphlonuridae, Gomphidae, Plecoptera	Smock (1994), Reese and Batzer (2007), Bright et al. (2010), Galatowitsch and Batzer (2011), Starr et al. (2014)
Group 3: Lentic aquatic organisms that spend low water periods in permanent water habitats on the floodplain proper (e.g., oxbow lakes, permanently flooded backwaters)	Asellidae, Corixidae, Notonectidae, Dytiscidae, Hydrophilidae	Benke et al. (1984), Reese and Batzer (2007), Tronstad et al. (2007), Starr et al. (2014)
Groups 4: Terrestrial organisms that can tolerate periodic flooding	Acarina, Araneae, Myriapoda, Collembola, Curculionidae, Formicidae	Braccia and Batzer (2001), Tronstad et al. (2005a, b), Reese and Batzer (2007), Bright et al. (2010)

In the Altamaha/Oconee River watershed (Georgia), Reese and Batzer (2007) found that invertebrate communities varied dramatically among floodplains in different locations (lower-river, mid-river, headwater-streams). Our reanalysis of that data (Fig. 13.3) suggests that desiccation-resistant aquatic organisms (Group 1) and flood-tolerant terrestrial organisms (Group 4) comprised most of the taxa on floodplains, regardless of their position. However, the relative contributions of lotic organisms from the river channels (Group 2) and lentic aquatic organisms from permanent water wetlands (Group 3) varied greatly among floodplains across the watershed (Fig. 13.3). These kinds of taxa were diagnostic for the different kinds of floodplain: (a) large lower-river floodplains; (b) mid-sized river floodplains; and (c) small headwater floodplains.

Large Lower-River Floodplains. Reese and Batzer (2007) collected 51 invertebrate families from larger, lower-river floodplains of the Altamaha River (Appendix). Despite only employing aquatic sampling, 17 families were flood-tolerant terrestrials (Group 4). Of the aquatics, 20 were desiccation-resistant (Group 1), 12 were lentic taxa from permanent water habitats (Group 3), and only 3 had migrated in from the river (Group 2) (Fig. 13.3). Group 1 and Group 3 organisms were by far the most abundant, with only a handful of Group 2 individuals being collected. Indicator analyses found that Dytiscidae beetles were the primary diagnostic taxon for lower river floodplains (Reese and Batzer 2007), and primarily spend dry periods in permanent oxbow lakes (i.e., use a Group 3 strategy).

Working in these same Altamaha River habitats, plus floodplains along the lower Savannah River, Bright et al. (2010) tested the hypothesis that invertebrate community structure would vary laterally across large floodplains, with lotic riverine taxa (Group 2) dominating habitat close to river channels, lentic aquatic taxa dominating floodplain interiors (Groups 1 and 3), and terrestrial taxa dominating habitat near upland areas (Group 4). However, they instead found that Group 1 and 3 organisms, termed the "obligate wetland fauna," dominated every habitat. Lotic riverine taxa (Group 2) were extremely rare throughout, even next to the river channels. Terrestrial taxa (Group 4) were more common, but were distributed across the entire floodplain rather than just near uplands, suggesting that they are also "obligate" wetland taxa, and not invading from uplands. Although the kinds of invertebrates were similar across the floodplain, overall abundance and biomass was highest close to the channels, suggesting that habitat near the river was higher quality (Bright et al. 2010).

Benke (2001), working in the Ogeechee River, also assessed lateral patterns of invertebrates in river-floodplain complexes, this time contrasting the channel itself with the adjacent floodplain. Over a given section of a river, aquatic invertebrate biomass on the floodplain exceeded that found in the channel, mostly due to the greater area of aquatic habitat in the floodplain (as terrestrial invertebrates were not considered, this contrast is conservative). He also confirmed that lotic taxa (Ephemeroptera, Plecoptera) were rare or absent on the floodplain, although abundant in river channels (Benke et al. 1984; Benke 2001).

The apparent lack of any significant migrations of lotic invertebrate taxa from river channels onto large Southeastern floodplains was unexpected. Why such migrations do not occur remains unknown. However, the same pattern seems to

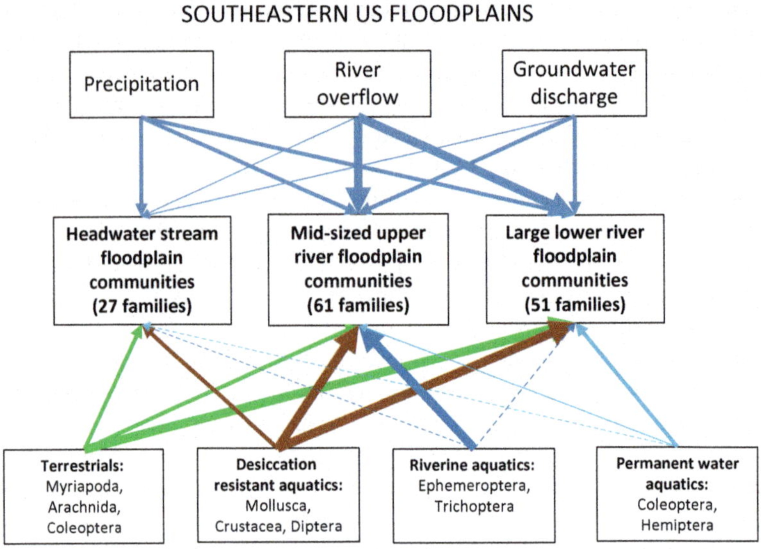

Fig. 13.3 Sources of invertebrate colonizers and water for headwater (Piedmont), upper river, (Piedmont) and lower river (Coastal Plain) floodplains of the Altamaha/Oconee watershed, Georgia, based on data from Reese and Batzer (2007). For colonizers, *thick arrows* denote ≥16 invertebrate families, *medium arrows* 10–15 families, *thin arrows* 4–9 families, and *dashed arrows* ≤3 families

develop for the resident fishes. Garnett and Batzer (2014) found that fish communities on large floodplains of the Altamaha and Savannah Rivers were dominated not by lotic fishes from the channels, but by lentic fishes (e.g., *Esox, Amia, Aphredoderus*) migrating from oxbow lakes and other permanent water bodies on the floodplain itself. In combination with the large numbers of predaceous dytiscid beetles typical of these habitats, this means that large numbers of invertivorous predators, all well adapted for wetland conditions, are full-time residents of large Southeastern floodplains, possibly making them inhospitable places for channel invertebrate species to enter.

Mid-sized River Floodplains. Higher up in the Altamaha watershed along mid-sized rivers (Oconee River and tributaries), Reese and Batzer (2007) collected 61 families of invertebrates (Appendix) of which 21 were desiccation-resistant aquatics (Group 1), 18 had migrated in from the river (Group 2) (some of which were very abundant), only 7 were lentic aquatic species from permanent water habitat (Group 3) (none of which were abundant), and 15 were terrestrial (Group 4) (Fig. 13.3). Group 3 colonizers from oxbow lakes and other permanent water bodies may be uncommon simply because these habitats are rare on mid-river floodplains. However, aerial dispersal of aquatic invertebrates can still become significant into summer (Tronstad et al. 2007), and notonectid and corixid populations can explode on floodplain areas during unusually wet summers.

Unlike on lower river floodplains, massive numbers of lotic taxa, particularly mayflies, migrate into some mid-river sites (Fig. 13.3; Reese and Batzer 2007; Galatowitsch and Batzer 2011). For leptophlebiid mayflies (the primary indicator taxon for these habitats), movements onto the floodplain involve active swimming, rather than passively being swept into the floodplains with flood waters (Galatowitsch and Batzer 2011). What motivates these mayfly migrations is not clear. Thermal conditions, food quality and quantity, growth rates, and even predation pressure are not necessarily superior on the floodplains compared to the channels, and migration undoubtedly confers considerable risk (Galatowitsch and Batzer 2011).

Landscape setting for mid-sized floodplains appears to affect invertebrate community composition. In the study by Reese and Batzer (2007), the mid-sized Oconee River sites were set in the Piedmont, where gradients are relatively high, floods are flashy, and permanently flooded habitat on the floodplain itself is rare. In contrast, Starr et al. (2014) worked on the mid-sized Sipsey River on Alabama's upper Coastal Plain. Here, gradients are low, floods are fairly predictable and extended, and permanent-water habitats on the floodplain are common. Unlike the Oconee in the Piedmont, no major influx of taxa from the river channels was detected here (mayflies were rare, except for some Caenidae). However, flood pulses that connected the Sipsey River channel to floodplain habitats were still important ecologically because flushing probably affected dissolved oxygen dynamics and, in turn, invertebrates. On low-gradient areas of the Southeastern US Coastal Plain, both mid-sized and larger river floodplains may share certain characteristics despite differences in discharge, with Group 1 and 3 taxa dominating the overall assemblages of aquatic invertebrates, and Group 2 taxa being rare.

Small Headwater Floodplains. The taxa of invertebrates that exploit floodplains of headwater streams of the Southeastern USA also appear to depend on landscape setting. In the Piedmont (Reese and Batzer 2007), headwater streams have high gradients, and floodplains rarely receive water from the channel, being inundated only briefly by direct precipitation and groundwater discharge. Here, invertebrate communities are dominated by terrestrial organisms (Fig. 13.3) and aquatic organisms that can both tolerate drying and develop rapidly after floods (e.g., mosquito larvae, harpacticoid microcrustaceans). On the Coastal Plain, however, low-gradient headwater streams and their floodplains are more connected hydrologically. The invertebrate communities here share characteristics with those of mid-sized river floodplains of the Piedmont, with migrations of mayflies being important, and a broader range of desiccation-resistant aquatic organisms occurring (Gladden and Smock 1990; Smock 1994).

Trophic Relationships

Southeastern US floodplains are among the only wetland habitats where secondary production statistics have been calculated for overall aquatic invertebrate communities (see also Platte River below). For two headwater coastal plain streams of Virginia, secondary production of aquatic floodplain invertebrates ranged from 1.1 to 6.6 g m^{-2} year^{-1}, with short hydroperiod sites being on the low end, and sites flooded longer being on the high end of this range (Gladden and Smock 1990; Smock et al. 1992). In permanently inundated floodplain of a larger coastal plain river of Georgia, aquatic invertebrate production was estimated at 22.0 g m^{-2} year^{-1} (Benke et al. 1984). Growth rates of resident Chironomidae (Tronstad et al. 2010) and microcrustaceans (Anderson et al. 1998) can be very rapid (generation times <30 days) and high turnover rates of these invertebrates may fuel production.

Given that floodplains of the Southeastern USA are largely forested, it is logical to assume that leaf litter and wood would be the primary trophic basis for invertebrates. However, the few studies that have addressed this issue have generated equivocal results. Taylor and Batzer (2010) used stable-isotope analyses to identify which basal resources were being assimilated by midge larvae (Chironomini) and amphipods (Crangonyctidae) on a South Carolina floodplain. Midges were generalist feeders, assimilating carbon from both plant leaf litter and algae. Amphipods turned out to be predators of midges, and not shredders or collectors as had been surmised. Libby (2013) assessed leaf litter breakdown on another South Carolina floodplain, finding few invertebrate shredders and scant evidence of an invertebrate role in the process.

Braccia and Batzer (2008) assessed invertebrate colonization and breakdown of dead wood on floodplains, contrasting patterns with wood in both the adjacent river channel and uplands. Wetland wood shared more similarities to upland than riverine wood; the trophic connection of floodplain wood to invertebrates is probably funneled primarily to the terrestrial rather than aquatic sector of the community. Floodplain wood is primarily colonized during floods, and used mostly as refugia by terrestrial invertebrates (Braccia and Batzer 2001).

Prairie River Floodplains of the Central USA (Platte River)

Rivers of the central prairie region of North America were historically wide, shallow, meandering systems, often with braided channels and sand or silt substrata. Extensive floodplain wetlands associated with these low-gradient systems would inundate during spring flooding, providing expansive habitat for wetland flora and fauna. Human activities in this region, mostly related to agriculture, have now greatly diminished floodplain wetland habitat and river-floodplain connectivity. As an example, wetland habitats along the central Platte River, a region recognized as a critical region for wetland-dependent wildlife, now constitute <5 % of land area in the region (Sidle et al. 1989; US Fish and Wildlife Service 1997); this is typical of the other river systems in the central prairie region. Loss of wetland habitats along the Platte and similar rivers is largely a result of direct conversion to agriculture, but groundwater extraction, impoundments, water diversion from the channels, navigation development, urbanization, and climate change have all contributed to the loss and degradation of floodplain wetlands (Sidle et al. 1989). While much has been lost, increasing awareness of the ecological and economic value of these systems has led to enhanced conservation and, in some regions, extensive restoration efforts. The central Platte River region, in particular, has been a focal area for wetland conservation and restoration as well as research on floodplain wetland communities.

The entire Platte River drainage area is 230,362 km^2, with land use dominated by agriculture (>90 %; Galat et al. 2005). While much of the headwaters are high-gradient mountain streams, lower prairie sections of the river are low gradient and turbid. Nutrient concentrations vary, depending upon discharge and seasonal agricultural activities (annual median NO_3-N concentrations range from 0.68 to 1.4 mg/L, total PO_4-P concentrations average 0.73 mg/L; Frenzel et al. 1998; Galat et al. 2005). The central Platte River region from Lexington to Chapman, Nebraska, includes some areas with significant intact floodplain wetland habitat. Mean annual precipitation here is ~650 mm, with cold winters and hot dry summers. Discharge in the central Platte River averages 39 m^3/s (Grand Island gauge), but has ranged from 0 to 147 m^3/s. Discharge volume and variation is influenced by upstream dams, canals, and power plants that provide water for irrigation and power generation. Flows now generally peak in May or June and then drop considerably during the late summer irrigation season.

The floodplain landscape of the central Platte is a series of ridges and swales covered with mesic prairie (in regions not converted to agriculture). Floodplain wetlands are mostly linear features formed in old river channels, locally called sloughs. When full, sloughs range from very shallow to nearly 1 m deep. Inorganic substrata are generally mixtures of sand and silt, usually with extensive organic materials ranging from very fine particulates to coarse detritus. Floodplain vegetation ranges from mixtures of grasses and forbs to dense forests of cottonwood (*Populus*) and willow (*Salix*), depending on local land use and river management. Systems with longer hydroperiods support stands of wetland macrophytes (*Typha*, *Scirpus*, *Sparganium*, *Eleocharis*, *Lemna*, *Potamogeton*), and copious filamentous algae (Whiles and Goldowitz 2001, 2005; Meyer et al. 2010). Most upland and lowland prairie habitat in this region has been converted to corn and soybean row crop systems, but conservation and restoration efforts are increasing.

The central Platte is a focal point of conservation and restoration activities, primarily because of its significance to birds (Sidle et al. 1989; US Fish and Wildlife Service 1997). The region is directly in the center of North America's central migration flyway, providing critical migration habitat for the federally endangered whooping crane (*Grus americana*), staging habitat for 80 % of the world's population of sandhill cranes (*Grus canadensis*), and wintering and migration habitat for myriad waterfowl.

Floodplain Invertebrate Community Variation and Assembly

Research on wetland invertebrates of Platte River floodplains has focused on the influence of hydrology on community structure and productivity, and invertebrate responses to wetland management and restoration. Results of these studies are often placed in the context of food availability for water birds, the primary management targets in the region. Across gradients from ephemeral, disconnected sites to permanent connected sites, invertebrate taxonomic richness follows a bell-shaped curve resembling predictions of the intermediate disturbance hypothesis (Fig. 13.4; Whiles and Goldowitz 2001). Taxa commonly encountered in drier sites are generally small-bodied groups with rapid life cycles (e.g., copepods, mosquitoes) or, if they are larger, are highly vagile (e.g., adult coleopterans) (Whiles and Goldowitz 2005). Fishless intermittent sites with annual hydroperiods in the range of ~300 days tend to harbor the highest richness, including taxa that can be found in both drier and wetter sites. Permanent sites lack some taxa that presumably are either poor competitors and/or intolerant of predation by larger invertebrates and fishes. Overall, the fauna is overwhelmingly dominated by obligate wetland taxa, either permanent water lentic taxa (Group 3; 28 families in Appendix) or desiccation-resistant forms (Group 1; 11 families), with only a single family, Baetidae, that might (or might not) have a connection to the channel.

While interactions between fishes and invertebrates in these systems are poorly studied, there is evidence that invertivorous fishes can reduce insect emergence production. Studies of benthic communities in sites ranging from fishless ephemeral ones to permanent sites with year-round fish indicated total benthic invertebrate biomass increased with hydroperiod (Whiles and Goldowitz 2005). However, a study examining insect emergence from these same sites showed highest emergence in intermittent sites and lower values in the permanent site with fish (Whiles and Goldowitz 2001). Contributions of non-insect groups accounted for some of the discrepancy between benthic biomass and emergence production, but even when considering only benthic biomass and emergence of insects, it appears that fish consume a considerable quantity of benthic insect production before adults emerge.

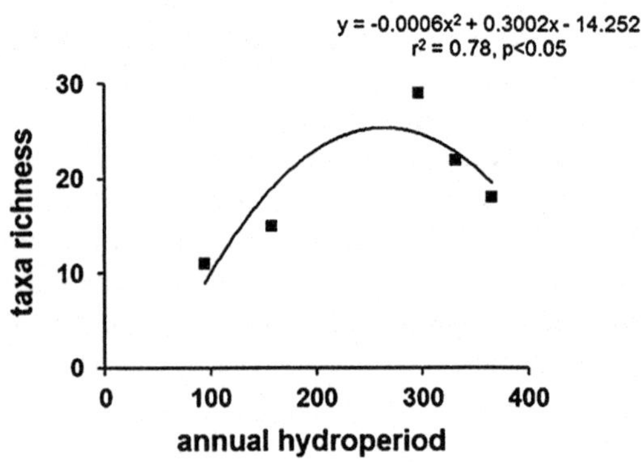

Fig. 13.4 Relationship between hydroperiod and insect taxon richness in floodplain slough habitats of the Platte River, Nebraska, USA

Endemism in the central prairie is rare, as much of the flora and fauna of the North American forests or deserts also occurs here. The Platte River caddisfly (*Ironoquia plattensis,* Alexander and Whiles 2000), however, is an interesting exception. Like other members of the genus, *I. plattensis* inhabits intermittent systems and migrates out of drying habitats in early summer. Final instar larvae then aestivate in the riparian litter layer, eventually pupating and emerging as adults in the late summer or early fall (Whiles et al. 1999). The Platte River caddisfly has been found in a few drainages beyond the Platte, but populations are few and far between and it is currently considered a "Tier I At-Risk Species" in Nebraska (Vivian et al. 2013). Habitat loss and modification are primary threats, including dewatering of intermittent habitats from drought and human water use, and cattle grazing in riparian areas (Harner and Geluso 2012; Vivian et al. 2013).

Wetland restoration projects are increasingly common in the floodplain of the central Platte River, but little is known about the overall success of these projects. Meyer and Whiles (2008) compared invertebrate communities in natural sloughs with a chronosequence (5–16 years post-restoration) of restored sites. Relatively coarse metrics, such as total abundance, biomass, and diversity, were similar among restored and natural sites. However, mollusks and amphipods were more abundant in natural sites, and leeches were only present in natural sites. These non-insect groups are less vagile and may colonize or recover slowly, and thus could be useful indicator taxa to monitor recovery following restoration. Meyer and Whiles (2008) also found that hydroperiod was linked to recovery and community structure, with restored sites generally holding water for shorter periods than natural sites, presumably because of reduced amounts of organic sediments in restored sites and/or less connectivity with groundwater. Wet-meadow restored sites in the region had less litter cover, lower root mass, lower soil moisture, and higher soil temperatures compared to native conditions (Riggins et al. 2009).

Most studies of Platte River wetland invertebrates have focused on communities inhabiting sloughs (Appendix), but moist-soil communities have also been examined. Davis et al. (2006) found that wet-meadow soil invertebrate communities were dominated by earthworms, isopods, and beetle larvae (mostly Scarabaeidae and Elateridae). As with slough communities, hydrology appears to be a primary driver of moist-soil communities; Scarabaeidae, Diplopoda, and Lepidoptera have higher biomass in drier soils, whereas biomass of Tipulidae and Isopoda is generally higher in wetter areas.

The combined results of the various wetland invertebrate studies in the central Platte River region suggest that a mosaic of wetland habitats with differing hydroperiods will maximize macroinvertebrate diversity and seasonal availability at the landscape scale, which should benefit management targets such as cranes and waterfowl. These results and patterns likely apply to wetland invertebrates in other river floodplain systems in the prairie region, but there is little information beyond the Platte basin. In one of the few studies of prairie floodplain wetland invertebrates from outside the Platte basin, Kohler et al. (1999) also found that fish predation was a dominant force shaping invertebrate communities in floodplain wetlands of the upper Mississippi and Sangamon Rivers. There, fish were present in both temporary and permanent wetlands, whereas in the central Platte fish may occasionally colonize temporary systems during high water periods, but are otherwise generally absent in nonpermanent habitat. Fish presence reduced abundances of large, active predaceous invertebrates (Kohler et al. 1999). Collectively, studies from the Platte and other prairie regions suggest that hydrology and predators, which are generally linked, strongly influence invertebrate diversity and productivity, and thus food availability to higher trophic levels.

Trophic Relationships

Wetlands of the central Platte River floodplain support high algal and plant production. Where macrophytes are productive, detritus accumulations can be extensive. However, large detritus-shredding invertebrates are not well represented in these habitats. The endemic caddisfly, *I. plattensis*, is an exception; populations of this shredder are only present in a few intermittent sites, but where they are present they can be abundant, and production and assimilation estimates suggest they can consume ~10 % of detrital standing stocks per year (Whiles et al. 1999). Invertebrate functional structure in central Platte River wetlands generally reflects the primarily autochthonous energy base of these systems. Collector-gatherers (e.g., most Chironomidae, Copepoda), scrapers (e.g., Physidae and Lymnaeidae), and predators (e.g., Odonata, Hemiptera) are abundant and diverse in these systems (Whiles and Goldowitz 2001, 2005; Meyer and Whiles 2008). In contrast, shredders and filter feeders are generally absent or poorly represented, except in more permanent habitats where active filter feeders such as bivalves and burrowing mayflies can be abundant.

In the central Platte region, fishes are the dominant predator of invertebrates. Their presence is linked to hydrology and connectivity to permanent water, with *Culea inconstans*, *Etheostoma exile*, *Lepomis cyanellus*, *Fundulus sciadicus*, and *Semotilus atromaculatus* being the more commonly encountered invertivorous species (Goldowitz and Whiles 1999). Anuran larvae can be abundant, but their interactions with invertebrates in this region are not well known; based on studies from other regions, tadpoles may compete with and/or consume some invertebrates. Tadpoles may also be an important food source for larger invertebrates such as odonates and hemipterans.

Management, conservation, and restoration activities in the region generally target cranes and other birds. However, there is increasing recognition of the linkages between wetland invertebrates and birds in this region. While cranes and other birds may acquire significant energy foraging in crop fields, wetland invertebrates represent protein-rich, and in the case of mollusks and crustaceans also calcium-rich, food resources. Hence, enhancing their availability may benefit the focal management targets in the region. Studies of restored wetlands in the region indicate that larger crustaceans such as amphipods, along with mollusks, may be the slowest to recolonize, which has implications for female birds preparing for breeding and in need of calcium (Meyer and Whiles 2008).

The insect component of wetland invertebrate communities can also be important to a variety of aerial predators such as bats and terrestrial birds that never enter the water. These and other terrestrial and riparian predators may exploit emerging insects as an abundant resource during emergence events (Baxter et al. 2005). Estimates from the central Platte region suggest this linkage is not trivial; Whiles and Goldowitz (2001) quantified insect emergence from a variety of sloughs along the central Platte River and estimated emergent production at >5 g dry mass m^{-2} $year^{-1}$ from their most productive fishless intermittent site.

Southern European Floodplains (Ebro River, Rhône River, Tagliamento River)

Europe has an extensive network of over 160 river-floodplains dominated by temperate and continental climates. Compared to temperate floodplains at similar latitudes in North America and Asia, those in Europe have milder air temperatures because of westerly winds warmed by the Gulf Stream (Tockner et al. 2009). Another distinctive feature of European floodplains is that they have been transformed for centuries. Today, over 6000 large dams store ~13 % of the mean annual runoff in Europe, and ~50 % of the total European human population lives on former floodplains (De Bono et al. 2004). As a consequence, 50 % of the original wetlands and 95 % of riverine floodplains have been lost (Tockner et al. 2009).

Three major foci of floodplain invertebrate research in Europe have been the Ebro River in Spain, the Rhône River in France, and the Tagliamento River in Italy. Mean annual precipitation in the Ebro catchment is ~600 mm, with high monthly and annual variability (Sabater et al. 2009). Groundwater inputs smooth its flow regime (Fig. 13.2c), producing the lowest inter-annual variation in flow among Iberian rivers (Sabater et al. 2009). Regulation in the 1960s substantially changed its discharge pattern by altering flow timing, particularly the flood peaks (Cabezas et al. 2009). Nevertheless, flooding still occurs in winter (beginning of the rainy season) and early summer (snow melt in the Pyrenees) (Sabater et al. 2009). The combination of disturbance pulses and natural river succession in the Ebro river-floodplain has historically created numerous wetlands that have varying degrees of hydrological connection (e.g., secondary channels, backwaters, oxbow lakes, and temporary pools, Fig. 13.5). However, the lack of intense erosive floods in recent decades has reduced the probability of creation of new wetlands. Consequently, >60 % of the floodplain area is used for agriculture whereas wetlands cover only 3.6 % of it (Cabezas et al. 2008).

Much of the invertebrate research on the Ebro floodplain has focused on its middle sector, where the Ebro meanders at 230 m^3/s across a floodplain averaging 5 km in width (Ollero 2007). Riparian woodlands in the Middle Ebro are

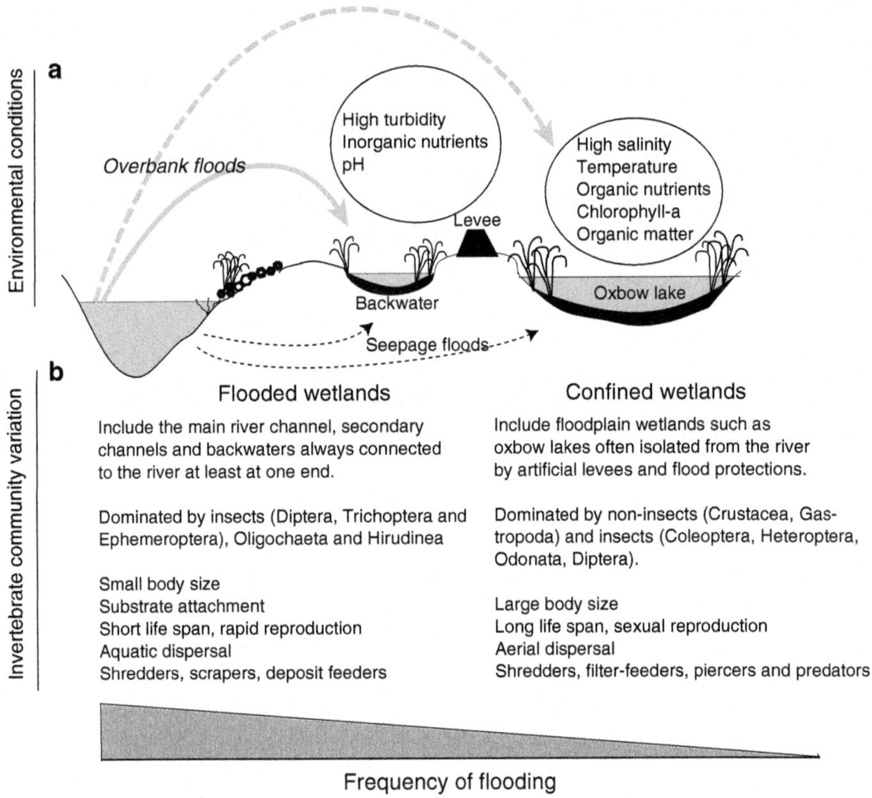

Fig. 13.5 Floodplain invertebrate community variation and assembly in the Ebro river-floodplain. (**a**) Wetlands are arranged along a gradient of hydrological connectivity. Flooded wetlands are characterized by a high frequency of flood disturbance that results in high turbidity and concentration of inorganic nutrients. Confined wetlands are usually isolated and are mainly affected by seepage from the main river channel, thus salinity, temperature and the concentration of organic matter and chlorophyll-a are high. (**b**) Macroinvertebrate assemblages respond to differences in flood frequency and water chemistry. Disturbance frequency in the river channel selects for those traits that allow organisms to resist flooding (small body size, attachment to substrate) and to recuperate faster after disturbance (short life span, rapid reproduction). Hydrological stability in confined wetlands enhanced traits that allow organisms to interact and compete with other organisms (large body size, sexual reproduction) and to efficiently use habitat and resources (diverse feeding strategies). Extracted from Gallardo et al. (2009a, b)

dominated by black poplar (*Populus nigra*), white willow (*Salix alba*), and several species of saltcedar (Tamaricaceae spp.), and are limited to a narrow band running parallel to the main channel (González et al. 2010). Aquatic vegetation is dominated by reeds (*Phragmites australis*) and cattails (*Typha latifolia*), with submerged vegetation being restricted to the main river channel (Gallardo et al. 2008).

The Rhône River is one of the major rivers in Europe, rising in Switzerland and running through southeastern France. Along its 813 km length, the Rhône is dynamic with a great diversity of floodplain habitats, despite being affected by dams, water extraction, and lateral barriers to flow (Olivier et al. 2009). In its French section, the river has an annual average discharge of 430 m^3/s through braided channels. Its hydrological regime is nivo-glacial, with annual floods normally occurring in spring from snow melt and low flows from late autumn to winter (Olivier et al. 2009). Hydroelectric development in the Rhône has severely changed the floodplain riparian vegetation, causing the regression of softwoods (e.g., *Salix, Alnus, Populus*) and progression of hardwoods (e.g., *Fraxinus, Acer, Quercus*) (Olivier et al. 2009).

The Tagliamento River in Italy rises in the Southern Alps and flows 172 km into the Adriatic Sea. Because of its relatively natural hydrology and morphology, the Tagliamento has been promoted as an especially useful reference system (Ward et al. 1999a). The catchment is divided into upper, middle, and lower sub-basins, with average slopes of 10 % in the upper section, 1 % in the middle section, and only 0.1 % in meandering lower section (Ward et al. 1999a). Flooding develops during the spring from snowmelt and in autumn from seasonal rains. Hydrology is complex because water levels are influenced not only by runoff patterns, but also by interactions with groundwater, with aquifer recharge and discharge patterns changing along the river. Because of the dynamic nature of the system, the active floodplain is largely bare gravel, apart from some vegetated islands (*Populus*, *Salix*).

Floodplain conditions differ among the sub-basins of the Tagliamento, with floodplain width and extent of channel braiding being greatest in the middle section and least in the upper and especially the lower-most sections (Ward et al. 1999a). Therefore, investigations of floodplain invertebrates have been stratified to address this longitudinal gradient (e.g., Arscott et al. 2005).

Floodplain Invertebrate Community Variation and Assembly

Hydrological connectivity has been identified as one of the most important drivers of aquatic communities on floodplains of the Ebro, Rhône, and Tagliamento rivers (Arscott et al. 2005; Paillex et al. 2007; Gallardo et al. 2009a, b, 2014). Hydrological connectivity for floodplains operates on the four dimensions of all fluvial systems: longitudinal (upstream-downstream), lateral (river-floodplain), vertical (groundwater-surface water), and temporal (Ward 1989). Consequently, it integrates a wide range of environmental processes such as water scouring and turbulence, reorganization of habitats, and transport of sediments, salts, nutrients, seeds and organisms, factors that all can directly or indirectly affect the presence of aquatic organisms (Amoros and Bornette 2002). An increasing body of literature deals with the critical influence of hydrological connectivity in structuring aquatic assemblages including macrophytes (e.g., Amoros and Bornette 1999), fish (e.g., Roach et al. 2009), macroinvertebrates (e.g., Arscott et al. 2005; Gallardo et al. 2014), and plankton (e.g., Frisch et al. 2005).

In recent years, changes in richness, composition, and traits of invertebrate assemblages in the Ebro River (Appendix) have been widely investigated to identify the main threats to biodiversity conservation (e.g., river regulation, land-use change, eutrophication, heavy metal pollution, invasive species), as well as to plan and monitor restoration activities (Cid 2011; Sabater et al. 2011; Gallardo et al. 2012). Lateral river-floodplain connectivity has been predominantly studied in the Middle Ebro (Gallardo et al. 2009a, b), whereas research in the lower reaches of the Ebro River has focused on longitudinal patterns (Muñoz and Prat 1994; Cid et al. 2010).

In the Middle Ebro, flooded (e.g., secondary channels and backwaters) and confined (e.g., oxbow lakes) wetlands significantly differ in invertebrate taxon composition (Fig. 13.5). Invertebrate assemblages in flooded wetlands are dominated by insect taxa (Gallardo et al. 2008). This reflects the functional characteristics of insects, as they are often considered generalist species capable of resisting unfavorable conditions, recuperating quickly after disturbances and aerially colonizing new habitats (Townsend et al. 1997). Flooding influences feeding behavior, favoring shredders, scrapers, and piercers that benefit from organic detritus and attached algae (Heino 2008). In contrast, non-insect taxa (including crustaceans, bivalves, and gastropods; Appendix) dominate confined wetlands of the Middle Ebro (Fig. 13.5; Gallardo et al. 2008), because these groups generally specialize in habitat and resource exploitation in stable habitats (Townsend et al. 1997). Overall, the fauna is split among taxa that rely on desiccation (Group 1, 9 families), lotic taxa from the channel (Group 2, 17 taxa), and especially permanent-water lentic taxa (Group 3, 28 families) (Appendix; terrestrials, Group 4, were not monitored from the Middle Ebro).

The large concentration of fine particulate organic matter characteristic of the lower courses of large rivers support filter- and collector-gatherers, the dominant feeding groups in the Lower Ebro, representing 48–81 % of the invertebrate community (Muñoz and Prat 1994). Representative species of these groups include insects of the families Chironomidae, Caenidae, and Hydropsychidae. Higher densities of macroinvertebrates were associated with higher water velocities, partly explained by the high abundance of submerged plants (*Cladophora* sp.) providing food and refuge for invertebrates, particularly scrapers and filter-feeders (Muñoz and Prat 1994; Cid 2011).

The lateral hydrological connectivity also has been acknowledged as a major driver of the Rhône floodplain's taxonomic biodiversity and functional structure. A series of papers examined distributions of various insect orders—Ephemeroptera and Plecoptera (Usseglio-Polatera and Tachet 1994), Coleoptera (Richoux 1994), and Trichoptera (Tachet et al. 1994)—across the upper Rhône River and floodplain in France. Plecoptera occurred only in the channel, and not on the floodplain. Distinct groups of Ephemeroptera and Trichoptera were associated with channel and floodplain habitats, with only a few families spanning both (Baetidae, Leptophlebiidae, Leptoceridae). Coleoptera also had distinct lotic channel and lentic floodplain faunas, plus a third semi-aquatic floodplain group. Paillex et al. (2007), also working in the upper Rhône floodplain, examined overall invertebrate community patterns across the floodplain. Riverine aquatic organisms (e.g., Ephemeroptera, Gastropoda) were found in the most connected channels, whereas lentic species (e.g., Coleoptera, Hemiptera) predominated in the most disconnected wetlands. A number of taxa (e.g., Oligochaeta, Chironomidae, Hydrachnida) were widespread along the hydrological gradient.

Along the Tagliamento River, Arscott et al. (2005) explored variation in floodplain and channel invertebrate communities among headwater, middle, and lower sections of the catchment. They found that invertebrate richness and densities tended to increase along the corridor, with differences primarily being driven by non-insect taxa (e.g., crustaceans, oligochaetes) associated with lentic habitats. Plecoptera and non-chironomid dipterans showed a reverse pattern, with downstream declines in richness and abundance. However, Karaus et al. (2013), focusing only on EPT (Ephemeroptera, Plecoptera, Trichoptera) taxa, found that cumulative numbers of EPT species increased continuously down the Tagliamento, especially in floodplain habitats (312 % increase).

Variation in invertebrate communities within individual floodplain sections of the Tagliamento River has also been addressed. The greatest overall density of invertebrates (193,567 individuals/m^2) occurred in a backwater habitat in the middle sub-basin of the river (Arscott et al. 2005). As in the Ebro floodplain, communities differed between lentic (backwaters, pools) and lotic (main and side channels) habitats. In headwater areas, Arscott et al. (2003) found trichopteran communities differed between forested and non-forested habitat. In lower sections of the floodplain where lentic taxa abounded, their densities tended to be higher in pools or back waters connected or close to lotic habitats (Arscott et al. 2005). Unexpectedly, more EPT taxa appear specialized for floodplain habitats of the Tagliamento than for the main channel (Karaus et al. 2013). Looking at terrestrial invertebrates (e.g., arachnids, ants, collembolans), Steward et al. (2011) found that floodplains of the Tagliamento supported floodplain-specific assemblages whose compositions tended to differ between vegetated and bare-gravel habitats.

Overall, Arscott et al. (2005) were able to categorize aquatic invertebrate communities along the Tagliamento based on (1) landscape position (headwater, middle, lower) and (2) habitat type (lentic, lotic). Distinct assemblages developed in mid- and lower reach lentic habitat, lower-reach lotic habitat, mid-reach lotic habitat, and headwater habitats (Table 13.2). In terms of overall diversity, both within-habitat (α) and among-habitat (β) diversities were typically high across all reaches, suggesting that longitudinal and lateral variation in diversity contributed equally to create heterogenous communities along the Tagliamento.

Table 13.2 Invertebrate taxa indicative of headwater, mid-reach lotic, lower-reach lotic, and mid- and lower-reach lentic habitat along the floodplain corridor of the Tagliamento River, Italy

Headwater streams and floodplains	Mid-reach lotic elements	Lower-reach lotic elements	Mid- and lower-reach lentic elements
Athericidae	Baetidae	Gammaridae	Cladocera
Nemouridae	Chloroperlidae	Gyrinidae	Corixidae
Perlodidae	Heptageniidae		Gastropoda
Simuliidae (*Prosimulium*)	Simuliidae (*Simulium*)		Nematoda
			Oligochaeta
			Ostracoda

Trophic Relationships

In the Ebro, invertebrate assemblage variation among floodplain positions and sub-habitats (Fig. 13.3) suggests some trophic control over assemblage compositions, with lentic habitats being dominated by gathering organisms and those that feed on submersed plant substrate (Muñoz and Prat 1994; Gallardo et al. 2008; Cid 2011). Higher plants growing on floodplains can provide copious supplies of organic matter as food and habitat for floodplain invertebrates. At the Rhône River, predators, scrapers, and filter-collectors were predominant in the more disconnected floodplain sites. The proportion of predators decreased along the connectivity gradient, a pattern that suggests a decrease in top-down control of invertebrate assemblages with increasing disturbance by floods (Paillex et al. 2007).

On the headwater floodplain of the Tagliamento, Arscott et al. (2003) found that Trichoptera abundance and species richness was greater in forested than non-forested portions. This pattern was driven by several species of Limnephilidae shredders being positively correlated with benthic organic matter (tree leaves and wood). However, further down the Tagliamento at a mid-reach floodplain, Langhans and Tockner (2006) found very limited evidence of an invertebrate role in leaf litter decomposition, with it only being important in permanently flooded areas.

Paetzold and colleagues studied predation by terrestrial arthropod predators (spiders, carabid ground beetles, staphylinid rove beetles, ants) on aquatic floodplain insects (see Paetzold and Tockner 2005; Paetzold et al. 2005, 2006). Large numbers of these predators inhabit the open gravel areas of the Tagliamento floodplain, despite the paucity of a terrestrial food base. However, aquatic insects often enter the floodplain via drift or to emerge. Using descriptive studies, stable isotope analyses, and manipulations of aquatic insect subsidies, Paetzold and Tockner (2005) showed that 45 % of emerging aquatic insects were consumed by these predators, and Paetzold et al. (2005) demonstrated that these aquatic subsidies accounted for up to 80 % of the diets of some predators. Manipulations of aquatic insect subsidies (Paetzold et al. 2006) indicated that terrestrial floodplain predators actively aggregate to this food source.

Southeastern Australian Floodplains (Murray-Darling River, Murrumbidgee River)

Most of the Australian continent is flat (<500 m above sea level) and semi-arid, with two-thirds of the mainland receiving less than 500 mm of rainfall annually (Boulton et al. 2014). The major river system of south-eastern Australia, the Murray-Darling, flows south and west from the inland and off the Great Dividing Range to the east. The Darling River, flowing down from the north, drains about two-thirds of the Murray-Darling Basin but except in flood years, contributes only about 10 % of the Basin's discharge and much of the flow remains in its incised channel. In contrast, the Murray River above its confluence with the Darling contributes some 90 % of the discharge and, along its lower reaches, is bordered by a broad floodplain supporting numerous wetlands, anabranches, and ox-bow lakes (colloquially termed billabongs) (Sheldon and Walker 1998). Extensive floodplains also characterize sections of the upper eastern reaches of the Murray, such as the Murrumbidgee River (Chessman and Hardwick 2014).

Along the lower Murray, the Chowilla floodplain has been a major focus for invertebrate research. In 1987, this floodplain was listed as a Wetland of International Significance under the Ramsar Convention because its water-dependent ecosystem is one of the last remnants of near-natural floodplain of the Murray that has not been irrigated or cleared of native vegetation (MDBA 2012). Significantly, it supports the largest remaining area of naturally occurring river red gum (*Eucalyptus camaldulensis*) forest in the lower Murray, and is crucial habitat for many native waterbirds, fish, and other biota that have disappeared from much of their former range along the river. Consequently, Chowilla is perhaps the best representative of a near-natural floodplain ecosystem of a large lowland river in southeastern Australia.

The Chowilla floodplain supports a mosaic of anabranch creeks, wetlands, temporary lagoons, and billabongs, variably connected at different flooding regimes across a floodplain whose vegetation composition is largely dictated by inundation frequency and duration (Overton and Doody 2008). The extent and depth of floodplain inundation is governed by the magnitude of floods down the lower Murray; flows of 460–700 m^3/s are needed to inundate low-lying aquatic habitats and those near the channel, at 925 m^3/s about half the Chowilla floodplain is inundated, and flows exceeding 1150 m^3/s will flood virtually the whole area (MDBC 2006). Upstream storages and a series of weirs have significantly modified the Murray's natural flow regime, reducing overall flow volume, prolonging periods of low flows, and reducing the magnitude of medium-sized floods (Walker 2006). Flows >900 m^3/s that occurred once every 2 years on average now occur only once every 8 years, the average period between spring-summer floods has more than tripled (from 2.4 to 9.3 years), and flood volumes have been cut by more than half (CSIRO 2008). These changes have led to engineered efforts since 2004 to "environmentally water" the Chowilla floodplain with appropriate volumes and timing to sustain river red gum and black box (*Eucalyptus largiflorens*) forests (MDBA 2012) as well as their associated floodplain biota.

Mean annual rainfall on the Chowilla floodplain is 260 mm, with most falling in late winter and spring. However, evaporation exceeds rainfall every month (Fig. 13.2d), assisted by low relative humidity, high temperatures, and strong winds. Thus, the floodplain's water budget relies heavily on inputs from occasional high river flows to sustain wetlands and anabranches (MDBA 2012). Groundwater inputs (Fig. 13.2d) are also important, although the regional groundwater is saline, contributing some 130 tonnes of salt per day to the Chowilla (Yan et al. 2005). Historically, evaporation from anabranch creeks matched inputs of saline groundwater, but the stored salts were flushed into the main river during floods. Recent river regulation has reduced flooding and flushing, resulting in daily storage of approximately 75 tonnes of salt per day (Yan et al. 2005). Now, when floods do occur, large salt loads are delivered to the main channel of the river.

In the upper Murray, the Murrumbidgee floodplain has been another focus of invertebrate research (Chessman and Hardwick 2014). The Murrumbidgee River's extensive floodplain has numerous wetlands occupying abandoned channels. Water diversions for agriculture and flow regulation have reduced the natural extent of Murrumbidgee floodplain inundation by 40 %, but flooding from the river remains a frequent event. Here, local rainfall and runoff contributes significantly to overall water budgets. Much of the floodplain has been cleared for agriculture, but portions are still covered by river red gum forest.

Floodplain Invertebrate Community Variation and Assembly

Despite the anthropogenic alterations to the hydrological regime of the Chowilla floodplain, its invertebrate community composition (Appendix) is reasonably representative of near-natural Southeastern Australian floodplains. A survey of aquatic invertebrates that included temporary and permanent billabongs, backwaters, anabranch channels, and two sites on the main river channel, collected 95 taxa in at least 43 families (Boulton and Lloyd 1991). Subsequent sampling at Chowilla, including collections from other wetlands and reflooding dried floodplain sediments (Boulton and Lloyd 1992; Goonan et al. 1992; Sheldon and Walker 1998; Skinner et al. 2001), has yielded a total of at least 65 families (Appendix).

Much of this diversity arises from the rich array of aquatic macrohabitats on the floodplain (Fig. 13.1). Comparisons of invertebrate diversity among the temporary and permanent billabongs, backwaters, and anabranches reveal substantial spatial beta-diversity with marked differences in invertebrate community composition among macrohabitats (Boulton and Lloyd 1991; Sheldon and Walker 1998). For example, temporary billabongs have a very different fauna from nearby permanent ones, and lentic macrohabitats support families of some taxa such as gastropods that are virtually absent from lotic macrohabitats dominated by amphipods and palaemonid and atyid shrimps. Slow-flowing anabranches consistently support the fewest taxa (typically only shrimps, oligochaetes, corixids, and amphipods), perhaps reflecting the low physical complexity of the largely unvegetated littoral zone. In the other macrohabitats, the wide array of microhabitats (e.g., emergent vegetation, submerged vegetation, leaf litter accumulations, woody debris) often harbor different assemblages of aquatic fauna, collectively contributing to high within-macrohabitat invertebrate diversity (Boulton and Lloyd 1991; Sheldon and Walker 1998).

Unfortunately, there has been no repeated sampling of aquatic habitats on the Chowilla floodplain during and after a flood to assess the extent of temporal beta-diversity. However, studies of temporal changes in diversity and invertebrate composition of permanent and temporary floodplain wetlands elsewhere along the Murray after flooding reveal marked differences over time, especially in the first 6–8 weeks after flooding (Hillman and Quinn 2002). It is likely that similar trends occur on the Chowilla floodplain, especially where differences in hydrological connectivity occur among wetlands (Sheldon et al. 2002), and that temporal beta-diversity is also high. Floodplain invertebrate diversity is probably enhanced by the spatial variability among aquatic habitats at several scales interacting, perhaps synergistically, with temporal changes in water permanence, flow, and level (and hence connectivity).

There is also a rich reserve of biodiversity within the floodplain sediments. When dry sediments from different parts of the Chowilla floodplain are inundated experimentally, a diverse soup of protozoans, nematodes, rotifers, microcrustaceans (e.g., copepods, cladocerans, and ostracods), notostracans, and anostracans soon develops (Boulton and Lloyd 1991; Skinner et al. 2001). Diversity is typically highest from sediments inundated annually and declines with increasing time since last flood, a pattern also observed in floodplain sediments of the Darling River (Jenkins and Boulton 2003, 2007) and increasing salinity (Skinner et al. 2001).

As in the case-study for floodplains of Southeastern USA, invertebrates of the Chowilla floodplain can be grouped according to four different sources: (1) desiccation-resistant aquatic organisms; (2) riverine aquatic organisms; (3) lentic water body aquatic organism; and (4) flood-tolerant terrestrial organisms (Table 13.1). Most of the taxa collected from the Chowilla floodplain listed in Appendix are from Groups 1 (10 families) and 3 (45 families) although the relative paucity of representation by Group 4 taxa (4 families) probably reflects the limited sampling for these organisms in studies focusing on aquatic invertebrates. Few taxa apparently entered from the adjacent channel of the Murray (i.e., Group 2; 6 families) because in this part of the lower Murray, invertebrate diversity is reduced by poor water quality, modified flow regimes, limited physical habitat, and alien fish such as carp. A survey of the invertebrates in the main channel of the lower Murray near Chowilla as part of a Basin-wide river health audit (Davies et al. 2008) corroborates this, reporting extremely low proportions of expected invertebrate families.

Some families in the group of taxa that spend low-water periods in permanent aquatic habitats on the floodplain (Group 3) disperse actively among the remnant aquatic habitats either by flying (e.g., notonectid and corixid bugs, dytiscid and hydrophilid beetle adults) or swimming (e.g., palaemonid and atyid shrimps). Others tend to be less mobile (e.g., hydrobiid and ancylid gastropods) and likely are dispersed passively by moving floodwater. As water recedes from the inundated floodplain, wetlands and some backwaters become isolated while flow slows and sometimes stops in anabranches. Movement by invertebrates among these fragmented aquatic habitats is restricted to aerial dispersal, so the resident invertebrate assembly in each isolated wetland is a function of which taxa are trapped when the hydrological connection is lost, which taxa fly in or out, and the outcomes of differential survival to deteriorating water quality and intensifying biotic interactions (e.g., competition and predation) as the wetlands dwindle in size (Boulton et al. 2006).

Further up river on the Murrumbidgee River floodplain (Chessman and Hardwick 2014), the aquatic invertebrate fauna (Appendix) is also dominated by permanent-water lentic taxa (Group 3, 36 families), followed by desiccation-resistant taxa (Group 1, 11 families), and finally by channel-associated lotic taxa (Group 2, 8 families). As would be expected, permanent water wetlands had different and richer assemblages than seasonally flooded habitats.

Trophic Relationships

Representation of the different trophic groups of invertebrates varies spatially among the aquatic macrohabitats on the Chowilla floodplain as a function of water permanence, flow, and temporal changes in organic matter bioavailability. As in many other temporary waters, mobile predators such as dytiscids, notonectids, and odonates are abundant in drying billabongs (Boulton and Lloyd 1991). In the final stages of drying, terrestrial invertebrates such as ants and spiders also likely prey heavily on stranded aquatic invertebrates (Steward et al. 2012), potentially representing an important pathway for organic carbon transfer on the Chowilla floodplain. Conversely, predatory aquatic invertebrates are uncommon in macrohabitats that are hydrologically connected to the main channel (Boulton and Lloyd 1991) where fish may dominate this trophic role.

Opportunistic omnivores such as shrimps and amphipods are extremely common in anabranches and the main channel, and probably play some role in processing the abundant eucalypt litter produced by the river red gum forest. However, this organic material is notoriously refractory, and takes a long time to break down (Watkins et al. 2010). Instead, it is likely that seston (suspended organic matter, <45 μm) comprising fine detritus and phytoplankton is a more important carbon source, especially during and soon after floodplain inundation. Using stable isotopes to compare the use of basal organic carbon sources by different consumers in a lowland river of the Murray upstream of Chowilla, Hladyz et al. (2012) showed that seston was the primary source of carbon during high flows whereas terrestrial vegetation and, for some consumers, filamentous algae increased in importance during low flows. This temporal change in organic matter bioavailability appears to reflect the suitability of conditions for algal growth (Hladyz et al. 2012) and, along with water permanence and movement, likely governs the trophic relationships of aquatic invertebrates on the Chowilla floodplain. Many of these invertebrates have flexible diets (Boulton et al. 2014), enabling them to opportunistically exploit different carbon sources.

Lentic macrohabitats on the Chowilla floodplain support high densities of invertebrate grazers such as aquatic snails that are virtually absent in lotic habitats and the main channel (Boulton and Lloyd 1991). Numbers of grazers are highest in permanent billabongs where low turbidity contributes to a greater photic zone and higher primary production, especially of algal-dominated biofilms on leaf litter, wood, and other benthic surfaces (Burns and Walker 2000). Emergent aquatic plants around the fringes of these lentic habitats not only provide shelter for aquatic invertebrates but may also be a significant source of carbon, especially after flood waters have receded and the floodplain is drying (Baldwin et al. 2013). Again, there is likely an important temporal component to the relative contributions to floodplain foodwebs by autochthonous carbon from biofilms and emergent aquatic plants versus allochthonous inputs from leaf litter of eucalypts and other terrestrial plants (Hladyz et al. 2012), potentially tied in with the "boom-bust cycle" of flooding and drying in semi-arid river ecosystems (Baldwin et al. 2013).

Synthetic Analysis of Invertebrate Ecological Controls in Temperate-Zone Floodplains

Across the different temperate-zone floodplains of the USA, Europe, and Australia featured in this chapter, there are numerous parallels in how resident invertebrate communities are structured and ecologically controlled.

Hydrologic and Climatic Controls

As would be expected, water budgets in temperate-zone floodplains are overwhelmingly dominated by inputs of water from the river (Fig. 13.2). However, inputs of groundwater, subsurface seepage of river water, and direct rainfall all appear to significantly influence the duration of inundation and water chemistry of low-lying areas of floodplains (Figs. 13.1 and 13.5), and these areas are also probably the most crucial sub-habitats for aquatic floodplain invertebrates. As such, more than just over-bank flow needs to be considered when evaluating hydrologic controls of floodplain invertebrates.

For the lowest-lying seasonally flooded areas such as sloughs and temporary ponds, longer-duration flooding leads to greater taxon richness of invertebrates (Fig. 13.4), up to a point (see below). Invertebrate production also tends to increase from upper to lower watershed floodplains, which is also likely related to lower-watershed habitats retaining water longer and providing more time for aquatic invertebrates to grow. Climate influences floodplain hydrology by regulating size and timing of floods, the primary control on floodplain invertebrate assemblage compositions. Additionally, in arid and semi-arid areas, evapotranspiration becomes very important (Fig. 13.2c, d), and here, extended dry phases can stress invertebrates aestivating in exposed soils.

Invertebrate Assemblage Compositions

Over 130 different families of invertebrates were recorded from the floodplains highlighted in this chapter (Appendix). Because different sampling approaches were used in each, we chose not to contrast assemblages quantitatively. However, it is clear that assemblages varied considerably among and within regions. Only seven families (Table 13.3) were ubiquitous across the floodplains we inventoried (Appendix), and most of these are widespread in wetlands in general (Batzer and Ruhí 2013). Despite the associations with rivers, invertebrate communities in floodplains were dominated by lentic aquatic organisms (all seven ubiquitous taxa are lentic). Typical lotic invertebrates (most mayflies and some caddisflies) were only common in a few of the floodplains studied (e.g., middle Ebro, upper Tagliamento River, Oconee; Appendix). The lentic invertebrates represented an obligate wetland fauna that persist on floodplains year-round, while the lotic invertebrates represented an ecotonal riverine fauna that only enter floodplains during flood events.

Invertebrates of floodplains employ a plethora of ecological strategies to exploit these highly variable habitats. Desiccation resistance is widely used, especially by flightless invertebrates (non-insects). However, many aquatic invertebrates of floodplains are active dispersers (flying, swimming) that readily migrate into newly inundated floodplain habitats, either from the river or from permanent water refugia on the floodplain. Early in a flood cycle, invertebrates that emerge from the soils may dominate, to be subsequently replaced by active dispersers. If a habitat is disturbed, passive dispersers may be slow to re-establish, and thus these organisms may be especially useful indicators.

We found ample evidence that a terrestrial component, generally overlooked by researchers, comprises much of the invertebrate fauna of floodplains. Contributions of these organisms to biodiversity and ecosystem function are both potentially important. Like lentic aquatic forms, these terrestrial invertebrates represented an obligate wetland fauna (i.e., were full-time residents, tolerant of flooding) rather an ecotonal influx of invertebrates from adjacent uplands.

Table 13.3 Families of aquatic macroinvertebrates collected at every temperate-zone floodplain habitat inventoried in Appendix

	Ubiquitous families
Odonata	Coenagrionidae (15)
Hemiptera	Corixidae (3)
	Notonectidae (9)
Coleoptera	Dytiscidae (2)
Diptera	Ceratopogonidae (7)
	Chironomidae (1)
	Culicidae (8)

Numbers in *parentheses* indicate the ranking of each family among the 40 most common macroinvertebrate families in wetlands overall, according to Batzer and Ruhi (2013)

A lentic-lotic dichotomy in aquatic community structure was evident spatially among and across most floodplain habitats. In general, lotic taxa are restricted to habitats near and/or strongly connected to the adjacent river channel. However, if lateral connections of sub-habitats to channels are weak or the sub-habitats are only flooded seasonally, invertebrate assemblages tended to be dominated by lentic organisms. Where lateral channel-floodplain connection is disrupted artificially by human activities, such as construction of levees (Fig. 13.5) or regulation of flood pulses, lentic forms may become dominant.

Besides lateral variation across floodplains, a lentic-lotic dichotomy was evident longitudinally down some floodplain corridors (i.e., from the headwaters downstream). Small headwater floodplains tended to be dominated by lotic organisms entering from the channel, while larger lower-river floodplains tended to be dominated by lentic invertebrates. For invertebrates, it appears that headwater floodplains function mostly as extensions of the river. In headwater areas, the development of a lentic invertebrate fauna may be hindered by the short duration of floods and a lack of extensive permanent-water ponding on those floodplains to provide aquatic refugia between floods.

In contrast, in large lower-river floodplains, habitat quality for lentic forms is enhanced by long duration and predictable flooding and the presence of extensive permanent-water lakes and ponds. Lentic invertebrates are well adapted to tolerate stagnant wetland conditions that develop on floodplains during all but the largest floods. However, the paucity of lotic invertebrates in larger floodplains is somewhat surprising given that large and long-lasting flooding should give lotic invertebrates ample opportunity to exploit the habitats. Our collective research suggests multiple possible mechanisms for this rarity of lotic organisms on large floodplains:

1. The presence of lentic wetland conditions, especially low oxygen concentrations, may make floodplains inhospitable places for many lotic invertebrates to live.
2. The occurrence on large floodplains of a robust lentic invertebrate and fish fauna, many of which are predators, may inhibit incursion of lotic invertebrates from the channel onto the floodplain.
3. The channels of large rivers may not support many typical lotic organisms, reducing the pool of lotic species to colonize lower-river floodplains.
4. As temperate-zone rivers get larger, the influence of human activities may increase cumulatively, disrupting connections between the channels and the floodplains.

If natural, the limited influx of lotic invertebrates onto floodplains seems inconsistent with the Flood Pulse Concept (Junk et al. 1989), which has strong channel-floodplain interactions as a foundational premise. The FPC was developed primarily for tropical areas and for very large rivers, most many times larger than the temperate-zone river systems covered in this chapter. Differences in climate, flooding patterns, and spatial scale of temperate- versus tropical-zone rivers may mean that the FPC must be modified to consider the unique conditions in each area. One such modification for temperate-zone invertebrates, at least in lower-river floodplains, might be to emphasize the link between permanent-water lentic habitats on the floodplain (oxbow lakes, billabongs, deep sloughs) and the vast expanse of seasonally inundated floodplain habitat rather than the link between the river channel and the floodplain. This internal floodplain link would still be controlled largely by flood pulses connecting permanent and temporary lentic habitat, and is equally prone to disruption by many human activities (e.g., agriculture, levee banks, urbanization) on lower-river floodplains.

Trophic Relationships

Floodplain invertebrates appear to be mostly opportunistic generalist feeders, flexibly exploiting the range of foods that become available in these spatially and temporally diverse habitats. However, a few generalizations can be made about feeding functions of floodplain invertebrates.

1. Collector-gatherers (e.g., midges) and predators (e.g., water bugs and beetles) are particularly well represented across most floodplain habitats, especially in lower-river floodplains.
2. Shredders of leaf litter and wood are generally rare, except where Limnephilidae caddisflies are abundant. Additionally, evidence is scant that aquatic invertebrate shredders are important to the breakdown of leaf litter or wood on floodplains.
3. Filterers (e.g., hydropsychid caddisflies) can sometimes become abundant where high water flows deliver suspended seston, or in soft-bottomed substrates conducive to burrowing by bivalves.
4. Grazers (e.g., snails) can become abundant in heavily vegetated, low-flow water bodies on the floodplain where filamentous algae accumulate.

Distributions of primary consumers tend to track organic matter availability, with gathering and filtering collectors concentrating in areas with copious small particles, shredders (if present) in areas with ample leaf litter, and grazers in areas where algae abound. As such, plant communities on floodplains exert considerable control on floodplain invertebrates via food supplies, and additionally provide them with structural habitat.

Predation emerged as a pervasive control in most floodplain habitats. Floodplain fishes occurring in permanent water bodies on the floodplain seem especially influential. They strongly structure the invertebrate communities in the long-hydroperiod lakes and ponds that serve as primary habitats, but also influence invertebrates across the floodplain during high flows when they migrate out of the lakes onto the main floodplain to feed. Predation by invertebrates, both from aquatic (see Table 13.1, Appendix) and terrestrial (e.g., spiders) taxa, is an equally pervasive influence. In some locations, amphibian predation might be important (Galatowitsch and Batzer 2011).

Conclusions

We conclude that in temperate-zone floodplains, multiple ecological gradients, both lateral and longitudinal, largely control the structure and function of resident invertebrate communities (Fig. 13.6). From a series of case studies of temperate-zone floodplains in the USA, Europe, and Australia, we hypothesize the following:

1. Increased connectivity of floodplain habitats with river channels increases access for lotic invertebrates from the channel, although other natural and anthropogenic factors may restrict their incursion across the floodplain. River-floodplain connectivity also promotes influx of nutrients from the river and reduces stagnation in lentic floodplain habitats, both being additional important controls on aquatic invertebrates.
2. Variation in duration of flooding (hydroperiod) controls lentic invertebrate communities, with short hydroperiods limiting taxa lacking resistance to desiccation, and long hydroperiods limiting taxa lacking resistance to fish predation.
3. Variation in plant communities across floodplains controls invertebrate taxa reliant on leaf litter (shredders) or algae (grazers) as food, as well as invertebrate taxa reliant on plants as habitat.

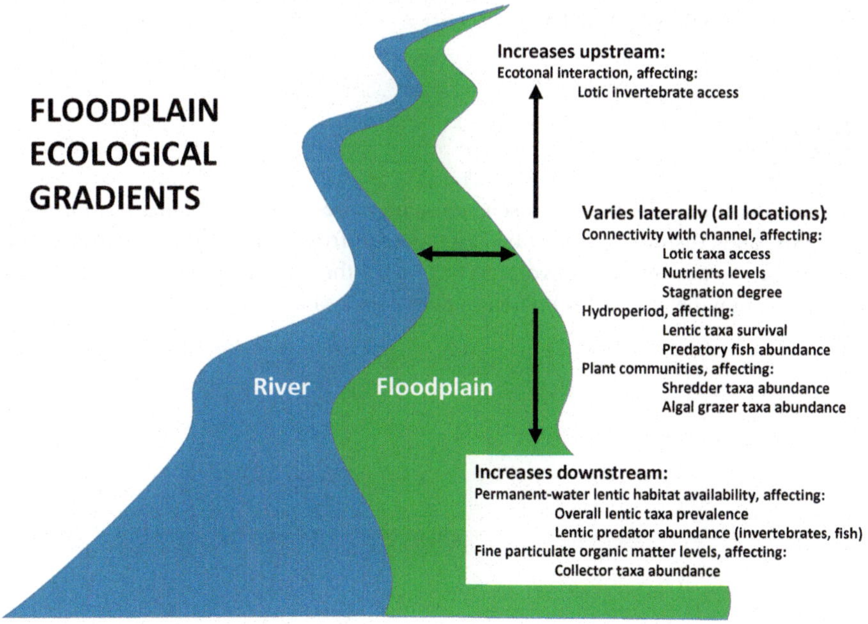

Fig. 13.6 Focal lateral and horizontal gradients that influence invertebrate community structure and function in temperate-zone floodplains

4. In upstream floodplains, lotic invertebrates tend to dominate because of the close proximity of floodplains to river channels (and perhaps because a lentic invertebrate and fish community is poorly developed).
5. In downstream habitats, lentic invertebrates tend to dominate because more permanent water lentic habitat is available. These lentic invertebrates, and lentic fishes, may limit incursions of lotic taxa from the channel.
6. In downstream habitats, collector taxa become dominant due to the copious supplies of small particles of organic matter (seston) suspended in lower-river flood waters.

Human activities can profoundly influence these ecological controls, and negatively affect floodplain invertebrates. To sustain healthy invertebrate communities, natural river flows that generate flood pulses should be maintained, and barriers to lateral floods (dikes, levees) should be minimized or, if possible, removed. Conservation and management efforts should ensure maintenance of the high heterogeneity in hydrology, sediment regimes, and plant communities characteristic of natural floodplains. Because invertebrates are crucial foods for most floodplain fishes, birds, and other wildlife, efforts to maintain healthy invertebrate communities will provide broader conservation benefits. Invertebrates also mediate many ecosystem processes such as nutrient regeneration and organic matter decomposition, and are an integral part of all functioning floodplains, their river channels, and their diverse aquatic habitats in temperate and, as we shall see in the next chapter, tropical floodplain ecosystems.

Appendix

Invertebrate taxa collected from selected temperate-zone floodplains from Europe, the USA, and Australia. (Some taxa not recorded from a site may simply reflect that the organism was not effectively collected by the sampling program employed; other taxa may be absent for biogeographic reasons).

	Southwestern Europe	Southeastern USA	Southeastern USA	US Central Prairie	Eastern Australia	Southeastern Australia
	Ebro River[a]	Oconee River[b]	Altamaha River[c]	Platte River[d]	Murrumbidgee River[e]	Murray River[f]
NEMATODA		X	X	X		X
PORIFERA			X			X
CNIDARIA				X	X	X
TURBELLARIA				X	X	X
NEMERTEA					X	X
ROTIFERA	X			X		X
MOLLUSCA						
Gastropoda						
Ancylidae	X					X
Hydrobiidae						X
Lymnaeidae	X			X	X	X
Neritidae	X					
Physidae	X			X	X	
Planorbidae		X	X	X	X	X
Bivalvia						
Corbiculidae	X	X				X
Sphaeriidae		X	X	X		
Unionidae	X					
ANNELIDA						
Oligochaeta						
Lumbriculidae				X		X
Naididae	X				X	
Tubificidae	X	X	X		X	X

(continued)

(continued)

	Southwestern Europe	Southeastern USA	Southeastern USA	US Central Prairie	Eastern Australia	Southeastern Australia
	Ebro River[a]	Oconee River[b]	Altamaha River[c]	Platte River[d]	Murrumbidgee River[e]	Murray River[f]
Hirudinea						
Erpobdellidae	X			X		
Glossiphoniidae	X			X	X	X
CRUSTACEA						
Branchiopoda						
Bosminidae	X					
Chirocephalidae			X			
Chydoridae	X			X		
Cyclestheridae						X
Cyzicidae					X	
Daphniidae	X	X	X	X		
Lynceidae					X	
Notostraca					X	X
Thamnocephalidae						X
Branchiura						
Arguliidae	X		X			
Copepoda						
Calanoida	X	X	X			X
Cyclopoida	X	X	X			X
Harpacticoida		X				X
Ostracoda	X	X	X		X	X
Amphipoda						
Ceinidae						X
Crangonyctidae		X	X	X		

Dogielinotidae		X	X	X		
Gammaridae	X		X			
Isopoda						
Asellidae	X	X	X			
Corallanidae						X
Janiridae						X
Decapoda						
Atyidae	X				X	X
Cambaridae	X	X	X			
Palaemonidae			X		X	X
Parastacidae					X	X
ARACHNIDA						
Araneae		X	X			X
Hydrachnidia				X	X	X
Oribatida		X	X		X	X
Pseudoscorpionida			X			
MYRIAPODA						
Chilopoda		X	X			
Diplopoda			X			
INSECTA						
Collembola	X	X	X			X
Ephemeroptera						
Baetidae	X			X	X	X

(continued)

(continued)

	Southwestern Europe	Southeastern USA	Southeastern USA	US Central Prairie	Eastern Australia	Southeastern Australia
	Ebro River[a]	Oconee River[b]	Altamaha River[c]	Platte River[d]	Murrumbidgee River[e]	Murray River[f]
Caenidae	X	X			X	X
Ephemerellidae		X				
Ephemeridae	X	X				
Heptageniidae	X					
Leptophlebidae	X	X			X	
Metretopodidae		X				
Polymitarcyidae	X					
Siphlonuridae		X				
Odonata						
Aeshnidae		X	X		X	X
Coenagrionidae	X	X	X	X	X	X
Corduliidae	X				X	X
Gomphidae	X	X			X	X
Isostictidae					X	
Lestidae				X	X	X
Libellulidae	X			X	X	
Plecoptera						
Gripopterygidae						X
Nemouridae		X				
Hemiptera						
Belostomatidae				X	X	X
Corixidae	X	X	X	X	X	X
Gerridae	X	X	X		X	
Hebridae					X	

Hydrometridae					X	
Mesoveliidae	X				X	X
Naucoridae					X	X
Nepidae					X	
Notonectidae	X	X	X	X	X	X
Ochteridae					X	
Pleidae					X	
Saldidae					X	
Veliidae	X				X	X
Neuroptera						
Corydalidae		X				
Sialidae		X				
Coleoptera						
Cantharidae		X	X			
Chrysomelidae				X		
Curculionidae		X	X			X
Dytiscidae	X	X	X	X	X	X
Elateridae		X	X			
Elmidae	X	X			X	X
Gyrinidae	X	X	X		X	X
Haliplidae		X			X	
Heteroceridae					X	
Hydraenidae					X	
Hydrophilidae/Helophoridae	X		X		X	X

(continued)

(continued)

	Southwestern Europe	Southeastern USA	Southeastern USA	US Central Prairie	Eastern Australia	Southeastern Australia
	Ebro River[a]	Oconee River[b]	Altamaha River[c]	Platte River[d]	Murrumbidgee River[e]	Murray River[f]
Hygrobiidae					X	
Lampyridae		X				
Noteridae	X					
Phengodidae			X			
Ptilodactylidae		X	X			
Scarabaeidae			X			
Scirtidae						X
Staphylinidae		X	X			
Trichoptera						
Ecnomidae	X				X	X
Glossosomatidae	X					
Hydropyschidae		X				
Hydroptilidae	X				X	X
Lepidostomatidae	X					
Leptoceridae				X	X	X
Limnephilidae		X		X		
Psychomyiidae	X					
Lepidoptera						
Pyralidae/Crambidae		X	X			
Diptera						
Bibionidae		X	X			
Cecidiomyiidae		X	X			
Ceratopogonidae	X	X	X	X	X	X

Chaoboridae		X	X			
Chironomidae						
(Chironominae)	X	X	X	X	X	X
(Tanypodinae)	X	X	X	X	X	X
(Orthocladiinae)	X	X	X	X	X	X
Culicidae	X	X	X	X	X	X
Dolichopodidae		X	X			X
Empididae		X	X			X
Ephydridae	X	X	X	X		X
Muscidae				X		X
Psychodidae		X	X			X
Sciaridae		X	X	X		
Sciomyzidae				X		
Simuliidae	X				X	
Stratiomyidae		X		X	X	
Syrphidae		X				
Tabanidae		X	X	X		
Tipulidae/Limoniidae	X	X	X	X		X

[a] Gallardo (unpublished)
[b] Reese and Batzer (2007)
[c] Reese and Batzer (2007), Bright et al. (2010)
[d] Whiles (unpublished)
[e] Chessman and Hardwick (2014)
[f] Boulton and Lloyd (1991, 1992), Goonan et al. (1992), Sheldon and Walker (1998)

References

Alexander KD, Whiles MR (2000) A new species of *Ironoquia* (Trichoptera: Limnephilidae) from an intermittent slough of the central Platte River, Nebraska. Entomol News 111:1–7

Amoros C, Bornette G (1999) Antagonistic and cumulative effects of connectivity: a predictive model based on aquatic vegetation in riverine wetlands. Archiv Hydrobiol 115:311–327

Amoros C, Bornette G (2002) Connectivity and biocomplexity in waterbodies of riverine floodplains. Freshw Biol 47:761–776

Anderson DH, Darring SD, Benke AC (1998) Growth of crustacean meiofauna in a forested floodplain swamp: implications for biomass turnover. J North Am Benthol Soc 17:21–36

Arscott DB, Keller B, Tockner K, Ward JV (2003) Habitat structure and Trichoptera diversity in two headwater flood plains, N.E. Italy. Internat Rev Hydrobiol 88:255–273

Arscott DB, Tockner K, Ward JV (2005) Lateral organization of aquatic invertebrates along the corridor of a braided floodplain river. J North Am Benthol Soc 24:934–954

Baldwin DS, Rees GN et al (2013) Provisioning of bioavailable carbon between the wet and dry phases in a semi-arid floodplain. Oecologia 172:539–550

MDBA (Murray–Darling Basin Authority) (2012) Assessment of environmental water requirements for the proposed basin plan: Riverland-Chowilla Floodplain. Murray–Darling Basin Authority, Canberra, Publ 26/12. http://www.mdba.gov.au/sites/default/files/archived/proposed/EWR-Riverland-Chowilla.pdf. Accessed Aug 2014

Batzer DP, Baldwin AH (eds) (2012) Wetland habitats of North America: ecology and conservation concerns. University of California Press, BerkeleyBatzer DP, Ruhí A (2013) Is there a core set of organisms that structure macroinvertebrate assemblages in freshwater wetlands? Freshw Biol 88:1647-1659

Batzer DP, Wissinger SA (1996) Ecology of insect communities in nontidal wetlands. Annu Rev Entomol 41:75–100

Baxter CV, Fausch KD, Saunders WC (2005) Tangled webs: reciprocal flows of invertebrate prey link streams and riparian zones. Freshw Biol 50:201–220

Benke AC (2001) Importance of flood regime to invertebrate habitat in an unregulated river-floodplain ecosystem. J North Am Benthol Soc 20:225–240

Benke AC, Cushing CE (eds) (2005) Rivers of North America. Elsevier, New York

Benke AC, Van Arsdall TC, Gillespie DM, Parrish FK (1984) Invertebrate productivity in a subtropical blackwater river: the importance of habitat and life history. Ecol Monogr 54:25–63

Benke AC, Chaubey I, Ward GM, Dunn EL (2000) Flood pulse dynamics of an unregulated river floodplain in the Southeastern U.S. coastal plain. Ecology 81:2730–2741

Boulton AJ, Lloyd LN (1991) Macroinvertebrate assemblages in floodplain habitats of the lower River Murray, South Australia. Regul Riv Res Manage 6:183–201

Boulton AJ, Lloyd LN (1992) Flooding frequency and invertebrate emergence from dry floodplain sediments of the River Murray, Australia. Regul Riv Res Manage 7:137–151

Boulton AJ, Sheldon F, Jenkins KM (2006) Natural disturbance and aquatic invertebrates in desert rivers. In: Kingsford RT (ed) Ecology of desert rivers. Academic, New York, pp 133–153

Boulton AJ, Brock MA et al (2014) Australian freshwater ecology: processes and management, 2nd edn. Wiley-Blackwell, Chichester

Braccia A, Batzer DP (2001) Invertebrates associated with woody debris in a southeastern forested floodplain wetland. Wetlands 21:18–31

Braccia A, Batzer DP (2008) Breakdown and invertebrate colonization of dead wood in wetland, upland, and river habitats. Can J Forest Res 38:2697–2704

Bright EG, Batzer DP, Garnett JA (2010) Variation in invertebrate and fish communities across floodplain ecotones of the Altamaha and Savannah Rivers. Wetlands 30:1117–1128

Brinson MM, Malvárez AI (2002) Temperate freshwater wetlands: types, status, and threats. Environ Conserv 29:115–133

Bunn SE, Arthington AH (2002) Basic principles and ecological consequences of altered flow regimes for aquatic biodiversity. Environ Manage 30:492–507

Burns A, Walker KF (2000) Effects of water level regulation on algal biofilms in the River Murray, South Australia. Regul Riv Res Manage 16:433–444

Cabezas A, González E et al (2008) Effects of hydrological connectivity on the substrate and understory structure of riparian wetlands in the Middle Ebro River (NE Spain): implications for restoration and management. Aquat Sci 70:361–376

Cabezas A, Comín FA, Beguería S, Trabucchi M (2009) Hydrologic and landscape changes in the Middle Ebro River (NE Spain): implications for restoration and management. Hydrol Earth Syst Sci 13:1–12

Chessman BC, Hardwick L (2014) Water regimes and macroinvertebrate assemblages in floodplain wetlands of the Murrumbidgee River, Australia. Wetlands 34:661–672

Cid N (2011) Ecology of the benthic macroinvertebrates in the Lower Ebro River: community characterization, population dynamics and bioaccumulation of pollutants in response to environmental factors. Ph.D. Dissertation, Universitat de Barcelona

Cid N, Ibáñez C, Palanques A, Prat N (2010) Patterns of metal bioaccumulation in two filter-feeding macroinvertebrates: exposure distribution, inter-species differences and variability across developmental stages. Sci Total Environ 408:2795–2806

Cole CA, Somerville DE (2014) United States wetland regulation, policy and assessment. In: Batzer DP, Sharitz RR (eds) Ecology of freshwater and estuarine wetlands, 2nd edn. University of California Press, Berkeley, pp 203–224

CSIRO (2008) Water availability in the Murray. A report to the Australian government from the CSIRO Murray–Darling Basin Sustainable Yields Project. CSIRO, Canberra, Australia

Davies PE, Harris JH, Hillman TJ, Walker KF (2008) SRA report 1: a report on the ecological health of rivers in the Murray–Darling Basin, 2004–2007. Murray–Darling Basin Commission, Canberra, Publ 16/08

Davis CA, Austin JE, Buhl DA (2006) Factors influencing soil invertebrate communities in riparian grasslands of the central Platte River floodplain. Wetlands 26:438–454

De Bono A, Del Pietro D et al (2004) Freshwater in Europe—facts, figures and maps. DEWA Europe/UNEP, Geneva

Frenzel SA, Swanson RB et al (1998) Water quality in the Central Nebraska Basins, Nebraska, 1992–95. US Geological Survey, Washington, DC, Circular 1163

Frisch D, Libman BS, D'Surney SJ, Threlkeld ST (2005) Diversity of floodplain copepods (Crustacea) modified by flooding: species richness, diapause strategies and population genetics. Arch Hydrobiol 162:1–17

Galat DL, Berry CRJ, Peters EJ, White RG (2005) The Missouri River basin. In: Benke AC, Cushing CE (eds) Rivers of North America. Elsevier Academic Press, San Diego, pp 427–480

Galatowitsch ML, Batzer DP (2011) Benefits and costs of *Leptophlebia* (Ephemeroptera) mayfly movements between river channels and floodplain wetlands. Can J Zool 89:714–723

Gallardo B, Garcia M et al (2008) Macroinvertebrate patterns along environmental gradients and hydrological connectivity within a regulated river-floodplain. Aquat Sci 70:248–258

Gallardo B, Gascón S, García M, Comín FA (2009a) Testing the response of the macroinvertebrate functional structure and biodiversity to flooding and confinement. J Limnol 68:315–326

Gallardo B, González-Sanchís M et al (2009b) Modelling the response of floodplain aquatic communities across the lateral hydrological gradient. Mar Freshw Res 60:924–935

Gallardo B, Cabezas Á, Gonzalez E, Comín FA (2012) Effectiveness of a newly created oxbow lake to mitigate habitat loss and increase biodiversity in a regulated floodplain. Restor Ecol 20:387–394

Gallardo B, Dolédec S et al (2014) Response of benthic macroinvertebrates to gradients in hydrological connectivity: a comparison of temperate, subtropical, Mediterranean and semiarid river floodplains. Freshw Biol 59:630–648

Garnett JA, Batzer DP (2014) Longitudinal variation in community structure of floodplain fishes along two rivers of the Southeastern USA. Can J Fish Aquat Sci 71:1291–1302

Gladden JE, Smock LA (1990) Macroinvertebrate distribution and production on the floodplains of two headwater streams. Freshw Biol 24:533–545

Goldowitz BS, Whiles MR (1999) Investigations of fishes, amphibians, and aquatic invertebrate species within the middle Platte River system. Final report, Platte Watershed Program Cooperative Agreement X99708101. USEPA Region VII, Kansas City, MO

González E, González-Sanchis M et al (2010) Recent changes in the riparian forest of a large regulated Mediterranean River: implications for management. Environ Manage 45:669–681

Goonan PM, Beer JA, Thompson TB, Suter PJ (1992) Wetlands of the River Murray floodplain, South Australia. 1. Preliminary survey of the biota and physico-chemistry of ten wetlands from Chowilla to Mannum. Trans R Soc South Australia 116:81–94

Harner MJ, Geluso K (2012) Effects of cattle grazing on Platte River caddisflies (*Ironoquia plattensis*) in central Nebraska. Freshw Sci 31:389–394

Heino J (2008) Patterns of functional biodiversity and function-environment relationships in lake littoral macroinvertebrates. Limnol Oceanogr 53:1446–1455

Hillman TJ, Quinn GP (2002) Temporal changes in macroinvertebrate assemblages following experimental flooding in permanent and temporary wetlands in an Australian floodplain forest. Riv Res Appl 18:37–154

Hladyz S, Nielsen DJ, Suter PJ, Krull ES (2012) Temporal variations in organic carbon utilization by consumers in a lowland river. Riv Res Appl 28:513–528

Jenkins KM, Boulton AJ (2003) Connectivity in a dryland river: short-term aquatic microinvertebrate recruitment following floodplain inundation. Ecology 84:2708–2723

Jenkins KM, Boulton AJ (2007) Detecting impacts and setting restoration targets in arid-zone rivers: aquatic microinvertebrate responses to reduced floodplain inundation. J Appl Ecol 44:823–832

Junk WJ, Wantzen KM (2006) Flood pulsing and the development and maintenance of biodiversity in floodplains. In: Batzer DP, Sharitz RR (eds) Ecology of freshwater and estuarine wetlands. University of California Press, Berkeley, pp 407–435

Junk WJ, Bailey PB, Sparks RE (1989) The flood-pulse concept in river-floodplain systems. Special Publ Can J Fish Aquat Sci 106:110–127

Karaus U, Larsen S, Guillong H, Tockner K (2013) The contribution of lateral aquatic habitats to insect diversity along river corridors in the Alps. Landscape Ecol 28:1755–1767

King SL, Battaglia LL et al (2012) Floodplain wetlands of the Southeastern Coastal Plain. In: Batzer DP, Baldwin AH (eds) Wetland habitats of North America: ecology and conservation concerns. University of California Press, Berkeley, pp 253–266

Kingsford RT (2000) Ecological impacts of dams, water diversions and river management on floodplain wetlands in Australia. Austral Ecol 25:109–127

Kohler SL, Corti D, Slamecka MC, Schneider DW (1999) Prairie floodplain ponds: mechanisms affecting invertebrate community structure. In: Batzer DP, Rader RB, Wissinger SA (eds) Invertebrates in freshwater wetlands of North America: ecology and management. Wiley, New York, pp 711–732

Langhans SD, Tockner K (2006) The role of timing, duration, and frequency of inundation in controlling leaf litter decomposition in a river-floodplain ecosystem (Tagliamento, northeastern Italy). Oecologia 147:501–509

Libby B (2013) Impacts of macroconsumers on leaf breakdown and detritivores in wet and dry wetlands of a Southeastern US Coastal Plain floodplain. M.S. thesis, Coastal Carolina University, Conway, SC

Malard F, Uehlinger U, Zah R, Tockner K (2006) Flood-pulse and riverscape dynamics in a braided glacial river. Ecology 87:704–716

MDBC (Murray–Darling Basin Commission) (2006) The Chowilla Floodplain and Lindsay–Wallpolla Islands icon site environmental management plan. Murray–Darling Basin Commission, Canberra

Meyer CK, Whiles MR (2008) Macroinvertebrate communities in restored and natural Platte River slough wetlands. J North Am Benthol Soc 27:626–639

Meyer CK, Whiles MR, Baer SG (2010) Plant community recovery following restoration in temporally variable riparian wetlands. Restor Ecol 18:52–64

Muñoz I, Prat N (1994) Macroinvertebrate community in the lower Ebro River (NE Spain). Hydrobiologia 286:65–78

Olivier JM, Carrel G et al (2009) The Rhône river basin. In: Tockner K, Uehlinger U, Robinson CT (eds) Rivers of Europe. Academic, London, pp 247–295

Ollero A (2007) Channel adjustments, floodplain changes and riparian ecosystems of the middle Ebro River: assessment and management. Internat J Water Resour Devel 23:73–90

Overton I, Doody T (2008) Groundwater, surface water, salinity and vegetation response to a proposed regulator on Chowilla Creek. CSIRO, Glen Osmond, South Australia

Paetzold A, Tockner K (2005) Effects of riparian arthropod predation on the biomass and abundance of aquatic insect emergence. J North Am Benthol Soc 24:395–402

Paetzold A, Schubert CJ, Tockner K (2005) Aquatic terrestrial linkage along a braided-river: riparian arthropods feeding on aquatic insects. Ecosystems 8:748–759

Paetzold A, Bernet JF, Tockner K (2006) Consumer-specific responses to riverine subsidy pulse in a riparian arthropod assemblage. Freshw Biol 51:1103–1115

Paetzold A, Yoshimura C, Tockner K (2008) Riparian arthropod responses to flow regulation and river channelization. J Appl Ecol 45:894–903

Paillex A, Castella E, Carron G (2007) Aquatic macroinvertebrate response along a gradient of lateral connectivity in river floodplain channels. J North Am Benthol Soc 26:779–796

Persson Vinnersten TZ, Östman Ö, Schäfer ML, Lunström JO (2014) Insect emergence in relation to floods in wet meadows and swamps in the River Dalälven floodplain. Bull Entomol Res 104:453–461

Poff NL, Allan JD et al (1997) The natural flow regime. BioScience 47:769–784

Reese EG, Batzer DP (2007) Do invertebrate communities in floodplains change predictably along a river's length? Freshw Biol 52:226–239

Richoux P (1994) Theoretical habitat templets, species traits, and species richness: aquatic Coleoptera in the Upper Rhône River and its floodplain. Freshw Biol 31:377–395

Riggins JJ, Davis CA, Hoback WW (2009) Biodiversity of belowground invertebrates as an indicator of wet meadow restoration success (Platte River, Nebraska). Restor Ecol 17:495–505

Roach KA, Thorp JH, Delong MD (2009) Influence of lateral gradients of hydrologic connectivity on trophic positions of fishes in the Upper Mississippi River. Freshw Biol 54:607–620

Sabater S, Feio M et al (2009) The Iberian rivers. In: Tockner K, Uehlinger U, Robinson CT (eds) Rivers of Europe. Academic, London, pp 113–150

Sabater S, Muñoz I et al (2011) Aquatic and riparian biodiversity in the Ebro watershed: prospects and threats. In: Barceló D, Petrovic M (eds) The Ebro River Basin. Springer, Berlin Heidelberg, pp 121–138

Sheldon F, Walker KF (1998) Spatial distribution of littoral invertebrates in the lower Murray-Darling River system, Australia. Mar Freshw Res 49:171–182

Sheldon F, Boulton AJ, Puckridge JT (2002) Conservation value of variable connectivity: aquatic invertebrate assemblages of channel and floodplain habitats of a central Australian arid-zone river, Cooper Creek. Biol Conserv 103:13–31

Sidle JG, Miller ED, Currier PJ (1989) Changing habitats in the Platte River valley of Nebraska USA. Prairie Naturalist 21:91–104

Skinner R, Sheldon F, Walker KF (2001) Animal propagules in dry wetland sediments as indicators of ecological health: effects of salinity. Regul Riv Res Manage 17:191–197

Smock LA (1994) Movements of invertebrates between stream channels and forested floodplains. J North Am Benthol Soc 13:524–531

Smock LA, Gladden JE et al (1992) Lotic macroinvertebrate production in three dimensions: channel surface, hyporheic, and floodplain environments. Ecology 73:876–886

Söderström O, Nillson AN (1987) Do nymphs of *Parameletus chelifer* and *P. minor* (Ephemeroptera) reduce mortality from predation by occupying temporary habitats? Oecologia 74:39–46

Starr SM, Benstead JP, Sponseller RA (2014) Spatial and temporal organization of macroinvertebrate assemblages in a lowland floodplain ecosystem. Landscape Ecol 29:1017–1031

Steward AL, Marshall JC et al (2011) Terrestrial invertebrates of dry river beds are not simply subsets of riparian assemblages. Aquat Sci 73:551–566

Steward AL, von Schiller D et al (2012) When the river runs dry: human and ecological values of dry rivers. Frontiers Ecol Environ 10:202–209

Tachet H, Usseglio-Polatera P, Roux C (1994) Theoretical habitat templets, species traits, and species richness: Trichoptera in the Upper Rhône River and its floodplain. Freshw Biol 31:397–415

Taylor AN, Batzer DP (2010) Spatial and temporal variation in invertebrate consumer diets in forested and herbaceous wetlands. Hydrobiologia 651:145–159

Tockner K, Malard F, Ward JV (2000) An extension of the flood pulse concept. Hydrol Process 14:2861–2883

Tockner K, Uehlinger U, Robinson CT (2009) Rivers of Europe. Academic, New York

Townsend CR, Doledec S, Scarsbrook MR (1997) Species traits in relation to temporal and spatial heterogeneity in streams: a test of habitat templet theory. Freshw Biol 37:367–387

Tronstad LM, Tronstad BP, Benke AC (2005a) Invertebrate seedbanks: rehydration of soils from an unregulated river floodplain of the south-eastern U.S. Freshw Biol 50:646–655

Tronstad LM, Tronstad BP, Benke AC (2005b) Invertebrate responses to decreasing water levels in a subtropical river floodplain wetland. Wetlands 25:583–593

Tronstad LM, Tronstad BP, Benke AC (2007) Aerial colonization and growth: rapid invertebrate responses to temporary aquatic habitat in a river floodplain. J North Am Benthol Soc 26:460–471

Tronstad LM, Tronstad BP, Benke AC (2010) Growth rates of chironomids collected from an ephemeral floodplain wetland. Wetlands 30:827–831

US Fish and Wildlife Service (1997) Biological opinion on the Federal Energy Regulatory Commission's preferred alternative for the Kingsley dam project and North Platte/Keystone dam project. US Fish and Wildlife Service, Grand Island, NE

Usseglio-Polatera P, Tachet H (1994) Theoretical habitat templets, species traits, and species richness: Plecoptera and Ephemeroptera in the Upper Rhône River and its floodplain. Freshw Biol 31:357–375

Vivian LA, Cavallaro M et al (2013) Current known range of the Platte River caddisfly, *Ironoquia plattensis*, and genetic variability among populations from three Nebraska rivers. J Insect Conserv 17:885–895

Walker KF (2006) Serial weirs, cumulative effects: the Lower River Murray, Australia. In: Kingsford RT (ed) Ecology of desert rivers. Academic, New York, pp 248–279

Ward JV (1989) The 4-dimensional nature of lotic ecosystems. J North Am Benthol Soc 8:2–8

Ward JV, Tockner K et al (1999a) A reference river system for the Alps: the 'Fiume Tagliamento'. Reg River Res Manage 15:63–75

Ward JV, Tockner K, Sciemer F (1999b) Biodiversity of floodplain river ecosystems: ecotones and connectivity. Regul Riv Res Manage 15:125–139

Ward JV, Tockner K, Arscott DB, Claret C (2002) Riverine landscape diversity. Freshw Biol 47:517–539

Watkins SC, Quinn GP, Gawne B (2010) Changes in organic-matter dynamics and physicochemistry, associated with riparian vegetation loss and river regulation in floodplain wetlands of the Murray River, Australia. Mar Freshw Res 61:1207–1217

Whiles MR, Goldowitz BS (2001) Hydrologic influences on insect emergence production from central Platte River wetlands. Ecol Appl 11:1829–1842

Whiles MR, Goldowitz BS (2005) Macroinvertebrate communities in Central Platte River wetlands: patterns across a hydrologic gradient. Wetlands 25:462–472

Whiles MR, Goldowitz BS, Charlton RE (1999) Life history and production of a semi-terrestrial limnephilid caddisfly in an intermittent Platte River wetland. J North Am Benthol Soc 18:533–544

Wissinger SA (1999) Ecology of wetland invertebrates: synthesis and applications for conservation and management. In: Batzer DP, Rader RD, Wissinger SA (eds) Invertebrates in freshwater wetlands of North America: ecology and management. Wiley, New York, pp 1043–1086

Yan W, Howles SR, Marsden Z (2005) Chowilla floodplain numerical groundwater model. South Australian Department of Water, Land and Biodiversity Conservation, Adelaide Rep 2004/65. https://www.waterconnect.sa.gov.au/Content/Publications/DEWNR/dwlbc_2004_65_screen.pdf. Accessed Aug 2014

Chapter 14
Invertebrates in Neotropical Floodplains

Karl M. Wantzen, Mercedes R. Marchese, Marinez I. Marques, and Leandro D. Battirola

Introduction: Environmental Conditions in Tropical Floodplain Wetlands and Their Implications for Invertebrates

Hydrology

The term "Tropical floodplain wetlands" covers a very wide range of ecosystem types. This chapter focuses mainly on systems studied by the authors, in Latin America (both tropical and subtropical floodplains, see Fig. 14.1). Even within these units, the variability is enormous. The overarching control in floodplains is the flood pulse (Junk et al. 1989; Neiff 1996; Junk and Wantzen 2004; Wantzen et al. 2008a), i.e., the seasonal wet-and-dry-cycle provoking habitat changes in a short-term, annual or multi-annual pattern, by the presence of water deriving from river or lake overflow, groundwater rise, direct rainfall, or a mixture of these. The interplay between these sources makes the hydrology of the systems extremely complex; specifically in coarse-grained alluvial river floodplains, the patchiness of hydrological conditions may vary across very small scales (Tockner et al. 2010).

The term "flood plain" is not restricted to large rivers, as often assumed. Rather, the occurrence of floodplains is defined by the local hydrogeomorphological factors such as slope and sediment dynamics; therefore floodplains can be found even in small streams. In mountainous areas they may have a "bead on a string" pattern (Ward et al. 2002), i.e., floodplain zones alternate with confined, steep river channels. The hydrodynamics vary with the catchment size. While small streams have a short, frequent, and intensive flood period according to rainfall events (Lewis et al. 1995; Wantzen 2003; Wantzen and Junk 2000), the overlapping hydrographs of an increasing number of tributaries have a tendency to smoothen and slow the inundation phase, resulting in mono- or bi-modal flood pulses, according to the seasonality of the precipitation in the catchment (Junk et al. 1989). Increased water levels of the main stem river during the flood period

D. Batzer, D. Boix (eds.), *Invertebrates in Freshwater Wetlands*,
DOI 10.1007/978-3-319-24978-0_14

may block the discharge of tributaries situated downstream, creating "back flooding" even if there was no intensive rainfall in local catchments, e.g., in the Southern Pantanal (Hamilton 2002). The type of flooding (speed of change, frequency, amplitude, duration, size, and type of the area covered, etc.) strongly influences how flood-borne resources become available for use by floodplain biota, and how biota may be disturbed by a flood event. Short floods (and subsequent moist phases) in forested stream floodplains enhance the reciprocal exchange of organic matter (and drifting animals) between aquatic and terrestrial phases, enabling e.g., stream detritivores to shred leaves outside the stream (Wantzen and Wagner 2006). Short and weakly synchronized emergence patterns of aquatic insects safeguard that a single rainstorm event will not eradicate an entire generation of offspring, rather many different larval stages of the same species (and winged adults) exist across a rainstorm-prone period (Wantzen and Junk 2000; Wantzen 2003).

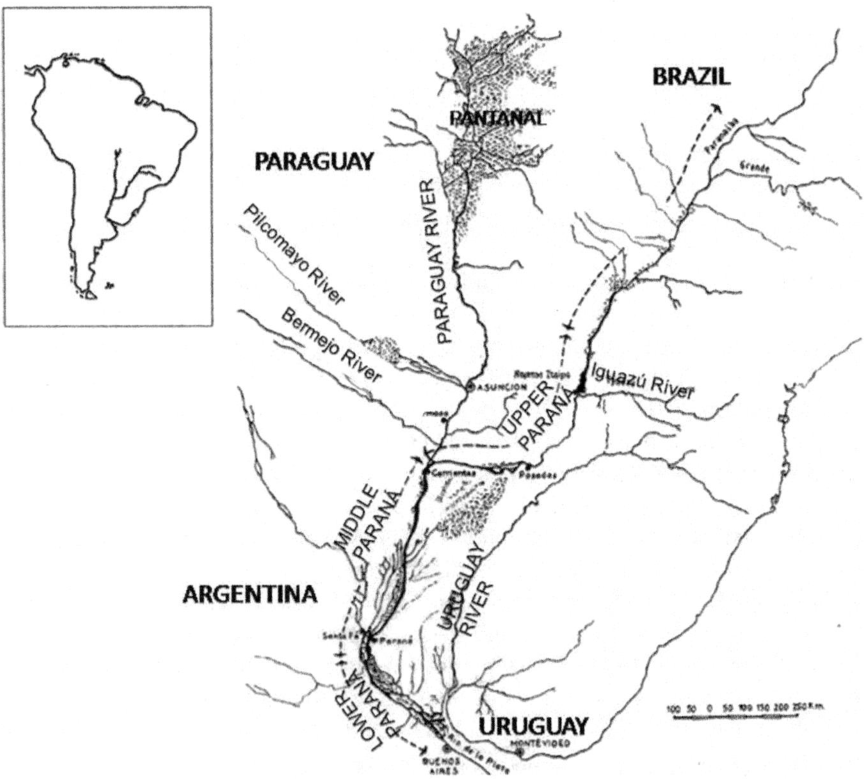

Fig. 14.1 Map showing the Paraguay and Paraná River floodplains in Latin America © M. Marchese

The longer the flood period, the more the foodweb depends on the productivity within the floodplain itself ("flood-borne resources": phytoplankton, macrophytes) rather than from productivity from river channels (Junk and Wantzen 2004). The floodpulse drives the development of a "shifting habitat mosaic" (Stanford et al. 2005; Tockner et al. 2010) of different successional stages of geomorphological and vegetational features, allowing the use of different resources during each phase of the flood pulse. The drawdown of the water during the beginning of drought creates a disconnection of floodplain water bodies, increases concentrations of nutrients (Heckman 1994; Furch and Junk 1997; Nogueira et al. 2011), and opens opportunities for terrestrial colonization of the aquatic-terrestrial transition zone (ATTZ) (Junk et al. 1989). High rates of evapotranspiration and the endorheic position of the floodplains may increase salinity and make osmoregulatory adaptations by the invertebrates necessary (Verschuren et al. 2000). Chemical substances leached from organic matter during rising and falling water may strongly influence the invertebrate fauna due to oxygen consumption during mineralization leading to anoxia (Hamilton et al. 1997) and by direct effects of dissolved substances (Baldwin 1999) or of nutrients that trigger algal growth as a food source (which may be more important than leaves for many invertebrates, especially in floodplains of dry regions; Bunn et al. 2003).

Geology and Water Chemistry

Floodplains are formed by the deposition of sediments eroded from highland areas of the catchment. The closer the floodplain to the source of these materials, the coarser the sediment size usually tends to be, and consequently the hydraulic conductance (porosity) of the sediments, sometimes enabling the colonization of the hyporheic zone far beyond the streams (see Marmonier et al. 2012 for a recent review). Sediment size is crucial for the biodiversity (Hynes 1970) and biogeochemical processes (Fischer et al. 2005) of river-floodplain systems. Most large Neotropical floodplains such as the Pantanal in Brazil (Irion et al. 2011) or the Paraná floodplains in Argentina (Drago 1990) (Fig. 14.1) receive sediments from highly weathered plains that already have a fine grain size before erosion begins. These "re-worked" sediments are generally very fine and release few nutrients into the water, and thus the existence of a hyporheic zone is very limited. Rocky outcrops may then be doubly important for the invertebrate fauna, as solid substrate for clingers, and as ion sources; they are often composed of chemically unique rock, such as the "carboniferous strips" in Central South America, that benefit calcium-dependent taxonomic groups such as crustaceans and mollusks (Junk and Robertson 1997; Wantzen et al. 2011). Tributaries may also introduce coarser sediments into otherwise fine-grained floodplains, such as the Río Bermejo that brings Andean materials into the Paraguay floodplain (Argentina), with important consequences for the invertebrate fauna (Ezcurra de Drago et al. 2004;

Amsler and Drago 2009). In terms of invertebrate biodiversity, the floodplains situated between highlands and lowlands can be unique: not only are they colonized by the faunal assemblages from both lower- and upper-section floodplains, but they also have overlapping physical habitat characteristics (sediment structure) from both of these sections (Wantzen et al. 2011). The presence of natural barriers such as large waterfalls may limit fish as predators and thereby allow different invertebrate faunas to develop (Flecker 1992).

Climate and Vegetation

To describe life conditions for floodplain invertebrates in the Tropics, we simplify discussion by depicting only large biomes, and focus only on freshwater systems. Climate and soil chemistry define vegetation patterns and thus have a strong influence on organic matter processing (Wantzen et al. 2008b; Graça et al. 2015), and in turn on the invertebrate community of floodplains. Plant biodiversity in riparian zones of tropical streams is generally very high, enabling a large array of floodplain invertebrate to use strategies including flower and fruit consumption, leaf and wood detritivory, or scavenging on terrestrial invertebrates that accidentally fall from the trees (Wantzen and Junk 2000). The reciprocal subsidies (Nakano and Murakami 2001) between aquatic and terrestrial food webs are very strong. In seasonal tropical climates, plant phenology tends to follow the wet-and-dry rhythm, however organic matter inputs are not as strongly pulsed as seasonal leaf-fall in temperate climates, promoting a different schedule of resource use, including rapid exploitation of short-term resources, such as green leaves after storms, or flowers and fruits (Wantzen et al. 2008b).

Tropical floodplains in evergreen lowland rainforest have humid conditions, although water availability may still be limited by seasonal rainfall patterns. The enormous peat swamp forests of tropical South East Asia (Wantzen et al. 2008d; Yule 2010) accumulate deep layers of organic matter due to permanent inundation and the presence of secondary compounds in the vegetation that reduces shredder activity (Haase and Wantzen 2008; Wantzen et al. 2008b; Lim et al. 2014). In Amazonian floodplain wetlands, large water-level fluctuations (up to 12 m) select for specifically adapted herbaceous and woody plants (herbaceous plants: rapid growth or the ability to float, woody plants: flood tolerance, biochemical adaptations to anoxia (Junk 1997; Parolin 2009)). Similarly, wetland invertebrates are selected to cope with fast-changing conditions including anoxic water and rapid inundation of terrestrial habitats.

In seasonal savanna wetlands such as in the Okavango Delta (Curtis et al. 1998; Davidson et al. 2012), in Kenya (Verschuren et al. 2000), or in the Pantanal, woody vegetation needs to be adapted to both flood and drought conditions, including fire (Nunes da Cunha and Junk 2004; de Oliveira et al. 2014), resulting in unique features such as xeromorphic leaves, high concentrations of secondary compounds, and thick bark, which may constrain their use by herbivorous or detritivorous invertebrates. The often luxurious aquatic macrophyte vegetation of these wetlands (in the Pantanal, more than 280 species are known (Pott et al. 2011) and in the Esteros del Iberá of the Paraná River, 161 species (Neiff et al. 2011)) can shift from terrestrial (dwarfish, hard-leaved) growth to fast growing and flowering aquatic forms. Once the floods recede, the ATTZ is rapidly recolonized by terrestrial grassy and herbaceous species. The floodplain sediments are seed banks for both aquatic and terrestrial plants that display long-term germination strategies to cope with variable hydrological situations (Pagotto et al. 2011). The hydrologically driven interannual variation in vegetation assemblages provides opportunities for a large diversity of herbivorous and detritivorous invertebrates to exist, and affects their interactions with overall foodwebs.

Major Ecological Controls on Invertebrates in Tropical Floodplains

The Hydrological Cycle (Flood Pulse)

Contrary to the common view of river floods being a nuisance, the flood pulse contributes nutrients and food sources, and creates valuable habitats for both aquatic and terrestrial species. Adaptations in life cycles to cyclic, seasonal, or irregular flood and drought events control the presence or absence of invertebrate species on floodplains. The same habitat can be subsequently colonized by both terrestrial and aquatic organisms, making floodplains doubly important to biodiversity. Moreover, there is strong interannual and multi-year variability in the hydrographs of natural systems, permitting certain species traits to be favored in 1 year, and others in subsequent years so that the resulting biotic community is rarely identical over time, termed "sequential biodiversity" by Junk and Wantzen (2004).

By wetting and drying the floodplain, the flood pulse drives the "Aquatic-Terrestrial-Transition Zone" (ATTZ *sensu* Junk et al. 1989) across the ecosystem. Terrestrial species need to cope with flooded conditions, and aquatic species to dry conditions. Diverse strategies to use highly variable flood-borne food resources are common among tropical floodplain invertebrates (Adis and Junk 2002). Compared with temperate water bodies, the bacterial activity in tropical systems is much higher, resulting in fast turnover rates and rapid changes in water quality. Anoxic conditions due to rapid mineralization or solution of carbon dioxide may occur within hours, sometimes leading to mass mortalities (e.g., the *dequada* (anoxia) phenomenon in the Pantanal (Calheiros and Hamilton 1998)). Under anoxic conditions, high concentrations of toxic hydrogen sulfide (H_2S) can develop, limiting all metazoan life (e.g., in stagnant floodplain water bodies of Central Amazonia during peak flooding season (Junk and Robertson 1997)).

Changing conditions of oxygenation and food resource availability are affected by water movements across the floodplain. Neiff (1996) distinguishes between the “limnophase” of stagnant water in permanently flooded lakes during the low-water phases, and the “potamophase” when river water flows across the floodplain. In large shallow river-floodplains such as the Pantanal, regional differences in evapotranspiration, and exchange with groundwater and river channels causes stagnant conditions to only occur locally/periodically. Water is typically flowing in most parts of the floodplain, although velocities can vary from few cm/s to >1 m/s.

Connectivity to and Distance from the Main Stem River

Connectivity is another important control on the composition of the floodplain invertebrate community (Marchese et al. 2005; Zilli and Marchese 2010; Davidson et al. 2012; Wantzen et al. 2014). Floodplains are often associated with permanently wetted water bodies such as lakes, rivers, and streams, with variable degrees of connectivity (Amoros and Bornette 2002; Wantzen et al. 2005; Drago et al. 2008a). This variation in the hydrological regime affects the exchange of water, nutrients, organic matter, and organisms at the aquatic-terrestrial, benthic-pelagic, and surface water–groundwater interfaces. Flow rates within these complex hydrological networks, and the residence times of the water are highly variable, and depend on the location of the habitat across the floodplain topography and the degree and type of connection (Drago 1990; Drago et al. 2008a, b). The chemical qualities of the water may be determined by direct access to river channel water (characterized by riverine-suspended solids and nutrients), indirect connection via secondary floodplain channels (i.e., suspended substances have settled or been filtered), or by groundwater inputs that may have high concentrations of dissolved organic and inorganic substances but are often anoxic. This suite of characteristics affects the structure of invertebrate assemblages and the persistence of species compositions. At the same time they determine the degree to which benthic macroinvertebrates contribute to the ecological functioning of the system, contributing to the cycling of nutrients and organic matter decomposition and providing a flux of energy from benthic organic matter to higher trophic levels. Studies on the Paraguay-Paraná river-floodplain system suggest that benthic invertebrate diversity increases laterally from the mainstream to floodplain habitats, as certain controlling physical factors (turbulence, flow velocity, hydraulic parameters, turbidity, etc.) decrease, strongly structuring benthic assemblage patterns (Figs. 14.2 and 14.3) (Marchese et al. 2002; Ezcurra de Drago et al. 2004; Marchese et al. 2005; Ezcurra de Drago et al. 2007; Wantzen et al. 2014).

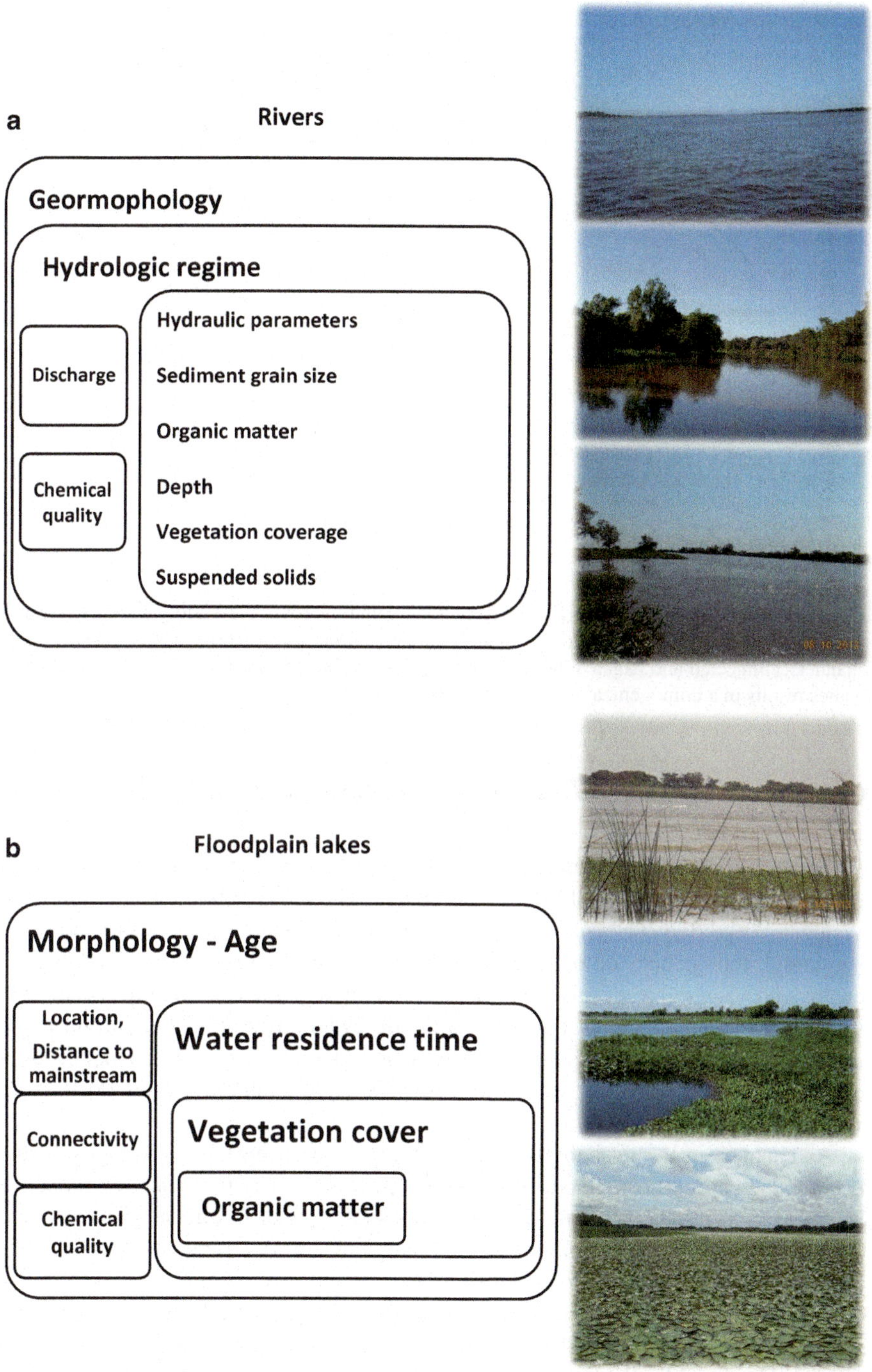

Fig. 14.2 Factors controlling the macroinvertebrates assemblage structure in riverine (*above*) and lacustrine (*below*) habitats of the Middle Paraná River system. © M. Marchese

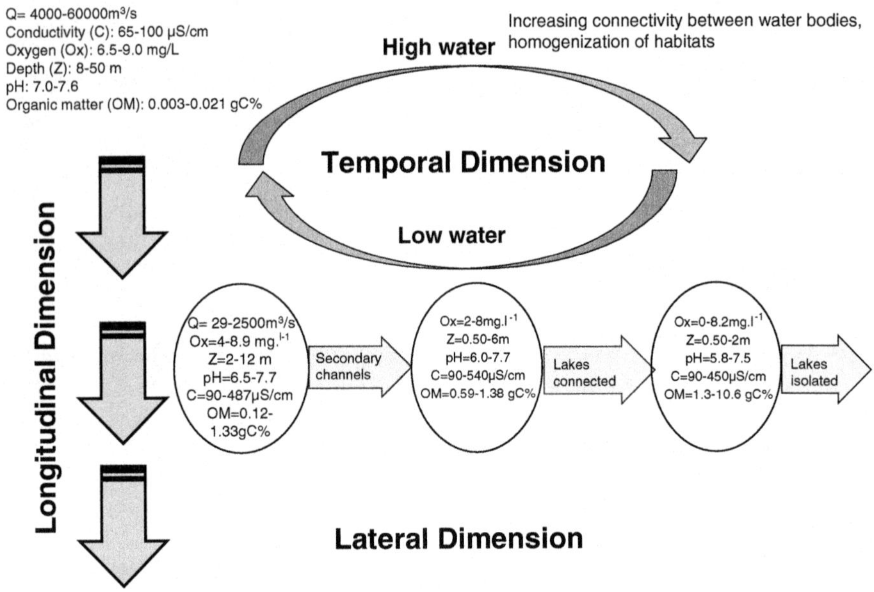

Fig. 14.3 Environmental conditions in a river-floodplain transect in the Middle Paraná River system. *Upper left*: range of values for the mainstem river, *middle row from left*: values for secondary channels, connected and isolated lakes. During high water, increasing connectivity between water bodies results in a homogenization of habitats. © M. Marchese

Congruence Between Life Cycles and Hydrological Cycles

Floodplain organisms are selected to profit from beneficial resources and to avoid adverse conditions provided by the flood pulse. The flood pulse concept (Junk and Wantzen 2004) depicts "*windows of opportunity*," i.e., when life cycles of an organism match the occurrence of an exploitable resource provided by the hydrological cycle, and "*windows of susceptibility*" when hydrological situations occur that impact vulnerable stages of the organism's life-cycle phase (Fig. 14.4). Man-made intervention to the hydrological regime may provoke non-natural flood and drought events that can profoundly impact the survival of floodplain species (Poff et al. 1997; Sparks et al. 1998).

As changes in the availability of a given resource during the hydrological cycle occur rather abruptly in floodplains, the development of populations, reproductive stages, or growth phases often occur over very short periods (*biotic hot moments*) and at restricted sites (*biotic hot spots*) (Wantzen and Junk 2006). Overall, a shifting mosaic of sites creates intensive biological activity in time and space (Stanford et al. 2005).

Metamorphosis of some floodplain insects tracks the switch between aquatic and terrestrial phases. The larvae of the chrysomelid beetle *Lysathia sioli,* discovered from the Pantanal by Wantzen and recently described by Medvedev (2001), live on

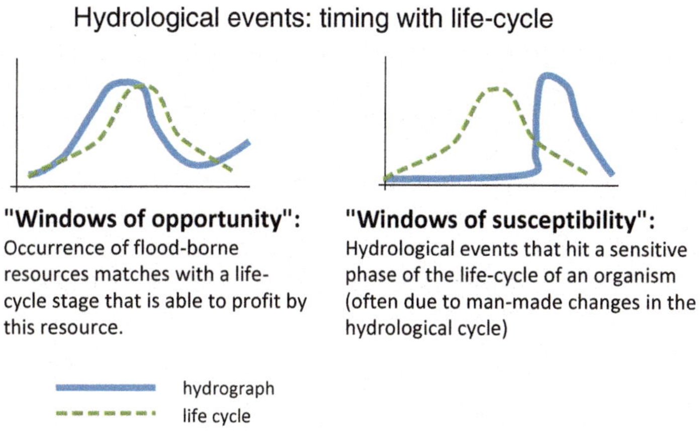

Fig. 14.4 Match and mismatch of the life cycle of the organisms with the flood pulse (Windows of opportunity, windows of susceptibility). Modified from Junk and Wantzen (2004)

the roots of *Ludwigia natans* during the inundation phase. As the wetland dries, the adults emerge in large numbers and feed on decaying leaves and reproduce on the rosettes of *L. natans*. Thereby, this species is optimally adapted to exploit the temporally changing resources provided by its plant host.

The high temperatures, constantly changing environmental conditions, and the brief availability of resources in tropical floodplains favor the development of specific life-cycle strategies: short life cycles, early reproduction, migration, and dormancy strategies (see also case studies). Aquatic animals with very short life cycles can employ a mixed strategy between migration and aestivation called "pond-hopping" (Wantzen et al. 2011); they reproduce and dwell in a pond until it dries out, then move to the next aquatic habitat by flight. This phenomenon is known for the chironomid *Apedilum elachistum* that can complete its entire life cycle in only 1 week (Nolte 1995).

Reproducing earlier and faster allows animals to exploit beneficial life conditions that occur only briefly. However, a high energy investment in reproduction may inhibit the survival of the progeny (Roff 1992). This pattern was found for the diplopod *Poratia salvator* in the Pantanal (Pinheiro et al. 2009). When comparing populations of *P. salvator* from the floodplain with a nearby non inundated habitat, the reproductive period of the floodplain population occurred much earlier. Similarly, the semiaquatic grasshopper *Cornops aquaticum,* associated with *Eichhornia azurea,* has ontogenic shifts over the flood pulse. The first two of six nymphal stages are associated with low-water periods, seemingly benefitting from the highly abundant food sources supporting high density levels, whereas later-stage nymphs, which have lower densities due to natural mortality, dominate on residual plant tips as water levels rise (Silva et al. 2014; da Silva et al. 2010). Although Orthoptera are usually not considered aquatic insects, some species cannot develop without aquatic plants for oviposition and nymphal development (Amédégnato and Devriese 2008). As such, in the Paraná River-floodplain, 7 of a total of 17 Orthoptera species were considered aquatic due to their associations with the freshwater environment, especially during high water levels (Capello et al. 2013). High water levels concentrate the abundance and species richness of orthopterans on the macrophytes they use as food and shelter (Capello et al. 2013).

Organisms that cannot maintain desirable environmental conditions by migrating or by morphological adaptations such as plastron respiration need to adapt to drought and re-wetting (aquatic species) or to flooding and re-drying (terrestrial species) by reducing their physiological activities to a minimum to survive the adverse conditions (see Fig. 14.4). During this dormancy phase, respiration is strongly reduced; there is no food uptake, growth, or egg production. The organisms live on stored energy and may be forced to accumulate metabolic end products until the adverse conditions have passed.

Invertebrate Case Studies

Aquatic Invertebrates in the Pantanal

The Pantanal is a huge floodplain in central South America (Fig. 14.1). It is fed by the tributaries of the upper section of the Paraguay River, with a seasonal monomodal flood pattern caused mostly by rainfall in its northern section, and back flooding of the larger rivers in its southern section. The interactions between different sources of flood water are particularly complex due to very flat terrain. For example, Baia Sinhá Mariana, a floodplain lake near Barão de Melgaço, is fed by a blackwater river (which crosses the lake), by a lateral channel from another floodplain lake, or by the Cuiabá River (mixed whitewater). Regional rainfall and water levels define the relative proportions from each of the sources, and even the flow directions of the water in floodplain channels. As a result, the lake benthos is either dominated by fauna drifting in from the blackwater tributary, or by filter-feeding chironomids and mayflies, profiting from the food particles from the whitewater tributary, or it is impoverished when the nutrient-poor lake is unconnected during stagnant, low-water periods (Wantzen et al. 2011). Near the Paraguay River, most floodplain waterbodies are oxbows with variable degrees of connectivity to the river mainstream; connectivity affects nutrient and biomass accumulation, and oxygenation of the habitats (Wantzen et al. 2005), which all affect the benthic assemblages. Leeward ends of oxbows tend to accumulate organic matter, mostly recalcitrant fiber from floating macrophytes, that reduces density and diversity of invertebrates to those few species adapted to low oxygen concentrations. Floodplain channels in close contact to fresh river water have a high biodiversity and biomass of invertebrates (Marchese et al. 2005). The high beta diversity of connected floodplain lakes stands in stark contrast to the low-diverse, low-biomass assemblages of the mobile sandy

Fig. 14.5 Floating or partially rooted aquatic macrophytes (*Eichhornia azurea*) provide important habitats for aquatic and terrestrial invertebrates. © K. M. Wantzen

substrates in the river mainstream (Wantzen et al. 2014). Recently flooded waterbodies have very steep gradients in environmental conditions such as organic matter and oxygen, although habitats become "homogenized" during later flood phases (Thomaz et al. 2007).

In the shallow floodplain habitats, most "benthic" invertebrates are not restricted to the sediments, rather, they also colonize the root systems of the floating macrophytes (*Eichhornia, Salvinia, Pistia*), called the "interrhizon" by Heckman (1997) (Fig. 14.5). This habitat supports a high biodiversity because it harbors invertebrates that typically colonize the sediment, the roots, and open-water zone. The root systems provide fresh and decomposing organic matter, biofilms, and accumulate organic particles in interrhizal pore space, which all are useful foods for invertebrates. Roots provide shelter from fish predators, and predators also profit because these zones attract zooplankton prey (Hamilton et al. 1990). Putting a floating macrophyte into a white plastic tray is a more efficient way to collect invertebrates than dragging and sieving large volumes of organic mud (see Wantzen and Rueda-Delgado 2009 for a review of methods for invertebrate studies).

Aquatic invertebrates of the Pantanal have developed a series of remarkable adaptations to survive drought conditions (Wantzen et al. 2011). In analogy to the late Joachim Adis' seminal work on terrestrial Amazonian invertebrates, survival strategies can be subdivided in non-migrating and migrating strategies with different substrategies and physiological adaptations for each (Adis 1997; Fig. 14.6).

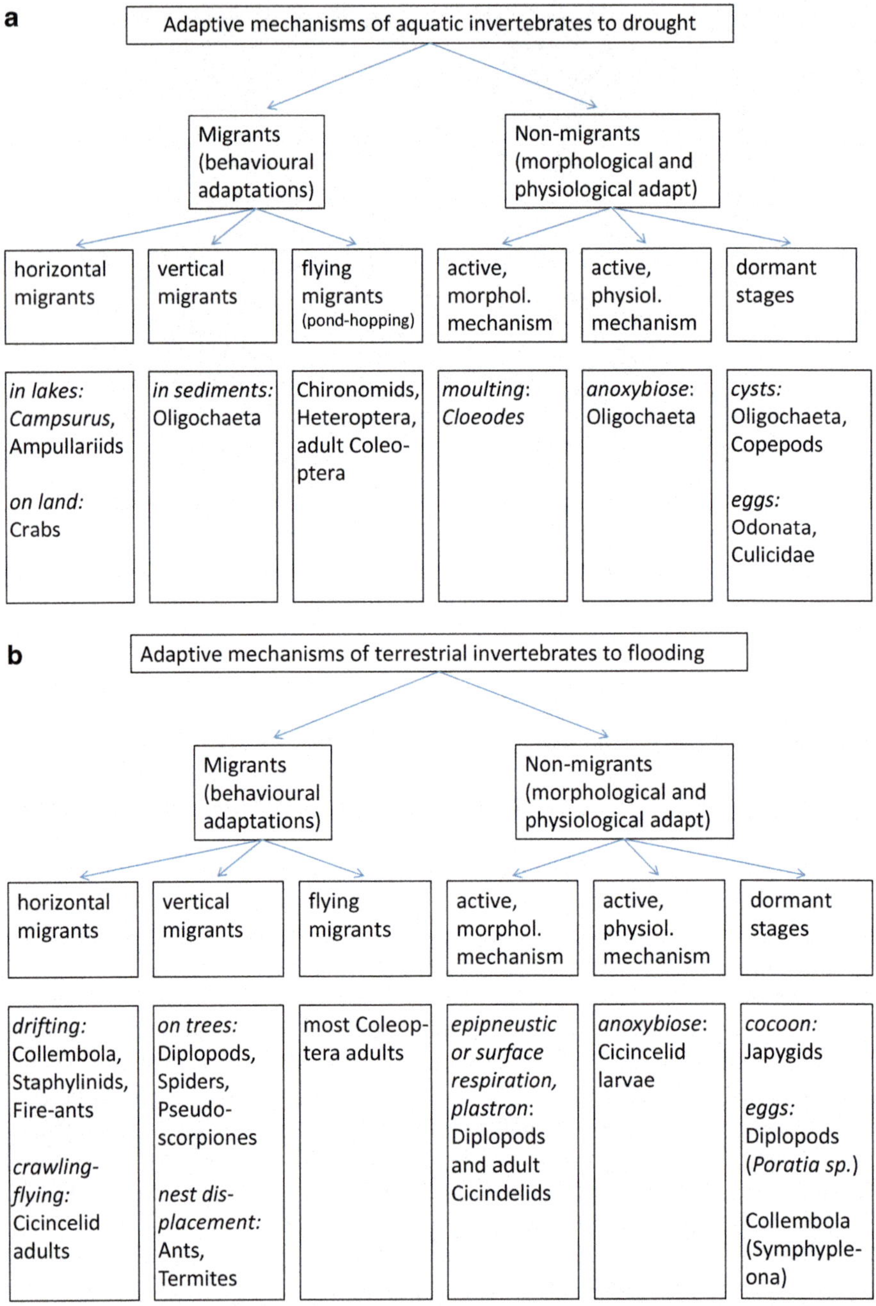

Fig. 14.6 Survival strategies of aquatic and terrestrial invertebrates from floodplains. Adaptive mechanisms of aquatic invertebrates to drought (**a**), and adaptive mechanisms of terrestrial invertebrates to flooding (**b**). Redrawn from Adis (1997) and Wantzen et al. (2011)

Adaptive Strategies of Aquatic Invertebrates to Survive Desiccation

In the Pantanal, only a few waterbodies remain wetted during the peak of the dry season. These lower zones accumulate large amounts of organic matter and nutrients (Nogueira et al. 2002) while the rest of the dry floodplain has rather hostile conditions for aquatic invertebrates. The dry phase exerts powerful selection on the distribution and abundance of aquatic invertebrates in these wetlands (Fig. 14.6a). As solubility of oxygen in water is much lower than in air, the respiratory organs of aquatic species often have large surfaces for oxygen uptake that are not protected against drought by impermeable cuticles. Once dried out, these respiratory surfaces cannot be reactivated, only exchanged by molting. An impressive example for this strategy is the mayfly *Cloeodes hydation* (Nolte et al. 1996) that molted up to nine times within 2 days when experimentally dried and re-wetted. Freshwater oligochaetes form cysts that resist desiccation (Montalto and Marchese 2005). Sponges such as *Drulia browni* can survive extreme drought for months, even while exposed on dry, dead wood (Junk and Robertson 1997). Less drought-tolerant species migrate with the receding water line. Larvae of the mayfly genus *Campsurus* migrate within the lake littoral as water levels rise and fall (Takeda and Grzybkowska 1996). Freshwater crabs in the Pantanal have very large brood pouches in which they carry their offspring over land. Crabs and large ampullariid snails migrate and accumulate at the lowest parts of lakes and flooded meadows to aestivate, and may provoke "hot moments" when they return to their original habitats when the waters return. Migrating kites (*Rostrhamus sociabilis*) flock by the thousands to prey upon the returning crabs and snails (Wantzen et al. 2011).

Habitat choice may enhance desiccation avoidance. General ecological life strategies of invertebrates such as r and K strategies may match the dynamics of their habitats, i.e., there are r and K habitats (Marchese et al. 2002). Reliable and stable microhabitats will favor longer-lived species with higher parental investment in offspring. Unstable, frequently and/or unpredictably disturbed habitats will instead be colonized by short-lived, small bodied r-strategists. For example, long-lived, large-bodied mussels are typically found in backwaters at the lower end of sand banks of Neotropical rivers, where flow conditions and supplies of organic matter are relatively stable (Wantzen et al. 2005; Colle and Callil 2012).

Adaptations of Aquatic Invertebrates to Flood Conditions

Floodplain sediments are "propagule banks" releasing large numbers of invertebrates, either when re-wetted or re-dried. Rising water levels in seasonal floodplains of the Pantanal mobilizes nutrients and organic matter that were stored during the dry phase. Bacteria, algae, and (later on) macrophytes (De-Lamonica-Freire and Heckman 1996; Loverde-Oliveira et al. 2011) experience rapid growth that provides support for aquatic invertebrate (Fellerhoff 2002; Marchese et al. 2014) and vertebrate (Wantzen et al. 2002) consumers. The earliest colonizers benefit most from these resources because their vertebrate predators have not yet arrived during the early phases of flooding. Mass emergence and reproduction of *Campsurus* mayflies at the beginning of the flood season ensures that larval development is timed to use these flood-borne resources (Wantzen and Junk 2006). Ceratopogonids hatch soon after flooding, allowing their larvae to develop rapidly by filter-feeding on planktonic bacteria, and to avoid predation by fish. Densities of chironomid larvae in the Coqueiro lake (Northern Pantanal) also peak at the beginning of the flood period (Sobreira et al. 2012). Invertebrate predators with long-distance flying capacity, however, quickly colonize these habitats (many heteropterans and odonates). Over time, invertebrate communities develop that are similar to those of permanent lakes, displaying "classical" sediment-associated benthic and planktonic assemblages, combined with epiphytic (benthos-like) organisms that make up the largest part of the biomass in most floodplain lakes (Wantzen et al. 2011).

As water levels decrease, many invertebrate consumers, including ampullariid snails, live on detritus of decaying macrophytes (Fellerhoff et al. 2003). In the Pantanal de Nabileque (southern Pantanal), high evapotranspiration rates induce saline lakes dominated by blue-green algae to develop. A specific invertebrate community colonizes this kind of habitat, and the lake shores are littered with dead bodies of corixids, witness to the high productivity of these lakes.

Terrestrial Floodplain Invertebrates in the Pantanal

Terrestrial habitats of the Pantanal such as rocks, soils, organic litter, trees, and the emergent parts of macrophytes provide habitat and resources for a rich terrestrial invertebrate fauna. The soil fauna of the Pantanal, however, seems less rich than that of Central Amazonia (taxonomic groups such as Collembola, Diplopoda, and Pseudoscorpiones are less diverse). The adaptive strategies of the terrestrial invertebrates are inversely complementary to those of the aquatic species, i.e., they require refugia during flood and profit from the drought season (see Fig. 14.6b). Our knowledge on strategies how terrestrial invertebrates survive adverse (flood) conditions largely originates from the extensive studies by Joachim Adis in the Central Amazonian Floodplain (Adis 1997; Adis and Junk 2002). Taxa previously known as being purely terrestrial were discovered surviving flooding by using specific respiratory organs that allowed them to remain under water. Japygids (Diplura) spin cocoons in which they survive (Adis et al. 1989). Diplopods or larvae of arctiid moths use plastron respiration and can freely inhabit flooded roots (Messner and Adis 1987; Adis et al. 1996). Larvae of tiger beetles (Carabidae) can survive flooded conditions by employing anaerobic metabolism (Zerm et al. 2004). However, most terrestrial invertebrates of the floodplain utilize impressive migration strategies. Vertical migrators climb to higher ground, such as the fire ant *Solenopsis saevissima* (Myrmicinae) that moves from soils to grass stems, often forming living carpets of ants on stems of *Setaria geniculata* (Gramineae) extending above the water-level. The leaf cutting ant *Acromyrmex lundi carli* (Myrmicinae) has been observed moving their subterranean

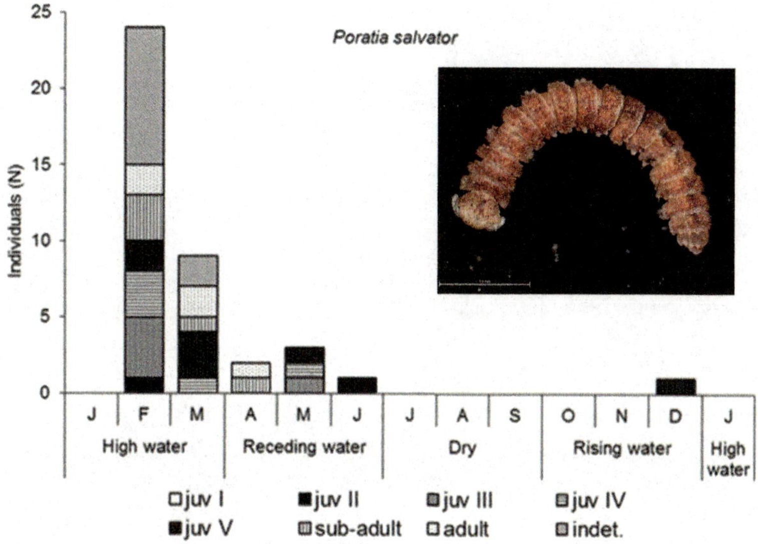

Fig. 14.7 Occurrence of larval and adult stages of *Poratia salvator* in the northern Pantanal of Mato Grosso. *Insert*: photograph of *P. salvator* © M. Marques

nests, typically located at the bases of tree trunks, to non-flooded tree axes. Immatures of *Plusioporus salvadorii* (Spirostreptidae) were located up to 1 m above the water-level on the bark of *Vochysia divergens* (Vochysiaceae, Adis et al. 2001). Most terrestrial invertebrates escape from the flooding by lateral migrations. These migrators represent an important prey for army ants that incorporate rising floods as a part of their hunting strategy (Adis et al. 2001). The migratory ability of tiger beetles may vary between populations, with beetles being winged in the lower (longer flooded) zones of the floodplain gradient in Amazonia, while wingless in the upper zone (Zerm and Adis 2002), presumably diverting saved energy for other purposes.

The millipedes *Poratia salvator* (Fig. 14.7) and *Promestosoma boggianii* (Polydesmida: Diplopoda), and some Cerambycidae (Coleoptera) synchronize their life cycles with the changing seasons of inundation and drought (Battirola 2007; Pinheiro et al. 2009, 2011; Marques et al. 2014). Non-migrating species of Collembola (Symphypleona) survive flooding as dormant eggs in soils (Battirola 2007). Diplopoda, Diplura, Chilopoda, Pseudoscorpiones, and Araneae inhabit plant leaves or tree trunks to remain dry. Some Araneae can escape inundated areas by crossing the water surface. *Attalea phalerata* palm associations in the Pantanal serve as refuges during the wet season with ground-dwelling ants and spiders occupying all vertical strata of these palms (Santos et al. 2003; Battirola et al. 2004, 2014). The architecture of this palm has leaf sheaths that accumulate organic debris and provide important microhabitats for numerous arthropod species, e.g., Coleoptera, Collembola, Formicidae, Diptera, and Psocoptera (Marques et al. 2009).

Fig. 14.8 Flooded grassy wetland with islands of termite mounds and individual trees (*Curatella americana*) in the Pantanal © K. M. Wantzen

A conspicuous feature of floodplain savannahs of the Pantanal are large termite mounds that are up to 3 m high. *Cornitermes cumulans* (Termitidae, Nasutitermitinae) build these mounds during the dry season but these termites increase their height during the wet season to survive the period of inundation flooding. Termite mounds are important refuges for terrestrial invertebrates during flood (Fig. 14.8). Impressive, at night, are abandoned termite mounds that have been colonized by larvae of the luminescent beetles *Pyrearinus termitiluminans* (Elateridae, Agrypninae), giving the impression of Christmas trees (Costa and Vanin 2010).

Some vegetation communities are dominated by single-tree species, e.g., *Vochysia divergens* (Vochysiaceae), *Attalea phalerata* (Arecaceae), or *Calophyllum brasiliense* (Clusiaceae). These single-tree species associations are known for their unique communities of ants, beetles, spiders, and millipedes, which exhibit strong seasonal dispersal among vertical strata during the inundation period (Battirola et al. 2009; Marques et al. 2001, 2006, 2011, 2014; see also Table 14.1). Pantanal wetlands are also known for monodominant stands of aquatic macrophytes, e.g., *Eichhornia* (Pontederiaceae) and *Salvinia* (Salviniaceae). Semiaquatic beetles (Curculionidae Staphylinidae) and aquatic beetles (Hydraenidae, Noteridae, Hydrophilidae) are common on these plants.

Table 14.1 Diversity of arthropods (ind./m²) in canopies of three monodominant vegetation stands in the dry and high water seasons in the Pantanal of Mato Grosso, Brazil

	Calophyllum brasiliense		*Vochysia divergens*		*Attalea phalerata*	
Taxa	Dry season	High water	Dry season	High water	Dry season	High water
Arachnida						
Acari	10.4	3.9	6.6	12.5	1.8	257.0
Araneae	8.7	5.3	20.9	10.2	15.3	13.4
Opiliones	–	–	<0.1	0.1	<0.1	<0.1
Pseudoscorpiones	<0.1	–	1.9	0.4	4.8	7.3
Crustacea						
Isopoda	–	–	–	–	–	0.1
Hexapoda						
Blattodea	<0.1	≤0.1	1.6	1.5	5.4	2.1
Coleoptera	18.0	12.2	13.2	17.1	65.5	77.5
Collembola	27.9	13.8	1.5	2.0	32.4	53.4
Dermaptera	–	–	–	–	0.2	0.3
Diptera	4.2	21.8	35.7	61.2	21.6	54.0
Embioptera	–	–	–	–	–	0.2
Ephemeroptera	–	–	0.1	0.2	–	0.2
Hemiptera	14.1	6.4	45.1	9.8	3.2	9.8
Hymenoptera	151.8	47.3	86.3	34.0	51.0	50.6
(Formicidae)	(131.8)	(41.7)	(64.1)	(13.2)	(45.4)	(25.2)
Isoptera	0.3	–	<0.1	≤0.1	1.5	2.7
Lepidoptera	0.3	<0.1	1.3	0.6	1.9	12.3
Mantodea	–	<0.1	0.1	–	<0.1	–
Neuroptera	<0.1	<0.1	0.3	<0.1	0.1	<0.1
Odonata	–	–	<0.1	0.1	<0.1	<0.1
Orthoptera	0.4	0.4	0.2	0.3	0.4	4.9
Phasmatodea	–	–	<0.1	–	–	–
Psocoptera	1.2	3.1	3.2	5.4	25.5	59.0
Thysanoptera	337.8	8.4	54.5	3.2	7.0	37.2
Thysanura	–	–	–	<0.1	–	–
Trichoptera	<0.1	–	0.1	0.1	0.8	0.4
Myriapoda						
Polydesmida	–	–	–	<0.1	–	<0.1
Polyxenida	–	–	0.2	–	–	–
Scolopendromorpha	–	–	–	<0.1	<0.1	–
Total	511.8	116.3	272.8	158.7	238.7	643.0

Overview of Pantanal Invertebrates

Overall, the knowledge of the Pantanal invertebrate fauna remains incomplete and requires a sustained taxonomic and ecological research effort to provide better estimates of species numbers, distributions, and evolutionary histories. The Pantanal is clearly a highly diverse wetland (Junk et al. 2006; Appendix), but taxonomic research has not adequately described the invertebrate fauna. Notably, taxonomists have described only six new terrestrial arthropod species, and one new genus, for the northern region of the Pantanal since 2005 (*Attaleachernes thaleri* Mahnert, 2009 (Pseudoscorpiones: Chernetidae); *Pantanalodesmus marinezae* Hoffmann, 2009 (Diplopoda: Polydesmida); *Schendylops inquilinus* Pereira et al., 2007 (Chilopoda: Geophilomorpha); *Hanseniella chapadensis* Scheller, 2007 (Symphyla); and *Allopauropus* (*Allopauropus*) *pantanalicus* Scheller, 2007 (Pauropoda)). This number of new species descriptions is negligible compared to the vast number of morphospecies still waiting to be described. This statement is also true for the aquatic invertebrates; many studies cite morphospecies awaiting their description. Recently a new shrimp has been described from the Pantanal (Santos et al. 2013).

Aquatic Invertebrates in the Paraná Floodplain

Benthic invertebrates in large river-floodplain systems like the Paraná River (Fig. 14.1) occupy a wide range of physical habitats. While the benthic assemblages in the main channel are remarkably homogeneous, specifically in sand-dominated stretches (Blettler et al. 2008), their diversity in the lateral floodplains is high (Marchese et al. 2002, 2005; Wantzen et al. 2014; Appendix). In the Paraná River system, low-flow secondary channels and floodplain lakes with longer water residence times have more favorable conditions for the development of benthic communities than the central channel (Ezcurra de Drago et al. 2004, 2007). Here there is a gradient of increasing diversity of species and functional feeding groups of the benthos from the main channel to secondary channels of higher flow to floodplain lakes (some) and their peripheral temporary wetlands (Marchese and Ezcurra de Drago 1992; Marchese et al. 2002; Zilli et al. 2008, Fig. 14.9).

The Oligochaeta and Chironomidae are the dominant taxonomic groups in the Middle Paraná River system and its floodplains. More than 80 species of oligochaetes and approximately 30 genera of chironomids are reported (Marchese and Paggi 2003; Zilli and Montalto 2011). In addition, Coleoptera is important in littoral wetlands with many species of different families.

The macrophyte cover has a strong influence on the floodplain habitats of the Paraná River system, where it reaches up to 100 % in hydrologically isolated lakes. This produces an abundant supply of detritus on the bottom, which can induce states of hypoxia or even anoxia to develop that decrease the diversity of benthic invertebrates, while at the same time the densities and dominance of individual tolerant

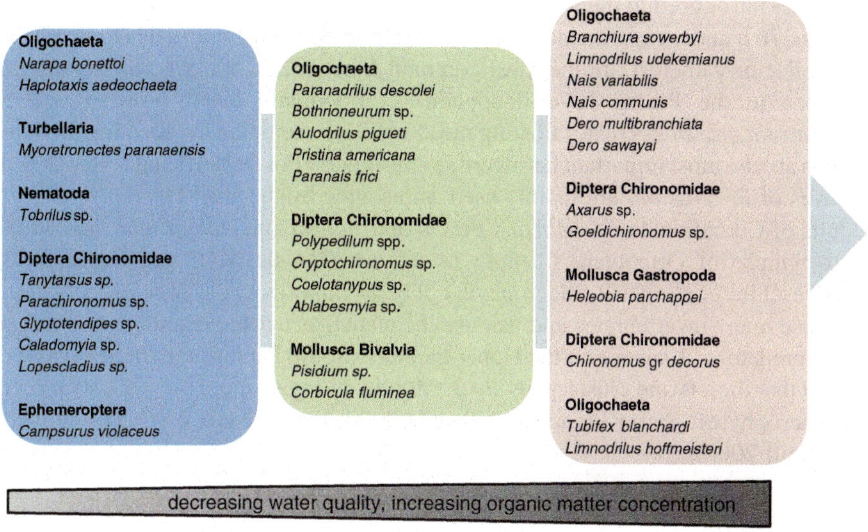

Fig. 14.9 Macroinvertebrates assemblages in relationship to enrichment gradients of organic matter in the Middle Paraná River system © M. Marchese

species can increase. The benthic diversity and species evenness are higher in lakes connected to the main channel or secondary channels (Zilli and Marchese 2010). This phenomenon may be explained by reduced competitive exclusion in the connected sites experiencing regular hydraulic disturbances, while competition is higher in permanent isolated habitats, following the Intermediate Disturbance Hypothesis (Ward and Stanford 1983). Additionally, the connectivity may directly interact with the diversity patterns, as the colonization by taxa with entirely aquatic life cycles are limited to aerial dispersion in isolated lakes, whereas the fauna of connected lakes may colonize surrounding aquatic habitats via multiple means (flying, swimming, drifting) (Montalto and Paggi 2006; Zilli and Marchese 2010; Zilli and Montalto 2011).

As with the Pantanal, large parts of the floodplain of the Paraná River dry completely and become re-wetted with the flood season, triggering a range of adaptive strategies to survive drought, colonize temporary wetlands, and disperse into other freshwater habitats. Aquatic oligochaetes such as *Dero multibranchiata* (Naidinae) and *Trieminentia corderoi* (Opistocystidae) were reported to form dehydration-resistant cysts (Montalto and Marchese 2005). Montalto and Paggi (2006) reported different genera of chironomids of the Paraná River-floodplain capable of surviving for more than 28 days in temporary wetlands under harsh desiccation conditions, where they built tubes or live in decaying macrophytes. Some larvae of the genera *Coelotanypus, Ablabesmyia, Polypedilum, Chironomus, Cryptochironomus,* and *Tanytarsus* were found in sediments up to 15 cm deep during drought conditions (Montalto pers. com. to MRM). These adaptations give floodpulse-adapted species a clear advantage over invasive species. *Limnoperna fortunei* (Bivalvia: Mytilidae), a mussel introduced to the Río de la Plata basin, has a limited tolerance to desiccation (<276 h under experimental conditions) (Montalto and Ezcurra de Drago 2003).

Collector-gatherers are the most common functional feeding group among the benthos in the Paraná River-floodplain. In floodplain lakes, however, where *Eichhornia crassipes* creates floating meadows, large scraping or shredding invertebrates are the most important herbivorous consumers (Poi de Neiff and Casco 2003). Leaves of *E. crassipes* are mainly used as resource by the snail *Omalonyx unguis,* adults of Curculionidae *Neochetina bruchi* and *Neochetina eichhorniae*, and adults and nymphs of Orthoptera *Cornops aquaticum* (Poi de Neiff and Casco 2003; Franceschini et al. 2010). Macrophytes play an important role in aquatic systems because many invertebrates and fish use them such as trophic resources and refuge from predators. A survey on the Upper Paraná, Middle Paraná, and Paraguay rivers found that insects and crustaceans are the dominant organisms associated with roots of macrophytes (Souza-Franco and Takeda 2000; Poi de Neiff 2003; Poi de Neiff and Neiff 2006).

Conservation and Management Issues

Conservation Issues

Invertebrates are still rarely considered for conservation activities (Agostinho et al. 2005; Loyola et al. 2006), especially in the tropics. Due to shorter life cycles and higher reproduction rates, invertebrates are presumed "resistant" to human activities, which is definitively not the case. As with vertebrates, invertebrate species with specific traits are most vulnerable (large territories, large body size, desirability for human harvesting, low reproduction rate, slow growth, low mobility, need to migrate). For example, migratory neretid snails and shrimps have similar restrictions as migratory fishes (Covich et al. 1996; Blanco and Scatena 2006). Large bivalves are perhaps the most threatened aquatic invertebrates in Neotropical floodplain, as they are among the first species to suffer from river regulation, and are also vulnerable to secondary colonization by invasive *Limnoperna* mussels (Wantzen et al. 2011).

The most severe environmental problems currently threatening tropical floodplain invertebrates are linked to changes of the flood pulse pattern (Hamilton 2002), either by reducing the size of the flood pulse (e.g., by damming and diking of the river, Bonetto et al. 1989; Zarfl et al. 2015) or by changing the natural rhythm of the flood pulse (e.g., by releasing water during the drought period in order to produce hydro-energy, Zeilhofer and de Moura 2009). Flood pulse-adapted species, not only invertebrates, either lose their habitats, or they lose their capacity to compete with other, non-flood pulse-adapted species. The latter phenomenon is very often observed in the context of invasive species such as *Limnoperna fortunei* (see below). A growing human population leads to huge water diversion and irrigation projects, but very little is known on the effects of ephemerization of water bodies in tropical wetlands. Comparisons between longer inundated Llanos lakes with the ephemeral floodplain lakes in the Pantanal have shown a strong influence of flood duration on the organic carbon dynamics (Vega et al. 2014), and thus a dramatic change in the dynamics of detritus-based foodwebs. Studies from naturally intermittent systems (Boulton and Suter 1986; Bunn and Davies 1990; Maltchik and Silva-Filho 2000) may serve as a baseline for future studies. Specifically, the plasticity of survival strategies (maximum length of aestivation, maximum temperature tolerance, etc.) is unknown for most species.

Floodplains are generally low-energy sites in terms of hydraulic forces, and they naturally accumulate sediments. The organisms living in floodplains are adapted to deal with a certain rate of sediment accumulation. If, however, this accumulation rate is increased, e.g., by erosion in catchments used for agriculture or mining, the capacity of the organisms to deal with new sediments is quickly overcome. Sediment dynamics in tropical floodplain wetlands can be influenced by numerous human impacts: dam reservoirs act as sediment traps, but water releases create erosional conditions below the dam. Mobilization of large amounts of eroded material due to unsustainable agriculture, road construction, or mining can lead to filling of the interstitial pore space of substrates (Marques Couceiro et al. 2010). The erosion problem in the upper catchments due to poor agricultural and mining practices (Wantzen et al. 2006; Ramirez et al. 2008; Wantzen and Mol 2013) can result in huge downstream deposits along the internal deltas of floodplain tributary rivers, e.g., the Taquari in the Pantanal (Pott and Pott 2004; Assine 2005). We have observed a transition of well-defined channels and lateral wetlands with highly diverse biota, into highly braided rivers where high sediment turnover rates reduce the invertebrate diversity and abundance. Only low numbers of scavengers or detritivores exist that are able to consume terrestrial organisms or organic matter that falls into the streams, and productivity of both the streams and their wetlands is minimal (Wantzen 2006).

Management Issues

When it comes to management, invertebrates are often considered as nuisances, especially in tropical floodplain wetlands that in some cases can support disease vector species (e.g., *Anopheles* mosquitoes that transmit malaria parasites). Therefore, management of floodplain invertebrates has primarily focused on control rather than on supporting invertebrates and the important ecosystem services they provide. Knowing the factors shaping the communities of invertebrates, the management of floodplain environments should set the highest priority in maintaining a natural hydrological regime (Poff et al. 1997; Hamilton 2002; Junk and Wantzen 2004; Wantzen et al. 2008c). Recurrent and timely floods and droughts preserve biodiversity and ecosystem functions. However, shallow landscape profiles, abundant water, and high productivity make tropical floodplains targets for large-scale "development" projects, which include isolating the floodplain from the main stem river by damming, diking or channel incision (Hamilton 1999). Along with multiple ecosystem services (Ricaurte et al. 2014; Vega et al. 2014), impacted floodplains lose their invertebrate diversity. Sustainable use scenarios are urgently needed (e.g., Wantzen et al. 2008c), but long-term benefits are often less appreciated than short-term returns from agriculture such as oil palm plantations (Yule 2010). In terms of governance, local groups are generally paid less attention than large well-financed groups (Ricaurte et al. 2014). A dramatic example for this phenomenon is the current discussion in Brazil: agricultural lobby groups are trying to force a definition of wetland size at its lowest extension (i.e., dry season ground water level) and to add a defined buffer width around them, rather than using the Ramsar definition which defines the upper flood level as the floodplain border (Ramsar Convention Secretariat 2006). Such a change would have deleterious consequences for biodiversity and ecosystem functions of the Pantanal and Paraná floodplains (Junk et al. 2014).

Many freshwater invertebrates (not only wetland species) are actively or passively introduced into areas beyond their natural biogeographic realm, and become a nuisance for native species or important ecosystem functions and services. A classic example is the invasion of Latin American Rivers and their floodplains by bivalve species originating from Asia, the golden mussel (*Limnoperna fortunei*) and the clams *Corbicula fluminea* and *C. langhillerti* (Darrigran 2002; Oliveira et al. 2006; Pereira et al. 2014). As mentioned above, the natural floodpulse regime helps to control *L. fortunei* fixed to solid benthic structures, but it cannot prevent colonization of floating vegetation in permanent lakes (Oliveira et al. 2011). Fortunately, predation by the diversity of fish in Neotropical floodplain rivers exerts an additional control on mussels (Montalto et al. 1999). However, we have observed that there is a time lag between initial invasion and the acceptance of mussels as fish prey, as predation effects in lower areas (where the invasion began) are higher than in upper sections of the Paraguay-Paraná system (Ezcurra de Drago et al. 2004). The consequences of introduced mollusks imply changes in habitat structure, food web alterations, changes in community structure, competition with native species, and bioaccumulation and magnification of toxic substances in tissues. Therefore, it is very important to know the effects of introduced species and their capacity to spread to threaten native biodiversity. Species that host parasites may affect various domestic and wildlife species, and occasionally even humans.

Appendix

Preliminary taxa list for wetland invertebrates of the Pantanal and Paraná floodplains. (*Abbreviations*: hab. = habitat, lot = lotic, running waters (may include tributaries), lac = lacustrine, stagnant waters, ubi = ubiquitous, occurring in both, lotic and lentic waters (indicating the main feeding types) her = herbivorous, fil = filter-feeding (organic particles or living plankton), det = detritivorous (including shredding and gather-collecting), pre = predacious). Due to the high taxonomic level (order or family), these attributions must be used cautiously, as individual genus or species may have different preferences. PTN = Pantanal, PAR = Paraná River. For terrestrial habitats, refer to Table 14.1

Taxon	hab.	FFG	PTN	PAR
HYDROZOA	lot	pre	x	x
PORIFERA		fil	x	x
PLATYHELMINTHES				
Planariidae	lot	pre	x	x
NEMATODA	ubi	det	x	x
ANNELIDA				
HIRUDINEA	ubi	pre	x	x
OLIGOCHAETA	ubi	det	x	x
Alluroididae	lot	det		x
Haplotaxidae	lot	pre	x	x
Lumbricidae	ubi	det	x	x
Lumbriculidae	ubi	det	x	
Naidinae	ubi	det	x	x
Narapidae	lot	det	x	x
Opistocystidae	ubi	det		x
Pristininae	ubi		x	x
Rhyacodrilinae	ubi	det	x	x
Tubificinae	ubi	det	x	x
MOLLUSCA			**x**	
GASTROPODA				
Ampullariidae	lac	det	x	x
Ancylidae	lot	her	x	x
Hydrobiidae	lac		x	x
Planorbidae	ubi	det	x	x
BIVALVIA				
Corbiculidae	ubi	fil	x	x
Hyriidae	lot	fil	x	x
Mycetopodidae	lac	fil	x	x
Mytilidae	ubi	fil	x	x
Psammobidae	lot	fil	x	
Sphaeridae	ubi	fil	x	x
ARACHNIDA		pre	x	
Arrenuridae	lot	pre	x	
Eylaidae	lot	pre	x	
Hydracarina	ubi	pre	x	x
CRUSTACEA				
AMPHIPODA				
Dogielinotidae	ubi	det		x
BRANCHIOPODA		fil	x	x
Bosminidae	lac	fil	x	x
Chydoridae	lot	fil	x	x
Conchostraca	lac	det	x	x
Daphnidae	lac	fil	x	x

(continued)

(continued)

Taxon	hab.	FFG	PTN	PAR
Ilyocryptidae	lac	fil	x	x
Macrotricidae	lot	fil	x	x
Moinidae	lac	fil	x	x
Sididae	lac	fil	x	x
COPEPODA			x	x
Cyclopidae	ubi		x	x
Diaptomidae	ubi		x	x
OSTRACODA	ubi	det	x	x
DECAPODA			x	
Palaeomonidae	ubi	omn	x	x
Trichodactylidae	lac	omn	x	x
INSECTA				
COLLEMBOLA	ubi	det	x	x
EPHEMEROPTERA				
Baetidae	lot	det	x	x
Caenidae	lot	det	x	x
Leptohyphidae	ubi	det	x	x
Leptophlebiidae	lot	det	x	x
Polymitarcyidae	ubi	det	x	x
ODONATA				
Aeshnidae	ubi	pre	x	x
Calopterygidae	lot	pre	x	x
Coenagrionidae	ubi	pre	x	x
Gomphidae	ubi	pre	x	x
Libellulidae	ubi	pre	x	x
Megapodagrionidae	lot	pre	x	
Polythoridae	lot	pre	x	
Protoneuridae	lot	pre	x	
THYSANOPTERA	ubi	det	x	
HEMIPTERA				
Belostomatidae	ubi	pre	x	x
Corixidae	lac	her	x	x
Gelastocoridae	ubi	pre	x	
Gerridae	ubi	pre	x	x
Naucoridae	lot	pre	x	
Pleidae	lot	pre	x	x
Veliidae	lot	pre	x	
NEUROPTERA	–	pre	x	
Corydalidae	lot	pre	x	
Sialidae	ubi	pre	x	
Sisyridae	lot	pre	x	
COLEOPTERA				
Carabidae	ubi	pre	x	x
Chrysomelidae	ubi		x	x

Taxon	hab.	FFG	PTN	PAR
Curculionidae	ubi	pre	x	x
Dryopidae	lot		x	x
Dytiscidae	ubi	pre	x	x
Elmidae	ubi	det	x	x
Eucnemidae	lot		x	
Georissidae	lot		x	
Gyrinidae	lot	pre	x	x
Heteroceridae	ubi			x
Hydraenidae	lot		x	x
Hydrochidae	lot		x	x
Hydrophilidae	ubi	pre	x	x
Hydroporinae	lot	pre	x	
Hydroscaphidae	lot		x	
Laccophilini	lot	pre	x	
Lampyridae	ubi		x	x
Noteridae	lot		x	x
Platypodidae	lot		x	
Ptilodactylidae	lot		x	
Scarabaeidae	ubi		x	x
Scirtidae	lot		x	x
Staphylinidae	ubi	pre	x	x
TRICHOPTERA				
Calamoceratidae	lot	det	x	
Glossosomatidae	lot		x	x
Helicopsychidae	lot	her	x	
Hydropschidae	lot	fil	x	x
Hydroptilidae	lot	det	x	x
Leptoceridae	ubi	det	x	x
Limnephilidae	lot		x	x
Odontoceridae	lot		x	
Philopotamidae	lot		x	
Polycentropodidae	lot	pre	x	x
Rhyacophilidae	lot		x	x
LEPIDOPTERA				
Arctiidae	lac	her	x	
Pyralidae	lot	her	x	
DIPTERA			x	
Athericidae	ubi		x	
Ceratopogonidae	ubi	omn	x	x
Chaoboridae	ubi	pre	x	x
Chironomidae	ubi		x	x
Culicidae	ubi	fil	x	x
Dolichopodidae	lot		x	x
Dixidae	ubi		x	x

(continued)

(continued)

Taxon	hab.	FFG	PTN	PAR
Ephydridae	ubi		x	x
Empididae	ubi		x	x
Psychodidae	ubi		x	x
Simuliidae	lot	fil	x	x
Sciomyzidae	lac		x	
Syrphidae	ubi		x	x
Sphaeroceridae	lac		x	
Stratiomyidae	lot		x	x
Tabanidae	lot		x	x
Tipulidae	lot		x	x

References

Adis J (1997) Terrestrial invertebrates: survival strategies, group spectrum, dominance and activity patterns. In: Junk WJ (ed) The Central Amazonian floodplain: ecology of a pulsing system. Springer, Berlin, pp 299–318

Adis J, Junk WJ (2002) Terrestrial invertebrates inhabiting lowland river floodplains of Central Amazonia and Central Europe: a review. Freshw Biol 47:711–731

Adis J, Messner B, Groth I (1989) Zur Überflutungsresistenz und zum Spinnvermögen von Japygiden (Diplura). Zool Jb Anat 119:371–382

Adis J, Golovatch SI, Hamann S (1996) Survival strategy of the terricolous millipede Cutervodesmus adisi Golovatch (Fuhrmannodesmidae, Polydesmida) in a Blackwater Inundation Forest of Central Amazonia (Brazil) in response to the flood pulse. Acta Myriapodol 169:23–532

Adis J, Marques MI, Wantzen KM (2001) First observations on the survival strategies of terricolous arthropods in the northern Pantanal wetland of Brazil. Andrias 15:127–128

Agostinho AA, Thomaz SM, Gomes LC (2005) Conservation of the biodiversity of Brazil's inland waters. Conserv Biol 19:646–652

Amédégnato C, Devriese H (2008) Global diversity of true and pygmy grasshoppers (Acridomorpha, Orthoptera) in freshwater. Hydrobiologia 595:535–543

Amoros C, Bornette G (2002) Connectivity and biocomplexity in waterbodies of riverine floodplains. Freshw Biol 47:761–776

Amsler ML, Drago EC (2009) A review of the suspended sediment budget at the confluence of the Paraná and Paraguay Rivers. Hydrol Process 23:3230–3235

Assine ML (2005) River avulsions on the Taquari megafan, Pantanal wetland, Brazil. Geomorphology 70:357–371

Baldwin DS (1999) Dissolved organic matter and phosphorus leached from fresh and 'terrestrially' aged River Red Gum leaves: implications for assessing river-floodplain interactions. Freshw Biol 41:675–685

Battirola LD (2007) Estratificação vertical e distribuição temporal da comunidade de artrópodes terrestres em uma floresta monodominante, sazonalmente inundável, na região norte do Pantanal de Mato Grosso, Brasil. In: Programa de Pós-Graduação em Ciências Biológicas (Entomologia). Universidade Federal do Paraná, Curitiba, p 613

Battirola LD, Marques MI, Adis J, Brescovit AD (2004) Aspectos ecológicos da comunidade de Araneae (Arthropoda: Arachnida), em copas da palmeira Attalea phalerata Mart. (Arecaceae), durante o período de Cheia no Pantanal de Mato Grosso, Brasil. Revista Brasil Entomol 48:421–430

Battirola LD, Marques MI et al (2009) Vertical and time distribution of Diplopoda (Arthropoda, Myriapoda) in a monodominant forest in Pantanal of Mato Grosso, Brazil. Revista Brasil Zool 26:479–487

Battirola LD, Santos GB, Rosado-Neto GH, Marques MI (2014) Coleoptera (Arthropoda, Insecta) associados às copas de *Attalea phalerata* Mart. (Arecaceae) no Pantanal de Mato Grosso, Brasil. Entomobrasilis (Vassouras) 7:20–28

Blanco JA, Scatena FN (2006) Hierarchical contribution of river-ocean connectivity, water chemistry, hydraulics and substrate to the distribution of diadromous snails in Puerto Rico streams. J North Am Benthol Soc 25:82–98

Blettler M, Amsler ML, Ezcurra de Drago I, Marchese MR (2008) Effects of stream hydraulics and other environmental variables on density of *Narapa bonettoi* (Oligochaeta) in the Paraná River system. Riv Res Appl 24:1124–1140

Bonetto AA, Wais JR, Castello HP (1989) The increasing damming of the Parana basin and its effects on the lower reaches. Regul Riv Res Manage 4:333–346

Boulton AJ, Suter PJ (1986) Ecology of temporary streams—an Australian perspective. In: Deckker PD, Williams WD (eds) Limnology in Australia. CSRIO Melbourne, Australia, pp 313–326

Bunn SE, Davies PM (1990) Why is the stream fauna of South-Western Australia so impoverished? Hydrobiologia 194:169–176

Bunn SE, Davies PM, Winning M (2003) Sources of organic carbon supporting the food web of an arid zone floodplain river. Freshw Biol 48:619–635

Calheiros DF, Hamilton SK (1998) Limnological conditions associated with natural fish kills in the Pantanal wetland of Brazil. Verh Int Ver Limnol 26:2189–2193

Capello S, Marchese M, de Wysiecki ML (2013) Orthoptera assemblages associated with macrophytes of floodplain lakes of the Paraná River. Revista Brasil Entomol 57:1–8

Colle AC, Callil CT (2012) Environmental influences on the composition and structure of the freshwater mussels in shallow lakes in the Cuiaba River floodplain. Brazil J Biol 72:249–256

Costa C, Vanin SA (2010) Coleoptera larval fauna associated with termite nests (Isoptera) with emphasis on the bioluminescent termite nests from Central Brazil. Psyche 2010:1–13, http://dx.doi.org/10.1155/2010/723947

Covich AP, Crowl TA, Sherri J, Pyron M (1996) Distribution and abundance of tropical freshwater shrimp along a stream corridor: response to disturbance. Biotropica 28:484–492

Curtis B, Roberts KS et al (1998) Species richness and conservation of Namibian freshwater macro-invertebrates, fish and amphibians. Biodiv Conserv 7:447–466

da Silva FRJ, Marques MI, Battirola LD, Lhano MG (2010) Phenology of *Cornops aquaticum* (Bruner) (Orthoptera: Acrididae) in *Eichhornia azurea* (Pontederiaceae) in the Northern Region of Pantanal of Mato Grosso, Brazil. Neotrop Entomol 39:535–542

Darrigran G (2002) Potential impact of filter-feeding invaders on temperate inland freshwater environments. Biol Invasions 4:145–156

Davidson TA, Mackay AW et al (2012) Seasonal and spatial hydrological variability drives aquatic biodiversity in a flood-pulsed, sub-tropical wetland. Freshw Biol 57:1253–1265

de Oliveira MT, Damasceno-Junior GA et al (2014) Regeneration of riparian forests of the Brazilian Pantanal under flood and fire influence. Forest Ecol Manage 331:256–263

De-Lamonica-Freire E, Heckman CW (1996) The seasonal succession of biotic communities in wetlands of the tropical wet-and-dry climatic zone III: the algal communities in the Pantanal of Mato Grosso, Brazil, with a comprehensive list of the known species and revision of two desmid taxa. Int Rev Ges Hydrobiol 81:253–280

Drago EC (1990) Hydrological and geomorphological characteristics of the hydrosystem of the middle Paraná River. Acta Limnol Bras 111:907–930

Drago EC, Paira AR, Wantzen KM (2008a) Channel-floodplain geomorphology and connectivity of the Lower Paraguay hydrosystem. Ecohydrol Hydrobiol 8:31–48

Drago EC, Wantzen KM, Paira AR (2008b) The Lower Paraguay river-floodplain habitats in the context of the Fluvial Hydrosystem Approach. Ecohydrol Hydrobiol 8:49–66

Ezcurra de Drago IE, Marchese M, Wantzen KM (2004) Benthos of a large neotropical river: spatial patterns and species assemblages in the Lower Paraguay and its floodplains. Archiv Hydrobiol 160:347–374

Ezcurra de Drago I, Marchese M, Montalto L (2007) Benthic invertebrates. In: Iriondo MH (ed) The Middle Paraná River: limnology of a subtropical wetland. Springer, Berlin, pp 251–275
Fellerhoff C (2002) Feeding and growth of apple snail *Pomacea lineata* in the Pantanal wetland, Brazil—a stable isotope approach. Isotopes Environ Health Stud 38:227–243
Fellerhoff C, Voss M, Wantzen KM (2003) Stable carbon and nitrogen isotope signatures of decomposing tropic macrophytes. Aquat Ecol 37:361–375
Fischer H, Kloep F, Wilczek S, Pusch M (2005) A river's liver—microbial processes within the hyporheic zone of a large lowland river. Biogeochemistry 76:349–371
Flecker AS (1992) Fish predation and the evolution of invertebrate drift periodicity evidence from neotropical streams. Ecology 73:438–448
Franceschini MC, Poi de Neiff A, Galassi ME (2010) Is the biomass of water hyacinth lost through herbivory in native areas important? Aquat Bot 92:250–256
Furch K, Junk WJ (1997) Physicochemical conditions in floodplains. In: Junk WJ (ed) The Central Amazonian floodplain: ecology of a pulsing system. Springer, Berlin, pp 69–108
Graça MAS, Ferreia M et al (2015) A conceptual model of litter breakdown in low order streams. Internat Rev Hydrobiol 100:1–12
Haase K, Wantzen KM (2008) Analysis and decomposition of condensed tannins in tree leaves. Environ Chem Lett 6:1610–1653
Hamilton SK (1999) Potential effects of a major navigation project (Paraguay-Parana Hidrovia) on inundation in the Pantanal floodplains. Reg Riv Res Manage 15:298–299
Hamilton SK (2002) Human impacts on hydrology in the Pantanal wetland of South America. Water Sci Technol 45:35–44
Hamilton SK, Sippel SJ, Lewis WM, Saunders Ii JF (1990) Zooplankton abundance and evidence for its reduction by macrophyte mats in two Orinoco floodplain lakes. J Plankton Res 12:345–363
Hamilton SK, Sippel SJ, Calheiros DF, Melack JM (1997) An anoxic event and other biogeochemical effects of the Pantanal wetland on the Paraguay River. Limnol Oceanogr 42:257–272
Heckman CW (1994) The seasonal succession of biotic communities in wetlands of the tropical wet-and-dry climatic zone I: physical and chemical causes and biological effects in the Pantanal of Mato Grosso, Brazil. Int Revue Ges Hydrobiol 79:397–421
Heckman CW (1997) Description of a dynamic ecotone in the Pantanal of Mato Grosso, Brazil. Ecotropica 3:1–76
Hynes HBN (1970) The ecology of running waters. University of Toronto Press, Toronto
Irion G, Buchas H et al (2011) Aspects of geological and sedimentological evolution of the Pantanal plain. In: Junk WJ, Da Silva CJ et al (eds) The Pantanal: ecology, biodiversity and sustainable management of a large neotropical seasonal wetland. Pensoft, Sofia-Moscow, pp 47–69
Junk WJ (1997) The Central Amazon floodplain—ecology of a pulsing system. Springer, Berlin
Junk WJ, Robertson BA (1997) Aquatic invertebrates. In: Junk WJ (ed) The Central Amazon floodplain—ecology of a pulsing system. Springer, Berlin, pp 279–298
Junk WJ, Wantzen KM (2004) The flood pulse concept: new aspects, approaches, and applications—an update. In: Welcomme R, Petr T (eds) Proceeding of the second international symposium on the management of large rivers for fisheries, vol 2. FAO & Mekong River Commission, FAO Regional Office Asia and the Pacific, Bangkok, pp 117–149, RAP Publ 2004/16
Junk WJ, Bayley PB, Sparks RE (1989) The flood pulse concept in river-floodplain systems. Can Spec Publ Fish Aquat Sci 106:110–127
Junk WJ, Nunes da Cunha C et al (2006) Biodiversity and its conservation in the Pantanal of Mato Grosso, Brazil. Aquat Sci 68:278–309
Junk WJ, Piedade MTF et al (2014) Brazilian wetlands: their definition, delineation, and classification for research, sustainable management, and protection. Aquat Conserv Mar Freshw Ecosys 24:5–22
Lewis WM, Hamilton SK, Saunders JF (1995) Rivers of Northern South America. In: Cushing CE, Cummins KW, Minshall GW (eds) River and stream ecosystems. Elsevier, Amsterdam, pp 219–256

Lim TY, Lim YY, Yule CM (2014) Bioactivity of leaves of Macaranga species in tropical peat swamp and non-peat swamp environments. J Trop Forest Sci 26:134–141

Loverde-Oliveira SM, Adler M, Pinto Silva V (2011) Phytoplankton, periphyton, and metaphyton of the Pantanal floodplains: species composition and richness, density, biomass, and primary production. In: Junk WJ, Da Silva CJ, Nunes da Cunha C, Wantzen KM (eds) The Pantanal: ecology, biodiversity and sustainable management of a large neotropical seasonal wetland. Pensoft, Sofia-Moscow, pp 235–255

Loyola RD, Brito SL, Ferreira RL (2006) Ecosystem disturbances and diversity increase: implications for invertebrate conservation. Biodiver Conserv 15:25–42

Maltchik L, Silva-Filho M (2000) Resistance and resilience of the macroinvertebrate community to disturbance by flood and drought in a Brazilian semiarid ephemeral stream. Acta Biol Leopoldensia 22:171–184

Marchese MR, Ezcurra de Drago I (1992) Benthos of the lotic environments in the middle Parana River system: transverse zonation. Hydrobiologia 237:1–13

Marchese MR, Paggi AC (2003) Diversidad de Oligochaeta (Annelida) y Chironomidae (Diptera) del Litoral. INSUGEO, Miscelánea 12:5–12

Marchese MR, Escurra de Drago I, Drago EC (2002) Benthic macroinvertebrates and physical habitat relationships in the Paran river-floodplain system. In: McClain ME (ed) The ecohydrology of South American rivers and wetlands. International Association of Hydrological Sciences, Oxfordshire, pp 111–132, Spec Publ 6

Marchese MR, Wantzen KM, Ezcurra de Drago I (2005) Benthic invertebrate assemblages and species diversity patterns of the upper Paraguay River. Riv Res Appl 21:485–499

Marchese MR, Saigo M et al (2014) Food webs of the Parana River floodplain: assessing basal sources using stable carbon and nitrogen isotopes. Limnologica 46:22–30

Marmonier P, Archambaud G et al (2012) The role of biodiversity in hyporheic processes: gaps in knowledge and needs for applications. Internat J Limnol (Annal Limnol) 48:253–266

Marques MI, Santos GB, Battirola LD, Tissiani ASO (2009) Entomofauna associada à matéria orgânica em bainhas foliares de *Attalea phalerata* Mart. (Arecaceae) na região norte do Pantanal de Mato Grosso. Acta Biológica Paranaense 38:93–112

Marques Couceiro SR, Hamada N, Forsberg BR, Padovesi-Fonseca C (2010) Effects of anthropogenic silt on aquatic macroinvertebrates and abiotic variables in streams in the Brazilian Amazon. J Soils Sediments 10:89–103

Marques MI, Adis J, Nunes da Cunha C, Brizzola dos Santos G (2001) Arthropod biodiversity in the canopy of *Vochysia divergens* (Vochysiaceae), a forest dominant in the Brazilian Pantanal. Stud Neotrop Fauna Environ 36:205–210

Marques MI, Adis J, Santos GB, Battirola LD (2006) Terrestrial arthropods from tree canopies in the Pantanal of Mato Grosso, Brazil. Revista Brasil Entomol 50:257–267

Marques MI, Adis J et al (2011) Arthropods associated with a forest of *Attalea phalerata* Mart. (Arecaceae) palm trees in the northern Pantanal of the Brazilian state of Mato Grosso. In: Junk WJ, Da Silva CJ, Nunes da Cunha C, Wantzen KM (eds) The Pantanal: ecology, biodiversity and sustainable management of a large neotropical seasonal wetland. Pensoft, Sofia-Moscow, pp 431–468

Marques MI, Santos GB, Battirola LD (2014) Cerambycidae (Insecta, Coleoptera) associados à *Vochysia divergens* Pohl (Vochysiaceae) na região norte do Pantanal de Mato Grosso, Brasil. EntomoBrasilis (Vassouras) 7:159–160

Medvedev LN (2001) A new species of *Lysathia* Bechyne, 1959 from the Pantanal, Brazil (Insecta, Coleoptera, Chrysomelidae, Alticinae). Amazoniana 16:417–420

Messner B, Adis J (1987) Die Plastronstrukturen der submers lebenden neotropischen Bärenraupe von Palustra laboulbeni Bar, 1873 (Lepidoptera: Arctiidae). Zoologische Jahrbücher Abteil Anatom Ontogen Tiere (Springer Berl) 115:531–538

Montalto L, Ezcurra de Drago I (2003) Tolerance to desiccation of an invasive mussel, *Limnoperna fortunei* (Dunker, 1857) (Bivalvia, Mytilidae), under experimental conditions. Hydrobiologia 498:161–167

Montalto L, Marchese M (2005) Cyst formation in Tubificidae (Naidinae) and Opistocystidae (Annelida, Oligochaeta) as an adaptive strategy for drought tolerance in fluvial wetlands of the Parana River, Argentina. Wetlands 25:488–494

Montalto L, Paggi A (2006) Diversity of chironomid larvae in a marginal fluvial wetland of the Middle Paraná River floodplain, Argentina. Ann Limnol Int J Limnol 42:289–300

Montalto L, Oliveros OB, Ezcurra de Drago I, Demonte LD (1999) Peces del rio Paraná Medio predadores de una espécie invasora: Limnoperna fortunei (Bivalvia, Mytilinae). Revista Facult Bioquim Cien Biol Unive Nacional Litoral (Santa Fé, Argent) 3:85–101

Nakano S, Murakami M (2001) Reciprocal subsidies: dynamic interdependence between terrestrial and aquatic food webs. Proc Natl Acad Sci U S A 98:166–170

Neiff JJ (1996) Large rivers of South America: towards the new approach. Verh Int Verein Limnol 26:167–180

Neiff JJ, Casco SL et al (2011) Vegetation diversity in a large neotropical wetland during two different climatic scenarios. Biodiver Conserv 20:2007–2025

Nogueira F, Couto EG, Bernardi CJ (2002) Geostatistics as a tool to improve sampling and statistical analysis in wetlands: a case study on dynamics of organic matter distribution in the Pantanal of Mato Grosso, Brazil. Brazil J Biol 62:861–870

Nogueira FMB, Silveira RML et al (2011) Hydrochemistry of lakes, rivers and groundwater. In: Junk WJ, Da Silva CJ, Nunes da Cunha C, Wantzen KM (eds) The Pantanal: ecology, biodiversity and sustainable management of a large neotropical seasonal wetland. Pensoft, Sofia-Moscow, pp 167–198

Nolte U (1995) From egg to imago in less than seven days: *Apedilum elachistum* (Chironomidae). In: Cranston PS (ed) Chironomids, from genes to ecosystems. CSIRO, Melbourne, Australia, pp 177–184

Nolte U, Tietböhl RS, McCafferty WP (1996) A mayfly from tropical Brazil capable of tolerating short-term dehydration. J North Am Benthol Soc 15:87–94

Nunes da Cunha C, Junk WJ (2004) Year-to-year changes in water level drive the invasion of Vochysia divergens in Pantanal grasslands. Appl Vegetat Sci 7:103–110

Oliveira MD, Takeda AM et al (2006) Invasion by *Limnoperna fortunei* (Dunker, 1857) (Bivalvia, Mytilidae) of the Pantanal wetland, Brazil. Biol Invasions 8:97–104

Oliveira MD, Calheiros DF, Jacobi CM, Hamilton SK (2011) Abiotic factors controlling the establishment and abundance of the invasive golden mussel *Limnoperna fortunei*. Biol Invasions 13:717–729

Pagotto MA, Silveira RDL, da Cunha CN, Fantin-Cruz I (2011) Distribution of herbaceous species in the soil seed bank of a flood seasonality area, Northern Pantanal, Brazil. Internat Rev Hydrobiol 96:149–163

Parolin P (2009) Submerged in darkness: adaptations to prolonged submergence by woody species of the Amazonian floodplains. Annal Bot 103:359–376

Pereira D, Dreher Mansur MC et al (2014) Bivalve distribution in hydrographic regions in South America: historical overview and conservation. Hydrobiologia 735:15–44

Pinheiro TG, Marques MI, Battirola LD (2009) Life cycle of *Poratia salvator* Sierwald & Golovatch, 2000 (Diplopoda, Polydesmida, Pyrgodesmidae). Zoologia (Curitiba) 26:658–662

Pinheiro TG, Battirola LD, Marques MI (2011) Fertility tables of two populations of the parthenogenetic species *Poratia salvator* (Diplopoda, Polydesmida, Pyrgodesmidae). Brazilian Journal of Biology 71(2):501–510

Poff NL, Allan JD et al (1997) The natural flow regime—a paradigm for river conservation and restoration. BioScience 47:769–784

Poi de Neiff ASG (2003) Macroinvertebrates living on *Eichhornia azurea* Kunth in the Paraguay River. Acta Limnol Bras 15:55–63

Poi de Neiff A, Casco SL (2003) Biological agents that accelerate winter decay of *Eichhornia crassipes* (Mart.) Solms in northeast Argentina. In: Thomaz SM, Bini LM (eds) Ecologıa e manejo de macrofitas aquaticas. Editora da Universidade Estadual de Maringa, Maringa, Brasil, pp 128–144

Poi de Neiff ASG, Neiff JJ (2006) Riqueza de especies y similaridad de los invertebrados que viven en plantas flotantes de la planicie de inundación del Río Paraná (Argentina). Interciencia 31:220–225

Pott A, Pott VJ (2004) Features and conservation of the Brazilian Pantanal wetland. Wetl Ecol Manage 12:547–552

Pott VJ, Pott A et al (2011) Aquatic macrophyte diversity of the Pantanal wetland and upper basin. Braz J Biol 71:255–263

Ramirez A, Pringle CM, Wantzen KM (2008) Tropical stream conservation. In: Dudgeon D (ed) Tropical stream ecology. Elsevier, London, pp 286–304

Ramsar Convention Secretariat (2006) The Ramsar Convention Manual: a guide to the convention on wetlands (Ramsar, Iran, 1971), 4th edn. Ramsar Convention Secretariat, Gland, Switzerland

Ricaurte LF, Wantzen KM et al (2014) Participatory rural appraisal of ecosystem services of wetlands in the Amazonian Piedmont of Colombia: elements for a sustainable management concept. Wetl Ecol Manage 22:343–361

Roff DA (1992) The evolution of life histories: theory and analysis. Chapman Hall, London

Santos GB, Marques MI, Adis J, Musis CR (2003) Artrópodos associados a copa de *Attalea phalerata* Mart. (Arecaceae) na região do Pantanal de Poconé-MT. Revista Brasil Entomol 47:211–224

Santos AD, Hayd L, Anger K (2013) A new species of *Macrobrachium* Spence Bate, 1868 (Decapoda, Palaemonidae), *M. pantanalense*, from the Pantanal, Brazil. Zootaxa 3700(4):534–546

Silva FRJ, Battirola LD et al (2014) Morphometry of *Cornops aquaticum* (Orthoptera: Acrididae: Leptysminae) in the Pantanal of Mato Grosso, Brazil. Brazil J Biol 74:730–738

Sobreira E, Butakka CMM, Wantzen KM (2012) Dinâmica de larvas de Chironomidae na Baía do Coqueiro, Pantanal Matogrossense, MT. Revista Brasil Zoociên 14:9–25

Souza-Franco GM, Takeda AM (2000) Invertebrates associated with *Paspalum repens* (Poaceae) at the mouth of Caracu Stream (1991-1992), affluent of the Paraná River, Porto-Rico (PR) Brazil. Braz Arch Biol Technol 43:317–325

Sparks RE, Nelson JC, Yin Y (1998) Naturalization of the flood regime in regulated rivers. BioScience 48:706–720

Stanford JA, Lorang MS, Hauer FR (2005) The shifting habitat mosaic of river ecosystems. Verh Internat Verein Limnol 29:123–126

Takeda AM, Grzybkowska M (1996) Seasonal dynamics and production of *Campsurus violaceus* nymphs (Ephemeroptera, Polymitarcyidae) in the Baia River, upper Parana River floodplain, Brazil. Hydrobiologia 356:149–155

Thomaz SM, Bini LM, Bozelli RL (2007) Floods increase similarity among aquatic habitats in river-floodplain systems. Hydrobiologia 579:1–13

Tockner K, Lorang MS, Stanford JA (2010) River flood plains are model ecosystems to test general hydrogeomorphic and ecological concepts. Riv Res Appl 26:76–86

Vega LF, da Cunha CN et al (2014) Does flood pulsing act as a switch to store or release sediment-bound carbon in seasonal floodplain lakes? Case study from the Colombian Orinoco-Llanos and the Brazilian Pantanal. Wetlands 34:177–187

Verschuren D, Tibby J, Sabbe K, Roberts N (2000) Effects of depth, salinity, and substrate on the invertebrate community of a fluctuating tropical lake. Ecology 81:164–182

Wantzen KM (2003) Cerrado streams—characteristics of a threatened freshwater ecosystem type on the tertiary shields of South America. Amazoniana 17:485–502

Wantzen KM (2006) Physical pollution: effects of gully erosion on benthic invertebrates in a tropical clear-water stream. Aquat Conserv Mar Freshw Ecosys 16:733–749

Wantzen KM, Junk WJ (2000) The importance of stream-wetland-systems for biodiversity: a tropical perspective. In: Gopal B, Junk WJ, Davies JA (eds) Biodiversity in Wetlands: assessment, function and conservation. Backhuys, Leiden, Netherlands, pp 11–34

Wantzen KM, Junk WJ (2006) Aquatic-terrestrial linkages from streams to rivers: biotic hot spots and hot moments. Arch Hydrobiol Suppl 158:595–611

Wantzen KM, Mol J (2013) Soil erosion from agriculture and mining: a threat to tropical stream ecosystems. Agriculture 3:660–683

Wantzen KM, Rueda-Delgado R (2009) Técnicas de mustreo de invertebrados acuáticos. In: Dominguez E, Fernandez HR (eds) Macroinvertebrados Bentonicos Sudamericanos. Editorial Fundacion Miguel Lillo, San Miguel de Tucumán, Argentina, pp 17–45

Wantzen KM, Wagner R (2006) Detritus processing by invertebrate shredders: a neotropical-temperate comparison. J North Am Benthol Soc 25:216–232

Wantzen KM, Machado FA et al (2002) Seasonal isotopic shifts in fish of the Pantanal wetland, Brazil. Aquat Sci 64:239–251

Wantzen KM, Drago E, da Silva CJ (2005) Aquatic habitats of the Upper Paraguay River-Floodplain-System and parts of the Pantanal (Brazil). Ecohydrol Hydrobiol 21:1–15

Wantzen KM, Ramirez A, Winemiller KO (2006) New vistas in neotropical stream ecology—preface. J North Am Benthol Soc 25:61–65

Wantzen KM, Junk WJ, Rothhaupt KO (2008a) An extension of the floodpulse concept (FPC) for lakes. Hydrobiologia 613:151–170

Wantzen KM, Mathooko J, Yule C, Pringle CM (2008b) Organic matter processing in tropical streams. In: Dudgeon D (ed) Tropical stream ecology. Elsevier, London, pp 43–64

Wantzen KM, Nunes da Cunha C et al (2008c) Towards a sustainable management concept for ecosystem services of the Pantanal wetland. Ecohydrol Hydrobiol 8:77–100

Wantzen KM, Yule C, Tockner K, Junk WJ (2008d) Riparian wetlands of tropical streams. In: Dudgeon D (ed) Tropical stream ecology. Elsevier, London, pp 199–218

Wantzen KM, Callil C, Butakka CMM (2011) Benthic invertebrates of the Pantanal and its tributaries. In: Junk WJ, da Silva CJ, Nunes da Cunha C, Wantzen KM (eds) The Pantanal: ecology, biodiversity and sustainable management of a large neotropical seasonal wetland. Pensoft, Sofia, pp 393–430

Wantzen KM, Blettler MCM et al (2014) Sandy rivers: a review on general ecohydrological patterns of benthic invertebrate assemblages across continents. Internat J River Basin Manage 12:163–174

Ward JV, Stanford JV (1983) The intermediate-disturbance hypothesis: an explanation for biotic diversity patterns in lotic ecosystems. In: Fontaine TD, Bartell SM (eds) Dynamics of lotic ecosystems. Ann Arbor Science, Ann Arbor, pp 347–356

Ward JV, Tockner K, Arscott DB, Claret C (2002) Riverine landscape diversity. Freshw Biol 47:517–539

Yule CM (2010) Loss of biodiversity and ecosystem functioning in Indo-Malayan peat swamp forests. Biodiver Conserv 19:393–409

Zarfl C, Lumsdon AE et al (2015) A global boom in hydropower dam construction. Aquat Sci 77:161–170

Zeilhofer P, de Moura RM (2009) Hydrological changes in the northern Pantanal caused by the Manso dam: impact analysis and suggestions for mitigation. Ecol Eng 35:105–117

Zerm M, Adis J (2002) Flight ability in nocturnal tiger beetles (Coleoptera: Carabidae: Cicindelinae) from Central Amazonian Floodplains (Brazil). Coleopt Bull 56:491–500

Zerm M, Walenciak O, Val AL, Adis J (2004) Evidence for anaerobic metabolism in the larval tiger beetle, *Phaeoxantha klugii* (Col. Cicindelidae) from a Central Amazonian floodplain (Brazil). Physiol Entomol 29:483–488

Zilli FL, Marchese MR (2010) Patterns in macroinvertebrate assemblages at different spatial scales. Implications of hydrological connectivity in a large floodplain river. Hydrobiologia 663:245–257

Zilli FL, Montalto L (2011) Benthic invertebrates in the Middle Paraná River Floodplain (Argentina). In: Alvarez MA (ed) Floodplains. Nova Science, New York, pp 99–126

Zilli FL, Montalto L, Marchese MR (2008) Benthic invertebrates diversity patterns and functional feeding groups in the Middle Paraná River Floodplain. Limnologica 38:159–171

Chapter 15
Invertebrates in Created and Restored Wetlands

Albert Ruhí, G. Winfield Fairchild, Douglas J. Spieles, Gustavo Becerra-Jurado, and David Moreno-Mateos

Introduction

A Single Term for a Variety of Systems

As natural wetlands have declined in many parts of the world (Dahl 1990; Gibbs 2000; Daniels and Cumming 2008), created wetlands and ponds have augmented, and in some regions dominated, total habitat available to lentic invertebrates. The terms *created* (or constructed, or man-made) *wetlands* and *ponds* are sometimes used to designate distinct ecosystems, especially with regard to their size and morphology (with created wetlands generally being comparatively bigger, shallower, and with more naturalized shores than ponds). However, both created wetlands and ponds present the interconnected three-component basis of lentic ecosystems (hydrology, physicochemical environment, and biota); and they can both serve a variety of purposes: they may be designed to reduce flooding (Persson et al. 1999), to retain sediments and organic carbon (Smith et al. 2002; Brainard and Fairchild 2012), to sequester and transform nutrients (Brix 1994; Sundaravadivel and Vigneswaran 2001), to intercept toxic pollutants in runoff during storm events (Herrmann 2012), to enhance biodiversity (Williams et al. 2004; Oertli et al. 2005; Ruhí et al. 2009, 2012b), or to fulfill several of these objectives simultaneously (Hansson et al. 2005; Becerra-Jurado et al. 2010; Ruhí et al. 2011). The frequent consideration of created ponds and wetlands as separate entities is likely unjustified from an ecological standpoint, since it ignores their joint contribution to biodiversity and ecosystem functioning at the landscape scale.

Throughout this chapter, the term *created wetland* (hereafter, CW) will be used to generically designate any lentic habitat with hydrophytes, undrained hydric soils, and shallow water at least during the growing season (after

D. Batzer, D. Boix (eds.), *Invertebrates in Freshwater Wetlands*,
DOI 10.1007/978-3-319-24978-0_15

Cowardin et al. 1979) that has been created *de novo* (by humans) in a place without wetland history. Contrary to recent papers that circumscribe the use of CW to treatment wetlands, here it is considered that given the variety of existing CWs, this practice can be misleading. Thus, throughout this chapter the term *CW* will remain generic and CW type will be specified in each case. Likewise, the term *restored wetland* (hereafter, RW) will refer to any wetland habitat reclaimed in former wetland habitat, without any assumption on the purpose and recovery of previous or reference physical, chemical, and biological processes.

Origins and History of Created and Restored Wetlands

Created and restored wetlands are numerically important in Europe and the United States, and still precursory but undergoing a steep increase in popularity in Asia (especially in China), Central and South America, some African countries, and Australia and New Zealand (Denny 1997; Kivaisi 2001; Kadlec and Wallace 2008; Liu et al. 2008; Ruhí 2012). The currently high rates of wetland construction are partly explained by wastewater treatment wetlands. However, creation and restoration of wetlands occurs for a variety of reasons, with its popularization dating back to the beginning of the twentieth century and being spurred by both economic interest and regulatory enforcement. In the United States, in direct response to devastating soil losses during the Dust Bowl of 1934–1937, approximately two million farm ponds were built during the period 1945–1975, with financial and technical encouragement by the US Soil Conservation Service, as a means of improving soil retention (Tuttle 2008). In 1972, the passage of the federal Clean Water Act represented another inflection point. This Act partly aimed at reducing or reversing substantial losses of wetlands to agriculture, road construction, and urbanization, which had exceeded 80 % of existing wetland habitat in many regions during the previous century (Zedler 2004). Under Section 404 of the legislation, Congress gave the primary responsibility of wetland protection to the US Army Corps of Engineers, which evaluates permit requests from applicants whose actions might impair or destroy existing wetlands. If a permit involving unavoidable wetland loss is issued, the applicant is required to mitigate the damage through construction of new wetlands, restoration of former wetlands, enhancement of existing wetlands, or protection of wetlands threatened with future impairment. Of these options, wetland creation and restoration have most frequently been selected; as a consequence, implementation of "no-net-loss" (NNL) policies in the United States have implied widespread creation and "relocation" of wetlands through a wetland mitigation banking system (Brown and Lant 1999). In Canada, since 1986 a homologous NNL is applied when Fisheries and Oceans Canada (DFO) issues an authorization for development activity affecting fish habitat. In the North American countries combined (including

Mexico) during the last 20 years more than 3,000,000 ha of wetlands have been created or restored (representing an investment of >US$70 billion; Copeland 2010). More recently, wetland creation and restoration has increasingly been motivated by urban/suburban needs for recreation, water and pollutant retention, and aesthetic amenities. In contrast to formally designated wetlands, the rapid proliferation of ponds has been largely determined by local rather than federal legislation.

Outside North America, despite a lack of a large-scale NNL policy, a wide range of human activities have been also responsible for the creation and restoration of lentic habitats, most remarkably in Europe. In the United Kingdom, declines in wetland abundance continued from the late nineteenth century until the 1980s (Boothby and Hull 1997), when new construction reversed the trend (Jeffries 2012). The Environment Agency of this country, Natural England, and county-based Wildlife Trusts have even developed a strategy to restore existing wetlands and double the number of small lentic waterbodies by 2050 (Hume 2008). Extensive wetland creation and restoration is taking place across Europe over the twenty-first century, often enhanced by EU directives for nature protection (especially the "Habitats Directive," Council Directive 92/43/EEC) and the transposed national legislation implemented in the various states (Verhoeven 2014). A considerable proportion of wetland creation and restoration in the EU is therefore biodiversity-focused, and often funded by Life programs that are aimed at preserving particular habitats and species of conservation priority (Silva et al. 2007). This is, in turn, offering opportunities for applied research (Rhazi et al. 2004; Louette et al. 2008; Rannap et al. 2009; Ruhí et al. 2012b) that could fill knowledge gaps on how to restore structure and functioning of lentic ecosystems, a critical need for the still emerging discipline of ecological restoration (Montoya et al. 2012; Moreno-Mateos et al. 2012). As shown in Fig. 15.1a, terms related to wetland creation, construction, and restoration first appeared in English-language books in approximately 1980, and increased rapidly in frequency during the following decade.

A New Landscape with Particular Conditions

Although not all wetlands are created equal from an ecological perspective (i.e., highly modified wastewater treatment wetlands tend to present a lower ecological potential than those designed for biodiversity enhancement), in some landscapes the proliferation of CWs and RWs has produced a dramatic change in the number and type of habitat patches available to aquatic biota. For instance, in six protected areas in S and SE Estonia, in 3 years (2005–2007) a total of 208 CWs for biodiversity enhancement (targeting amphibians) were created and 22 were restored, increasing by 56 % the total number of waterbodies regionally present (Rannap et al. 2009). In Doñana National Park (S Spain), the restoration of a former marshland area (also for biodiversity enhancement) involved the construction of 96 ponds (of several sizes and depths). In that area, the only previously existing waterbodies were ditches and some small rain pools (Frisch and Green 2007). CWs aimed at recreation and biodiversity enhancement in SE Pennsylvania increased 18-fold during the period 1937–2005, and eventually represented 97 % of all lentic waterbodies in the study area (Fairchild et al. 2013). In that county, an important proportion of lentic systems are currently "young" due to wetland creation (Fig. 15.1b). Although these examples illustrate strong interventions at local to regional spatial scales and do not necessarily represent global trends, they show how historical and ongoing episodes of wetland loss and creation (regardless of the driver) can alter wetland habitat density and provide a changing array of habitats to biota. Increases in density of new habitats, and the concomitant decrease in average ecosystem age at the regional scale, has the potential to impact invertebrate communities through changes in their dispersal and successional patterns (Bohonak and Jenkins 2003; Ruhí et al. 2011, 2013; Johnson et al. 2013). Therefore, research on wetland invertebrates needs to consider CWs and RWs as an integral component of freshwater habitats at the landscape level.

Most of the major environmental controls influencing invertebrate composition, structure, and dynamics in CWs and RWs (Appendix) also occur in natural wetlands, and include a wide range of hierarchical filters operating from ecoregion to microhabitat levels (Cañedo-Argüelles and Rieradevall 2011; Ruhí et al. 2012a, 2013; Moreno-Mateos et al. 2012). However, some controls—either environmental or anthropogenic—may affect CWs to a greater extent than natural wetlands, and others may be particularly important in some (but not all) CW types (Table 15.1): climate (Ruhí et al. 2012a), hydrology (Ruhí et al. 2013), and vegetation structure (Fairchild et al. 2000; Ruhí et al. 2011) widely control colonization and primary succession rates, either directly (by influencing dispersal and effective establishment of pioneer colonizers) or indirectly (by mediating biotic interactions). Nutrient loadings are particularly high in CWs aimed at treating wastewater (Vymazal et al. 2006); accordingly, this control has been consistently identified as a major driver of macroinvertebrate composition and diversity in these systems (Spieles and Mitsch 2000; Becerra-Jurado et al. 2009; Hsu et al. 2011). Likewise, CWs for the treatment of acid mine drainage are generally affected by metals (Mitsch and Wise 1998; Mays and Edwards 2001); wetlands created in agricultural landscapes often experience water quality stresses related to nutrient loadings, salinity, and pesticides (Schulz and Peall 2001; Moore et al. 2009; Moreno-Mateos et al. 2010); and nutrient and metal pollution can be important in stormwater CWs (Karouna-Renier and Sparling 2001; Scher and Thièry 2005; Herrmann 2012). Wetlands created for aesthetic values in urban settings generally present low connectivity to other waterbodies, high susceptibility to invasion and associated predation pressure (e.g., from fish and wildfowl), and public overuse (Zedler and Leach 1998). As a consequence of both environmental and anthropogenic controls in CWs and RWs, most of the conditions that invertebrates experience are specific to each wetland type (Table 15.1).

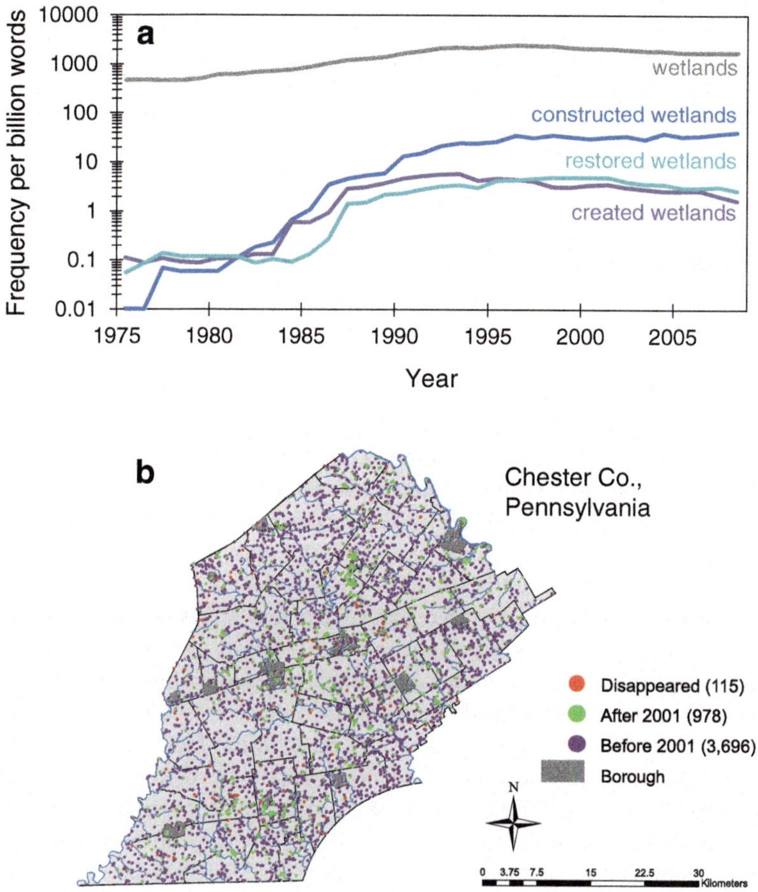

Fig. 15.1 (**a**) The use of three terms describing CWs and RWs, but not the more general term "wetland," has increased drastically since 1985. Terms were compiled using *Google Ngrams* based on 5.2 million books published in English up to 2008 and data points are 3-year moving averages. (**b**) Example of how wetland creation impacts the mean age of lentic ecosystems at the landscape level: in the Chester County, SE Pennsylvania, ≥21 % of the lentic waterbodies are younger than 15 years (modified from Fairchild et al. 2013)

Overview of Invertebrate Research in Created and Restored Wetlands

Research on (or including) invertebrates in CWs and RWs has flourished during the last 20 years, following the rapid increase in popularity of these systems in Europe and the United States (Kivaisi 2001; Liu et al. 2008). Research themes can be divided into the study of (1) colonization, succession, and food-web assembly, from both a fundamental and an applied perspective; (2) the use of invertebrates as indicators of ecological conditions across the wide variety of CW and RW types; and (3) conflicting demands on CW ecosystem services, and particularly between nutrient retention and biodiversity.

Colonization, Succession, and Food-Web Assembly

Invertebrate Colonization and Succession: Phases and Patterns

Newly created aquatic habitats are useful model systems to study primary succession (e.g., Milner et al. 2008; Matthews et al. 2009; Ruhí et al. 2013). In particular, CWs offer quasi-experimental conditions (as they are often relatively small, highly replicable systems) and well-defined boundaries (ideal to test distance effects, species-area curves, and metapopulation/metacommunity theory). According to the classical definition by Connell and Slatyer (1977), succession refers to "changes observed in an ecological community following a perturbation that opens up a relatively large space." Current views reject succession as a linear, directional, and deterministic process (Odum 1969), and suggest it is instead a nonlinear process that rarely reaches equilibrium (Walker and Del Moral 2003). Indeed, potential successional trajectories are multiple; successional rates depend on a variety of local and regional factors, often unpredictable; and primary and secondary succession can be difficult to disentangle (Walker and Del Moral 2003). Despite this ambiguity, some commonalities have been identified, namely the existence of an initial colonization phase preceding any consistent change in community structure (Fisher 1983), and the dichotomy between allogenic phases (i.e., driven by exogenous, abiotic controls) and autogenic phases (i.e., driven by in-pond, biotic interactions). In CWs and RWs, invertebrate communities tend to show three successional periods that have been named as the *colonization*, *mid succession*, and *advanced succession* phases (Fig. 15.2; Ruhí 2012).

The Colonization Phase

The first phase or colonization phase is characterized by a sudden increase in species richness and a dominance of some pioneer active dispersers, shared with source (nearby) natural wetlands (Fig. 15.2) (Herrmann et al. 2001; Ruhí et al. 2009). Colonization of created wetlands by invertebrates implies (1) dispersal from source populations to sink habitats, and (2) successful establishment. On the one hand, invertebrate dispersal between noncontiguous habitat patches is dependent on emigration rates (controlled by both biotic interactions and abiotic conditions

Table 15.1 Environmental and anthropogenic controls of invertebrate composition, structure, and dynamics in different types of CWs organized by wetland type where each particular control is likely to be most relevant

Wetland type where control is relevant	Control	Rationale	Reference
All CW types (including CWs for biodiversity enhancement)	Climate	During early stages after wetland construction, invertebrate colonization rates and changes in community structure depend on the amount of effective time that insects have had to disperse	Ruhí et al. (2012a)
	Hydroperiod	Extreme hydrological variability (e.g., intermittence) slows down primary succession and prolongs pioneer invertebrate communities	Ruhí et al. (2013)
	Submersed aquatic vegetation	Establishment of aquatic macrophytes has a positive feedback on succession by increasing habitat structure and complexity, and by reducing the foraging efficiency of predators	Gee et al. (1997), Fairchild et al. (2000), Ruhí et al. (2011)
CWs for domestic and municipal wastewater treatment; stormwater CWs	Water quality	Invertebrate richness and diversity tends to respond negatively to high nutrient loadings via differential loss of oligosaprobic organisms	Spieles and Mitsch (2000), Becerra-Jurado et al. (2009), Hsu et al. (2011), Herrmann (2012)
CWs for treatment of acid mine drainage	Water quality	Precipitation of metals (e.g., iron hydroxides) interferes with oxygen uptake, although rarely represents direct toxicity	Mays and Edwards (2001)
CWs for agricultural landscape restoration	Water quality	Nonpoint-source pesticide pollution via agricultural runoff increases mortality of wetland insect larvae	Schulz and Peall (2001)
Urban CWs	Connectivity	Habitat fragmentation may limit dispersal and hence prevent "natural" community dynamics	Zedler and Leach (1998)
	Invasion	A high propagule pressure may facilitate invasion by alien predators and cause top-down effects in wetland food webs	Zedler and Leach (1998)
	Public overuse	Trampling and vandalism may cause physical disturbance of wetland habitat	Zedler and Leach (1998)

that stimulate departure, Baines et al. 2014) and on the interaction between habitat connectivity (i.e., the spatial template of aquatic habitats at the landscape level) and species dispersal capacity (which differs widely among invertebrate groups, Bilton et al. 2001; Bohonak and Jenkins 2003). Active dispersers (especially flying adult hemipterans, dipterans, and coleopterans) may dominate during the colonization stage. In turn, passive dispersers may travel long distances and start to arrive shortly after, creating distinct spatial patterns of diversity due to the stochastic processes that govern their dispersal (Ruhí et al. 2013). Moreover, research examining the effects of habitat isolation on the abundance of active early colonists (the backswimmer *Notonecta*) has shown that it may be inaccurate to consider colonization probability as a monotonically decreasing function of habitat isolation, since colonizers may be behaviorally adapted to bypass habitats they find early in the dispersal process to settle later on (McCauley et al. 2009). This advances the notion that successional trajectories may be more complex than predicted by models that focus on habitat template and species' dispersal abilities alone. Habitat connectivity does not only determine local composition and structure, but also beta diversity or variation in composition among local communities (Forbes and Chase 2002) and susceptibility to invasion. On the other hand, when propagules arrive at the vacant habitat probabilities of successful colonization can be further controlled by water and habitat factors like emergent plant cover, light reflectance from the water surface, water chemistry, or fish presence (De Szalay and Resh 2000; Resetarits and Binckley 2009, 2014). Overall, dispersal capacity and habitat connectivity largely mediate the linkage between emigration and colonization rates, but much of the subsequent trajectory depends on local environmental characteristics and biotic interactions.

The Mid- and the Advanced Succession Phases

The colonization phase in CWs shares many characteristics with the *flooding* phase observed in natural temporary wetlands (after Lake et al. 1989), namely a distinct community, an increasing species richness over time, and a dominance of aerial colonizers (Moorhead et al. 1998; Schneider 1999; Boix et al. 2004). However, subsequent successional phases observed in CWs differ in several ways from the phases described under the temporary wetland paradigm. Whereas the mid- and the advanced succession phases have been described as autogenic in CWs (Fig. 15.2; modified from Ruhí 2012), in natural temporary wetlands undergoing secondary succession the final phase is considered allogenic (Lake et al. 1989; Boix et al. 2004).

The mid-succession phase in CWs is characterized by the arrival of passive dispersers that increase idiosyncrasy among localities (Ruhí et al. 2013), and by profound changes in community structure (Barnes 1983; Solimini et al. 2003; Ruhí et al. 2013). Besides, the advanced succession phase in CWs has been usually studied by means of space-for-time substitution approaches (e.g., by contrasting

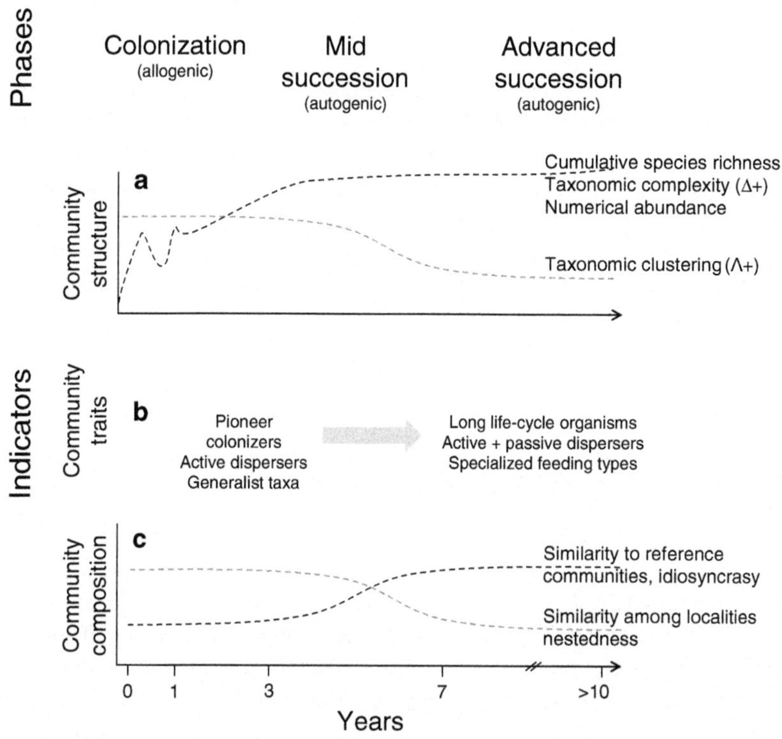

Fig. 15.2 Synthesis of the observed temporal changes in (**a**) community structure (species richness, abundance, taxonomic distinctness), (**b**) dominant biological traits (life cycle, dispersal, feeding specialization), and (**c**) composition trajectories (relative to reference or natural communities) that invertebrate communities in CWs and RWs tend to display over time (modified from Ruhí 2012)

community patterns over CWs differing in age); as a consequence, hypotheses and constructs on how communities may change beyond the 3-year horizon are currently very limited. Classical succession theory suggests that during the advanced succession phase, invertebrate abundance and cumulative species richness should reach an asymptote. Abundance should respond to increasing ecosystem production, attaining maximum values towards the advanced successional phase (Margalef 1968; Odum et al. 1971; Legendre et al. 1985). In turn, the arrival of species new to the system should decrease as succession proceeds and ecological niches become occupied (Margalef 1968). Because species replacing early colonizers may be taxonomically diverse (spread across several taxonomic groups), the relative importance of the few species-rich pioneer groups should decrease over time (Barnes 1983; Christman and Voshell 1993; Herrmann et al. 2001; Solimini et al. 2003; Ruhí et al. 2009). If this is true, assemblage relatedness (for example, measured as taxonomic-tree complexity and clustering, $\Delta+\Lambda+$respectively, Warwick and Clarke 1995;

Clarke and Warwick 2001) should also change over time. These predictions have been empirically corroborated in some cases (see Ruhí et al. 2011 for a 13-year study in a Swedish CW), but they would need to be tested in a variety of CW types, climates, and time horizons. Finally, it is interesting to note that in some cases communities characterizing the *advanced succession* phase may be richer than those found in younger man-made wetlands or in natural wetlands, due to a compositional overlap between these two (Ruhí et al. 2011). Trophic specialists may nevertheless be restricted to natural sites (Fairchild et al. 2000), hence composition similarities, rather than absolute numbers of species, should be used when comparing CWs and RWs to natural wetland communities.

The Importance of the Local Habitat

Throughout primary succession, invertebrate communities are influenced by a rapidly changing environment associated with pond age (Marchetti et al. 2010). Sediment in-filling and redistribution, often accelerated by riparian soil disturbance during the construction process, create microhabitats differing in substrate particle size and organic content (Angélibert et al. 2004). The chemical composition of the organic fraction is likewise variable, particularly when comparing detritus of in-pond vs. terrestrial origin (Spieles and Mitsch 2003). Gradual changes in water depth, in light reaching the sediments, in water temperature, and in oxygen may further affect the conditions that invertebrates experience over the succession phases. Nevertheless, macrophyte establishment is likely the most important single factor facilitating succession (Van de Meutter et al. 2008; Kim et al. 2014). Plants provide structural habitat (e.g., protective cover, adult emergence, and oviposition sites), surface for epiphytic algae, leaf/stem tissues consumed by specialist herbivores, and detrital resources. Therefore, invertebrate food-web assembly highly depends on the successful establishment of macrophytes, both directly (as a resource or a physical support for resource growth) and indirectly (as a way for potential prey to hide and decrease predation pressure; Warfe and Barmuta 2004). This is in agreement with the important role that submersed macrophytes play in lake ecosystem processes (Carpenter and Lodge 1986).

Food-Web Assembly and Controls

Food webs in CWs are often considered three-tiered, with primary producers and detritus as basal sources, and two levels of consumers. As these food webs assemble, the number of biotic interactions increases and particular energetic pathways change in relative importance. A prevalent view is that bottom-up processes may generally dominate (Spieles and Mitsch 2003), with food-web assembly following an approximately predictable pattern of sequential dominance of functional feeding groups (Spieles and Mitsch 2003; Ruhí et al. 2011). Within this context, scrapers of benthic biofilms, filter-feeders that consume detritus, and deposit feeders that collect settled particles, are all likely to benefit from the early arrival and growth of algae. In contrast, shredder abundance and diversity may track availability of plant tissues of terrestrial and in-pond origin (Spieles and Mitsch 2000; Ruhí et al. 2011). Phytoplankton proliferation in shallow ponds can intercept light needed for macrophyte growth, whereas early macrophyte establishment may reduce phytoplankton abundance and potentially produce an alternative stable state (Scheffer et al. 1993). These stable-state changes can have very different bottom-up implications for invertebrates; whereas phytoplankton dominance should favor filter-feeders and collectors, macrophyte dominance may benefit a much wider range of invertebrates based on both habitat modification and increased food availability. Extensive cover by free-floating plants and metaphyton (common in nutrient-rich ponds) also intercepts light at the water surface and suppresses growth of both submersed macrophytes and phytoplankton (Irfanullah and Moss 2005). Additionally, free-floating plants and metaphyton provide habitat and food resources invertebrates adapted for life near the surface (Fontanarrosa et al. 2013), and the settling of relatively short-lived metaphyton scums (Saunders et al. 2012) may also contribute to sediment collectors' diets (Spieles and Mitsch 2003). Whereas filter-feeders and sediment collectors are expected early in succession, more specialized herbivores may depend on the prior establishment of particular plant species. For example, a study by Fairchild et al. (2000) comparing aquatic beetle assemblages of recently created mitigation marshes (age 1–8 years) to those inhabiting nearby reference marshes (age >35 years) found that although overall species richness at the mitigation sites rapidly approached that of the reference wetlands, species within the phytophagous families Curculionidae and Chrysomelidae were typically restricted to the much older reference marshes.

Top-down controls also occur in CW invertebrate food webs, and are often exerted by nonnative vertebrate predators (Zedler and Leach 1998). Fish predation is known to reduce invertebrate density, biomass, size, foraging behavior, and diversity (Fairchild et al. 2000; McPeek 2004; Chase et al. 2009). Fish are often size-selective and prey on multiple trophic levels, and their impacts on the invertebrate community strongly depend on fish identity and abundance (Lazzaro et al. 2009). Consumption by waterfowl (Marklund et al. 2002) can similarly reduce invertebrate abundances. Since game fish are often stocked, and waterfowl encouraged, in recreational CWs (often for "aesthetic" reasons), vertebrate predation pressure may be greater in these systems than in most natural wetlands (Zedler and Leach 1998). Invertebrate community assembly, and successional trajectories as a whole, may highly depend on whether and when these vertebrates enter the local food web.

Invertebrate predators can similarly cause top-down effects in created wetland food webs. In a controlled mesocosm experiment, Chase (2003) found that predation by the heteropteran *Belostoma* suppressed overall herbivore (snail) abundance, thereby augmenting primary production at low-nutrient levels. However, at higher nutrient levels predation modified the relative abundance of the herbivore species but did not exert top-down control of primary production. Given the importance of colonization in CWs, dispersal mass effects may be especially important in these systems, hence variation in invertebrate predator:prey abundance may present food-web implications via top-down controls (Hein and Gillooly 2011). Jeffries (2002) found that after experimentally manipulating relative numbers of predator:prey species in earthen ponds, subsequent dispersal caused a return to the original richness ratios. Although many predators are strong fliers and may occur in small numbers in marginally suitable sites early in succession (Bie et al. 2012; Miguel-Chinchilla et al. 2014), the need for prior establishment of food resources may accentuate apparent dispersal limitation of predator species (Fig. 15.3). Consistent with this prediction, Shulman and Chase (2007) found that mosquitoes were most abundant in isolated ponds, whereas predatory beetles reduced mosquito densities in ponds situated closer to source habitats. Microinvertebrate prey (zooplankton and meiofauna), with their high rates of biomass-specific production (P/B; Stead et al. 2005), could subsidize unexpectedly high predator abundances during early food-web assembly.

Other Considerations: Invasive Plants, Priority Effects, and Compartmentalization

CWs and RWs, owing to their initially high resource availability, are particularly prone to habitat preemption by invasive plants (Spieles 2005; Matthews and Spyreas 2010). A range of effects have been documented, with some invasive plants reducing invertebrate diversity and modifying food-web structure (Zedler and Kercher 2004; Cañedo-Argüelles and Rieradevall 2011). This, combined with priority effects (effects of the order in which species initially enter the community) can potentially cause divergent successional trajectories to develop, relative to a desired or expected reference. The importance of submersed macrophytes may thus be even higher in CWs and RWs than in natural systems (Carpenter and Lodge 1986), since they can affect not only concurrent food-web structure but also future successional trajectories. Within this context, it is important to note that although "local" food-web assembly is typically described for CWs and RWs as a single entity (based on integrated sampling of a wide range of microhabitats), many food-web interactions in lentic ecosystems occur within-microhabitats (Lodge et al. 1988; Burks et al. 2006). This scale-dependency may not only bias estimates of food-web connectance and prey:predator ratios (as observed in streams, Thompson and Townsend 2005), but may also hinder an accurate understanding of how food webs assemble in CWs (Fig. 15.3). Food-web topology may be relatively simple during early succession, and increasingly modular as microhabitats form and differentiate. Models predict that modularity (or compartmentalization) is positively associated with food-web persistence (Stouffer and Bascompte 2011); therefore, increasing habitat complexity should theoretically increase the robustness of the assembling food web. On the whole, CWs and RWs provide a useful arena to study invertebrate primary succession, but a greater understanding of the interplay between habitat succession and food-web assembly based on empirical evidence (currently very limited) would be beneficial if we are to predict successional trajectories and target outcomes.

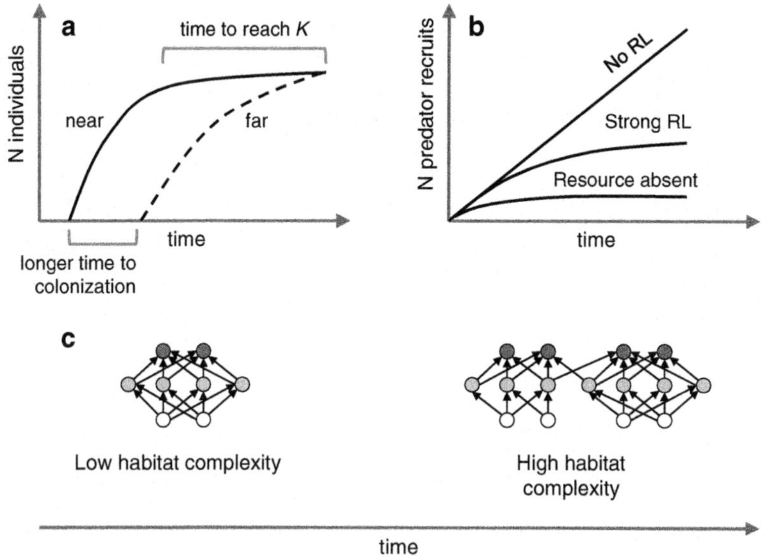

Fig. 15.3 (**a**) Both predators and prey experience dispersal limitation (*x*-axis), which influences the time required to achieve local carrying capacity (*K*); (**b**) Despite early arrival, predator abundances may be strongly resource limited (*RL*) by available prey (adapted from Hein and Gillooly 2011); (**c**) As succession occurs, habitat complexity (physical heterogeneity, macrophyte structure) increases and this has a positive feedback on food-web assembly: more habitat allows for a higher *K*, more niches allow for a higher diversity of ecological traits, and a higher habitat complexity allows for more compartmentalized food webs. Whereas some predators may link the several compartments, most of the interactions occur at the microhabitat scale (*white circles* represent resources, *light gray circles* represent primary consumers, and *dark gray circles* represent secondary consumers)

Invertebrates as Bioindicators of Ecological Conditions in Created and Restored Wetlands

Indicator Potential and Examples of Their Use

Another important focus of research on invertebrate communities in created wetlands has been that of exploring their usefulness as indicators of wetland status and function. Aquatic macroinvertebrates have long been used as bioindicators in streams, dating back to at least Kolkwitz and Marsson's (1908) saprobien system of relating organic pollution to stream communities. The concept has since evolved into many configurations of pollutants and indicator metrics (Goodnight 1973; Cairns and Pratt 1993). Application of the bioindicator approach to lentic systems was spurred in the United States by federal regulation of wetlands, but direct application of indices designed for lotic systems (e.g., Spieles and Mitsch 2000) has been complicated by the hydrogeomorphic diversity of wetland ecosystems, by the unique biotic and abiotic conditions of each system, and by stressors that can differ greatly from those of streams (Adamus and Brandt 1990). Given the wide variety of wetlands and the necessary specificity of site, condition, and objective, there is no single standard approach to the use of macroinvertebrates as wetland bioindicators. That being said, a number of researchers have proposed specific wetland integrity indices based on a local macroinvertebrate community (Table 15.2). In the Great Lakes region, for example, several studies have evaluated the potential of stream invertebrate metrics for use in a biotic index for coastal wetlands (Burton et al. 1999; Kashian and Burton 2000; see Great Lakes marshes chapter). Aimed particularly at seasonal domestic wastewater and stormwater impacts, these studies found the proportions of three indicator taxa (Ephemeroptera, Isopoda, and Trichoptera) and four trophic guilds (predators, collector-filterers, herbivores, and detritivores) to be useful metrics in these wetlands. Notably, 28 of 38 stream metrics tested were not found to be useful in this particular context. In semiarid Mediterranean wetlands of Spain, where stressors include hydrologic, nutrient, and salinity fluctuation, the ratio of Coleoptera to Heteroptera and the overall family richness of aquatic macroinvertebrates were included in a wetland integrity index (Ortega et al. 2004). In urban wetlands of northern California (Lunde and Resh 2012), a screen of 56 macroinvertebrate metrics yielded eight with utility; these included the relative abundance of indicator taxa (Ephemeroptera, Odonata, Trichoptera, Tanypodinae and Chironomidae, Oligochaeta, and Coleoptera) and feeding guilds (scrapers and predators). Ethiopian marshlands subjected to hydrologic and agricultural stress were categorized by macroinvertebrate taxon richness, indicator taxa, and tolerant taxa in an index proposed by Yimer and Mengistou (2010). In the Himalayan foothills of India, where silt and organic suspended solids are stressors, the overall density of aquatic macroinvertebrates was found to be the strongest predictor of wetland condition (Sharma and Rawat 2009). Awal and Svozil (2010) suggested that macroinvertebrate species richness can be used as a primary indicator of wetland integrity in Melbourne, Australia. Clearly, there is wide interest in the use of the invertebrate community as an indicator of wetland condition. It seems, however, that there may be as many macroinvertebrate integrity indices as there are wetlands.

One useful application of wetland macroinvertebrate bioindicators is in the evaluation of CWs and RWs, as compared with reference wetlands (Table 15.3). CWs for mitigation, wastewater treatment, education, habitat provision, or other reasons are typically evaluated by the development of vegetation (Breaux and Serefiddin 1999) or by the efficacy of particular functions, such as water quality improvement (Hosomi et al. 1994). Detailed evaluation of a biotic community can supplement these standard metrics and broaden the understanding of CW development and performance. Research along these lines has shown that wetland macroinvertebrate communities do not necessarily develop at the same rate as wetland vegetation. In an evaluation of mitigation wetlands, Spieles et al. (2006) found that macroinvertebrate taxon richness and diversity did not differ between 10-year old created wetlands and natural reference wetlands, though the proportion of tolerant invertebrates and the trophic structure of the respective macroinvertebrate communities did differ. In this study, the 10-year old macroinvertebrate community was dominated by detri-

Table 15.2 Selected applications of macroinvertebrates as biological indicators of ecological condition in wetlands

Location	Wetland type	Primary stressors	Focal invertebrate metrics	Source
Brazil	Natural Palustrine	Eutrophication Dissolved solids	Odonata abundance, richness Hydrologic	Maltchik et al. (2010)
Ethiopia	Natural Palustrine	Eutrophication Dissolved solids	Overall richness Tolerant taxa richness	Yimer and Mengistou (2010)
India	Created Riverine	Hydrologic	Community density	Sharma and Rawat (2009)
Iran	Natural Lacustrine	Eutrophication Dissolved solids	Taxon richness Diversity index Coleoptera abundance, richness Chironomidae abundance, richness	Ahmadi et al. (2011)
Spain	Natural/created Riverine/palustrine	Eutrophication Dissolved solids Hydrologic	Coleoptera abundance, richness Heteroptera abundance, richness Overall richness	Ortega et al. (2004)
Spain	Natural Palustrine	Eutrophication Habitat alteration	Chironomidae abundance, richness Dytiscidae abundance, richness Odonata abundance, richness Tanypodinae abundance, richness Macropelopini abundance, richness Diversity index	Trigal et al. (2009)
USA: California	Natural/created Palustrine	Land use Dissolved solids	Ephemeroptera abundance, richness Odonata abundance, richness Trichoptera abundance, richness Tanypodinae abundance, richness Oligochaeta abundance, richness Coleoptera abundance, richness Feeding guilds	Lunde and Resh (2012)

(continued)

Table 15.2 (continued)

Location	Wetland type	Primary stressors	Focal invertebrate metrics	Source
USA: Michigan	Natural	Eutrophication	Ephemeroptera abundance, richness	Kashian and Burton (2000)
	Lacustrine	Hypoxia	Trichoptera abundance, richness	
		Dissolved solids	Isopoda abundance, richness	
			Feeding guilds	
USA: North Carolina	Natural Riverine	Highway proximity	Overall richness Taxon dominance Feeding guilds	King et al. (2000)

tivores, while the reference community was herbivore dominated. Other studies have similarly found equal or greater macroinvertebrate richness and diversity in CWs of 4–20 years in age as compared with natural reference wetlands (Levin et al. 1996; Craft et al. 1999; Balcombe et al. 2005), but dissimilar trophic guild distribution among created and reference systems (Craft et al. 1999). As noted before in this chapter (Fig. 15.2), Stewart and Downing (2008) observed that macroinvertebrates are quick to colonize newly created wetlands and rapidly achieve biomass, density, and richness that are equivalent to much older wetlands, but they also noted that macroinvertebrate communities are highly influenced by the plant community and by the availability and quality of coarse particular organic matter. Thus, trophic representation in macroinvertebrate communities may be a useful indicator of vegetation and detrital development in CWs.

Caveats

There are also some important caveats to the use of macroinvertebrates as wetland bioindicators, the first of them being temporal. Macroinvertebrate populations can be highly variable by season and year. In a comparison of two CWs with hydrogeomorphic similarity but different ages (15 and 45 years) and successional stages, Spieles and Horn (2009) found similar macroinvertebrate communities in terms of richness, diversity, biomass, and trophic distribution despite wide differences in water chemistry and soil quality, but community measures within wetlands differed significantly by season. This seasonal variation can make system-wide signals of age, location, or condition difficult to discern, and it means that sampling regimes must be temporally diverse and consistent among sites. The second caveat is spatial; the same sampling technique in different locations of the same wetland can yield very different results. For instance, Balcombe et al. (2005) found differences in the

Table 15.3 Selected applications of macroinvertebrates as biological indicators of ecological condition in created wetlands (CWs) as compared with reference wetlands

Location	CW type	General comparison with reference wetland(s)	Source
Australia	Palustrine	Similar macroinvertebrate communities among three created metropolitan wetlands of similar age	Awal and Svozil (2010)
USA: West Virginia	Palustrine	Similar macroinvertebrate communities among created and natural wetlands; spatial variability within wetlands	Balcombe et al. (2005)
USA: North Carolina	Estuarine	Rapid equivalence with natural system in terms of density and species richness, but not diversity or trophic structure	Levin et al. (1996)
USA: North Carolina	Estuarine	Created wetland macroinvertebrate equivalence with natural system by year 15, 5–10 years after plant community	Craft et al. (1999)
USA: Iowa	Lacustrine	Macroinvertebrate communities vary by wetland, strongly influenced by CPOM, vegetation and turbidity	Stewart and Downing (2008)
USA: Ohio	Palustrine	Richness similar in created and natural wetlands, but trophic structure differs by construction method	Spieles et al. (2006)
USA: Ohio	Palustrine	Trophic differences in wetlands of different successional stages; seasonal community variability within wetlands	Spieles and Horn (2009)

macroinvertebrate communities in benthic and water column samples in vegetated vs. open water habitat. As with temporal variability, this spatial variability could easily confound assessments. A third caveat deals with sampling and identification effort. Excessive sampling and species-level identification could easily become prohibitively laborious, and yet minimal sampling and overgeneralization may not provide adequate resolution. A useful discussion of this problem is provided by King and Richardson (2002), who evaluated the efficacy of different levels of sampling intensity. Finally, any macroinvertebrate-based assessment of wetlands needs to account for differences in landscape and geographic setting. To this point, Hall et al. (2004) found landscape characteristics to be significant factors for wetland macroinvertebrate communities in Texas, while Davis et al. (2006) identified the need for regional specificity as the primary impediment to widespread use of macroinvertebrate bioindicators for wetlands in Australia. Together, these caveats suggest limitations to the utility of macroinvertebrate bioindicators in wetlands. However, when used as part of a suite of biological and physicochemical measurements, the macroinvertebrate community can be an important tool in the understanding and evaluation of CW and RW condition.

Conflicting Demands on Wetland Ecosystem Services: Integrated Constructed Wetlands as a Paradigm

Nutrient Retention and Biodiversity Enhancement, an Assumed but Rarely Tested Concert

Eutrophication continues to be a global pressing issue for freshwater ecosystems and biodiversity (Smith and Schindler 2009), and in the light of the European Water Framework Directive (EU 2000/60/EEC) and the Convention for Biological Diversity, management approaches that simultaneously address water pollution and biodiversity should be emphasized. Increasing evidence supports that surface-flow CWs could potentially provide a wide range of services in agricultural landscapes, including water treatment, biodiversity enhancement, and recreational–educational functions (Knight et al. 2001; Hansson et al. 2005; Becerra-Jurado et al. 2010; Boets et al. 2011; Hsu et al. 2011). Over the last five decades, numerous studies have focused on the water treatment function, and design options are now well advanced (Kadlec and Wallace 2008; Vymazal 2010). However, few studies have focused on the potential of these systems to enhance regional biodiversity, and their biodiversity potential has only recently garnered more attention.

Integrated Constructed Wetlands

Integrated Constructed Wetlands (ICWs) are a type of surface-flow CW consisting of a series of interconnected ponds with emergent macrophytes and controlled water depths (Scholz et al. 2007). Despite being mainly designed to reduce nutrients and pollutants from agricultural, livestock, municipal, industrial, and domestic wastewaters, suitable habitat for aquatic macroinvertebrates can be also integrated in their design. Recognized features influencing macroinvertebrate diversity are the total number of ponds and their overall size (the more and the bigger, the better; Hsu et al. 2011), the wetland system profile and shape (Hansson et al. 2005; Scholz et al. 2007), and shore sloping and habitat complexity (Williams et al. 1999; Hansson et al. 2005) (Table 15.4; Fig. 15.4). Research has shown that these systems can actually host communities with a high a proportion of unique species (Becerra-Jurado et al. 2010), advancing the notion that wetlands (regardless of type) may importantly contribute to biodiversity at the landscape level (Williams et al. 2004). In Ireland, five interconnected ponds were compared to five nearby natural wetlands and nine river sites (Becerra-Jurado et al. 2010, 2014). A total of 135 taxa were collected from the ICWs, with 38 % of the taxa being unique to these systems (see characteristic groups in Table 15.5 and Appendix). Ponds located near the ICW system outflow contributed disproportionally to overall diversity (72 % of the taxa). In general, excessive nutrient loadings (through a negative impact on some sensitive taxa that are not able to complete their life cycles; Hawkes 1998) and variation in pH due to differences in microbial respiration (Feldman and Connor 1992; Becerra-Jurado et al. 2009) may be particularly important controls in ICW systems.

Table 15.4 Key studies including invertebrate communities in Integrated Constructed Wetlands (ICWs), organized by wastewater type

Wastewater type	Country	Focal groups	Main findings on macroinvertebrate diversity	Reference
Agricultural wastewater	Sweden	Macroinvertebrates, macrophytes, amphibians and birds	• Relatively shallow, large created wetlands with high shoreline complexity are more likely to yield higher macroinvertebrate diversity values	Hansson et al. (2005)
Agricultural wastewater	Ireland	Macroinvertebrates	• Macroinvertebrate communities are dominated by Coleoptera and Hemiptera taxa, followed by Diptera, Gastropoda, and Trichoptera • Ponds near the outflow have similar numbers of taxa as natural ponds • Contributions of the ponds near the outflow to macroinvertebrate diversity at the catchment level are high	Becerra-Jurado et al. (2010)
Primarily treated pig manure	Belgium	Macroinvertebrates	• Diversity of macroinvertebrates increases along the water treatment process • Several pollution-sensitive taxa occur near the outflow	Boets et al. (2011)
Municipal, industrial, and livestock wastewater, as well as storm and river water	USA	Macroinvertebrates, macrophytes, fish, amphibians, reptiles, birds, and mammals	• Surface-flow constructed wetlands constitute an important habitat for macroinvertebrates • Average benthic macroinvertebrate diversity for constructed wetlands is low when compared to natural treatment wetlands	Knight et al. (2001)
Domestic wastewater (secondarily treated)	USA	Macroinvertebrates	• Highest macroinvertebrate diversity occurs amongst Coleoptera, Diptera, Hemiptera, and Odonata • Highest diversity occurs near the outflow	Spieles and Mitsch (2000)
Domestic wastewater	Taiwan	Macroinvertebrates, macrophytes, algae, fish, and birds	• Richness and density of aquatic macroinvertebrates increases with the cover of aquatic macrophytes, while diversity increases with wetland area • Wetland area, cover of aquatic macrophytes, and water quality are the most important factors governing the diversity in the constructed wetlands • Variations in the macroinvertebrate community structure is best explained by species richness of fish	Hsu et al. (2011)

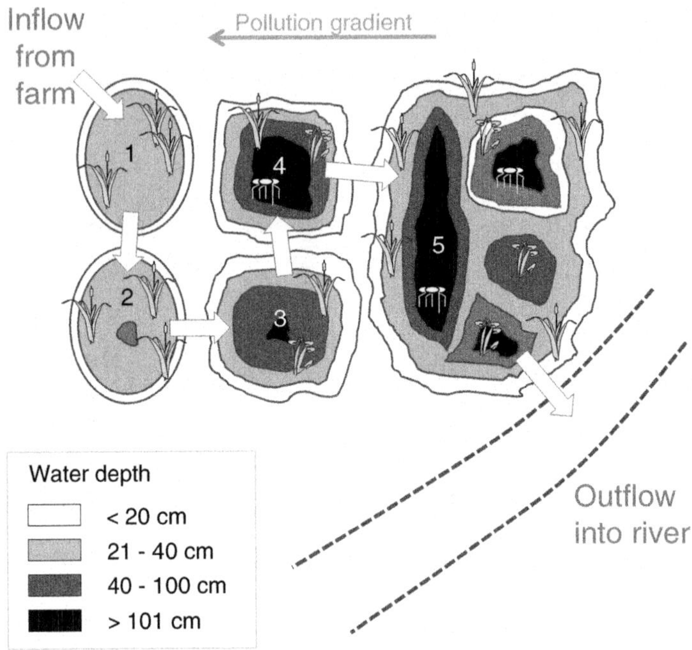

Fig. 15.4 Schematization of an ICW with a design that integrates wastewater treatment from farms and macroinvertebrate diversity enhancement (plant symbols represent helophytes, hygrophytes, and hydrophytes)

Although water treatment constitutes the primary function of ICWs and similar surface-flow wetlands, research is increasingly focusing on how to combine this demand with biodiversity enhancement. Several critical questions should be further investigated: (1) Are there optimal trade-offs between the inherently conflicting demands of these systems? (2) Can their ecological (or biodiversity) potentials be accurately predicted?; and (3) Are there features that lead to an optimized design under a multifunctional perspective? If answered positively, these avenues could inform guidelines on ICW construction, and their potential to mitigate pollution and biodiversity erosion in agricultural landscapes could be adequately appraised.

Conservation and Management Issues

Wetland Construction and Restoration: An Evolving Science

The ability of CWs and RWs to recover biodiversity and ecosystem functions is, up to the present time, limited (Ballantine and Schneider 2009; Hossler et al. 2011; Moreno-Mateos et al. 2012; Kovalenko et al. 2013). CWs and RWs of all types (including depressional, tidal, and riverine wetlands) from all over the world tend to be ~25 % less biodiverse and functional than reference systems in pre-disturbance

Table 15.5 Most important taxa characterizing ICW vs. natural wetlands in Ireland (up to 50 % cumulative community dissimilarity)

Taxa	Tolerance	ICW ab.	Natural ab.	Diss.	Diss./SD	Contrib.%
Asellus aquaticus	High	34.47	9.99	6.55	1.42	10.54
Chironomidae	High	13.76	18.08	3.52	1.18	5.66
Planorbis spp.	High	4.90	16.49	3.51	1.00	5.65
Plea minutissima	Medium	10.91	2.98	2.18	0.88	3.51
Coleoptera larvae	Medium	10.96	5.39	1.96	0.80	3.15
Cloeon dipterum	Medium/low	10.20	10.14	1.77	1.25	2.84
Noterus clavicornis	Medium	5.88	7.15	1.49	1.41	2.39
Oligochaeta	High	8.99	6.31	1.38	0.82	2.22
Pisidium spp/*Sphaerium* spp	Medium	5.01	2.73	1.29	1.05	2.07
Chaoborus crystallinus	Medium/low	5.66	3.42	1.26	0.86	2.02
Radix labiata	Medium	2.34	5.21	1.20	0.78	1.93
Tricladida	High	4.94	3.36	1.19	1.36	1.91
Hydracarina	Medium	0.47	4.19	1.05	0.86	1.68
Coenagrionidae	Medium/low	7.18	5.04	1.01	1.30	1.63
Erpobdella octoculata	High	1.15	4.11	0.95	1.05	1.52
Haliplus ruficollis	Medium	4.98	1.98	0.91	1.02	1.46
Helobdella stagnalis	High	3.40	3.34	0.86	1.20	1.38

Tolerance = Tolerance to organic pollution, *ICW ab* = Average abundance in ICW, *Natural ab* = Average abundance in natural wetlands, *Diss.* = Average dissimilarities between wetland types, *Diss./SD* = measure of consistency in the contribution to dissimilarity, *Contrib. %* = percentage contributed to the overall Bray-Curtis dissimilarity between communities (adapted from Becerra-Jurado et al. 2010)

states, decades or centuries after construction or restoration (Moreno-Mateos et al. 2012) (Fig. 15.5). However, diversity of macroinvertebrate communities in CWs and RWs tends to converge to that from reference "undisturbed" wetlands within 5–10 years after construction or restoration (Moreno-Mateos et al. 2012). Another recent meta-analysis on worldwide wetlands showed nonsignificant differences in aquatic invertebrate diversity between restored and natural wetlands (Meli et al. 2014). As discussed previously, the frequently described "quick" development or recovery of CWs and RWs is generally based on macroinvertebrate abundance and species richness data, and could be explained by aerial active dispersal, by passive aquatic dispersal (when wetlands are reconnected to stream or tidal flow), and even by aerial passive dispersal when propagules are carried in by birds (Badosa et al. 2010). Fast life cycles may also accelerate the recovery of their populations (Levin and Talley 2002). Other studies have also shown that macroinvertebrate communities can recover in 15–25 years in coastal ecosystems (Borja et al. 2010). However, recovery is less obvious in studies measuring species similarity. Some studies have shown recovery of the benthic invertebrate community in less than 5 years (Stanczak and Keiper 2004), while others report that only 56 % of species similarity is reached more than 20 years after wetlands were created (Hartzell et al. 2007). Using

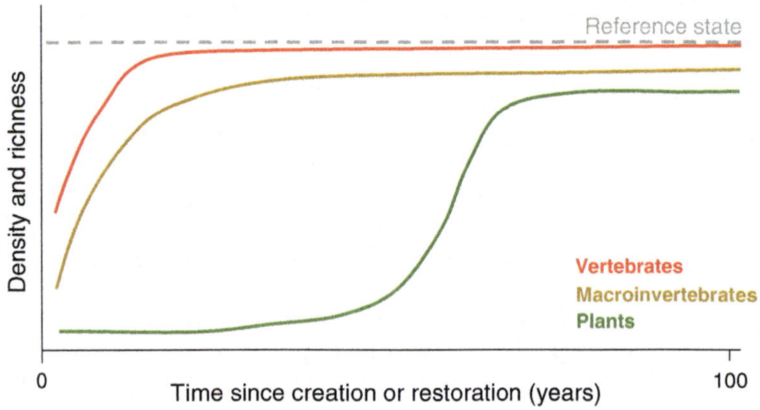

Fig. 15.5 Recovery of density and richness of some of the main biological components of wetlands after restoration or construction. Reference state refers to a condition similar to that existing before those wetlands were degraded due to human actions. Adapted from Moreno-Mateos et al. (2012)

ordination methods, clear groupings in community structure and composition are consistently observed when comparing macroinvertebrate communities from natural and created wetlands (Fig. 15.6). Even when conspicuous change in plant communities and the physical environment occurs (Fig. 15.7), macroinvertebrate communities inhabiting CWs and RWs generally differ significantly from those found in natural nearby or reference communities (Ruhí et al. 2013). More research is thus necessary to understand the long-term recovery of macroinvertebrate communities, including the frequently neglected aspect of biotic interactions.

Context-Dependent Success, Uncertainties, and the Risk of No-Net-Loss Policies

The recovery of wetland structure and functioning is highly context-dependent (Meli et al. 2014), and can be affected by multiple abiotic factors. For example, large wetlands (especially those over 10–100 ha), wetlands in temperate climates, and hydrologically connected wetlands may recover faster than small, isolated, or cold-climate wetlands (Moreno-Mateos et al. 2012). Small wetlands may be more isolated and may act as habitat islands (O'Connell et al. 2013), and the chances of plant propagules arriving spontaneously, and then persisting, are lower than in larger or more connected ecosystems. A similar situation may exist in wetlands with low hydrological connectivity (like some depressional wetlands), where the effectiveness of hydrochory and other propagule dispersal mechanisms is limited (Nilsson et al. 2010). Finally, higher temperature might accelerate the nutrient turnover of soils of created and restored wetlands (Rustad et al. 2001) and also,

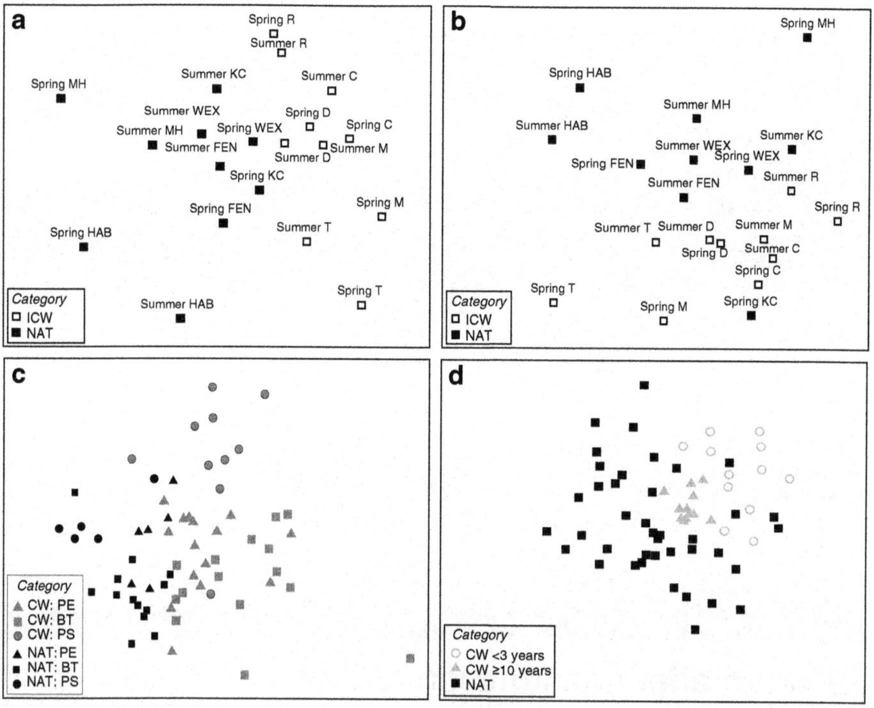

Fig. 15.6 Nonmetric multidimensional scaling ordinations of different studies on composition and structure of invertebrate assemblages in natural vs. CWs. (**a**) and (**b**) show communities in five ICWs in Ireland and reference (nearby natural) systems (Becerra-Jurado et al. 2009) sampled in Spring and Summer; whereas (**a**) shows similarities in macroinvertebrate structure (Bray-Curtis similarities on relative abundances), (**b**) shows similarities in composition (Sorensen on presence–absence data). Acronyms represent the different sites (ICW sites: FEN, HAB, KC, MH, WEX; Reference sites (NAT): C, D, M, R, T). In both cases, differences between ICW and CW communities were smaller than seasonal effects. (**c**) shows similarities in macroinvertebrate structure between pioneer communities (sampled seasonally during the first year after construction) in CWs designed for biodiversity enhancement (CW) and in reference wetlands (NAT), in three different areas (PE, BT, PS) in Catalonia (Ruhí et al. 2009). Finally, (**d**) shows aquatic coleopteran compositional similarities between young CWs (<3 years), old CWs (≥10 years), and reference wetlands (NAT) in Sweden (Ruhí et al. 2011)

potentially, the number of reproductive cycles of aquatic macroinvertebrates, thus affecting community dynamics—and hence successional trajectories—of the overall macroinvertebrate assemblage (Ruhí et al. 2012a). Taken together, these findings suggest that wetland creation and restoration is highly context-dependent and must be more place-based than has been practiced so far. For example, wetland designs facilitating immigration and the rapid establishment of critical habitat (e.g., revegetation) may be particularly important in cold climates.

Fig. 15.7 CW for biodiversity enhancement close to Banyoles, Catalonia. The goal of this CW was to increase amphibian populations within the framework of the LIFE-Nature project LIFE03 NAT/E/000067. The *upper picture* shows bare shores and scarcity of macrophytes 0.5 years after wetland construction; the *lower picture* shows the dense macrophyte community (dominated by common reed *Phragmites australis* and cattail *Typha latifolia*) that established shortly after (still present)

Beyond spatial and temporal contingency, CW and RW responses to human interventions are also highly uncertain today (Zedler and West 2008; Matthews and Spyreas 2010). Even the need for the most commonly used restoration actions, such as revegetation, is questioned due to unclear benefits in most cases (Moreno-Mateos et al. in press). While some studies report a need for revegetation (Klimkowska et al. 2007; Kiehl et al. 2010), in efforts aiming to restore or create salt marshes (Morzaria-Luna and Zedler 2007; Garbutt and Wolters 2008), other studies show the lack of positive effects of this intervention (Wolters et al. 2008). In many cases, regardless of the creation or restoration approach used, environmental factors such as invasive species, or extreme climatic events, might force recovery trajectories to converge or diverge over short (~10 years) periods of time (Collinge and Ray 2009). Thus, a combination of environmental and anthropogenic factors, often stochastic, may limit success, or significantly delay the time to full success, in the recovery trajectories of CW and RW ecosystems. Overall, the still limited knowledge on these environmental and anthropogenic influences underpins the need for place-based, adaptive approaches.

For these reasons, knowledge is limited at this point to support ecosystem trading policies, like biodiversity offset policies (including "no-net-loss" (NNL) policies). In terrestrial ecosystems, models have shown that when metrics more complex than richness are considered (e.g., species composition and community structure), recovery may lag over centuries (Curran et al. 2014). Similar results could be expected in aquatic ecosystems, which is in agreement with the fact that even communities in the *advanced* succession phase generally differ in composition (often by missing rare species) from reference ones. Under that scenario, NNL regulations could be actually creating a net loss, due to the long-term interim loss caused during the recovery process until the reference is supposedly reached (Gutrich and Hitzhusen 2004); and also due to the loss of uniqueness of the ecosystem that has been degraded, which may or may not recover to pre-disturbance conditions (Gardner et al. 2013; Curran et al. 2014). Because NNL of wetland habitat acreage does not translate into NNL of biodiversity, before offset policies can be considered safe, research will need to shed light into best construction and restoration practices (to increase and accelerate recovery rates), and into ecologically-meaningful ways of measuring recovery.

Conclusions

Creation and restoration of lentic waterbodies in altered landscapes can certainly have positive effects on freshwater biodiversity, particularly for faunal groups that suffer habitat fragmentation (e.g., Lichko and Calhoun 2003; Rannap et al. 2009;

Ruhí et al. 2012b). In such cases, the recovery of "stepping-stone" habitat patches, by increasing habitat availability at the landscape level, may have positive effects on metacommunity dynamics even if the CWs or RWs are not structurally and functionally analogous to natural wetlands. Additionally, CWs and RWs provide many societal values, including important aesthetic and recreational benefits (Fig. 15.7). However, precisely because both connectivity and aesthetic values are often high in CWs and RWs, and these systems are often used by emblematic fauna and enjoyed by humans, a frequent problem has been to assume that they may be equivalent to natural wetlands. Research has shown that structural and functional differences between CWs-RWs and natural wetlands persist for long periods, with successional trajectories depending on many factors that are difficult to predict. Additionally, CWs and RWs are being increasingly designed to fulfill several objectives simultaneously (Hansson et al. 2005; Becerra-Jurado et al. 2010; Moreno-Mateos and Comín 2010; Ruhí et al. 2011), and because the virtues of multifunctionality in CWs and RWs are still uncertain, this topic will likely receive increasing attention. The scientific advancement of these two key topics (anticipation of outcomes and combination of services) will determine, to a large extent, the impact that CWs and RWs can make to freshwater biodiversity conservation.

Acknowledgements Albert Ruhi researched CWs and RWs thanks to a FPU doctoral fellowship grant and a mobility grant from the Spanish Ministry of Education (AP2006-00807), and is currently supported by the US National Science Foundation project "WSC-Category 3: Collaborative Research: Water sustainability under near-term climate change: a cross-regional analysis incorporating socioecological feedbacks and adaptations" (award #1204478). We would like to thank Miquel Campos (Consorci de l'Estany, www.consorcidelestany.org) for pictures and Chris Robinson for assistance with GIS.

Appendix

Macroinvertebrates reported from five constructed wetlands for wastewater treatment in Ireland (WW; Becerra-Jurado et al. 2009, 2014), in a created wetland for stormwater treatment in Sweden (SW; Ruhí et al. 2011, 2012a), and in three created wetlands for biodiversity enhancement in Catalonia (B1, B2, B3; Ruhí et al. 2013). Taxonomic hierarchy and names after Fauna Europaea (http://www.faunaeur.org)

Phylum	Class	Order	Family	Taxa	WW	SW	B1	B2	B3
Cnidaria	Leptolida	Capitata	Hydridae	*Hydra*				+	
Annelida	Oligochaeta	–	–	Oligochaeta indet.	+	+			
		Lumbriculida	Lumbriculidae	Lumbriculidae indet.				+	
		Tubificida	Enchytraeidae	Enchytraeidae indet.				+	+
			Naididae	Naididae indet.			+		
			Tubificidae	Tubificidae indet.			+	+	+
	Hirudinea	Arhynchobdellida	Erpobdellidae	*Dina*		+			
				Erpobdella	+	+			
			Haemopidae	*Haemopis*	+				
		Rhynchobdellida	Glossiphoniidae	*Helobdella*	+	+			
				Hemiclepsis		+			
				Theromyzon		+			
Platyhelminthes	Turbellaria	Tricladida	Dendrocoelidae	*Dendrocoelum*		+			
			Dugesiidae	*Dugesia*	+	+			
Mollusca	Gastropoda	Neotaenioglossa	Hydrobiidae	*Potamopyrgus*	+				
		Pulmonata	Gastrodontidae	*Zonitoides*	+				
			Lymnaeidae	*Galba*		+			
				Lymnaea	+				
				Radix	+	+			
			Physidae	*Aplexa*	+				
				Haitia	+		+	+	
				Physa		+			
			Planorbidae	*Ferrissia*				+	
				Gyraulus		+			
				Planorbis	+	+			
	Bivalvia	Veneroidea	Sphaeriidae	*Pisidium*	+	+		+	

(continued)

(continued)

Phylum	Class	Order	Family	Taxa	WW	SW	B1	B2	B3
Arthropoda	Malacostraca	Amphipoda	Gammaridae	*Gammarus*	+				
		Isopoda	Asellidae	*Asellus*	+	+			
				Proasellus			+		
		Decapoda	Atyidae	*Atyaephyra*				+	
			Cambaridae	*Procambarus*			+	+	
	Arachnida	Prostigmata	–	Hydrachnidia indet.		+			
			Arrenuridae	*Arrenurus*				+	+
			Hydrachnidae	*Hydrachna*				+	+
			Hydrodromidae	*Hydrodroma*			+	+	+
			Hydryphantidae	*Hydryphantes*				+	
			Pionidae	*Piona*				+	
	Insecta	Ephemeroptera	Baetidae	*Cloeon*	+	+	+	+	+
				Centroptilum		+			
			Caenidae	*Caenis*		+	+	+	+
			Ephemeridae	*Ephemera*				+	
			Leptophlebiidae	*Leptophlebia*		+			
		Odonata	Aeshnidae	Aeshnidae indet.	+				
				Aeshna			+	+	+
				Anax			+	+	+
				Boyeria					+
				Brachytron		+			
			Coenagrionidae	Coenagrionidae indet.	+				
				Coenagrion		+			+
				Ischnura		+	+	+	+
			Lestidae	*Lestes*		+		+	
			Libellulidae	Libellulidae indet.					
				Leucorrhinia		+			
				Libellula		+			+
				Orthetrum		+		+	+
				Sympetrum		+	+	+	+

Phylum	Class	Order	Family	Taxa	WW	SW	B1	B2	B3
		Heteroptera	Corixidae	*Callicorixa*	+	+			
				Corixa	+	+		+	+
				Cymatia				+	
				Hesperocorixa		+			
				Micronecta			+	+	
				Paracorixa		+	+	+	
				Sigara		+	+	+	+
			Gerridae	*Aquarius*			+	+	+
				Gerris	+	+	+	+	+
			Hydrometridae	*Hydrometra*	+				+
			Nepidae	*Nepa*	+				
				Ranatra		+			+
			Notonectidae	*Anisops*			+	+	+
				Notonecta		+	+		+
			Pleidae	*Plea*	+			+	
			Veliidae	*Microvelia*			+	+	
				Velia		+			
		Coleoptera	Curculionidae	*Nanophyes*	+				
				Tanysphyrus	+				
			Dryopidae	*Dryops*	+				
			Dytiscidae	*Acilius*	+				
				Agabus	+	+		+	+
				Colymbetes	+	+			
				Copelatus	+				
				Dytiscus	+	+			
				Eretes			+	+	+

(continued)

(continued)

Phylum	Class	Order	Family	Taxa	WW	SW	B1	B2	B3
				Graptodytes	+				
				Hydaticus	+				
				Hydroglyphus			+	+	+
				Hydroporus	+	+			
				Hygrotus	+	+			+
				Hyphydrus	+	+			
				Ilybius	+	+			
				Laccophilus	+	+		+	+
				Nebrioporus		+			
				Porhydrus	+	+			
				Rhantus	+	+	+		+
				Scarodytes		+			
				Stictonectes	+				
				Yola					+
			Elmidae	*Elmis*	+				
				Esolus	+				
			Gyrinidae	*Gyrinus*	+	+	+	+	
			Haliplidae	*Haliplus*	+	+			
			Helophoridae	*Helophorus*	+		+	+	+
			Hydraenidae	*Aulacochthebius*					+
				Hydraena	+	+			
				Limnebius	+				
				Ochthebius	+		+	+	+

Phylum	Class	Order	Family	Taxa	WW	SW	B1	B2	B3
			Hydrophilidae	*Anacaena*	+				
				Berosus		+			+
				Cercyon	+				
				Coelostoma	+				
				Cymbiodita	+				
				Enochrus	+	+	+	+	
				Helochares			+	+	+
				Hydrobius	+				
				Laccobius	+				
				Limnoxenus					+
			Hygrobiidae	*Hygrobia*	+				
			Noteridae	*Noterus*	+	+		+	
			Scirtidae	*Contacyphon*	+				
		Megaloptera	Sialidae	*Sialis*		+			
		Trichoptera	Ecnomidae	*Ecnomus*				+	
			Hydropsychidae	*Hydropsyche*	+				
			Leptoceridae	*Oecetis*		+			
				Trianodes		+			
			Limnephilidae	*Glyphotaelius*		+			
				Grammotaulius		+			
				Limnephilus	+	+			
				Nemotaulius		+			
			Phryganeidae	*Agrypnia*		+			
				Phryganea		+			
			Polycentropodidae	*Holocentropus*	+	+			

(continued)

(continued)

Phylum	Class	Order	Family	Taxa	WW	SW	B1	B2	B3
		Diptera	Ceratopogonidae	Ceratopogonidae indet.		+			
				Bezzia			+	+	+
				Culicoides			+	+	+
			Chaoboridae	Chaoboridae indet.		+			
				Chaoborus	+			+	+
			Chironomidae	Chironomidae indet.	+	+			
				Ablabesmyia			+	+	
				Chironomus/Einfeldia				+	
				Cladopelma				+	
				Cladotanytarsus			+	+	
				Corynoneura				+	
				Cricotopus			+	+	+
				Cryptochironomus				+	
				Endochironomus				+	
				Kiefferulus			+	+	+
				Labrundinia				+	
				Meropelopia				+	
				Microchironomus				+	
				Parachironomus				+	
				Paramerina				+	
				Paratanytarsus			+	+	+
				Polypedilum			+	+	+
				Procladius			+	+	+
				Psectrocladius			+	+	+
				Tanytarsus			+	+	

Phylum	Class	Order	Family	Taxa	WW	SW	B1	B2	B3
			Culicidae	Culicidae indet.	+	+			
				Aedes			+		
				Anopheles			+	+	+
				Culex			+	+	
			Dixidae	Dixidae indet.	+	+			
			Dolichopodidae	Dolichopodidae indet.		+			
			Empididae	Empididae indet.		+			
			Ephydridae	Ephydridae indet.		+			
				Scatella			+	+	
			Limoniidae	Limoniidae indet.	+	+		+	
			Psychodidae	Psychodidae indet.	+				
			Ptychopteridae	Ptychopteridae indet.		+			
			Stratiomyidae	Stratiomyidae indet.		+			
				Stratiomys					+
			Syrphidae	Syrphidae indet.		+		+	
			Tabanidae	Tabanidae indet.		+			
			Tipulidae	Tipulidae indet.		+			

References

Adamus PR, Brandt KH (1990) Impacts on quality of inland wetlands of the United States: a survey of indicators, techniques, and applications of community-level biomonitoring data. EPA Report 600/3–90/073 (1990). USEPA Environ Research Laboratory

Angélibert S, Marty P, Céréghino R, Giani N (2004) Seasonal variations in the physical and chemical characteristics of ponds: implications for biodiversity conservation. Aquat Conserv Mar Freshw Ecosys 14:439–456

Awal S, Svozil D (2010) Macro-invertebrate species diversity as a potential universal measure of wetland ecosystem integrity in constructed wetlands in South East Melbourne. Aquat Ecosys Health Manage 13:472–479

Badosa A, Frisch D et al (2010) Recovery of zooplankton diversity in a restored Mediterranean temporary marsh in Doñana National Park (SW Spain). Hydrobiologia 654:67–82

Baines CB, McCauley SJ, Rowe L (2014) The interactive effects of competition and predation risk on dispersal in an insect. Biol Lett 10:20140287

Balcombe CK, Anderson JT, Fortney RH, Kordek WS (2005) Aquatic macroinvertebrate assemblages in mitigated and natural wetlands. Hydrobiologia 541:175–188

Ballantine K, Schneider R (2009) Fifty-five years of soil development in restored freshwater depressional wetlands. Ecol Appl 19:1467–1480

Barnes LE (1983) The colonization of ball-clay ponds by macroinvertebrates and macrophytes. Freshw Biol 13:561–578

Becerra-Jurado G, Callanan M et al (2009) Comparison of macroinvertebrate community structure and driving environmental factors in natural and wastewater treatment ponds. Hydrobiologia 634:153–165

Becerra-Jurado G, Johnson J et al (2010) The potential of integrated constructed wetlands (ICWs) to enhance macroinvertebrate diversity in agricultural landscapes. Wetlands 30:393–404

Becerra-Jurado G, Foster G, Harrington R, Kelly-Quinn M (2014) Integrated constructed wetlands: hotspots for freshwater coleopteran diversity in the landscape of Ireland. Biology and environment: Proceedings of the Royal Irish Academy, p 271–279

Bie T, Meester L et al (2012) Body size and dispersal mode as key traits determining metacommunity structure of aquatic organisms. Ecol Lett 15:740–747

Bilton DT, Freeland JR, Okamura B (2001) Dispersal in freshwater invertebrates. Annu Rev Ecol Syst 32:159–181

Boets P, Michels E et al (2011) Integrated constructed wetlands (ICW): ecological development in constructed wetlands for manure treatment. Wetlands 31:763–771

Bohonak AJ, Jenkins DG (2003) Ecological and evolutionary significance of dispersal by freshwater invertebrates. Ecol Lett 6:783–796

Boix D, Sala J, Quintana XD, Moreno-Amich R (2004) Succession of the animal community in a Mediterranean temporary pond. J North Am Benthol Soc 23:29–49

Boothby J, Hull AP (1997) A census of ponds in Cheshire, North West England. Aquat Conserv Mar Freshw Ecosys 7:75–79

Borja A, Dauer DM, Elliott M, Simenstad CA (2010) Medium- and long-term recovery of estuarine and coastal ecosystems: patterns, rates and restoration effectiveness. Estuar Coasts 33:1249–1260

Brainard AS, Fairchild GW (2012) Sediment characteristics and accumulation rates in constructed ponds. J Soil Water Conserv 67:425–432

Breaux A, Serefiddin F (1999) Validity of performance criteria and a tentative model for regulatory use in compensatory wetland mitigation permitting. Environ Manag 24:327–336

Brix H (1994) Use of constructed wetlands in water pollution control: historical development, present status, and future perspectives. Water Sci Technol 30:209–224

Brown PH, Lant CL (1999) The effect of wetland mitigation banking on the achievement of no-net-loss. Environ Manage 23:333–345

Burks RL, Mulderij G et al (2006) Center stage: the crucial role of macrophytes in regulating trophic interactions in shallow lake wetlands. Springer, New York

Burton TM, Uzarski DG et al (1999) Development of a preliminary invertebrate index of biotic integrity for Lake Huron coastal wetlands. Wetlands 19:869–882

Cairns J, Pratt JR (1993) A history of biological monitoring using benthic macroinvertebrates. In: Rosenberg DM, Resh VH (eds) Freshwater biomonitoring and benthic macroinvertebrates. Kluwer, Boston, pp 10–27

Cañedo-Argüelles M, Rieradevall M (2011) Early succession of the macroinvertebrate community in a shallow lake: response to changes in the habitat condition. Limnologica—Ecol Manage Inland Waters 41:363–370

Carpenter SR, Lodge DM (1986) Effects of submersed macrophytes on ecosystem processes. Aquat Bot 26:341–370

Chase JM (2003) Experimental evidence for alternative stable equilibria in a benthic pond food web. Ecol Lett 6:733–741

Chase JM, Biro EG, Ryberg WA, Smith KG (2009) Predators temper the relative importance of stochastic processes in the assembly of prey metacommunities. Ecol Lett 12:1210–1218

Christman VD, Voshell JR (1993) Changes in the benthic macroinvertebrate community in two years of colonization of new experimental ponds. Internat Rev Gesam Hydrobiol Hydrogr 78:481–491

Clarke KR, Warwick RM (2001) A further biodiversity index applicable to species lists: variation in taxonomic distinctness. Mar Ecol Progr Ser 216:265–278

Collinge SK, Ray C (2009) Transient patterns in the assembly of vernal pool plant communities. Ecology 90:3313–23

Connell JH, Slatyer RO (1977) Mechanisms of succession in natural communities and their role in community stability and organization. Am Nat 111(982):1119–1144

Copeland C (2010) Wetlands: an overview of issues. Congressional Research Service, Washington, DC, RL33483

Cowardin LM, Carter V, Golet FC, LaRoe ET (1979) Classification of wetlands and deepwater habitats of the United States. US Department of the Interior, Fish and Wildlife Service, Washington, DC

Craft C, Reader J, Sacco JN, Broome SW (1999) Twenty-five years of ecosystem development of constructed Spartina alterniflora (Loisel) marshes. Ecol Appl 9:1405–1419

Curran M, Hellweg S, Beck J (2014) Is there any empirical support for biodiversity offset policy? Ecol Appl 24:617–32

Dahl TE (1990) Wetlands losses in the United States 1780's to 1980's. US Department of the Interior, Fish and Wildlife Service, Washington, DC

Daniels AE, Cumming GS (2008) Conversion or conservation? Understanding wetland change in northwest Costa Rica. Ecol Appl 18:49–63

Davis J, Horwitz P et al (2006) Are river bioassessment methods using macroinvertebrates applicable to wetlands? Hydrobiologia 572:115–128

De Szalay FA, Resh VH (2000) Factors influencing macroinvertebrate colonization of seasonal wetlands: responses to emergent plant cover. Freshw Biol 45:295–308

Denny P (1997) Implementation of constructed wetlands in developing countries. Water Sci Technol 35:27–34

Fairchild GW, Faulds AM, Matta JF (2000) Beetle assemblages in ponds: effects of habitat and site age. Freshw Biol 44:523–534

Fairchild GW, Robinson C, Brainard AS, Coutu GW (2013) Historical changes in the distribution and abundance of constructed ponds in response to changing population density and land use. Landsc Res 38:593–606

Feldman RS, Connor EF (1992) The relationship between pH and community structure of invertebrates in streams of the Shenandoah National Park, Virginia, USA. Freshw Biol 27:261–276

Fisher SG (1983) Succession in streams. In: Barnes JR, Minshall CW (eds) Stream ecology. Springer, New York, pp 7–27

Fontanarrosa MS, Chaparro GN, O'Farrell I (2013) Temporal and spatial patterns of macroinvertebrates associated with small and medium-sized free-floating plants. Wetlands 33:47–63
Forbes AE, Chase JM (2002) The role of habitat connectivity and landscape geometry in experimental zooplankton metacommunities. Oikos 96:433–440
Frisch D, Green AJ (2007) Copepods come in first: rapid colonization of new temporary ponds. Fund Appl Limnol 168:289–297
Garbutt A, Wolters M (2008) The natural regeneration of salt marsh on formerly reclaimed land. Appl Veg Sci 11:335–344
Gardner TA, von Hase A et al (2013) Biodiversity offsets and the challenge of achieving No Net Loss. Conserv Biol 27:1254–1264
Gibbs JP (2000) Wetland loss and biodiversity conservation. Conserv Biol 14:314–317
Goodnight CJ (1973) The use of aquatic macroinvertebrates as indicators of stream pollution. Trans Am Microsc Soc 1:1–13
Gutrich J, Hitzhusen F (2004) Assessing the substitutability of mitigation wetlands for natural sites: estimating restoration lag costs of wetland mitigation. Ecol Econ 48:409–424
Hall DL, Willig MR et al (2004) Aquatic macroinvertebrate diversity of playa wetlands: the role of landscape and island biogeographic characteristics. Wetlands 24:77–91
Hansson L, Brönmark C, Anders Nilsson P, Åbjörnsson K (2005) Conflicting demands on wetland ecosystem services: nutrient retention, biodiversity or both? Freshw Biol 50:705–714
Hartzell D, Bidwell JR, Davis CA (2007) A comparison of natural and created depressional wetlands in central Oklahoma using metrics from indices of biological integrity. Wetlands 27:794–805
Hawkes HA (1998) Origin and development of the biological monitoring working party score system. Water Res 32:964–968
Hein AM, Gillooly JF (2011) Predators, prey, and transient states in the assembly of spatially structured communities. Ecology 92:549–555
Herrmann J (2012) Chemical and biological benefits in a stormwater wetland in Kalmar, SE Sweden. Limnologica—Ecol Manage Inland Waters 42:299–309
Herrmann J, Boström A, Bohmann I (2001) Invertebrate colonisation into the man-made Kalmar Daemme wetland dam system. Verh Int Ver Theor Angew Limnol 27:1653–1656
Hosomi M, Murakami A, Sudo R (1994) A four-year mass balance for a natural wetland system receiving domestic wastewater. Water Sci Technol 30:235–244
Hossler K, Bouchard V, Fennessy M (2011) No-net-loss not met for nutrient function in freshwater marshes: recommendations for wetland mitigation policies. Ecosphere 2:1–36
Hsu CB, Hsieh HL et al (2011) Biodiversity of constructed wetlands for wastewater treatment. Ecol Eng 37:1533–1545
Hume C (2008) Wetland vision technical document [electronic resource]: overview and reporting of project philosophy and technical approach. Wetland Vision Partnership
Irfanullah H, Moss B (2005) A filamentous green algae-dominated temperate shallow lake: variations on the theme of clear-water stable states? Arch Hydrobiol 163:25–47
Jeffries MJ (2002) Evidence for individualistic species assembly creating convergent predator: prey ratios among pond invertebrate communities. J Anim Ecol 71:173–184
Jeffries MJ (2012) Ponds and the importance of their history: an audit of pond numbers, turnover and the relationship between the origins of ponds and their contemporary plant communities in south-east Northumberland, UK. Hydrobiologia 689:11–21
Johnson PTJ, Hoverman JT et al (2013) Urbanization and wetland communities: applying metacommunity theory to understand the local and landscape effects. J Appl Ecol 50:34–42
Kadlec RH, Wallace S (2008) Treatment wetlands. CRC Press, Boca Raton, FL
Karouna-Renier NK, Sparling DW (2001) Relationships between ambient geochemistry, watershed land-use and trace metal concentrations in aquatic invertebrates living in stormwater treatment ponds. Environ Pollut 112:183–192
Kashian DR, Burton TM (2000) A comparison of macroinvertebrates of two Great Lakes coastal wetlands: testing potential metrics for an index of ecological integrity. J Great Lakes Res 26:460–481

Kiehl K, Kirmer A et al (2010) Species introduction in restoration projects—evaluation of different techniques for the establishment of semi-natural grasslands in Central and Northwestern Europe. Basic Appl Ecol 11:285–299

Kim DG, Kang HJ et al (2014) Analyses of benthic macroinvertebrate colonization during the early successional phases of created wetlands in temperate Asia. Fund Appl Limnol 184:35–49

King RS, Richardson CJ (2002) Evaluating subsampling approaches and macroinvertebrate taxonomic resolution for wetland bioassessment. J North Am Benthol Soc 21:150–171

Kivaisi AK (2001) The potential for constructed wetlands for wastewater treatment and reuse in developing countries: a review. Ecol Eng 16:545–560

Klimkowska A, Van Diggelen R, Bakker JP, Grootjans AP (2007) Wet meadow restoration in Western Europe: a quantitative assessment of the effectiveness of several techniques. Biol Conserv 140:318–328

Knight R, Clarke R, Bastian R (2001) Surface flow (SF) treatment wetlands as a habitat for wildlife and humans. Water Sci Technol 44:27–37

Kolkwitz R, Marsson M (1908) Ökologie der pflanzlichen Saprobien. Borntraeger

Kovalenko KE, Ciborowski JJH et al (2013) Food web structure in oil sands reclaimed wetlands. Ecol Appl 23:1048–60

Lake PS, Bayly IAE, Morton DW (1989) The phenology of a temporary pond in western Victoria, Australia, with special reference to invertebrate succession. Arch Hydrobiol 115:171–202

Lazzaro X, Lacroix G et al (2009) Predator foraging behaviour drives food-web topological structure. J Anim Ecol 78:1307–1317

Legendre P, Dallot S, Legendre L (1985) Succession of species within a community: chronological clustering, with applications to marine and freshwater zooplankton. Am Nat 125:257–288

Levin LA, Talley TS (2002) Natural and manipulated sources of heterogeneity controlling early faunal development of a salt marsh. Ecol Appl 12:1785–1802

Levin LA, Talley D, Thayer G (1996) Succession of macrobenthos in a created salt marsh. Mar Ecol Progr Ser Oldendorf 141:67–82

Lichko LE, Calhoun AJK (2003) An evaluation of vernal pool creation projects in New England: project documentation from 1991–2000. Environ Manage 32:141–151

Liu D, Ge Y et al (2008) Constructed wetlands in China: recent developments and future challenges. Front Ecol Environ 7:261–268

Lodge DM, Barko JW et al (1988) Spatial heterogeneity and habitat interactions in lake communities. In: Complex interactions in lake communities. Springer, New York, p 181–208

Louette G, De Meester L, Declerck S (2008) Assembly of zooplankton communities in newly created ponds. Freshw Biol 53:2309–2320

Lunde KB, Resh VH (2012) Development and validation of a macroinvertebrate index of biotic integrity (IBI) for assessing urban impacts to Northern California freshwater wetlands. Environ Monitor Assess 184:3653–3674

Marchetti MP, Garr M, Smith ANH (2010) Evaluating wetland restoration success using aquatic macroinvertebrate assemblages in the Sacramento Valley, California. Restor Ecol 18:457–466

Margalef R (1968) Perspectives in ecological theory. Chicago University Press, Chicago

Marklund O, Sandsten H, Hansson L, Blindow I (2002) Effects of waterfowl and fish on submerged vegetation and macroinvertebrates. Freshw Biol 47:2049–2059

Matthews JW, Spyreas G (2010) Convergence and divergence in plant community trajectories as a framework for monitoring wetland restoration progress. J Appl Ecol 47:1128–1136

Matthews JW, Spyreas G, Endress AG (2009) Trajectories of vegetation-based indicators used to assess wetland restoration progress. Ecol Appl 19:2093–2107

Mays PA, Edwards GS (2001) Comparison of heavy metal accumulation in a natural wetland and constructed wetlands receiving acid mine drainage. Ecol Eng 16:487–500

McCauley SJ, Davis CJ, Nystrom J, Werner EE (2009) A hump-shaped relationship between isolation and abundance of Notonecta irrorata colonists in aquatic mesocosms. Ecology 90:2635–2641

McPeek MA (2004) The growth/predation risk trade-off: so what is the mechanism? Am Nat 163:E88–E111

Meli P, Rey Benayas JM, Balvanera P, Martínez Ramos M (2014) Restoration enhances wetland biodiversity and ecosystem service supply, but results are context-dependent: a meta-analysis. PLoS One 9, e93507

Miguel-Chinchilla L, Boix D, Gascón S, Comín FA (2014) Taxonomic and functional successional patterns in macroinvertebrates related to flying dispersal abilities: a case study from isolated manmade ponds at reclaimed opencast coal mines. Hydrobiologia 732:111–122

Milner AM, Robertson AL et al (2008) Colonization and development of an Alaskan stream community over 28 years. Front Ecol Environ 6:413–419

Mitsch WJ, Wise KM (1998) Water quality, fate of metals, and predictive model validation of a constructed wetland treating acid mine drainage. Water Res 32:1888–1900

Montoya D, Rogers L, Memmott J (2012) Emerging perspectives in the restoration of biodiversity-based ecosystem services. Trends Ecol Evol 27:666–672

Moore MT, Cooper CM et al (2009) Mitigation of two pyrethroid insecticides in a Mississippi Delta constructed wetland. Environ Pollut 157:250–256

Moorhead DL, Hall DL, Willig MR (1998) Succession of macroinvertebrates in playas of the Southern High Plains, USA. J North Am Benthol Soc 17:430–442

Moreno-Mateos D, Comín FA (2010) Integrating objectives and scales for planning and implementing wetland restoration and creation in agricultural landscapes. J Environ Manage 91:2087–2095

Moreno-Mateos D, Pedrocchi C, Comín FA (2010) Effects of wetland construction on water quality in a semi-arid catchment degraded by intensive agricultural use. Ecol Eng 36:631–639

Moreno-Mateos D, Power ME, Comín FA, Yockteng R (2012) Structural and functional loss in restored wetland ecosystems. PLoS Biol 10, e1001247

Moreno-Mateos D, Meli P, Vara-Rodríguez M, Aronson J (in press) Ecosystem response to interventions: lessons from restored and created wetland ecosystems. J Appl Ecol

Morzaria-Luna HN, Zedler JB (2007) Does seed availability limit plant establishment during salt marsh restoration? Estuar Coasts 30:12–25

Nilsson C, Brown RL, Jansson R, Merritt DM (2010) The role of hydrochory in structuring riparian and wetland vegetation. Biol Rev 85:837–858

O'Connell JL, Johnson LA et al (2013) Predicting dispersal-limitation in plants: optimizing planting decisions for isolated wetland restoration in agricultural landscapes. Biol Conserv 159:343–354

Odum EP (1969) The strategy of ecosystem development. Science 164:262–270

Odum EP, Odum HT, Andrews J (1971) Fundamentals of ecology. Saunders, Philadelphia

Oertli B, Biggs J et al (2005) Conservation and monitoring of pond biodiversity: introduction. Aquat Conserv Mar Freshw Ecosys 15:535–540

Ortega M, Velasco J, Millán A, Guerrero C (2004) An ecological integrity index for littoral wetlands in agricultural catchments of semiarid Mediterranean regions. Environ Manage 33:412–430

Persson J, Somes NLG, Wong THF (1999) Hydraulics efficiency of constructed wetlands and ponds. Water Sci Technol 40:291–300

Rannap R, Lõhmus A, Briggs L (2009) Restoring ponds for amphibians: a success story. Hydrobiologia 634:87–95

Resetarits WJ, Binckley CA (2009) Spatial contagion of predation risk affects colonization dynamics in experimental aquatic landscapes. Ecology 90:869–876

Resetarits WJ, Binckley CA (2014) Species responses of colonising beetles to variation in patch quality, number, and context in experimental aquatic landscapes. Ecol Entomol 39:226–235

Rhazi M, Grillas P, Charpentier A, Médail F (2004) Experimental management of Mediterranean temporary pools for conservation of the rare quillwort *Isoetes setacea*. Biol Conserv 118:675–684

Ruhí A (2012) Primary succession in man-made wetlands: biodiversity, structure and dynamics of macrofaunal assemblages. Ph.D. Dis., University of Girona, Catalonia

Ruhí A, Boix D et al (2009) Spatial and temporal patterns of pioneer macrofauna in recently created ponds: taxonomic and functional approaches. Hydrobiologia 634:137–151

Ruhí A, Herrmann J et al (2011) Change in biological traits and community structure of macroinvertebrates through primary succession in a man-made Swedish wetland. Freshw Sci 31:22–37

Ruhí A, Herrmann J et al (2012a) How do early successional patterns in man-made wetlands differ between cold temperate and Mediterranean regions? Limnologica—Ecol Manage Inland Waters 42:328–339

Ruhí A, San Sebastian O et al (2012b) Man-made Mediterranean temporary ponds as a tool for amphibian conservation. Ann Limnol 48:81–93

Ruhí A, Boix D et al (2013) Nestedness and successional trajectories of macroinvertebrate assemblages in man-made wetlands. Oecologia 171:545–556

Rustad LE, Campbell JL et al (2001) A meta-analysis of the response of soil respiration, net nitrogen mineralization, and aboveground plant growth to experimental ecosystem warming. Oecologia 126:543–562

Saunders LL, Kilham SS, Fairchild GW, Verb R (2012) Effects of small-scale environmental variation on metaphyton condition and community composition. Freshw Biol 57:1884–1895

Scheffer M, Hosper SH et al (1993) Alternative equilibria in shallow lakes. Trends Ecol Evol 8:275–279

Scher O, Thièry A (2005) Odonata, Amphibia and environmental characteristics in motorway stormwater retention ponds (Southern France). Hydrobiologia 551:237–251

Schneider DW (1999) Snowmelt ponds in Wisconsin. In: Batzer DP, Rader RB, Wissinger SA (eds) Invertebrates in freshwater wetlands of North America: ecology and management. Wiley, New York, pp 299–318

Scholz M, Harrington R, Carroll P, Mustafa A (2007) The integrated constructed wetlands (ICW) concept. Wetlands 27:337–354

Schulz R, Peall SKC (2001) Effectiveness of a constructed wetland for retention of nonpoint-source pesticide pollution in the Lourens River catchment, South Africa. Environ Sci Technol 35:422–426

Sharma RC, Rawat JS (2009) Monitoring of aquatic macroinvertebrates as bioindicator for assessing the health of wetlands: a case study in the Central Himalayas, India. Ecol Indicat 9:118–128

Shulman RS, Chase JM (2007) Increasing isolation reduces predator: prey species richness ratios in aquatic food webs. Oikos 116:1581–1587

Silva JP, Phillips L, Jones W (2007) LIFE and Europe's wetlands: restoring a vital ecosystem. Public Office European Union, Luxembourg

Smith VH, Schindler DW (2009) Eutrophication science: where do we go from here? Trends Ecol Evol 24:201–207

Smith SV, Renwick WH, Bartley JD, Buddemeier RW (2002) Distribution and significance of small, artificial water bodies across the United States landscape. Sci Total Environ 299:21–36

Solimini AG, Ruggiero A, Bernardini V, Carchini G (2003) Temporal pattern of macroinvertebrate diversity and production in a new man made shallow lake. Hydrobiologia 506:373–379

Spieles DJ (2005) Vegetation development in created, restored, and enhanced mitigation wetland banks of the United States. Wetlands 25:51–63

Spieles DJ, Horn JD (2009) Macroinvertebrate community structure in created wetlands of different successional stage. Aquat Ecosys Health Manage 12:320–329

Spieles DJ, Mitsch WJ (2000) Macroinvertebrate community structure in high- and low-nutrient constructed wetlands. Wetlands 20:716–729

Spieles DJ, Mitsch WJ (2003) A model of macroinvertebrate trophic structure and oxygen demand in freshwater wetlands. Ecol Model 161:183–194

Spieles DJ, Coneybeer M, Horn J (2006) Community structure and quality after 10 years in two central Ohio mitigation bank wetlands. Environ Manage 38:837–852
Stanczak M, Keiper JB (2004) Benthic invertebrates in adjacent created and natural wetlands in northeastern Ohio, USA. Wetlands 24:212–218
Stead TK, Schmid-Araya JM, Hildrew AG (2005) Secondary production of a stream metazoan community: Does the meiofauna make a difference? Limnol Oceanogr 50:398–403
Stewart TW, Downing JA (2008) Macroinvertebrate communities and environmental conditions in recently constructed wetlands. Wetlands 28:141–150
Stouffer DB, Bascompte J (2011) Compartmentalization increases food-web persistence. Proc Natl Acad Sci U S A 108:3648–3652
Sundaravadivel M, Vigneswaran S (2001) Constructed wetlands for wastewater treatment. Crit Rev Environ Sci Technol 31:351–409
Thompson RM, Townsend CR (2005) Food-web topology varies with spatial scale in a patchy environment. Ecology 86:1916–1925
Tuttle RW (2008) Farm ponds. In: Trimble SW, Stewart BA, Howell TA (eds) Encyclopedia of water science, 2nd edn. Marcel Dekker, New York, pp 359–362
Van de Meutter F, Cottenie K, De Meester L (2008) Exploring differences in macroinvertebrate communities from emergent, floating-leaved and submersed vegetation in shallow ponds. Fund Appl Limnol 173:47–57
Verhoeven JTA (2014) Wetlands in Europe: perspectives for restoration of a lost paradise. Ecol Eng 66:6–9
Vymazal J (2010) Constructed wetlands for wastewater treatment: five decades of experience. Environ Sci Technol 45:61–69
Vymazal J, Greenway M et al (2006) Constructed wetlands for wastewater treatment. In: Verhoeven JTA, Beltman B, Bobbink R, Whigham DF (eds) Wetlands and natural resource management. Springer, New York, pp 69–96
Walker LR, Del Moral R (2003) Primary succession and ecosystem rehabilitation. Cambridge University Press, Cambridge
Warfe DM, Barmuta LA (2004) Habitat structural complexity mediates the foraging success of multiple predator species. Oecologia 141:171–178
Warwick RM, Clarke KR (1995) New 'biodiversity' measures reveal a decrease in taxonomic distinctness with increasing stress. Mar Ecol Progr Ser 129:301–305
Williams P, Biggs J et al (1999) The Pond book: a guide to the management and creation of ponds. Ponds Conservation Trust, Oxford
Williams P, Whitfield M et al (2004) Comparative biodiversity of rivers, streams, ditches and ponds in an agricultural landscape in Southern England. Biol Conserv 115:329–341
Wolters M, Garbutt A et al (2008) Restoration of salt-marsh vegetation in relation to site suitability, species pool and dispersal traits. J Appl Ecol 45:904–912
Yimer HD, Mengistou S (2010) Water quality parameters and macroinvertebrates index of biotic integrity of the Jimma wetlands, Southwestern Ethiopia. J Wetl Ecol 3:79–99
Zedler JB (2004) Compensating for wetland losses in the United States. Ibis 146:92–100
Zedler JB, Kercher S (2004) Causes and consequences of invasive plants in wetlands: opportunities, opportunists, and outcomes. Crit Rev Plant Sci 23:431–452
Zedler JB, Leach MK (1998) Managing urban wetlands for multiple use: research, restoration, and recreation. Urban Ecosys 2:189–204
Zedler JB, West JM (2008) Declining diversity in natural and restored salt marshes: a 30-year study of Tijuana Estuary. Restor Ecol 16:249–262

Chapter 16
Invertebrates in Managed Waterfowl Marshes

Joshua D. Stafford, Adam K. Janke, Elisabeth B. Webb, and Steven R. Chipps

Introduction

Wetlands provide critical habitats for breeding, migrating, and wintering waterfowl. Accordingly, management agencies and conservation organizations have long-sought to improve habitats for waterfowl during key life phases through active management of wetland ecosystems. The tools used for wetland management are diverse, though most focus on the manipulation of hydrology and (or) vegetation (e.g., disking, mowing), sowing of annual plants to provide high-energy foods, or controlling vertebrate populations, such as fish or exotic mammals, that can have negative effects on management goals. Managed waterfowl marshes are unique environments, because unlike other systems explored in this volume, active wetland management for waterfowl, and hence a "waterfowl marsh" may be found in most wetland ecosystems and are ubiquitous in the northern hemisphere. Accordingly, we have adapted a broad definition of managed waterfowl marshes and review literature on a diversity of ecosystems ranging from large coastal wetlands to isolated systems, such as playa lakes or prairie potholes. In this context, we define managed wetlands as those that receive direct manipulations intended to alter the naturally occurring hydrology, vegetation, or biotic communities of the wetland (hereafter active management) with the goal of providing habitat for waterfowl during at least one phase of their annual life cycle.

There is considerable variability in the objectives, intensity, and approaches to managing wetlands as waterfowl habitat. Such management may not explicitly focus on promotion of invertebrate populations and may occur on highly variable time intervals ranging from days to decades. Combinations of many wetland management techniques are often used in an integrated strategy, and impacts of such practices are often anecdotal, particularly as they relate to aquatic invertebrate populations. For example, the widespread practice of water-level manipulation for wintering waterfowl in the southern United States is

D. Batzer, D. Boix (eds.), *Invertebrates in Freshwater Wetlands*,
DOI 10.1007/978-3-319-24978-0_16

intended to promote access to plant-based food for waterfowl, but only recently have researchers begun to understand how invertebrates respond to those manipulations and their importance for wintering waterfowl (Anderson and Smith 2000; Foth et al. 2014; Tapp and Webb 2015). Wetland management for waterfowl often conveys multiple benefits to fish, wildlife, and plant communities (Baldassarre 2014) and likely reflects practices and principles detailed in previous chapters. In this chapter, we review wetland management practices used to promote waterfowl habitat and their impacts on aquatic invertebrates in those systems.

Waterfowl Classification and Foraging Ecology

Understanding the impetus behind wetland management for waterfowl first requires a general understanding of the diversity of life-history strategies and functional morphology among waterfowl. For the purposes of this review, we focused our discussion of waterfowl to those within the family Anatidae. Anatidae is a diverse family comprising 5 subfamilies (Dendrocygninae, Anserinae, Stictonettinae, Tadorninae, Anatinae) and 171 extant species occupying all continents except Antarctica (Baldassarre and Bolen 2006). Detailed phylogenies of the group have been described elsewhere (e.g., Livezay 1997); the relevant discussion of these phylogenies for our review is to identify taxa that are likely to benefit from invertebrates in management wetlands. The first criterion for inclusion in our discussion is that the species consume invertebrates to successfully complete some part of their life-cycle. Herbivory, or more generally a plant-based diet (i.e., plant seeds and vegetation), is ubiquitous among Anatidae. The subfamily Anserinae (geese and swans) is comprised of herbivores, which exclusively forage on plant material, even during periods of high protein demand during ontogeny. Plant-dependent species also occur throughout other subfamilies within Anatidae (e.g., whistling ducks within Dendrogygninae) and many species rely on plant-dominated diets throughout most of their life cycle. The second criterion for inclusion in our discussion is that the species or taxa must consume invertebrates in habitats that are subject to management during a phase of their life cycle. Species that fall within the Mergini tribe of the Anatinae subfamily (sea ducks) for example consume considerable amounts of invertebrates in their diets but forage primarily in pelagic systems that are not subject to management. With these two criteria, our discussion focuses primarily on duck species in two subfamilies: Tadorinae (shellducks and torrent ducks) and Anatinae (pochards, stiff-tailed ducks, and surface-feeding ducks). Further, our discussion will be primarily constrained to species and examples within this group that occupy ranges in western Europe and North America, where most active management for invertebrates and research occurs and is relatively well documented.

Among the species meeting the above criteria, there remains considerable variability in their reliance on invertebrates. The extent of aquatic invertebrate use among waterfowl species is driven by two main factors: interspecific adaptations for foraging and interseasonal variation in nutrient demands. There are two general foraging strategies used by ducks in aquatic systems: diving and surfacing feeding or dabbling (Fig. 16.1). Diving is characteristic of sea ducks, stiff-tailed ducks, and pochards and facilitates foraging within the water column and along the benthos. Long-tailed ducks (*Clangula hyemalis*) are the deepest-documented diving ducks, having achieved diving depths in excess of 50 m (Schorger 1947), although most diving ducks likely restrict foraging to depths of 0.5 to 3 m (Baldassarre 2014). Dabbling ducks are constrained to foraging by skimming the surface, submerging their head and neck, or tipping up to submerse their head and upper body to reach foods at deeper depths - up to approximately 40 cm (Pöysä 1983). Location of foraging within the water column can have considerable influence on invertebrate use and availability among waterfowl species using varying foraging strategies. For example, midge-larvae that emerge along the water surface are consumed extensively by surface-feeding ducks, whereas benthic invertebrates such as amphipods are consumed extensively by diving ducks such as lesser scaup (*Aythya affinis*; Afton and Hier 1991).

Interspecific morphological variation among ducks has a well-documented influence on the structuring of waterfowl communities in wetlands through influences on foraging efficiency or functional foraging depths (Siegfried 1976; Pöysä 1983; Nudds and Bowlby 1984; Torrence and Butler 2006). Bill shape and structure, neck length, and body length are the primary morphological factors influencing this differentiation. The role of body and neck length and bill structure in determining foraging depths is particularly pronounced among

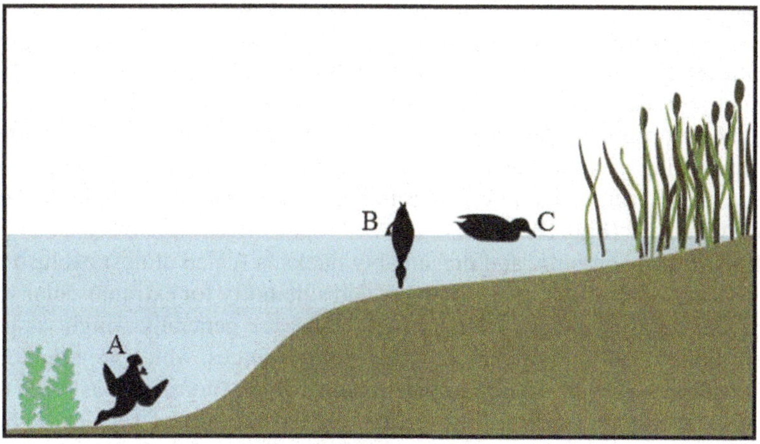

Fig. 16.1 Diving (**a**) and surface feeding by tipping up (**b**) or skimming (**c**) are the primary foraging habits used by waterfowl that consume invertebrates in wetlands

surface-feeding ducks, which are all characterized by elongated necks and bodies and flat bills lined with variable densities of lamellae to facilitate filtering food. Neck and body lengths influence functional feeding depths of surface-feeding ducks and facilitate depth-dependent segregation of conspecific foraging guilds in wetlands (Pöysä 1983; Pöysä et al. 1994; Isola et al. 2000). Lamellar densities among surface-feeding ducks facilitate considerable variability in invertebrate consumption; species such as mallards (*Anas platyrhynchos*) with relatively coarse lamellae densities (8 lamellae/cm) consume larger macroinvertebrates (Batzer et al. 1993), whereas northern shoveler (*Anas clypeata*) can capture and consume microinvertebrates such as rotifers because of their high-density lamellae (21 lamellae/cm) (Euliss et al. 1991). Less variability in lamellar density has been documented among diving ducks (6.7–8.3 lamellae/cm; Lagerquist and Ankney 1989) though interspecific variation in lamellar densities has been reported to contribute to structuring of conspecific foraging guilds of both surface-feeding and diving ducks in wetlands (Nudds and Bowlby 1984; Lagerquist and Ankney 1989).

Because they migrate and reproduce in highly seasonal environments, waterfowl face many energetic demands throughout their annual life cycle. Accordingly, diet can vary considerably throughout the year, depending on motivations for foraging related to major life-cycle events (e.g., ontogenesis, migration, clutch formation, feather molt) and composition and availability of prey in foraging habitats. The annual progression of the life cycle of waterfowl starting in spring is: breeding (territory establishment, clutch formation, incubation, and brood rearing); remigial feather molt in late summer; fall migration; overwintering; and finally spring migration to the breeding grounds (Fig. 16.2). Most species in our discussion are seasonally monogamous, forming pair bonds on the wintering grounds and maintaining them through spring migration and breeding. Pairs subsequently dissolve after clutches are completed and females are responsible for brood rearing independent of their mates. This life-history trait is important in considering the relative importance of invertebrates in wetlands throughout the annual cycle of males and females, which have highly variable nutrient demands within a year. In general, protein demands increase throughout spring into the breeding season; with peak invertebrate consumption on the breeding grounds and followed by a transition to plant-based (seeds and vegetative material) diets late in the breeding season and throughout the non-breeding periods (Fig. 16.2).

Migration among birds and presumably ducks is fueled almost exclusively by lipids because of its high-energy density and suitability for extramuscular storage (Jenni and Jenni-Eiermann 1998). Invertebrates are generally sought as protein sources rather than lipid or carbohydrate energy sources, which are often derived from plant-based foods. Therefore invertebrates are not consumed as much during migration as during the breeding season, except in cases of species with more carnivorous diets (e.g., lesser scaup, Anteau and Afton 2008b; Anteau et al. 2014), or those seeking to build or maintain protein reserves during migration for use in clutch formation. The tendency to transition from plant-based diets to animal-based diets late in winter and throughout spring migration in anticipation of

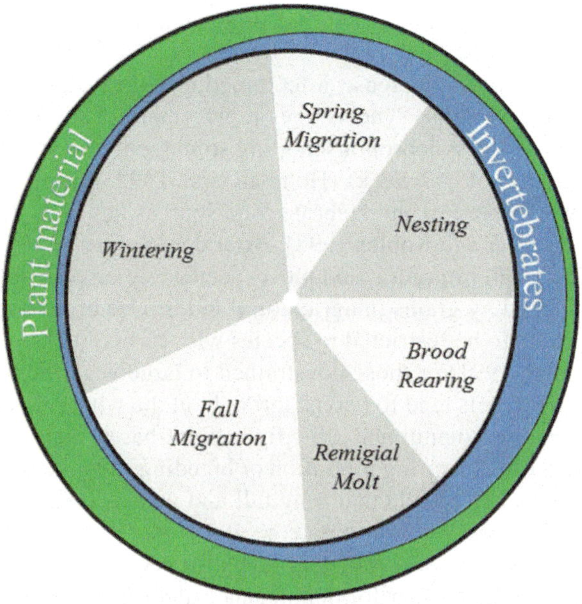

Fig. 16.2 Generalized life-cycle and diet requirements of omnivorous waterfowl. The relative size of the circle outside the life cycle illustrates the relative reliance on plant-based foods or invertebrates in waterfowl diets during each phase in their life cycle. Although many exceptions and deviations on this pattern exist, the general pattern for increasing reliance on invertebrates during the breeding season and plants-based foods during the non-breeding season holds across many omnivorous waterfowl species

nutrient demands for reproduction has been documented in many species (e.g., Euliss and Harris 1987), with some dabbling ducks selectively foraging on invertebrates disproportionate to their availability in wetlands during spring migration (Tidwell et al. 2013).

Waterfowl lay relatively large, nutrient-rich eggs and therefore incur substantial protein and lipid costs during clutch formation, which are often satisfied by invertebrate consumption on the breeding grounds or reserve accumulation prior to arrival on the breeding grounds (Ankney et al. 1991; Alisauskas and Ankney 1992). Duck diet studies ranging from generally herbivorous species, such as gadwalls (*Anas strepera*; Ankney and Alisauskas 1991) or canvasbacks (*Aythya valisineria*; Bartonek and Hickey 1969), to carnivorous species, such as lesser scaup (Afton and Hier 1991) or northern shovelers (Ankney and Afton 1988), have consistently illustrated the ubiquity of invertebrate consumption during the breeding season. Protein, and therefore invertebrate demands, remain throughout the breeding season associated with ontogenic growth of precocial young, which depart the nest and begin feeding on invertebrates (aquatic and emerging adults) within 24 h of hatching (Sedinger 1992). Distribution and abundance of aquatic invertebrates during brood rearing periods can influence habitat use, behavior and growth rates of broods and individual ducklings and, thus, are often the focus of wetland management strategies in breeding habitats (Hunter et al. 1984; Cox et al. 1998).

Protein demands also increase after breeding when adult (>second year) males and females undergo synchronous replacement of flight feathers during the remigial molt. This synchronous molting strategy of waterfowl renders them flightless for periods of 2–7 weeks (Hohman et al. 1992) during which selection for wetlands with abundant invertebrate food sources has been documented for some species (Kohler and Kohler 1998). After completion of the remigial molt, waterfowl initiate fall migration and most species rely on plant-based diets that are often subsidized by grains in agricultural fields. Use of invertebrates during this period appears to be restricted to species with carnivorous diets (e.g., lesser scaup; Afton et al. 1991) or those constrained to large water bodies (Ross et al. 2005). Plant-based diets tend to persist throughout the wintering period in many species until the aforementioned shift from plant-based diets to invertebrates occurs in anticipation of spring migration or breeding. Despite the prevalence of plant material in waterfowl diets during fall and winter, invertebrate consumption during the period is ubiquitous (e.g., Baldassarre et al. 1983; Euliss and Harris 1987; Ballard et al. 2004) and is likely necessary to derive essential amino acids for metabolism or for maintenance and synthesis of somatic tissues (Loesch and Kaminski 1989; Heitmeyer and Fredrickson 1990; Richardson and Kaminski 1992).

Goals of Waterfowl Management in Wetlands

The three principle motivations for wetland management for waterfowl are (1) provision of space for non-foraging activities (i.e., courtship and pair formation, or roosting) (2) facilitating waterfowl-based recreation (i.e., observing or hunting) and (3) provision of food for foraging waterfowl. These three motivations are rarely mutually exclusive and the focus of this chapter is on the final motivation (food) so we only briefly mention the first two (space for waterfowl and (or) recreation) to provide context for some wetland-management programs targeting waterfowl that may not emphasize food resources as a desired management outcome. Management focused on provision of space for waterfowl seeks to provide either roosting habitats, refuges from hunting pressure during fall and winter, or nesting cover. As alluded to earlier, many waterfowl shift from wetland food resources to terrestrial food resources, generally in the form of waste agricultural seeds, throughout the non-breeding period. However, these species still seek out and use wetlands for roosting and supplemental foraging habitats, which can become limiting in some landscapes (Fox et al. 1994; Anteau et al. 2011b). Accordingly, some wetland management strategies aim to provide suitable roosting cover for waterfowl without any explicit concern for food within those habitats. Another common objective for wetland management during the non-breeding season is to provide refuge from anthropogenic disturbances (primarily hunting) for migrating or wintering waterfowl (Madsen 1995; Knapton et al. 2000). In general, most ducks included in our discussion nest in upland habitats around wetlands and use wetlands only for foraging or roosting. However, some species nest over water in wetland vegetation and therefore have specific nesting habitat requirements that can be the focus of wetland management practices (e.g., redheads, *Aythya americana*; Low 1945).

Many waterfowl-based wetland management strategies aim to increase abundance of food resources for waterfowl during at least one period of their annual life cycle. Wetland management for invertebrates and waterfowl is the focus of this chapter, so we will only briefly mention the context of that management, and provide more detail on methods and results later. In general, wetland management practices for waterfowl foods overwhelmingly focus on the manipulation of vegetation or hydrologic regimes to promote early successional plant communities capable of producing large quantities of carbohydrate-rich seeds for waterfowl during the non-breeding period. Primary waterfowl breeding habitats in northern temperate and arctic latitudes, where invertebrates are most important in the diet, are rarely subject to management control, whereas wetland complexes used during non-breeding periods at mid- and southern latitudes, where waterfowl consume mostly plant-based diets, are often subject to substantial annual management control. Despite the overall focus on plant-based foods in many waterfowl management practices, invertebrate community responses to plant-focused manipulations are increasingly recognized at southern latitudes (e.g., Gray et al. 1999; Davis and Bidwell 2008; Foth et al. 2014; Tapp and Webb 2015) and some management and conservation efforts are ongoing at northern latitudes with the intention to increase invertebrate abundance in wetlands used by migrating or breeding waterfowl (e.g., Hornung and Foote 2006; Anteau and Afton 2008a; Anteau et al. 2011a).

An additional motivation for wetland management relevant to waterfowl aims to attract waterfowl or influence their distribution to satisfy wetland and waterfowl-based recreationists, such as observers, photographers, or hunters. Such management engages a broad constituency, can foster strong sociopolitical and financial support for wetland conservation (e.g., Enck et al. 1993; Adams et al. 1997; Vrtiska et al. 2013), and may provide positive local economic impacts that strengthen support for wetland conservation (Gren et al. 1994; Grado et al. 2001). Nongovernmental organizations with recreation-based constituencies, such as Ducks Unlimited, Inc. in North America, have made substantial contributions to wetland conservation and in raising public awareness and support for wetland conservation issues beyond those relevant exclusively to recreational endeavors (Tori et al. 2002). Widespread public interest in waterfowl conservation, driven strongly by waterfowl hunters in the case of North America (Organ et al. 2010), along with the international migratory behavior of waterfowl across the world has led to the development of many international conservation strategies with net benefits for wetlands and aquatic invertebrates therein. For example, since its inception in 1986, the North American Waterfowl Management Plan (NAWMP) and associated funding support mechanisms in the United States, Canada, and Mexico have expended over $4 billion (USD) on conservation of over 63,000 km^2 of wetland habitats throughout North America (NAWMP 2012). A comparable international agreement and conservation strategy exists in Eurasia (African-Eurasian Waterbird Agreement) to identify conservation priorities for Eurasian waterfowl and other wetland-dependent birds.

Where Wetlands Important to Waterfowl Occur

Wetlands of importance to waterfowl occur in many regions of the Northern Hemisphere, and in North America in particular. Their geographic location may partly relate to their importance in providing invertebrate forage. In the United States, wetlands most subject to active waterfowl management tend to occur in the mid- and southern latitudes. Key regions include California's Central Valley, wetlands of the Great Basin region in Oregon, Nevada, and Utah, the Playa Lakes region of Texas, the Mississippi Alluvial Valley, the Upper Mississippi River and its associated pools and wetlands, and Nebraska's Rainwater Basin. Within most of these areas large wetland complexes (e.g., Bear River Migratory Bird Refuge in Utah) are managed primarily for waterfowl and other migratory birds and, therefore, are well suited to the provision of plant and animal foods. Other regions, such as the Mississippi Alluvial Valley, comprise predominantly small, privately owned wetlands (e.g., rice fields) that provide habitat for migrating and wintering waterfowl. Breeding habitats at more northern latitudes in the United States and Canada, including the Prairie Pothole Region, the Boreal Forest, and Arctic region of Canada and Alaska, are keystone habitats for waterfowl, but because of their geographic isolation and extreme climatic variability, receive little direct active management for waterfowl.

Hagy et al. (2014) reviewed wetland conservation issues of relevance to waterfowl in Latin America. They concluded that most priorities involved inventorying existing wetland areas and bird populations; management of wetlands at meaningful scales in this region (for invertebrates or otherwise) was not a common priority. Numerous important wetland systems exist for waterfowl outside of North America, but many are not subject to management consideration. Much of coastal United Kingdom and continental Europe and Africa contain thousands of areas deemed critical wetland habitats for migratory waterfowl, most of which are accounted for and may be investigated using the Critical Site Network Tool and associated publications (http://wow.wetlands.org/INFORMATIONFLYWAY/CRITICALSITENETWORKTOOL/tabid/1349/language/en-US/Default.aspx). Notable regions include the Camargue Region of France, the Ebro Delta of Spain, and the rice-growing regions of Australia and Japan. Wetland areas in Argentina are often subject to waterfowl hunting, but the extent of wetland management on these sites is poorly documented (Ferreyra et al. 2014). Many of these regions are associated with rice or other agriculture subject to flooding, although managed marshes exist in some regions, especially the Camargue, that provide important foraging habitats for migratory waterfowl (Arzel et al. 2009).

Unfortunately, little has been written on management of wetlands outside of North America for invertebrate production for waterfowl. Most studies of invertebrates and waterfowl in these other regions describe associations between duck use and invertebrate abundances (e.g., Phillips 1991), foraging ecology (e.g., Giles 1990), or diet (e.g., Mouronval et al. 2007). Similarly, critical networks of wetlands for waterfowl exist throughout Asia, Australia and New Zealand, Africa, and the Indian Subcontinent, but the extent of management for invertebrates in these regions is largely undocumented. Detailing the international scope of wetlands managed for waterfowl would be untenable for this chapter; therefore, we focus on general ecological principles of wetland management that should broadly apply to invertebrate communities and waterfowl, acknowledging that the majority of examples and insights are of North American studies.

Wetland Management Strategies

Wetland management for waterfowl takes many forms that can generally be classified into three strategies; hydrologic manipulations, direct-vegetation manipulations, and biotic manipulations (of vertebrates). Use of these strategies is rarely exclusive and they are often implemented simultaneously in systems ranging from geographically isolated wetlands such as Playa Lakes or prairie wetlands to coastal or riverine wetlands. Similarly, a wide gradient of management intensity exists with two extremes spanning from systems that may receive management on decades-long return intervals to those that receive active manipulations or management on intraannual (seasonal) or perhaps even daily timescales. For example, as we discuss in detail below, efforts to eradicate fish populations in prairie wetlands over a few years can have long-lasting positive impacts that favor improved waterfowl habitat for many years after active management has occurred. In contrast many wetland management strategies for waterfowl use seasonal or even monthly/weekly management of vegetation or water levels based on fine-tuned understanding of seasonal habitat needs of waterfowl. Our discussion focuses primarily on active management strategies applied on annual time scales, but we also discuss longer-term management strategies, specifically related to biotic manipulation of vertebrates or invasive species in managed wetlands.

Hydrologic Manipulations

Hydrologic manipulations in wetlands are typically achieved using levees containing water control structures with the capacity to introduce or remove water from the wetland (Fredrickson and Taylor 1982; Gray et al. 1999). Examples of water control structures include screw and flap gates and flash-board (also referred to as drop-board) risers (Gray et al. 2013). Water can be introduced to manage wetlands through gravity-flow water source from a higher elevation or through use of gas or electric pumps capable of moving water across elevation gradients and into managed wetlands (Gray et al. 2013).

Water-level manipulations in wetlands are often done for a variety of management objectives related to providing favorable water depths for foraging water birds, eliciting a vegetative response, or promoting dry soil conditions that allow for additional management activities such as direct-vegetation or soil manipulations (Fredrickson and Taylor 1982; Isola et al. 2000; Taft et al. 2002; Greer et al. 2007). Within seasonal wetlands managed for early successional moist-soil plant communities, water levels are typically reduced during spring, to provide suitable conditions for germination and growth of annual grasses, sedges, rushes, and forbs. These plants are fast growing, reach sexual maturity quickly (i.e., <3 months), and produce large quantities of seeds with relatively high nutritional quality for dabbling ducks (Haukos and Smith 1993; Gray et al. 1999). To reduce monocultures of perennially emergent plants, including cattail (*Typha* spp.) and phragmites (*Phragmites australis*), which are also poor producers of seeds for waterfowl, managers of prairie wetlands often periodically draw down water levels in late spring every 4–6 years (Merendino et al. 1990; Gray et al. 2013). This water-level manipulation disrupts the life-history needs of these two wetland species, allowing other plants to recolonize.

Within managed wetlands, the duration and timing of inundation, or hydroperiod, as well as depth of flooding, can influence macroinvertebrate abundance, species richness, community composition, and production (Batzer and Resh 1992). However the relationship between hydroperiod and invertebrate community response is not consistent among wetlands types and has resulted in contradictory results (Neckles et al. 1990; Schneider and Frost 1996; Hall et al. 2004; Batzer 2013). For example, temporary and seasonal wetlands in California with longer hydroperiods had greater invertebrate diversity and abundance for some taxa (Batzer and Resh 1992), whereas hydroperiod explained minimal variation in invertebrate metrics within seasonal woodland wetlands in northern Minnesota and Wisconsin (Palik et al. 2001; Batzer et al. 2004; Hanson et al. 2010). Alternatively, emergent invertebrate taxonomic richness, biomass, and production peaked in riparian wetlands along the Platte River with intermediate hydroperiods, compared to sites with longer or shorter hydroperiods (Whiles and Goldowitz 2001). The contradictory results of numerous studies examining the relationship between wetland hydrology and invertebrate communities led Batzer (2013) to conclude that while hydrology, and specifically hydroperiod, was clearly important in structuring invertebrate communities within seasonally flooded wetlands, generalization about the direction and strength of this relationship was likely confounded by the sensitivity of invertebrates to other environmental factors (e.g., water and air temperature, water depth and chemistry, predators, etc.).

Hydroperiod can indirectly influence wetland invertebrate diversity and production by supporting populations of vertebrate predators, which are often implicated as a primary factor structuring wetland invertebrate communities (Schneider and Frost 1996; Corti et al. 1997; Wissinger 1999). Depending on the water source, inundating wetlands and maintaining water levels for longer time periods allows for the colonization and survival of fish, which prey on invertebrates (Batzer and Resh 1992). Wetlands with shorter hydroperiods or those that are drawn down more frequently typically support lower vertebrate predator densities, especially fish (Batzer and Wissinger 1996). In a study that simultaneously evaluated effects of hydroperiod and predation on Mississippi River floodplain invertebrate communities, total invertebrate taxa richness, abundance, and biomass were lower in sites with more predators and the strength of this relationship increased with duration of floodplain inundation (Corti et al. 1997). Schneider and Frost (1996) reported greater abundance and diversity of predators in vernal ponds with longer hydroperiods and subsequent increased effects of biotic interactions, particularly predation, on invertebrate populations.

The timing of spring water-level drawdowns in managed wetlands influences the macrophyte community composition, with earlier drawdowns often resulting in germination of moist-soil plant species capable of producing large biomasses of seeds (Fredrickson and Taylor 1982; Gray et al. 2013). However, there is also potential for early drawdowns to preclude invertebrate taxa with longer life-history stages (Wilbur 1987; Neckles et al. 1990). Wetlands with shorter hydroperiods, or those drawn down earlier in the spring, may restrict presence of some species because they are unable to complete the aquatic portion of their life cycle before water is removed from the wetland or they lack capacity to survive longer dry periods in desiccation-resistant life stages (Schneider and Frost 1996; Corti et al. 1997). In addition, managed wetlands with differing hydrology or spring drawdown dates exhibit distinct temporal patterns of insect emergence, which can result in varying peaks of adult insect biomass availability along a temporal gradient (Whiles and Goldowitz 2001).

Late-summer inundation of managed seasonal wetlands can influence invertebrate communities: Anderson and Smith (2000) reported the greatest invertebrate abundance and biomass in playas flooded in September (compared to later November inundation) and Batzer et al. (1993) found greater densities of benthic midge larvae in moist-soil wetlands inundated earlier in autumn. Although both studies noted that earlier flooding schedules did not necessarily reflect natural conditions, early fall flooding is sometimes implemented by wetland managers to create inundated wetland habitat for autumn migrating shorebirds and early migrating dabbling ducks (Batzer et al. 1993; Anderson and Smith 2000).

Wetland hydrology is often manipulated to directly alter macrophyte distribution, structure, and community composition, which in turn can influence aquatic invertebrate communities, making it challenging to separate the effects of manipulating hydroperiod and vegetation on invertebrates (Fredrickson and Taylor 1982; Batzer and Resh 1992; Hall et al. 2004; Batzer 2013). Hydrologic variation is the driving force structuring diversity and production of wetland plants, which increase structural habitat for invertebrates, serve as refugia from predators, provide attachment sites, and produce detritus consumed by many invertebrates (Batzer and Wissinger 1996). Whiles and Goldowitz (2001) reported greatest plant species richness and production at Platte River wetlands with intermediate levels of flood duration and speculated that the high invertebrate species richness and production at these same sites could be linked to macrophyte communities. In an experimental study that manipulated wetland plant density, de Szalay and Resh (2000) found greater macroinvertebrate diversity in areas with high densities of emergent plants, but also reported greater overall invertebrate abundance and lower diversity in open water areas. Given that invertebrate communities vary with wetland plant species composition, as well as varying structure or density within the same plant species, it is challenging to understand and differentiate the effects of various wetland management practices on aquatic macroinvertebrates (Olson et al. 1995; Streever et al. 1995; Batzer 2013).

Direct-Vegetation Manipulations

Wetland vegetation may be manipulated to achieve habitat objectives for migrating waterbirds, and these techniques strive to set back plant succession to enhance overall seed production and improve water-cover ratios. Such techniques also may influence abundance and diversity of aquatic invertebrates. Common manipulations include prescribed burning (Fig. 16.3a), grazing, mowing, and disking of soils, sometimes in conjunction with planting agricultural grains. Davis and Bidwell (2008) reported that wetlands experiencing these different types of vegetation manipulation yielded similar aquatic invertebrate richness and diversity, although these metrics were generally greatest in grazed wetlands and least in disked wetlands. Notably, plant biomass in mowed areas was less than unmanipulated areas, supporting the need to leave some vegetation intact (Davis and Bidwell 2008). Although direct manipulations didn't offer a distinct advantage in enhancing invertebrate communities, the authors suggested direct manipulations were reasonable given management objectives of controlling nuisance plant species and enhancing seed production (Davis and Bidwell 2008). In a study evaluating autumn invertebrate communities in managed, seasonally inundated wetlands in the Lower Mississippi Alluvial Valley, Tapp and Webb (2015) reported no effects of mowing on invertebrate biomass or family richness but did find invertebrate production was approximately ten times greater on sites that were mowed.

Fig. 16.3 These photos illustrate two types of active wetland management for waterfowl habitat conducted by the US Fish and Wildlife service at the 130 km^2 Horicon National Wildlife Refuge in southeastern Wisconsin, USA. Figure (**a**) depicts the use of prescribed fire as a direct-vegetation manipulation intended to increase availability of open foraging habitats and macrophyte diversity in a 1200 ha wetland impoundment that was completely dominated by cattail vegetation. Figure (**b**) shows a trap full of carp in the Rock River that feeds the wetland complex. The trap uses an electronic barrier and vertical bars in the river to capture and remove carp moving upstream to spawn (Photos courtesy of USFWS)

Gray et al. (1999) reported that aquatic invertebrate biomass and diversity was greatest in mowed and control plots in moist-soil units in Mississippi, with tilled and disked plots generally being least in these metrics. In a study of an experimentally manipulated cattail marsh, Murkin et al. (1982) reported invertebrate populations were unaffected by various treatments to manipulate vegetation cover-water ratios, but noted that dabbling duck pair densities were correlated positively with invertebrate biomass on treated plots. Overall, direct manipulations of vegetation appears to have mostly neutral influences on invertebrate biomass and diversity, with potentially positive effects on production, and thus appear to be a reasonable practice when such manipulations obtain other desirable management objectives.

Biomanipulation

Wetland management strategies designed to manipulate vertebrate populations to improve wetland functioning or suitability for waterfowl are often termed biomanipulation. The most common of these practices focus on the management of fishes because the presence of fishes can have a strong influence on invertebrate composition and nutrient dynamics in wetland food webs. Aquatic invertebrates are generally less abundant in wetlands that harbor fish populations owing to direct predation by fish, changes in phytoplankton and macrophyte communities, or a combination of both (Mallory et al. 1994; Hanson and Riggs 1995; Parkos et al. 2003). In wetlands of the Prairie Pothole Region (PPR), the presence of fish had a greater influence on macroinvertebrate composition than surrounding land-use attributes (Tangen et al. 2003). It has even been suggested that the absence of fishes can promote greater richness of aquatic birds, plants, amphibians and invertebrates in small, isolated wetlands (Scheffer and van Geest 2006). Thus, the introduction and (or) management of fishes in wetlands raises important concerns among wildlife biologists because of potential interactions between fishes and waterfowl.

Wetland macroinvertebrates represent important food resources for both fish and waterfowl. In a South Dakota study, diets of spring-migrating lesser scaup were similar (92%) to those of yellow perch (*Perca flavescens*), an intensively managed sport fish (Strand et al. 2008). Fishes can also have important indirect effects on wetland invertebrate communities. The presence of fish in a New York marsh resulted in reduced abundance of snails and midge predators, but greater abundance of epiphytic midges owing to indirect effects of fish on competitive/predatory release from other invertebrates (Batzer et al. 2000). Although a variety of fishes are known to inhabit wetlands and shallow lakes, two species in particular—fathead minnow (*Pimphales promelas*) and common carp (*Cyprinus carpio*)—have been well studied with regard to their effects on food web structure and nutrient cycling in shallow lakes and wetlands and provide a good case-study of the role of biotic manipulations in wetland management for waterfowl.

Fathead Minnow

The fathead minnow is widely distributed in central North America (Scott and Crossman 1973) and has been introduced to many areas along the Atlantic and Pacific coast drainages in the United States (Andreasen 1975). Commonly found in wetlands, fathead minnow densities can reach over 350,000/ha (Duffy 1998). Because fathead minnows are a valued baitfish, they are easily moved among waterbodies via "bait bucket" introductions by anglers. Fathead minnow are also commonly harvested from wetlands and sold to commercial bait dealers (Duffy 1998). In palustrine wetlands of eastern South Dakota, the annual wholesale value of fathead minnows averaged about $232/ha in the early 1990s (Carlson and Berry 1990).

The effects of fathead minnow on wetland invertebrates have been well documented (see Zimmer et al. Chap. 8 of this volume). Early work in Minnesota showed that invertebrate abundance, biomass, and taxon richness were lower in wetlands containing fathead minnows (Hanson and Riggs 1995). Similarly, a study of 19 semi-permanent wetlands in west-central Minnesota showed that invertebrate community structure was affected by fathead minnow abundance, wetland depth, and abundance of aquatic macrophytes, with fathead minnow being the most influential parameter measured (Zimmer et al. 2000). In addition to reducing invertebrate abundance, fathead minnows are known to affect other important biotic and abiotic attributes in wetlands (Anteau and Afton 2008a; Anteau et al. 2011a). In the eastern PPR, wetlands that contained fathead minnows had fewer cladocerans, calanoid copepods, ostracods, and larval tiger salamanders but greater turbidity, total phosphorus, and chlorophyll-*a* concentrations (Zimmer et al. 2001, 2002). Because fathead minnows often represent a dominant species in wetlands where fish occur, they can have strong ecosystem affects that influence plant and invertebrate abundance (Zimmer et al. 2002)—that ultimately affects waterfowl use. Mallard ducklings, for example, grew and survived better in fishless wetlands compared to wetlands that contained fathead minnow (Cox et al. 1998).

Common Carp

Native to Europe and Asia, the common carp has been introduced worldwide and is found on every continent except Antarctica (Weber and Brown 2009). In North America, common carp are often a dominant species in shallow lakes and riverine wetlands and are considered by biologists as a nuisance species with regard to waterfowl management in wetlands. A survey of National Wildlife Refuge managers in the United States revealed that 80% of refuges with wetland impoundments identified common carp as a management challenge (Ivey et al. 1998). The benthivorous feeding behavior of adult carp can have direct, "bottom-up" effects on aquatic systems that results in increased nutrient concentration, greater turbidity, and reduced submerged macrophyte abundance (Parkos et al. 2003). Similarly, nutrient enhancement via excretion and (or) increased decomposition of aquatic macrophytes can increase turbidity and have profound effects on ecosystem state (i.e., clear- or turbid-water state; Scheffer et al. 1993). In Great Lakes wetlands, carp-related turbidity above 20 nephelometric turbidity units was associated with fewer submerged plant species (mean = 2.6 plants) than observed in wetlands with lower turbidity (mean = 10.5 plants; Lougheed et al. 1998).

Life-history attributes of common carp, such as rapid growth and early maturation, contribute to their success in introduced environments, where they can rapidly expand to reach high biomass (>3000 kg/ha; Weber and Brown 2009). In experimental wetlands, total phosphorus, turbidity, and suspended solids were positively related to carp biomass (175–476 kg/ha), whereas macrophyte and macroinvertebrate abundance decreased with increasing carp biomass (Parkos et al. 2003). Early life stages of common carp (<100 mm total length) are known to consume zooplankton (Meijer et al. 1990), but effects of common carp on zooplankton densities are often equivocal. Common carp have been associated with increases (Drenner et al. 1998; Parkos et al. 2003) and decreases (Lougheed et al. 1998) in zooplankton biomass. At sizes greater than 100 mm, common carp switch to foraging on benthic macroinvertebrates and detritus and can reduce benthic invertebrate abundance and diversity. In a recent review, Weber and Brown (2009) found that in 94% of the surveyed literature, benthic macroinvertebrates decreased in response to common carp. Invertebrate taxa that are negatively influenced by common carp include amphipods, annelids, chironomids, and odonates (Miller and Crowl 2006; Weber and Brown 2009; Anteau et al. 2011a). Both predation (Kloskowski 2011) and reduction of macrophytes (Matsuzaki et al. 2007) are believed to be responsible for the reduction of invertebrate biomass in carp-invaded communities.

The establishment of common carp has long been implicated in the disappearance of aquatic macrophytes and reduction of waterfowl use in shallow lakes and wetlands (Bajer et al. 2009). At biomass concentrations of <30 kg/ha, carp were shown to have no effect on macrophytes or waterfowl abundance in Illinois wetlands. However, when carp biomass exceeded 250 kg/ha, vegetative cover declined by 17% and waterfowl use was only 10% of that documented before carp became established (Bajer et al. 2009). Similarly, Ivey et al. (1998) estimated that carp invasion in Malheur National Wildlife Refuge in Oregon reduced waterfowl production there to 25% of the level observed prior to carp invasion. Because fishes such as carp and fathead minnow can have dramatic effects on wetland plants, invertebrates, and waterfowl, a variety of management approaches have been developed to aid in control or removal of these wetland fishes (Fig. 16.3b).

Control/Manipulation of Wetland Fish Populations

Biomanipulation is a common technique used to improve water quality and ecosystem functioning in eutrophic lakes and wetlands. In a review of biomanipulation techniques commonly applied to fish assemblages, Drenner and Hambright (1999) found that partial fish removal was the most successful technique (90%) in improving water quality and habitat conditions for plants and aquatic invertebrates. In a related study of 18 Dutch shallow lakes, removing less than 70% of fish biomass had no impact on water turbidity, but in lakes where fish biomass was reduced below 30% of their initial abundance, water clarity improved appreciably (Drenner and Hambright 1999). Other techniques, such as piscivorous fish stocking (26% of studies reviewed), piscivore stocking with partial fish removal (60%), and elimination of fish followed by re-stocking (67%) generally appear to be less successful at combating eutrophication (Drenner and Hambright 1999).

In efforts to restore waterfowl populations in Lake Christina, Minnesota, the lake was chemically treated with rotenone in the late 1980s to reduce the abundance of benthivorous fish (primarily carp). Early responses to fish removal showed that the cladoceran community shifted from small-bodied taxa (e.g., *Bosmina* spp.) to larger cladoceran taxa (e.g., *Daphnia* spp.) within 6 months (Hanson and Butler 1990). Concomitant with changes in zooplankton composition, chlorophyll-*a* concentrations declined and water transparency and submergent macrophytes increased dramatically (Hanson and Butler 1990). Three years post-treatment, use of Lake Christina by migrating waterfowl increased significantly, presumably due to changes in macroinvertebrate abundance and submerged plants (Hanson and Butler 1994). Although reductions in fish biomass can have profound effects on plant-invertebrate-waterfowl interactions in wetlands, they are generally not permanent because over time, because fish populations increase via recruitment, invasion, and (or) introductions. For example, Lake Christina has been chemically treated three times in the last 40 years to reduce effects of benthivorous fish. Similar efforts to reduce carp populations through direct removal or through rotenone treatments in systems managed for waterfowl are ubiquitous across the United States and report variable long-term success (Fig. 16.4; e.g., Cahoon (1953) at Lake Mattamuskeet in North Carolina or Weier and Starr (1950) in coastal marshes in Ohio).

Stocking of piscivorous fish, such as walleye (*Sander vitreus*), has been used to control fathead minnow populations in wetlands. Fisheries managers in the PPR often use semi-permanent and permanent wetlands to temporarily rear juvenile walleye, thus providing an opportunity to evaluate effects of walleye stocking on wetland attributes. In Minnesota wetlands, the stocking of walleye fry (12,000 fry/ha) resulted in significant reductions in fathead minnow populations (Herwig et al. 2004; Ward et al. 2008). In addition, walleye stocking resulted in increases in water clarity, cladoceran abundance, and some macroinvertebrates whereas phytoplankton biomass declined (Reed 2006; Potthoff et al. 2008; Fig. 16.4). Changes in biotic and abiotic wetland attributes following stocking of walleye fry was attributed to their rapid diet shift from invertebrates to small fish (i.e., fathead minnows). As with fish removal techniques, however, effects of walleye stocking can be short-lived, requiring repeated re-stocking to maintain desired wetland conditions (Potthoff et al. 2008).

Other Biomanipulations

Fish are the most commonly manipulated vertebrates in wetlands managed for waterfowl because of their direct and indirect impacts on food web structure. However, other vertebrates are often targeted for reduction (or removal) in efforts

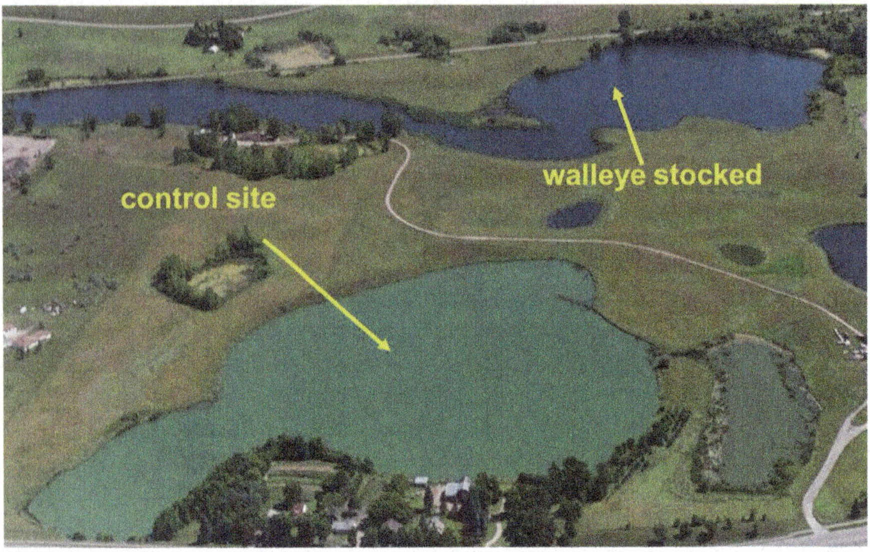

Fig. 16.4 Stocking walleye fingerlings to reduce fathead minnow abundance has proven to be an effective management strategy for improving wetland habitat and water quality in Prairie Pothole wetlands. By reducing fathead minnow abundance, walleyes impose a trophic cascade, resulting in increased water clarity, submerged macrophytes, and greater macroinvertebrate abundance. Below are two wetlands—one stocked with walleye to reduce fathead minnow density (walleye stocked) and one that contains only fathead minnows (control site). Photo courtesy of B. Herwig, Minnesota Department of Natural Resources

to improve wetland conditions for waterfowl. Aquatic mammals have received considerable attention in this respect where a variety of strategies have been used to manage their populations in waterfowl marshes. Muskrats (*Ondatra zibethicus*) and beavers (*Castor canadensis*) for example can be "friend or foe" to wetland managers and are often a focus of habitat management strategies. Beavers and muskrats have well-documented roles in improving waterfowl habitat in wetlands through their impacts on water levels (Beard 1953; see beaver wetlands Chap. 12 of this volume) and vegetation density and composition (Clark 2000; de Szalay and Cassidy 2001), respectively. Further, habitat heterogeneity created by beavers may influence aquatic invertebrates, such as increasing diversity (Hood and Larson 2014). However, conflicts between these species and wetland management objectives often exist and can become the focus of population reduction efforts. The most common conflicts between wetland managers and beavers relate to construction of dams in areas that interfere with water control structures—beavers often respond to drawdowns in managed systems by constructing dams on outlet pipes. Similarly, muskrat populations can be of management concern in cases where their abundance has negative impacts on desired vegetation within a wetland or because of their tendency to construct burrows in the steep slopes of dikes around wetland impoundments. Considerable design and engineering efforts go into abating the latter of these issues related to the optimal slope and dimensions of dikes around impoundments or through burying fence to prevent burrowing into dikes.

Nutria (*Myocastor coypus*) is another aquatic mammal garnering increased attention among wetland managers in North America. Nutria were imported to several states for fur production during the 1930s, and were also deemed useful as a means to control undesirable aquatic vegetation (Carter and Leonard 2002). Since this time, nutria have become well established in the wild in several North American states, and are most abundant along the Gulf Coast and in Oregon (Carter and Leonard 2002). Nutria can cause crop damage, destroy wetland infrastructure via burrowing (e.g., undermining levees), and have been implicated in the loss of coastal wetlands in Louisiana though direct vegetation damage. Therefore, nutria likely indirectly influence wetland invertebrate populations through the removal of important vegetation and substrates, and destruction of wetland infrastructure. Fortunately, control techniques, such as trapping and toxicants, exist to reduce nutria populations where they have become problematic.

Ironically, another emerging threat for waterfowl habitat in North America comes from an exotic species of waterfowl, the mute swan (*Cygnus olor*) which was introduced from its native range in Europe (Petrie and Francis 2003). Mute swans are increasingly a concern for wetland and waterfowl managers across a diversity of ecosystems because of their impacts on aquatic plant productivity and potential indirect effects on invertebrate populations. Many studies have shown reductions in aquatic macrophytes in wetlands associated with grazing by mute swans (e.g., Allin and Husband 2003; Tatu et al. 2007; Stafford et al. 2012); however, the direct effects of such grazing on macroinvertebrates in these wetlands are equivocal (e.g., Allin and Husband 2003) and understudied. Direct control efforts on mute swans have been conducted in some systems though such efforts are often met with controversy from the general public (Ellis and Elphick 2007).

Controversies and Challenges

Of controversies facing wetland management for waterfowl, we suggest the foremost is the notion that active waterfowl management, for invertebrates and otherwise, promotes static, artificial situations intended to maintain consistent hydrology and cover-water ratios that approximate 50:50 (i.e., hemi-marsh; Weller and Spatcher 1965). Various studies support the notion that waterfowl abundance in breeding, migratory, and wintering regions is greatest in wetlands where cover-water interspersion approaches even ratios (Kaminski and Prince 1984; Smith et al. 2004; Webb et al. 2010). However, these intended recommendations may work against normal wetland hydrologic cycles, resulting in wetlands that are less productive or functional over the long-term, which may impede ecosystem health and services. Recent clarifications to wetland management concepts have encouraged more natural manipulations (e.g., use of fire) and hydrology under controlled situations. Nonetheless, the mandate to manage migratory birds may create a mismatch in wetland quality and productivity, and we suggest this topic requires specific study on the costs and benefits of such practices. We note, however, that wetland complexes, properly managed, may rotate through various stages; thus, in some situations static conditions may be more easily avoided.

Another important and contemporary challenge to waterfowl management for invertebrates is accessibility of water for active wetland management. This is of recent and particular importance in the western United States, where many states can experience drought conditions that lead to conflicts for water use. Regions that actively manage wetlands (e.g., in the Klamath Basin; Burke et al. 2004) face pressure on providing water for wetland management given urban and agriculture requirements in the area and downstream. In arid areas, wetlands that remain unflooded during breeding or migration may result in considerable losses of waterfowl habitat, and prolonged dry periods exacerbate the need for solutions to water shortages that address ecological and anthropogenic demands (Burke et al. 2004). Water banks, where water is temporarily purchased from willing sellers for specific uses, may provide one mitigation option, but long-term solutions are evasive (Burke et al. 2004).

Perhaps the greatest challenge to effective wetland management to provide invertebrates for waterfowl is the lack of consistent and conclusive trends identified by previous research on the environmental factors and ecological processes that influence wetland invertebrate communities (Batzer 2013). Inability to detect consistent patterns in invertebrate response to wetland management practices may be because invertebrates, which are tremendously diverse taxa, are simultaneously influenced by numerous ecological and environmental factors including detritus, hydrology, water chemistry, and predation (Batzer 2013). Many studies have evaluated effects of wetland management on invertebrate biomass and community composition, however invertebrate secondary production, or the formation of animal biomass over time, is less commonly evaluated and the dynamics of aquatic invertebrate populations are likely an important element to consider when determining the implications of wetland management for waterfowl (Anderson and Smith 2000; Whiles and Goldowitz 2001; Tapp and Webb 2015). Secondary production of invertebrate communities incorporates invertebrate abundance, biomass, growth, and even colonization of different invertebrate taxa, which may make it a useful metric to evaluate overall invertebrate response to wetland habitat management strategies (Benke and Huryn 2010). Studies of aquatic invertebrates are challenging, time-consuming, and often costly; however, this frontier in wildlife and wetland research needs to be explored more fully if best management practices are to be science-driven. Research on waterfowl selection of different invertebrate groups relative to different life-history events may provide more insight into when and where invertebrates as a food resource are most limiting, and which should be the focus of management. In addition, variation inherent in invertebrate communities make active management for invertebrates difficult, and recommendations, management strategies, and best practices are lacking. The dearth of information may be apparent from our literature review, and we believe new emphasis be focused on bottom-up processes that relate to invertebrate communities with respect to active and passive wetland management practices (Box 16.1).

Box 16.1: Neonicotinoids: An Emerging Threat to Invertebrate Communities in Managed Wetlands?

Neonicotinoids are a class of pesticide developed in the 1980s that have been widely adopted for agricultural use throughout North America and Europe, in large part because they are selectively more toxic to insects than vertebrates (Kollmeyer et al. 1999; Tomizawa and Casida 2005), and their use has grown considerably since 2000 promoted by their flexible use, and systemic (uptake by the plant) activity (Goulson 2013). However, >90 % of active ingredients found in neonicotinoids applied as seed treatments enter the soil, where half-life can range from 200 to >1000 days (Goulson 2013). Given the relatively long half-life and high degree of water solubility of many neonicotinoids, repeated application has the potential to accumulate high concentrations in soils, which can then leach to groundwater or surface water following precipitation events (Gupta et al. 2008; Starner and Goh 2012). Recently, neonicotinoids were detected in >90 % of wetlands sampled within the prairie pothole region of Saskatchewan (Main et al. 2014), whereas Hladik et al. (2014) reported neonicotinoid occurrence in all Midwestern streams sampled, many with concentrations exceeding acute and chronic exposure levels for aquatic macroinvertebrates. In addition to receiving neonicotinoids indirectly through hydrologic inputs from the surrounding watershed, managed wetlands also have the potential for increased neonicotinoid concentrations through direct application as part of moist-soil management practices. Portions of managed wetlands are sometimes planted with agricultural grains (usually corn) as part of the soil disturbance to reset vegetative successional patterns (though disking action) and provide an alternative, high carbohydrate food source for dabbling ducks, however if managers use neonicotinoid-treated seed, direct application of these insecticides has the potential to influence wetland macroinvertebrate communities.

Beketov et al. (2008) demonstrated that a single-pulse application of the neonicotinoid thiacloprid altered long-term community structure of stream macroinvertebrates in a mesocosm, with populations of longer-lived species less likely to recover. A study in the Netherlands reported a significant negative correlation between imidacloprid concentrations in surface waters and macroinvertebrate abundance, including orders Diptera and Ephemeroptera, which commonly occur in wetland ecosystems (Van Dijk et al. 2013). However, this study relied on disparate, previously collected data, where invertebrate and neonicotinoid samples were collected in different locations (up to 1 km apart) and times (up to 160 days apart) (Van Dijk et al. 2013). Prolonged exposure in water to the neonicotinoid imidacloprid was linked to changes in growth, persistence, and community composition of aquatic invertebrates, particularly to individuals in the functional group known as shredders, potentially reducing ecosystem functions related to decomposition, nutrient cycling, and water quality (Kreutzweiser et al. 2009; Agatz et al. 2014; Chagnon et al. 2015). Currently, little is known on neonicotinoid concentrations in managed wetlands and the potential to impact on aquatic invertebrate communities, however if results of lab studies translate to field conditions, direct application of neonicotinoid-treated seeds in managed wetlands could pose an emerging threat to wetland invertebrates.

Conclusions

Invertebrates in managed wetlands (Appendix) clearly provide valuable food resources for waterfowl during all portions of the annual cycle. Because many species rely on proteinaceous foods prior to and during nesting, and young feed almost exclusively on aquatic invertebrates, management to promote these resources may be particularly valuable. However, it appears that invertebrate production for waterfowl is often a consequence of other wetland management practices intended to promote desirable vegetation and carbohydrate-rich foods. Correspondingly, few studies have involved invertebrate production in managed waterfowl marshes, although this topic is beginning to receive more attention.

Managed wetlands for waterfowl occur globally, but we have focused on those in North America and Europe. However, wetland management strategies vary considerably even within close geographic areas. Hydrologic manipulations are common, which involve controlling hydroperiods and extent of inundation to achieve desired results. These techniques may influence macroinvertebrates, but results are inconsistent. Direct manipulations of vegetation may also affect invertebrate abundances and diversity, but likely provide results similar to unmanaged wetlands. Vertebrates may also be managed to "biomanipulate" wetlands. Fish, in particular fathead minnows and common carp, can negatively influence invertebrate abundance, biomass, and richness in wetlands of importance to waterfowl. These species may be subject to control, thereby improving water quality, ecosystem function, and invertebrate populations. Finally, other vertebrate species, such as mute swans and nutria, may alter wetland habitats in a fashion that negatively impacts invertebrate foods for waterfowl. It may also be desirable to control these species.

Wetland management for waterfowl may create long-term artificially static hydrologic patterns with the potential to negatively impact wetland ecosystem processes. Efforts are ongoing to improve hydrologic practices for waterfowl. Issues of availability of water resources for wetland management affect many regions, especially in western North America. Climate change could exacerbate these issues and present important socioeconomic and ecological challenges. Finally, the influence of waterfowl habitat management on invertebrates remains poorly understood, and a unified approach, perhaps in the context of adaptive resource management, would improve understanding and efficiency of management strategies for waterfowl.

Appendix

Aquatic invertebrates collected from waterfowl marshes across North America.

	Manitoba, Canada	California, USA	California, USA	Prairie Pothole Region, USA and Canada	US Central Prairie	Mississippi Alluvial Valley, USA	Mississippi Alluvial Valley, USA	Rainwater Basin, Nebraska, USA
	Lake Manitoba Experimental Cells[a]	Suisun Marsh[b]	Sacramento National Wildlife Refuge[c]	Throughout[d]	Platte River[e]	Greentree reservoirs and flooded forests[f]	Seasonally inundated wetlands with herbaceous vegetation[g]	Playa wetlands[h,i]
NEMATODA	X				X			
CNIDARIA					X			
TURBELLARIA					X			
ROTIFERA				X	X			
MOLLUSCA								
Gastropoda								
Hydrobiidae			X	X				
Lymnaeidae	X			X	X			X
Physidae	X		X	X	X	X	X	X
Planorbidae	X			X	X	X	X	X
Valvatidae				X				
Bivalvia								
Sphaeriidae					X	X		X
ANNELIDA								
Oligochaeta	X	X						
Lumbriculidae				X	X		X	X
Naididae				X			X	X
Tubificidae				X				
Hirudinea				X				
Erpobdellidae				X	X			X

(continued)

(continued)

	Manitoba, Canada	California, USA	California, USA	Prairie Pothole Region, USA and Canada	US Central Prairie	Mississippi Alluvial Valley, USA	Mississippi Alluvial Valley, USA	Rainwater Basin, Nebraska, USA
	Lake Manitoba Experimental Cells[a]	Suisun Marsh[b]	Sacramento National Wildlife Refuge[c]	Throughout[d]	Platte River[e]	Greentree reservoirs and flooded forests[f]	Seasonally inundated wetlands with herbaceous vegetation[g]	Playa wetlands[h,i]
Glossiphoniidae	X			X	X			X
CRUSTACEA								
Branchiopoda								
Artemiidae				X				X
Bosminidae				X				
Branchinectidae							X	
Chirocephalidae				X				
Chydoridae					X			
Daphniidae			X	X	X	X	X	X
Lynceidae				X				·
Thamnocephalidae								X
Branchiura								
Copepoda	X	X		X			X	X
Calanoida			X					
Cyclopoida			X					
Ostracoda	X	X	X	X				X
Amphipoda	X							X
Crangonyctidae					X	X		
Dogielinotidae	X			X	X			X
Gammaridae	X			X			X	X

Isopoda								X
Asellidae						X	X	
Decapoda								
Cambaridae						X	X	
Palaemonidae							X	
ARACHNIDA								
Hydrachnidia	X				X			
Arrenuridae								X
Hygrobatidae						X		
MYRIAPODA								
Chilopoda							X	
INSECTA								
Ephemeroptera								
Baetidae	X		X	X	X		X	
Caenidae	X		X	X			X	
Ephemerellidae								X
Leptophlebidae								X
Metretopodidae							X	
Siphlonuridae							X	X
Odonata								
Aeshnidae	X		X	X		X	X	X
Calopterygidae				X				
Coenagrionidae	X		X	X	X		X	X
Corduliidae			X	X			X	
Gomphidae			X	X			X	
Lestidae	X			X	X		X	X

(continued)

(continued)

	Manitoba, Canada	California, USA	California, USA	Prairie Pothole Region, USA and Canada	US Central Prairie	Mississippi Alluvial Valley, USA	Mississippi Alluvial Valley, USA	Rainwater Basin, Nebraska, USA
	Lake Manitoba Experimental Cells[a]	Suisun Marsh[b]	Sacramento National Wildlife Refuge[c]	Throughout[d]	Platte River[e]	Greentree reservoirs and flooded forests[f]	Seasonally inundated wetlands with herbaceous vegetation[g]	Playa wetlands[h,i]
Libellulidae	X			X	X		X	X
Petaluridae							X	
Plecoptera								
Chloroperlidae			X					
Hemiptera								
Belostomatidae	X		X	X	X		X	X
Corixidae	X	X	X	X	X	X	X	X
Gerridae			X	X		X	X	X
Hebridae							X	
Hydrometridae							X	
Macroveliidae			X					
Mesoveliidae				X			X	
Naucoridae			X					
Nepidae				X		X	X	X
Notonectidae	X		X	X	X		X	X
Ochteridae			X					
Saldidae				X			X	
Pleidae				X				
Veliidae				X			X	
Coleoptera								

Chrysomelidae					X		X	
Curculionidae			X				X	X
Dytiscidae	X		X	X	X	X	X	X
Elmidae			X				X	X
Gyrinidae	X			X			X	X
Haliplidae	X			X			X	X
Hydrophilidae/ Helophoridae	X	X	X	X		X	X	X
Psephenidae								X
Scirtidae	X							
Staphylinidae							X	
Trichoptera								
Hydroptilidae	X			X				
Leptoceridae	X			X	X			
Limnephilidae				X	X		X	
Molannidae				X				
Psychomyiidae				X				
Phryganeidae	X			X				
Polycentropodidae	X			X				
Rhyacophilidae							X	
Lepidoptera							**X**	
Pyralidae/ Crambidae							X	X
Diptera								
Canacidae			X					
Cecidiomyiidae								X

(continued)

(continued)

	Manitoba, Canada	California, USA	California, USA	Prairie Pothole Region, USA and Canada	US Central Prairie	Mississippi Alluvial Valley, USA	Mississippi Alluvial Valley, USA	Rainwater Basin, Nebraska, USA
	Lake Manitoba Experimental Cells[a]	Suisun Marsh[b]	Sacramento National Wildlife Refuge[c]	Throughout[d]	Platte River[e]	Greentree reservoirs and flooded forests[f]	Seasonally inundated wetlands with herbaceous vegetation[g]	Playa wetlands[h,i]
Ceratopogonidae	X		X	X	X		X	X
Chaoboridae			X	X				X
Chironomidae	X	X		X		X	X	X
(Chironominae)		X	X	X	X			
(Tanypodinae)			X	X	X			
(Orthocladiinae)		X	X	X	X			
Culicidae	X	X	X	X	X	X	X	X
Dolichopodidae			X					
Dixidae			X	X			X	X
Empididae			X					
Ephydridae	X	X	X		X		X	
Muscidae			X		X			
Psychodidae								X
Sciaridae					X			
Sciomyzidae			X		X			
Simuliidae			X					
Stratiomyidae			X	X	X	X	X	X
Syrphidae						X		

Tabanidae			X		X	X	X	X
Tipulidae/ Limoniidae	X		X	X	X	X	X	
Hymenoptera								
Pteromalidae				X				

[a]Murkin and Ross (2000)
[b]de Szalay and Resh (1997)
[c]Hicks et al. (1997)
[d]Euliss et al. (1999)
[e]Whiles (unpublished)
[f]Foth (2011)
[g]Tapp and Webb (2015)
[h]Riens et al. (2013)
[i]T. Schepker and E. Webb (unpublished data)

References

Adams CE, Jason AL, Herron JSC (1997) Understanding wildlife constituents: birders and waterfowl hunters. Wildl Soc Bull 25:653–660

Afton AD, Hier RH (1991) Diets of lesser scaup breeding in Manitoba. J Field Ornithol 62:325–334

Afton AD, Hier RH, Paulus SL (1991) Lesser scaup diets during migration and winter in the Mississippi flyway. Can J Zool 69:328–333

Agatz A, Ashauer R, Brown CD (2014) Imidacloprid perturbs feeding of *Gammarus pulex* at environmentally-relevant concentrations. Environ Toxicol Chem 33:648–653

Alisauskas RT, Ankney CD (1992) The cost of egg laying and its relationship to nutrient reserves in waterfowl. In: Batt BDJ (ed) Ecology and management of breeding waterfowl. University of Minnesota Press, Minneapolis, pp 30–61

Allin CC, Husband TP (2003) Mute Swan (*Cygnus olor*) impact on submerged aquatic vegetation and macroinvertebrates in a Rhode Island coastal pond. Northeast Nat 10:305–318

Anderson JT, Smith LM (2000) Invertebrate response to moist-soil management of playa wetlands. Ecol Appl 10:550–558

Andreasen JK (1975) Occurrence of the fathead minnow, *Pimephales promelas*, in Oregon. Calif Fish Game 61:155–156

Ankney CD, Afton AD (1988) Bioenergetics of breeding northern shovelers: diet, nutrient reserves, clutch size, and incubation. Condor 90:459–472

Ankney CD, Alisauskas RT (1991) Nutrient-reserve dynamics and diet of breeding female gadwalls. Condor 93:799–810

Ankney CD, Afton AD, Alisauskas RT (1991) The role of nutrient reserves in limiting waterfowl reproduction. Condor 93:1029–1032

Anteau MJ, Afton AD (2008a) Amphipod densities and indices of wetland quality across the upper-Midwest, USA. Wetlands 28:184–196

Anteau MJ, Afton AD (2008b) Diets of lesser scaup during spring migration throughout the upper-Midwest are consistent with the spring condition hypothesis. Waterbirds 31:97–106

Anteau MJ, Afton AD, Anteau ACE, Moser EB (2011a) Fish and land use influence *Gammarus lacustris* and *Hyalella azteca* (Amphipoda) densities in large wetlands across the upper Midwest. Hydrobiologia 664:69–80

Anteau MJ, Sherfy MH, Bishop AA (2011b) Location and agricultural practices influence spring use of harvested cornfields by cranes and geese in Nebraska. J Wildl Manage 75:1004–1011

Anteau MJ, DeVink J-M, Koons DM, Austin JE, Custer CM, Afton AD (2014) Lesser scaup (*Aythya affinis*). In: Poole A (ed) The birds of North America Online. Cornell Lab of Ornithology, Ithaca, http://bna.birds.cornell.edu/bna/species/338

Arzel C, Elmberg J et al (2009) A flyway perspective on food resource abundance in a long-distance migrant, the Eurasian teal (*Anas crecca*). J Ornithol 150:61–73

Bajer PG, Sullivan G, Sorensen PW (2009) Effects of a rapidly increasing population of common carp on vegetative cover and waterfowl in a recently restored Midwestern shallow lake. Hydrobiologia 632:235–245

Baldassarre GA (2014) Ducks, geese, and swans of North America. Johns Hopkins University Press, Baltimore

Baldassarre GA, Bolen EG (2006) Waterfowl ecology and management. Wiley, New York

Baldassarre GA, Whyte RJ, Quinlan EE, Bolen EG (1983) Dynamics and quality of waste corn available to postbreeding waterfowl in Texas. Wildl Soc Bull 11:25–31

Ballard BM, Thompson JE et al (2004) Diet and nutrition of northern pintails wintering along the southern coast of Texas. J Wildl Manage 68:371–382

Bartonek JC, Hickey JJ (1969) Food habits of canvasbacks, redheads, and lesser scaup in Manitoba. Condor 71:280–290

Batzer DP (2013) The seemingly intractable ecological responses of invertebrates in North American wetlands: a review. Wetlands 33:1–15

Batzer DP, Resh VR (1992) Macroinvertebrates of a California seasonal wetland and responses to experimental habitat manipulation. Wetlands 12:1–7

Batzer DP, Wissinger SA (1996) Ecology of insects in nontidal wetlands. Annu Rev Entomol 41:75–100

Batzer DP, McGee M, Resh VH, Smith RR (1993) Characteristics of invertebrates consumed by mallards and prey response to wetland flooding schedules. Wetlands 13:41–49

Batzer DP, Pusateri CP, Vetter R (2000) Impacts of fish predation on marsh invertebrates: direct and indirect effects. Wetlands 20:307–312

Batzer DP, Palik BJ, Buech R (2004) Relationships between environmental characteristics and macroinvertebrate communities in seasonal woodland ponds in Minnesota. J North Am Benthol Soc 23:50–68

Beard EB (1953) The importance of beaver in waterfowl management at the Seney National Wildlife Refuge. J Wildl Manage 17:398–436

Beketov MA, Schafer RB et al (2008) Long-term stream invertebrate community alterations induced by the insecticide thiacloprid: effect concentrations and recovery dynamics. Sci Total Environ 405:96–108

Benke AC, Huryn AD (2010) Benthic invertebrate production—facilitating answers to ecological riddles in freshwater ecosystems. J North Am Benthol Soc 29:264–285

Burke SM, Adams RM, Wallender WW (2004) Water banks and environmental water demands: case of the Klamath Project. Water Resour Res 40:W09S02. doi:10.1029/2003WR002832

Cahoon WG (1953) Commercial carp removal at Lake Mattamuskeet, North Carolina. J Wildl Manage 17:312–317

Carlson BN, Berry CR (1990) Population size and economic value of aquatic bait species in palustrine wetlands of eastern South Dakota. Prairie Natur 22:119–128

Carter J, Leonard BP (2002) A review of the literature on the worldwide distribution, spread of, and efforts to eradicate the Coypu (*Myocastor coypus*). Wildl Soc Bull 30:162–175

Chagnon M, Kreutzweiser D et al (2015) Risks of large-scale use of systemic insecticides to ecosystem functioning and services. Environ Sci Pollu Res 22:119–134

Clark RG (2000) Ecology of muskrats in prairie wetlands. In: Murkin HR, van der Valk AG, Clark RG (eds) Prairie Wetland Ecology: the contribution of the Marsh Ecology Research Program. Iowa State University Press, Ames, pp 287–313

Corti D, Kohler SL, Sparks RE (1997) Effects of hydroperiod and predation on a Mississippi River floodplain invertebrate community. Oecologia 109:154–165

Cox RR, Hanson MA et al (1998) Mallard duckling growth and survival in relation to aquatic invertebrates. J Wildl Manage 62:124–133

Davis CA, Bidwell JR (2008) Response of aquatic invertebrates to vegetation management and agriculture. Wetlands 28:793–805

de Szalay FA, Cassidy W (2001) Effects of muskrat (*Ondatra zibethicus*) lodge construction on invertebrate communities in a Great Lakes coastal wetland. Am Midl Natur 146:300–310

de Szalay F, Resh V (1997) Responses of wetland invertebrates and plants important in waterfowl diets to burning and mowing of emergent vegetation. Wetlands 17:149–156

de Szalay FA, Resh VH (2000) Factors influencing macroinvertebrate colonization of seasonal wetlands: responses to emergent plant cover. Freshw Biol 45:295–308

Drenner RW, Hambright KD (1999) Biomanipulation of fish assemblages as a lake restoration technique. Archiv Hydrobiol 146:129–165

Drenner RW, Gallow KL, Baca RM, Smith JD (1998) Synergistic effects of nutrient loading and omnivorous fish on phytoplankton biomass. Can J Fish Aquat Sci 55:2087–2096

Duffy WG (1998) Population dynamics, production, and prey consumption of fathead minnows (*Pimephales promelas*) in prairie wetlands: a bioenergetics approach. Can J Fish Aquat Sci 54:15–27

Ellis MM, Elphick CS (2007) Using a stochastic model to examine the ecological, economic and ethical consequences of population control in a charismatic invasive species: mute swans in North America. J Appl Ecol 44:312–322

Enck JW, Swift BL, Decker DJ (1993) Reasons for decline in duck hunting: insights from New York. Wildl Soc Bull 21:10–21

Euliss NH, Harris SW (1987) Feeding ecology of northern pintails and green-winged teal wintering in California. J Wildl Manage 51:724–732

Euliss NH, Jarvis RL, Gilmer DS (1991) Feeding ecology of waterfowl wintering on evaporation ponds in California. Condor 93:582–590

Euliss NH, Mushet DM, Wrubleski DA (1999) Wetlands of the prairie pothole region: invertebrate species composition, ecology, and management. In: Batzer DP, Rader RB, Wissinger SA (eds) Invertebrates in freshwater wetlands of North America: ecology and management. Wiley, New York, pp 471–514

Ferreyra H, Romano M et al (2014) Lead gunshot pellet ingestion and tissue lead levels in wild ducks from Argentine hunting hotspots. Ecotoxicol Environ Saf 103:47–81

Foth JR (2011) Aquatic invertebrate biomass and community composition in greentree reservoirs and naturally flooded forests in the Mississippi Alluvial Valley and Interior Flatwoods. Thesis. Mississippi State University

Foth JR, Straub JN et al (2014) Aquatic invertebrate abundance and biomass in Arkansas, Mississippi, and Missouri bottomland hardwood forests during winter. J Fish Wildl Manage 5:243–251

Fox AD, Mitchell C et al (1994) Winter movements and site-fidelity of Pink-footed Geese *Anser brachyrhynchus* ringed in Britain, with particular emphasis on those marked in Lancashire. Bird Study 41:221–234

Fredrickson LH, Taylor ST (1982) Management of seasonally flooded impoundments for wildlife. US Fish and Wildlife Service, Washington, DC, Resource Publication 148

Giles N (1990) Effects of increasing larval chironomid densities on the underwater feeding success of downy Tufted ducklings *Aythya fuligula*. Wildfowl 41:99–106

Goulson D (2013) An overview of the environmental risks posed by neonicotinoid insecticides. J Appl Ecol 50:977–987

Grado SC, Kaminski RM, Munn IA, Tullos TA (2001) Economic impacts of waterfowl hunting on public lands and at private lodges in the Mississippi Delta. Wildl Soc Bull 29:846–855

Gray MJ, Kaminski RM et al (1999) Aquatic invertebrate and plant responses following mechanical manipulations of moist-soil habitat. Wildl Soc Bull 27:770–779

Gray MJ, Hagy HM, Nyman JA, Stafford JD (2013) Management of wetlands for wildlife. In: Anderson JT, Davis CA (eds) Wetland techniques: vol 3: applications and management. Springer, New York, pp 121–180, 10.1007/978-94-007-6907-6_4

Greer AK, Dugger BD, Graber DA, Petrie MJ (2007) The effects of seasonal flooding on seed availability for spring migrating waterfowl. J Wildl Manage 71:1561–1566

Gren I-M, Folke C, Turner K, Batemen I (1994) Primary and secondary values of wetland ecosystems. Environ Resource Econ 4:55–74

Gupta S, Gajbhiye VT, Gupta RK (2008) Soil dissipation and leaching behavior of a neonicotinoid insecticide thiamethoxam. Bull Environ Contam Toxicol 80:431–437

Hagy HM, Yaic SC, et al. (2014) Wetland issues affecting waterfowl conservation in North America. Wildfowl Spec Issue 4:343–367

Hall DL, Willig MR et al (2004) Aquatic macroinvertebrate diversity of playa wetlands: the role of landscape and island biogeographic characteristics. Wetlands 24:77–91

Hanson MA, Butler MG (1990) Early responses of plankton and turbidity to biomanipulation in a shallow prairie lake. Hydrobiologia 200:317–327

Hanson MA, Butler MG (1994) Responses to food web manipulation in a shallow waterfowl lake. Hydrobiologia 279:457–466

Hanson MA, Riggs MR (1995) Potential effects of fish predation on wetland invertebrates: a comparison of wetlands with and without fathead minnows. Wetlands 15:167–175

Hanson MA, Palik BJ, Church JO, Miller AT (2010) Influences of upland timber harvest on aquatic invertebrate communities in seasonal ponds: efficacy of forested buffers. Wetlands Ecol Manage 15:167–175

Haukos DA, Smith LM (1993) Moist-soil management of playa lakes for migrating and wintering ducks. Wildl Soc Bull 21:288–298

Heitmeyer ME, Fredrickson LH (1990) Fatty acid composition of wintering female mallards in relation to nutrient use. J Wildl Manage 54:54–61

Herwig BR, Hanson MA et al (2004) Walleye stocking as a tool to suppress fathead minnows and improve habitat quality in semipermanent and permanent wetlands in the Prairie Pothole Region of Minnesota. Minnesota Department of Natural Resources, Little Falls, Spec Publ 159

Hicks JM, Euliss NH, Harris SW (1997) Aquatic invertebrate ecology during a simulated botulism epizootic in a Sacramento Valley wetland. Wetlands 17:157–162

Hladik ML, Kolpin DW, Kuivila KM (2014) Widespread occurrence of neonicotinoid insecticides in streams in a high corn and soybean producing region, USA. Environ Pollut 193:189–196

Hohman WL, Ankney CD, Gordon DH (1992) Ecology and management of postbreeding waterfowl. In: Batt BDJ, Afton AD et al (eds) Ecology and management of breeding waterfowl. University of Minnesota Press, Minneapolis, pp 128–189

Hood GA, Larson DG (2014) Beaver-created habitat heterogeneity influences aquatic invertebrate assemblages in boreal Canada. Wetlands 34:19–29

Hornung J, Foote AL (2006) Aquatic invertebrate responses to fish presence and vegetation complexity in Western Boreal wetlands, with implications for waterbird productivity. Wetlands 26:1–12

Hunter ML, Witham JW, Dow H (1984) Effects of a carbaryl-induced depression in invertebrate abundance on the growth and behavior of American black duck and mallard ducklings. Can J Zool 62:452–456

Isola C, Colwell M, Taft O, Safran R (2000) Interspecific differences in habitat use of shorebirds and waterfowl foraging in managed wetlands of California's San Joaquin Valley. Waterbirds 23:196–203

Ivey GL, Cornely JE, Ehlers BD (1998) Carp impacts on waterfowl at Malheur National Wildlife Refuge, Oregon. Transactions of the North American Wildlife Natural Resources conference. No. 63, pp 66–74

Jenni L, Jenni-Eiermann S (1998) Fuel supply and metabolic constraints in migrating birds. J Avian Biol 29:521–528

Kaminski RM, Prince HH (1984) Dabbling duck-habitat associations during spring in Delta Marsh, Manitoba. J Wildl Manage 48:37–50

Kloskowski J (2011) Impact of common carp *Cyprinus carpio* on aquatic communities: direct trophic effects versus habitat deterioration. Fund Appl Limnol 178:245–255

Knapton RW, Petrie SA, Herring G (2000) Human disturbance of diving ducks on Long Point Bay, Lake Erie. Wildl Soc Bull 28:923–930

Kohler P, Kohler U (1998) Considerable increase of moulting waterfowl in fishponds without carp in the Ismaninger Teichgebiet. Sylvia 34:27–32

Kollmeyer WD, Flattum RF et al (1999) Discovery of the nitromethylene heterocycle insecticides. In: Yamamoto I, Casida J (eds) Nicotinoid insecticides and the nicotinic acetylcholine receptor. Springer, Tokyo, pp 71–89

Kreutzweiser DP, Thompson DG, Scarr TA (2009) Imidacloprid in leaves from systemically treated trees may inhibit litter breakdown by nontarget invertebrates. Ecotoxicol Environ Saf 72:1053–1057

Lagerquist BA, Ankney CD (1989) Interspecific differences in bill and tongue morphology among diving ducks (*Aythya* spp., *Oxyura jamaicensis*). Can J Zool 67:2694–2699

Livezay BC (1997) A phylogenetic classification of waterfowl (Aves: Anseriformes), including selected fossil species. Ann Carneigie Mus 66:457–496

Loesch CR, Kaminski RM (1989) Winter body-weight patterns of female mallards fed agricultural seeds. J Wildl Manage 53:1081–1087

Lougheed VL, Crosbie B, Chow-Fraser P (1998) Predictions on the effect of common carp (*Cyprinus carpio*) exclusion on water quality, zooplankton, and submergent macrophytes in a great lakes wetland. Can J Fish Aquat Sci 55:1189–1197

Low JB (1945) Ecology and management of the redhead, *Nyroca americana*, in Iowa. Ecol Monogr 15:35–69

Madsen J (1995) Impacts of disturbance on migratory waterfowl. Ibis 137:S67–S74

Main AR, Headley JV et al (2014) Widespread use and frequent detection of neonicotinoid insecticides in wetlands of Canada's Prairie Pothole Region. PLoS One 9(3), e92821

Mallory ML, Blancher PJ, Weatherhead PJ, McNicol DK (1994) Presence or absence of fish as a cue to macroinvertebrate abundance in boreal wetlands. Hydrobiologia 279(280):345–351

Matsuzaki SS, Nisikawa U, Noriko T, Izumi W (2007) Effects of common carp on nutrient dynamics and littoral community composition: roles of excretion and bioturbation. Fund Appl Limnol 168:27–38

Meijer ML, De Hamm MW, Breukelaar AW, Buiteveld H (1990) Is reduction of the benthivorous fish an important cause of high transparency following biomanipulation in shallow lakes? Hydrobiologia 200(201):303–315

Merendino MT, Smith LM, Murkin HR, Pederson RL (1990) The response of prairie wetland vegetation to seasonality of drawdown. Wildl Soc Bull 18:245–251

Miller SA, Crowl TA (2006) Effects of common carp (*Cyprinus carpio*) on macrophytes and invertebrate communities in a shallow lake. Freshw Biol 51:85–94

Mouronval J-B, Guillemain M, Canny A, Poirier F (2007) Diet of non-breeding wildfowl Anatidae and Coot *Fulica atra* on the Perthois gravel pits, northeast France. Wildfowl 57:68–97

Murkin HR, Ross LCM (2000) Invertebrates in prairie wetlands. In: van der Valk A (ed) Prairie wetland ecology. Iowa State University Press, Ames, pp 201–248

Murkin HR, Kaminski RM, Titman RD (1982) Responses by dabbling ducks and aquatic invertebrates to an experimentally manipulated cattail marsh. Can J Zool 60:2324–2332

Neckles HA, Murkin HR, Cooper JA (1990) Influences of seasonal flooding on macroinvertebrate abundance in wetland habitats. Freshw Biol 23:311–322

North American Waterfowl Management Plan [NAWMP] (2012) North American waterfowl management plan 2012: people conserving waterfowl and wetlands. US Fish and Wildlife Service, Department of Interior, Washington, DC

Nudds TD, Bowlby JN (1984) Predator–prey size relationships in North American dabbling ducks. Can J Zool 62:2002–2008

Olson EJ, Engstrom ES, Doeringsfield MR, Bellig R (1995) Abundance and distribution of macroinvertebrates in relation to macrophyte communities in a prairie marsh, Swan Lake, Minnesota. J Freshw Ecol 10:325–335

Organ JF, Mahoney SP, Geist V (2010) Born in the hands of hunters. Wildl Prof 4:22–27

Palik B, Batzer DP et al (2001) Seasonal pond characteristics across a chronosequence of adjacent forest ages in northern Minnesota. Wetlands 21:532–542

Parkos JJ, Santucci VJ, Wahl DH (2003) Effects of adult common carp (*Cyprinus carpio*) on multiple trophic levels in shallow mesocosms. Can J Fish Aquat Sci 60:182–192

Petrie SA, Francis CM (2003) Rapid increase in the lower Great Lakes population of feral mute swans: a review and recommendation. Wildl Soc Bull 31:407–416

Phillips VE (1991) Pochard *Aythya ferina* use of chironomid-rich feeding habitat in winter. Bird Study 38:118–122

Potthoff AJ, Herwig BR et al (2008) Cascading food-web effects on piscivore introductions in shallow lakes. J Appl Ecol 45:1170–1179

Pöysä H (1983) Morphology-mediated niche organization in a guild of dabbling ducks. Ornis Scand 14:317–326

Pöysä H, Elmberg J, Nummi P, Sjöberg K (1994) Species composition of dabbling duck assemblages: ecomorphological patterns compared with null models. Oecologia 98:193–200

Reed JR (2006) Effects of landscape-scale factors on wetland biomanipulation. Minnesota Department of Natural Resources, Investigational Report 533

Richardson DM, Kaminski RM (1992) Diet restriction, diet quality, and prebasic molt in female mallards. J Wildl Manage 56:531–539

Riens JR, Schwartz MS, Mustafa F, Hoback WW (2013) Aquatic macroinvertebrate communities and water quality at buffered and non-buffered wetland sites on federal waterfowl production areas in the Rainwater Basin, Nebraska. Wetlands 33:1025–1036

Ross RK, Petrie SA, Shannon SB, Mullie A (2005) Autumn diet of greater scaup, lesser scaup, and long-tailed ducks on eastern Lake Ontario prior to zebra mussel invasion. Wildl Soc Bull 33:81–91

Scheffer M, van Geest GJ (2006) Small habitat size and isolation can promote species richness: second-order effects on biodiversity in shallow lakes and ponds. Oikos 121:227–231

Scheffer M, Hosper SH et al (1993) Alternative equilibria in shallow lakes. Trends Ecol Evol 8:275–279

Schneider DW, Frost TM (1996) Habitat duration and community ecology of temporary ponds. J North Am Benthol Soc 15:64–86

Schorger AW (1947) The deep diving of the loon and old-squaw and its mechanism. Wilson Bull 59:151–159

Scott WB, Crossman EJ (1973) Freshwater fishes of Canada. Bull Fish Res Board Can 184:1–966

Sedinger JS (1992) Ecology of prefledging waterfowl. In: Batt BDJ, Afton AD et al (eds) Ecology and management of breeding waterfowl. University of Minnesota Press, Minneapolis

Siegfried WR (1976) Segregation in feeding behaviour of four diving ducks in southern Manitoba. Can J Zool 54:730–736

Smith LM, Haukos DA, Prather RM (2004) Avian response to vegetative pattern in playa wetlands during winter. Wildl Soc Bull 32:474–480

Stafford J, Eichholz M, Phillips A (2012) Impacts of mute swans (*Cygnus olor*) on submerged aquatic vegetation in Illinois River Valley backwaters. Wetlands 32:851–857

Starner K, Goh KS (2012) Detection of the neonicotinoid insecticide imidacloprid in surface waters of three agricultural regions of California, USA, 2010–2011. Bull Environ Contam Toxicol 88:316–321

Strand KA, Chipps SR et al (2008) Patterns of prey use by lesser scaup *Aythya affinis* and diet overlap with fishes during spring migration. Hydrobiologia 598:389–398

Streever WJ, Evans DL, Keenan CM, Crisman TL (1995) Chiromidae (Diptera) and vegetation in a created wetland and implications for sampling. Wetlands 15:285–289

Taft OW, Colwell MA, Isola CR, Safran RJ (2002) Waterbird responses to experimental drawdown: implications for the multispecies management of wetland mosaics. J Appl Ecol 39:987–1001

Tangen BA, Butler MG, Ell MJ (2003) Weak correspondence between macroinvertebrate assemblages and land use in Prairie Pothole Region wetlands, USA. Wetlands 23:104–115

Tapp JL, Webb EB (2015) Aquatic invertebrate food base for waterbirds at Wetland Reserve Program easements in the Lower Mississippi Alluvial Valley. Wetlands 35:183–192

Tatu KS, Anderson JT, Hindman LJ, Seidel G (2007) Mute swans' impact on submerged aquatic vegetation in Chesapeake Bay. J Wildl Manage 71:1431–1439

Tidwell PR, Webb EB, Vrtiska MP, Bishop AA (2013) Diets and food selection of female mallards and blue-winged teal during spring migration. J Fish Wildl Manage 4:63–74

Tomizawa M, Casida JE (2005) Neonicotinoid insecticide toxicology: mechanisms of selective action. Annu Rev Pharmacol Toxicol 45:247–268

Tori GM, McLeod S et al (2002) Wetland conservation and Ducks unlimited: real world approaches to multispecies management. Waterbirds Int J Waterbird Biol 25:115–121

Torrence SM, Butler MG (2006) Spatial structure of a diving duck (*Aythya*, *Oxyura*) guild: how does habitat structure and competition influence diving duck habitat use within northern prairie wetlands? Can J Zool 84:1358–1367

Van Dijk TC, Van Staalduinen MA, Nan der Sluijs JP (2013) Macro-invertebrate decline in surface water polluted with imidacloprid. PLoS One 8(5), e62374

Vrtiska MP, Gammonley JH, Naylor LW, Raedeke AH (2013) Economic and conservation ramifications from the decline of waterfowl hunters. Wildl Soc Bull 37:380–388

Ward MC, Willis DW et al (2008) Consumption estimates of walleye stocked as fry to suppress fathead minnow populations in west-central Minnesota wetlands. Ecol Freshw Fish 17:59–70

Webb EB, Smith LM, Vrtiska M, LaGrange T (2010) Effects of local and landscape variables on wetland bird use during migration through the Rainwater Basin. J Wildl Manage 74:109–119

Weber MJ, Brown ML (2009) Effects of common carp on aquatic ecosystems 80 years after "Carp as a dominant": ecological insights for fisheries management. Rev Fish Sci 17:524–537

Weier JL, Starr DF (1950) The use of rotenone to remove rough fish for the purpose of improving migratory waterfowl refuge areas. J Wildl Manage 14:203–205
Weller MW, Spatcher CF (1965) Role of habitat in the distribution and abundance of marsh birds. Iowa Agricultural Home Economic Experiment Station, Ames, Spec Rep 43
Whiles MR, Goldowitz BS (2001) Hydrologic Influences on insect emergence production from Central Platte River wetlands. Ecol Appl 11:1829–1842
Wilbur HM (1987) Regulation of structure in complex systems: experimental temporary pond communities. Ecology 68:1437–1452
Wissinger SA (1999) Ecology of wetland invertebrates: synthesis and application for conservation and management. In: Batzer DP, Rader RB, Wissinger SA (eds) Invertebrates in freshwater wetlands in North America. Wiley, New York, pp 1043–1086
Zimmer KD, Hanson MA, Butler MG (2000) Factors influencing invertebrate communities in prairie wetlands: a multivariate approach. Can J Fish Aquat Sci 57:76–85
Zimmer KD, Hanson MA, Butler MG (2001) Effects of fathead minnow colonization and removal on a prairie wetland ecosystem. Ecosystems 2001:346–357
Zimmer KD, Hanson MA, Butler MG (2002) Effects of fathead minnows and restoration on prairie wetland ecosystems. Freshw Biol 47:2017–2086

Chapter 17
Invertebrate Assemblages and Their Ecological Controls Across the World's Freshwater Wetlands

Dani Boix and Darold Batzer

Introduction

The 15 habitat-based chapters in this book (Table 17.1) provide a unique opportunity for cross-wetland comparisons of invertebrate assemblages. Because of the global coverage and the enormous natural variation among the habitats covered, it permits a "macro" scale perspective on wetland invertebrate distribution and ecology, never before attempted. For example, the influence of habitat size on invertebrate taxonomic richness has been examined within several wetland types (e.g., rock pools, Vanschoenwinkel et al. 2009; alpine ponds, Wissinger et al. 1999). However, in this book, we were dealing with habitats like rock pools that can be <1 m across to the Pantanal at almost 200,000 km^2 in area. We deal with small habitats with very simplistic water budgets (e.g., precipitation in temporary alpine ponds, groundwater in springs), to large habitats that concurrently receive water from precipitation, groundwater, and surface flows and also concurrently have temporary, semipermanent, and permanent hydroperiods (e.g., floodplains). Thus, we examine variation in wetlands and their invertebrates through a very broad macroscopic rather than a narrow habitat-specific perspective, and we feel this yields some new and insightful ways to view wetland invertebrate ecology.

Invertebrate Assemblages Across Different Types of Wetlands

To begin our macroscopic tour of wetland invertebrates, we first examine assemblage structure among the different wetland types. We focus on the family level because this is the lowest taxon that was reliably reported by authors across all habitats, and this level eliminates, to some extent, biogeographic considerations (a great

D. Batzer, D. Boix (eds.), *Invertebrates in Freshwater Wetlands*,
DOI 10.1007/978-3-319-24978-0_17

Table 17.1 Fifteen habitat-based chapters in this book, with the abbreviated habitat names we used for habitat labeling in summary tables

Book chapter titles	Abbreviated habitat names
2. Invertebrates in Rock Pools	Rock pools
3. Invertebrate Communities of Alpine Ponds	Alpine ponds
4. Invertebrates in Temporary Wetland Ponds of the Temperate Biomes	Temporary ponds
5. Invertebrates of Freshwater Temporary Ponds in Mediterranean Climates	Mediterranean ponds
6. Invertebrates in Irish Turloughs	Turloughs
7. Peatland Invertebrates	Peatlands
8. Invertebrates in Permanent Wetlands (Long-hydroperiod Marshes and Shallow Lakes)	Permanent marshes
9. Invertebrates in Great Lakes Marshes	Great Lakes marshes
10. Invertebrates in the Florida Everglades	Everglades
11. Invertebrates in Groundwater Springs and Seeps	Springs
12. Invertebrate in Beaver-Created Wetlands and Ponds	Beaver ponds
13. Invertebrates in Temperate-zone River Floodplains	Temperate floodplains
14. Invertebrates in Neotropical Floodplains	Neotropical floodplains
15. Invertebrates in Created and Restored Wetlands	Created wetlands
16. Invertebrates in Managed Waterfowl Marshes	Waterfowl marshes

number of genera only occur in specific areas or continents, but this is the case for relatively few families). However, family level classification was only comprehensively covered in chapter appendices for certain higher taxonomic groups, mostly macroinvertebrates. Thus, for our analysis, we focus on mollusks, leeches, large branchiopods, malacostracans, odonates, the so-called EPT taxa (Ephemeroptera, Plecoptera, and Trichoptera), hemipterans, coleopterans, and dipterans, compiling cross-habitat summary tables for each.

In each table, we list all families of a group reported across all of the wetland types (see those appendices), ranked by percent occurrence (vertical axis). We also rank the different kinds of wetlands by how many families of a group were reported (least to most; horizontal axis). For each taxonomic group, a small number of families were ubiquitous across all wetland types (occurred in virtually every wetland type; e.g., Chironomidae). We considered these families as *generalists*, able to exploit any wetland regardless of its environmental conditions (acknowledging that more specialization undoubtedly occurs for many genera or species in these groups). The families that only occurred in a single wetland type were considered *rare*. These rare taxa were mostly incidental colonizers, have very specific habitat needs, or in some cases occurred in very restrictive locations (e.g., only Australia or South America). While individual species in these rare families may be noteworthy (highly specialized), we felt these rare families would provide minimal information regarding cross-habitat ecological controls on invertebrates.

Those families that were neither ubiquitous nor rare, were deemed the most likely to suggest how invertebrate assemblages might be controlled across environmental gradients or transitions. We refer to these families as being *constrained*; i.e., they were clearly adapted for wetland conditions, but some environmental factor (or factors) controlled whether they occurred in some wetland types and not others. After identifying constrained families in a group, we then looked for gradients and/or transitions in the richness of these moderately widespread families across the set of 15 wetland types, focusing on which habitats supported large numbers of otherwise constrained taxa and which habitat supported few or none. We further provided gross characterizations of the environmental conditions that occur in the different wetlands, focusing on relative size, relative hydroperiod lengths, primary sources of water input, whether the water flowed (was lotic) or not (was lentic), how the vegetation was structured, and whether fish occurred or not. These six different environmental factors and the criteria we used to identify variation for each factor are summarized in Table 17.2. Criteria by intention only reflect gross change. Note also that several wetland types defy even a relatively gross categorization (in sizes, hydroperiods, water sources, or flow patterns). Thus, we often used multiple values or ranges in values to describe some wetlands (e.g., Mediterranean ponds range from small to medium in size, floodplains support a range of vegetation types).

For every taxonomic group, we discovered a large range in the numbers of families supported by the different wetland types, with some types supporting only a handful (or even none) and others supporting a large number of families (>20 for the larger insect orders and mollusks). This indicates that considerable variation among wetland types is captured using family-level classification. In the following subsections, we use the generated Tables 17.3, 17.4, 17.5, 17.6, 17.7, 17.8, 17.9, 17.10, and 17.11 to discuss each of the different macroinvertebrate taxonomic groups, looking for patterns in family richness and possible mechanisms of environmental control. We then follow that analysis with an attempt to integrate effects of multiple taxa on overall wetland change, and then finally to develop an overarching model that could help explain the enormous variation in invertebrate assemblages across wetlands globally.

Mollusca

Generalist mollusks include pulmonate snails (Planorbidae, Physidae) and Sphaeriidae fingernail clams (Table 17.3), which all can tolerate both drying and low levels of dissolved oxygen (pulmonates air breathe; Brown and Lydeard 2010). Wetland conditions pose few constraints for these generalists. Table 17.3 shows that

Table 17.2 Labels used to categorize environmental conditions in the 15 habitat types applied in summary Tables 17.3–17.11

Environmental variables[a]	Wetland examples
Size	
XS—<10 m diameter (long axis)	Rock pools, some temporary ponds
S—10–100 m diameter	Some temporary ponds, springs
M—100–1000 m diameter	Permanent marshes
L—1000–10,000 m diameter	Peatlands, Great Lakes marshes
XL—>10,000 m diameter	Everglades, Neotropical floodplains
Hydroperiod	
Sh—short (<3 months of inundation)	Most rock pools
Me—medium (3–12 months)	Mediterranean ponds
Lo—long (1–10 years)	Permanent marshes
Pe—permanent (or nearly so)	Springs
Water source	
P—precipitation	Temporary ponds, peatlands
GW—ground water	Springs, turloughs
R—runoff (lateral) of surface water	Floodplains, Great Lakes marshes
Flow	
Le—lentic (standing water)	Rock pools, permanent marshes
Lo—lotic (flowing water)	Neotropical floodplains, springs
Vegetation	
E—emergent macrophytes	Permanent marshes
F—floating, submersed macrophytes	Permanent marshes
W—woody trees and shrubs	Neotropical floodplains
O—mostly open water (acknowledging peripheral emergent vegetation occurs)	Alpine ponds
Fish	
(+)—widespread	Everglades, Neotropical floodplains
(−)—rare or absent	Rock pools, temporary ponds
(+/−)—usually occur, but sometimes absent	Permanent marshes
(−/+)—usually absent, but sometimes occur	Alpine ponds

[a]For any measure, a habitat may be assigned more than one label (e.g., XS-S for temporary ponds)

numbers of constrained taxa increased steadily as overall molluscan family richness increases among the 15 wetland types, with a surge in constrained family richness for Irish turloughs (Chap. 6), Great Lakes marshes (Chap. 9), and permanent marshes (Chap. 8). This change in constrained taxa occurs as hydroperiods lengthen and water sources other than precipitation (especially groundwater) begin to dominate water budgets. Thus, drying and a lack of calcium for shell formation are likely strong constraints for non-generalist mollusks. The large increase of constrained mollusks in permanent marsh and Great Lakes marsh habitats coincides with a prevalence of lush emergent, submersed, and floating macrophyte growth (Table 17.3)

Table 17.3 **Mollusca** families in 15 wetland types (Table 17.1), arranged from the habitats with the fewest reported families (*left*) to the most (*right*). Families are listed from the most widespread (top) to the least (bottom). The solid lines across the midportions of the matrix identify ubiquitous, generalist taxa (above the top line, i.e, only missing from at most two wetland types), and rare families (below the second line, i.e., singletons). Families falling between these two lines represent constrained taxa (i.e., reported from 2 to 12 wetland types) that appeared to be adapted for wetland environments but are restricted in their distributions. Environmental conditions in each wetland type are listed in the top five rows (size, hydroperiod, water source, flow, vegetation, fish), with criteria used to assign broad categorizations being listed in Table 17.2. If more than one category label is provided for a cell, the most important descriptor is listed first, with others listed after in descending order of importance; for hydroperiod, two labels indicate the range of inundation. (If an influence is trivial, it is not listed, even if present; e.g., direct precipitation contributes to every wetland, but is often trivial compared to other water sources). The bottom rows indicate the total numbers of families reported for each wetland type, broken down by relative numbers of generalist, constrained, and rare families

	Rock pools	Beaver ponds	Alpine ponds	Created wetlands	Waterfowl marshes	Peatlands	Springs	Everglades	Temperate floodplains	Temporary ponds	Mediterranean ponds	Neotropical floodplains	Turloughs	Great Lakes marshes	Permanent marshes
Size	XS	S	S-M	M	M	M-L	S	XL	XL	S	S-M	XL	M	L	M-L
Hydroperiod	Sh	Lo	Sh-Pe	Me-Pe	Me-Lo	Lo-Pe	Pe	Sh-Lo	Sh-Lo	Sh-Me	Sh-Me	Sh-Pe	Me	Lo-Pe	Lo-Pe
Water source	P	R	P	R	R	P-GW	GW	R	R-GW	P	P-GW	P-R	GW	R-GW	R-GW
Flow	Le	Lo	Le	Le	Le	Le	Lo	Le-Lo	Lo-Le	Le	Le	Lo-Le	Le	Le-Lo	Le
Vegetation	O	E-O-F-W	O	O-E-F	E-O-F	E-W-F-O	O-E	E-O-F	E-W-O-F	O	O-E	W-E-F-O	O-E	E-F-O	F-O-E
Fish	–	+	–/+	+	+/–	+/–	+/–	+	+	–	–	+	–	+	+/–
Planorbidae	X	X	X	X	X	X	X	X	X	X	X	X	X	X	X
Sphaeriidae[a]		X	X	X	X	X	X	X	X	X	X	X	X	X	X
Physidae	X	X	X	X	X		X	X	X	X	X		X	X	X
Lymnaeidae		X	X	X	X	X		X	X	X	X		X	X	X
Hydrobiidae[b]			X	X	X		X	X	X		X	X	X	X	X
Ancylidae[c]						X		X	X	X	X	X		X	X
Valvatidae					X	X				X			X	X	X
Succineidae						X				X	X		X	X	
Bithyniidae						X							X	X	X
Pomatiopsidae							X				X			X	X
Unionidae								X	X					X	X
Ampullariidae								X				X			X
Corbiculidae									X			X			X
Acroloxidae													X		X

(continued)

Table 17.3 (continued)

	Rock pools	Beaver ponds	Alpine ponds	Created wetlands	Waterfowl marshes	Peatlands	Springs	Everglades	Temperate floodplains	Temporary ponds	Mediterranean ponds	Neotropical floodplains	Turloughs	Great Lakes marshes	Permanent marshes
Chilinidae										X	X				
Ellobiidae													X		X
Gastrodontidae				X									X		
Hyriidae										X		X			
Melanopsidae							X								X
Mycetopodidae												X			X
Neritidae									X						X
Pleuroceridae														X	X
Vertiginidae										X			X		
Viviparidae														X	X
Assimineidae							X								
Cochliopidae															X
Dreissenidae														X	
Glacidorbidae											X				
Mytilidae												X			
Pharidae															X
Psammobidae												X			
Stenothyridae															X
Thiaridae								X							
TOTAL	2	4	5	6	6	7	7	9	9	10	10	10	12	14	22
Generalist	2	3	3	3	3	2	3	3	3	3	3	2	3	3	3
Constrained	0	1	2	3	3	5	3	5	6	7	6	6	9	10	16
Rare	0	0	0	0	0	0	1	1	0	0	1	2	0	1	3

[a]Sphaeriidae category includes *Pisidium* spp. that are included in Pisidiidae in some taxonomies
[b]Hydrobiidae category includes the family Tateidae that was recently split
[c]Ancylidae is now considered in some taxonomies as being part of Planorbidae. However, most authors identified the group, and thus we do the same

Table 17.4 **Hirudinea** families in 14 wetland types, arranged from the habitats with the fewest reported families (*left*) to the most (*right*)

	Springs	Rock pools	Alpine ponds	Peatlands	Beaver ponds	Temperate floodplains	Waterfowl marshes	Everglades	Turloughs	Temporary ponds	Created wetlands	Great Lakes marshes	Mediterranean ponds	Permanent marshes
Size	S	XS	S-M	M-L	S	XL	M	XL	M	S	M	L	S-M	M-L
Hydroperiod	Pe	Sh	Sh-Pe	Lo-Pe	Lo	Sh-Lo	Me-Lo	Sh-Lo	Me	Sh-Me	Me-Pe	Lo-Pe	Sh-Me	Lo-Pe
Water source	GW	P	P	P-GW	R	R-GW	R	R	GW	P	R	R-GW	P-GW	R-GW
Flow	Lo	Le	Le	Le	Lo	Lo-Le	Le	Le-Lo	Le	Le	Le	Le-Lo	Le	Le
Vegetation	O-E	O	O	E-W-F-O	E-O-F-W	E-W-O-F	E-O-F	E-O-F	O-E	O	O-E-F	E-F-O	O-E	F-O-E
Fish	+/–	–	–/+	+/–	+	+	+/–	+	+/–	–	+	+	–	+/–
Glossiphoniidae			X	X	X	X	X	X	X	X	X	X	X	X
Erpobdellidae			X	X	X	X	X	X		X	X	X	X	X
Haemopidae									X		X			X
Hirudinidae										X			X	
Piscicolidae												X		X
Richardsonianidae													X	
TOTAL	0	0	2	2	2	2	2	2	2	3	3	3	4	4
Generalists	0	0	1	1	1	1	1	1	1	1	1	1	1	1
Constrained	0	0	1	1	1	1	1	1	1	2	2	2	3	3
Rare	0	0	0	0	0	0	0	0	0	0	0	0	1	0

See caption for Table 17.3 to interpret information in cells. Neotropical floodplains are not included because although it has leeches, the family is unknown

Table 17.5 **Large Branchiopoda (LB)** families in 15 wetland types, arranged from the habitats with the fewest reported families (*left*) to the most (*right*)

	Great Lakes marshes	Peatlands	Springs	Beaver ponds	Created wetlands	Everglades	Permanent marshes	Alpine ponds[a]	Neotropical floodplains	Turloughs	Temporary ponds[a]	Waterfowl marshes	Temperate floodplains	Rock pools	Mediterranean ponds
Size	L	M-L	S	S	M	XL	M-L	S-M	XL	M	S	M	XL	XS	S-M
Hydoperiod	Lo-Pe	Lo-Pe	Pe	Lo	Me-Pe	Sh-Lo	Lo-Pe	Sh-Pe	Sh-Pe	Me	Sh-Me	Me-Lo	Sh-Lo	Sh	Sh-Me
Watersource	R-GW	P-GW	GW	R	R	R	R-GW	P	P-R	GW	P	R	R-GW	P	P-GW
Flow	Le-Lo	Le	Lo	Lo	Le	Le-Lo	Le	Le	Lo-Le	Le	Le	Le	Lo-Le	Le	Le
Vegetation	E-F-O	E-W-F-O	O-E	E-O-F-W	O-E-F	E-O-F	F-O-E	O	W-E-F-O	O-E	O	E-O-F	E-W-O-F	O	O-E
Fish	+	+/–	+/–	+	+	+	+/–	–/+	+	–	–	+/–	+	–	–
Chirocephalidae											X	X	X	X	X
Branchinectidae								X			X	X		X	X
Lynceidae												X	X	X	X
Thamnocephalidae												X	X	X	X
Triopsidae											X		X	X	X
Cyclestheridae									X		X		X		
Cyzicidae													X	X	X
Branchipodidae														X	X
Leptestheriidae														X	X
Limnadiidae														X	X
Streptocephalidae														X	X
Tanymastigidae										X					X
Artemiidae												X			
TOTAL	0	0	0	0	0	0	0	1	1	1	4	5	6	10	11
Constrained	0	0	0	0	0	0	0	1	1	1	4	4	6	10	11
Rare	0	0	0	0	0	0	0	0	0	0	0	1	0	0	0

See caption for Table 17.3 to interpret information in cells. (Generalist families did not occur)

[a]Appendices for the chapters on alpine ponds and temperate-zone temporary ponds were based on collections from representative habitats, and list only the LB families actually encountered. Neither list was intended to be an exhaustive accounting of all LB families that might occur in these habitat types, and it is established that other families do occur. Thus readers should not overinterpret this table as reflecting the global distributions of LB taxa across all wetlands. For example, species of the families Tanymastigidae and Chirocephalidae are known in ponds located on a high altitude in the Pyrenees and/or Apennines

Table 17.6 **Malacostraca** families in 15 wetland types, arranged from the habitats with the fewest reported families (*left*) to the most (*right*)

	Rock pools	Alpine ponds	Temporary ponds	Turloughs	Neotropical flooplains	Beaver ponds	Created wetlands	Waterfowl marshes	Everglades	Peatlands	Great Lakes marshes	Mediterranean ponds	Springs	Temperate floodplains	Permanent marshes
Size	XS	S-M	S	M	XL	S	M	M	XL	M-L	L	S-M	S	XL	M-L
Hydroperiod	Sh	Sh-Pe	Sh-Me	Me	Sh-Pe	Lo	Me-Pe	Me-Lo	Sh-Lo	Lo-Pe	Lo-Pe	Sh-Me	Pe	Sh-Lo	Lo-Pe
Water source	P	P	P	GW	P-R	R	R	R	R	P-GW	R-GW	P-GW	GW	R-GW	R-GW
Flow	Le	Le	Le	Le	Lo-Le	Lo	Le	Le	Le-Lo	Le	Le-Lo	Le	Lo	Lo-Le	Le
Vegetation	O	O	O	O-E	W-E-F-O	E-O-F-W	O-E-F	E-O-F	E-O-F	E-W-F-O	E-F-O	O-E	O-E	E-W-O-F	F-O-E
Fish	–	–/+	–	–	+	+	+	+/–	+	+/–	+	–	+/–	+	+/–
Asellidae			X	X		X	X	X	X	X	X	X	X	X	X
Crangonyctidae			X			X		X	X	X	X	X	X	X	X
Gammaridae	X	X		X			X	X		X	X		X	X	X
Dogielinotidae[a]		X	X		X	X		X	X		X	X		X	X
Cambaridae						X	X	X	X	X	X	X	X	X	
Palaemonidae					X			X	X	X	X		X	X	X
Atyidae							X			X				X	X
Ceinidae												X		X	X
Parastacidae												X		X	X
Sphaeromatidae									X				X		X
Anthuridae													X		X
Janiridae														X	X
Scyphacidae												X			X
Niphargidae				X											
Trichodactylidae					X										
Aeglidae												X			

(continued)

Table 17.6 (continued)

	Rock pools	Alpine ponds	Temporary ponds	Turloughs	Neotropical flooplains	Beaver ponds	Created wetlands	Waterfowl marshes	Everglades	Peatlands	Great Lakes marshes	Mediterranean ponds	Springs	Temperate floodplains	Permanent marshes
Amphisopidae												X			
Perthiidae												X			
Stenasellidae													X		
Leptocheliidae															X
Koonungidae															X
Mysidae									X						
Bogidiellidae													X		
Hymenosomatidae															X
Corophiidae															X
Corallanidae														X	
Potamidae													X		
TOTAL	1	2	3	3	3	4	4	6	7	6	6	10	10	11	16
Constrained	1	2	3	2	2	4	4	6	6	6	6	7	7	10	12
Rare	0	0	0	1	1	0	0	0	1	0	0	3	3	1	4

See caption for Table 17.3 to interpret information in cells (Generalist families did not occur)

[a]Dogielinotidae category includes *Hyalella*, which is considered Hyalellidae in some recent taxonomies

Table 17.7 **Odonata** families in 15 wetland types, arranged from the habitats with the fewest reported families (*left*) to the most (*right*)

	Temporary ponds	Created wetlands	Turloughs	Everglades	Alpine ponds	Rock pools	Peatlands	Beaver ponds	Temperate floodplains	Neotropical flooplains	Waterfowl marshes	Great Lakes marshes	Springs	Permanent marshes	Mediterranean ponds
Size	S-M	M	M	XL	S-M	XS	M-L	S	XL	XL	M	L	S	M-L	S-M
Hydroperiod	Sh-Pe	Me-Pe	Me	Sh-Lo	Sh-Pe	Sh	Lo-Pe	Lo	Sh-Lo	Sh-Pe	Me-Lo	Lo-Pe	Pe	Lo-Pe	Sh-Me
Water source	P	R	GW	R	P	P	P-GW	R	R-GW	P-R	R	R-GW	GW	R-GW	P-GW
Flow	Le	Le	Le	Le-Lo	Le	Le	Le	Lo	Lo-Le	Lo-Le	Le	Le-Lo	Lo	Le	Le
Vegetation	O	O-E-F	O-E	E-O-F	O	O	E-W-F-O	E-O-F-W	E-W-O-F	W-E-F-O	E-O-F	E-F-O	O-E	F-O-E	O-E
Fish	–	+	–	+	–/+	–	+/–	+	+	+	+/–	+	+/–	+/–	–
Coenagrionidae	X	X	X	X	X	X	X	X	X	X	X	X	X	X	X
Libellulidae	X	X	X	X	X	X	X	X	X	X	X	X	X	X	X
Aeshnidae	X	X		X	X	X	X	X	X	X	X	X		X	X
Lestidae	X	X	X		X	X	X	X	X		X	X	X	X	X
Corduliidae				X	X	X	X		X		X	X	X	X	X
Gomphidae				X				X	X	X		X	X	X	X
Calopterygidae			X							X	X	X	X	X	X
Megapodagrionidae										X			X		X
Petaluridae											X		X		X
Dicteriadidae											X			X	
Protoneuridae										X				X	

(continued)

Table 17.7 (continued)

	Temporary ponds	Created wetlands	Turloughs	Everglades	Alpine ponds	Rock pools	Peatlands	Beaver ponds	Temperate floodplains	Neotropical flooplains	Waterfowl marshes	Great Lakes marshes	Springs	Permanent marshes	Mediterranean ponds
Austrocorduliidae															X
Cordulegastridae													X		
Hemicorduliidae													X		
Isostictidae									X						
Macromiidae												X			
Perilestidae														X	
Platycnemididae															X
Polythoridae										X					
Synthemistidae															X
Telephlebiidae															X
TOTAL	4	4	4	5	5	5	5	5	7	8	8	8	10	10	13
Generalists	4	4	3	3	4	4	4	4	4	3	4	4	3	4	4
Constrained	0	0	1	2	1	1	1	1	2	4	4	3	5	5	5
Rare	0	0	0	0	0	0	0	0	1	1	0	1	2	1	4

See caption for Table 17.3 to interpret information in cells

Table 17.8 EPT (Ephemeroptera, Plecoptera, Trichoptera) families in 15 wetland types, arranged from the habitats with the fewest reported families (*left*) to the most (*right*)

	Rock pools	Everglades	Turloughs	Peatlands	Created wetlands	Temporary ponds	Alpine ponds	Water-fowl marshes	Neotropical flooplains	Mediterranean ponds	Beaver ponds	Temperate floodplains	Permanenet marshes	Great Lakes marshes	Springs
Size	XS	XL	M	M-L	M	S	S-M	M	XL	S-M	S	XL	M-L	L	S
Hydroperiod	Sh	Sh-Lo	Me	Lo-Pe	Me-Pe	Sh-Me	Sh-Pe	Me-Lo	Sh-Pe	Sh-Me	Lo	Sh-Lo	Lo-Pe	Lo-Pe	Pe
Water source	P	R	GW	P-GW	R	P	P	R	P-R	P-GW	R	R-GW	R-GW	R-GW	GW
Flow	Le	Le-Lo	Le	Le	Le	Le	Le	Le	Lo-Le	Le	Lo	Lo-Le	Le	Le-Lo	Lo
Vegetation	O	E-O-F	O-E	E-W-F-O	O-E-F	O	O	E-O-F	W-E-F-O	O-E	E-O-F-W	E-W-O-F	F-O-E	E-F-O	O-E
Fish	–	+	–	+/–	+	–	–/+	+/–	+	–	+	+	+/–	+	+/–
Baetidae	X	X	X	X	X	X	X	X	X	X	X	X	X	X	X
Leptoceridae	X	X	X	X	X	X	X	X	X	X	X	X	X	X	X
Limnephilidae	X		X	X	X	X	X	X	X	X	X	X	X	X	X
Caenidae		X	X	X	X	X		X	X	X	X	X	X	X	
Hydroptilidae		X		X		X	X	X	X	X	X	X	X	X	X
Leptophlebiidae			X	X	X	X	X	X	X	X	X	X	X	X	
Polycentropodidae				X	X	X		X	X	X	X		X	X	X
Phryganeidae				X	X		X	X		X	X		X	X	X
Hydropyschidae					X				X	X	X	X		X	X
Ephemerellidae								X			X	X		X	X
Nemouridae			X				X				X	X			X
Siphlonuridae			X					X			X	X	X		
Chloroperlidae								X			X			X	X
Ecnomidae					X					X		X	X		
Ephemeridae					X						X	X		X	

(continued)

Table 17.8 (continued)

	Rock pools	Everglades	Turloughs	Peatlands	Created wetlands	Temp orary ponds	Alpine ponds	Wate rfowl marshes	Neotropical flooplains	Mediterr anean ponds	Beaver ponds	Temperate floodplains	Permanenet marshes	Great Lakes marshes	Springs
Helicopsychidae									X				X	X	X
Heptageniidae											X	X	X	X	X
Molannidae								X					X	X	X
Rhyacophilidae								X	X		X				X
Apataniidae													X	X	X
Brachycentridae													X	X	X
Glossosomatidae									X			X			X
Gripopterygidae						X				X		X			
Lepidostomatidae												X		X	X
Leptohyphidae									X				X	X	
Metretopodidae								X				X		X	
Perlidae										X	X			X	
Philopotamidae		X							X						X
Psychomyiidae						X		X				X			
Polymitarcyidae									X			X	X		
Leuctridae							X								X
Nesameletidae						X	X								
Notonemouridae							X			X					
Odontoceridae									X						X
Calamoceratidae									X				X		
Sericostomatidae										X					X
Ameletidae														X	
Atriplectidae													X		
Austroperlidae							X								

Beraeidae															X
Baetiscidae														X	
Capniidae										X					
Coloburiscidae										X					
Conoesucidae							X								
Dipseudopsidae														X	
Goeridae															X
Hydrobiosidae						X									
Isonychiidae														X	
Neoephemeridae														X	
Oeconesidae							X								
Oniscigastridae						X									
Peltoperlidae															X
Perlodidae															X
Philorheithridae													X		
Rossianidae															X
Stenopsychidae										X					
Tricorythidae														X	
Uenoidae															X
Xiphocentronidae															X
TOTAL	3	5	7	8	10	12	13	15	16	17	17	19	20	27	29
Generalists	3	2	3	3	3	3	3	3	3	3	3	3	3	3	3
Constrained	0	3	4	5	7	7	7	12	13	11	14	16	14	18	19
Rare	0	0	0	0	0	2	3	0	0	3	0	0	3	6	7

See caption for Table 17.3 to interpret information in cells

Table 17.9 **Hemiptera** families in 15 wetland types, arranged from the habitats with the fewest reported families (*left*) to the most (*right*)

	Alpine ponds	Turloughs	Rock pools	Temporary ponds	Created wetlands	Springs	Neotropical floodplains	Peatlands	Everglades	Mediterranean ponds	Temperate floodplains	Great Lakes marshes	Permanent marshes	Waterfowl marshes	Beaver ponds
Size	S-M	M	XS	S	M	S	XL	M-L	XL	S-M	XL	L	M-L	M	S
Hydroperiod	Sh-Pe	Me	Sh	Sh-Me	Me-Pe	Pe	Sh-Pe	Lo-Pe	Sh-Lo	Sh-Me	Sh-Lo	Lo-Pe	Lo-Pe	Me-Lo	Lo
Water source	P	GW	P	P	R	GW	P-R	P-GW	R	P-GW	R-GW	R-GW	R-GW	R	R
Flow	Le	Le	Le	Le	Le	Lo	Lo-Le	Le	Le-Lo	Le	Lo-Le	Le-Lo	Le	Le	Lo
Vegetation	O	O-E	O	O	O-E-F	O-E	W-E-F-O	E-W-F-O	E-O-F	O-E	E-W-O-F	E-F-O	F-O-E	E-O-F	E-O-F-W
Fish	−/+	−	−	−	+	+/−	+	+/−	+	−	+	+	+/−	+/−	+
Corixidae	X	X	X	X	X	X	X	X	X	X	X	X	X	X	X
Gerridae	X	X	X	X	X		X	X	X	X	X	X	X	X	X
Veliidae	X	X	X	X	X		X	X	X	X	X	X	X	X	X
Notonectidae	X	X	X	X	X	X		X	X	X	X	X	X	X	
Belostomatidae			X	X		X	X	X	X	X	X	X	X	X	X
Naucoridae			X			X	X	X	X	X	X	X	X	X	
Nepidae		X		X	X			X	X	X	X	X	X	X	X
Pleidae			X	X	X		X	X		X	X	X	X	X	
Mesoveliidae						X		X	X	X	X	X	X	X	X
Hydrometridae					X			X	X	X	X	X	X	X	X
Saldidae	X	X				X				X	X	X	X	X	
Hebridae									X	X	X	X	X	X	X
Macroveliidae									X			X		X	
Ochteridae						X					X			X	
Gelastocoridae							X								X

Aphidae															X
Cercopidae															X
Cicadellidae															X
Coccoidea															X
Delphacidae															X
Miridae															X
Reduviidae															X
Rhyparochromidae													X		
TOTAL	5	6	7	7	7	7	7	10	11	12	13	13	13	14	16
Generalist	4	4	4	4	4	2	3	4	4	4	4	4	4	4	3
Constrained	1	2	3	3	3	5	4	6	7	8	9	9	8	10	6
Rare	0	0	0	0	0	0	0	0	0	0	0	0	1	0	7

See caption for Table 17.3 to interpret information in cells

Table 17.10 **Coleoptera** families in 15 wetland types, arranged from the habitats with the fewest reported families (*left*) to the most (*right*)

	Rock pools	Peatlands	Alpine ponds	Everglades	Waterfowl marshes	Temporary ponds	Created wetlands	Springs	Turloughs	Great Lakes marshes	Permanent marshes	Temperate floodplains	Neotropical floodplains	Beaver ponds	Mediterranean Ponds
Size	XS	M-L	S-M	XL	M	S	M	S	M	L	M-L	XL	XL	S	S-M
Hydroperiod	Sh	Lo-Pe	Sh-Pe	Sh-Lo	Me-Lo	Sh-Me	Me-Pe	Pe	Me	Lo-Pe	Lo-Pe	Sh-Lo	Sh-Pe	Lo	Sh-Me
Water source	P	P-GW	P	R	R	P	R	GW	GW	R-GW	R-GW	R-GW	P-R	R	P-GW
Flow	Le	Le	Le	Le-Lo	Le	Le	Le	Lo	Le	Le-Lo	Le	Lo-Le	Lo-Le	Lo	Le
Vegetation	O	E-W-F-O	O	E-O-F	E-O-F	O	O-E-F	O-E	O-E	E-F-O	F-O-E	E-W-O-F	W-E-F-O	E-O-F-W	O-E
Fish	–	+/–	–/+	+	+/–	–	+	+/–	–	+	+/–	+	+	+	–
Dytiscidae	X	X	X	X	X	X	X	X	X	X	X	X	X	X	X
Hydrophilidae/ Helophoridae	X	X	X	X	X	X	X	X	X	X	X	X	X	X	X
Haliplidae	X	X	X	X	X	X	X	X	X	X	X	X		X	X
Scirtidae	X	X	X	X	X	X	X	X		X	X	X	X	X	X
Gyrinidae	X	X	X	X	X	X	X			X	X	X	X	X	X
Chrysomelidae			X	X	X			X	X	X	X	X	X	X	X
Hydraenidae	X	X				X	X	X	X	X	X	X	X	X	X
Noteridae		X		X		X	X		X	X	X	X	X	X	X
Elmidae						X	X	X	X	X	X	X	X	X	X
Staphylinidae			X			X			X	X	X	X	X	X	X
Curculionidae					X		X		X	X	X	X	X	X	X
Dryopidae				X			X	X	X		X		X		X
Hydrochidae								X		X	X		X		X
Ptilodactylidae										X		X	X	X	X
Hygrobiidae							X		X		X	X			X
Lampyridae										X	X	X	X	X	

Carabidae									X				X	X	X
Heteroceridae											X	X	X		X
Limnichidae	X										X				X
Anthicidae										X				X	
Elateridae												X		X	
Georissidae										X			X		X
Hydroscaphidae								X					X		
Psephenidae				X	X										
Scarabaeidae												X	X		X
Alleculidae															X
Cantharidae												X			
Coccinellidae														X	
Eucnemidae													X		
Erirhinidae															X
Latridiidae														X	
Lutrochidae								X							
Melolonthidae											X				
Phalacridae														X	
Phengodidae												X			
Ptilidae															X
Silvanidae														X	
Spercheidae															X
TOTAL	7	7	7	9	9	10	11	11	13	16	18	19	20	20	24
Generalist	5	5	5	5	5	5	5	4	3	5	5	5	4	5	5
Constrained	2	2	2	4	3	5	6	6	10	11	12	11	15	11	15
Rare	0	0	0	0	1	0	0	1	0	0	1	3	1	4	4

See caption for Table 17.3 to interpret information in cells

Table 17.11 Diptera families in 15 wetland types, arranged from the habitats with the fewest reported families (*left*) to the most (*right*)

	Alpine ponds	Turloughs	Everglades	Rock pools	Created wetlands	Beaver ponds	Temporary ponds	Peatlands	Permanent marshes	Great Lakes marshes	Neotropical floodplains	Temperate floodplains	Waterfowl marshes	Springs	Mediterranean ponds
Size	S-M	M	XL	XS	M	S	S	M-L	M-L	L	XL	XL	M	S	S-M
Hydroperiod	Sh-Pe	Me	Sh-Lo	Sh	Me-Pe	Lo	Sh-Me	Lo-Pe	Lo-Pe	Lo-Pe	Sh-Pe	Sh-Lo	Me-Lo	Pe	Sh-Me
Water source	P	GW	R	P	R	R	P	P-GW	R-GW	R-GW	P-R	R-GW	R	GW	P-GW
Flow	Le	Le	Le-Lo	Le	Le	Lo	Le	Le	Le	Le-Lo	Lo-Le	Lo-Le	Le	Lo	Le
Vegetation	O	O-E	E-O-F	O	O-E-F	E-O-F-W	O	E-W-F-O	F-O-E	E-F-O	W-E-F-O	E-W-O-F	E-O-F	O-E	O-E
Fish	–/+	–	+	–	+	+	–	+/–	+/–	+	+	+	+/–	+/–	–
Ceratopogonidae	X	X	X	X	X	X	X	X	X	X	X	X	X	X	X
Chironomidae	X	X	X	X	X	X	X	X	X	X	X	X	X	X	X
Culicidae	X	X	X	X	X	X	X	X	X	X	X	X	X	X	X
Tipulidae/ Limoniidae	X	X	X	X	X	X	X	X	X	X	X	X	X	X	X
Stratiomyidae	X	X	X		X	X	X	X	X	X	X	X	X	X	X
Dolichopodidae			X	X	X	X	X	X	X	X	X	X	X	X	X
Ephydridae	X		X	X	X	X	X		X	X	X	X	X	X	X
Psychodidae		X	X		X	X	X	X	X	X	X	X	X	X	X
Tabanidae		X	X	X	X	X	X	X	X	X	X	X	X		X
Chaoboridae	X				X	X	X	X	X	X	X	X	X	X	X
Dixidae					X	X	X	X	X	X	X		X	X	X
Empididae					X	X	X		X	X	X	X	X	X	X
Sciomyzidae		X		X				X	X	X	X	X	X	X	X
Syrphidae					X		X	X	X		X	X	X	X	X
Simuliidae						X	X	X			X	X	X	X	X
Muscidae				X			X	X	X			X	X		

Ptychopteridae					X	X		X	X	X				X	
Athericidae										X	X				X
Cecidomyiidae												X	X		
Scatopsidae									X						X
Sciaridae												X	X		
Thaumaleidae														X	X
Anthomyiidae															X
Bibionidae												X			
Canacidae													X		
Chloropidae															X
Corethrellidae						X									
Deuterophlebiidae														X	
Nymphomyiidae														X	
Pelecorhynchidae														X	
Phoridae										X					
Rhagionidae															X
Sarcophagidae										X					
Sphaeroceridae											X				
TOTAL	7	8	9	9	14	15	15	15	17	17	17	18	19	19	21
Generalist	6	7	9	7	9	9	9	8	9	9	9	9	9	8	9
Constrained	1	1	0	2	5	5	6	7	8	6	7	8	9	8	9
Rare	0	0	0	0	0	1	0	0	0	2	1	1	1	3	3

See caption for Table 17.3 to interpret information in cells

that creates a complex three-dimensional habitat for snails, which primarily graze algae and biofilms from surfaces (Brown and Lydeard 2010). Thus, a lack of this kind of habitat is likely another constraint on mollusks. Molluscan family richness is high in the habitats where fish predation is also a pervasive force (see permanent marshes Chap. 8), but this constraint is probably minimized by protections conferred by molluscan shells.

Endemic and threatened species of mollusks exist, including hydrobiid snails (see Chap. 11), ampullariid (apple) snails (see Chap. 10), vertiginid snails (see Chap. 6), and unionid clams (see Chap. 9). Several mollusks have become important exotic invasives that threaten wetlands, including zebra mussels (Dreissenidae; see Chap. 9), Asiatic clams (Corbiculidae), some apple snails (Ampullariidae, see Chap. 10), and the golden mussel (Mytilidae; see Chap. 14). Mollusks are arguably the most important invertebrates in freshwater wetlands, overall, given that they can be sensitive to multiple environmental constraints, snails are perhaps the most important herbivores in wetlands, and several mollusks are species of special concern.

Hirudinea

Relatively few families of leeches (6) were reported from wetlands globally (Table 17.4). None were completely ubiquitous, with leeches being absent from rock pools (probably due to short hydroperiods and poor access; see Chap. 2) and springs (which to us was a surprise, as neither conditions nor access should be problematic across all springs; see Chap. 11). Glossiphoniidae occur in every other kind of habitat, and Erpobdellidae were also fairly widespread, regardless of wetland conditions. A few wetland types supported additional leech families. Some of those leech families with restricted distributions are parasitic on fish (Govedich et al. 2010), and occurrence may reflect fish presence, or at least a connection to fish bearing waters.

Large Branchiopoda

No generalist families were observed for large branchiopods, and this taxonomic group was not recorded from seven wetland types (Table 17.5). Branchinectidae and Chirocephalidae were the most widespread families, and yet were only recorded from 5 of 15 wetland types each. Only two wetland types recorded high family richness, rock pools (Chap. 2) and Mediterranean ponds (Chap. 5), although the lists for some wetlands types were based on representative habitats and were not intended to be exhaustive.

Wetland types where large branchiopods were absent are characterized by permanent or long hydroperiods, whereas the two taxon-rich wetland types had medium to short hydroperiods. Large branchiopods have long been recognized as a group of crustaceans that inhabits almost exclusively temporary ponds (Hartland-Rowe 1972). However, besides hydrology, patterns for large branchiopods could be also related to fish predation. The wetland types where large branchiopods were lacking generally also supported fish; rock pools and Mediterranean ponds are fishless habitats. Both environmental factors, permanence and fish predation, are known to control invertebrate community structure in many wetlands (Wellborn et al. 1996), and it is likely that richness patterns of large branchiopods are directly or indirectly related to these factors. However, large branchiopods are known to occasionally occur in permanent waters without fish, and in some arctic and alpine ponds with fish (Jeppesen et al. 2001; Beladjal et al 2007). The relatively high large branchiopod richness in temperate floodplains (Table 17.5) was somewhat of a surprise given frequent inundations by river water and high fish abundances.

Malacostraca

Most wetland types supported only a handful of families of Amphipoda, Decapoda, or Isopoda. Family richness was low across small, short-hydroperiod, precipitation-based wetland types, especially those with simplistic vegetative community structures (Table 17.6). No families were truly ubiquitous, and we did not classify any as generalists. Numerous families occurred in only a single wetland type, mostly due to restricted biogeographic ranges.

Despite being a small habitat type, springs supported a rich assortment of malacostracan families, probably because the group frequently inhabits groundwater (see Chap. 11). Family richness was also very high in permanent marshes (Chap. 8) and temperate floodplains (Chap. 13) (Table 17.6). Here malacostracans can be among the most abundant invertebrates, and given their large body sizes are undoubtedly crucially important ecologically (as consumers, and as fish and duck foods). They are similarly important ecologically in Great Lakes marshes and the Everglades (see Chaps. 9 and 10). Although typically considered to be restricted to permanent waters (Wiggins et al. 1980), there are numerous accounts of various malacostracans successfully persisting in dry wetland.

Odonata

The dragonfly and damselfly fauna of most wetland types was composed mostly of generalist families (Table 17.7). When wetlands supported numerous non-generalist families, it was often due to biogeographic reasons, i.e., the surveys included habitats outside of Europe and North America. At a macroscopic level, the Odonata seem to lack much specialization. Strong flying and long-lived adults should be able to access most wetland habitats, and virtually all species are generalist predators able to exploit most any prey resource (Tennessen 2008). However, ample evidence exists that more specialization occurs among genera and species within families (especially Coenagrionidae and Libellulidae; see beaver pond Chap. 12).

EPT Taxa (Ephemeroptera + Plecoptera + Trichoptera)

We examined the mayflies, stoneflies, and caddisflies together because their pooled richness is often a metric used to assess the environmental health of streams and rivers (Rosenberg et al. 2008). Despite the large numbers of EPT taxa in wetlands worldwide, only three families, the Baetidae, Leptoceridae, and Limnephilidae were ubiquitous generalists (Table 17.8). No plecopterans were generalists, although a few were reasonably widespread constrained taxa. Plecoptera is often discounted completely in wetlands, but this is probably not warranted for some wetland types.

Constrained EPT families were best represented in long hydroperiod habitats and in those with lotic characteristics (springs, beaver ponds, floodplains). Springs (Chap. 11) had the most EPT families, but many of them only occurred there. Long hydroperiod, lentic habitats such as permanent marshes (Chap. 8) and Great Lakes marshes (Chap. 9) had very high EPT family richness. Great Lakes marshes are where invertebrates have been most successfully used in bioassessment (see Chap. 9), and perhaps the prevalence of EPT families in those habitats makes the practice efficacious (as it is in EPT-rich streams). Alternatives to EPT taxa have been sought for bioassessment in many wetlands (e.g., Ruhí and Batzer 2014), but this may not be necessary in those wetland types where EPT taxa thrive.

Limnephilid caddisflies are probably the only EPT family that plays a major ecological role in wetlands. In some alpine (Chap. 3) and temporary ponds (Chap. 4), limnephilids can be very speciose and abundant, and can be important large-bodied shredders of coarse organic matter (leaf litter) (e.g., Klemmer et al. 2012). Baetid mayflies can be occasionally abundant in some wetlands, and are likely important prey items for predators in those locations (see Chap. 12).

Hemiptera and Coleoptera

We discuss both of these orders together because they varied across wetland types in fairly similar ways. Several aquatic families of Hemiptera and Coleoptera comprised the generalist component, including many of the families most familiar to researchers (e.g., Corixidae, Dytiscidae, Gyrinidae, Hydrophilidae, Notonectidae). It seems that these families can successfully exploit wetlands regardless of conditions, likely because they are all strong flyers able to colonize any available habitat (Batzer and Wissinger 1996).

The constrained taxa also included some typical aquatic families (e.g., Belostomatidae), but increasingly semiaquatic and terrestrial families (e.g., surface and shore bugs, nonaquatic herbivorous and predaceous beetles and bugs) were reported as overall family richness increased across the habitats (Tables 17.9 and

17.10). Those habitats with numerous semiaquatic and terrestrial families tended to support complex vegetative communities, and it is likely that many of the nonaquatic families were associated with these plants.

However, the plethora of terrestrial Hemiptera and especially Coleoptera families in Mediterranean ponds (Chap. 5) was perplexing, given that vegetative complexity is not unusually high there (see Tables 17.9 and 17.10). However, one factor that was unique to Mediterranean ponds was the arid to semiarid climate (that defines those habitats). Researchers that study wetland invertebrates are well aware that drying is a significant constraint on the aquatic fauna (Wiggins et al. 1980). However, water-loss is also a major threat to terrestrial arthropods (Schowalter 2000). It seems plausible that wetland habitats may be very attractive to terrestrial invertebrates of arid climates because the perimeters and vegetation of the wetlands would be humid, lush refugia. Even the terrestrial invertebrates associated with vegetation in other wetland types (e.g., beaver ponds, Chap. 12) may be exploiting the plants because of the humid conditions near the water surface, and not simply because plants are food.

When considering the terrestrial fauna, however, we need to recognize that some of the cross-habitat variation reflects sampling and processing protocols. In virtually all instances, the researchers were using equipment (nets, large cores) designed to sample the aquatic fauna, and they happened to also collect large numbers of terrestrials. In some cases, researchers may simply disregard the terrestrial fauna, even if present, because of a solely aquatic focus to their research questions. Thus, the terrestrial fauna is likely underrepresented in our summary tables. Even so, it is obvious that the terrestrial fauna is a very important component of many wetland habitats (see Chaps. 12, 13, and 14). If terrestrial invertebrates of wetlands were specifically targeted, using devices that would collect them efficiently (e.g., pitfall, malaise, or sticky traps), we would not be surprised to find that the terrestrial fauna was richer taxonomically than the aquatic fauna (there are more species of nonaquatic Coleoptera than all aquatic invertebrate species combined; Triplehorn and Johnson 2005). It seems obvious that to truly understand how invertebrates function ecologically in wetland ecosystems, the terrestrial fauna, in both wet and dry phases, needs to be addressed.

Diptera

For the true flies (Table 17.11), an unusually large number of generalist families occurred (9). The constrained families were relatively few, and we could find no consistent pattern of richness related to any particular environmental factor. While dipteran family richness in the Everglades (Chap. 10) was relatively small (Table 17.11), chironomid midges have received considerable attention there, with very large numbers of genera and species being recorded (see Appendix in Chap. 10). Some Everglades midge genera and species have proved environmentally sensitive (see King and Richardson 2002, Chap. 10). For the Diptera, like the Odonata, family level classification may yield meagre amounts of information regarding environmental controls, and finer taxonomic resolution may be required to acquire significant knowledge.

Cross-Wetland Comparison Using Multiple Taxonomic Groups

In order to compare family richness patterns among wetland types using all of the major taxonomic groups, we compiled the information from Tables 17.3–17.11 into a single summary Table 17.12. In this table we ranked the family richness of each taxonomic group according with its maximum richness value observed across all wetland types. Thus, category 0 includes wetland types without any family for a specific taxonomic group, and category 1 includes wetland types with <25 % of the richness of the richest wetland type for a specific taxonomic group. Category 2, category 3, and category 4 include wetland types with 25–50 %, 51–75 %, and >75 % of maximum richness, respectively. We ranked the 15 wetland types in terms of both constrained family richness and total family richness, but prioritize constrained taxa for rankings as they should provide the most ecological insight (Table 17.12).

Richness patterns of the nine analyzed taxonomic groups clearly distinguished between high- and low-richness wetland types (Table 17.12). A group of five wetland types, Alpine ponds, Rock pools, the Everglades, Turloughs, and Created wetlands had low overall richness of constrained families. Another small group, Permanent marshes, Mediterranean ponds, and Temperate floodplains, had high overall family richness. When comparing the wetland types with low richness to the wetland types with high richness according to whether the families they supported were generalist, constrained, or rare, another obvious pattern emerged (Table 17.13a). Whether wetlands were high richness or low richness, they supported fairly similar numbers of generalist families (from 24 to 29). However, high-richness wetlands supported over 2–3 times the number of constrained families as low-richness wetlands (Table 17.13a). Rare families were few or absent from low-richness wetland types, while they were common in high-richness wetland types.

Focusing on the compositions of constrained families in low-richness wetlands (Table 17.13b), we found certain unique taxonomic groups were favored in each. In Alpine ponds, EPT taxa seemed to do well, and the authors of Chap. 3 spend considerable time discussing cold-water adaptations for the resident fauna, a feature of many EPT taxa. In Rocks pools, large branchiopods overwhelmingly dominated the constrained taxa (53 %), and the authors of Chap. 2 emphasize why this group does well there (well adapted for desiccation, and benefit from an absence of certain predators). In the Everglades, hemipterans, malacostracans, and molluscans did well, and the authors of Chap. 10 note how crayfish and certain snails play critical roles in Everglades ecology. In Created wetlands, various insect groups do well, and the authors of Chap.

Table 17.12 Constrained family richness of each taxonomic group (total family richness values in parenthesis) in each wetland type

	Alpine ponds	Rock pools	Everglades	Turloughs	Created wetlands	Peatlands	Temp orary ponds	Springs	Beaver ponds	Neotropical floodplains	Waterfowl marshes	Great Lakes marshes	Temperate floodplains	Permanent marshes	Mediterranean ponds
Size	S-M	XS	XL	M	M	M-L	S	S	S	XL	M	L	XL	M-L	S-M
Hydroperiod	Sh-Pe	Sh	Sh-Lo	Me	Me-Pe	Lo-Pe	Sh-Me	Pe	Lo	Sh-Pe	Me-Lo	Lo-Pe	Sh-Lo	Lo-Pe	Sh-Me
Water source	P	P	R	GW	R	P-GW	P	GW	R	P-R	R	R-GW	R-GW	R-GW	P-GW
Flow	Le	Le	Le-Lo	Le	Le	Le	Le	Lo	Lo	Lo-Le	Le	Le-Lo	Lo-Le	Le	Le
Vegetation	O	O	E-O-F	O-E	O-E-F	E-W-F-O	O	O-E	E-O-F-W	W-E-F-O	E-O-F	E-F-O	E-W-O-F	F-O-E	O-E
Fish	–/+	–	+	–	+	+/–	–	+/–	+	+	+/–	+	+	+/–	–
Mollusca	1 (1)	0 (1)	2 (2)	3 (3)	1 (2)	2 (2)	2 (2)	1 (2)	1 (1)	2 (2)	1 (2)	3 (3)	2 (2)	4 (4)	2 (2)
Hirudinea	2 (2)	0 (0)	2 (2)	2 (2)	3 (3)	2 (2)	3 (3)	0 (0)	2 (2)	–	2 (2)	3 (3)	2 (2)	4 (4)	4 (4)
Lg Branchiopoda	1 (1)	4 (4)	0 (0)	1 (1)	0 (0)	0 (0)	2 (2)	0 (0)	0 (0)	1 (1)	2 (2)	0 (0)	3 (3)	0 (0)	4 (4)
Malacostraca	1 (1)	1 (1)	3 (2)	1 (1)	2 (2)	3 (2)	2 (1)	3 (3)	2 (2)	1 (1)	3 (2)	3 (2)	4 (3)	4 (4)	3 (3)
Odonata	2 (2)	2 (2)	2 (2)	2 (2)	0 (2)	2 (2)	0 (2)	4 (3)	2 (2)	3 (3)	3 (3)	3 (3)	2 (3)	4 (3)	4 (4)
EPT taxa	2 (2)	0 (1)	1 (1)	1 (2)	2 (2)	2 (2)	2 (2)	4 (4)	3 (3)	3 (3)	3 (3)	4 (4)	4 (3)	3 (3)	3 (3)
Hemiptera	1 (2)	2 (2)	3 (3)	1 (2)	2 (2)	3 (3)	2 (2)	2 (2)	3 (4)	2 (2)	4 (4)	4 (4)	4 (4)	3 (4)	3 (3)
Coleoptera	1 (2)	1 (2)	2 (2)	3 (3)	2 (2)	1 (2)	2 (2)	2 (2)	3 (4)	4 (4)	1 (2)	3 (3)	3 (4)	4 (3)	4 (4)
Diptera	1 (2)	2 (2)	0 (2)	1 (2)	3 (3)	3 (3)	3 (3)	4 (4)	3 (3)	3 (4)	4 (4)	3 (4)	4 (4)	4 (4)	4 (4)
MEAN RANK (constrained families)	**1.3**	**1.3**	**1.7**	**1.7**	**1.7**	**2.0**	**2.0**	**2.2**	**2.2**	**2.4**	**2.6**	**2.9**	**3.2**	**3.3**	**3.4**
MEAN RANK (total families)	*1.7*	*1.7*	*1.8*	*2.0*	*2.0*	*2.0*	*2.1*	*2.2*	*2.3*	*2.5*	*2.7*	*2.9*	*3.1*	*3.2*	*3.4*

Family richness was ranked using the maximum richness observed for each taxonomic group as 100 % and then using percentiles. Thus, category 0 includes wetland types without any family for a specific taxonomic group, and category 1 includes wetland types with <25 % of the richness of the richest wetland type for a specific taxonomic group. Category 2, category 3, and category 4 include wetland types with 25 %–50 %, 51 %–75 %, and >75 %, respectively

Table 17.13 Kinds of invertebrate families that characterize the five low-richness and three high-richness wetland types, based on rankings in Table 17.12 and data from Tables 17.3–17.11

(a) Numbers of generalist, constrained, and rare families in the 5 low-richness and 3 high-richness wetland types

	Generalist	Constrained	Rare
Alpine ponds	26	18	3
Rock Pools	25	19	0
Everglades	27	28	2
Created wetlands	29	30	1
Turloughs	24	31	0
Permanent marshes	29	78	13
Mediterranean ponds	29	75	19
Temperate floodplains	29	69	6

(b) Taxonomic groups that dominate (>15 % of total) the constrained set of invertebrate families in each of the five low-richness wetland habitat types (n = total number of constrained families reported from Tables 17.3–17.11)

Alpine ponds (n=18)	Rock Pools (n=19)	Everglades (n=28)	Created wetlands (n=30)	Turloughs (n=31)
39 % EPT taxa	53 % Lg branchiopods	25 % Hemipterans	23 % EPT taxa	32 % Coleopterans
	16 % Hemipterans	21 % Malacostracans	20 % Coleopterans	29 % Molluscans
		18 % Molluscans	17 % Dipterans	

15 emphasize the importance of aerial colonization in these newly created habitats. EPT taxa, which are considered environmentally sensitive, appeared to do well in Created wetlands, which was a surprise given they are the most human-impacted habitats covered in this book. In Turloughs, coleopterans and molluscans did well, with at least the latter group probably benefitting from the strong groundwater connection. Most importantly, Table 17.13b clearly demonstrates that completely different factors are responsible for the low richness in each of the five habitats (i.e., a universal constraint on richness is not operating), and further establishes that the constraints being imposed on overall assemblages are concurrently benefitting certain taxa (i.e., a handful of taxa still thrive, with this group being unique in each wetland type).

Taxa Sorting (or Taxa Filtering): Assemblage Models

One initial finding of our macroscopic examination of taxon richness among wetland types was that some of the factors that we thought might be most important in controlling richness, e.g., habitat size, hydroperiod, and the presence or absence of fish, did not seem to be controlling overall richness patterns. Indeed, among the low-richness wetland types, Rock pools are very small, have temporary hydroperiods,

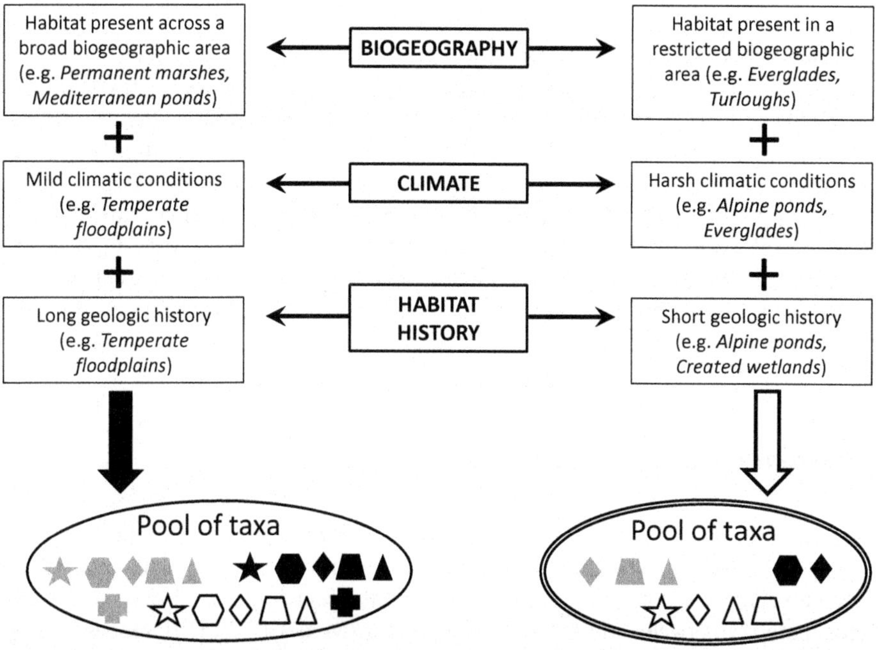

Model 17.1 Diagram showing three mechanisms (biogeography, climate, and history) that determine the available pool of taxa that can colonize wetland habitats of a region. Symbols (shapes for taxa are arbitrary) correspond to generalists (*open*), constrained (*grey*), and rare taxa (*black*). Habitats located across a limited biogeographic region, with harsh climatic conditions, and a short geologic history are characterized by a reduced pool of taxa

and lack fish, while the Everglades is very large, mostly permanent, and supports a plethora of fish. Among the high-richness wetlands types, Mediterranean temporary ponds are fairly small, have mostly short hydroperiods, and lack fish, while Permanent marshes are fairly large, almost always hold water, and usually support fish. Other kinds of constraints on richness were clearly operating, but what were they?

We think the answer is fairly obvious, but entails three mechanisms that ecologists working on wetlands may simply forget to consider (see Model 17.1). The first is biogeography. The pool of taxa available to colonize a wetland type that occurs across a broad geographic range, such as Permanent marshes, will be larger than the pool available to colonize a wetland type that has a restricted (or isolated) geographic range, such as the Everglades or Irish Turloughs. The second is climate. The pool of taxa available to colonize a wetland type that occurs in a mild climate, such as Temperate floodplains, will be larger than the pool available to colonize a wetland type set in a harsh climate, such as cold Alpine ponds (the very hot climate in the Everglades may also impose constraints). The third is related to history. The pool of taxa available to colonize a wetland type that is geologically old (such Temperate floodplains) will be larger than the pool available to colonize a wetland type that is geologically young (such as Alpine ponds or the Everglades).

If geologic age is relevant, it is not surprising that invertebrate family richness in Created wetlands, that may be only decades old, is also low (see Chap. 15).

These larger constraints will dictate the size and composition of the taxa pool available to a specific wetland or set of wetlands, and depending on the characteristics of the available taxa may modify the relative importance of traditional environmental constraints (habitat size, hydrology, vegetation, or predation). This innate variation may be why finding overarching mechanisms for the environmental control of wetland invertebrates have been so elusive (Batzer 2013). None-the-less, the authors of virtually every chapter in this book provide compelling evidence that various abiotic and biotic constraints strongly affect wetland invertebrate communities, and we now try to model how this control might occur.

Environmental Filters

The family richness patterns observed across wetland types and among taxonomic groups suggest the existence of a variety of constraints on each wetland type and each taxonomic group. Most of these constraints occur across environmental gradients that vary between and within wetland types (Model 17.2). Differences in these abiotic and biotic conditions in each wetland type imply a more or less severe constraint for a specific taxonomic group, while also offering opportunities for some groups, i.e., constraints on some taxa may confer benefits for others (see Table 17.13b). For wetland invertebrates, one of the most known examples is the conflicting relationship between hydroperiod lengths and predation rates (Schneider and Frost 1996; Wellborn et al. 1996; Wissinger 1999; Zokan and Drake 2015).

Thus, several and different environmental filters (abiotic and biotic controls) operate in each wetland type and will explain the community characteristics of each. Following this rationale, we use Model 17.3 to show how low-richness communities are characterized by environmental gradients that create a severe filter (i.e., low numbers of taxa are adapted), and we use Model 17.4 to show how high-richness communities are less filtered by environmental factors. The filtering mechanisms for abiotic controls act in a hierarchical manner selecting organisms that have matching biological traits in a sequential way for the different environmental factors (Poff 1997; Ripley and Simovich 2009). However, if abiotic factors do not control a system (i.e., benign environmental conditions exist), the potentially rich community will then become filtered by biotic controls (i.e., competition, predation). The relative importance of abiotic versus biotic control, in turn, will act as an assembly rule determining the community composition. Thus, communities in habitats with strong abiotic control (i.e., more filtered) will be clustered together phylogenetically (i.e., taxa forming these communities will probably share more traits than would be expected by chance; see Table 17.13b), whereas communities in habitats with weak abiotic control will be controlled by biotic factors (i.e., competition and predation) and be characterized by a phylogenetic overdispersion (i.e., taxa forming these communities will be composed by organisms that share fewer traits than would be expected by chance) (Ruhí et al. 2013). The main abiotic and biotic filters identified

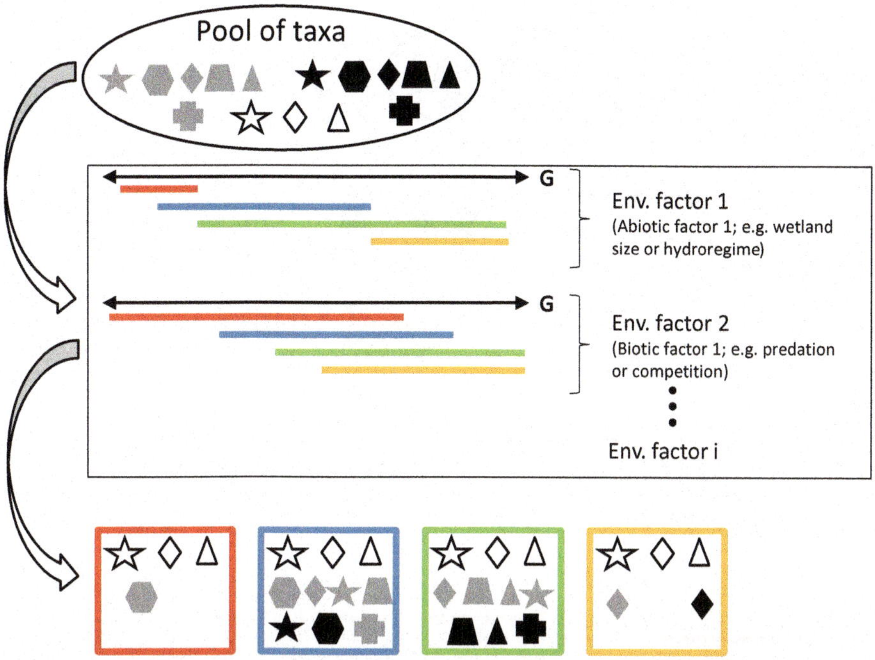

Model 17.2 Diagram showing how assemblage composition might develop in four hypothetical wetland habitats (*colored boxes* at the bottom) from the available pool of taxa (upper ellipse of open, black, and grey taxa; see Model 17.1). Environmental filters (abiotic and biotic mechanisms; *central boxes*) differentially sieve the pool of taxa in each wetland type. Environmental gradients (G) for each filter are represented by *colored bars*, with the environmental gradients and resulting habitat assemblages for each of the four habitats indicated by the same color. Two of these hypothetical habitats host taxa-rich assemblages (*blue* and *green boxes*), while, due to more truncated environmental gradients, the other two habitats host taxa-poor assemblages (*red* and *orange boxes*). The compositions of the two taxa-poor assemblages differ because they are each constrained by different environmental gradients. Similarly, compositions of the two taxa-rich assemblages also differ because they are subject to different gradients for each environmental factor. More detail on how environmental filtering mechanisms operate in taxa-poor (Model 17.3) and taxa-rich (Model 17.4) wetlands is provided below

for wetland invertebrates have been discussed elsewhere (hydroregime, habitat heterogeneity, water quality, macrophytes, predation, and competition; e.g., Declerck et al. 2005; Williams 2006; Vanschoenwinkel et al. 2009; Quintana et al. 2015), including, to some extent, by authors of every chapter of this book.

Human Impacts Altering Assemblage Models

Human impacts can alter both the compositions of taxa pools (Model 17.5) and the actions of environmental filters (Model 17.6). Global drivers (Model 17.5) such as biogeography can change due to movements of taxa among biogeographic regions,

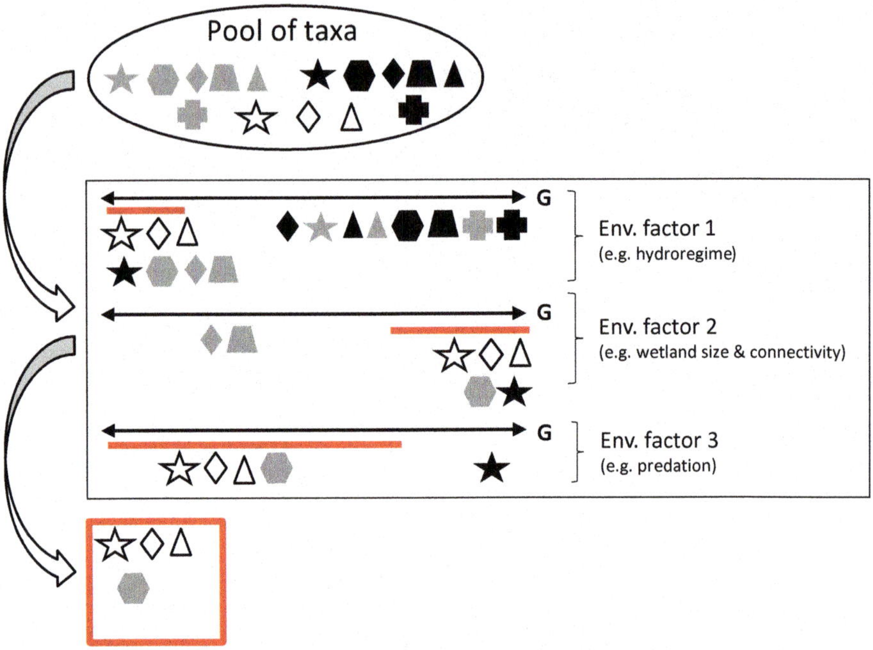

Model 17.3 A focused example of how environmental filtering might create a low-richness assemblage (e.g., Rock pools; Table 17.12). Starting with the available pool of taxa (Model 17.1), different environmental gradients sequentially reduce the assemblage from 15 to 4 taxa, with the degree of filtering indicated by the range of each gradient (i.e., short-ranges imply a greater reduction than long-ranges). Symbols located under the environmental gradient (*red bar*) represent taxa adapted for that range (i.e., these taxa are not excluded by this environmental factor). Symbols outside of the bar are nonadapted taxa and are excluded from this wetland by this environmental factor. For example, the first environmental factor causes a major reduction in taxa (from a possible 15 down to 8), while the third environmental factor only excludes a single taxon from the pool of 5 taxa that were adapted to the two first environmental factors

which nowadays occur mostly from human actions (Sax et al. 2005). Species originate in a specific evolutionary context, occurring in specific bioregions isolated from other species by borders that can't easily be crossed naturally. Although generalizations about why certain human-introduced taxa establish successfully are being debated (Simberloff et al. 2013), some tendencies are shared (Williamson 1996; Roy and Kauffman 2001). For example, low-diversity communities tend to be more invasion prone, and successfully introduced taxa tend to have generalist rather than specialist traits (McKinney and Lockwood 2001). Climate, another global driver (Model 17.5), is also changing, and will likely strongly modify potential pools of taxa for wetlands. With new ranges in climatic conditions (temperature, precipitation, etc.), native species not adapted to cope will be eliminated, native species able to adapt will persist, and new species previously unable to live in the region may successfully colonize. It will be difficult to predict how a pool of taxa will be affected by climate change, in terms of numbers of species. First, the ratio between the numbers of species whose geographical ranges will expand versus

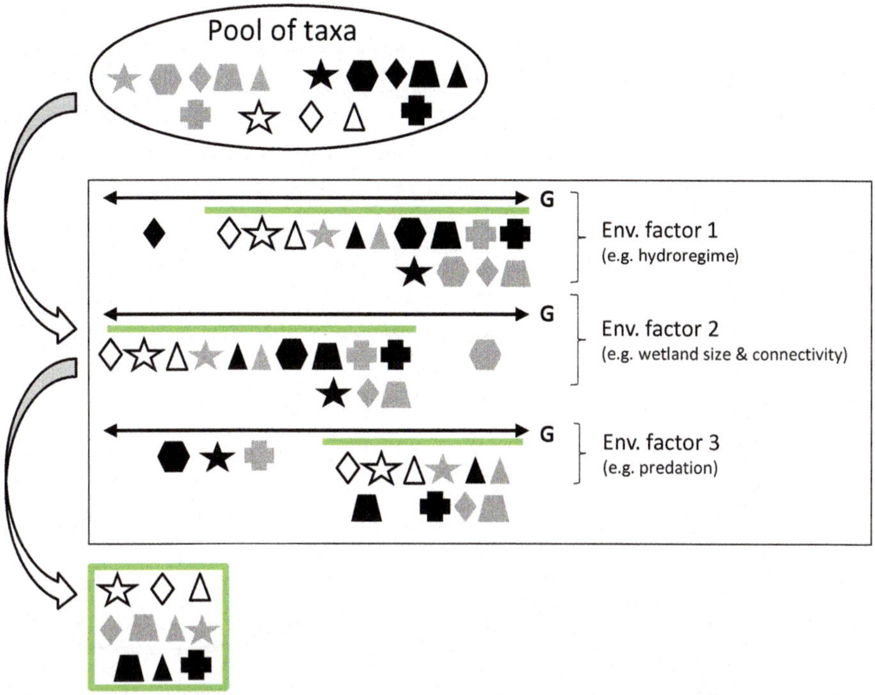

Model 17.4 A focused example of how environmental filtering might create a high-richness assemblage (e.g., Permanent marshes; Table 17.12). Starting with the available pool of taxa (Model 17.1), different environmental gradients sequentially reduce the assemblage from 15 to 10 taxa, with the degree of filtering indicated by the range of each gradient (i.e., short-ranges imply a greater reduction than long-ranges). Symbols located under the environmental gradient (*green bar*) represent taxa adapted for that range (i.e., these taxa are not excluded by this environmental factor). Symbols outside of the bar are nonadapted taxa and are excluded from this wetland by this environmental factor. For example, the first environmental factor imposes a minimal reduction (from a possible 15 taxa down to 14). The third environmental factor exerts the most filtering but still only excludes 3 of the 13 taxa that were adapted to the two first environmental factors

contract will vary among taxonomic groups (Rosset and Oertli 2011). Second, contradictory predictions are generated by different analytic approaches. The metabolic theory of ecology predicts an increase in species richness with climate change (Meerhoff et al. 2012), while bioclimatic approaches predict that current species ranges will contract or become fragmented, resulting in reduced species richness with climate change (Hughes 2003). Moreover, changes in land-use patterns may inhibit migration and adaptation of species due to climate change (IPCC 2007).

Invasive species and climate change may have both direct and indirect effects over the environmental filtering process (Model 17.6). Introduced taxa may modify predation impacts, food-web functioning, habitat structure (i.e., presence of macrophytes), and water quality (e.g., impacts of *Procambarus clarkii* crayfish introductions; Rodríguez et al. 2005; Geiger et al. 2005). Climate change will mod-

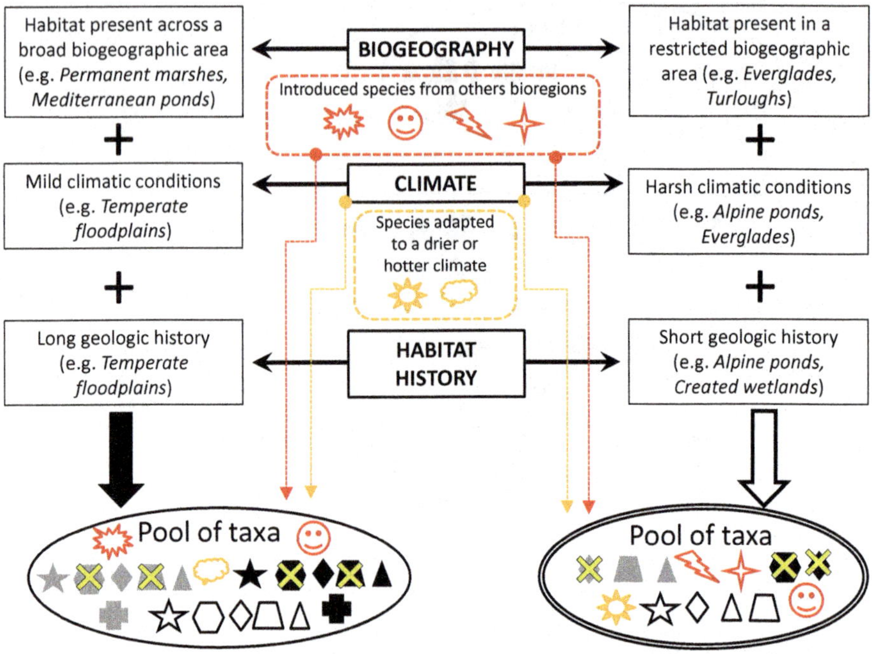

Model 17.5 Diagram showing how human impacts can alter the configuration of the available pool of taxa (see Model 17.1). Symbols correspond to generalist taxa (*open*), constrained taxa (*grey*), and rare taxa (*black*). Generalist taxa (*open symbols*) are further divided into three types by color: native taxa (*black*), new taxa introduced by humans (*red*), and new taxa able to naturally invade due to climate change (*orange*) (taxa that newly colonize will likely have generalist tendencies). Introduced taxa probably establish more easily in nonsaturated communities (i.e., taxa-poor communities). *Yellow crosses* indicate taxa that climate change removes from the pool of taxa. Climate change probably creates a reduction in taxa richness due to habitat fragmentation or an inability of taxa to change their geographic distributions (e.g., surrounding human land uses prevent it). However, climate change may also add some new taxa to the pool of taxa. Climate change should mainly affect rare and constrained taxa

ify hydroregimes, perhaps the key environmental control on invertebrate community structures in wetlands (as indicated by virtually every chapter in this book). Climate change may also disrupt trophic linkages (Winder and Schindler 2004; Brucet et al. 2012), or decouple processes controlling phenological and behavioral responses (Adrian et al. 2006; Hassall et al. 2007). In aquatic environments, species introductions and climate change can act synergistically, since climate change can modify the ecological impacts of invasive species by enhancing their competitive and predatory effects on native species, and by increasing the virulence of some diseases (Rahel and Olden 2008). Other human activities, such as agriculture, livestock grazing, or urbanization, may also alter environmental filtering processes in wetlands by modifying hydroregimes, habitat structures (landscape changes, fire frequency and intensity, etc.), or water qualities (pollution, eutrophication, etc.) (Holland et al. 1995; Batzer and Wissinger 1996; Euliss and Mushet 1999). As

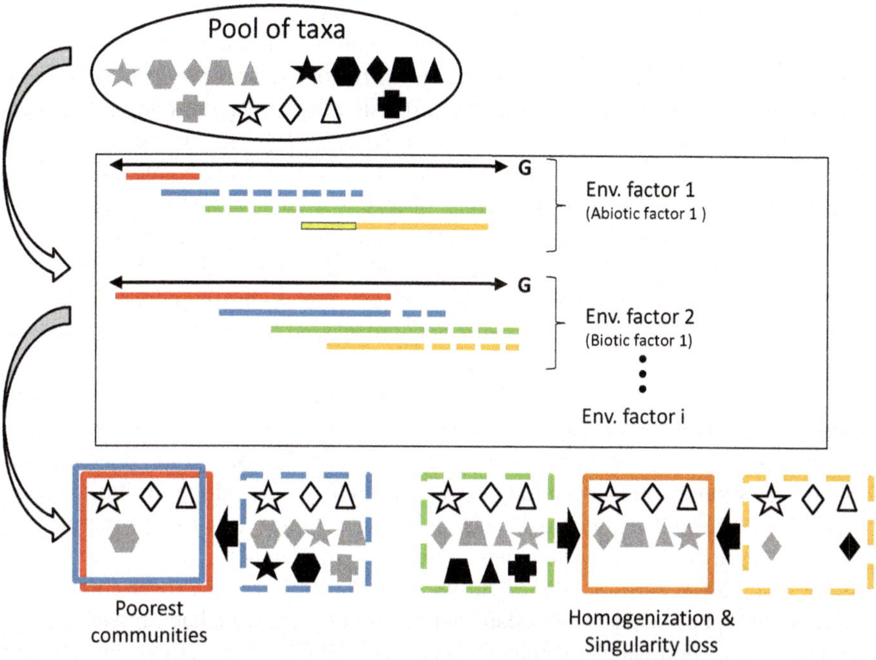

Model 17.6 Diagram showing how human impacts can alter how environmental filtering mechanisms operate (see Models 17.2, 17.3, and 17.4). The original range of an environmental gradient can be either reduced or extended by humans. *Solid lines* indicate the residual environmental gradients from their original ranges. *Dashed lines* indicate the portion of a gradient that was truncated; a color change indicates a gradient that was extended. For example, the red habitat was not human altered. The *blue* and *green* habitats had their environmental gradients reduced by human impacts. In the *orange* habitat, human activity extended the gradient of the first environmental factor (*yellow-bar fragment*), but reduced the second gradient (*dashed-line fragment*). For the *blue* habitat, the reduced gradients induced it to become a taxa-poor assemblage, similar to the *red* habitat (e.g., an invading taxa eliminates numerous native wetland species). In the case of the *orange* and *green* habitats, human impacts modify each wetland in such a way that they now support similar assemblages (e.g., humans alter hydroperiods to become more similar, and invertebrate assemblages also homogenize). Thus, *dashed boxes* at the bottom of the diagram are the old communities, whereas *solid boxes* are the newly created ones

human activities intensify, small, shallow, and/or temporary wetlands that are easily drained or filled by sediment will be most affected (Grillas et al. 2004; Boix et al. 2012; see Chaps. 4 and 5). However, even large, permanent wetland areas will become degraded (e.g., Dimentman et al. 1992; Richardson et al. 2005; see Great Lakes marshes Chap. 9).

In general, human impacts will have two main effects on environmental filtering (Model 17.6). First, human activities can increase the intensity of environmental filtering (reduce the environmental gradients for the habitat), and habitats will support fewer taxa. For example, if natural hydroperiods are truncated, that filter will operate more broadly across more wetland habitats and more intensively in individual wetland habitats (Zokan and Drake 2015). Second, habitat and biotic homogenization can occur at regional scales. For example, introductions of *P. clarkii* crayfish into Europe has caused wetlands to shift from clear stable states to consistently turbid states (with low macrophyte density, high plankton chlorophyll content, low zooplankton abundance, etc.; Scheffer et al. 1993, see Chap. 8) that host similar invertebrate communities (Rodríguez et al. 2005). Other examples of habitat homogenization at regional scales can be observed in areas of intensive agriculture, where temporary waterbodies are converted into more permanent ones to provide water for livestock or irrigation, reducing the range of available habitat types. Furthermore, habitat homogenization can also occur within individual wetlands, since waterbodies transformed by human activities tend to have fewer depth irregularities and more simplistic littoral zones.

Global biodiversity of wetlands is high (Gopal and Junk 2000), but the vulnerability of these habitats is also high. To conserve high invertebrate biodiversity, priorities must be the preservation of existing natural wetlands, and the prevention of homogenization (biotic or habitat) across different scales (global biotic, regional biotic, among habitats, within habitats). Even wetlands in close proximity can exhibit very different environmental gradients, and support different taxa (e.g., Smith et al. 2003; Jeffries 2003; Escrivà et al. 2010). For some invertebrates, life cycles require the existence of wetland heterogeneity; e.g., taxa that migrate between temporary and permanent wetlands (Wissinger 1997). Management efforts may reverse, or at least mitigate past human impacts. Wetland restoration or creation is undoubtedly a valuable strategy to mitigate wetland loss, although biological recovery can be slow (see Moreno-Mateos et al. 2012; Chap. 15). To conserve biodiversity, it is necessary to maintain the natural heterogeneity in environmental gradients across and within wetlands (Biggs et al. 1994; Gee et al. 1997; Moss et al. 2009), especially ecologically key gradients—i.e., natural hydrological functioning (Wissinger 1999; Euliss et al. 2004).

Acknowledgements We thank Xavier Quintana, Stéphanie Gascón, Jordi Sala, Courtney Holt, Kelly Murray, and Bryana Bush for their valuable reviews of this chapter.

References

Adrian R, Wilhelm S, Gerten D (2006) Life-history traits of lake plankton species may govern their phenological response to climate warming. Global Change Biol 12:652–661

Batzer DP (2013) The seemingly intractable ecological responses of invertebrates in North American wetlands: a review. Wetlands 33:1–15

Batzer DP, Wissinger SA (1996) Ecology of insect communities in nontidal wetlands. Annu Rev Entomol 41:75–100

Beladjal L, Dierckens K, Mertens J (2007) Dispersal of fairy shrimp *Chirocephalus diaphanus* (Branchiopoda: Anostraca) by the trout (*Salmo trutta*). J Crust Biol 27:71–73

Biggs J, Corfield A et al (1994) New approaches to the management of ponds. Br Wildl 5:273–287

Boix D, Biggs J et al (2012) Pond research and management in Europe: "Small is Beautiful". Hydrobiologia 689:1–9

Brown KM, Lydeard C (2010) Mollusca: Gastropoda. In: Thorp JH, Covich AP (eds) Ecology and classification of North American freshwater invertebrates, 3rd edn. Elsevier, Amsterdam, pp 277–308

Brucet S, Boix D et al (2012) Effects of temperature, salinity and fish in structuring the macroinvertebrate community in shallow lakes: implications for effects of climate change. PLoS One 7(2), e30877

Declerck S, Vandekerkhove J et al (2005) Multi-group biodiversity in shallow lakes along gradients of phosphorus and water plant cover. Ecology 86:1905–1915

Dimentman C, Bromley H, Por FD (1992) Lake Hula: reconstruction of the fauna and hydrobiology of a lost lake. Israel Academy of Sciences and Humanities, Jerusalem

Escrivà A, Armengol X, Mezquita F (2010) Microcrustacean and rotiferan communities of two close Mediterranean mountain ponds, lagunas de Bezas and Rubiales (Spain). J Freshw Ecol 25:427–435

Euliss NH, Mushet DM (1999) Influence of agriculture on aquatic invertebrate communities of temporary wetlands in the prairie pothole region of North Dakota, USA. Wetlands 19:578–583

Euliss NH, LaBaugh JW et al (2004) The wetland continuum: a conceptual framework for interpreting biological studies. Wetlands 24:448–458

Gee JH, Smith BD, Lee KM, Griffiths SW (1997) The ecological basis of freshwater pond management for biodiversity. Aquat Conserv Mar Freshw Ecosys 7:91–104

Geiger W, Alcorlo P, Baltanás A, Montes C (2005) Impact of an introduced crustacean on the trophic webs of Mediterranean wetlands. Biol Invasions 7:49–73

Gopal B, Junk WJ (2000) Biodiversity in wetlands: an introduction. In: Gopal B, Junk WJ, Davis JA (eds) Biodiversity in wetlands: assessment, function and conservation. Backhuys, Leiden, pp 1–10

Govedich FR, Bain BA et al (2010) Annelida (Clitellata): Oligochaeta, Branchiobdellida, Hirudininida, and Acanthobdellida. In: Thorp JH, Covich AP (eds) Ecology and classification of North American freshwater invertebrates, 3rd edn. Elsevier, Amsterdam, pp 385–436

Grillas P, Gauthier P, Yavercovski N, Perennou C (2004) Mediterranean temporary pools. Issues relating to conservation, functioning and management, vol 1. Station Biologique de la Tour du Valat, Arles

Hartland-Rowe R (1972) The limnology of temporary waters and the ecology of Euphyllopoda. In: Clark RB, Wootton RJ (eds) Essays in hydrobiology. University of Exeter, Exeter, pp 15–31

Hassall C, Thompson DJ, French GC, Harvey IF (2007) Historical changes in the phenology of British Odonata are related to climate. Global Change Biol 13:933–941

Holland CC, Honea J, Gwin SE, Kentula ME (1995) Wetland degradation and loss in the rapidly urbanizing area of Portland, Oregon. Wetlands 15:336–345

Hughes L (2003) Climate change and Australia: trends, projections and impacts. Austral Ecol 28:423–443

IPCC (2007) Climate change 2007: impacts, adaptation and vulnerability. In: Contribution of working group II to the fourth assessment report of the intergovernmental panel on climate change. Cambridge University Press, Cambridge

Jeffries M (2003) Idiosyncratic relationships between pond invertebrates and environmental, temporal and patch-specific predictors of incidence. Ecography 26:311–324

Jeppesen E, Christoffersen K et al (2001) Fish and crustaceans in northeast Greenland lakes with special emphasis on interactions between Arctic char (*Salvelinus alpinus*), *Lepidurus arcticus* and benthic chydorids. Hydrobiologia 442:329–337

King RS, Richardson CJ (2002) Evaluating subsampling approaches and macro invertebrate taxonomic resolution for wetland bioassessment. J North Am Benthol Soc 21:150–171

Klemmer AJ, Wissinger SA, Greig HS, Ostrofsky ML (2012) Nonlinear effects of consumer density on multiple ecosystem processes. J Anim Ecol 81:770–780

McKinney ML, Lockwood JL (2001) Biotic homogenization: a sequential and selective process. In: Lockwood JL, McKinney ML (eds) Biotic homogenization. Kluwer, New York, pp 1–17
Meerhoff M, Teixeira-de-Mello F et al (2012) Environmental warming in shallow lakes: a review of potential changes in community structure as evidenced from space-for-time substitution approaches. Adv Ecol Res 46:259–349
Moreno-Mateos D, Power ME, Comín FA, Yockteng R (2012) Structural and functional loss in restored wetland ecosystems. PLoS Biol 10:e1001247
Moss B, Hering D et al (2009) Climate change and the future of freshwater biodiversity in Europe: a primer for policy-makers. Freshw Rev 2:103–130
Poff NL (1997) Landscape filters and species traits: towards mechanistic understanding and prediction in stream ecology. J North Am Benthol Soc 16:391–409
Quintana XD, Arim M et al (2015) Predation and competition effects on the size diversity of aquatic communities. Aquat Sci 77:45–57
Rahel FJ, Olden JD (2008) Assessing the effects of climate change on aquatic invasive species. Conserv Biol 22:521–533
Richardson CJ, Reiss P et al (2005) The restoration potential of the Mesopotamian marshes of Iraq. Science 307(5713):1307–1311
Ripley BJ, Simovich MA (2009) Species richness on islands in time: variation in ephemeral pond crustacean communities in relation to habitat duration and size. Hydrobiologia 617:181–196
Rodríguez CF, Bécares E, Fernández-Aláez M, Fernández-Aláez C (2005) Loss of biodiversity and degradation of wetlands as a result of introducing exotic crayfish. Biol Invasions 7:75–85
Rosenberg DM, Resh VH, King RS (2008) Use of aquatic insects in biomonitoring. In: Merritt RW, Cummins KW, Berg MB (eds) An introduction to the aquatic insects of North America. Kendall/Hunt Publications, Dubuque, pp 123–138
Rosset V, Oertli B (2011) Freshwater biodiversity under climate warming pressure: identifying the winners and losers in temperate standing waterbodies. Biol Conserv 144:2311–2319
Roy K, Kauffman JS (2001) Biotic homogenization: lessons from the past. In: Lockwood JL, McKinney ML (eds) Biotic homogenization. Kluwer, New York, pp 19–32
Ruhí A, Batzer DP (2014) Assessing congruence and surrogacy among wetland macroinvertebrate taxa towards efficiently measuring biodiversity. Wetlands 34:1061–1071
Ruhí A, Boix D et al (2013) Functional and phylogenetic relatedness in temporary wetland invertebrates: current macroecological patterns and implications for future climatic change scenarios. PLoS One 8:e81739
Sax DF, Stachowicz JJ, Gaines SD (2005) Species invasions: insights into ecology, evolution and biogeography. Sinauer Associations, Sunderland
Scheffer M, Hosper SH et al (1993) Alternative equilibria in shallow lakes. Trends Ecol Evol 8:275–279
Schneider DW, Frost TM (1996) Habitat duration and community structure in temporary ponds. J North Am Benthol Soc 15:64–86
Schowalter TD (2000) Insect ecology: an ecosystem approach. Academic, New York
Simberloff D, Martin JL et al (2013) Impacts of biological invasions: what's what and the way forward. Trends Ecol Evol 28:58–66
Smith GR, Vaala DA, Dingfelder HA (2003) Distribution and abundance of macroinvertebrates within two temporary ponds. Hydrobiologia 497:161–167
Tennessen KJ (2008) Odonata. In: Merritt RW, Cummins KW, Berg MB (eds) An introduction to the aquatic insects of North America. Kendall/Hunt Publications, Dubuque, pp 237–294
Triplehorn CA, Johnson NF (2005) Borror and Delong's introduction to the study of insects. Thomson Brooks/Cole, Belmont
Vanschoenwinkel B, Hulsmans A et al (2009) Community structure in temporary freshwater pools: disentangling the effects of habitat size and hydroregime. Freshw Biol 54:1487–1500
Wellborn GA, Skelly DK, Werner EE (1996) Mechanisms creating community structure across a freshwater habitat gradient. Annu Rev Ecol Syst 27:337–363

Wiggins GB, Mackay RJ, Smith IM (1980) Evolutionary and ecological strategies of animals in annual temporary pools. Arch Hydrobiol Suppl 58:97–206

Williams DD (2006) The biology of temporary waters. Oxford University Press, Oxford

Williamson MH (1996) Biological invasions. Chapman Hall, London

Winder M, Schindler DE (2004) Climate change uncouples trophic interactions in an aquatic ecosystem. Ecology 85:2100–2106

Wissinger SA (1997) Cyclic colonization in predictably ephemeral habitats: a template for biological control in annual crop systems. Biol Control 10:4–15

Wissinger SA (1999) Ecology of wetland invertebrates. Synthesis and applications for conservation and management. In: Batzer DP, Rader RB, Wissinger SA (eds) Invertebrates in freshwater wetlands of North America. Wiley, New York, pp 1043–1086

Wissinger S, Bohonak AJ, Whiteman HH, Brown WS (1999) Subapline wetlands in Colorado: habitat permanence, salamander predation and invertebrate communities. In: Batzer DP, Rader RB, Wissinger SA (eds) Invertebrates in freshwater wetlands of North America. Wiley, New York, pp 757–790

Zokan M, Drake JM (2015) The effect of hydroperiod and predation on the diversity of temporary pond zooplankton communities. Ecol Evol 5(15):3066–3074

Index

D. Batzer, D. Boix (eds.), *Invertebrates in Freshwater Wetlands*,
DOI 10.1007/978-3-319-24978-0

CPSIA information can be obtained
at www.ICGtesting.com
Printed in the USA
LVOW02*2042120216
474873LV00001B/3/P

9 783319 249766